Discrete Mathematics

DISCRETE MATHEMATICS

AN INTRODUCTION TO CONCEPTS, METHODS, AND APPLICATIONS

JERROLD W. GROSSMAN

Oakland University
Rochester, Michigan

Macmillan Publishing Company

New York

Collier Macmillan Publishers

London

Book team

Acquisition Editor: Robert W. Pirtle
Production Supervisor: Elaine W. Wetterau
Production Manager: Richard C. Fischer
Text Designer: Eileen Burke
Cover Designer: Eileen Burke
Cover photograph: FPG International Corporation
Illustrations: Hadel Studio

This book was set in Times Roman by ETP Services, Inc.,
printed and bound by Von Hoffmann Press, Inc.
The cover was printed by Lehigh Press Lithographers.

Macmillan Publishing Company
866 Third Avenue, New York, New York 10022

Collier Macmillan Canada, Inc.

Library of Congress Cataloging-in-Publication Data

Grossman, Jerrold W.
    Discrete mathematics / Jerrold W. Grossman.
       p.   cm.
    Includes index.
    ISBN 0-02-348331-8
    1. Mathematics—1961- 2. Electronic data processing—Mathematics.
    I. Title.
    QA39.2.G7749   1990
    510—dc20   89-2552
                                                       CIP

Printing: 1 2 3 4 5 6 7 8   Year: 9 0 1 2 3 4 5 6 7 8

PREFACE

TO THE INSTRUCTOR

This book is designed for an introductory course in what has come to be called "discrete mathematics" for undergraduates majoring in mathematics, computer science, and related disciplines. In content, level, emphasis, and spirit, it follows most of the guidelines of the *1986 Report of the Committee on Discrete Mathematics in the First Two Years* of the Mathematical Association of America. In particular, the primary themes of this book are the notions of **proof**, **recursion**, **induction**, **modeling** and **algorithmic thinking**. These themes are developed as subjects in themselves, and they are applied to the two major content areas of discrete mathematics: **combinatorics** and **graph theory**.

The reader will find unity in the content of this book. Logic and set theory are developed in the first two chapters and are then used throughout, to provide the preciseness of thinking that is emphasized. Proofs are discussed early, and the reader is expected to follow proofs throughout the book and construct proofs in doing many of the exercises. There is a detailed discussion of algorithms fairly early, and much of the emphasis in the graph theory portion of the book is on the algorithmic aspects of the subject. Induction and recursion are seen extensively after their introduction. Themes recur in discrete mathematics, and the reader will see the same general topics from different points of view recurring in this book.

The writing style is somewhat formal but very leisurely and conversational; it is intense, intellectual, and serious but also relaxed and light-hearted. Theorems are stated formally (and careful, understandable proofs given when appropriate), but definitions are given informally in the text (with a complete glossary at the end of each section). Much of the text is centered around substantive examples, many involving applications from computer science, mathematics, or other areas. In this book, the reader learns to see discrete mathematics as a mathematician or theoretical computer scientist sees it, to understand what is important and why. I do not hesitate to discuss topics at the forefront of current research, for it lets the reader know that this is very much a living and exciting subject. Top students will find the book exciting and challenging, while at the same time average students will be able to understand mathematics in a way they have not understood it before.

The book is intended primarily for sophomores. It is assumed that the reader has had a course in calculus, not because the results or ideas of calculus are heavily used (only rarely are there references to such things as differentiation or L'Hôpital's rule), but rather because the mathematical maturity obtained in a calculus course prepares the student for the next step in abstraction and mathematical sophistication that this subject matter presents. I want the reader to progress beyond the point of thinking of mathematics as a subject that can be made simple if only the right tricks and procedures are learned. In studying from this book, the student is forced to confront mathematics as the rich, ambiguous, and complex subject it really is. I also assume—and do not expect to lose any readers thereby—that the reader has been exposed to the idea of a computer program. Otherwise, the book is complete and self-contained.

The book may be used in a variety of courses, from a brief one-quarter course concentrating on the foundational material (logic, proofs, sets, functions, relations, algorithms, and induction), to a one-semester course treating these topics as well as combinatorics and graph theory, to a two-quarter course in which these latter topics are explored more deeply. By omitting many optional sections and subsections along the way and varying the order of presentation, the instructor can also structure a course emphasizing combinatorics and/or graph theory. A detailed discussion of the contents, section dependencies, and possible course outlines appears below.

Overview of the Contents

The book is organized into four parts, each part containing two or three related chapters. Part One (Chapters 1, 2, and 3), Foundations of Discrete Mathematics, deals with logic, proof, Boolean functions, sets, functions, and relations. Part Two (Chapters 4 and 5), Tools of Discrete Mathematics, deals with algorithms, recursion, and induction. Part Three (Chapters 6 and 7), Combinatorics, deals with permutations, combinations, the pigeonhole principle, combinatorial identities, setting up and solving recurrence relations, the inclusion–exclusion principle, and generating functions. Part Four (Chapters 8, 9, and 10), Graph Theory, deals with graphs, trees and networks.

The choice of these broad areas, as well as the particular topics that are included, has been influenced primarily by what I, as well as the MAA committee mentioned above, consider to be important in a first course in discrete mathematics. (The committee envisions a year-long course in which linear algebra and either abstract algebra or probability and statistics are also taught; as there are several good books on these subjects, I have not attempted to include them here.) Some topics that I feel to be less important, peripheral, or more advanced have been included for those instructors who may wish to teach this material to some classes. They are marked as optional sections and subsections, and results from these sections are not needed in the mainline material that follows. Throughout the entire book, substantial examples and applications make this much more than just an exposition of the mathematical topics at hand.

Chapter 1 lays the logical foundations upon which all else is based. No prior knowledge of logic is assumed. I begin in Section 1.1 by introducing propositions and the various ways of combining them to make meaningful, precise statements. The need for quantifiers immediately becomes apparent in Section 1.2, and care is taken to explain what a quantified proposition really means. This leads into Section 1.3, where the reader sees what it means to establish the truth of—or to refute—a proposition; the discussion here touches on philosophical issues such as Gödel's Incompleteness Theorem and computer-aided proofs, as well as covering standard proof techniques, applied mostly to elementary facts from number theory. Finally, optional Section 1.4 looks at propositional logic from the algebraic viewpoint of Boolean functions and switching circuits.

Chapter 2 introduces sets as formal mathematical objects. Since the reader has probably seen some of this material before, I move quickly through the elementary definitions and include many subtler concepts. Set membership, subsets, the power set, cardinality, and (optionally) infinite sets are discussed in Section 2.1; different ways to describe sets are investigated. Section 2.2 takes a unified look at how a set can be endowed with structure to turn it into more than just a set; depending on the structure one can talk about sets of sets, ordered pairs rm or n-tuples, finite or infinite sequences, strings, matrices, or arbitrary indexed collections of sets. These ideas are applied in Section 2.3, where I discuss various operations on sets, not just for two or three sets, but also over arbitrary collections; a discussion of the various proof techniques for establishing the Boolean identities gives the reader a second taste of proofs in mathematics.

The most important kinds of sets with structure are functions and relations, and a complete treatment of these topics appears in Chapter 3. The reader will have worked extensively but informally with functions before (and to a lesser extent, with relations), but here they are developed from scratch in a mathematically precise way. In Section 3.1, I give the basic definition of function as a set (with a discussion of why this is necessary), along with many examples of numerical and nonnumerical functions that will be used later in the book. Section 3.2 is more abstract, and the reader gets a third taste of proving mathematical truths by confronting issues such as injectivity, surjectivity, images and inverse images, and compositions. Relations, thought of as generalizations of functions, are introduced in Section 3.3; matrix and digraph representations of relations and notions such as symmetry and transitivity are the key ideas, again with an emphasis on nontrivial examples and applications. Finally, there is an extensive treatment of partial orders (with Hasse diagrams) and equivalence relations and partitions in Section 3.4; general theorems about and important examples of these relations receive equal emphasis. Throughout Chapter 3, the reader is asked to follow and write many proofs of the important facts about functions and relations. Also, both here and in Chapter 4 some elementary number theory is covered.

Much of the foundational material up to this point is important to all of mathematics. However, in Chapter 4, on algorithms, the content and flavor of discrete mathematics really begin to predominate. The reader has certainly dealt with algorithms before (from the grade-school arithmetic algorithms to complex procedures possibly encountered in a computer science or programming class), but he or she has probably not studied algorithms per se. Using the Euclidean algorithm as a paradigm, I give a wide-ranging exposition of algorithms in Section 4.1. In order to state algorithms clearly, I have chosen to use an easy-to-understand structured pseudocode, emphasizing input and output; a quick "programming course" is given in Section 4.2. A study of the efficiency of algorithms is a vital part of any discrete mathematics course, and a careful explanation of the basis for this analysis is given in Section 4.3; I adopt the modern viewpoint of $O(f)$ as a class of functions and explain how to compare algorithms in big-oh terms. Optional Section 4.4 explains that some easy-to-state problems cannot be solved algorithmically, while others are apparently hard to solve algorithmically; for the reader who has tackled this section, I point out in the rest of the book whenever an NP-complete problem arises. Finally, Section 4.5 looks at some specific arithmetic and other algorithms, including some interesting applications of modular arithmetic.

Chapter 5 is perhaps the most important in the book, for it is here that the reader learns about induction and recursion. These topics are postponed to this point to let the reader gain some experience with abstraction and to understand algorithms first. No prior knowledge of these topics is assumed, and the treatment is extremely thorough. I have found that it is best to proceed from the concrete to the abstract. Therefore I begin, in Section 5.1, with recursive definitions of sequences, sets, and functions. Recursive algorithms are treated in Section 5.2, with the towers of Hanoi and merge sort as prime examples. Finally, in Section 5.3, I describe proof by mathematical induction, as a natural extension of these other two, more concrete, settings of recursive processes.

Elementary counting techniques are treated in Chapter 6; no prior exposure to these topics is assumed. Section 6.1 shows how to solve counting problems by looking at them in the right way, using the multiplication principle, the addition principle, the overcounting principle, or some combination of all three. I have taken extra care to confront and thereby avoid ambiguity in the statement of combinatorial problems. A systematic discussion of permutations and combinations is given in Section 6.2, with applications to probability and other areas. More complex combinatorial problems are solved in Section 6.3, especially combinations with repetitions allowed. Finally, optional Section 6.4 on the pigeonhole principle explores some simple and not-so-simple results that follow from this most obvious yet surprisingly powerful mathematical truth; this exploration goes even to the boundaries of current research in Ramsey theory.

Chapter 7 deals with additional standard topics in combinatorics, building on the results in the first two or three sections of Chapter 6. Combinatorial proofs are illustrated in Section 7.1 on combinatorial identities. Section 7.2 returns to the theme of recursively defined sequences introduced in Section 5.1 and shows the wide-ranging application of recurrence relations in discrete modeling. Then in Section 7.3, I present the standard elementary techniques for solving recurrence relations. There is an optional Section 7.4 on the inclusion–exclusion principle and its applications to combinatorial problems; and optional Section 7.5 introduces generating functions as a tool for solving combinatorial problems.

Topics in graph theory occupy Chapter 8. The emphasis is algorithmic, and again I proceed from scratch. Section 8.1 lays the foundations, emphasizing precise definitions and many examples and applications of graphs and digraphs. Paths in graphs are the subject of Section 8.2, culminating in a thorough comparison and contrast of Euler tours and Hamilton cycles. The question of how best to represent a graph is discussed in Section 8.3, and the important topic of isomorphism between graphs is introduced. Optional Section 8.4 treats a beautiful, self-contained topic in graph theory: the theory of planar graphs; Euler's formula is proved and then used to establish the nonplanarity of K_5 and $K_{3,3}$. In a similar vein, optional Section 8.5 discusses graph coloring and, building on Section 8.4, the four- and five-color theorems for planar graphs.

The algorithmic emphasis continues in Chapter 9 on trees. This chapter (which builds on the first three sections of Chapter 8) has a decidedly computer science flavor. Definitions, examples, and applications are laid out in Section 9.1, and the distinction between different types of trees is made clear. In Section 9.2, I show how graphs can be systematically traversed as their depth-first or breadth-first spanning trees are constructed. In Section 9.3 preorder, postorder, and inorder traversals of (mainly binary) trees are discussed, with application to expressions. Additional substantial applications of trees are given in two optional sections at the end of this chapter. Section 9.4 deals with two topics from computer science: binary search trees and Huffman codes. In Section 9.5 game trees provide a wonderful application of trees and recursion.

Finally, Chapter 10 deals with various algorithmic questions regarding graphs and digraphs in which the edges have weights. Portions of Chapters 8 and 9 are prerequisite. Section 10.1 deals with algorithms for finding shortest paths, methods for finding longest paths in acyclic digraphs, and the traveling salesperson problem. Section 10.2 discusses the problem of finding minimum spanning trees. Optional Section 10.3 presents the most complex algorithm in the book, the Ford–Fulkerson algorithm for finding maximum flows and minimum cuts in a capacitated network.

Section Dependencies and Course Outlines

The instructor has wide latitude in constructing a course around this book. A digraph giving the logical dependencies between sections is shown here (the sections marked optional are shown in lighter boxes). Note that the numerical order of the sections is consistent with the partial order implied by this digraph.

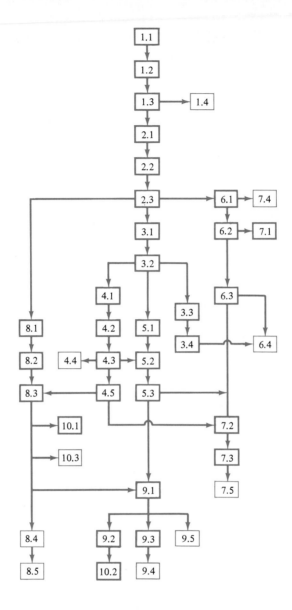

I would consider a basic one-term discrete mathematics course to consist of all of Chapters 1 through 8, *omitting the sections and subsections marked optional*, together with Section 9.1. If there is time at the end, additional topics from Part Four could be included. Well-prepared students could be taken through some of the foundational material more rapidly, allowing more time for topics from combinatorics and graph theory. A course could emphasize combinatorics by including more sections from Part Three and less graph theory, or it could emphasize graphs and trees and omit much or all of Part Three.

A "bridge to advanced mathematics course" (with a bit of a discrete flavor) could be constructed by going very thoroughly through Parts One and Two, and then perhaps using graph theory as a vehicle in which students perfect their proof techniques. A brush-up course for prospective computer science graduate students might touch lightly on the entire book.

Exercises

The heart of any textbook is its exercises, for only by doing mathematics can a student learn mathematics. There must be straightforward exercises to begin with, but much more important, if there is to be any real growth and understanding, there must be exercises in which the student is forced to think about the new concepts and apply them in a variety of different settings. The extensive exercise sets in this book have both kinds of exercises.

In fact, each exercise set contains four categories of exercises:

- The first few exercises in each section are mostly fairly routine problems on the ideas and techniques introduced in the section. Some of these easier exercises are similar to worked-out examples in the text; others require merely a simple application of a definition or theorem. A typical section will have about six such exercises, give or take a few; the exact dividing lines between these straightforward exercises and the more substantive ones in each section are given in the *Instructors' Manual*.

- The remaining **regular exercises** force the student to go a bit further, to deal with less straightforward examples, perhaps to recall some key ideas from previous sections, perhaps to look at some new topic that could not be discussed in the text itself. It is my strong belief that the majority of the learning that the student will retain from the course will be generated by working such exercises.

- In addition there are several **challenging exercises** in each section. Sometimes they are not too much more difficult than the hardest of the regular exercises, but most students will find most of them quite hard. Good students seem to be excited by working on problems like these, and I am often pleasantly surprised and gratified at the results some students are able to obtain.

- Finally, each exercise set ends with several **exploratory exercises**. These are really not exercises at all, but instead open-ended suggestions for further exploration. Some are discussion questions (Do you think computer-assisted proofs are as valid as totally human ones?). Others are hard challenging exercises that the student is not expected to solve completely or for which there is no good solution (How many lines might be determined by n points some of which might be collinear?); these exercises usually include references to the literature where solutions or further information can be found. Finally, many of the exploratory exercises are suggestions for directed reading and term projects; there are references to over 300 books and articles, ranging from more advanced textbooks to pieces in *Scientific American*, *Mathematics Magazine*, or *The American Mathematical Monthly* to essays on famous mathematicians.

Many exercises have more than one part (there are **about 3000 exercises**, when parts are counted). In some cases the parts are simply several different examples to which a technique or idea is to be applied. More often than not, an exercise with parts forms a progression, leading the student to explore the topic at hand in depth, looking at all of its facets and really coming to grips with it.

Solutions to all the exercises (except for the exploratory ones) have been worked out in full both by the author and by his colleague Jon Froemke. Our answers to the odd-numbered exercises—the straightforward ones at the beginning of each section, the standard ones, and the challenging ones—are given in the back of the book. The *Students' Solutions Manual* contains worked-out solutions to these odd-numbered exercises. The *Instructors' Manual* contains the solutions to both the odd- and even-numbered exercises (this supplement also has sample exam questions and complete solutions to all the exercises).

Acknowledgments

Much in this book is owed to other people. I thank Jack Tsui and Stuart Wang at Oakland University and Gary Shannon at California State University at Sacramento, who taught from a preliminary version of the book. They and their students, as well as my own students, provided many helpful suggestions. I gratefully acknowledge the comments of the following people who reviewed the book in various stages of its development: Thomas Dowling, Ohio State University; Robert Feinerman, Lehman College; Alejandro Garcia, University of California, Los Angeles; Jerrold Griggs, University of Minnesota; Gordon Hughes, California State University, Chico; Richard Molnar, Macalester College; William Nico, California State University, Hayward; Gary Shannon, California State University, Sacramento; and Alan Tucker, State University of New York, Stony Brook. Jon Froemke performed an extensive and extremely helpful line by line review of the entire manuscript and worked all of the exercises to double-check my work; his insightful comments, suggestions, and corrections are very much appreciated. My colleagues Don Malm, Ernie Schochetman, and Bob Stern also offered helpful advice on various sections. The staff at Macmillan has made the task of turning a manuscript into a book surprisingly satisfying in itself. Finally, my wife Suzanne deserves the most thanks for giving me her unfailing support and for allowing me the time away from my other duties to devote to this project.

TO THE STUDENT

All of pure and applied mathematics can be roughly divided into two parts: the **continuous** and the **discrete**. The distinction can best be appreciated by contrasting examples. The real number line is continuous; the integers and finite sets are discrete. Measurements of physical properties—such as length, area, force, or elasticity—are continuous; tabulation of individual entities—such as the number of people who voted for the latest Republican presidential candidate, the number of electrons in a carbon atom, or the size of a personal computer's main memory—are discrete. Finding an approximate solution to $x^3 + 3x - 2 = 0$ is a continuous problem ($x \approx 0.596$); determining whether there are any rational number solutions to this equation is a discrete problem (there are none). Worrying about the slope of the curve $y = \sqrt{x}$ at $x = 5$ is a continuous problem (it is $\sqrt{5}/10$); looking for the smallest prime number larger than 1,000,000 is a discrete problem (it is 1,000,003). Computing the average of a list of numbers is a continuous problem; sorting these numbers into increasing order and finding their median is a discrete problem. The mathematics motivated by physics and engineering is primarily continuous; the mathematics behind computer science (and many social sciences, as well) is primarily discrete. This book focuses mainly on discrete mathematics, although much of the first half of the book is relevant to all parts of mathematics.

In studying from this book, you will find that mathematics is much more than what you have seen in algebra or calculus courses. For one thing, the raw material of mathematics goes well beyond numbers; logical propositions, sets, strings of characters, functions, relationships, computer programs, and figures of points and lines in the plane will all be the objects of our study. For another, the emphasis here is somewhat more on the abstract. If in calculus you worried about how to differentiate a given function from real numbers to real numbers, here the problem might be to determine how many different functions there are from one finite set to another. Much less emphasis here is placed on manipulating symbols in a routine way to get the right answer to a problem; a more likely concern would be an analysis of the symbol-manipulating procedure itself to see whether it can be executed by a computer in a reasonable length of time. There will be more of an emphasis on stating very precisely exactly what we are talking about (giving complete, rigorous definitions of the objects we are studying) and proving that the claims we make are true. It is impossible to give you a real understanding of the flavor of this material without actually presenting the material, but when you have finished, I think you will agree that things have been a lot different from what you were used to.

Here is some specific advice about how to use this book and take a discrete mathematics course.

- Be sure to approach discrete mathematics with an open mind as to what it is about, what is important, and what you will need to do to learn it. No mathematics can be "made easy," and discrete mathematics is no exception; you will not find a cookbook here. Revel in the subtleties, ambiguities, novelty, and richness of the subject as you gradually come to understand it. You will get from this book a feeling for the subject as a mathematician or a computer scientist sees it, not as it might be oversimplified and handed to a student.

- Do not be disturbed if there are things in this book that you do not understand completely in this introductory course. Discrete mathematics is a new, exciting, and evolving field (unlike, say, calculus, which was discovered over 300 years ago), and occasionally you will read here of recent developments and unsolved problems at the center of current research. Take advantage of the suggestions for further reading and exploration in the "exploratory exercises" at the end of each section to learn more about this fascinating area.

- You must read a mathematics book totally differently from the way in which you read a magazine article, novel, or history textbook. Do not expect to read it in linear order and understand everything the first time through. Skim over passages that seem a little confusing at first and come back to them after you have read a little further. In many cases it will make sense to skim the statement of a definition or theorem, then study in detail the examples that follow it, then return to the definition or theorem to appreciate how it applies to the examples. Furthermore, read with pencil in hand and a pad of paper beside you, so that you can draw pictures of the things you are learning about, construct your own examples, and work out details in proofs, derivations, or calculations that are presented.

- Many of the examples in this book are in the form of a problem with a solution. After reading the problem, try to solve it before reading on. There are two advantages to this approach. First, you may be able to solve it, in which case you can be sure that you understand the concepts or methods being discussed. (Read the solution anyway, because it may give you more insight into the problem or describe an interesting alternative approach.) Second, if you have worked on a problem yourself, then it is much easier to get into and appreciate the solution that someone else provides—you are already familiar with the setting, and you may find that your ideas had taken you almost to the solution.

- In mathematics, and especially in this course, you need to pay careful attention to being precise. Just as a computer demands that you write your programs with exactly the correct syntax in order for it to know what you mean, so in mathematics you need to be careful to say exactly what you mean. In particular, you must learn the words we define—not just have a vague idea of what the words mean, but be able to give a rigorous definition. Sometimes words that are very close in form (*maximum* and *maximal*, for instance) mean two different things, and the subtle differences are important. In a similar vein, always be aware of the level of abstraction at which a concept is operating; it is a serious error to let the level become confused in your mind. (For example, there is a big difference between a function and a set of functions.)

- Since learning the definitions of words is so important in this subject, you should make up your own dictionary as you study. The glossaries at the end of each section should be helpful in this regard. Either make your dictionary on 3 by 5 cards, or use a word processor, inserting things into alphabetical order. Write out the definitions precisely, and then make up your own examples to go with them. Often it is useful to give yourself an example of something that does *not* satisfy the definition, as well as something that does.

- Use the index of this book often. In it you will usually find all the pages on which you can find out something useful about a topic. The most important reference (where the topic is discussed most completely) is in boldface type (unless it is the first one). The page on which a formal definition is given in the summary of definitions at the end of each section is shown in italics. References to exercises are indicated with parentheses. As you read, when you come to a concept that you have forgotten, look it up in the index and refresh your memory. You will also find the symbol index on the endpapers (inside covers) of the book useful for quick reference.

- If all of the words of advice had to be condensed into just one item, it would be this one: **do the exercises**. They are the heart of this book. Your learning of discrete mathematics will take place primarily through your working the exercises. The first several exercises in each section are meant to be fairly straightforward and routine, and you should do as many of them as is necessary to make sure you understand the basics. The majority of the exercises, however, are meant to be more than just practice, and you should not approach them as if they were just drill. Do as many as you have time for, even more than your instructor assigns. (Choose ones that look interesting to you.) Spend some time, also, attempting some of the challenging exercises grouped near the end of each exercise set. Even if you do not solve them, the effort will have been worthwhile. Most mathematicians work on many more problems than they ever actually end up solving.

- Think of each exercise as an adventure, in which you are looking into a particular facet of the discrete mathematical concepts at hand. Solve the exercises completely. Do not be satisfied just to get an answer (even the right answer). Make sure that you understand why it is the right answer. Devise ways to check your answer to make sure that it is right (maybe come up with two different ways to solve the same problem). Ask yourself how the problem or the answer might generalize, or what related problems you can ask and answer. Many of the exercises are already laid out to encourage you to do this, with several parts that build on each other. Remember that exercises are not obstacles to be gotten through but rather vehicles to learn from. Squeeze as much from each exercise as you can.

- Write down your answers in a form in which you would wish someone else to read them. Doing mathematics is a form of communication, and it is almost as important to communicate your mathematical ideas well as it is to come up with the ideas. Use complete sentences and well-written explanations; very rarely is a correct solution to an exercise just a string of symbols. In particular, the standing rule for all exercises is that you must **justify your answers**. For example, if the exercise asks you to determine whether every zibble has a nook, then "yes" can never be a correct solution; you must show (i.e., prove) *why* every zibble has a nook. Think of most solutions to exercises in this book as mathematical essays. It is just as important to write a clear and convincing essay in mathematics as it is in literature, history, or political science.

- **Answers** to the odd-numbered exercises (except for the exploratory ones) are given in the back of the book. These are answers, not solutions; they are just there to help you determine whether you have done the problem correctly. **Solutions** to the odd-numbered exercises appear in a supplement, the *Students' Solutions Manual*. There the solutions are as complete as you are expected to write them. It is extremely important that you do not cheat yourself by misusing either the answers in the back of the book or the solutions manual. Under no circumstances should you ever look at an answer or a solution to a problem until you have spent a great deal of time working on it yourself (preferably having worked it all the way to the point of obtaining a final answer and convincing yourself that it is right). What constitutes "a great deal of time" depends on the problem. Certainly you should never "give up" before working on a problem for at least 20 minutes (longer if you seem to be making some progress or if the problem is difficult). If you do not know where to begin, or come to a point where you cannot continue, then go back and review what you know rather than turn to someone else's solution. See if you can find something that can be applied to solve the problem at hand. Expect to try two or more approaches that fail before being able to solve a problem successfully (indeed, mathematical problem solving, even at the highest research levels, consists mostly of discovering that the approach being tried does not work). On challenging exercises allow yourself many hours of concentrated work, spread over several days, before getting help from the answer section or solutions manual. (In real life, mathematicians and computer scientists will often spend months working on the same problem.) You will be amazed as you watch the problem-solving process at work in your brain. The feeling of satisfaction at having obtained a beautiful solution to a difficult intellectual problem is an emotional high that cannot be described to one who has not felt this thrill first-hand.

- If your instructor permits it, there is much to be gained from working with your fellow students on exercises. Each of you will contribute different insights as the solution evolves. After the group of you has satisfied yourselves that you have solved the problem, then each one should go away and write up the complete solution alone. That way you will be sure that you understand it.

There are probably many more bits of advice that could be added to this list. Ask your instructor to provide some, or try to come up with two or three of your own. Review this list every now and then, and make sure that you are making the most of your study of discrete mathematics.

J.W.G.

BRIEF CONTENTS

DETAILED CONTENTS

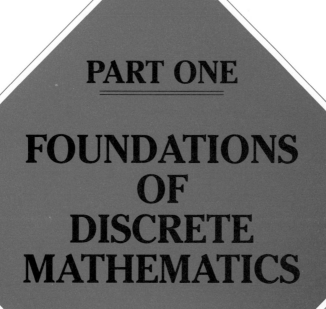

PART ONE

FOUNDATIONS
OF
DISCRETE
MATHEMATICS

LOGIC

Mathematics, unlike most other disciplines, deals with indisputable truth. The basic building blocks of mathematical discourse are declarative sentences, called propositions. One goal of the mathematician is to determine which propositions are true and which are false. Mathematical propositions may deal with integers, functions of a real variable, points and lines in the plane, computer programs, or any of a vast number of other specific areas; but the framework in which propositions are stated, and their truth or falsity determined, is common to all.

The study of propositions—what they mean and how the true ones are proved to be true—is a part of a subject called mathematical logic. In this chapter we will give a brief introduction to the content and symbolism of mathematical logic, enough to serve us throughout the remainder of the book. Section 1.1 deals with propositions and how they are combined to form more complex propositions. In Section 1.2 we study propositions that contain the quantifiers "for every" and "there exists"; these propositions can express ideas too complex to state without such quantifiers. In Section 1.3 we discuss ways to prove that true propositions are true. Finally, in Section 1.4 we look at logic from a more abstract and computational point of view; this material is tangential to the main goals of this book and may comfortably be omitted.

Logic is an indispensable tool of mathematics and computer science. With logic a mathematician can prove that a conjecture is true, and a computer scientist can prove that a computer program will work the way it is intended to work, for all possible inputs. The study of logic dates back at least to ancient Greece, with mathematical logic coming into its own in the late nineteenth century. During the present century many profound insights have been gained, and mathematical logic remains a fruitful area of research at major universities today.

Good general references for mathematical logic and proofs include Enderton [77], Fletcher and Patty [85], Jeffrey [163], Kneebone [173], Mendelson [210], Monk [214], and Suppes [297]. Friedman and Menon [92] and Sloan [274] contain an abundance of material relevant to Section 1.4.

SECTION 1.1
PROPOSITIONS

"When I use a word," Humpty said, in a rather scornful tone, "it means just what I choose it to mean—neither more nor less."

—Lewis Carroll, *Through the Looking Glass*

A **proposition** is a declarative sentence, with a subject and a verb. A proposition may be true or false as it stands, or it may contain variables, called **free variables**, which need to be assigned values before the truth or falsity of the proposition is determined. We assume that these values will be ones that make the proposition meaningful. For instance, in the proposition "x is married," we would assume that the free variable x stands for a human being. The **truth value** of a proposition is **true** (abbreviated T) if the sentence is true and **false** (abbreviated F) if the sentence is false.

EXAMPLE 1. The following are all propositions. For this example, and throughout the book, recall that the **integers** are the numbers $\ldots, -2, -1, 0, 1, 2, \ldots$, and that the **natural numbers** are the nonnegative (i.e., either **positive** or zero) integers (0, 1, 2, ...). A nonzero integer d is said to be a **factor** or **divisor** of an integer x if x is a **multiple** of d, that is, $x = dm$ for some integer m; we also say that x is **divisible** by d in this case. Finally, a **prime number** is a natural number greater than 1 whose only natural number divisors are 1 and itself.

    1. $4 + 5 = 6$.
    2. 7 is a prime number.
    3. 2 is a prime number, but 4 is not.
    4. $x + y$ is divisible by 3.

4

5. Every natural number can be written as the sum of the squares of four natural numbers.
6. Every even number (an integer that is a multiple of 2) greater than 4 can be written as the sum of two prime numbers.
7. A computer running the following computer program will eventually halt:

> **procedure** *halt_or_not*
> $\quad i \leftarrow 1$
> $\quad$ **while** $i < 10$ **do**
> $\quad\quad i \leftarrow i + 1$

8. He is over 6 feet tall.

Note that each proposition shown here has a subject and a verb. For example, in (1) the subject is $4 + 5$ and the verb is $=$ ("equals"). In (2) the subject is 7 and the verb is "is." Proposition 3 is a compound sentence and has two subject–verb pairs. Proposition 7 is the sentence declaring that a computer running a certain program (the subject) will eventually halt (the verb).

It is clear that propositions 2 and 3 are true and (1) is false. Proposition 4 contains free variables (x and y) and hence does not have a truth value until specific values (presumably, integers, so that the notion of divisibility makes sense) are substituted for x and y. If, for example, 2 is substituted for x and 6 for y, then the proposition will be false, since 8 is not a multiple of 3; but if 0 is substituted for both x and y, then the proposition will be true, since $3 \cdot 0 = 0$. Similarly, (8) contains the free variable "he" and may be either true or false, depending on who "he" is; presumably, in this context any living male human being is a legitimate value for "he" to be assigned.

Proposition 5 is a theorem from number theory: The French mathematician J. L. Lagrange proved in the eighteenth century that it is, indeed, true. Proposition 6, on the other hand, is a famous conjecture in number theory, dating back to C. Goldbach in the eighteenth century, whose truth value is still not known. Note that the fact that we imperfect mortals do not *know* whether (6) is true or false does not prevent it from being a proposition; it clearly *is* either true or false. Neither (5) nor (6) contains free variables.

Finally, let us see why proposition 7 is also true. [The program shown here is expressed in a "pseudocode" similar to the programming language Pascal. We discuss our method of writing computer programs in some detail in Section 4.2, but this simple example should be easy to understand once the reader is told that the left-pointing arrow ($\leftarrow$) means that the variable on the left is to be set to the value computed on the right.] The variable i is set equal to 1 initially and then repeatedly incremented by 1 as long as its value remains less than 10. When its value reaches 10, the **while** loop is finished, and since there are no more steps in the program, the computer halts. ◆

Not everything one can say or write down is a proposition. Meaningless strings of words such as "discrete should mathematics" or sentence fragments such as "since $3 < 2$" or phrases such as "$n(n-1)/2$" are not propositions. Questions such as "Is x even?" are interrogative sentences, not declarative ones and hence are not propositions. Program statements such as "$i \leftarrow i + 1$" are not propositions since they are commands (imperative sentences), not assertions (declarative ones). The student of mathematics should get into the habit of speaking—both orally and in writing—in complete mathematical sentences (i.e., in propositions), and pay careful attention to *exactly* what these sentences mean.

In mathematical logic, we study propositions in somewhat the same way that we study real numbers in algebra. We will use letters (variables) to represent propositions, just as we use variables to represent numbers in algebra. We will usually use letters such as P, Q, and R to stand for propositions, and call them **propositional variables**. (Do not confuse propositional variables with the numerical or other variables that may occur *within* a proposition.) Just as numerals stand for constants in algebra (like "25"), so we will use the constants T and F to stand for a constantly true proposition (say, "$1 = 1$") and a constantly false proposition (say, "$1 = 2$"), respectively. Also, just as operations (such as addition) and relations (such as "is less than") allow us to form more complex arithmetic and algebraic expressions and sentences from simpler ones, so we will need to define logical operations with which to combine propositions. Note that algebra has a coherent symbolic notation—a shorthand way of expressing the necessary operations and relations (like $+$ for addition and $<$ for "is less than"). In mathematical logic, too, we will use special symbols for operations and relations. Finally, recall that algebra has certain laws—key statements that capture the way the system behaves, such as the commutative law for addition; we will find (and prove true) similar laws in logic. Propositional logic, studied in this way, is sometimes called the **propositional calculus**.

LOGICAL CONNECTIVES AND COMPOUND PROPOSITIONS

Given two propositions, we can, by using the word "and," make the assertion that they are both true simultaneously. This assertion is again a proposition, a **compound proposition**. For example, "$4 < 7$ and 8 is an even number" is the compound proposition formed from the propositions "$4 < 7$" and "8 is an even number" by combining them with the word "and." Similarly, we can connect these two propositions with the word "or," obtaining the compound proposition "$4 < 7$ or 8 is an even number." A somewhat subtler way to combine these propositions is with the construction "if...then..."; we then obtain the compound proposition "If $4 < 7$, then 8 is an even number." Similarly, we can form a compound proposition by using the word "not"; for example, we can assert that "8 is not an even number" (which happens to be a false proposition).

Each of these four **logical connectives** ("and," "or," "if...then....," and "not") has a meaning, or sometimes one of several meanings, in everyday English. We need to give precise meanings to them in mathematical logic, meanings we hope will correspond in most cases to their everyday ones. The need for precision in the meaning of propositions—and in particular, the interpretation of these logical connectives—cannot be overemphasized. If we are careless as to exactly what a mathematical utterance means, we risk producing ambiguity, mistakes, confusion, or just plain nonsense. Just as computer programming languages have a precise syntax, and each statement in a program has a precise meaning, so too in mathematics, every statement must have a precise meaning.

Since we are concerned with the truth values of propositions (distinguishing between true propositions and false ones), we will define these four connectives solely in terms of the way in which they treat truth values. For example, under what conditions should we define a proposition "P or Q" to be true? We will accomplish these definitions by the use of **truth tables**, which give the truth values of compound propositions for all possible combinations of truth values of the propositional variables that appear in them.

We begin with the simplest connective, whose meaning should not be controversial: "not." If P is a proposition, then the **negation** of P, written $\overline{P}$ or $\neg P$, and read "not P," is false when P is true, and true when P is false (see Table 1.1).

Table 1.1
Truth table
for negation.

P	$\overline{P}$
T	F
F	T

For example, the negation of the false proposition "7 is an even number" is the true proposition "7 is not an even number," which can be written as $\overline{7 \text{ is an even number}}$ or $\neg(7 \text{ is an even number})$. We tend to prefer the notation $\overline{P}$ to the notation $\neg P$.

There should also be no debate about the way the logical connective "and" is defined. If P and Q are propositions, then the **conjunction** of P and Q, written $P \wedge Q$ and read "P and Q," is true when P and Q are both true, and false otherwise (see Table 1.2).

For example, the compound proposition "$4 < 7$ and 8 is an odd number" is false. We see this by consulting the second line of Table 1.2: P ("$4 < 7$") is true and Q ("8 is an odd number") is false.

As used in English, the word "or" is somewhat ambiguous. For example, if a child is offered a chocolate chip cookie *or* a peanut butter cookie, is she allowed

Table 1.2 Truth table for conjunction.

P	Q	$P \wedge Q$
T	T	T
T	F	F
F	T	F
F	F	F

to choose to have both? Mathematicians usually will say that she is; they resolve the ambiguity in favor of the *inclusive* use of the word "or." Thus we make the following definition. If P and Q are propositions, then the **disjunction** of P and Q, written $P \vee Q$ and read "P or Q," is true when either P is true, or Q is true, or both P and Q are true, and false otherwise (see Table 1.3).

Table 1.3 Truth table for disjunction.

P	Q	$P \vee Q$
T	T	T
T	F	T
F	T	T
F	F	F

This inclusive sense of the word "or" is what some people might call "and/or." If P and Q are both true, then $P \vee Q$ is true; the only time $P \vee Q$ is false is when both P and Q are false. For example, the compound proposition "$4 < 7$ or 8 is an even number" is true, as we see by consulting the first line of Table 1.3.

We can form more complex compound propositions by using combinations of logical connectives. Just as in algebra, we use parentheses for grouping, to avoid ambiguity. (In Section 9.3 we look at notations that avoid ambiguity without using parentheses.)

EXAMPLE 2. Let P be the proposition "$x > 2$"; let Q be "$x < 7$"; and let R be "x is odd." We assume in this example that x can take on values that are natural numbers.

(a) Write in symbols the proposition "Either $x < 7$ and $x > 2$, or x is not odd."

(b) Express in English $R \wedge \overline{P \wedge Q}$.

(c) Construct a truth table for $R \wedge \overline{P \wedge Q}$.

Solution.   (a) The given sentence is a disjunction of two propositions, the first of which is the conjunction of Q and P, and the second of which is the negation of R. Thus we write this proposition in symbols as $(Q \wedge P) \vee \overline{R}$ or $(Q \wedge P) \vee (\neg R)$. Note that the English words "either…or" still mean the inclusive "or" in mathematics. Note also that the parentheses are needed to distinguish between the proposition we wrote and the proposition $Q \wedge (P \vee \overline{R})$, whose meaning and truth value for some choices of x are quite different (see Exercise 17).

   (b) The proposition $R \wedge \overline{P \wedge Q}$ is the conjunction of R and the negation of the conjunction of P and Q. Note that the long bar over $P \wedge Q$ already implies a grouping, so that parentheses are not necessary here; it would not be wrong, though, to write this as $R \wedge \overline{(P \wedge Q)}$. We may render this in English as "x is odd, but it is not the case that $2 < x < 7$." Note that "but" is a synonym for "and." Note also that a string of inequalities in mathematics is the conjunction of the inequalities in the string; in this case, "$2 < x < 7$" means "$2 < x$ and $x < 7$." Finally, note that it would be incorrect to render the last part of the sentence as "$x \not> 2$ and $x \not< 7$." This would have been the translation of $\overline{P} \wedge \overline{Q}$, not $\overline{P \wedge Q}$. They mean different things and have different truth tables, as the reader should verify (see Exercise 14).

   (c) The truth table requires eight lines to take care of all the possible truth values for P, Q, and R (see Exercise 13, however). We work from the inside out, using the definitions to find the truth values of $P \wedge Q$, $\overline{P \wedge Q}$, and finally $R \wedge \overline{P \wedge Q}$.

P	Q	R	$P \wedge Q$	$\overline{P \wedge Q}$	$R \wedge \overline{P \wedge Q}$
T	T	T	T	F	F
T	T	F	T	F	F
T	F	T	F	T	T
T	F	F	F	T	F
F	T	T	F	T	T
F	T	F	F	T	F
F	F	T	F	T	T
F	F	F	F	T	F

   We turn next to the most important—and most confusing—of the logical operations. If P and Q are propositions, then the **implication** "P implies Q," also read "If P, then Q" and written $P \rightarrow Q$, is false in case P is true and Q is false, and true in the other three cases (see Table 1.4). We call P the **hypothesis** of the implication, and we call Q the **conclusion**.

   Although there should be no disagreement that our definitions of "not" ($^-$ or $\neg$), "and" ($\wedge$), and (inclusive) "or" ($\vee$) agree with their everyday meanings, this definition of "implies" ($\rightarrow$) seems rather arbitrary and not necessarily in agreement

Table 1.4 Truth table for implication.

P	Q	$P \rightarrow Q$
T	T	T
T	F	F
F	T	T
F	F	T

with nonmathematical usage. We will motivate this definition by appealing to the following example.

EXAMPLE 3. Let P be the proposition "$x < 3$." Let Q be the proposition "$x < 7$." Thus $P \rightarrow Q$ is the proposition "If $x < 3$, then $x < 7$." Now we would all agree that this implication is a true statement: If x is less than 3, then certainly x is also less than 7. Thus we want our definition to show that $P \rightarrow Q$ is given the truth value T for all possible values that might be substituted for the free variable x. If we let $x = 2$, then $P \rightarrow Q$ becomes "If $2 < 3$, then $2 < 7$." Here both P and Q are true. Thus we must define T $\rightarrow$ T to be T, as we have done in the first line of the body of Table 1.4. Next, if we let $x = 5$, then the proposition reads "If $5 < 3$, then $5 < 7$." Here the hypothesis ($5 < 3$) is false, but the conclusion ($5 < 7$) is true. This necessitates the third line of the truth table: F $\rightarrow$ T has to be defined to be T. Finally, if we let $x = 10$, then $P \rightarrow Q$ is "If $10 < 3$, then $10 < 7$." This is of the form F $\rightarrow$ F. Again, since we agreed that the statement was a true one for all values of x, we are forced to define this last proposition to be true, as we have done in the last line of Table 1.4. ◆

Note, then, that we have defined $P \rightarrow Q$ to be true in every case except the case in which P is true and Q is false. As a further rationale for this definition, consider a mother's promise to her child, "If it is sunny tomorrow, then we will go on a picnic." The only case in which we can accuse the mother of lying (not keeping her promise) is that in which the sun shines but they do not go on the picnic.

There are many ways to say, in English, "P implies Q." We have already indicated the most common form, "If P, then Q." Other equivalent forms are "P only if Q"; "P is a sufficient condition for Q"; "Q is a necessary condition for P"; and "Q is a necessary consequence of P." For example, consider the true proposition "If it is raining outside, then there are clouds in the sky." We can render this in any of the following ways:

Rain outside implies clouds in the sky.
It is raining outside only if there are clouds in the sky.
Rain outside is a sufficient condition for there to be clouds in the sky.
Clouds in the sky are a necessary condition for rain.

Note that the order in which we combine the propositions is important. The sentence "If there are clouds in the sky, then it is raining outside" is not true on a dry overcast day; equivalently, having clouds in the sky is not a sufficient condition for rain, nor is rain a necessary condition for (or consequence of) cloudiness.

In mathematics we assign a truth value to statements of the form $P \rightarrow Q$ even if there is no cause and effect relationship between the hypothesis and the conclusion. The implication is true as long as P is false or Q is true (or both— note our inclusive use of the word "or"), regardless of whether the truth of P is responsible for the truth of Q. For example, "If $x = 3$, then $1 + 1 = 2$" is true for all values of x, because the conclusion is true; it is irrelevant that the value of x has no bearing on the truth of "$1 + 1 = 2$."

Although there seems to be symmetry between the propositional variables P and Q in our definitions of conjunction and disjunction, we have noted that the order in which an implication is written influences its meaning and its truth value. The proposition $Q \rightarrow P$ is called the **converse** of the proposition $P \rightarrow Q$. For instance, the converse of the true proposition "If James Garfield is the President of the United States, then the year is 1881" is the false proposition "If the year is 1881, then James Garfield is the President of the United States" (Hayes and Arthur also held that office during portions of 1881).

Sometimes we want to assert both an implication and its converse, and there are many ways to do so. Thus $(P \rightarrow Q) \wedge (Q \rightarrow P)$ could be verbalized as "P implies Q, and conversely" or as "If P, then Q, and conversely" or as "P precisely when Q." The most concise way to express this idea, however, and the one that is used most commonly in mathematical texts (including this one), is "P if and only if Q." For instance, "Nixon is the Vice President of the United States if and only if Eisenhower is the President of the United States" is a true proposition: Both clauses were true from noon on January 20, 1953, to noon on January 20, 1961, and both were false at all other times.

EXAMPLE 4. The following are equivalent ways to express the fact that $x + 3 = 5$ precisely when $x = 2$. Note that this compound proposition is true for all values of x.

> If $x + 3 = 5$, then $x = 2$; and if $x = 2$, then $x + 3 = 5$.
> $x + 3 = 5$ if and only if $x = 2$.
> If $x + 3 = 5$, then $x = 2$, and conversely.
> $x = 2$ is a necessary and sufficient condition for $x + 3 = 5$.
> $x + 3 = 5$ is a necessary and sufficient condition for $x = 2$.   ◆

We formally define a fifth logical connective to embrace this notion. If P and Q are propositions, then the **equivalence** of P and Q, written $P \leftrightarrow Q$ and read "P is equivalent to Q" or "P if and only if Q," is true when P and Q have the same truth value, and false when P and Q have different truth values (see Table 1.5).

Table 1.5 Truth table for equivalence.

P	Q	$P \leftrightarrow Q$
T	T	T
T	F	F
F	T	F
F	F	T

TAUTOLOGIES AND LOGICAL EQUIVALENCE

Some compound propositions are always true, regardless of the truth values of the propositions in them. The simplest example is known in philosophy as the **law of the excluded middle:** $P \vee \overline{P}$ (in words, "P or not P"). Whether P itself is true or false, $P \vee \overline{P}$ is always true, as the reader can easily see by looking at its truth table. A proposition that is always true is called a **tautology**. Similarly, some propositions are always false; they are called **contradictions**. The simplest example of a contradiction is $P \wedge \overline{P}$. The truth table for a tautology, then, will consist of an entire column of T's, and the truth table for a contradiction will consist of an entire column of F's.

EXAMPLE 5. We can verify that $(P \wedge Q) \rightarrow (P \vee \overline{R})$ is a tautology by constructing the following truth table and noting that the last column contains only T's.

P	Q	R	$P \wedge Q$	$P \vee \overline{R}$	$(P \wedge Q) \rightarrow (P \vee \overline{R})$
T	T	T	T	T	T
T	T	F	T	T	T
T	F	T	F	T	T
T	F	F	F	T	T
F	T	T	F	F	T
F	T	F	F	T	T
F	F	T	F	F	T
F	F	F	F	T	T

We can also prove that this proposition is a tautology by the following ad hoc argument. Since an implication is true unless the hypothesis is true and the conclusion is false, let us see under what conditions the hypothesis can be true. Now $P \wedge Q$ is true if and only if both P and Q are true. But if P is true, then by the definition of $\vee$, we know that $P \vee \overline{R}$ is also true. Since we have just shown that the conclusion $P \vee \overline{R}$ is true whenever the hypothesis $P \wedge Q$ is true, we know that the implication $(P \wedge Q) \to (P \vee \overline{R})$ is always true, as desired. ◆

More generally, it often happens that two distinct (usually compound) propositions, say S_1 and S_2, have truth tables that agree, line for line; in other words, for every assignment of truth values to the propositional variables in S_1 and S_2, the truth value of S_1 is the same as the truth value of S_2. In this case we say that the two propositions are **logically equivalent**. We write $S_1 \iff S_2$ to indicate that S_1 and S_2 are logically equivalent.

EXAMPLE 6.   To show that $\overline{P \wedge Q}$ is logically equivalent to $\overline{P} \vee \overline{Q}$ (i.e., that $\overline{P \wedge Q} \iff \overline{P} \vee \overline{Q}$), we note that the last two columns of the following truth table are identical. (Note that we have not explicitly indicated in the table some intermediate calculations that were needed to complete the third and fourth columns—a column showing the truth values of $\overline{P}$, for instance.)

P	Q	$\overline{P \wedge Q}$	$\overline{P} \vee \overline{Q}$
T	T	F	F
T	F	T	T
F	T	T	T
F	F	T	T

Using the alternative notation for negation, we can write this logical equivalence as $\neg(P \wedge Q) \iff (\neg P) \vee (\neg Q)$. ◆

LOGICAL IDENTITIES

In mathematical logic there are also deductive and manipulative ways to verify that two propositions involving propositional variables are logically equivalent, rather than by comparing columns in truth tables, but for our purposes truth table verification is sufficient. Our first theorem (a proposition that we prove to be true) contains a list of important and commonly used logical equivalences (also called **logical identities**). The reader is asked to supply proofs of several of them, by constructing truth tables, as we have done for the first of DeMorgan's laws (part (i) of Theorem 1) in Example 6, or by ad hoc reasoning, as in the second part of Example 5 (see Exercises 28 and 30).

You should think about what each of these logical equivalences means, and make sure that you believe them all. For the most part they are a part of the thought processes that we have grown up with from childhood. Their truth seems to be at the very heart of what we all consider logical thinking (at least when we are being careful). For example, our basic understanding of the meaning of the words "and," "or," and "not" is surely reflected in DeMorgan's laws (part (i) of this theorem): If it is not the case that P and Q both hold, then either P fails to hold or Q fails to hold (or both fail to hold); if it is not the case that either P or Q holds, then P does not hold and Q does not hold.

THEOREM 1.   *The following pairs of propositions are logically equivalent. Here P, Q, and R stand for any propositions, T stands for any true proposition, and F stands for any false proposition.*

(a)   *Associative laws*

$$P \wedge (Q \wedge R) \Longleftrightarrow (P \wedge Q) \wedge R$$
$$P \vee (Q \vee R) \Longleftrightarrow (P \vee Q) \vee R$$

(b)   *Commutative laws*

$$P \wedge Q \Longleftrightarrow Q \wedge P$$
$$P \vee Q \Longleftrightarrow Q \vee P$$

(c)   *Distributive laws*

$$P \wedge (Q \vee R) \Longleftrightarrow (P \wedge Q) \vee (P \wedge R)$$
$$P \vee (Q \wedge R) \Longleftrightarrow (P \vee Q) \wedge (P \vee R)$$

(d)   *Identity laws*

$$P \wedge T \Longleftrightarrow P$$
$$P \vee F \Longleftrightarrow P$$

(e)   *Dominance laws*

$$P \wedge F \Longleftrightarrow F$$
$$P \vee T \Longleftrightarrow T$$

(f)   *Idempotent laws*

$$P \wedge P \Longleftrightarrow P$$
$$P \vee P \Longleftrightarrow P$$

(g)   *Complement laws*

$$P \wedge \overline{P} \Longleftrightarrow F$$
$$P \vee \overline{P} \Longleftrightarrow T$$

(h)   *Double negative law*

$$\overline{\overline{P}} \Longleftrightarrow P$$

(i)   *DeMorgan's laws*

$$\overline{P \wedge Q} \Longleftrightarrow \overline{P} \vee \overline{Q}$$
$$\overline{P \vee Q} \Longleftrightarrow \overline{P} \wedge \overline{Q}$$

(j)   *Implication*

$$P \to Q \Longleftrightarrow \overline{P} \vee Q$$

(k) *Negation of implication* $\overline{P \to Q} \iff P \wedge \overline{Q}$

(l) *Contrapositive* $P \to Q \iff \overline{Q} \to \overline{P}$

(m) *If and only if* $P \leftrightarrow Q \iff (P \to Q) \wedge (Q \to P)$

As we employ propositional logic when talking about the material in this book, we will be using implicitly many of these identities. For example (and a very important one), we may express a certain fact as an implication $(P \to Q)$, and then use the **contrapositive** of the implication $(\overline{Q} \to \overline{P})$ in an application. Since these two propositions are logically equivalent (part (l) above), such use is justified. For instance, suppose that we conducted a thorough search of an unabridged dictionary and found that if a string of letters is an English word, then that string does not contain three consecutive a's. It follows that if a string of letters *does* contain three consecutive a's, then that string is not an English word. Here P is the proposition "the string is an English word," and Q is the proposition "the string does not contain three consecutive a's" (so that $\overline{Q}$ is the proposition "the string does contain three consecutive a's").

EXAMPLE 7. Because of the associative law for conjunction (the first half of part (a) of Theorem 1), we may omit parentheses in a string of three or more "and" operations. The truth value of $P \wedge Q \wedge R$ is the same, whether we interpret it as $P \wedge (Q \wedge R)$ or as $(P \wedge Q) \wedge R$. Furthermore, because of the commutative law, we can reorder the propositions in the conjunction without changing the truth value of the conjunction. For instance, $P \wedge Q \wedge R$ is logically equivalent to $Q \wedge P \wedge R$. The corresponding statements for "or" also hold. (This last remark is an instance of a general phenomenon called "duality"; it is discussed in Exercise 37.) ◆

LOGICAL IMPLICATIONS

We conclude this section by introducing a notion somewhat weaker than, but just as important as, the notion of logical equivalence: logical implication. As we will see in Section 1.3, the notion of logical implication is what allows us to construct valid proofs. Let S_1 and S_2 be (usually compound) propositions. Then S_1 is said to **logically imply** S_2, written $S_1 \implies S_2$, if every assignment of truth values to the propositional variables in S_1 and S_2 that results in S_1 being true also results in S_2 being true.

EXAMPLE 8. To see that $P \wedge (P \to Q) \implies Q$, we construct the following truth table and compare the column headed $P \wedge (P \to Q)$ to the column headed Q.

P	Q	$P \to Q$	$P \wedge (P \to Q)$
T	T	T	T
T	F	F	F
F	T	T	F
F	F	T	F

Now $P \wedge (P \to Q)$ is true only in the case in which P is true and Q is true (line 1), and in this case Q is indeed also true. Note that $P \wedge (P \to Q)$ is not logically equivalent to Q, however, since in one case (line 3), $P \wedge (P \to Q)$ is false, but Q is true. ◆

We collect the following cases of logical implication for future use. Again we leave their verification, by truth tables or ad hoc arguments, to the reader (Exercises 29 and 30).

THEOREM 2. *The following logical implications hold. Here P, Q, and R stand for any propositions,* T *stands for any true proposition, and* F *stands for any false proposition.*

$$(a) \qquad P \wedge (P \to Q) \Longrightarrow Q$$

$$(b) \qquad P \Longrightarrow P \vee Q$$

$$(c) \qquad P \wedge Q \Longrightarrow P$$

$$(d) \qquad \overline{P} \to P \Longrightarrow P$$

$$(e) \qquad \overline{P} \to F \Longrightarrow P$$

$$(f) \qquad F \Longrightarrow P$$

$$(g) \qquad P \Longrightarrow T$$

$$(h) \qquad (P \to Q) \wedge (Q \to R) \Longrightarrow P \to R$$

Astute readers may have noticed that the logical connective $\to$ (which we called "implication") and the notion of logical implication (which we denoted by $\Longrightarrow$) are intricately related, as are equivalence ($\leftrightarrow$) and logical equivalence ($\Longleftrightarrow$). We make this precise in the following theorem.

THEOREM 3. *Let P and Q stand for any propositions.*

(a) $P \Longrightarrow Q$ if and only if the proposition $P \to Q$ is a tautology.
(b) $P \Longleftrightarrow Q$ if and only if the proposition $P \leftrightarrow Q$ is a tautology.

Proof. We will indicate why part (a) is true and leave part (b) to the reader (Exercise 35). In a sense we are getting ahead of ourselves, because we will not be formally discussing proofs until Section 1.3. As we will see there, however, a proof is really nothing more than a convincing argument that a proposition is true, and that is what we will give now for statement (a).

There are two things to prove here, the "if" part and the "only if" part of the sentence. First we consider the "only if" part; in other words, we show that $P \Longrightarrow Q$ implies that $P \rightarrow Q$ is a tautology. By definition $P \Longrightarrow Q$ means that every assignment of truth values to the propositional variables in the propositions P and Q that results in P being true also results in Q being true. For no assignment of truth values to the propositional variables, then, can it happen that P is true and Q is false. But by the definition of implication (Table 1.4), this means that $P \rightarrow Q$ will be true for every assignment of truth values; in other words, $P \rightarrow Q$ is a tautology.

Finally, we prove the "if" part of our sentence, namely that if $P \rightarrow Q$ is a tautology, then $P \Longrightarrow Q$. Now if $P \rightarrow Q$ is a tautology, then it must be true for every assignment of truth values to its propositional variables. In particular, there can be no assignment for which P is true and Q is false (see Table 1.4). But this is precisely the definition that $P \Longrightarrow Q$. ∎

One of the themes of this book is that mathematical concepts operate at various *levels of abstraction*, and it is important for the reader to keep the levels straight. We see this theme illustrated in part (b) of Theorem 3. The notion of equivalence is being used at three distinct levels there. It is being asserted that the logical connective "is equivalent to" (first level) and the notion of "is logically equivalent to" (second level) are in some sense equivalent mathematically (third and highest level).

SUMMARY OF DEFINITIONS

compound proposition: a proposition formed by combining other propositions with logical connectives (e.g., $P \vee \overline{Q}$).

conclusion of the implication $P \rightarrow Q$: Q.

conjunction of two propositions P and Q: the proposition $P \wedge Q$, read "P and Q," which is true if both P and Q are true, and false if at least one of P and Q is false.

contradiction: a proposition that is false for all possible assignments of truth values to the propositional variables appearing in it (e.g., $P \wedge \overline{P}$)

contrapositive of the implication $P \rightarrow Q$: the implication $\overline{Q} \rightarrow \overline{P}$; it is logically equivalent to the original implication.

converse of the implication $P \rightarrow Q$: the implication $Q \rightarrow P$.

disjunction of two propositions P and Q: the proposition $P \vee Q$, read "P or Q," which is true if at least one of P and Q is true, and false if both P and Q are false; thus "or" is being used in the inclusive sense.

divisible: An integer x is divisible by a nonzero integer y if y is a divisor of x (e.g., 91 is divisible by 7).

divisor of the integer x: a nonzero integer y such that x is a multiple of y (e.g., 7 is a divisor of 91).

equivalence of two propositions P and Q: the proposition $P \leftrightarrow Q$, read "P if and only if Q," which is true if both P and Q are true or both P and Q are false, and false if P and Q have opposite truth values.

factor of the integer x: a divisor of x.

false: truth value for a proposition such as "$1 + 1 = 3$."

free variable: a variable appearing in a proposition about which the proposition makes a statement (e.g., x in "$x^2 - 4 = 0$").

hypothesis of the implication $P \to Q$: P.

implication: a proposition $P \to Q$, read "if P, then Q" or "P implies Q," which is false if P is true and Q is false, and true in the other three cases.

integers: the positive and negative whole numbers and zero ($0, 1, -1, 2, -2, \dots$).

law of the excluded middle: $P \vee \overline{P}$ is a tautology.

logical connective: a word or words used to combine propositions into compound propositions, including "and," "or," "not," "if...then...," and "if and only if."

logical equivalence: Two propositions S_1 and S_2 are logically equivalent (written $S_1 \Longleftrightarrow S_2$) if for every assignment of truth values to the propositional variables in them, the truth value of S_1 is the same as the truth value of S_2 (e.g., $P \to Q \Longleftrightarrow \overline{P} \vee Q$).

logical identity: a logical equivalence.

logical implication: A proposition S_1 logically implies a proposition S_2 (written $S_1 \Longrightarrow S_2$) if every assignment of truth values to the propositional variables in them that makes S_1 true also makes S_2 true (e.g., $P \wedge Q \Longrightarrow P \vee Q$).

multiple of the integer x: an integer of the form mx for some integer m (e.g., 91 is a multiple of 7, since $91 = 13 \cdot 7$).

natural numbers: the nonnegative integers, that is, the positive integers and 0 ($0, 1, 2, 3, \dots$).

negation of the proposition P: the proposition $\overline{P}$ (also written $\neg P$), read "not P," which is false if P is true, and true if P is false.

positive number: a number strictly greater than 0 (e.g., the positive integers are $1, 2, 3, \dots$).

prime number: a natural number greater than 1 whose only natural number factors are 1 and itself (e.g., 17 is prime, but $18 = 3 \cdot 6$ is not).

proposition: a declarative sentence (e.g., "p is a prime number").

propositional calculus: the study of logical propositions and the ways in which they can be combined using logical connectives.

propositional variable: a letter such as P or Q used to stand for any proposition.

tautology: a proposition that is true for all possible assignments of truth values to the propositional variables appearing in it (e.g., $P \vee \overline{P}$).

true: truth value for a proposition such as "$1 + 1 = 2$."

truth table: a table giving the truth values of a compound proposition for all possible assignments of truth values to the propositional variables appearing in the proposition (e.g., Table 1.5).

truth value of a proposition: true if the proposition expresses a true statement, false if the proposition expresses a false statement.

EXERCISES

1. Determine which of the following are propositions.
    (a) $x^2 - 4 = 0$.
    (b) 6 is divisible by 7.
    (c) Divide 26 by 2.
    (d) p is a prime number.
    (e) $x = x + |y| + 1$.
    (f) Some prime number is even.
    (g) He was the Democratic party candidate for Vice President in 1984.
    (h) Did George Bush win the 1984 election for Vice President?
    (i) George Washington was over 5 feet tall.
    (j) The sum of all the integers from 1 to n.
    (k) 121 is a perfect square.
    (l) 121 is equal to N^2.

2. Identify all free variables in the propositions in Exercise 1.

3. Determine the truth value of each proposition in Exercise 1 that does not have free variables.

4. Determine whether each of these propositions is true or false.
    (a) $4 < 7$ or $3 < 2$.
    (b) If 6 is divisible by 7, then 7 is divisible by 6.
    (c) 7 is not divisible by 3.
    (d) 7 is prime, and 4 is not prime.
    (e) If 7 is prime, then 100 is not prime.
    (f) If 100 is prime, then 7 is prime.
    (g) If 100 is prime, then 7 is not prime.

5. Let P be "$4 < 7$"; let Q be "13 is prime"; and let R be "Paris is the capital of Spain." Translate the following propositions into English.
    (a) $\overline{R}$.
    (b) $P \vee Q$.
    (c) $P \to (Q \wedge R)$.
    (d) $\overline{P} \vee \overline{Q}$.
    (e) $\overline{P \wedge Q}$.
    (f) $(P \to Q) \vee (Q \to R)$.

(g)   The conjunction of Q and R.

(h)   The negation of P.

(i)   The disjunction of Q and the negation of R.

6. Determine the truth value of each proposition in Exercise 5.

7. Using truth tables, show that the following propositions are tautologies.

   (a) $(P \wedge (P \to Q)) \to Q$.

   (b) $(P \to Q) \leftrightarrow (Q \vee \overline{P})$.

   (c) $(P \wedge \overline{P}) \to Q$.

8. Rewrite the following propositions using each of the following forms: "if...
    then...," "only if," "is a sufficient condition for," and "is a necessary condition
    for."

   (a)   The car's being out of gas implies that it will not run.

   (b) $x < 3$ implies $x < 7$.

9. For each proposition in Exercise 1 that does have free variables, find values
    of the variables that make the proposition true and find values of the variables
    that make the proposition false; or explain why this cannot be done.

10. Let P be "$4 < 7$"; let Q be "13 is prime"; and let R be "Paris is the capital
    of Spain." Translate the following propositions into symbols.

   (a)   If Paris is the capital of Spain, then $4 < 7$.

   (b) $4 \geq 7$, and 13 is prime.

   (c)   Either 13 is not prime, or Paris is the capital of Spain.

   (d)   Either 13 is not prime, or Paris is the capital of Spain, but not both.

   (e)   13 is prime only if $4 < 7$.

   (f)   A necessary condition for 13 to be prime is that $4 < 7$.

   (g)   A sufficient condition for 13 to be prime is that $4 < 7$.

   (h)   A necessary and sufficient condition for 13 to be prime is that $4 < 7$.

   (i)   Paris is the capital of Spain implies that 13 is prime.

11. Determine the truth value of each proposition in Exercise 10.

12. Consider the statement "If it's cold and rainy, then we won't go swimming."

   (a)   Express this proposition in symbols, letting C, R, and S stand for
         the obvious simple propositions here whose key words start with these
         letters.

   (b)   Make a truth table for the proposition obtained in part (a).

   (c)   Assume that this proposition is true. Determine under how many dif-
         ferent conditions we can go swimming.

13. In Example 2 we listed eight possible combinations of truth values for P,
    Q, and R. Are they all really possible in this case? For each of the eight
    possibilities, either find a natural number that can be substituted for x so that
    that particular set of truth values applies, or show that no such number exists.

14. Construct truth tables for each of the following compound propositions.
    (a) $\overline{P} \wedge \overline{Q}$.
    (b) $\overline{P \wedge Q}$.
    (c) $(P \wedge Q) \rightarrow R$.
    (d) $((P \wedge Q) \rightarrow R) \rightarrow (P \wedge \overline{Q})$.

15. Classify each of the following propositions as a tautology, a contradiction, or neither. Note that if you claim that a proposition is a tautology, then you must argue (by using truth tables or otherwise) that it is true for every assignment of truth values to the propositional variables; if you claim that it is a contradiction, then you must argue (by using truth tables or otherwise) that it is false for every assignment of truth values to the propositional variables; and if you claim that it is neither a tautology nor a contradiction, then you must find an assignment of truth values to the propositional variables that makes it true and another assignment that makes it false.
    (a) $P \rightarrow \overline{P}$.
    (b) $P \rightarrow P$.
    (c) $(P \wedge Q) \rightarrow (P \vee Q)$.
    (d) $\overline{P} \vee (P \rightarrow Q)$.
    (e) $P \wedge \neg(P \vee Q)$.
    (f) $(P \vee Q) \rightarrow P$.
    (g) $(P \vee Q) \wedge (\overline{P} \vee \overline{Q})$.
    (h) $(P \wedge Q) \vee (\overline{P} \wedge \overline{Q})$.

16. Consider the proposition $P \rightarrow (P \wedge Q)$.
    (a) Construct the truth table for this proposition.
    (b) Find a simpler proposition logically equivalent to this proposition.

17. Show that $(Q \wedge P) \vee \overline{R}$ is not logically equivalent to $Q \wedge (P \vee \overline{R})$.

18. Show that $(P \vee Q) \wedge (P \rightarrow Q) \Longrightarrow Q$.

19. Show that $(P \rightarrow Q) \wedge (\overline{P} \rightarrow Q)$ logically implies Q. Explain this fact in nontechnical, everyday terms.

20. Determine whether $(P \wedge Q) \rightarrow R \Longrightarrow (P \rightarrow R) \wedge (Q \rightarrow R)$.

21. Determine whether the proposition $(P \rightarrow (Q \rightarrow R)) \leftrightarrow (\overline{P \wedge Q} \vee R)$ is a tautology.

22. "Clean up your room or you'll not get any dessert today," said the father to the child. Let P be "The child cleans up her room," and let Q be "The child gets dessert."
    (a) Write the father's statement in symbols, and explain how his statement can be viewed as a proposition, even though it is being used as a command.

(b)   Rewrite this proposition in symbols as a logically equivalent implication.

(c)   Translate the implication in part (b) into English in three different ways that make sense in this context.

23. A lifeguard at a swim club was heard to make the following statement to some 8-year-olds swimming in the baby pool: "To swim in this pool you need to be less than 6 years old and your parent has to be present. If you are not less than 6 years old and your parent isn't here, then you may not swim in this pool."

(a)   Render each of the lifeguard's two statements symbolically as implications (your answers should have three propositional variables).

(b)   Show that the lifeguard's first statement logically implies the second, but is not (contrary to what she probably thought) logically equivalent to it.

(c)   What did the lifeguard really mean to say for her second sentence? Use Theorem 1 to show that the first statement and the intended second statement are logically equivalent.

24. The following proposition is true for all values of the (implied) free variable: "Being at least 35 years old is a necessary condition for serving as President of the United States."

(a)   Formulate this proposition as an implication using the word "if."

(b)   State the converse of the proposition in part (a); determine whether the converse is always true.

(c)   State the contrapositive of the proposition in part (a); determine whether the contrapositive is always true.

25. Let P be the following proposition about real numbers: "For x and y both to be positive, it is necessary that $x \cdot y$ be positive."

(a)   Determine whether P is true for all possible values of the free variables.

(b)   Rewrite P using the word "sufficient."

(c)   State the converse of P in two different ways, once using the word "necessary" and once using the word "sufficient."

(d)   Determine whether the converse of P is true for all possible values of the free variables.

26. The following proposition (which we will encounter as Theorem 3 in Section 4.5) is true for all positive integers p: "That p is an odd prime number is a sufficient condition for the number $2^{p-1} - 1$ to be divisible by p."

(a)   Rewrite this proposition as an implication using the word "if."

(b)   Verify that this proposition is true for $p = 7, 3, 6$, and 2.

(c)   Since 101 is prime, can we conclude that $2^{100} - 1$ is divisible by 101? Explain.

(d)   Since 341 is not prime, can we conclude that $2^{340} - 1$ is not divisible by 341? Explain.

27. Consider the following proposition, which is true for all natural numbers: "A natural number is divisible by 3 if and only if the sum of its digits is divisible by 3."
    (a) Verify that this proposition is true for 42548 and for 121551.
    (b) Rewrite this proposition in English using the words "necessary" and/or "sufficient."

28. Using truth tables, verify the following parts of Theorem 1.
    (a) The associative law for $\vee$.
    (b) The second distributive law.
    (c) The second of DeMorgan's laws.
    (d) The logical equivalence between the negation of an implication and the conjunction of the hypothesis and the negation of the conclusion.
    (e) The logical equivalence between an implication and its contrapositive.

29. Using truth tables, verify Theorem 2.

30. Use an ad hoc prose argument (as in the second half of Example 5) to verify the following parts of Theorems 1 and 2.
    (a) $P \wedge (Q \vee R) \iff (P \wedge Q) \vee (P \wedge R)$.
    (b) $P \implies P \vee Q$.
    (c) $(P \to Q) \wedge (Q \to R) \implies P \to R$.

31. Determine which, if any, of the logical implications given in Theorem 2 are actually logical equivalences.

32. Comment on the logical structure and meaning of the following sentence: "If you win that contest, I'll be a monkey's uncle."

33. Construct compound propositions involving variables P and Q satisfying each of the following conditions.
    (a) The proposition is true if and only if P and Q have different truth values.
    (b) The proposition is true if and only if P and Q are both false.
    (c) The proposition is true if and only if at least one of P and Q is false.

34. Construct compound propositions involving variables P, Q, and R satisfying each of the following conditions.
    (a) The proposition is true if and only if exactly one of P, Q, and R is true.
    (b) The proposition is true if and only if exactly two of P, Q, and R are true.
    (c) The proposition is true if and only if P is false or at least one of Q and R is true.
    (d) The proposition is true if a majority of the three propositions are true.

35. Explain why part (b) of Theorem 3 is true.

36. Determine whether the following statement about logical implication and logical equivalence is true, where P and Q are any propositions.

$$P \iff Q \text{ if and only if both } P \implies Q \text{ and } Q \implies P.$$

37. The **dual** of a proposition involving only propositional variables, $\wedge$, $\vee$, $^-$, and the constants T and F is obtained by replacing every occurrence of $\wedge$ with $\vee$, every occurrence of $\vee$ with $\wedge$, every occurrence of T with F, and every occurrence of F with T.

    (a)   Find the dual of $(P \vee Q) \wedge R$ and the dual of $\overline{P \wedge Q} \vee (F \vee Q)$.

    (b)   Explain why the dual of the dual of S is S for every compound proposition S of this form.

    (c)   It is a theorem of mathematical logic that if the duals of compound propositions S_1 and S_2 are defined, then S_1 is logically equivalent to S_2 if and only if the dual of S_1 is logically equivalent to the dual of S_2. Notice that each of the parts of Theorem 1 with two equivalences contains such a dual pair. Find the duals of the logically equivalent propositions $(P \wedge Q) \vee (P \wedge \overline{Q})$ and P, and verify that both the originals and the duals are indeed logically equivalent.

Challenging Exercises

38. Write "P unless Q" using the symbols we have defined. In essence, this exercise is asking for a mathematical (truth table) definition of the word "unless."

39. A logical connective called **nand** and denoted by $\uparrow$ is defined by the following truth table.

P	Q	$P \uparrow Q$
T	T	F
T	F	T
F	T	T
F	F	T

    (a)   Construct a proposition logically equivalent to $\overline{P}$ using only the connective $\uparrow$.

    (b)   Construct a proposition logically equivalent to $P \vee Q$ using only the connective $\uparrow$.

    (c)   Construct a proposition logically equivalent to $P \wedge Q$ using only the connective $\uparrow$.

40. A logical connective called **nor** and denoted by $\downarrow$ is defined by the following truth table.

P	Q	$P \downarrow Q$
T	T	F
T	F	F
F	T	F
F	F	T

(a) Construct a proposition logically equivalent to $\overline{P}$ using only the connective $\downarrow$.

(b) Construct a proposition logically equivalent to $P \vee Q$ using only the connective $\downarrow$.

(c) Construct a proposition logically equivalent to $P \wedge Q$ using only the connective $\downarrow$.

41. A set of logical connectives is called **complete** if, given any truth table involving any number of propositional variables, we can construct a compound proposition that has the given truth table and uses only the connectives in the set.

(a) Explain why the set $\{\wedge, \vee, ^-\}$ is a complete set of logical connectives. [*Hint*: For each line of the truth table with a T as the desired truth value, construct, using only conjunction and negation, a proposition that is true for only that combination of truth values of the propositional variables; then take an appropriate disjunction.] The proposition constructed in this manner is said to be in **disjunctive normal form**; this topic is discussed further in Section 1.4.

(b) Using part (a) and DeMorgan's laws, explain why each of the sets $\{\wedge, ^-\}$ and $\{\vee, ^-\}$ is complete.

(c) Using part (b) and Exercises 39 and 40, explain why each of the sets $\{\uparrow\}$ and $\{\downarrow\}$ is complete.

(d) Construct a proposition logically equivalent to $P \rightarrow Q$ using only the logical connective "nand."

(e) Construct a proposition logically equivalent to $P \rightarrow Q$ using only the logical connective "nor."

Exploratory Exercises

42. If the plus sign in proposition 7 of Example 1 is replaced by a minus sign, will the proposition be true or false? Explain fully. (Either answer may be considered correct here, depending on the assumptions you make. This exercise illustrates the fact that mathematics is often not totally clear cut and can generate interesting discussion.)

43. Consider the following sentences. They appear to be propositions, since they appear to be declarative sentences. Are they, or are they meaningless jumbles

of words? Discuss. [*Hint*: If a proposition contains no free variables, then it must be either true or false, but not both.]
(a)   This sentence is false.
(b)   This sentence is true.
(c)   Every rule has an exception.

44. Some prime numbers can be expressed as the sum of two squares. For example, $13 = 3^2 + 2^2$. Some, such as 7, cannot.
(a)   Experiment until you can formulate a conjecture as to which primes can be so expressed.
(b)   For a discussion of this problem, consult Frei [90].

45. Investigate (perhaps with a computer) the problem of expressing natural numbers as the sums of cubes of natural numbers. For example, $10 = 2^3 + 1^3 + 1^3$.
(a)   What is the largest number of cubes that seems to be required?
(b)   Find out more about expressing natural numbers as sums of powers of natural numbers (this is known as Waring's problem) by consulting Gardner [95] and Small [277].

46. Find out about logical puzzles and paradoxes by consulting Gardner [110] and Smullyan [281–283].

47. Find out about the master of logical puzzles, Raymond Smullyan (see Exercise 46), by consulting Albers and Alexanderson [4].

48. Find out about the importance of mathematical logic in artificial intelligence by consulting Genesereth and Nilsson [113].

49. Find out about Joseph Lagrange (see proposition 5 of Example 1) by consulting Bell [15].

SECTION 1.2
LOGICAL QUANTIFIERS

You may fool all the people some of the time;
you can even fool some of the people all the time,
but you can't fool all of the people all the time.

—Abraham Lincoln

The propositional logic discussed in Section 1.1 is inadequate for expressing all the propositions we need to deal with in mathematics and computer science. Something as simple as "Every nonzero real number has a reciprocal" contains in its logical structure more than simple propositions and logical connectives. To express propositions like this, we need the notion of quantified variables, and they are the subject of this section.

Recall that a proposition with one or more free variables is neither true nor false until values are substituted for the free variables. Consider the proposition "If $x \neq 0$, then $xy = 1$." If we substitute 2 for x and $\frac{1}{2}$ for y, then we obtain the true proposition "If $2 \neq 0$, then $2 \cdot \frac{1}{2} = 1$." Similarly, if we substitute 0 for x and any number at all for y, then the proposition becomes a true one (since the hypothesis is false in that case). On the other hand, if we substitute 2 for x and 4 for y, then we obtain the false proposition "If $2 \neq 0$, then $2 \cdot 4 = 1$."

We substituted real numbers for x and y in this proposition. We did not, for instance, try to let x equal King Henry VIII or y equal an equilateral triangle 1 inch on a side. We were discussing the issue of reciprocals of real numbers, so we had the implicit understanding that the variables were to range over the collection (or set) of real numbers. (We discuss sets more fully in Chapter 2.) In other contexts, the variables might range over people or geometric figures, in which case substitutions of Henry VIII or a triangle for a variable would be perfectly legitimate. When we consider propositions with variables, then, we will always have in mind a **domain under discussion** for the variables to range over; we will usually make this domain explicit.

Let us return to our example. Suppose that what we actually wanted this proposition to express was the idea that every nonzero real number has a reciprocal. Let us analyze what this really means. To say that a number x has a reciprocal is to say, "There is a number y such that $xy = 1$." The most important words in this sentence are "there is"; this proposition is asserting the *existence* of a certain number (one whose product with x yields the number 1).

We will use a special symbol to indicate this idea of mathematical existence. Suppose that P is a proposition with a free variable v. Then $\exists v \colon P$ is a proposition, read "there exists a v such that P" or "for some v, P." The variable v is called a **bound** or **quantified** (as opposed to free) **variable** in $\exists v \colon P$. The proposition $\exists v \colon P$ is true if P is true for *at least one* value of v chosen from whatever domain is under discussion; it is false if P is false no matter what value from this domain is substituted for v. The symbol $\exists$ is called the **existential quantifier**, and $\exists v \colon P$ is said to be an **existentially quantified proposition**. (Some authors omit the colon from the notation.)

EXAMPLE 1. Suppose that P is the proposition $x^2 + 3x + 2 = 0$. Note that P has a free variable x, and the truth value of P depends on x. If $x = 2$, then P is false, since $2^2 + 3 \cdot 2 + 2 \neq 0$. If $x = -2$, then P is true, since $(-2)^2 + 3 \cdot (-2) + 2 = 0$. Let us assume that we are talking about integer values of x; in other words, integers are the domain under discussion. We can existentially quantify P, obtaining the proposition $\exists x \colon x^2 + 3x + 2 = 0$, which asserts the existence of a solution to a particular quadratic equation. This proposition contains no free variables, so it must be either true or false as it stands. In fact it is a true proposition, since there does indeed exist an integer x (there are two of them, in fact) such that $x^2 + 3x + 2 = 0$.

If the domain under discussion were the positive real numbers, then the proposition $\exists x: x^2 + 3x + 2 = 0$ would be false, since there does not exist any positive real number satisfying the equation (the left-hand side is always bigger than 2 if x is positive).   ◆

Propositions with free variables can be thought of as functions of these variables, and we sometimes write them to suggest this interpretation. Thus $P(v)$ is a statement about v. For example, we might write the proposition $x^2 + 3x + 2 = 0$ as $P(x)$. Then we would write $P(-5)$ for the proposition $(-5)^2 + 3 \cdot (-5) + 2 = 0$, which happens to be false, and $P(-2)$ for the true proposition $(-2)^2 + 3 \cdot (-2) + 2 = 0$. Also, $P(t)$ is $t^2 + 3t + 2 = 0$, so that $\exists t: P(t)$ is another way to assert the existence of a solution to the quadratic equation discussed in Example 1. As another example, we might write $F(x, y)$ for the proposition "x is the father of y"; then $F(\text{Henry VIII}, \text{Elizabeth I})$ is the true proposition about English monarchs "Henry VIII is the father of Elizabeth I," and $F(\text{Elizabeth I}, \text{Henry VIII})$ is a false proposition.

Let us now look at a more complicated example.

EXAMPLE 2.   The proposition "100 can be written as the sum of two even integers" can be expressed in symbols as follows, assuming that the domain under discussion is integers:

$$\exists x: \exists y: (x + y = 100 \,\wedge\, (\exists a: x = 2a) \,\wedge\, (\exists b: y = 2b)).$$

Let us see why. To say that 100 can be written as the sum of two even integers is to say that *there exist* even integers—let us call them x and y—whose sum is 100. Furthermore, to say that x is even is to say that it is a multiple of 2 (this is the *definition* of an even number—we will have more to say about even and odd numbers in Section 1.3); in other words, x is equal to $2a$ for *some* number a. In the same way, we can express the statement that y is even by $\exists b: y = 2b$. Since we want all three conditions to hold—$x + y = 100$ *and* x is even *and* y is even—we join these propositions with the logical connective "and." Thus the proposition displayed above is the desired symbolic translation. We should note that the proposition is true, since such x, y, a, and b do in fact exist: for example, $x = 20$, $y = 80$, $a = 10$, and $b = 40$; or $x = y = 50$, $a = b = 25$. The latter solution illustrates the important point that *distinct variables need not receive distinct values*.

What we have written is certainly not the only way, nor even the simplest way, to express the given proposition in logical notation. Exercise 20 asks for a simplification.   ◆

Once a variable has been quantified, it no longer has an existence of its own in the proposition. The proposition in Example 2 does not speak about variables

x and y. Indeed, x and y can be thought of merely as **dummy variables** in that proposition. The proposition could be written equivalently as

$$\exists w\colon \exists a\colon (w + a = 100 \ \wedge \ (\exists q\colon w = 2q) \ \wedge \ (\exists q\colon a = 2q)).$$

(Note that the two q's here do not need to be the same number.) The situation is analogous to that of dummy variables in definite integrals. The expression $\int_1^2 \sqrt{x^3 + 1}\,dx$ has nothing to do with x; it represents a number—the area under a certain curve. This number is not a function of x, and $\int_1^2 \sqrt{t^3 + 1}\,dt$ represents exactly the same number.

Returning to our original motivating example, where we wanted to express the idea that x had a reciprocal, we can write $\exists y\colon xy = 1$ and read this as "There exists a y such that x times y equals 1." This proposition does speak about a variable x (i.e., x is a free variable in this proposition), but it does not speak about a variable y (the bound, quantified variable y is just a dummy variable being used to make a statement about the existence of a reciprocal for x). Now if we combine this proposition with the restriction we want to impose on x, then we have the compound proposition $x \neq 0 \rightarrow \exists y\colon xy = 1$, which we would read, "If x is not equal to 0, then there exists a y such that x times y equals 1." It is important to use the English language with precision when stating mathematical sentences such as this, and to understand what all the words mean—especially the logical words like "if," "there exists," and "such that."

Unfortunately, we still have not expressed in our proposition the sentiment originally intended, that *every* nonzero real number has a reciprocal, because we have not yet incorporated the concept of "every"; the proposition still speaks about a particular (and unspecified) x. We need another logical quantifier, to whose definition we now turn.

Suppose that P is a proposition with a free variable v. Then $\forall v\colon P$ is a proposition, read "for every v, P" or "for each v, P" or "for all v, P." The variable v is a bound or quantified variable in $\forall v\colon P$. The proposition $\forall v\colon P$ is true if P is true for *every* value of v chosen from whatever domain is under discussion; it is false if P is false for at least one value from this domain. The symbol $\forall$ is called the **universal quantifier**, and $\forall v\colon P$ is said to be a **universally quantified proposition**.

We can now complete our ongoing example. To express the fact that every nonzero real number has a reciprocal, we universally quantify the proposition as we left it above, obtaining

$$\forall x\colon (x \neq 0 \rightarrow \exists y\colon xy = 1).$$

The proper way to read this proposition is "For every x, if x is not equal to 0, then there exists a y such that x times y equals 1." Here we are assuming that the domain under discussion is the real numbers: The quantified variables (or, more simply, the quantifiers) range over the real numbers. Explicitly, we are saying that the implication holds for every *real* number x, and we are asserting (as long as $x \neq 0$) the existence of a *real* number y with a certain property. If we were to change the domain under discussion, then this proposition, which is true over the real numbers, might become false. If we let the quantifiers range over the integers, for example, then the proposition is false, since not every integer has an integer reciprocal; in fact, none of them do except for 1 and -1.

Sometimes a restriction on a quantified variable is incorporated into the symbolism more succinctly, by writing the restriction along with the name of the variable immediately following the quantifier. Thus we can write our assertion that every nonzero real number has a reciprocal (assuming that the quantifiers are meant to range over the real numbers) as

$$\forall x \neq 0 \colon \exists y \colon xy = 1.$$

We read this as "For every x not equal to 0, there exists a y such that x times y equals 1."

There is one more thing we might do with our original example. Not only does every nonzero real number have a reciprocal, but that reciprocal is unique. For instance, $\frac{1}{2}$ is the *only* number that can be multiplied by 2 to give the product 1. In other words, for every nonzero x, there exists a *unique* y such that $xy = 1$. A notation for the phrase "there exists a unique" is $\exists!$. Thus we can write our proposition as $\forall x \neq 0 \colon \exists! y \colon xy = 1$. Exercise 35 asks you to rewrite this without the special notation.

Let us consider some more examples.

EXAMPLE 3.   The domain under discussion is natural numbers. Consider the two propositions

(1) $$\forall x \colon \exists y \colon x < y$$

and

(2) $$\exists y \colon \forall x \colon x < y.$$

Proposition 1 states, "For every x, there exists a y such that x is less than y." Because of the logical (and grammatical) structure of this proposition, the value of y whose existence is being asserted can depend on x. Proposition 1 is true: Given any natural number x, we can find a larger natural number y (by letting $y = x + 1$, for example). Proposition 2 states, "There exists a y such that for every x, x is less than y." Note the grammatical and logical structure of this sentence, and note how it differs from (1). Here the existence of y is mentioned first, and the assertion is that there exists a value of y that is greater than any value of x that might thereafter be chosen. In this case the value of y *cannot* depend on x. Clearly, proposition 2 is false: There is no such "largest" natural number, no "infinity." In fact, even if our domain were restricted to a finite set of numbers, so that there would be a largest number in the domain, the proposition would still be false, because the proposition $x < y$ is false if x is assigned the same value as y.

In general, then, the two propositions $\forall x\colon \exists y\colon P(x, y)$ and $\exists y\colon \forall x\colon P(x, y)$ mean quite different things. In this example, $P(x, y)$ was the proposition $x < y$, and we found that the first of these quantified propositions was true, whereas the second was false. ◆

EXAMPLE 4. Express Lagrange's four square theorem (proposition 5 of Example 1 in Section 1.1) in logical notation.

Solution. We want to assert that every number (universal quantifier needed, ranging over all natural numbers) can be written as (existential quantifier needed) the sum of four squares. In words, we want to say that for every natural number x, there exist natural numbers a, b, c, and d such that x is equal to the sums of the squares of a, b, c, and d. Translating this into symbols, we obtain

$$\forall x\colon \exists a\colon \exists b\colon \exists c\colon \exists d\colon x = a^2 + b^2 + c^2 + d^2. \qquad ◆$$

We repeat that the most important words in mathematical utterances are often the logical words. It is a totally different thing to say that "$f(x) = 0$ *and* $f(-x) = 0$" than it is to say that "*If* $f(x) = 0$, *then* $f(-x) = 0$." The assertion $\forall x\colon x^2 = x$ is quite a different statement from $\exists x\colon x^2 = x$. Furthermore, even innocuous words like "such that" must be used very carefully. When we verbalize the existential quantifier ($\exists$) as "there exists," the following colon (:) can be thought of as corresponding to the words "such that." For example, $\exists u\colon u + 1 = 7$ is read "There exists a u such that u plus 1 equals 7." The proposition is asserting the existence of a number having a certain property—a number "such that" the property holds. On the other hand, the colon after the universal quantifier ($\forall$) does not correspond to the English words "such that." For example, $\forall z\colon z + 0 = z$ is read "For every z, z plus 0 equals z." It would be incorrect to read this as "For every z such that z plus 0 equals z." For one thing, the latter quote is not even an English sentence (it has no main verb) and so cannot be the translation of a proposition. For another, the universally quantified proposition is asserting that *every* number has the property in question; it is not referring to a number that happens to have the property. Similarly, if we choose to read $\exists$ as "for some," then it makes no grammatical sense to include the words "such that." For example, $\exists u\colon u + 1 = 7$ can be read "For some u, u plus 1 equals 7," but not "For some u such that u plus 1 equals 7." If the reader keeps in mind that propositions must be complete, meaningful English sentences, these pitfalls can often be avoided.

EXAMPLE 5. It is not a correct verbalization of $\forall x\colon \exists y\colon x < y$ to read "For every x such that there exists a y such that x is less than y." Similarly, it is not correct to read $\exists y\colon \forall x\colon x < y$ as "There exists a y for every x such that x is less than y." ◆

We need to mention one convention and one technical consideration at this point. The convention is that universal quantifiers are often understood, even when words like "for all" are not used explicitly. For example, suppose that we want to state that the collection of odd integers is closed under multiplication; this means that the product of two integers is odd whenever the two integers are both odd. We could write this last utterance as (x is odd $\land$ y is odd) $\rightarrow$ xy is odd. This proposition (both in English and in symbols) appears to contain free variables x and y, but the intent is that x and y are universally quantified. After all, we were trying to say that the property of being odd was preserved by the multiplication operation, and objects x and y should have nothing to do with it. Thus the proposition we mean is really $\forall x : \forall y : ((x$ is odd $\land$ y is odd) $\rightarrow$ xy is odd).

The technical consideration is that there are some restrictions that apply to the construction of meaningful compound propositions involving quantified and free variables. For example, the "proposition" $\forall x : ((\exists x : x < x) \land x \neq 0)$ is ambiguous at best, since the variable x is being used in several overlapping contexts. (The issue here is related to the notion of the scope of variables in computer programming languages.) We will not concern ourselves with these restrictions in this book, and instead warn the reader to choose variable names that will avoid such problems.

DETERMINING TRUTH VALUES
OF QUANTIFIED PROPOSITIONS

How can we determine whether a quantified proposition is true or false? The truth value of such a proposition is determined by the definitions. Let us in effect restate the definitions by pointing out explicitly what is involved in justifying a quantified proposition (i.e., demonstrating that it is true) or in refuting it (i.e., demonstrating that it is false).

Rules for Justifying or Refuting Quantified Propositions

1. To show that a **universally** quantified proposition $\forall x : P$ is **true**, you must show that P holds for *all* values of x, not just for a specific x that you choose. An example—an x that makes P true—proves nothing. At best, examples can only show that a universally quantified proposition is plausible.

2. To show that a **universally** quantified proposition $\forall x : P$ is **false**, you need only find a **counterexample**, that is, you need only find one value of x for which P fails to hold.

3. To show that an **existentially** quantified proposition $\exists x : P$ is **true**, you need only find an **example**, that is, you need only find one value of x for which P holds.

4. To show that an **existentially** quantified proposition $\exists x : P$ is **false**, you must show that *no* value of x makes P true, that is, that P is false for *all*

values of x, not just for a specific x that you choose. An example—an x that makes P false—proves nothing.

EXAMPLE 6.   To show that the statement "Every prime number is odd" (a universally quantified proposition) is false, you need only point out that 2 is a prime number that is not odd. In symbols, the proposition is

$$\forall x \colon (x \text{ is prime } \to x \text{ is odd}).$$

When x is assigned the value 2, we obtain the proposition "2 is prime $\to$ 2 is odd," which is false because the hypothesis is true and the conclusion is false. Thus the universally quantified statement is false. It is irrelevant that there are other prime numbers that *are* odd (in fact, all prime numbers except for 2 are odd).   ◆

EXAMPLE 7.   To show that the statement "The product of two odd numbers is always odd" is true, it is useless to point out, for example, that 5 times 7 is 35. The proposition is the universally quantified

$$\forall x \colon \forall y \colon ((x \text{ is odd } \wedge y \text{ is odd}) \to xy \text{ is odd}),$$

and noting that the implication is true for $x = 5$ and $y = 7$ does nothing toward justifying the entire, quantified statement. (We will see in Section 1.3 how to prove that this proposition is indeed true.)   ◆

EXAMPLE 8.   A proof of Lagrange's four square theorem (Example 4) is far beyond the scope of this book. For any natural number x, it is fairly clear how to look systematically for natural numbers a, b, c, and d so that $x = a^2 + b^2 + c^2 + d^2$. For example, for $x = 18$ we can find without too much effort that $18 = 4^2 + 1^2 + 1^2 + 0^2$. The fact that we have written a particular x, namely 18, as the sum of four squares does nothing toward showing that the theorem is true, nor does the fact that we have a method for *looking for* the summands show that our search will always be successful. Indeed, 18 can be written as the sum of *three* squares ($18 = 4^2 + 1^2 + 1^2$), and for any x one can look systematically for solutions to $x = a^2 + b^2 + c^2$; but the proposition $\forall x \colon \exists a \colon \exists b \colon \exists c \colon x = a^2 + b^2 + c^2$ (a "three square theorem") is false, as the reader is asked to show in Exercise 36.   ◆

All this is not to say that looking at examples is a poor way to explore mathematical problems. On the contrary: *When faced with trying to decide whether a given assertion is true or false, it is usually very helpful to experiment with some examples*. If the examples you choose allow you to refute a universally quantified assertion, or to justify an existentially quantified one, then your problem is solved. On the other hand, if the examples lead you to conjecture that a universally quantified assertion is true, or that an existentially quantified one is false, then the understanding of what is going on gained by having worked with the examples may help you construct a valid proof.

NEGATIONS OF QUANTIFIED PROPOSITIONS

We end this section by considering the relationships between logical quantifiers and the negation operation. Given a quantified proposition, can its negation be rewritten as a quantified proposition? In other words, is there a quantified proposition that is logically equivalent to the negation of the original quantified proposition? The answer is yes, and a method for writing the negation as a quantified proposition follows immediately from rules 2 and 4 above, namely those for refuting quantified propositions. Since these rules were essentially just restatements of the definitions, the following theorem really follows immediately from the definitions of the quantifiers. Rule 2 implies that the negation of $\forall x\colon P$ is the assertion that there exists an x such that P is false—in other words, the proposition $\exists x\colon \overline{P}$. Rule 4 implies that the negation of $\exists x\colon P$ is the assertion that P is false for all x—in other words, the proposition $\forall x\colon \overline{P}$. Let us formally summarize these facts.

THEOREM 1.   *Let P be a proposition. Then*

$$(a) \quad \overline{\forall x\colon P} \iff \exists x\colon \overline{P}$$

and

$$(b) \quad \overline{\exists x\colon P} \iff \forall x\colon \overline{P}.$$

We can restate this theorem as an operative rule: To negate a quantified proposition, you change the quantifier (from universal to existential, and vice versa) and negate the rest. If there are several quantifiers in a row, then the rule needs to be applied successively, from left to right. If we write negations with the symbol $\neg$, then our rule says that to slide that symbol past a quantifier, from left to right, we must change the quantifier. In other words, $\neg\forall x\colon$ is the same as $\exists x\colon \neg$, and $\neg\exists x\colon$ is the same as $\forall x\colon \neg$.

EXAMPLE 9.   The negation of Lagrange's four square theorem

$$\forall x\colon \exists a\colon \exists b\colon \exists c\colon \exists d\colon x = a^2 + b^2 + c^2 + d^2$$

is

$\neg \forall x \colon \exists a \colon \exists b \colon \exists c \colon \exists d \colon x = a^2 + b^2 + c^2 + d^2$

$$\Longleftrightarrow \quad \exists x \colon \neg \exists a \colon \exists b \colon \exists c \colon \exists d \colon x = a^2 + b^2 + c^2 + d^2$$

$$\Longleftrightarrow \quad \exists x \colon \forall a \colon \neg \exists b \colon \exists c \colon \exists d \colon x = a^2 + b^2 + c^2 + d^2$$

$$\Longleftrightarrow \quad \exists x \colon \forall a \colon \forall b \colon \neg \exists c \colon \exists d \colon x = a^2 + b^2 + c^2 + d^2$$

$$\Longleftrightarrow \quad \exists x \colon \forall a \colon \forall b \colon \forall c \colon \neg \exists d \colon x = a^2 + b^2 + c^2 + d^2$$

$$\Longleftrightarrow \quad \exists x \colon \forall a \colon \forall b \colon \forall c \colon \forall d \colon \neg x = a^2 + b^2 + c^2 + d^2$$

$$\Longleftrightarrow \quad \exists x \colon \forall a \colon \forall b \colon \forall c \colon \forall d \colon x \neq a^2 + b^2 + c^2 + d^2.$$

Since there are five quantifiers at the beginning of this proposition, we applied the rule five times, changing all the quantifiers; after the last application of the rule, we simply rewrote the negation of an equality as an inequality. In words, the negation of Lagrange's theorem says that there exists a natural number such that no matter what four numbers are chosen, their squares do not add up to that number. Since Lagrange's theorem is true, its negation is false: No such natural number exists. ◆

EXAMPLE 10. The negation of the assertion that every nonzero number has a reciprocal is the assertion that some nonzero number does not have a reciprocal. Note carefully the steps below, and note in particular the second line, whose justification is part (k) of Theorem 1 from Section 1.1. In symbols, we have

$$\overline{\forall x \colon (x \neq 0 \rightarrow \exists y \colon xy = 1)} \quad \Longleftrightarrow \quad \exists x \colon \overline{x \neq 0 \rightarrow \exists y \colon xy = 1}$$

$$\Longleftrightarrow \quad \exists x \colon (x \neq 0 \wedge \overline{\exists y \colon xy = 1})$$

$$\Longleftrightarrow \quad \exists x \colon (x \neq 0 \wedge \forall y \colon xy \neq 1).$$

We can read this last proposition as "There exists a nonzero number x such that no number is the reciprocal of x."

To negate the version of this proposition in which the restriction on x is built into the quantifier is even simpler:

$$\neg(\forall x \neq 0 \colon \exists y \colon xy = 1) \quad \Longleftrightarrow \quad \exists x \neq 0 \colon \forall y \colon xy \neq 1. \qquad ◆$$

The logical apparatus that we have introduced in this section and the preceding one—propositions, logical connectives, and quantifiers—enables us to express, with clarity and complete precision, most of the mathematical facts and definitions we will encounter in discrete mathematics. Remember that in mathematics one must say exactly what one means and mean exactly what one says. At this point we have the tools for achieving this rigor.

SUMMARY OF DEFINITIONS

bound variable: a quantified variable.

counterexample to a false universally quantified proposition $\forall v \colon P(v)$: a value v_0 such that $P(v_0)$ is false (e.g., 11 is a counterexample to the proposition $\forall p \colon (p$ is prime $\to 2^p - 1$ is prime), since $2^{11} - 1 = 2047 = 23 \cdot 89$).

domain under discussion: the collection of objects that the free or bound variables in a proposition might stand for (e.g., in the proposition $\forall x \colon x + 0 = x$, if the variable x is meant to range over real numbers, then the domain under discussion is real numbers).

dummy variable: a variable that has no existence of its own in an expression; quantified variables are dummy variables in quantified propositions.

example for a true existentially quantified proposition $\exists v \colon P(v)$: a value v_0 such that $P(v_0)$ is true (e.g., the triple of numbers 5, 12, and 13 is an example for $\exists x \colon \exists y \colon \exists z \colon x^2 + y^2 = z^2$, since $25 + 144 = 169$; here we are thinking of the three existential quantifiers as together asserting the existence of the triple x, y, z).

existential quantifier: the logical symbol $\exists$.

existentially quantified proposition: a proposition of the form $\exists v \colon P(v)$, which asserts that $P(v)$ holds for at least one value of the variable v (e.g., $\exists x \colon x + 3 = 5$).

quantified variable: a variable in a proposition appearing immediately after a quantifier (e.g., x in $\forall x \colon 0 + x = x$).

quantifier: an existential or universal quantifier.

universal quantifier: the logical symbol $\forall$.

universally quantified proposition: a proposition of the form $\forall v \colon P(v)$, which asserts that $P(v)$ holds for every value of the variable v (e.g., $\forall x \colon \exists y \colon x + y = 0$; this proposition is not an existentially quantified proposition).

EXERCISES

1. Express the following true propositions in symbols. The variables range over the integers.
   - (a) For every x, $x^2 > 4$ if and only if $x > 2$ or $x < -2$.
   - (b) There exists a number that equals its own square.
   - (c) For every number x greater than 1, there exists a number y that is greater than x and less than $2x$.
   - (d) There exists a number x such that for every y, $y^2 > x$.

2. Let $F(x, y)$ be the proposition that x is the father of y, and let $M(x, y)$ be the proposition that x is the mother of y. Express each of the following propositions symbolically.
   - (a) Fred is Ann's father.
   - (b) A is the child of B.

(c)   Everybody has a mother and a father.
(d)   Tom and Harry have the same mother.
(e)   Nobody is his or her own mother.
(f)   Peter has no children.
(g)   Charles and Diana have at least one child.

3. Express each of the following propositions in English, using the language properly and concisely. The quantifiers are meant to range over the real numbers.
   (a) $\exists x \colon \forall y \colon x + y = y.$
   (b) $\forall x \colon \exists y \colon x + y = y.$
   (c) $\forall x \colon \forall y \colon x + y = y.$
   (d) $\exists x \colon \exists y \colon x - y = y.$
   (e) $\exists x \colon \forall y \colon x - y = y.$
   (f) $\forall x \colon \exists y \colon x - y = y.$

4. Determine for each part of Exercise 3 whether the proposition is true or false.

5. Express each of the following propositions in English, using the language properly and concisely. The quantifiers are meant to range over the integers.
   (a) $\exists a \colon \exists b \colon (x = ab \wedge a > 1 \wedge b > 1).$
   (b) $\forall x \colon \exists y \colon x < y^2.$
   (c) $\forall x \colon \exists y \colon (x = 3y \vee x = 3y + 1 \vee x = 3y + 2).$
   (d) $\forall x \colon (x < 2 \rightarrow x^2 < 4).$
   (e) $\forall x \colon (x^2 < 4 \rightarrow x < 2).$
   (f) $\exists x \colon (x < 5 \rightarrow x < 3).$
   (g) $\exists x \colon x^2 - 2x - 120 = 0.$
   (h) $\forall x \colon x^2 > a.$

6. Consider the proposition $\forall x \colon \forall y \colon (x < y \rightarrow \exists z \colon (x < z \wedge z < y)).$ Determine whether this proposition is true in each of the following cases.
   (a)   The domain under discussion is the real numbers.
   (b)   The domain under discussion is the natural numbers.

7. Let M and F be as defined in Exercise 2. Express the following propositions symbolically.
   (a)   Charles and Diana have at least two children.
   (b)   Jerry and Suzanne have exactly one child.
   (c)   Sam is Pam's maternal grandfather.
   (d)   Pamela and Conrad have the same paternal grandmother.
   (e)   Some people have no children.

8. Let $Q(x)$ denote "x is even," $P(x)$ denote "x is prime," $R(x)$ denote "x is divisible by 6," $G(x)$ denote $x \leq 5$, and $L(x, y)$ denote $x < y$. Determine what each of the following propositions asserts and whether it is true or false. The domain under discussion is the natural numbers.
   (a) $\exists x \colon (R(x) \wedge P(x)).$

(b) $(\exists x \colon R(x)) \wedge (\exists x \colon P(x))$.

(c) $\exists x \colon (G(x) \to \overline{P(x)})$.

(d) $\forall x \colon (P(x) \to \overline{Q(x)})$.

(e) $(\forall x \colon P(x)) \to \forall x \colon \overline{Q(x)}$.

(f) $(\exists x \colon P(x)) \to \exists x \colon (P(x) \wedge R(x))$.

(g) $\exists x \colon (P(x) \to (P(x) \wedge R(x)))$.

(h) $\forall x \colon (R(x) \to \exists y \colon (L(x, y) \wedge R(y)))$.

(i) $\forall x \colon \exists y \colon (L(x, y) \wedge L(y, x))$.

9. Express each of the following propositions in symbols, using only numerals, variables, logical symbols, and the mathematical symbols $=$ and $\cdot$ (times). The domain under discussion is the integers.

   (a) 100 is a multiple of 5.

   (b) 1000 is not a multiple of 8.

   (c) Every multiple of 6 is also a multiple of 2, and conversely.

   (d) Every multiple of 3 is not even.

   (e) Some number has no square root.

10. Express each of the following propositions in symbols, using only numerals, variables, logical symbols, and the mathematical symbols $=$, $>$, $<$, and ^{2} (squared). Quantifiers range over the natural numbers.

    (a) Every perfect square is less than 500.

    (b) Some perfect square is less than 500.

    (c) No perfect square is less than 500.

    (d) Not every perfect square is less than 500.

    (e) Every perfect square is not less than 500.

    (f) Some perfect square is not less than 500.

11. Determine the truth value of each of the propositions in Exercise 9.

12. Determine the truth value of each of the propositions in Exercise 10.

13. Determine for each part of Exercise 5 whether the proposition is true, false, or neither (because it contains free variables).

14. Write the negation of each proposition in Exercise 5 as a quantified proposition, in symbols.

15. Write the negation of each proposition in Exercise 9 as a quantified proposition, in English.

16. Write the negation of each proposition in Exercise 9 as a quantified proposition, in symbols.

17. Write the negation of each proposition in Exercise 10 as a quantified proposition, in English.

18. What facts about the real numbers are expressed by the following propositions? (Express your answers succinctly in English.)

(a) $\forall x \colon \forall y \colon \forall z \colon (x(y+z) = xy + xz)$.
(b) $\exists z \colon \forall x \colon xz = x$.
(c) $\exists z \colon ((\forall x \colon x + z = x) \wedge (\forall y \colon [(\forall x \colon x + y = x) \rightarrow y = z]))$.
(d) $\exists z \colon ((\forall x \colon x + z = x) \wedge (\forall x \colon \exists y \colon x + y = z))$.

19. Let $P(x, y)$ be the proposition $x^2 = y$, where x and y range over the positive integers. Determine the truth value of each of the following propositions.
    (a) $P(1, 2)$.
    (b) $P(4, 2)$.
    (c) $P(3, 9)$.
    (d) $\exists x \colon P(x, 6)$.
    (e) $\exists y \colon P(6, y)$.
    (f) $\exists x \colon P(6, x)$.
    (g) $\forall x \colon \exists y \colon P(x, y)$.
    (h) $\forall y \colon \exists x \colon P(x, y)$.
    (i) $\exists y \colon \forall x \colon P(x, y)$.

20. Express the proposition in Example 2 using only two quantified variables, rather than four.

21. Express each of the following propositions in symbols (using only numerals, variables, logical symbols, $=$, $>$, and $\cdot$), and determine its truth value. Assume that the domain under discussion is the natural numbers. [*Hint*: Recall the definition of a prime number.]
    (a) 7921 is a prime number.
    (b) Every number that has a square root is not prime.

22. Determine the truth value of the following propositions. The quantifiers range over the integers.
    (a) $\forall x \colon \forall y \colon (x^2 = y^2 \rightarrow x = -y)$.
    (b) $\exists x \colon \exists y \colon (x = y \rightarrow x > y)$.
    (c) $\forall x \colon \exists y \colon (x = y \rightarrow x > y)$.
    (d) $\exists x \colon \forall y \colon (x = y \rightarrow x > y)$.

23. If the domain under discussion is finite, then a universally quantified proposition can be expressed as a conjunction rather than as a quantified proposition; similarly, an existentially quantified proposition can be expressed as a disjunction. Assume that the domain consists of just the two numbers 1 and 2. Express each of the following propositions without quantifiers.
    (a) $\forall x \colon \forall y \colon P(x, y)$.
    (b) $\exists x \colon \exists y \colon P(x, y)$.
    (c) $\forall x \colon \exists y \colon P(x, y)$.
    (d) $\exists x \colon \forall y \colon P(x, y)$.

24. Let $P(x, y)$ be the proposition $x \geq y$. Determine the truth value of each of the propositions in Exercise 23. Remember that the domain under discussion consists of just the two numbers 1 and 2.

25. Translate into symbols the statement "Everybody loves somebody sometime" using $L(x, y, t)$ to stand for "x loves y at time t."

26. Let $P(x, y)$ be the proposition $x \leq y$, where the variables range over the real numbers. Determine the truth value of each of the following propositions. (These particular propositions will be important in Section 3.3.)
  (a) $\forall x \colon P(x, x)$.
  (b) $\forall x \colon \overline{P(x, x)}$.
  (c) $\forall x \colon \forall y \colon (P(x, y) \to P(y, x))$.
  (d) $\forall x \colon \forall y \colon ((P(x, y) \land P(y, x)) \to x = y)$.
  (e) $\forall x \colon \forall y \colon ((P(x, y) \land x \neq y) \to \overline{P(y, x)})$.
  (f) $\forall x \colon \forall y \colon \forall z \colon ((P(x, y) \land P(y, z)) \to P(x, z))$.
  (g) $\forall x \colon \forall y \colon (P(x, y) \lor P(y, x))$.

27. Let $C(x, m, f)$ be the proposition "m is the mother of x and f is the father of x." Express each of the following propositions symbolically.
  (a) x and y are (full) siblings.
  (b) x is the paternal grandfather of y.
  (c) x has no grandchildren.
  (d) x is the aunt or uncle of y (directly, not by marriage).

28. Let $B(x)$ be "x is male," and let $P(x, y)$ be "x is a parent of y." Translate each of these propositions into the simplest possible English.
  (a) $P(s, t) \land B(s)$.
  (b) $P(s, t) \land \overline{B(t)}$.
  (c) $B(s) \land B(t) \land s \neq t \land \exists z_1 \colon \exists z_2 \colon (P(z_1, s) \land P(z_2, s) \land P(z_1, t) \land P(z_2, t) \land z_1 \neq z_2)$.
  (d) $\exists x \colon \forall y \colon \forall z \colon (P(x, y) \to \overline{P(y, z)})$.
  (e) $B(s) \land \exists v \colon \exists w \colon \exists z \colon (P(v, z) \land P(w, z) \land P(v, s) \land P(w, s) \land s \neq z \land v \neq w \land P(z, t))$.

29. In each case, determine whether the two propositions are logically equivalent.
  (a) $\forall x \colon (P \land Q)$ and $(\forall x \colon P) \land (\forall x \colon Q)$.
  (b) $\forall x \colon (P \lor Q)$ and $(\forall x \colon P) \lor (\forall x \colon Q)$.
  (c) $\exists x \colon (P \land Q)$ and $(\exists x \colon P) \land (\exists x \colon Q)$.
  (d) $\exists x \colon (P \lor Q)$ and $(\exists x \colon P) \lor (\exists x \colon Q)$.
  (e) $\forall x \colon (P \to Q)$ and $(\forall x \colon P) \to (\forall x \colon Q)$.
  (f) $\exists x \colon (P \to Q)$ and $(\exists x \colon P) \to (\exists x \colon Q)$.

30. Recall Goldbach's conjecture (proposition 6 of Example 1 in Section 1.1).
  (a)   Express Goldbach's conjecture entirely in symbols.
  (b)   Verify that the following numbers can be written as the sum of two primes: 4, 6, 8, 10, 20, 30, 100, and 1000.
  (c)   Does the verification in part (b) say anything about the truth of Goldbach's conjecture?

31. Pierre de Fermat, a seventeenth-century French mathematician, while reading an old book on number theory, noticed that there were many examples in which a square is the sum of two squares (e.g., $5^2 = 3^2 + 4^2$). He scribbled in his copy of the old book, "However, it is impossible to write a cube as the sum of two cubes, a fourth power as the sum of two fourth powers and in general any power beyond the second as the sum of two similar powers. For this I have discovered a truly wonderful proof, but the margin is too small to contain it." The domain under discussion for this exercise is the positive integers.

(a) Express this proposition, known as Fermat's last theorem, in symbols. (You may use exponentiation notation.)

(b) Despite Fermat's claim, no one has been able to supply a proof, so the proposition remains only a conjecture, and no one knows whether it is true or false. Exactly what would constitute a counterexample to Fermat's last theorem?

32. Let $M(x)$ be the proposition that x is a male. Assume that quantifiers range over the children in the McKay family. Express in symbols the proposition that every child in this family has at least one brother and at least one sister. (A simple brain-teaser asks for the smallest number of children the family can have if this proposition is true.)

33. Criticize the following statement: $\forall y: \exists x: x + y = y$ means that $x = 0$.

Challenging Exercises

34. Express the following propositions in symbols. State explicitly the domain over which the quantifiers range.

(a) There are an infinite number of twin primes. [*Hint*: Twin primes are pairs of prime numbers, like 101 and 103, which differ by 2. To say that there are an infinite number of such pairs is equivalent to saying that there are arbitrarily large ones. Incidentally, no one knows whether this proposition is true or false.]

(b) The only rational roots of the third-degree polynomial $ax^3 + bx^2 + cx + d$, where a, b, c, and d are integers and neither a nor d is zero, are of the form $\pm$(divisor of d)/(divisor of a).

(c) Every third-degree polynomial with real coefficients has at least one real root.

(d) The positive integer x is a power of 2. (Do not use a symbol for exponentiation; express this proposition using just multiplication. What makes this exercise feasible is the fact that 2 is a prime number. To express "x is a power of 6" is ridiculously hard and far, far beyond the scope of this book.)

(e) Every quadratic polynomial has at most two roots.

35. Suppose that $P(x)$ is a proposition with a free variable x. Express $\exists!x\colon P(x)$ using ordinary quantifiers. [*Hint*: You will need to use the "equals" symbol, as well as some logical connectives.]

36. Determine the truth value of the following propositions.
  (a) The analogue of Lagrange's four square theorem involving three squares instead of four.
  (b) The analogue of Lagrange's four square theorem involving five squares instead of four.

37. Determine the truth value of the proposition "$\forall n \geq 0\colon n^2 - n + 41$ is prime." Quantifier ranges over natural numbers. [*Hint*: Do not jump to conclusions.]

38. Some propositions cannot be expressed conveniently with a fixed number of quantified variables. For example, half of the fundamental theorem of arithmetic states that every natural number greater than 1 has a prime factorization (the other half is the uniqueness of this factorization). This means that for every natural number $x > 1$, there exist distinct prime numbers p_1, p_2, ..., p_n and positive integer exponents e_1, e_2, ..., e_n (for some $n \geq 1$), such that $x = p_1^{e_1} p_2^{e_2} \cdots p_n^{e_n}$. Here x is universally quantified and all the p_i's and e_i's are existentially quantified. Since n, the number of distinct prime factors of x, is not a fixed number but depends on x, we need to treat n as a quantified variable also. Using symbols and the "is prime" predicate, we could write this as

$$\forall x > 1\colon \exists n \geq 1\colon \exists p_1\colon \exists p_2\colon \cdots \exists p_n\colon \exists e_1\colon \exists e_2\colon \cdots \exists e_n\colon$$

$$((\forall i\colon (e_i \geq 1 \wedge p_i \text{ is prime})) \wedge x = p_1^{e_1} p_2^{e_2} \cdots p_n^{e_n}).$$

Express each of the following propositions using quantifiers in this manner.
  (a) Every finite collection of real numbers contains a largest number.
  (b) Every polynomial of odd degree has at least one real root.
  (c) Every natural number can be written as the sum of cubes of natural numbers. (In fact, it is a theorem of number theory that, in analogy with Lagrange's four square theorem, the number of cubes required is *not* arbitrarily large (see Exercise 45 in Section 1.1).)
  (d) Every nth-degree polynomial has at most n roots.

39. Express with logical symbols the following statements from various parts of mathematics, paying special attention to the use of quantifiers.
  (a) A function f of real numbers is **continuous** at a if and only if for every positive number ϵ there exists a positive number δ such that $|f(x) - f(a)| < \epsilon$ whenever $|x - a| < \delta$.
  (b) That an integer x is a **quadratic residue** modulo a positive integer p means that $x - y^2$ is a multiple of p for some integer y.

(c)   If a function f of real numbers is continuous on $[a, b]$ and differentiable on (a, b), and if $f(a) = f(b) = 0$, then $f'(c) = 0$ for some c strictly between a and b. (Do not translate the notions of continuity or differentiability into symbols here.)

(d)   Two distinct points determine one and only one line.

Exploratory Exercises

40. Translate Abraham Lincoln's famous quotation that appears at the beginning of this section into symbols, using $F(p, t)$ to mean "you can fool person p at time t." Explain why the correct answer here is not unique—in other words, why Lincoln's statement is ambiguous in English. [*Hint*: The issue is the order of quantifiers.] Argue that your translation is preferable.

41. Explain the relationship between Theorem 1 and DeMorgan's laws (part (i) of Theorem 1 in Section 1.1). [*Hint*: See Exercise 23.]

42. Find out more about prime numbers by consulting Dudley [72], Gardner [98], and Yates [334].

43. Find out more about Fermat's last theorem (Exercise 31) by consulting Edwards [74] and Ribenboim [245].

44. Find out about Pierre de Fermat (Exercise 31) by consulting Bell [15].

SECTION 1.3
PROOFS

Sir, $\dfrac{a + b^n}{n} = x$, *hence God exists; reply!*

—Leonhard Euler

In the preceding two sections we have indicated how to determine if a proposition is true. For example, an implication $P \to Q$ is true as long as the hypothesis P is false or the conclusion Q is true. As another example, the universally quantified proposition $\forall x \colon (P \lor Q)$ is true as long as for every choice of x either P is true or Q is true. Now suppose that a person wants to make sure that he or she has not made a mistake in determining the truth value of a proposition—that he or she has used the human reasoning process correctly. Or suppose that one person wants to convince a second person that a certain proposition is true. In fact, mathematicians regularly engage in both of these activities. It would be desirable if certain protocols were observed in the arguments that people used to prove the truth of propositions in mathematics (as well as other fields), rules of the game that we can all agree upon ahead of time. Ideally, these protocols would allow us to prove all the true propositions to everyone's satisfaction and avoid mistakenly "proving" any false ones. Unfortunately, this desired state of affairs is impossible to achieve completely, for at least two reasons, one practical and one profound. Before we find out how to go about constructing proofs, we should look at these limitations.

The practical problem is that proofs are human endeavors, and no matter how carefully we lay down rules for proofs, there will always be a chance that the person or persons involved in a particular proof will make a mistake. This is not just farfetched speculation. Consider the case of one of the most famous theorems in discrete mathematics, the four-color theorem. This proposition, which we discuss at length in Section 8.5, states, roughly, that the countries in any geographical map drawn on a flat sheet of paper can be colored, using only four colors, in such a way that no two countries with a common boundary have the same color. The four-color theorem was first "proved" in the late nineteenth century by a British lawyer and mathematician, A. B. Kempe, and his "proof" was published in a reputable mathematics research journal of the day. Everyone assumed that the proof was correct, because it seemed to follow the protocols of mathematical proof. About a dozen years later, however, another mathematician, P. J. Heawood, found a mistake in the proof. At one point in the published proof, there was a subtle flaw in the reasoning, and the whole proof fell apart. Heawood published, in another respected mathematics journal, his refutation of the proof. The second journal article did not show that the four-color theorem was false, merely that the purported proof published a dozen years earlier was incorrect. Everyone was convinced by the latter article, and so what had apparently been a theorem in 1885 was suddenly no longer a theorem in 1895. (The story does not end there, however. Mathematicians continued to work on proving the four-color theorem, and another proof was published in 1976. It is so long, however, that it needed the assistance of a high-speed computer, working for about 1200 hours, and the intermediate steps in the computations—performed by the computer—were not written down. Some philosophers today argue that the proof is therefore not valid, since it cannot readily be checked by human beings.)

The other reason that we cannot hope to lay down once and for all a nice set of rules enabling us to prove exactly all the true propositions of mathematics is that it has been proved that *any such consistent set of rules is incomplete*. To be more precise, no matter what reasonable set of rules and assumptions one makes for proving theorems in mathematics, there will always be true propositions that cannot be proved with these rules and assumptions. That this is so was discovered—and proved—in 1931 by the Austrian mathematician K. Gödel. His incompleteness theorem is one of the most profound discoveries in mathematical logic. Fortunately, we will not encounter (at least knowingly) any of these true but unprovable propositions in elementary discrete mathematics.

We will not dwell on the limitations any longer, for at the level of this book, the protocols of mathematical proof are well established and effective. Knowing and practicing them will enable the reader to recognize correct proofs, and, with practice, to construct them. Proving propositions is much more an art than a science, however, and there is no algorithm for coming up with the *idea* that can be turned into a correct mathematical proof. Creativity remains a remarkable and little understood human ability!

In this section we discuss two aspects of mathematical proofs—a *global* aspect and a *local* aspect. In the large, we need to agree on what statements need proof and what can be assumed without proof. At the local level we need to decide on what constitutes valid reasoning, that is, what makes each step in a proof correct.

AXIOM SYSTEMS AND DEFINITIONS

Just as the 4-year-old child can respond to every explanation with an inquisitive, or merely annoying, "Why?", so a skeptical person could counter, at every step of a colleague's proof, with a demand for further justification of that step. To avoid an endless regress, therefore, all mathematical proofs need to be based on certain assumptions that are themselves *not proved*. Formally, these assumptions are called **axioms** or **postulates** (the two words are synonymous). The axioms in a given setting should capture the fundamental truths that are self-evident about objects in that setting; they should be statements that everyone would be willing to accept as true (or at least agree to tentatively). In advanced courses in mathematical logic, abstract algebra, number theory, combinatorics, geometry, or other areas of mathematics, it is often useful to be very explicit as to what these assumptions are. For example, it is an axiom in most discussions of geometry that given two distinct points in the plane, we can find exactly one line containing both of the points— in short, *two points determine a line*. As another example, we could explicitly introduce the algebraic axiom

$$\forall x\colon \forall y\colon \forall a\colon \forall b\colon ((x = y \land a = b) \to x + a = y + b),$$

which states that *equals added to equals produce equals*. In this book we will be much more informal and very rarely make explicit our basic assumptions about the nature of numbers, sets, or the geometry of the plane. It is assumed that the reader has enough experience with and intuition about these entities to agree to the implicit assumptions we will make.

There is another set of statements that we will accept without proof—the **definitions** of technical terms. We have already encountered definitions in our discussions of logic. For example, we defined the logical connective "if...then..." ($\to$); the reason that $P \to Q$ is true when P is false and Q is false is that *we defined it that way*. Almost every proof that we present here, and almost every proof that the reader is likely to construct, will appeal, at one or more points, directly or indirectly, to definitions. This is not surprising, since it would be hard to prove a proposition without making it clear as to what we meant by the terms used in the proposition we were attempting to prove. In short, we cannot prove something if we don't know what we're talking about. (On the other hand, many students tend to overrate the power of a definition and will sometimes claim—perhaps grasping at straws—that a certain proposition is true simply "by definition." Such a claim is seldom correct; all but the most trivial propositions are proved true by a line of reasoning, not simply by an appeal to a definition. A definition can only tell us what a certain term *means*.)

Definitions, then, are really propositions whose truth we all agree to without proof; they are part of the rules of the game, and the person inventing the game gets to make up the rules. Often a person doing mathematics—writing a proof, say—will find it convenient to define a concept or notation; as long as the definition is precise and not contradictory, such a step is completely acceptable.

Most definitions are explicitly or implicitly in the form of an equivalence. To define a new concept N, we write down a proposition that says that N holds if and only if some other proposition is true. Thus a typical definition might look like $\forall x : (N(x) \leftrightarrow P(x))$, where $P(x)$ is some proposition involving only notions that have already been defined.

EXAMPLE 1.   Let us give a formal definition of even number and odd number, as applied to integers. We will say that a number is **even** if and only if it is a multiple of 2. In symbols, then,

$$x \text{ is even } \longleftrightarrow \exists a : x = 2a,$$

with the quantifiers ranging over the integers (note also the *implied* universal quantifier on x). This definition applies to all integers x; by applying the criterion given in the definition, we can determine that certain integers are even and certain integers are not even. Thus we know that -6 is even, since we can take $a = -3$ in this definition, and 0 is even, since we can take $a = 0$; but 1 is not even, since for no integer a is $2a = 1$. Similarly, we make the following definition of **odd** integer:

$$x \text{ is odd } \longleftrightarrow \exists a : x = 2a + 1.$$

For instance, 1 is odd, since $1 = 2 \cdot 0 + 1$.   ◆

We might summarize this subsection by saying that proofs of propositions are *relative*—dependent upon the definitions and axioms. Anyone who is willing to accept the meaning of terms as we define them, and who believes our axioms, must also accept the truth of the propositions we prove.

VALID REASONING

We now turn to the local question. How do we know when a sequence of pencil marks on a piece of paper or a flow of sound emanating from a human larynx constitutes a valid mathematical proof? A mathematical proof is a sequence of statements. Each statement in the sequence must either

1. Be an assumption, or
2. Be a proposition proved previously, or
3. Follow logically from one or more preceding statements in the proof.

The assumptions may be axioms or definitions, or they may be temporary assumptions made in the course of the proof, according to rules that we will discuss below. For example, a proof might contain sentences such as "By definition, since T is a tree, it has no cycles" or "Let us assume that the function f is not one-to-one" or "By Theorem 7, we know that T has a vertex with degree less than 6."

What about item 3: When does a statement follow logically from preceding statements? The answer is amazingly simple to state and is contained in the definitions of the logical connectives and the theorems about logical equivalence and logical implication that we discussed in Sections 1.1 and 1.2:

A proposition Q follows logically from propositions P_1, P_2, ..., P_n
if Q must be true whenever P_1, P_2, ..., P_n are true.

In particular, every logical implication (or logical equivalence), such as the ones stated in Theorems 1 and 2 of Section 1.1, is the basis for a valid step in a proof. For example, if we have a proposition P at some step in our proof, and if we have the proposition $P \rightarrow Q$ at some other step, then by the definition of the logical connective $\wedge$ and part (a) of Theorem 2 in Section 1.1, we can write Q as the next step in our proof. In words we are saying that if P is true, and if P implies Q, then Q must be true. Classical logicians accepted this because of an inherent belief that it represents valid human reasoning; they called it *modus ponens*, Latin for "method of affirmation." We accept it because the truth table definitions that we gave in Section 1.1 allowed us to prove part (a) of Theorem 2. To make this example concrete with a nonmathematical application, suppose that we look outside and observe that it is now raining. Suppose further that the meteorological proposition "If it is raining, then there are clouds in the sky" is a true one. By *modus ponens* we may conclude that there are now clouds in the sky.

On the other hand, certain modes of reasoning are not valid. For example, there is no logical implication $Q \wedge (P \rightarrow Q) \implies P$. Indeed, if Q is true, then $Q \wedge (P \rightarrow Q)$ is true, and we can make no inferences at all about P. Thus it would be incorrect to justify a statement in a proof by trying to invoke the line of reasoning given by this *nonlogical* implication.

EXAMPLE 2. Suppose that it has been established that all persons in political party X voted for proposal Y. In other words, if a person is a member of political party X, then that person voted for proposal Y. Suppose further that Ms. Jones voted for proposal Y. Can we infer that Ms. Jones belongs to political party X? Certainly not. Such reasoning is invalid, both in the informal context of political discussions and in the more formal setting of mathematical proofs. Here Q is "Ms. Jones voted for proposal Y," and P is "Ms. Jones belongs to political party X." We are told that Q is true and that $P \rightarrow Q$ is true, but we cannot thereby conclude that P is true. ◆

We will not dwell on such fallacious reasoning but rather focus on the various valid methods of proof that are available.

Constructing a mathematical proof consists of two major steps. First you have to figure out what is going on—what it is that is really forcing the proposition you are trying to prove to be true. This is the hard part, and it will take anywhere from a few seconds to a lifetime. You need to experiment with the situation facing you, try some special cases, make tentative assumptions, play around with equations or inequalities, or draw pictures. In general, you have to immerse yourself in the situation at hand in an attempt to get at its key ingredients. The second step is to write down a convincing and logically valid argument, in an acceptable style. Keep in mind that a proof is an essay that you write to explain why a mathematical proposition is true.

Do not be discouraged if at first you have difficulty giving valid proofs. It is important to have an experienced reader look over your proofs to point out any fallacies and slips in good style. Eventually, you will be writing good proofs yourself, and you will be able to tell whether a proof you read, whether yours or that of someone else, is correct.

Certain conventions have evolved for writing mathematical proofs, and you should follow them. We usually indicate the beginning of a proof by writing the word *Proof*, and we indicate the end with a special symbol. (In this book we use ■; another time-honored marking for the end of a proof is "QED," which stands for the Latin *quod erat demonstrandum*—"which was to have been proved.") The proof should use complete sentences and be grammatically correct; even mathematical formulas are sentences, with verbs such as "equals" or "is less than." Also, the proof should flow smoothly and prosaically. Appropriate use of such words as "therefore," "thus," or "hence" can make the proof read well. (In more formal settings, proofs can be constructed so that the statements constituting the proof are given entirely in symbols. Statements can be listed, one per line, with the reason for each step listed opposite to it. The reader has probably encountered this style for writing proofs in geometry. We prefer the prose paragraph form, except for proofs or parts of proofs that are pure calculations.) In general, there must be a reason behind each statement made in a proof. Sometimes, if it is obvious, a small step or a reason may be omitted, but the beginning student of proof-writing should definitely err on the side of writing too much rather than trying to streamline his or her proof. You should study mathematical proofs, in this book and elsewhere, and attempt to adopt their style.

Ultimately, a correct proof is simply an argument for the truth of a proposition that is convincing to a person with the proper mathematical and logical knowledge to understand it. In the rest of this section we will study various techniques for constructing such arguments.

METHODS OF PROOF

Having established what modes of reasoning we will accept in justifying steps in a proof (namely, the ones inherent in the definitions and theorems of Sections 1.1 and 1.2), we turn to a discussion of the various forms that our proofs typically take. First let us deal with the universal quantifier, since most propositions to be proved in mathematics are universally quantified.

As we saw in Section 1.2, a universally quantified proposition $\forall x\colon P(x)$ asserts that the proposition $P(x)$ is true *for every* value of the variable x. Thus to prove $\forall x\colon P(x)$, we must prove the proposition $P(x)$ for an arbitrary, unspecified x in the domain under discussion. Here is a simple example from college algebra.

THEOREM 1.   *For every positive real number x other than 1, $x^{1/\ln x} = e$, where $\ln x$ denotes the natural logarithm of x, and e is the base of the natural logarithm (≈ 2.7).*

Proof.   By the exponent law for logarithms ($\ln a^b = b\ln a$), we have

$$\ln(x^{1/\ln x}) = \frac{1}{\ln x} \cdot \ln x = 1.$$

The only number whose natural logarithm is 1 is e; therefore, $x^{1/\ln x} = e$.   ∎

Evaluating $5^{1/\ln 5}$ with a calculator and obtaining $2.71828\ldots$ as the "answer" does not prove this theorem, although performing such calculations is probably a reasonable thing to do when first confronted with this proposition, to get a feel for what it is saying and to convince ourselves that it is plausible. Similarly, proving that $e^{1/\ln e} = e$ (which is trivially true) does nothing toward proving this universally quantified proposition.

Many of the propositions that we wish to prove are in the form of an implication, $P \rightarrow Q$. Explicitly, there are hypotheses (P) and a conclusion (Q). The proposition is that *if* the hypotheses are met, *then* the conclusion must follow. (In fact, there are always some hypotheses around, whether or not we treat the proposition as an implication. In Theorem 1, the hypothesis was that x was a positive real number different from 1.) In the remainder of this section we consider several different methods for proving implications, all of them important.

Direct proof.   To prove $P \rightarrow Q$, prove Q, using P as an assumption.

Indirect proof.   To prove $P \rightarrow Q$, prove the contrapositive $\overline{Q} \rightarrow \overline{P}$ directly. In other words, prove the negation of P, using the negation of Q as an assumption.

Proof by contradiction.   To prove a proposition by contradiction, assume the negation of the proposition to be proved and derive a contradiction. In the case of an implication, to prove $P \rightarrow Q$, assume both P and $\overline{Q}$, and derive a contradiction.

Proof by cases. To prove $P \to Q$, find a set of propositions P_1, P_2, ..., P_n, $n \geq 2$, at least one of which has to hold if P holds—in other words, so that $P \to (P_1 \lor P_2 \lor \cdots \lor P_n)$. Then prove the n propositions $P_1 \to Q$, $P_2 \to Q$, ..., $P_n \to Q$.

Vacuous proof. To prove $P \to Q$, prove $\overline{P}$. In other words, show that the hypothesis cannot hold.

We now examine these methods more closely, by giving a justification and an example or two for each. The propositions we will be proving in these examples are simple, but the form is representative of proofs occurring throughout this book and throughout mathematics. All our quantifiers here are assumed to range over the integers (positive, negative, and zero), and when we say "number" in these examples we mean "integer." We also will assume the following basic fact as an axiom: *Each integer is either even or odd, but not both* ("even" and "odd" were defined in Example 1). We will also assume all the usual facts about arithmetic and algebra that one learns in high school.

Direct proof is justified by the truth table definition of implication ($\to$). Recall that $P \to Q$ is false only in the case that P is true and Q is false. Thus if we want to *prove* $P \to Q$ (i.e., to show that $P \to Q$ is always true), we do not need to concern ourselves with the possibility that P might be false, for in that case $P \to Q$ will necessarily be true. It is only in the case that P is true that we need to worry about the truth of Q. Thus it is legitimate to construct a proof that starts by *assuming* that P is true and reaches the conclusion that—in that case— Q must be true. (It may happen that Q is true regardless of whether or not P is true. In this case the direct proof is still valid, but the hypothesis P will probably not have been actually *used* to justify any of the steps in the proof. Ideally, the person constructing the proof should restate the proposition being proved to make it stronger: It says more if the unneeded hypothesis is eliminated. In practice, however, it is rare that a proposition a student is likely to encounter at this level has an unneeded hypothesis. Thus if you discover at the end of a proof that you did not use the hypothesis, then more likely than not you have made a mistake in your proof.)

We annotate our proof of the next theorem (and some others in this section) with bracketed comments; they are not part of the proof.

THEOREM 2. *The square of an odd number is always odd.*

Proof. Explicitly, what we are proving is the proposition

$$\forall x \colon (x \text{ is odd } \to x^2 \text{ is odd }).$$

Let x be an odd number. [Since this is a direct proof, we get to *assume* the hypothesis that x is odd. Note that we are proving a universally quantified proposition, so we need to treat x as an arbitrary, unknown odd number; we could not, for instance, let $x = 17$.] Then $x = 2n + 1$, for some integer n. [Here we appealed to the *definition* of odd number; note the existential quantifier on n.] By algebra, we have $x^2 = (2n + 1)^2 = 4n^2 + 4n + 1 = 2(2n^2 + 2n) + 1$. Thus, since we have written x^2 in the form $2(\text{something}) + 1$, it is an odd number by definition. ∎

THEOREM 3. *The square of an even number is always even.*

Proof. The proof is similar to the proof of Theorem 2 and is omitted. ∎

One might call the proof given here for Theorem 3 a "proof by intimidation." Obviously, this is not really a proof at all. Another example of intimidation is the use of the word "clearly" in place of an explicit reason for a step in a proof. Such devices are appropriate when used properly; the alternative to using them is a waste of natural resources (the trees used to make the paper a proof is written on) and time (of both the writer and the reader of the proof). However, it is important that when details of a proof are omitted there is no doubt how they could be supplied if requested (see Exercise 1).

Indirect proof is justified by part (l) of Theorem 1 in Section 1.1: The contrapositive of a proposition is logically equivalent to the proposition itself. Thus if we can prove that the contrapositive is true, then we know automatically that the proposition is true. Usually, we prove the contrapositive directly: Assume the negation of the original conclusion (i.e., the hypothesis of the contrapositive) and derive through a sequence of logically valid steps the negation of the original hypothesis (i.e., the conclusion of the contrapositive). An indirect proof is also sometimes called a proof by contradiction, because we conclude the proof by deriving a contradiction to the original hypothesis, namely its negation.

THEOREM 4. *If n^2 is even, then n is even.*

Proof. [Note the implicit universal quantifier in this proposition: The assertion is really that *for all* n, if n^2 is even, then n is even. Our proof must treat n as an arbitrary, unknown number.] Suppose that n is not even. [This is the negation of the conclusion; an indirect proof always begins by assuming the negation of the conclusion.] Then n is odd [we agreed to accept the axiom that every integer is either even or odd], so by Theorem 2, n^2 is also odd and hence not even [we are using a previously proved proposition here, as well as our axiom about even and odd numbers]. This contradicts the assumption that n^2 is even [in other words, we have proved the negation of the hypothesis that n^2 is even, so our indirect proof is finished]. ∎

The justification for **proof by contradiction** is part (e) of Theorem 2 in Section 1.1. If the assumption of $\overline{P}$ leads to a falsehood, then P must be true. As a variation on this theme, if we can derive P by assuming $\overline{P}$, then by part (d) of Theorem 2, we have proved that P is true. In the case of an implication ($P \rightarrow Q$), to assume the negation of the proposition we wish to prove is to assume $P \wedge \overline{Q}$, by part (k) of Theorem 1.

Students often have difficulty with proof by contradiction and sometimes misuse it. It is certainly *not* valid to assume the proposition you are trying to prove. A proof by contradiction works by *assuming the negation of what you want to prove to be true*, and showing that this assumption is untenable; hence the negation cannot possibly be true, so the original proposition itself has to be true.

Our example of proof by contradiction is a classic dating back to the ancient Greeks. Again, the bracketed comments, which are not part of the proof itself, explain the subtleties in this argument.

THEOREM 5. *The square root of 2 is not a rational number.*

Proof. Suppose, on the contrary, that $\sqrt{2}$ is rational. This means that $\sqrt{2} = a/b$ for some positive integers a and b [note the use of a definition here, in this case the definition of what a (positive) **rational number** is]. By reducing this fraction to lowest terms, we may assume that a and b have no common factors. [This step may seem a little obscure, but tricks like this come up often in proofs. Here is what this bit of rhetoric means. If a and b have no common factors, we leave them alone. If they do have common factors, we factor out all the common factors and obtain an equivalent fraction, say c/d, in lowest terms (i.e., such that c and d have no common factors and $a/b = c/d$). We then *rename* c and d as a and b, respectively; this is justifiable, since the letters were chosen arbitrarily to begin with. In either case, then, we end up with a/b in lowest terms.] Squaring both sides of $\sqrt{2} = a/b$, we obtain $2 = a^2/b^2$, or $2b^2 = a^2$. By definition, therefore, a^2 is even, and hence by Theorem 4, a is even. Writing $a = 2n$, we have $2b^2 = 4n^2$, whence, upon dividing both sides by 2, we have $b^2 = 2n^2$. At this point we see that b^2 is even and hence (again by Theorem 4) that b is even. We have now deduced a contradiction, namely, that a and b are both even (i.e., have the common factor 2) while at the same time a and b have no common factors. Hence our original assumption (that $\sqrt{2}$ was rational) is false, and hence $\sqrt{2}$ is not a rational number. ∎

The reader may have tried to read the proof given above much as one would read a newspaper article. It cannot be done. Reading mathematical proofs requires very careful attention to every word. The arguments must be followed exactly and the computations verified. The reader should stop after every sentence (or even more often) and be sure that he or she agrees with the argument up to that point. It is not unusual for a proof to require five or more readings before its validity becomes clear to the reader.

To justify a **proof by cases**, we appeal to the following logical implication, which holds for every $n \geq 2$:

$$((P \to (P_1 \vee P_2 \vee \cdots \vee P_n)) \wedge (P_1 \to Q) \wedge (P_2 \to Q) \wedge \cdots \wedge (P_n \to Q)) \implies (P \to Q).$$

If we let $n = 1$ here, then we obtain part (h) of Theorem 2 in Section 1.1, so in some sense this logical implication is a generalization of part (h) of that theorem. Intuitively, it says that if the hypothesis P can be broken down into cases P_1, P_2, ..., P_n, and if it can be verified that at least one of the cases has to hold if P holds (this part is usually clear), and if Q follows in each case, then Q follows from P.

THEOREM 6.   *For every nonzero integer x, $x^2 > 0$.*

Proof.   There are two cases to consider, since every nonzero integer is either positive or negative. If $x > 0$, then certainly $x^2 > 0$. On the other hand, if $x < 0$, then since the product of a negative number and a negative number is positive, again $x^2 > 0$.   ∎

In this proof there were two cases. It is not unusual to have proofs with three, four, or even more cases (the proof of the four-color theorem had over a thousand). Often it is helpful to organize the possibilities into cases and subcases. Sometimes there are two (or more) cases that are practically identical. This is an instance of a general phenomenon that appears over and over in mathematics: symmetry. If the symmetry of cases for a proposition is clear when we are constructing a proof, we need only point out that the cases are symmetric and give the proof for just one of them. We usually signal this technique by using the words "without loss of generality." Our next example deals with some of these complexities.

THEOREM 7.   *For all real numbers x and y, $|x + y| \leq |x| + |y|$.*

Proof.   Recall the definition of absolute value: $|a| = a$ if $a \geq 0$, and $|a| = -a$ if $a < 0$. There are three cases to consider.

Case 1.   If both x and y are nonnegative, then so is their sum, and we have $|x + y| = x + y = |x| + |y|$.

Case 2.   If both x and y are negative, then so is their sum, and we have $|x+y| = -(x + y) = (-x) + (-y) = |x| + |y|$.

Case 3.   Finally, suppose that one of x and y is nonnegative and the other is negative. By symmetry, without loss of generality, we may assume that $x \geq 0$ and $y < 0$.

There are two subcases to consider.

Subcase 3a. Suppose that $|x| \geq |y|$; since $|y| = -y$ in the case under consideration, this means that $x \geq -y$. Thus $x + y \geq 0$, and we have $|x + y| = x + y = x - (-y) = |x| - |y| \leq |x| \leq |x| + |y|$.

Subcase 3b. Suppose that $|x| < |y|$; again since $|y| = -y$, this means that $x < -y$. Thus $x + y < 0$, and we have $|x + y| = -(x + y) = (-x) + (-y) = (-|x|) + |y| \leq |y| \leq |x| + |y|$. ∎

Finally, to justify a **vacuous proof**, we need only refer to the truth table definition of implication (Table 1.4). If P is false, then $P \to Q$ is true. Thus we can prove an implication merely by showing that the hypothesis is false. As a simple example, suppose that we wish to verify that the proposition $P(n)$ holds for $n = 3$, where $P(n)$ is "If $n \geq 4$, then $n^3 < 3^n$." In other words, we wish to prove $P(3)$, which is "If $3 \geq 4$, then $27 < 27$." Since the hypothesis in this implication is false, the implication is vacuously true (it does not matter whether the conclusion is true or false—in this case it happens to be false). In fact, $\forall n\colon P(n)$ is a theorem, where the domain of discussion is the natural numbers (Exercise 14 in Section 5.3).

THEOREM 8. *Every natural number that cannot be expressed as the sum of four squares is exactly divisible by 5.*

Proof. This proposition is the universally quantified implication that if x cannot be expressed as the sum of four squares, then x is divisible by 5. By Lagrange's four-square theorem, there is no natural number that cannot be expressed as the sum of four squares. Hence the proposition is (vacuously) true. ∎

The conclusion of Theorem 8 could have been anything at all (e.g., "... is both even and odd" or "... is the year of your birth"). Since the hypothesis is never true, the implication is always true, regardless of the truth value of the conclusion. Admittedly, this proposition is rather unsatisfying—it does not say anything useful. We will see an example or two in this book, however, of important propositions or parts of important propositions with vacuous proofs.

There is one additional proof technique of utmost importance in the mathematical sciences, especially discrete mathematics and computer science: proof by mathematical induction. We postpone a discussion of it to Section 5.3.

We close with a word about terminology. Formal mathematics has been around for a couple of thousand years. A lot of terminology and traditions have developed, in many languages, in many countries, over many centuries. Some persist to this day, and the student of discrete mathematics may as well join in this heritage. There are several words used to mean "proposition that has been proved to be true," depending on the context. The most general is **theorem**. A **lemma** is a proposition that may or may not be of much interest in its own right but is useful in proving other propositions. A **corollary** is a proposition whose truth follows directly from a previous proposition. If someone thinks that a statement is true but has not yet developed a proof, the statement is called a **conjecture**.

SUMMARY OF DEFINITIONS

axiom: a self-evident statement that is accepted without proof (e.g., $\forall x\colon \forall y\colon x+y = y+x$ in the setting of the real numbers).

conjecture: a proposition believed to be true, for which no proof has yet been given (e.g., the statement that there are an infinite number of twin primes).

corollary: a proposition whose proof follows easily from a previous theorem.

definition: a proposition giving the meaning of a term or notation; the proposition is accepted as true (e.g., "A prime number is an integer greater than 1 that has no positive integer divisors other than 1 and itself").

direct proof of $P \to Q$: a proof that begins with the assumption that P is true and concludes that Q must be true (e.g., the proof of Theorem 2).

even integer: an integer that is a multiple of 2 (e.g., 42548, 0, and -2).

indirect proof of $P \to Q$: a proof that begins with the assumption that Q is false and concludes that P is false; in other words, a direct proof of the contrapositive $\overline{Q} \to \overline{P}$ (e.g., the proof of Theorem 4).

lemma: a proposition that has been proved to be true and is helpful in proving other propositions.

odd integer: an integer that is one greater than a multiple of 2 (e.g., 215, which is $2 \cdot 107 + 1$).

postulate: an axiom.

proof by cases: a proof that considers an exhaustive set of possibilities and produces the desired conclusion in each case (e.g., the proof of Theorem 7).

proof by contradiction: a proof of a proposition that begins with the assumption that the proposition is false and concludes with a contradiction (e.g., the proof of Theorem 5).

rational number: a real number that can be expressed as a/b, where a is an integer and b is a positive integer (e.g., $-0.6\overline{6}$, which is $-\frac{2}{3}$, and 17, which is $\frac{17}{1}$).

theorem: a proposition that has been proved to be true.

vacuous proof of $P \to Q$: a proof in which P is shown to be false (e.g., the proof of Theorem 8).

EXERCISES

Note that many of these exercises state a proposition and ask you either to prove that it is true or to prove that it is false. Before starting each proof, be sure you understand completely the proposition under consideration. You should write it out explicitly, with quantifiers, and then experiment with it a little to see whether you believe it is true (and, perhaps, give you some ideas for constructing the proof or counterexample). Unless otherwise stated, all explicit or implied quantifiers are assumed to range over the integers.

1. Prove Theorem 3.

2. Prove the following propositions. Note the implied universal quantifiers.
   - (a) The sum of two odd numbers is even.
   - (b) The sum of two even numbers is even.
   - (c) The sum of an even number and an odd number is odd.

3. State and prove a proposition that gives the **parity** (oddness or evenness) of the product of two integers in terms of the parity of the two integers.

4. Prove or disprove each of these propositions.
   - (a) If p and q are prime numbers, then $pq + 1$ is a prime number.
   - (b) If x and y are positive integers with $x > y + 1$, then $x^2 - y^2$ is not prime.
   - (c) If Tom has at most n brothers and Chris is Tom's sibling, then Chris has at most n brothers.
   - (d) If a and b are positive integers, then $\sqrt{a^2 + b^2} = a + b$.
   - (e) Every even number can be written as the sum of an odd number and a perfect square.
   - (f) The product of a multiple of 6 and a multiple of 10 is a multiple of 60.

5. Prove or disprove the following propositions.
   - (a) Every multiple of 6 is also a multiple of 3.
   - (b) Every multiple of 3 is also a multiple of 6.
   - (c) For every x, if x is not a multiple of 2, then x is not a multiple of 6.

6. Prove the following propositions.
   - (a) If n is an integer and a solution to $n^2 = n + 4$, then $n > 7$.
   - (b) If n is an integer and a solution to $n^2 = n + 4$, then $n < 7$.
   - (c) Every natural number between 9 and 14, inclusive, can be written as the sum of the squares of three natural numbers.

7. Prove that the average of any finite collection of real numbers must be less than or equal to at least one of the numbers in the collection. [*Hint*: Use an indirect proof.]

8. Prove the following propositions.
   - (a) There is no largest prime number. [*Hint*: Assume as given the fundamental theorem of arithmetic, which asserts that every natural number greater than 1 has a (unique) prime factorization. If there were only a finite number of primes, then one could look at their product plus 1.]
   - (b) $\forall x\colon \forall y\colon ((x > 0 \wedge y > 0 \wedge x^2 < y^2) \rightarrow x < y)$. Assume that the quantifiers range over the real numbers. Do *not* assume any properties of real numbers that involve the square root. [*Hint*: Try an indirect proof.]
   - (c) If n is not a multiple of 3, then $n^2 - 1$ is a multiple of 3. [*Hint*: Try a proof by cases. A number that is not a multiple of 3 will have one of two remainders when divided by 3, so it can be written in one of two forms.]

(d)   Every prime number greater than 3 is of the form $6n + 1$ or $6n + 5$.

9. What is wrong with the following "proofs"? In each case, find the mistake and then either correct the proof or show that the proposition is false.

(a)   Proposition: If n^2 is a multiple of 8, then n is a multiple of 8.
Proof: Let $n = 8m$. Then $n^2 = 64m^2 = 8(8m^2)$, which is a multiple of 8, as desired. QED.

(b)   Proposition: If n is not a multiple of 4, then $n^2 - 1$ is a multiple of 4.
Proof: Since n is not a multiple of 4, n must be odd, say $n = 2m + 1$. Then $n^2 - 1 = 4m^2 + 4m = 4(m^2 + m)$, as desired. QED.

(c)   Definition: A number N is "twinly" if for every prime number p that is a factor of N, the number $p + 2$ is a factor of N.
Proposition: 76 is not twinly.
Proof: Let $p = 17$. Then $17 + 2 = 19$ is a prime factor of 76, but 17 is not a factor of 76. Hence 76 is not twinly. QED.

(d)   Proposition: Suppose that the proposition $P(x, y)$, containing two free variables x and y, satisfies the following two axioms: $\forall a\colon \forall b\colon (P(a, b) \to P(b, a))$ and $\forall a\colon \forall b\colon \forall c\colon ((P(a, b) \land P(b, c)) \to P(a, c))$. Then P also satisfies the following: $\forall a\colon P(a, a)$.
Proof: Let a be given. By the first of the axioms, we have both $P(a, b)$ and $P(b, a)$. But if we now apply the second of the axioms, letting the a, b, and c there be a, b, and a, respectively, then we obtain $P(a, a)$, as desired. QED.

10. Prove that the equation $3x - 6 = 0$ has a unique solution. [*Hint*: Be sure to write down *exactly* what proposition you are trying to prove.]

11. Prove or disprove the following propositions.
(a)   If a is a multiple of b, and b is a multiple of c, then a is a multiple of c.
(b)   6083824773 is a multiple of 13.
(c) $n^2 + n$ is always an even number.

12. Prove or disprove the following propositions.
(a) $\forall x\colon \forall y\colon x < x + |y| + 1$.
(b)   If x is the product of four distinct prime numbers, then $x > 200$.
(c) $\forall x\colon \forall y\colon (x - y)^2 > 0$.

13. Prove or disprove the following propositions.
(a)   The difference of the squares of two odd numbers is divisible by 4.
(b)   The sum of the squares of two odd numbers is not divisible by 4.

14. Prove that the number 12345678903 cannot be written as the sum of two squares. [*Hint*: Consider three cases, based on the parity of the two numbers whose squares might add up 12345678903.]

15. Prove that strictly between every two distinct real numbers there is another real number.

16. Prove or disprove the following propositions.
  (a)   If n^2 is a multiple of 4, then n is a multiple of 4.
  (b)   If n^3 is a multiple of 2, then n is a multiple of 2.
  (c)   If n^3 is a multiple of 2, then n^2 is a multiple of 4.
  (d)   If n^2 is a multiple of 2, then n^2 is a multiple of 4.

17. Use the fact that $\sqrt{2}$ is not rational to prove that $5\sqrt{2}$ is not rational. [*Hint*: Try a proof by contradiction, using the fact that the product of rational numbers is rational.]

18. Suppose that we were to adopt the falsehood $0 = 1$ as an axiom.
  (a)   Show that we would then be able to prove *every* proposition.
  (b)   Explain why the fact that we would then be able to prove all the true propositions does not contradict Gödel's incompleteness theorem.

Challenging Exercises

19. Prove that $\sqrt{3}$ is not a rational number.

20. Prove that $\sqrt[3]{2}$ is not a rational number.

21. If r and s are irrational real numbers, can r^s be rational? Formulate your answer explicitly as a true proposition and prove it. [*Hint*: Contemplate $\left(\sqrt{2}^{\sqrt{2}}\right)^{\sqrt{2}}$.]

22. If r and s are irrational real numbers, can r^s be irrational? Formulate your answer explicitly as a true proposition and prove it. [*Hint*: Contemplate $\sqrt{2}^{1+\sqrt{2}}$.]

23. The **arithmetic mean** of two positive real numbers x and y is, by definition, $(x + y)/2$; and their **geometric mean** is $\sqrt{xy}$. For example, the arithmetic mean of 5 and 20 is 12.5 and their geometric mean is 10. Prove that the arithmetic mean is never less than the geometric mean, and that the two means are equal if and only if the two numbers are equal. [*Hint*: Note all that is being asked for. Try starting with the inequality $(\sqrt{x} - \sqrt{y})^2 \geq 0$.]

24. Prove that the square of an odd integer always leaves a remainder of 1 when divided by 8.

25. Prove that the product of any five consecutive positive integers (such as $17 \cdot 18 \cdot 19 \cdot 20 \cdot 21$) must be divisible by 120.

26. Prove or disprove each of the following propositions. [*Hint*: Review the fundamental theorem of arithmetic (see Exercise 8a).]
  (a)   A number that is both a multiple of 3 and a multiple of 5 is a multiple of 15.
  (b)   A number that is both a multiple of 6 and a multiple of 10 is a multiple of 60.
  (c)   The base 2 logarithm of 3 is irrational.

Exploratory Exercises

27. Do you agree that there are philosophical problems with a proof in which a high-speed computer is needed to carry out the computations involved? (Assume that the computations are so lengthy that no human being possibly could, or would want to, verify them by hand.)

    (a) For example, suppose that you program a computer to divide the number 76543 by all the odd numbers from 3 to $\sqrt{76543}$ and print out the message "not prime" if the remainder of any of these computations is 0. Suppose that the computer executes your program and halts without printing out this message. Do you feel any less certain that 76543 is prime than that 13 is prime?

    (b) Find out more about the controversy surrounding computer-assisted proofs by consulting Swart [301] and Tymoczko [308,309].

28. The philosophy of mathematics is an area in which there is no agreement, even among mathematicians.

    (a) In what sense do you believe mathematical objects *exist*? Does the number 7 exist? If so, what is it? Does the collection of all natural numbers exist?

    (b) Find out about the philosophical foundations of mathematics by consulting Benacerraf and Putnam [16], N. D. Goodman [121], Henkin [147], Kreisel [188], Lehman [195], Russell [262], Snapper [284,285], and Wilder [319].

29. Investigate the following problem (come up with a conjecture and then prove it): Which even positive integers are the sum of two odd composite natural numbers? (A composite number is an integer greater than 1 that is not prime.) See Just and Schaumberger [166] for the solution.

30. The quotation at the beginning of this section is due to a prolific eighteenth-century Swiss mathematician whose work laid the foundations for much of discrete (as well as continuous) mathematics. Find out the circumstances in which he said it and what he meant by it, by consulting Bell [15].

31. Find out more about Gödel's incompleteness theorem and unprovable true propositions by consulting Chaitin [37], DeLong [63], Hofstadter [152,153], Kolata [184], Monk [215], Nagel and Newman [219], and Smullyan [281,282].

32. Find out about the notion of proof in mathematics, from social, historical, and philosophical perspectives, by consulting DeMillo, Lipton, and Perlis [64], Galda [93], Lakatos [192], and Renz [244].

33. It has been proved that there exist theorems whose statements are short yet whose shortest proof are very long. Find out about this sad state of affairs by consulting Norwood [224] and J. Spencer [288,289].

34. In mathematics, unlike other fields, it is often possible to *prove* that something is impossible. Explore this topic by consulting Richards [246].

35. Find out more about the role of axioms in mathematics by consulting Montague and Montgomery [216].

36. Learn more about writing proofs in mathematics by consulting Bittinger [20], Morash [217], and Solow [286].

37. Find out more about Kurt Gödel by consulting Dawson [62] and Wang [316].

SECTION 1.4
BOOLEAN FUNCTIONS

"But if everybody obeyed that rule," said Alice, who was
always ready for a little argument, "and if you only spoke
when you were spoken to, and the other person always waited
for you to begin, you see nobody would ever say anything."

—Lewis Carroll, *Through the Looking Glass*

In Section 1.1 we constructed compound propositions that were true if and only if the propositional variables in the compound propositions had certain truth values. For example, the compound proposition $P \to Q$ is true in exactly three cases: that in which P is true and Q is true, that in which P is false and Q is true, and that in which P is false and Q is false. The truth value of $P \to Q$ depends only on the truth values of P and of Q. Let us look at what we did from a different point of view, by dispensing with the propositions and looking only at the truth values.

First we make a change in notation: We will denote truth by the number 1 and falsehood by the number 0, and we will call these values **Boolean constants**. Rather than looking at propositional variables, we will consider variables that can take on one of these Boolean values; we call such a variable a **Boolean variable**. Then a compound proposition can be thought of as defining a Boolean-valued *function* of Boolean variables, or a **Boolean function** for short. (We will study the concept of function more fully and formally in Section 3.1, but for now we rely on the reader's familiarity with functions from algebra and calculus.) For example, the compound proposition $P \to Q$ defines the following Boolean function f if we let the first argument be the Boolean variable representing the truth of P and let the second argument be the Boolean variable representing the truth of Q:

$$f(0,0) = 1, \qquad f(0,1) = 1, \qquad f(1,0) = 0, \qquad f(1,1) = 1.$$

In this section we study such functions. Our primary concern will be with how best to represent and manipulate them. We will emphasize the analogy between what we are doing and the representation and manipulation of polynomial functions of real number variables using algebra—in other words, we will be doing Boolean algebra. Some of the examples will provide concrete applications of Boolean functions. (We will see in Section 3.3 that Boolean algebra can be viewed as the natural setting in which to represent relations by matrices.) We will also briefly discuss a visual representation of Boolean functions that is the foundation for circuit theory, an important area of electrical and computer engineering.

In algebra, expressions involving real variables are built up from numbers and variables using a few basic operations (such as addition, subtraction, and multiplication). These expressions represent certain real-valued functions of these variables. For example, the function $f(x, y) = x^2 - 3xy + 5$ is defined by the expression $x^2 - 3xy + 5$ (meaning $x \cdot x - 3 \cdot x \cdot y + 5$), which is made up of the constants 3 and 5 and the variables x and y, combined in a certain way by multiplication, addition, and subtraction. Similarly, we will build up expressions representing Boolean functions from the Boolean constants 0 and 1 and Boolean variables, using just three operations, those corresponding to the logical connectives "or," "and," and "not." We only need to explain the notation we will use for these three operations, and define the way in which the two constants 0 and 1 behave with respect to them. We could continue to use the logical connective notation ($\wedge$, $\vee$, and $\bar{}$), and some authors do so, but instead, we will use the simpler algebraic notation.

Boolean addition, denoted by the ordinary plus sign, corresponds to the logical "or." Recalling that 0 and 1 are intended to represent false and true, respectively, we define the **Boolean sum** by

$$0 + 0 = 0, \qquad 0 + 1 = 1, \qquad 1 + 0 = 1, \qquad 1 + 1 = 1.$$

Boolean multiplication, denoted by the ordinary raised-dot multiplication sign or by juxtaposition, corresponds to the logical "and." Thus we define the **Boolean product** by

$$0 \cdot 0 = 0, \qquad 0 \cdot 1 = 0, \qquad 1 \cdot 0 = 0, \qquad 1 \cdot 1 = 1.$$

The **complement** of a Boolean value, just like the negation of a proposition, is denoted by an overbar, and its definition is

$$\overline{0} = 1, \qquad \overline{1} = 0.$$

Parentheses or the length of the overbar are used to indicate the order of operations, with multiplication otherwise taking precedence over addition. A **Boolean expression** is any expression constructed by applying any sequence of these operations to 0, 1, and variable names. For example, 0, 1, x, $0 \cdot x$, and $(x\,y)z$, are all Boolean expressions. We say that two Boolean expressions B_1 and B_2 are **equivalent** (and write $B_1 = B_2$) if they represent the same function of the variables appearing in either. Explicitly in terms of these variables, $B_1(x_1, x_2, \ldots, x_n)$ and $B_2(x_1, x_2, \ldots, x_n)$ are equivalent if and only if the values of $B_1(x_1, x_2, \ldots, x_n)$ and $B_2(x_1, x_2, \ldots, x_n)$ are the same for every choice of values for x_1, x_2, ..., x_n.

EXAMPLE 1. Find the value of the Boolean expression $(x + xy) \cdot \overline{y + 1}$ when $x = 1$ and $y = 1$. In other words, compute $f(1, 1)$, where f is the Boolean function of two variables defined by $f(x, y) = (x + xy) \cdot \overline{y + 1}$.

Solution. The term xy has the value $1 \cdot 1 = 1$. Therefore, $x + xy = 1 + 1 = 1$. Similarly, $y + 1 = 1 + 1 = 1$, so $\overline{y + 1} = \overline{1} = 0$. Therefore, the entire expression has the value $1 \cdot 0 = 0$. In other words, $f(1, 1) = 0$. Note that in fact this expression will always have the value 0 (no matter what values x and y have), because the second factor, $\overline{y + 1}$, has the value $\overline{1} = 0$, regardless of whether $y = 0$ or $y = 1$, and hence the product will be 0, regardless of the value of the first factor $(x + xy)$. Therefore, f is just the constant function $f(x, y) = 0$. In other words, $(x + xy) \cdot \overline{y + 1}$ and 0 are equivalent Boolean expressions, that is, $(x + xy) \cdot \overline{y + 1} = 0$. Note that two expressions can be equivalent even though they look quite different. ◆

Because of the way in which we defined Boolean addition, multiplication, and complementation—to correspond to logical disjunction, conjunction, and negation, respectively—the identities (a)–(i) given in Theorem 1 of Section 1.1 hold for Boolean algebra if we replace ∨ by +, ∧ by ·, F by 0, T by 1, ⟺ by =, and propositional variables by Boolean variables. We restate these identities here. Their proofs by truth table from Section 1.1 are valid in this setting, since the operations were defined in exactly the same way (only the notation has changed).

THEOREM 1. *The following identities (Boolean equivalences) hold for all Boolean variables x, y, and z.*

(a) *Associative laws*
$$x(y\,z) = (x\,y)z$$
$$x + (y + z) = (x + y) + z$$

(b) *Commutative laws*
$$x\,y = y\,x$$
$$x + y = y + x$$

(c) *Distributive laws*
$$x(y + z) = x\,y + x\,z$$
$$x + y\,z = (x + y)(x + z)$$

(d) *Identity laws*
$$x \cdot 1 = x$$
$$x + 0 = x$$

(e)	*Dominance laws*	$x \cdot 0 = 0$
		$x + 1 = 1$
(f)	*Idempotent laws*	$x \cdot x = x$
		$x + x = x$
(g)	*Complement laws*	$x \, \overline{x} = 0$
		$x + \overline{x} = 1$
(h)	*Double complement law*	$\overline{\overline{x}} = x$
(i)	*DeMorgan's laws*	$\overline{x \, y} = \overline{x} + \overline{y}$
		$\overline{x + y} = \overline{x} \, \overline{y}$

These identities can be used to manipulate and simplify Boolean expressions.

EXAMPLE 2. There are many ways to show algebraically that $(x + xy) \cdot \overline{y + 1} = 0$. We will give two proofs here, not necessarily the simplest ones but ones that use several of the parts of Theorem 1. Here is one using DeMorgan's law.

$$(x + xy) \cdot \overline{y + 1} = (x + xy) \cdot (\overline{y} \cdot \overline{1}) \quad \text{(by DeMorgan's law)}$$
$$= ((x + xy) \cdot \overline{y}) \cdot \overline{1} \quad \text{(by the associative law)}$$
$$= ((x + xy) \cdot \overline{y}) \cdot 0 \quad \text{(by definition)}$$
$$= 0 \quad \text{(by the dominance law)}.$$

Here is another proof.

$$(x + xy) \cdot \overline{y + 1} = (x \cdot 1 + xy) \cdot \overline{y + 1} \quad \text{(by the identity law)}$$
$$= (x(1 + y)) \cdot \overline{y + 1} \quad \text{(by the distributive law)}$$
$$= x((1 + y) \cdot \overline{y + 1}) \quad \text{(by the associative law)}$$
$$= x((y + 1) \cdot \overline{y + 1}) \quad \text{(by the commutative law)}$$
$$= x \cdot 0 \quad \text{(by the complement law)}$$
$$= 0 \quad \text{(by the dominance law)}. \quad \blacklozenge$$

DISJUNCTIVE NORMAL FORM

We seem to have restricted ourselves severely in constructing Boolean expressions by allowing only three operations. We will see in this subsection, however, that we can represent *every* Boolean function—of any number of variables—using Boolean expressions. This is certainly not the case with functions of a real variable: Using only the basic operations of addition, subtraction, and multiplication, we can construct only the polynomial functions; we cannot, for instance, define the logarithmic or trigonometric functions with such algebraic expressions.

Let us begin by looking at the simplest case, functions of just one variable. It is not hard to see that there are exactly four different Boolean functions of one variable:

x	$f_1(x)$	$f_2(x)$	$f_3(x)$	$f_4(x)$
0	0	1	0	1
1	0	1	1	0

Indeed, there is no way to fill in a column of values for $f(0)$ and $f(1)$ other than the four ways shown here. Now note that f_1 is the constant function $f_1(x) = 0$, f_2 is the constant function $f_2(x) = 1$, f_3 is the identity function $f_3(x) = x$, and f_4 is the complementation function $f_4(x) = \overline{x}$. In each case the function is represented by a Boolean expression.

The situation quickly becomes more complicated when we move to functions of two variables. It turns out that there are 16 different Boolean functions of two variables (we will learn how to count such things in Section 6.1). One such function was discussed at the beginning of this section, the function corresponding to the logical connective "if...then...":

x_1	x_2	$f(x_1, x_2)$
0	0	1
0	1	1
1	0	0
1	1	1

We will now show how to represent this function with a Boolean expression. Our approach will be to represent each line of the table in which the function value is 1 by a Boolean product, and then to take the Boolean sum of these products. The method we use will easily be seen to generalize to arbitrary Boolean functions of any number of variables.

Consider the first line of the table, which tells us that $f(0,0) = 1$. To capture this fact in our function, let us construct an expression that will have the value 1 exactly when $x_1 = 0$ and $x_2 = 0$. A little thought will show that $\overline{x}_1 \overline{x}_2$ will have this property. This product will be 1 if and only if each factor is 1 (by the definition of Boolean multiplication), and this occurs if and only if both x_1 and x_2 are 0 (by the definition of complementation). Similarly, the second line of the table is represented by the Boolean product $\overline{x}_1 x_2$, which has the value 1 if and only if $x_1 = 0$ and $x_2 = 1$. We do not write down an expression for the third line of the table, since we want to represent only those cases in which the function value is 1. The product for the fourth line of the table is $x_1 x_2$.

Now let us form the Boolean sum of these three Boolean products: $\overline{x}_1 \overline{x}_2 + \overline{x}_1 x_2 + x_1 x_2$. Since the sum has the value 1 if and only if at least one of the summands has the value 1 (this follows from the definition of Boolean sum), we see that this expression has the value 1 precisely in the three cases that we wish it to, and has the value 0 in the remaining case. In other words, we have shown that

$$f(x_1, x_2) = \overline{x}_1 \overline{x}_2 + \overline{x}_1 x_2 + x_1 x_2.$$

Note that each of the products in the sum contained either x_i or its complement, for each variable x_i. This representation of the function as a sum of such products is called the **disjunctive normal form** or **sum of products representation** of the function. Each of the products is called a **minterm**. By the commutativity of Boolean multiplication and addition (part (b) of Theorem 1), the order in which we list the factors and the summands does not matter. It makes sense to list the factors in their obvious order (i.e., with increasing subscripts) within each minterm. If there are more than two variables, we can and do omit parentheses indicating the grouping for multiplication, since by part (a) of Theorem 1, Boolean multiplication is associative. By convention, no parentheses are needed to remind us that the multiplications are meant to be performed before the additions.

Let us generalize to an arbitrary Boolean function of n variables, say $f(x_1, x_2, \ldots, x_n)$; we wish to construct a Boolean expression to represent this function. For each n-tuple $(a_1, a_2, \ldots, a_n)$ of 0's and 1's such that $f(a_1, a_2, \ldots, a_n) = 1$, we form the minterm $y_1 y_2 \cdots y_n$, where y_i is x_i if $a_i = 1$, and y_i is $\overline{x}_i$ if $a_i = 0$. The desired expression is the sum of these minterms. If there are no minterms, then the sum is 0.

EXAMPLE 3. Thinking of 1 as being greater than 0, express the following function with a Boolean expression:

$$f(x, y, z) = \begin{cases} 1 & \text{if } x \leq y \leq z \\ 0 & \text{otherwise.} \end{cases}$$

Solution. We need to list the values of x, y, and z that make $f(x, y, z) = 1$. Listing them as triples, we see that they are precisely $(0, 0, 0)$, $(0, 0, 1)$, $(0, 1, 1)$, and $(1, 1, 1)$, since as soon as one of the variables, in alphabetical order, is 1, all the subsequent ones must be 1. Applying the procedure for finding the disjunctive normal form of this function, we obtain

$$f(x, y, z) = \overline{x}\,\overline{y}\,\overline{z} + \overline{x}\,\overline{y}\,z + \overline{x}\,y\,z + x\,y\,z.$$ ◆

Disjunctive normal form provides a way to standardize the representation of Boolean functions, and such standardization allows us to compare Boolean functions. Given a Boolean expression representing a Boolean function, we can find the disjunctive normal form for this function with straightforward algebraic manipulation. First we "multiply out" the expression, using the distributive law, DeMorgan's laws, and the double complement law repeatedly, until we end up with an expression that is the sum of products. Next we use the fact that $x\,x = x$ (idempotent law) and $x\,\overline{x} = 0$ (complement law) to get rid of all repeated occurrences of the same variable within a product. Those terms with a 0 in them drop out by the dominance law. At this point each term has at most one occurrence of each variable. For those terms with one or more variables missing, we multiply by terms of the form $z + \overline{z}$ for each missing variable z (this does not change the value of the term, since $z + \overline{z} = 1$), and again multiply out. Finally, we eliminate any duplicate terms (by the fact that $A + A = A$). The result is then a Boolean expression in disjunctive normal form that is equivalent to the original Boolean expression.

EXAMPLE 4. Find the disjunctive normal form for $(x + y\,z)\overline{x\,\overline{\overline{y}}}$; assume that x, y, and z are the only variables.

Solution. We follow the plan outlined above. We use the associative law throughout; its use is not mentioned explicitly.

$(x + y\,z)\overline{x\,\overline{y}}$

$= (x + y\,z)(\overline{x} + y)$ (by DeMorgan's and double complement laws)

$= x\,\overline{x} + x\,y + \overline{x}\,y\,z + y\,y\,z$ (by the distributive and commutative laws)

$= 0 + x\,y + \overline{x}\,y\,z + y\,z$ (by the complement and idempotent laws)

$= x\,y + \overline{x}\,y\,z + y\,z$ (by the identity law)

$= x\,y(z + \overline{z}) + \overline{x}\,y\,z + (x + \overline{x})\,y\,z$ (by the identity and complement laws)

$= x\,y\,z + x\,y\,\overline{z} + \overline{x}\,y\,z + x\,y\,z + \overline{x}\,y\,z$ (by the distributive law)

$= x\,y\,z + x\,y\,\overline{z} + \overline{x}\,y\,z$ (by the idempotent and commutative laws).   ◆

SIMPLIFYING BOOLEAN EXPRESSIONS

Although the procedure described in the preceding subsection for finding the disjunctive normal form of a Boolean function always produces a Boolean expression to represent the function, it does not always produce the simplest one in any sense. In Example 4, for example, it certainly would have been simpler to stop with

$x\,y + \overline{x}\,y\,z + y\,z$, in the sense that this expression has fewer symbols in it. In fact, since $\overline{x}\,y\,z + y\,z = y\,z$, we can simplify this even further to $x\,y + y\,z$ or to $y(x + z)$, which involves only two operations. As a worst case, the disjunctive normal form of the constant function $f(x_1, x_2, \ldots, x_n) = 1$, which is represented by the Boolean expression 1, has 2^n minterms in its disjunctive normal form representation. For example, 1 is clearly a much simpler expression than

$$x\,y\,z + x\,y\,\overline{z} + x\,\overline{y}\,z + x\,\overline{y}\,\overline{z} + \overline{x}\,y\,z + \overline{x}\,y\,\overline{z} + \overline{x}\,\overline{y}\,z + \overline{x}\,\overline{y}\,\overline{z}.$$

Sometimes a simpler form for the Boolean expression is suggested by the definition of the function itself. Our next example gives an application in which this is the case.

EXAMPLE 5.   In a club's executive committee, which has five members including the president, a motion is deemed to have passed if the president and at least one other member of the committee are in favor of it, or if all four of the other members are in favor. Find a Boolean expression for the outcome of any motion before this committee; let x_i, for $i = 1, 2, 3, 4, 5$, represent the vote of the ith member (1 for aye, 0 for nay), where the first member is the president.

Solution.   Observe that the motion will pass (the function value will be 1) if $x_1 = 1$ and at least one of x_2, x_3, x_4, and x_5 is 1, or if $x_i = 1$ for all $i > 1$. We can translate this observation directly into our solution, recalling the parallel between Boolean sums and products and logical disjunctions and conjunctions, respectively: $x_1(x_2 + x_3 + x_4 + x_5) + x_2x_3x_4x_5$. The disjunctive normal form would be long and messy.   ◆

In other cases, we may have to start with the sum of products representation and try to simplify it. The problem is analogous to simplifying algebraic expressions in calculus, and in both contexts it is not always clear exactly what is "simplest." Unfortunately, the problem of simplifying Boolean expressions sometimes turns out to be very difficult if there are many minterms and a large number of variables (in the terminology of Section 4.4, the problem of obtaining the simplest Boolean expression is *NP*-complete). On the other hand, if the expressions involved are not too long, we can often perform simplifications by hand fairly easily. As was the case in the manipulations discussed at the end of the preceding subsection, the identities in Theorem 1 play the role of the algebraic identities learned in high school.

To simplify an expression, we try to apply the distributive law of multiplication over addition in such a way that after we have factored out a suitable subexpression from a sum, what remains equals 1 by the complement law. Then by the identity law, we can drop the 1. In other words, we look for an instance of $A\,B + A\,\overline{B}$. This subexpression can be factored to obtain $A(B + \overline{B}) = A \cdot 1 = A$. Obviously, we achieve a substantial simplification each time we are able to do this. The trick is to find the appropriate terms to attack and the appropriate pieces to factor out. We can repeat this process as long as such factoring is possible.

EXAMPLE 6.   We found that the disjunctive normal form for the Boolean function representing the logical connective "if...then..." was

$$\overline{x}_1\overline{x}_2 + \overline{x}_1 x_2 + x_1 x_2.$$

Suppose that we factor out $\overline{x}_1$ from the first two minterms; this is valid by part (c) of Theorem 1. We thus obtain the equivalent expression

$$\overline{x}_1(\overline{x}_2 + x_2) + x_1 x_2.$$

But now $\overline{x}_2 + x_2 = 1$ by part (g) of Theorem 1, and $\overline{x}_1 \cdot 1 = \overline{x}_1$ by part (d) of Theorem 1. Therefore, we obtain the simplified expression $\overline{x}_1 + x_1 x_2$ for this function. In fact, this can be simplified a bit further with the following sequence of steps:

$$\begin{aligned}
\overline{x}_1 + x_1 x_2 &= \overline{x}_1(1 + x_2) + x_1 x_2 \\
&= \overline{x}_1 + \overline{x}_1 x_2 + x_1 x_2 \\
&= \overline{x}_1 + (\overline{x}_1 + x_1)x_2 \\
&= \overline{x}_1 + x_2.
\end{aligned}$$

Note that this last expression corresponds to the logical expression $\overline{P} \vee Q$, and we saw in part (j) of Theorem 1 of Section 1.1 that $P \to Q \iff \overline{P} \vee Q$—in words, P implies Q if and only if P is false or Q is true. ◆

Other factoring techniques and ad hoc tricks can also be used in simplifying Boolean expressions, but we will not pursue them. Indeed, we have merely scratched the surface here, to give the reader an idea of what is involved. This topic lies more in computer engineering (as we will see in the final subsection) than in discrete mathematics. Various systematic geometric and algebraic procedures have been developed for finding terms to combine and keeping track of the bookkeeping, and the interested reader should consult a good circuit theory text for details (see Exercises 23 and 24).

SWITCHING CIRCUITS

One often thinks of functions as machines: They accept inputs (the arguments) and produce outputs (the function values). For example, the Boolean function $f(x_1, x_2) = \overline{x}_1 + x_2$ can be envisioned as a machine that accepts two inputs (the values of x_1 and x_2) and produces one output (the value of $\overline{x}_1 + x_2$). If we do

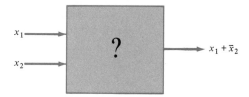

FIGURE 1.1   A function as a black box.

not care to look inside the machine to see *how* it performs the calculation, we can envision the function as a "black box" (see Figure 1.1).

In electrical and computer engineering, this model is much more than a convenient way to visualize a mathematical abstraction. Engineers must actually *build* such machines as part of the circuitry for computers and other electronic devices. We have seen that any Boolean function can be built up using the three basic operations of addition, multiplication, and complementation. If an electronics engineer can build a black box to perform each of these three basic operations, then he or she can construct arbitrarily complex circuits to represent complex Boolean functions, by combining them as the relevant Boolean expressions dictate. For example, one can construct circuits to perform arithmetic in computers (see Exercise 26).

We will not worry about the electronics behind circuits; instead, we simply think of them as visual models of Boolean functions. Furthermore, we will not rigorously define our model, but rather, think of it as a convenient way to "see" Boolean algebra in our minds.

Each circuit (or, more formally, **switching circuit**) will have one or more inputs and one output. We think of each input or output as a Boolean variable, but in the real world it might be given by a voltage level along a wire. We will visualize the values of these Boolean variables (0 or 1) as being carried along lines in our pictures, following the arrows. The lines represent wires in real circuits.

To begin, then, let us assume that the following three black boxes exist: an **OR gate** to perform Boolean addition (Figure 1.2), an **AND gate** to perform Boolean multiplication (Figure 1.3), and an **inverter** to perform the Boolean operation of complementation (Figure 1.4).

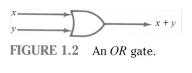

FIGURE 1.2   An *OR* gate.

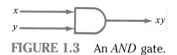

FIGURE 1.3   An *AND* gate.

FIGURE 1.4   An inverter.

The *OR* and *AND* gates shown here have two inputs, but we can assume the existence of such gates with an arbitrary number of inputs (see Figure 1.5 for an *AND* gate with three inputs); recall that the operations are commutative and associative, so no order of operations need be specified.

FIGURE 1.5   An *AND* gate with three inputs.

We can combine these elementary circuits (*AND* gates, *OR* gates, and inverters) to form more complex circuits by using the outputs of some gates as inputs to others. For example, the circuit shown in Figure 1.6 represents the Boolean expression $xy + \bar{x}\bar{y}$. Note that x and y are the inputs to this circuit and there is one output. The inputs are each used twice, so the lines split; the values are carried along each branch. When a gate is encountered, the output from the gate is determined by the input values. For example, suppose that $x = 1$ and $y = 0$ in the circuit in Figure 1.6. Then the output of the *AND* gate on top is 0, since $1 \cdot 0 = 0$. Similarly, the outputs of the two inverters are 0 and 1, and hence the output of the bottom *AND* gate is also 0. These two 0 values are carried as inputs to the *OR* gate, whose output is therefore $0 + 0 = 0$. Thus this entire circuit would produce output 0 if given the inputs $x = 1$ and $y = 0$. On the other hand, if the inputs were $x = 0$ and $y = 0$, then the output would be 1.

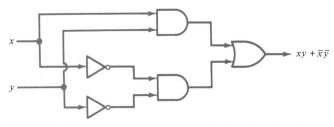

FIGURE 1.6   A circuit representing the Boolean function $f(x, y) = xy + \bar{x}\bar{y}$.

EXAMPLE 7.   Build a circuit to represent the Boolean function of two variables discussed at the beginning of this section, whose value is 1 except when the first argument is 1 and the second argument is 0.

Solution.   There are many circuits that will represent this function. The first step is to find a Boolean expression for this function, and we have already done so. One such expression is the disjunctive normal form $f(x_1, x_2) = \overline{x}_1\,\overline{x}_2 + \overline{x}_1\,x_2 + x_1\,x_2$. To draw the circuit for this expression would require two inverters, three *AND* gates, and an *OR* gate with three inputs. The reader is asked in Exercise 17 to draw such a circuit. We can draw a simpler circuit for this function, however, if we use the simpler Boolean expression for this function that we found in Example 6: $f(x_1, x_2) = \overline{x}_1 + x_2$. The circuit is shown in Figure 1.7.

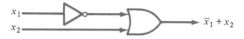

FIGURE 1.7   A circuit representing $f(x_1, x_2) = \overline{x}_1 + x_2$.

Thus we need only one inverter and one *OR* gate for this circuit.   ◆

It should now be clear why simplifying Boolean expression is important. A typical electronic device may contain a large number of circuits. The simpler these circuits can be made, the more elegantly and inexpensively the device can be built.

SUMMARY OF DEFINITIONS

***AND* gate**: a circuit that realizes the Boolean multiplication operation (see Figure 1.3).

Boolean addition: the operation on Boolean constants defined by $0 + 0 = 0$, $0 + 1 = 1 + 0 = 1 + 1 = 1$.

Boolean constant: 0 (representing false) or 1 (representing true).

Boolean expression: an expression constructed by applying any sequence of Boolean additions, Boolean multiplications, and complementations to Boolean constants and Boolean variables; parentheses are used to indicate grouping, with multiplication taking precedence over addition otherwise (e.g., $\overline{x} + y(z + \overline{x\,y + \overline{z}})$).

Boolean function: a function with one or more arguments, all of which are Boolean variables, that takes only the values 0 and 1 (e.g., the function $f(x, y) = x\,y + \overline{x}\,\overline{y}$, which is given by $f(0,0) = 1$, $f(0,1) = 0$, $f(1,0) = 0$, and $f(1,1) = 1$).

Boolean multiplication: the operation on Boolean constants defined by $0 \cdot 0 = 0 \cdot 1 = 1 \cdot 0 = 0$, $1 \cdot 1 = 1$.

Boolean product: the result of a Boolean multiplication.

Boolean sum: the result of a Boolean addition.

Boolean variable: a variable that takes the value 0 (false) or 1 (true).

(Boolean) **complement**: the result of a Boolean complementation.

(Boolean) **complementation**: the operation on Boolean constants defined by $\bar{0} = 1, \bar{1} = 0$.

disjunctive normal form of a Boolean function: an expression defining the function, consisting of a sum of distinct minterms (e.g., $f(x, y) = x y + x \bar{y}$).

equivalence of Boolean expressions: Two Boolean expressions are equivalent (written by placing an equals sign between them) if they represent the same Boolean function (e.g., $\overline{x + \bar{y}} = \bar{x} y$).

inverter: a circuit that realizes the Boolean complementation operation (see Figure 1.4).

minterm: a product $y_1 y_2 \cdots y_n$, where for each i either $y_i = x_i$ or $y_i = \bar{x}_i$; it is assumed that $x_1, x_2, \ldots, x_n$ are all the Boolean variables under consideration (e.g., $x_1 x_2 \bar{x}_3 x_4$).

OR **gate**: a circuit that realizes the Boolean addition operation (see Figure 1.2).

sum of products representation of a Boolean function: the disjunctive normal form of the function.

switching circuit: a diagram constructed from *AND* gates, *OR* gates, and inverters, which represents a Boolean function.

EXERCISES

1. Let a Boolean function be defined by $f(x, y, z) = x \bar{y} z + x(y + z) + \overline{x + y}$.
   (a) Compute $f(1, 1, 1)$.
   (b) Compute $f(0, 1, 0)$.

2. Prove that $x + xy = x$, where x and y are Boolean variables, by showing that the equality holds for each of the four possible values of the pair (x, y) (i.e., $x = y = 0$; $x = y = 1$; $x = 0$, $y = 1$; and $x = 1$, $y = 0$).

3. Repeat Exercise 2 for the identity $x(x + y) = x$.

4. Write down the disjunctive normal form for the Boolean function $f(x, y, z)$ that has the value 1 precisely when $x = y = z = 0$ or $x = y = 0$, $z = 1$ or $x = z = 1$, $y = 0$.

5. Write down the Boolean function that is represented by the circuit on p. 73.

6. Draw a switching circuit to represent each of the following Boolean expressions.
   (a) $x y + \bar{x} y$.
   (b) $x y + y z + \bar{z}$.

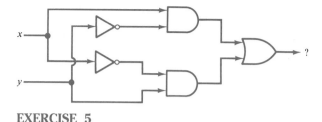

EXERCISE 5

7. Suppose that x, y, and z represent the votes of three people who make up a committee of three (the value 1 represents an affirmative vote, the value 0 represents a negative vote—abstaining is not allowed).
    (a) Write down the Boolean function in disjunctive normal form that represents the outcome of the committee's vote (majority rules); note that there are four cases in which a motion will pass.
    (b) Find a simpler expression for the function in part (a).

8. Repeat Exercise 7 if there are four members of the committee (on a tie vote, a motion does not pass).

9. By manipulating the expression algebraically, using the laws contained in Theorem 1, prove the following identities.
    (a) $x \cdot \overline{xy} = x\,\overline{y}$.
    (b) $x + \overline{x+y} = x + \overline{y}$.
    (c) $x + xy = x$.
    (d) $x(x+y) = x$.

10. Prove the dominance law $x + 1 = 1$ from the other parts of Theorem 1. [*Hint*: Write $x + 1$ as $x + (x + \overline{x})$.]

11. Consider the Boolean function $f(x, y) = \overline{xy}$.
    (a) Draw a circuit to represent this function.
    (b) Apply DeMorgan's law to obtain an equivalent expression and draw a circuit based on this expression.
    (c) Express this function in disjunctive normal form.

12. Repeat Exercise 11 for the function $f(x, y) = \overline{x + y}$.

13. Simplify the following Boolean expressions, which are given in disjunctive normal form.
    (a) $x\,y + x\,\overline{y}$.
    (b) $x\,y + x\,\overline{y} + \overline{x}\,y + \overline{x}\,\overline{y}$.
    (c) $x\,y\,\overline{z} + x\,\overline{y}\,z + \overline{x}\,y\,\overline{z} + \overline{x}\,\overline{y}\,z$.
    (d) $x\,y\,\overline{z} + x\,\overline{y}\,z + x\,\overline{y}\,\overline{z}$.

14. Convert the following simple Boolean expressions to disjunctive normal form by algebraic manipulation, using the identities in Theorem 1. Except in

part (e), the only variables in each case are the ones that appear in the expression.

(a) $x\,y + \overline{y}$.
(b) $x + \overline{y}$.
(c) $\overline{x}\,\overline{y} + y\,z$.
(d) $(\overline{x} + z)y$.
(e) x (the variables are x, y, and z).

15. Rewrite these expressions in disjunctive normal form, using algebraic manipulations. The only variables in each case are x, y, and z.

(a) $\overline{\overline{x}\,\overline{y}\,\overline{z}}$.
(b) $(x + y\,\overline{z})(x + \overline{y}\,\overline{z})$.
(c) $(x + y)(y + z)\overline{x}$.

16. Simplify the Boolean expressions in Exercise 15 as much as possible.

17. Draw a switching circuit that represents the disjunctive normal form for the function discussed in Example 7.

18. Consider the Boolean expression $x_1(\overline{\overline{x}_2 + x_3} + x_2 x_3) \cdot \overline{x_1\overline{x}_2 x_3} + \overline{x}_3$.

(a) How many inverters and (not necessarily binary) gates would be needed in a circuit that corresponds to this expression?
(b) Find a simpler expression (preferably as simple as possible) equivalent to this expression.
(c) Draw a circuit that corresponds to the simpler expression.

19. Show each of the following Boolean nonequivalences.

(a) $\overline{xy} \neq \overline{x}\,\overline{y}$.
(b) $\overline{x + y} \neq \overline{x} + \overline{y}$.

20. Consider the Boolean function represented by the following circuit.

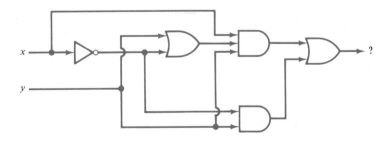

EXERCISE 20

(a) Write down the Boolean expression that corresponds to this circuit.
(b) Find a simplest Boolean expression for this function—simplest in the sense of using the smallest number of Boolean operations.

(c)   Express this function in disjunctive normal form.

Challenging Exercises

21. Develop a product of sums form for representing Boolean functions by interchanging the roles of 0 and 1 and the roles of addition and multiplication in the development of sum of products form. This form is also called the **conjunctive normal form.**

    (a)   Explain clearly in prose the algorithm for obtaining this form.

    (b)   Find the conjunctive normal form for the Boolean function that represents majority voting in a committee of three (Exercise 7).

    (c)   Explain a purely manipulative method for obtaining the conjunctive normal form for a Boolean expression, given the disjunctive normal form.

22. Consider the Boolean function of two variables that takes the value 1 if and only if the variables are not equal. We write the result of applying this function to x and y as $x \oplus y$.

    (a)   Express this function in disjunctive normal form.

    (b)   Determine whether $\oplus$ is commutative and associative.

    (c)   Prove that $x(y \oplus z) = xy \oplus xz$.

Exploratory Exercises

23. Find out about Karnaugh maps, a device used for simplifying Boolean expressions, by consulting Friedman and Menon [92] and Kohavi [183].

24. Find out about the Quine–McCluskey procedure for simplifying Boolean expressions, by consulting Kohavi [183] and Sloan [274].

25. Find out about Boolean algebra from a more abstract point of view (in which some of the identities in Theorem 1 are taken as axioms) by consulting Dornhoff and Hohn [71].

26. Find out how switching circuits can be combined to perform arithmetic operations such as addition and multiplication of n-digit binary numbers, by consulting Sloan [274].

27. Find out about George Boole by consulting Bell [15].

SETS

Nearly all of mathematics and theoretical computer science can be formulated in terms of set theory. The language, ideas, and results of set theory permeate these disciplines. Set theory allows us to formalize concepts and to be precise. With set theory, we can actually say exactly what something *is*; without it, confusion is likely to abound. In this chapter we lay the set-theoretic foundations for the rest of this book, just as we laid the logical foundations in Chapter 1.

Some of this material may be familiar to the reader of this book. Such notions as subset, intersection, and ordered pair are so fundamental to modern mathematics that most people who have studied calculus or high school algebra will probably have seen them and used them often. In this chapter we treat these topics more abstractly and also get into some deeper ideas that will be less familiar.

The basic notions of set theory are laid out in Section 2.1. Set theory can be used to formalize and make precise the notions of sequence, string, and matrix; we see how in Section 2.2. Finally, in Section 2.3 we look at the algebra of set theory based on such operations as intersection and union; we will find parallels there with the propositional calculus of Section 1.1.

The reader should keep two things in mind while working through this chapter (and the entire book): precision and level of abstraction. We will give exact, unambiguous definitions of the concepts we discuss. We want to avoid the trap of talking a lot about a concept without being able to say what it is we are talking about. Make sure that you can explain *precisely* what each concept is. One goal of this chapter is to get the reader used to thinking at more than one level of abstraction at the same time. For example, the set of subsets of a set is quite a different thing from—one level of abstraction higher than—the set of elements of the set. Do not yield to the temptation to gloss over such distinctions.

Good general references for set theory include Fletcher and Patty [85], Halmos [137], Kamke [167], Monk [213], and Suppes [296].

SECTION 2.1
BASIC DEFINITIONS IN SET THEORY

> *By a "set" we shall understand any collection into a whole,*
> *M, of definite, distinguishable objects m (which will be*
> *called the "elements" of M) of our intuition or thought.*

> **—Georg Cantor**

The notion of **set** is so fundamental in mathematics that a rigorous definition is impossible—there is essentially nothing more fundamental to define it in terms of. Rather, we will simply agree that a set is any well-defined collection of objects, called the **elements** or **members** of the set, and illustrate the concept with some examples. We also say that the elements of a set are "in" the set, and we use the words "set" and "collection" interchangeably. Now that we have agreed on what a set is, we can define everything else precisely.

Even mathematicians who do research in set theory do not define explicitly what a set is. Instead, they write down the axioms—self-evident statements—that they assume sets obey. (Some of the axioms for set theory are discussed in Exercises 29, 34, and 37.) The notion of a set seems to be fundamental to human thinking, so after a little discussion to clarify the finer points, we will probably all have the same implicit definition of set floating around in our brains.

EXAMPLE 1. The collection of letters in the English alphabet is a set. There are 26 elements in this set, for example, q and w. The number 4 is not an element of this set, nor is this book, nor is the word "book."

The set of **real numbers** is a set that the reader has worked with in algebra or calculus. Its elements are too numerous to list, but we presumably all agree on what is a real number and what is not. For example, 4 is an element of this set, as are 2.83493 and $\sqrt{2}$ and π; but the letter q, the Empire State Building, and the imaginary number $\sqrt{-1}$ are not.

There is a set consisting precisely of King Henry VIII of England together with all sets of natural numbers with exactly two elements. This example illustrates that sets can have other sets as elements, and that the elements of a set do not need to be meaningfully related to each other (although in practice they usually are).

There is a set consisting of all propositions with no free variables, written using numerals, variable names, symbols for equality, greater than, addition, and multiplication, and logical symbols (such as $\rightarrow$ and $\exists$), that are true when the quantifiers are assumed to range over the natural numbers. For example, $2+2=4$ and $\forall x\colon x+0=x$ are in this set, and $\exists x\colon (x>1 \wedge x=2\cdot x)$ and Henry VIII are not (the former because it is false, the latter because it is not a proposition of the specified type). A proposition such as Goldbach's conjecture (proposition 6 in Example 1 of Section 1.1 when written out in symbols) is either in this set or not in this set; at this point in human history, we do not know which is the case. Our own inability to determine whether certain objects are elements of this set does not keep the set from being well defined. We would have the same difficulty with the set of persons serving in the U.S. Senate on April 25, 2048, another perfectly valid set.

There is no such set as the set of uninteresting natural numbers. Although this string of words seems to define a collection, a little thought will convince us that it really does not. For one thing, the property of being "interesting" is not well defined; whereas 1729 is interesting to most number theorists (it is the smallest number that can be written as the sum of two cubes in two different ways: $1729 = 12^3 + 1^3 = 10^3 + 9^3$), it is probably not very interesting to most other people. Exercise 36 deals with another, more profound reason that this purported "set" is a sham. ◆

We usually use braces ("{" and "}") to describe sets. If a set is small enough to let us write down all its elements, then we can describe it by listing these elements, in any order, separated by commas and enclosed within braces. For example, $\{\{1,9\},5,\{3\}\}$ is a set with three elements, two of which are sets of numbers and one of which is a number. A set is determined by its elements, not by the way in which we list them, so, for example, this set could also be written as $\{\{9,1\},\{3\},5,5\}$ (note that writing the 5 twice was superfluous but harmless). A set with no elements is called **the empty set**. By our convention, this set could be written as $\{\ \}$, but we usually use the special symbol $\emptyset$ to denote it. We are justified in referring to *the* empty set, because, since a set is uniquely defined by its elements, there is *only one* set with no elements.

If it is inconvenient or impossible to list explicitly all the elements of a set, we have two choices. The informal way out, which is often, but not always appropriate, is to write down the beginning of the list of elements of a set to establish a pattern, write an ellipsis ($\ldots$) to indicate that some elements are missing, and write down the end of the list, if there is one. For example, the set of letters in the English alphabet can be written $\{a,b,c,\ldots,z\}$. Similarly, $\{2,4,6,8,\ldots\}$ is probably meant to be the set of even positive integers. From a rigorous point of view, of course, this is nonsense, since there are many sets four of whose elements are 2, 4, 6, and 8; we rely on common sense with this informal approach. More generally, multiple ellipses can be used to describe sets whose elements form more complicated patterns, such as $\{\ldots, -4, -1, 2, 5, 8, \ldots\}$ (the set of integers that are 1 less than a multiple of 3) or $\{10, 11, \ldots, 19, 30, 31, \ldots, 39, 50, 51, \ldots, 59, \ldots, 190, 191, \ldots, 199\}$ (the set of natural numbers less than 200 whose tens' digit is odd).

A more precise way to define a particular set is through the use of propositions. Suppose that we have agreed upon some set as the domain under discussion (real numbers, for instance). Let $P(x)$ be a proposition with free variable x. Then we can form the set S whose elements are precisely the objects x that satisfy $P(x)$— that make $P(x)$ a true proposition. This set is well defined because we insisted that propositions be either true or false, with no ambiguity allowed. In other words,

$$\forall x \colon [(x \text{ is in } S \longleftrightarrow P(x)].$$

We write the set so described as $\{x \mid P(x)\}$, read "the set of all x such that $P(x)$." Note that x has become a bound variable in this expression; in particular, $\{t \mid P(t)\}$ is exactly the same set as $\{x \mid P(x)\}$. For example, the set $\{\ldots, -4, -1, 2, 5, 8, \ldots\}$ mentioned in the preceding paragraph can be described as $\{q \mid \exists t \colon q = 3t - 1\}$, where we assume that t ranges over integers.

EXAMPLE 2.   Assume that we are talking about real numbers. Then $\{x \mid x^5 - 7x^4 + 1 = 0\}$ is a set containing three real numbers, approximately (but not exactly) -0.60, 0.63, and 7.00. (One can tell this by analyzing the graph of $y = x^5 - 7x^4 + 1$ using algebra and/or calculus; see Exercise 24.) In this case there is really no other good way to describe the set, since the three numbers do not have any obvious names.

The third set described in Example 1 can be written as follows:

$$\{A \mid A = \text{Henry VIII} \ \lor \ \exists x \colon \exists y \colon (x \neq y \land A = \{x, y\})\},$$

where we are assuming that the quantified variables x and y range over natural numbers.   ◆

More generally, we can describe a set by writing $\{E(x) \mid P(x)\}$, where $E(x)$ is some expression involving x. For example, we could write the set of nonnegative even integers as $\{2n \mid n \text{ is a natural number}\}$.

Just as we used letters as variables to denote propositions in Chapter 1, when our object of study was propositions, here we will use letters as variables to denote sets. More often than not, capital (uppercase) letters will denote sets, and lowercase letters will denote arbitrary elements of sets. Some sets are used so often that we make up standard names for them. In particular, we let $\mathbf{Z} = \{0, 1, -1, 2, -2, 3, -3, \ldots\}$ denote the set of integers; $\mathbf{N} = \{0, 1, 2, 3, \ldots\}$, the set of natural numbers; and $\mathbf{R}$, the set of real numbers.

Clearly, the basic notion underlying set theory is the notion of *being an element of*. We use the symbol $\in$ to indicate this relationship, and $\notin$ to indicate its negation. Thus $x \in A$ is read "x is an element of A," or "x is in A"; and $x \notin A$, which is the same as $\overline{x \in A}$, is read "x is not an element of A," or "x is not in A."

EXAMPLE 3. Let $A = \{\{1, 9\}, 5, \{3\}\}$. Then $5 \in A$ and $\{9, 1\} \in A$; but $1 \notin A$. Note the care we need to exercise to be precise and remain at the right level of abstraction. The number 3 is not an element of this set A (i.e., $3 \notin A$), but the *set* consisting of the single element 3 is an element of A (i.e., $\{3\} \in A$). The *number* 3 and the *set* consisting of the number 3 are two different things. Similarly $\{5\} \notin A$.

The empty set ($\emptyset$) is the unique set satisfying the property $\forall x \colon x \notin \emptyset$.

Perhaps the most concise way to describe $\{\ldots, -4, -1, 2, 5, 8, \ldots\}$ is as $\{3t - 1 \mid t \in \mathbf{Z}\}$.

We can use the symbol $\in$ to clarify the range of a quantifier. For example, the statement that every nonzero real number has an inverse can be written as

$$\forall x \in \mathbf{R} \colon (x \neq 0 \;\rightarrow\; \exists y \in \mathbf{R} \colon xy = 1).$$

Thus the "domain under discussion" in any given situation, as we have been calling it, is usually some (large) set. We call it the **universal set** for that situation.

Similarly, we can write $\{x \in \mathbf{N} \mid x^3 + x^2 + x = 0\}$ for the set of *natural number* solutions to $x^3 + x^2 + x = 0$. This set equals $\{0\}$. ◆

SUBSETS AND THE POWER SET

Much of mathematics can be phrased in terms of relations between and among sets. We begin with one of the simplest.

Suppose that A and B are sets. We say that A is a **subset** of B (written $A \subseteq B$), and that B is a **superset** of A, if and only if every element of A is also an element of B. For example, $\mathbf{N} \subseteq \mathbf{Z}$, $\mathbf{N} \subseteq \mathbf{R}$, and $\mathbf{Z} \subseteq \mathbf{R}$. On the other hand $\{\{1, 2\}, 3\} \nsubseteq \{1, 2, 3\}$ since the element $\{1, 2\}$ of the first set is not an element of the second. If every element of A is an element of B but, in addition, B has at least one element that A does not have, then we say that A is a **proper** subset of B. Every set B is trivially a subset of itself (since every element of B is an element of B), but not a proper subset; every other subset of B is a proper subset of B.

Note that we do not use the word *in* as a synonym for "a subset of," since we already reserved "in" as a synonym for "an element of." It is important to keep the two notions separate. Whereas every set is a subset of itself, no set is an element of itself. (The latter statement is discussed in Exercise 39—it is far from obvious that there might not be a situation in which a set is an element of itself.)

A little reflection will show that two sets are equal (i.e., have exactly the same elements) if and only if each is a subset of the other. In symbols,

$$A = B \longleftrightarrow (A \subseteq B \wedge B \subseteq A).$$

This observation provides a useful way to prove that two sets are equal. If we want to show that $A = B$, we need to show that $A \subseteq B$ and $B \subseteq A$. By definition, to do the first, we must show that every element in A is also in B; then for the second we have to go the other way, showing that every element in B is also in A. We will see some examples of this proof technique in Section 2.3.

Our first theorem about sets is a good exercise in vacuous proofs.

THEOREM 1.   *If A is any set, then $\emptyset \subseteq A$.*

Proof.   By the definition of subset, we need to prove the proposition $\forall x: (x \in \emptyset \rightarrow x \in A)$. Now for every x, the hypothesis of this implication is false, since $\emptyset$ has no elements. Therefore, the implication is itself true, by the truth table definition of implication. In other words, we have indeed shown that every element of $\emptyset$ (of which there are none) is also an element of A (see also Exercise 11).   ■

Sometimes we need to collect together *all* the subsets of a given set. Such a collection is again a set. We call the set of all subsets of A the **power set** of A, written $\mathcal{P}(A)$. In symbols,

$$\mathcal{P}(A) = \{ B \mid B \subseteq A \}.$$

For example, $\mathcal{P}(\{2,3\}) = \{\emptyset, \{2\}, \{3\}, \{2,3\}\}$ and $\mathcal{P}(\emptyset) = \{\emptyset\}$. Note in particular that $A \in \mathcal{P}(A)$ always holds, but A is not usually a *subset* of $\mathcal{P}(A)$ (see Exercise 19, however).

EXAMPLE 4.   Let P be the set of points in the plane. Then $\mathcal{P}(P)$ is the set consisting of all *sets* of points in the plane. For example, every line is in $\mathcal{P}(P)$, as is every drawing of the Empire State Building, if we think of lines and pictures simply as sets of points. Furthermore, if x is a point, then x is not in $\mathcal{P}(P)$, but $\{x\}$ is.   ◆

CARDINALITY OF FINITE SETS

We often care about how big a set A is, in other words, how many elements it has. If A has only a finite number of elements (i.e., zero elements, one element, two elements, etc.), then A is said to be a **finite set**; and the **cardinality** of A, written $|A|$, is the number of elements in A. Note that $|A|$ is always a natural number if A is a finite set. For this reason the natural numbers are sometimes called **cardinal numbers**; thus the cardinality of a set is a cardinal number indicating its size. A set with cardinality n is called an **n-set**.

A set that is not finite is said to be an **infinite set**. (We give an alternative way to characterize infinite sets in the next subsection. As it stands, one could complain that we have really begged the question here, and that our definition amounts to nothing more than saying that a finite set is a finite set. These questions get at the heart of the philosophy of mathematics (see Exercise 28 in Section 1.3). We avoid the issue by assuming that for any natural number n we all have the same notion of what it means for a set to have n elements.)

EXAMPLE 5. Since the set of letters of the English alphabet has 26 elements, we say that this set has cardinality 26 and write $|\{a, b, c, \ldots, z\}| = 26$.

The set of U.S. senators, when there are no vacancies in the Senate, is a 100-set.

The empty set is a finite set with no elements, so $|\emptyset| = 0$.

Clearly $|\{\{1, 3, 5, 7, 9\}, 5, \{3\}, 5\}| = 3$, even though there are five different numbers appearing in the symbolism and the list appears to have four items.

The sets $\mathbf{N}$, $\mathbf{Z}$, and $\mathbf{R}$ are all infinite; we have not yet defined cardinality for infinite sets. On the other hand, $|\{\mathbf{N}, \mathbf{Z}, \mathbf{R}\}| = 3$ and $|\{\{\mathbf{N}, \mathbf{Z}, \mathbf{R}\}\}| = 1$. ◆

EXAMPLE 6. Suppose that A and B are finite sets.

(a) What is the relationship between $|A|$ and $|B|$ if $A \subseteq B$?
(b) What is the relationship between $|A|$ and $|\mathcal{P}(A)|$?

Solution. Part (a) is clear: If $A \subseteq B$, then B has to have at least as many elements as A has, so $|A| \le |B|$.

Part (b) is more interesting. A little experimentation will show that a set with n elements seems to have 2^n subsets. For instance, the set $\{a, b, c\}$, which has cardinality 3, has $2^3 = 8$ subsets (as the reader will show in Exercise 4a). Thus we might conjecture that $|\mathcal{P}(A)| = 2^{|A|}$. In fact, this formula is correct, and we will prove it in Example 7 of Section 5.3, using the technique of mathematical induction. We will also prove there that $n < 2^n$ for all natural numbers n, so $|A| < |\mathcal{P}(A)|$.

Let us carry this quite a bit further. We can prove that $|A| < |\mathcal{P}(A)|$ for finite sets by noting first that each set of the form $\{a\}$, for $a \in A$, is an element of $\mathcal{P}(A)$. Thus $|\mathcal{P}(A)|$ is at least as large as $|A|$. Furthermore, there is at least one more set in $\mathcal{P}(A)$ that is not of this form, namely $\emptyset$. Therefore, $|\mathcal{P}(A)|$ must be at least one more than $|A|$. Note that this argument relies heavily on the fact that A is a finite set; we have shown that $|\mathcal{P}(A)| \ge |A| + 1 > |A|$.

Finally, we give a more conceptual, self-contained proof that $|A| < |\mathcal{P}(A)|$—a proof that does not rely on this finiteness assumption. The idea in this subtle proof by contradiction—called the diagonal argument—dates back to G. Cantor, the German mathematician who founded set theory at the end of the nineteenth century. (The quotation at the beginning of this section is from Cantor's pioneering work.)

Suppose, on the contrary, that A had as many elements as (or more elements than) $\mathcal{P}(A)$. Then we could pair off each element of $\mathcal{P}(A)$ (i.e., each subset of A) with an element of A, never using the same element of A more than once (maybe using some of them not at all). Now some elements of A might get paired with sets of which they are elements, and some might not. [For example, if c is paired with $\{a, c\}$ or with $\{c\}$, then c is an element that is paired with a set of which it is an element. On the other hand, if b is paired with $\{a\}$ or with $\emptyset$, or if b is not paired at all, then b is an element that is not paired with a set that contains it as an element.]

We now look at a particular subset of A: Let B consist of those elements of A that are not paired with sets that contain them as elements. In other words, $x \in B$ if and only if either x is not paired with any subset of A, or else x is paired with a subset of which it is not an element. [In our example in the preceding paragraph, we would have $b \in B$ but $c \notin B$.] Certainly, B is a well-defined subset of A. Therefore, B is paired with some particular element z of A. Now either $z \in B$ or $z \notin B$. We will see that each of these assumptions (one of which has to hold by the law of the excluded middle) leads to a contradiction.

Suppose that $z \in B$. Then, patently, z is an element that is paired with a set of which it is an element. That contradicts our definition of B, which was to contain only elements that were *not* paired with sets that contained them as elements. On the other hand, suppose that $z \notin B$. Then z has the property that it is paired with a subset of A (namely, B) of which it is not an element. But then $z \in B$ by our definition of B, again a contradiction. Thus in either case we arrive at a contradiction. Therefore, our assumption that A had as many elements as $\mathcal{P}(A)$ was false, and the result is proved. ◆

INFINITE SETS (*OPTIONAL*)

In the remainder of this section, we take a brief look at the fascinating notion of cardinality for infinite sets. Observe that two finite sets have the same cardinality if and only if their elements can be paired off, one by one. (In the language to be described in Section 3.2, two finite sets have the same cardinality if and only if there is a one-to-one and onto function from one of the sets to the other.) Cantor extended this observation and *defined* two *infinite* sets to have the same cardinality if such a pairing is possible. For example, the set **N** of natural numbers and the set of squares of natural numbers have the same cardinality because there is the requisite one-to-one correspondence between them:

$$
\begin{array}{cccccc}
0 & 1 & 2 & 3 & 4 & 5 \\
\updownarrow & \updownarrow & \updownarrow & \updownarrow & \updownarrow & \updownarrow \\
0 & 1 & 4 & 9 & 16 & 25
\end{array} \quad \cdots
$$

Cantor's ideas were controversial at the time. After all, it goes against one's intuition that a proper subset of **N** (namely, the set of natural numbers that are perfect squares) can be "as big as" all of **N**. (Indeed, an alternative definition of an infinite set is that it is a set that can be put into a one-to-one correspondence with a proper subset of itself.) Even more at odds with intuition, the set of rational numbers (which spread out over the entire real number line) has the same cardinality as the set of natural numbers (which occur only in discrete steps along the line); in Exercise 31 the reader is asked to find a pairing that demonstrates this.

Nevertheless, it gradually sunk in that Cantor's definition worked. He was able to define arithmetic operations such as addition and multiplication for his *infinite* cardinal numbers, analogous to arithmetic operations for the usual, finite cardinal numbers. He gave names to various infinite cardinal numbers. He called the cardinality of the natural numbers $\aleph_0$, read "aleph sub zero" or "aleph nought." (The symbol $\aleph$ is the first letter of the Hebrew alphabet.) Thus $|\mathbf{N}| = |\text{set of perfect squares}| = |\text{set of rational numbers}| = \aleph_0$. It is possible to show that $\aleph_0$ is the smallest infinite cardinal number. Sets with this cardinality are called **countably infinite sets**, and sets that are either finite or countably infinite are called **countable**.

The cardinality of the set of real numbers, which he showed was the same as the cardinality of $\mathcal{P}(\mathbf{N})$, Cantor called $\mathbf{c}$. Then he showed, using a diagonal argument (essentially, just the argument we gave in Example 6b—see Exercise 32), that $\aleph_0 \neq \mathbf{c}$. In other words, the set of real numbers is not countable. Informally, this says that there is no way to write the real numbers in an infinite list $r_0, r_1, r_2, \ldots$ without omitting some real numbers from the list. As we remarked above, such an infinite listing *is* possible for the rational numbers. Thus something much more subtle is going on than simply that there is never a "next" real or rational number after a given real or rational number in their usual numerical order (see Exercise 33). Cantor found other cardinal numbers bigger than $\aleph_0$ as well, all arranged in a nice order. He named them with subscripts on the $\aleph$. Thus we have

$$\aleph_0 < \aleph_1 < \aleph_2 < \cdots < \aleph_\omega < \aleph_{\omega+1} < \aleph_{\omega+2} < \cdots < \cdots$$

and on and on, further than the mind can imagine. (To explain how $\aleph_1$ is defined, or what $\aleph_\omega$ is, or how this sequence continues "beyond infinity" is beyond the scope of this book. For a taste of what is involved, see Exercises 35, 41, and 43.)

What Cantor could not show was whether $\mathbf{c}$ was the next-bigger cardinal number after $\aleph_0$, in other words, whether $\mathbf{c} = \aleph_1$. That question, known as the **continuum hypothesis**, was not settled until the middle of the twentieth century, and the answer is not entirely satisfying. The Austrian K. Gödel (whom we met in Chapter 1) showed that using the usual axioms for set theory, it is not possible to answer the question of whether $\mathbf{c} = \aleph_1$ in the negative. The American P. Cohen proved a few decades later that using the usual axioms for set theory, it is not possible to answer the question of whether $\mathbf{c} = \aleph_1$ in the affirmative. In other words, no contradiction would result from adding to the axioms of set theory either an axiom stating that $\mathbf{c} = \aleph_1$ or an axiom stating that $\mathbf{c} \neq \aleph_1$ (but not both at the same time, of course). In fact, no contradiction would result from postulating, for instance, that $\mathbf{c} = \aleph_{423}$ or $\mathbf{c} = \aleph_{\omega+2}$. Research continues to this day as mathematicians try to gain a better understanding of very large infinite sets.

SUMMARY OF DEFINITIONS

cardinal number: a number that is the cardinality of some set; the finite cardinal numbers are the natural numbers, whereas the infinite cardinal numbers include $\aleph_0$, **c**, and others.

cardinality of a finite set: the number of elements in the set; the cardinality of A is written $|A|$ (e.g., $|\{10, 11, \{1, 2, 3, 10\}\}| = 3$).

cardinality of an infinite set: an indication of the size of the set, relative to other infinite sets, as measured by one-to-one correspondences; the cardinality of A is written $|A|$ (e.g., $|\mathbf{Z}| = \aleph_0$).

continuum hypothesis: the statement that there is no cardinality strictly between $\aleph_0$ and **c**.

countable set: a set that is either finite or countably infinite (e.g., **R** is not countable).

countably infinite set: a set whose cardinality is $\aleph_0$ (the cardinality of **N**).

element of a set: any one of the objects in the collection; "the object x is an element of (or is in) the set A" is written $x \in A$.

empty set: the set with no elements.

finite set: a set with only a finite number of elements (e.g., $\emptyset$ or $\{3, \{10\}, \mathbf{N}\}$).

infinite set: a set that is not finite (e.g., **N**).

member of a set: an element of the set.

***n*-set**: a set whose cardinality is n (e.g., $\{a, b, e\}$ is a 3-set).

power set of the set A: the set of all subsets of A, written $\mathcal{P}(A)$ (e.g., $\mathcal{P}(\{1\}) = \{\emptyset, \{1\}\}$).

proper subset of the set A: a subset of A other than A itself.

real numbers: numbers corresponding to the points along a line, including integers, other rational numbers, and irrational numbers such as $\sqrt{2} - \sqrt{3}$.

set: a well-defined collection of objects (e.g., the set of compound propositions using no propositional variables other than P, Q, and R that are tautologies; $P \vee (P \rightarrow (Q \wedge \mathrm{T}))$ is in this set, but $3 + 2$ and $P \vee Q \vee R$ are not).

subset: A is a subset of B, written $A \subseteq B$, if every element of A is also an element of B (e.g., $\{1, 2, 3\} \subseteq \{-1, 0, 1, 2, 3\}$).

superset: A is a superset of B if and only if B is a subset of A.

universal set: the set consisting of all objects under discussion in a given situation (e.g., **N** if we are discussing natural numbers).

EXERCISES

1. Give an example of a set that has four elements, two of which are people, one of which is a set consisting precisely of three numbers, and one of which is a set with exactly five sets as its elements.

2. Determine whether each of the following propositions is true or false.
   (a) $\emptyset \subseteq \emptyset$.

    (b) $\emptyset \in \emptyset$.

    (c) $\emptyset \subseteq \{\emptyset\}$.

    (d) $\emptyset \in \{\emptyset\}$.

    (e) $\{1\} \subseteq \{1, 2, 3\}$.

    (f) $\{1\} \in \{1, 2, 3\}$.

    (g) $\{1, 2\} \subseteq \{1, 2, 3, \{1, 2\}\}$.

    (h) $\{1, 2\} \in \{1, 2, 3, \{1, 2\}\}$.

3. Which of the following are proper subsets of $\{1, 2, \{1, 2\}\}$?

    (a)   1.

    (b) $\{1, 2\}$.

    (c) $\{\{1, 2\}\}$.

    (d) $\{1, \{1, 2\}\}$.

    (e) $\{1, 2, \{1, 2\}\}$.

    (f) $\{\{\{1, 2\}\}\}$.

4. Write down explicitly the power set of the following sets.

    (a) $\{a, b, c\}$.

    (b) $\{a, \{b, c\}\}$.

    (c) $\{\{a\}, \{b\}\}$.

5. Describe each of the following sets using braces and ellipses.

    (a) $\{x \in \mathbf{Z} \mid |x| < 100\}$.

    (b)   The set of natural numbers between 1 and 100, inclusive, excluding 69.

    (c) $\{n^2 \mid n \in \mathbf{Z}\}$.

6. Take as true the assertion, made in Example 6b, that $|\mathcal{P}(A)| = 2^{|A|}$ for every finite set A. How many nonempty proper subsets does the set $\{a, b, c, d, e\}$ have?

7. List all the 2-sets that are subsets of $\{\text{red}, \text{blue}, \text{green}, \text{yellow}\}$.

8. Write down the definition of $A \subseteq B$ in symbols.

9. Determine which of the following equalities are true.

    (a) $\{\{1, 2\}, \{1, 2, 3\}\} = \{1, 2, 3\}$.

    (b) $\{\{1, 2\}, \{1, 2, 3\}\} = \{\{1, 2, 3\}\}$.

    (c) $\{1, 2, 1, 2, 3\} = \{1, 2, 3\}$.

10. Write down all correct statements of the form $A \in B$ and $A \subseteq B$, where A and B are chosen from 1, $\{1\}$, and $\{\{1\}\}$.

11. Consider the proposition $A \not\subseteq B$, that A is not a subset of B.

    (a)   Write out what this statement means, as a quantified proposition involving elements.

    (b)   Use part (a) to show that $\emptyset \not\subseteq A$ is false for all sets A. (This gives an alternative way to see why $\emptyset \subseteq A$ for all sets A.)

12. Write out explicitly $\mathcal{P}(\mathcal{P}(\{5\}))$.

13. Describe using the $\{\, x \mid P(x)\,\}$ or $\{\, E(x) \mid P(x)\,\}$ notation the sets apparently intended by the following descriptions.
    (a) $\{1, 2, \ldots, 9, 11, 12, \ldots, 19, \ldots, 91, 92, \ldots, 99\}$.
    (b) $\{\frac{1}{1}, \frac{1}{8}, \frac{1}{27}, \frac{1}{64}, \frac{1}{125}, \ldots\}$.
    (c) $\{2, 6, 12, 20, 30, 42, 56, 72, 90, 110, \ldots, 9900\}$.
    (d) $\{\ldots, -9, -6, -3, 0, 3, 6, 9, \ldots\}$.
    (e) $\{\ldots, -11, -5, 1, 7, 13, 19, 25, \ldots\}$.

14. Describe the following sets by listing their elements (or by listing some of their elements and using ellipses). The quantifiers range over **N**.
    (a) $\{\, x \mid \exists y\colon x = 5y\,\}$.
    (b) $\{\, p^2 + 1 \mid p \text{ is an odd prime}\,\}$.
    (c) $\{\, x \mid \exists p\colon \exists q\colon x = pq \land (p \text{ and } q \text{ are primes})\,\}$.
    (d) $\{\, x \mid 101 \leq x \leq 105\,\}$.
    (e) $\{\, \{x, y\} \mid x = y + 1\,\}$.
    (f) $\{\, \{x, y\} \mid |x - y| \leq 1\,\}$.

15. Write down a subset of cardinality four of the third set described in Example 1.

16. Write down the set of all sets X such that $\{1, 2, 3\} \subseteq X \subseteq \{1, 2, 3, 4, 5\}$.

17. Prove that a subset of a subset is a subset. In other words, if A, B, and C are sets, and $A \subseteq B$ and $B \subseteq C$, then $A \subseteq C$. [*Hint*: Your proof must make use of the definition of subset; your argument needs to involve elements.]

18. Describe the following sets in the form $\{\, x \mid P(x)\,\}$ or $\{\, E(x) \mid P(x)\,\}$. Do not use any ellipses. The obvious patterns are intended.
    (a) $\{\{0\}, \{1\}, \{2\}, \{3\}, \ldots\}$.
    (b) $\{\{0, 1\}, \{0, 2\}, \{0, 3\}, \ldots, \{1, 2\}, \{1, 3\}, \{1, 4\}, \ldots, \ldots\}$.
    (c) $\{\{0\}, \{0, 1\}, \{0, 1, 2\}, \{0, 1, 2, 3\}, \ldots\}$.

19. Give an example of each of the following.
    (a) A nonempty set A such that $A \subseteq \mathcal{P}(A)$.
    (b) A set A such that $A \not\subseteq \mathcal{P}(A)$.
    (c) Sets A and B such that both $A \in B$ and $A \subseteq B$.

20. Give a definition of the empty set in the form $\{\, x \mid P(x)\,\}$.

21. Prove that if $A \subseteq B$, then $\mathcal{P}(A) \subseteq \mathcal{P}(B)$.

22. Prove that if $\mathcal{P}(A) \subseteq \mathcal{P}(B)$, then $A \subseteq B$. [*Hint*: Try a proof by contradiction.]

23. Take as true the assertion, made in Example 6b, that $|\mathcal{P}(A)| = 2^{|A|}$ for every finite set A. How many elements does $\mathcal{P}(\mathcal{P}(\mathcal{P}(\mathcal{P}(\mathcal{P}(\emptyset)))))$ have?

24. Verify the claim made in Example 2 that the first set described there has cardinality 3.

25. Prove that if $A \subseteq \emptyset$, then $A = \emptyset$.

26. Prove—or disprove with a counterexample—each of the following statements about sets.
    (a) If $A \in \emptyset$, then $\emptyset \in A$.
    (b) If $A \in \emptyset$, then $\emptyset \notin A$.
    (c) If $A \in B$ and $B \in C$, then $A \in C$.
    (d) If $A \in B$ and $B \in C$, then $A \notin C$.

27. Show that the following sets are countable.
    (a) The set of even positive integers.
    (b) The set of all the grains of sand in the world.
    (c) The set of real numbers that are either integers or half-integers (such as $\frac{5}{2}$ or $-\frac{1}{2}$).

28. Consider the proposition "For every A and B, if A is a proper subset of B, then $|A| < |B|$."
    (a) Explain why this proposition is true or why it is false (as the case may be) if the variables range over finite sets.
    (b) Explain why this proposition is true or why it is false (as the case may be) if the variables range over infinite sets.

29. Using only logical connectives, variables, parentheses, quantifiers ranging over sets, the equal sign (=), and the "is an element of" symbol ($\in$), write the following propositions. (These are some of the axioms that are taken to be self-evident when studying set theory from a rigorous, formal point of view.)
    (a) For every pair of sets A and B there exists a set whose elements are precisely A and B.
    (b) The empty set exists.
    (c) For each set A, $\mathcal{P}(A)$ exists.

30. If A and B are sets of natural numbers, let their sum be defined by $A + B = \{ a + b \mid a \in A \wedge b \in B \}$.
    (a) Compute $\{1, 2, 7\} + \{5, 6\}$.
    (b) What is the sum of the set of even natural numbers and the set $\{0, 1\}$?
    (c) State Goldbach's conjecture (proposition 6 of Example 1 in Section 1.1) using this notation.
    (d) Find with proof a set of even positive integers that cannot be written as $A + B$ for any sets A and B of positive integers such that $|A| \geq 2$ and $|B| \geq 2$.

Challenging Exercises

31. Recall that a rational number is a real number of the form a/b, where a is an integer and b is a positive integer. Show that the set of rational numbers is countable. [*Hint*: Order the positive rational numbers by the sum of their numerator and denominator.]

32. Prove that $|A| < |\mathcal{P}(A)|$ for infinite sets A in the following sense.

(a)   Show that there is no one-to-one correspondence between A and $\mathcal{P}(A)$. [*Hint*: We really did this in Example 6b; you are to rewrite that proof, in your own words and in this setting. Do it with the book closed!]

(b)   Show that there is a one-to-one correspondence between A and a proper subset of $\mathcal{P}(A)$.

33. Prove, using a proof by contradiction, that the set of real numbers between 0 and 1 is not countable, using the following ideas.

(a)   Explain how real numbers between 0 and 1 can be represented uniquely using decimal notation $r = 0.d_1 d_2 d_3 \ldots$. There is one subtlety here, since, for example, $0.5 = 0.5000\ldots = 0.4999\ldots$.

(b)   Assume (hoping to derive a contradiction) that there is an enumeration $r_1, r_2, r_3, \ldots$ of *all* the real numbers between 0 and 1. Write each r_i in its decimal expansion, say as $0.d_{i1} d_{i2} d_{i3} \ldots$.

(c)   Use Cantor's diagonal trick: Define a real number $r = 0.d_1 d_2 d_3 \ldots$ so that r cannot possibly equal any of the numbers r_i in the purported list, by making sure that r disagrees with r_i in the ith decimal place.

(d)   Conclude that the set of real numbers between 0 and 1 is not countable.

34. Explain fully in English what the following proposition says. (It can be taken as one of the axioms of set theory.)

$$\exists I : [\emptyset \in I \wedge \forall x : (x \in I \rightarrow \exists y : (y \in I \wedge \forall z : (z \in y \leftrightarrow z = x)))]$$

35. Set theory can be used to model the natural numbers. We can actually think of *defining* the natural numbers by the following process. First 0 is the empty set. Then 1 is the set consisting of 0; thus 1 is $\{0\} = \{\emptyset\}$. Next 2 is the set consisting of 0 and 1, namely $\{0, 1\} = \{\emptyset, \{\emptyset\}\}$. Similarly, we take 3 to be the set $\{0, 1, 2\} = \{\emptyset, \{\emptyset\}, \{\emptyset, \{\emptyset\}\}\}$. We continue in this manner. Numbers formed in this way are called the **ordinal** numbers. In general, each ordinal number *is* the set whose elements are all the smaller ordinal numbers. The ordinal numbers actually go beyond the natural numbers. We can take the set of all the natural numbers, $\{0, 1, 2, \ldots\} = \{\emptyset, \{\emptyset\}, \{\emptyset, \{\emptyset\}\}, \ldots\}$, and define it to be the first infinite ordinal, denoted ω (the Greek lowercase letter *omega*). The next ordinal, $\omega + 1$, is formed by including ω as well; thus $\omega + 1 = \{0, 1, 2, \ldots, \omega\}$. We can continue this process forever, and beyond.

(a)   Write down the sets that are the next two ordinal numbers after $\omega + 1$ (called, not surprisingly, $\omega + 2$ and $\omega + 3$).

(b)   Write down the set that is the next ordinal after all the ordinals $\omega + n$ for natural numbers n, assuming that these are defined in the manner indicated in part (a). This is sometimes called $\omega \cdot 2$.

(c)   Continue the pattern from parts (a) and (b) and write down the set that is the ordinal number ω^2. Then write down the set and the name for the ordinal number after that.

Exploratory Exercises

36. Argue that there is no such set as the set of interesting natural numbers, unless all natural numbers are interesting. [*Hint*: Look at the smallest natural number that is not interesting.]

37. The following proposition is a version of the **axiom of choice**, an important but somewhat controversial axiom of set theory. Recall from Section 1.2 that $\exists!z$: means "there exists a unique z such that."

$$\forall A: [(\forall X \in A: [(\exists a: a \in X)$$
$$\wedge \ (\forall Y \in A: (X \neq Y \rightarrow \overline{\exists z: (z \in X \wedge z \in Y)}))])$$
$$\longrightarrow \exists B: ((\forall z \in B: \exists X \in A: z \in X) \wedge (\forall X \in A: \exists! z \in B: z \in X))]$$

   (a) Explain fully in English what the proposition is saying.
   (b) Give an example to illustrate what is going on.
   (c) Discuss whether you believe that the axiom is self-evidently true about your conception of sets.

38. The twentieth-century British mathematician and philosopher B. Russell came up with the following paradox that shows that we have to be careful about collecting objects into sets. All the sets we have encountered are not elements of themselves—indeed, it is hard to imagine how a set could be an element of itself (see Exercise 39). In any case, suppose that we try to look at the "set"

$$S = \{ A \mid (A \text{ is a set}) \wedge A \notin A \}$$

of *all* sets that are not elements of themselves. By the law of the excluded middle, either $S \in S$ or $S \notin S$. Show that $S \in S$ leads to a contradiction and that $S \notin S$ leads to a contradiction. Conclude that this "set" does not exist. (The problem with the existence of sets like this is that in some sense they are "too big." Russell and others tried to refine set theory so that paradoxes like this are specifically avoided.)

39. Try to write down a specific example of a set A containing exactly one element, A itself. Discuss the difficulty in doing so. In particular, what is wrong with just writing $A = \{A\}$? (In most axiom systems for set theory there is an axiom that explicitly forbids possibilities like this.)

40. Give one good explanation as to what is so "natural" about the natural numbers, from a set-theoretic point of view; in particular, why do we include 0 as a natural number?

41. Find out more about infinite sets by consulting Gardner [99], Maor [207], Rucker [260], and Zippin [335].

42. Find out more about Cantor, and the problems he had with the controversy over his discoveries, by consulting Dauben [56,57] and Bell [15].

43. Find out more about axiomatic set theory, the axiom of choice (Exercise 37), and the continuum hypothesis by consulting Cohen and Hersh [48], Gödel [116], and Halmos [137].

SECTION 2.2
SETS WITH STRUCTURE

Fresh as a lark mounting at break of day,
Festively she puts forth in trim array.

—William Wordsworth

Sets are certainly not the only important structure one encounters in mathematics. We also need to consider n-tuples, matrices, sequences, strings, and a host of other more complex structures that are the bread and butter of pure and applied mathematics. Let us see how some of these can be defined. Not surprisingly, we will use set theory in these discussions.

The reader has probably had some exposure to many of these concepts at an informal level, and an informal, intuitive understanding is probably sufficient for reading the remainder of this book. This section can be viewed both as a review of these concepts and as an explanation of how they can be made rigorous through the use of set theory.

ORDERED PAIRS AND TUPLES

An **ordered pair** consists of two (not necessarily distinct) objects, considered in order. If the objects are x and y, we write the ordered pair as (x, y) and sometimes call x and y the **coordinates** of the pair. Equality of ordered pairs is defined by saying that $(a, b) = (c, d)$ if and only if $a = c$ and $b = d$. If A and B are sets, then the **Cartesian product** $A \times B$ of A and B, which may be read "A cross B," is the set $\{(a, b) \mid a \in A \land b \in B\}$.

A set with two elements can be thought of as an *unordered* pair. Thus $\{a, b\} = \{b, a\}$, but the ordered pairs (a, b) and (b, a) are not equal unless $a = b$. This distinction will be important in Chapter 8, when we discuss directed and undirected graphs.

EXAMPLE 1.   If $A = \{1, 3, 5\}$ and $B = \{3, 4\}$, then

$$A \times B = \{(1, 3), (1, 4), (3, 3), (3, 4), (5, 3), (5, 4)\}.$$

The Cartesian product $\mathbf{R} \times \mathbf{R}$ is used in analytic geometry and calculus to represent the plane: Each point in the plane corresponds to an ordered pair of real numbers.

Ordered pairs are useful in referring to mathematical objects that consist of more than just sets—sets with additional structure imposed upon them. One such structure familiar to readers of this book is the set of positive real numbers together with the operation of multiplication. If we let $\mathbf{R}^+$ stand for the set of positive real numbers, then we might denote this structure by the ordered pair $(\mathbf{R}^+, \cdot)$. Note that the first object in this pair is a set, and the second is an operation (multiplication) on the set. Similarly, we could consider the structure $(\mathbf{R}, +)$ consisting of the set of real numbers and the operation of addition. A nontrivial statement we could make about these two structures is that they are **isomorphic** (meaning "having the same form"), under the pairing of each element of $\mathbf{R}^+$ with its natural logarithm (which is an element of $\mathbf{R}$), since taking the logarithm transforms the operation of multiplication into the operation of addition: $\ln(xy) = \ln x + \ln y$. Every property of the structure $(\mathbf{R}^+, \cdot)$ is reflected in the structure $(\mathbf{R}, +)$. For example, the number 1 is the identity for $(\mathbf{R}^+, \cdot)$, since $\forall x \in \mathbf{R}^+ : x \cdot 1 = x$, whereas 0 (which is $\ln 1$) is the identity for $(\mathbf{R}, +)$, since $\forall x \in \mathbf{R} : x + 0 = x$. On the other hand, the structures $(\mathbf{R}, \cdot)$ and $(\mathbf{R}, +)$ are not isomorphic, since the former has an element z with the property $\forall x : x \cdot z = z$ (namely, $z = 0$), whereas the latter has no element z such that $\forall x : x + z = z$. We will have more to say about isomorphic structures in later chapters; the notion is one of the cornerstones of advanced abstract mathematics. ◆

Suppose that we write an ordered pair as (a_1, a_2). Then this ordered pair can be looked at as two (not necessarily different) numbers, together with an order imposed on them by the **index set** $\{1, 2\}$. The indices are the subscripts. The index set tells which element comes first and which comes second. By generalizing the index set to $\{1, 2, \ldots, n\}$ we can define n-tuples for natural numbers n other than $n = 2$. We simply mimic the definition above. More precisely, an (**ordered**) **n-tuple** $(a_1, a_2, \ldots, a_n)$ consists of n (not necessarily distinct) objects, considered in order. Again each a_i in $(a_1, a_2, \ldots, a_n)$ is called a **coordinate**. We say that $(a_1, a_2, \ldots, a_n) = (b_1, b_2, \ldots, b_n)$ if and only if $\forall i : a_i = b_i$. If A_1, A_2, $\ldots$, A_n are sets, then their **Cartesian product** is the set

$$A_1 \times A_2 \times \cdots \times A_n = \{ (a_1, a_2, \ldots, a_n) \mid \forall i : a_i \in A_i \}.$$

The n-fold Cartesian product of the same set A (a special case that arises frequently), written $A \times A \times \cdots \times A$ (with n A's), is denoted A^n.

EXAMPLE 2. In calculus, physics, and engineering we use the n-fold Cartesian product of the set of real numbers, $\mathbf{R}^n$, to represent n-dimensional space.

To specify a complicated mathematical structure we often use an n-tuple. For instance, in Chapter 8 we will give a formal definition of a "multidigraph" as a

4-tuple. We will not make much use of the definition as such—it will be clear intuitively at the time what a multidigraph is—but this formalism gives us the ability to define such a creature rigorously. ◆

An n-tuple, then, is just a finite ordered list, also known as a **finite sequence**. If $n = 1$, an ordered n-tuple is just an object with a pair of parentheses around it—a list with one item. If $n = 0$, then we have the **empty list**, containing no items. In particular, if A is any set, then $A^0 = \{(\ )\}$, that is, the set whose only element is the empty list. An alternative notation for the finite sequence $(a_1, a_2, \ldots, a_n)$ is $\{a_i\}_{i=1}^n$; note that i is a bound variable in this notation, but n is not, since n tells us the length of the sequence.

EXAMPLE 3. The programming language LISP organizes everything as lists. If we needed to keep track of the capitals and the years of admission to the union of the states in the United States, then in LISP we could have a list of 50 ordered 3-tuples, perhaps in alphabetical order (in other words, a 50-tuple of 3-tuples):

((Alabama, Montgomery, 1819), (Alaska, Juneau, 1959),

$\ldots$, (Wyoming, Cheyenne, 1890)).

Suppose that we needed to find the capital of Michigan. Then we could search this list for a 3-tuple whose first coordinate was Michigan, and report the second coordinate of that 3-tuple (Lansing) as the answer. We will consider searching processes in more detail in Sections 4.3 and 9.4. ◆

We will spend the rest of this section discussing various interpretations, applications, and generalizations of sequences, for they play an important role in discrete mathematics and computer science.

SEQUENCES AND STRINGS

Suppose that we let n "become infinite" in our definition of n-tuples, and consider infinite lists (i.e., infinite sequences.) We will have need for infinite sequences especially in Chapters 5 and 7. Formally, then, an **infinite sequence** from a set A is an ordered list, indexed by the positive integers, of elements from A, with repetitions allowed. The a_i's are called the **terms** of the sequence. We sometimes abbreviate the sequence $(a_1, a_2, a_3, \ldots)$ by $\{a_i\}_{i=1}^\infty$, or simply by $\{a_i\}$. The term "sequence" can mean either a finite sequence or an infinite one. Also, the enclosing parentheses are sometimes omitted.

Note that the symbol ∞ (infinity) in the notation $\{a_i\}_{i=1}^\infty$ does not denote an integer—there is no such largest integer. In particular, the notation does not imply that the sequence ends with a term a_∞; it simply means that the sequence continues forever. Again note that i is not a free variable in the notation; $\{a_i\}_{i=1}^\infty$ says nothing about any number i.

Sometimes it is convenient to have the indexing for a sequence begin with 0 instead of 1. Thus an infinite sequence might be denoted $(a_0, a_1, a_2, \ldots)$ or $\{a_i\}_{i=0}^{\infty}$.

EXAMPLE 4. We can consider the sequence of prime numbers in increasing order, $(2, 3, 5, 7, 11, \ldots)$, perhaps denoting it by $\{p_n\}$. Thus p_n would denote the nth prime number, starting from $p_1 = 2$. There are an infinite number of prime numbers (see Exercise 8a in Section 1.3), so this really is an infinite sequence, not just a long finite one. ◆

Many times we need to pick out some of the terms of a (finite or infinite) sequence and use them—in the same order—to form a new sequence. We call the result a subsequence of the original. For example, the prime numbers in increasing order form a subsequence of the natural numbers in increasing order. The reader should try his or her hand at writing out a formal definition of "subsequence" before reading on. It is not an easy task, and our definition, which follows, is quite complicated. You should not worry too much about mastering this particular formal definition, because you can probably gain a satisfactory understanding of the concepts involved on an informal level, by looking at the examples. Nevertheless, it is important to realize that in mathematics concepts can be, and usually should be, rigorously defined.

Let $\{a_i\}$ be a finite or infinite sequence. Another sequence $\{b_i\}_{i=1}^{m}$ (or $\{b_i\}_{i=1}^{\infty}$) is a **subsequence** of the given sequence if there is an increasing sequence of positive integers $i_1 < i_2 < \cdots < i_m$ (or, in the second case, an increasing sequence of positive integers $i_1 < i_2 < \cdots$) such that each $b_j = a_{i_j}$. A subsequence is said to be a **consecutive subsequence** if, in this notation, $i_{j+1} = i_j + 1$ for all relevant j. In other words, we obtain a subsequence by selecting any terms we wish of the original sequence and listing them in the same order as they originally appeared. We obtain a consecutive subsequence by insisting that there be no "gaps": Whenever we select a_r and a_s for the consecutive subsequence, where $r < s$, we need to select all the terms occurring between a_r and a_s, as well (i.e., $a_{r+1}, a_{r+2}, \ldots, a_{s-1}$).

EXAMPLE 5. Consider the sequence of prime numbers in increasing order: $(2, 3, 5, 7, 11, \ldots)$. The sequence of odd primes less than 100, in increasing order, is a consecutive subsequence of this sequence; it is the sequence $(3, 5, 7, 11, \ldots, 97)$. The sequence of prime numbers greater than 1,000,000 is also a consecutive subsequence, as are the sequence (7) and the empty sequence. The sequence of primes that can be written as $4k + 1$ (for some integer k), in increasing order, is a subsequence, but not a consecutive one; it is the sequence $(5, 13, 17, 29, \ldots)$. The sequence $(11, 7, 5)$ is not a subsequence, since its terms do not appear in the same order as in the original sequence, nor is the sequence $(2, 3, 4)$.

Let S be the sequence $0, 0, 1, 0, 1, 2, 0, 1, 2, 3, 0, 1, 2, 3, 4, \ldots$. Then *every* sequence of natural numbers is a subsequence of S. For instance, the sequence

Then $A_{2,3} = 1957$.

We will use matrices for representing relations in Section 3.3 and for representing graphs in Chapter 8. Matrices also arise in multivariable calculus, linear algebra, and many other areas of pure and applied mathematics, both as structures in their own right and as convenient ways of representing information.

EXAMPLE 8.   Represent the following system of linear equations by a matrix:

$$10x_1 + 3x_2 + 3x_3 + 18x_4 = 1$$
$$6x_1 - 4x_2 \qquad + 18x_4 = 5$$
$$7x_2 + 3x_3 \qquad = 0.$$

Solution.   We put the coefficients and right-hand sides of the equations into a matrix A, so that $a_{i,j}$ is the coefficient of x_j (for $1 \le j \le 4$) and $a_{i,5}$ is the constant term on the right, in the ith equation. Thus we obtain the matrix

$$A = \begin{bmatrix} 10 & 3 & 3 & 18 & 1 \\ 6 & -4 & 0 & 18 & 5 \\ 0 & 7 & 3 & 0 & 0 \end{bmatrix}.$$

All the information in the system now appears in the matrix. Methods of linear algebra allow us to obtain the complete solution to the system of equations by doing operations on the matrix (see Exercise 39).   ◆

We have been talking about how elements of some universal set (the domain under discussion) could be indexed by certain index sets to produce lists or matrices. In fact, the index set can be any set at all, and in some applications the index set can get quite complicated. Furthermore, it is often useful to look at indexed collections of sets, rather than just indexed collections of elements.

Let us deal first with indexed collections of objects such as numbers. Formally, we set up the following notation. Let I be any set (finite or infinite). If for each $i \in I$ there is associated an object x_i, we say that the set

$$X = \{ x_i \mid i \in I \}$$

is **indexed** by I, or that I is an **index set** for X. We think of the index i as being a label for the object x_i. (The plural of "index" is "indices.") Note that it may happen that $x_i = x_j$ for two different elements i and j in the index set, so there is a subtle difference between X viewed as just a set (where repetitions of elements are ignored) and X viewed as an indexed set (where the same element may occur more than once, with different indices); see Exercise 23. In most cases, each element in an indexed collection is an element of some universal set; in other words, there is some universal set U under discussion such that $x_i \in U$ for each

$i \in I$. The following example discusses some notation for working with indexed sets.

EXAMPLE 9. Suppose that we have a finite sequence of positive real numbers $x_1, x_2, \ldots, x_n$. We can consider this as the set $\{x_1, x_2, \ldots, x_n\}$, indexed by the set I of integers from 1 to n. If we want to find the sum of the numbers in the sequence, then we write

$$\sum_{i=1}^{n} x_i \quad \text{or} \quad \sum_{i \in I} x_i$$

for the desired sum $x_1 + x_2 + \cdots + x_n$. (If $n = 1$, then $\sum_{i=1}^{1} x_i = x_1$, and if $n = 0$, then the "empty sum" is defined to be 0.) For example, one learns in high school algebra (and we will prove in Example 8 of Section 5.3) that the sum of a geometric series, indexed by the natural numbers from 0 to n, whose ith term is r^i (where $r \neq 1$ is a constant), is given by the following formula:

$$\sum_{i=0}^{n} r^i = \frac{1 - r^{n+1}}{1 - r}.$$

If the index set is infinite, say the set of natural numbers, we can still talk about the sum of all the elements in the sequence $x_0, x_1, x_2, \ldots$, using the notion of limit from calculus. In this case we write

$$\sum_{i=0}^{\infty} x_i \quad \text{or} \quad \sum_{i \in \mathbf{N}} x_i$$

to represent this limit (which will be either a finite real number or ∞, since we assumed that all the terms were positive real numbers). Note that the symbol ∞ at the top of the summation sign does not imply that there is a final term to be added, merely that "the sum is to go on forever," or, more precisely, that

$$\sum_{i=0}^{\infty} x_i = \lim_{n \to \infty} \sum_{i=0}^{n} x_i.$$

Again from high school algebra we have the following formula for the sum of an infinite geometric series, as long as $|r| < 1$:

$$\sum_{i=0}^{\infty} r^i = \frac{1}{1 - r}.$$

As a final example, suppose that we want to talk about the largest of the absolute values of the entries in an m by n matrix of real numbers. If $A = (a_{ij})$

is such a matrix, we refer to this value as

$$\max_{\substack{1 \le i \le m \\ 1 \le j \le n}} |a_{ij}|.$$

Here the index set is the set of pairs (i, j) with $1 \le i \le m$ and $1 \le j \le n$. ◆

Finally, we turn to the case in which the indexed collection is a collection of sets. The notation is similar to what we had before. Let I be any set (finite or infinite). If for each $i \in I$ there is associated a set A_i, we say that the collection of sets

$$C = \{\, A_i \mid i \in I \,\}$$

is indexed by I, or that I is an index set for C. Thus C is a set of sets. Again it may happen that $A_i = A_j$ for two different elements i and j in the index set. In most cases, each set in an indexed collection is a set of elements from some universal set; in other words, there is some universal set U under discussion such that $A_i \subseteq U$ for each $i \in I$.

EXAMPLE 10. For each natural number n we can form the set $A_n = \{0, 1, 2, \ldots, n\}$, consisting of all the natural numbers less than or equal to n. We can collect all of these sets into one indexed set $C = \{\, A_n \mid n \in \mathbf{N} \,\}$. Note that C is an infinite set, but that all the elements of C are finite sets.

For each real number x, we might be interested in the interval of real numbers greater than or equal to x, usually denoted $[x, \infty)$. Thus $[x, \infty) = \{\, y \in \mathbf{R} \mid y \ge x \,\}$. We could collect all of these intervals together into a set C, indexed by the real numbers. Thus we would have $C = \{\, [x, \infty) \mid x \in \mathbf{R} \,\}$.

For each positive integer n, we might be interested in the set of prime factors of n. This would give us an indexed collection of sets $\{\, F_n \mid n \in \mathbf{N} \wedge n > 0 \,\}$, indexed by the set of positive integers, where $F_n = \{\, p \mid p \text{ is a prime number } \wedge p \text{ is a factor of } n \,\}$. Thus, for example, $F_{12} = \{2, 3\}$, $F_{54} = \{2, 3\}$, and $F_1 = \emptyset$. As a set, stripped of the indexing structure, this collection is the set of all finite subsets of the set of prime numbers, as the reader should verify (Exercise 30). ◆

In the next section we will see how some operations on sets can be applied to collections of sets in the same way that operations on numbers (adding and finding the maximum) were applied to collections of numbers in Example 9.

SUMMARY OF DEFINITIONS

alphabet: a nonempty finite set, when used in the context of looking at strings; if V is an alphabet, then the set of all strings over V is denoted V^* (e.g., the alphabet is $\{0, 1, \ldots, 9\}$ when we are forming numerals for the natural numbers; $\{0, 1, \ldots, 9\}^*$ consists of all such numerals, including those with leading zeros, together with the empty string).

bit string: a string over the alphabet $\{0, 1\}$, that is, an element of $\{0, 1\}^*$ (e.g., 0010001111).

Cartesian product: If A and B are sets, then the Cartesian product of A and B, written $A \times B$, is $\{(a, b) \mid a \in A \wedge b \in B\}$; more generally, the Cartesian product of the sets A_1, A_2, ..., A_n is the set of ordered n-tuples $(a_1, a_2, \ldots, a_n)$ in which the ith coordinate is in A_i (e.g., $\{1, 2\} \times \{s\} \times \{t\} = \{(1, s, t), (2, s, t)\}$).

concatenation of strings u and v: the string obtained by writing v immediately following u (e.g., *houseboat*, the concatenation of *house* and *boat*).

consecutive subsequence: a sequence all of whose elements occur, in the same order and without gaps, in a given sequence (e.g., $(3, 4, 5)$ is a consecutive subsequence of $(2, 3, 4, \ldots)$).

coordinate of the n-tuple $(a_1, a_2, \ldots, a_n)$: any of the a_i's (e.g., 3 is the first coordinate of $(3, 1)$).

empty list: the finite sequence of length 0 (the list with no elements).

empty string: the string, denoted λ, of length 0 (having no symbols).

finite sequence of length n from the set A: an ordered n-tuple of elements from A (e.g., $(2, 2, 1)$ is a sequence of length 3 from the set of natural numbers).

index set: a set used to label the elements in an indexed collection (e.g., the set $\{1, 2, 3, 4\}$ in the set of 4-tuples of real numbers (a_1, a_2, a_3, a_4)).

indexed set: a set in which each element has a label or index; different indices may be associated with the same element (e.g., if $A_i = \{i + 3n \mid n \in \mathbf{Z}\}$, then $\{A_i \mid i \in \{1, 2, 3, 4\}\}$ is an indexed set of sets in which A_1 happens to equal A_4).

infinite sequence from the set A: an ordered infinite list of elements of A, usually indexed by the natural numbers or the positive integers (e.g., $(1, 0, 1, 0, \ldots)$).

initial substring: a consecutive subsequence of a string, beginning at the beginning of the string (e.g., *car* and λ are initial substrings of *carrie*).

isomorphic mathematical structures: structures having the same form; two structures are isomorphic if the elements of the first can be put into a one-to-one correspondence with the elements of the second in such a way that all relations and operations in the first structure behave exactly as the corresponding relations and operations in the second (e.g., the structure consisting of the positive integers and the "less than" relation is isomorphic to the structure consisting of the negative integers and the "greater than" relation, under the correspondence that matches each positive integer n with $-n$, since $n_1 < n_2$ if and only if $-n_1 > -n_2$).

m by n matrix (plural—**matrices**): a set indexed by $\{1, 2, \ldots, m\} \times \{1, 2, \ldots, n\}$, displayed in a rectangular array of m rows and n columns; the entry in the ith row and jth column of matrix A is usually denoted a_{ij}; two m by n matrices are equal if and only if their corresponding entries are equal.

(ordered) n-tuple, where n is a natural number: an ordered list of n not necessarily distinct objects, written between parentheses and separated by commas (e.g., $(2, 4, 6, 2)$ is a 4-tuple of integers).

ordered pair: two not necessarily distinct objects written in order, between parentheses and separated by a comma; two ordered pairs (a, b) and (c, d) are equal if and only if $a = c$ and $b = d$.

sequence: a finite or infinite sequence.

size of an m by n matrix: the pair (m, n).

string over the alphabet V: a finite sequence whose elements come from V, written without commas or parentheses (e.g., *carrie*, a string of length 6 over the set of English letters).

subsequence: a sequence all of whose elements occur, in the same order, in a given sequence (e.g., $(4, 5, 8, 892)$ is a subsequence of $(2, 3, 4, \ldots)$).

substring: a consecutive subsequence of a string (e.g., λ, c, and *he* are substrings of *headache*).

symbol: an element of an alphabet.

term of a sequence: an item in the list.

word: a string over some alphabet.

EXERCISES

1. Write down explicitly the following sets.
    (a) $\{1, 2\} \times \{1, 3\}$.
    (b) $\{a, b\}^3$.
    (c) $\{1, 2, 3, 4\} \times \{5\}$.
    (d) $\{1, 2, 3\} \times \{4\} \times \{5\}$.
    (e) $\{1, 2, 3, 4, 5\} \times \emptyset$.

2. Let $\{a_i\}_{i=1}^{\infty}$ be the sequence defined by $a_i = 2^i$ for each i; let $\{b_i\}_{i=1}^{\infty}$ be the sequence defined by $b_i = 2i$ for each i; and let $\{c_i\}_{i=1}^{\infty}$ be the sequence defined by $c_i = 2i + 100$ for each i. Determine all of the relationships "is a subsequence of" and "is a consecutive subsequence of" that hold among $\{a_i\}_{i=1}^{\infty}$, $\{b_i\}_{i=1}^{\infty}$, and $\{c_i\}_{i=1}^{\infty}$.

3. Let the variables u, v, and w stand for the strings r, *ove*, and *oll*, respectively. Express the string *rolloverrover* as a concatenation, using these variables.

4. Consider the sequence (p, d, q).
    (a) List the set of all subsequences of this sequence.
    (b) List the set of all consecutive subsequences of this sequence.

5. List the set of all substrings of *roll*, and indicate which ones are initial substrings.

6. List the set of all nonempty subsequences of the sequence (a, a, b, a), and indicate which one is not consecutive (there is only one).

7. Write down the following matrices.
   (a)   The 3 by 4 matrix whose (i, j)th entry is $i + j$.
   (b)   The 5 by 5 matrix whose (i, j)th entry is 1 if j is a multiple of i, and 0 otherwise.
   (c)   The 4 by 4 matrix whose (i, j)th entry is the substring of the string *papa* extending from the ith letter to the jth letter of the string if $i \leq j$, and is λ otherwise (e.g., the $(2, 4)$th entry is *apa*).

8. Write the following sums using the summation ($\sum$) notation. The obvious patterns are intended.
   (a) $1 + 2 + 4 + 8 + 16 + \cdots + 1024$.
   (b) $1 + 2 + 3 + 4 + \cdots + 100$.
   (c) $4 + 4 + 4 + \cdots + 4$ (there are 37 4's).
   (d) $1 + 3 + 5 + 7 + 9 + \cdots + 99$.

9. Find the following sums.
   (a) $\sum_{i=3}^{6}(3i - 4)$.
   (b) $\sum_{n=0}^{3} 2^{2^n}$ (recall that 2^{2^n} means $2^{(2^n)}$, not $(2^2)^n$).
   (c) $\sum_{i=1}^{100} 7$.
   (d) $\sum_{p \in P} p^2$, where $P = \{2, 3, 5, 7, 11\}$.
   (e) $\sum_{x \in A} |x|$, where $A = \{1, -1, 2, -2, 3, -3\}$.

10. Thinking of a line in the plane as a set of points, and thinking of points as ordered pairs of real numbers (their usual Cartesian coordinates), describe the set of all nonvertical lines in the plane, using the $\{ E(x) \mid P(x) \}$ notation. [*Hint*: First recall from high school algebra how to describe one line.]

11. Prove that if $A \subseteq B$ and $C \subseteq D$, then $A \times C \subseteq B \times D$.

12. Show that $A \times B = \emptyset \longleftrightarrow (A = \emptyset \lor B = \emptyset)$.

13. Show that if A and B are nonempty sets, then $A \times B = B \times A$ if and only if $A = B$.

14. Explicitly list the set of all strings over $\{a, d\}$ that satisfy each of the following properties.
   (a)   Having length 3.
   (b)   Having length at most 5 and not containing either aa or dd as a substring.
   (c)   Having length at most 4 and being of the form uv (concatenation of u and v), where u contains no d's and v contains no a's.
   (d)   Having length 7 and having *dadaa* as an initial substring.

15. Repeat the exercise in Example 7 for bit strings of length 8.

16. An exponent notation is sometimes used when talking about strings. If u is a string and n a natural number, then u^n is the concatenation of n copies of u: for example, $a^2 cb^3 = aacbbb$, and $(011)^0 (10)^2 1^1 = 10101$.

(a) Using this notation, write down all the substrings of a^2bc^2.

(b) Prove or disprove that for all strings u and v and natural numbers n, $(uv)^n = u^n v^n$.

(c) Find all the equalities that always exist among $u^{(mn)}$, $u^m u^n$, $u^{(m+n)}$, $(u^m)^n$, and $u^{(m^n)}$, where u is a string and m and n are natural numbers.

17. Write down explicitly the set $\{(i, j, k) \mid i + j + k = 3\}$, where the universal set under consideration is the set of natural numbers.

18. Construct a list of lists that records the name, sex, and date of birth of each member of your immediate family (yourself and your parents and siblings).

19. Use a matrix to represent each of the following.

(a) The coefficients and constants in the system of equations $x + y + z = 6$, $x - y = 5$, and $x - y - 3z = 4$.

(b) The results of a tournament among four teams, where the entry a_{ij} in the matrix is the number of points by which team i beat team j if, in fact, team i beat team j (and is 0 otherwise), assuming that team 1 beat team 2 by 3 points, team 1 beat team 3 by 2 points, team 4 beat team 1 by 1 point, team 3 beat team 2 by 3 points, team 2 beat team 4 by 7 points, team 4 beat team 3 by 4 points.

(c) All the distances between pairs of four points in the plane, where the entry a_{ij} in the matrix is the distance between point i and point j, assuming that the points are at positions $(0, 0)$, $(1, 0)$, $(0, 1)$, and $(2, 2)$.

20. Express in logical symbols the following notions. In each case you should have at least one existential quantifier.

(a) Two n-tuples are not equal; that is, $(a_1, a_2, \ldots, a_n) \neq (b_1, b_2, \ldots, b_n)$.

(b) Two m by n matrices A and B are not equal.

21. Describe the following sets as indexed sets, explicitly indicating the index set in each case.

(a) All nonempty open intervals (of finite length) of real numbers.

(b) All last names of people in the United States.

(c) $\{\{0, 1, 2, 3, \ldots\}, \{0, 2, 4, 6, \ldots\}, \{0, 3, 6, 9, \ldots\}, \ldots\}$ (the obvious pattern is intended).

22. Consider the set $\{\{x \in \mathbf{N} \mid i \leq x \leq j\} \mid i \in \mathbf{N} \land j \in \mathbf{N} \land 1 \leq i \leq j \leq 3\}$.

(a) Write out explicitly the elements of this set.

(b) Explain how this is an indexed collection of sets. What is the index set? What is the underlying universal set?

23. We defined an indexed set to be a set together with more structure, namely the structure that the indexing imparts. Give an example of two indexed sets that are different as indexed sets but equal as sets with the indexing structure stripped away.

24. Use the formulas given in Example 9 (together with algebra and common sense) to find the following sums. The obvious patterns are intended.
    (a) $1 + 2 + 4 + 8 + 16 + \cdots + 1024$.
    (b) $3^3 + 3^4 + 3^5 + \cdots + 3^{10}$.
    (c) $\frac{1}{3} + \frac{1}{9} + \frac{1}{27} + \cdots$.
    (d) $0.1 + 0.01 + 0.001 + \cdots$.
    (e) $0.9 + 0.09 + 0.009 + \cdots$.

25. Find the following sums. The indices range over natural numbers.
    (a) $\sum_{0 < i < j < 4} ij$.
    (b) $\sum_{0 < i, \, j < 4} ij$.

26. Translate into formulas with the summation notation the following theorems of mathematics.
    (a) The sum of the reciprocals of the positive integers diverges (i.e., equals ∞).
    (b) The sum of the cardinalities of all the subsets of a set with n elements is $n2^{n-1}$.

27. Translate into formulas with the summation notation the following definitions of mathematics.
    (a) $\sigma(n)$ is the sum of the proper divisors of the positive integer n. (A divisor of n is a positive integer d such that n is a multiple of d, and a divisor d is called proper if $d < n$. A useful notation for "d is a divisor of n" is $d \mid n$.)
    (b) The **trace** of a square matrix (one that has the same number of rows as it has columns) is the sum of the entries on the **main diagonal** (the entries along the line joining the upper left-hand corner of the matrix to the lower right-hand corner).

28. Translate into prosaic English the following beautiful theorems of mathematics.
    (a) $\sum_{p \in P} 1/p = \infty$, where $P = \{2, 3, 5, 7, 11, 13, 17, 19, 23, \ldots\}$.
    (b) $\sum_{n=1}^{\infty} 1/n^2 = \pi^2/6$.

29. Just as we can talk about the sum of the numbers in an indexed set, so we can talk about their product. The expression $\prod_{i=1}^{n} a_i$ denotes the product $a_1 a_2 \cdots a_n$. Find the following.
    (a) $\prod_{i=1}^{7} i$.
    (b) $\prod_{i=1}^{47} (47 - i)$.
    (c) $\prod_{i=1}^{3} \left(\sum_{j=1}^{i} ij \right)$.
    (d) $\sum_{i=1}^{3} \left(\prod_{j=1}^{i} ij \right)$.

30. Explain why the third collection of sets in Example 10, as just a set (rather than as an indexed set), is the set of all finite subsets of the set of prime numbers.

Challenging Exercises

31. Let E be the set of even integers $\{0, 2, -2, 4, -4, \ldots\}$. See Example 1 for the notion of isomorphic structures. Determine whether the following pairs of structures are isomorphic.
    (a) $(\mathbf{Z}, +)$ and $(E, +)$.
    (b) $(\mathbf{Z}, \cdot)$ and $(E, \cdot)$.
    (c) $(\mathbf{R}, +)$ and $(\mathbf{R}, -)$.

32. It is possible to define ordered pairs totally in terms of sets, as follows: The ordered pair (a, b) is the set $\{\{a\}, \{a, b\}\}$. Show that this definition works, by proving that $\{\{a\}, \{a, b\}\} = \{\{c\}, \{c, d\}\}$ if and only if $a = c$ and $b = d$. [*Hint*: There are a few cases to consider. The issue is basically this: Given a set that represents an ordered pair under this definition, how do you recover the first and second coordinates of the pair?]

33. A **palindrome** is a string (such as *noon* or *eve*) that reads the same forward and backward. Write down a simple, formal, self-contained definition of the set of all palindromes over an alphabet V.

Exploratory Exercises

34. Explain how we might define A^∞ if A is a finite set, and contrast this set to A^*. In both cases, is there any problem with letting A be infinite?

35. See Exercise 27 for the meaning of $\sigma(n)$. We call n **perfect** if $\sigma(n) = n$, **abundant** if $\sigma(n) > n$, and **deficient** if $\sigma(n) < n$.
    (a)   Classify all the integers from 1 to 20 on this basis.
    (b)   Find another perfect number besides the one you found in part (a).
    (c)   Find a formula for $\sigma(p^k)$, where p is prime, and prove that p^k is always deficient.
    (d)   Prove that if $2^p - 1$ is prime, then $2^{p-1}(2^p - 1)$ is perfect.
    (e)   Consult Wagon [312] to find out more about perfect numbers.

36. Find out about the problem of finding substrings in a string, an important problem in text editing, by consulting Baase [12] and Standish [292].

37. Find out how to prove the theorem stated in Exercise 28b by consulting Choe [43].

38. The study of sets with structure (including relations and operations) is called **abstract algebra**. Find out about abstract algebra and its applications by consulting Dornhoff and Hohn [71].

39. Find out more about the use of matrices in linear algebra by consulting Anton [9] and Kolman [186].

SECTION 2.3
OPERATIONS ON SETS

Haven't I met you somewhere before?

—Anonymous, overheard in a bar

In Section 1.1 we defined several operations on propositions that produced new (compound) propositions from given ones—conjunction, disjunction, and so on. In a similar way we will now define some operations on sets that produce other sets. The analogy runs very deep, because we will see that exactly the same identities that hold in propositional logic hold for sets, once we suitably translate from one context to the other. (Indeed, both propositions and sets, with their respective operations, form structures called abstract Boolean algebras; more advanced treatments would approach these topics from an abstract, axiomatic point of view.)

Many of the set-theoretic operations may well be familiar to a student who has progressed this far in mathematics, but just as in Sections 2.1 and 2.2, we will be including a few twists that should be new.

INTERSECTIONS AND UNIONS

We begin with some basic definitions. Let A and B be two sets. The **intersection** of A and B, written $A \cap B$, is the set of all objects that are simultaneously elements of A and elements of B. The **union** of A and B, written $A \cup B$, is the set of all objects that are elements of A or elements of B (or both—we always use the word "or" in the inclusive sense). In symbols,

$$A \cap B = \{ x \mid x \in A \wedge x \in B \} \qquad \text{and} \qquad A \cup B = \{ x \mid x \in A \vee x \in B \}.$$

In other words, the intersection of two sets is the set of elements common to both, whereas their union is the set of elements appearing in either.

EXAMPLE 1.   Let P be the set of prime numbers, and let M be the set of positive integers that are 1 greater than a multiple of 4. So, $P = \{2, 3, 5, 7, 11, 13, 17, 19, 23, 29, \ldots\}$, and $M = \{ 4n + 1 \mid n \in \mathbf{N} \} = \{1, 5, 9, 13, 17, 21, 25, 29, \ldots\}$. Then $P \cap M$ is the set of prime numbers that can be written in the form $4n + 1$:

$$P \cap M = \{ 4n + 1 \mid n \in \mathbf{N} \wedge 4n + 1 \text{ is prime} \} = \{5, 13, 17, 29, \ldots\}.$$

It is a theorem of number theory that this is an infinite set. Similarly, $P \cup M$ is the set of numbers that are either prime *or* of the form $4n + 1$:

$$P \cup M = \{\, x \mid (\exists n \in \mathbf{N} : x = 4n + 1) \vee x \text{ is prime}\,\}$$
$$= \{1, 2, 3, 5, 7, 9, 11, 13, 17, 19, 21, 23, 25, 29, \ldots\}. \qquad \blacklozenge$$

It often happens that two sets have no elements at all in common. In that case their intersection is the empty set. If $A \cap B = \emptyset$, we say that A and B are **disjoint**. For example, the set of prime numbers and the set of multiples of 10 are disjoint: No multiple of 10 is prime, since every multiple of 10 has at least 2 and 5 as nontrivial factors.

The intersection of two sets and the union of two sets are again sets, just as the conjunction of two propositions and the disjunction of two propositions are again propositions. The definitions of intersection and union, when written in symbols, were identical, except that in the case of $\cap$ we used the logical "and" ($\wedge$), and in the case of $\cup$ we used the logical "or" ($\vee$). The reader should keep this correspondence in mind, since there is a tendency to want to reverse it. (The correspondence is the one that preserves the position of the opening in the symbol: $\cap$ and $\wedge$ both have their openings at the bottom, while $\cup$ and $\vee$ both have their openings at the top.) We will have more to say about the analogy between propositions and sets later in this section.

Intersections and unions apply in a much more general setting than just the case of two sets. Recall from Section 2.2 that we can talk about large collections of sets, even infinite collections, by viewing them as indexed collections. We extend the definitions of intersection and union to these collections in the fairly obvious way. Let $C = \{\, A_i \mid i \in I \,\}$ be an indexed collection of sets. Then the **intersection** of this collection, written $\bigcap C$ or, more usually, $\bigcap_{i \in I} A_i$ or just $\bigcap A_i$, is the set of all objects that are simultaneously in every one of the sets in the collection:

$$\bigcap C = \bigcap_{i \in I} A_i = \{\, x \mid \forall i \in I : x \in A_i \,\}.$$

The **union** of the collection, written $\bigcup C$ or, more usually, $\bigcup_{i \in I} A_i$ or just $\bigcup A_i$, is the set of all objects that are in at least one of the sets in the collection:

$$\bigcup C = \bigcup_{i \in I} A_i = \{\, x \mid \exists i \in I : x \in A_i \,\}.$$

Actually, both $\bigcap C$ and $\bigcup C$ depend only on C as a set, not on the indexing structure. If C is any set of sets, then we can write

$$\bigcap C = \{\, x \mid \forall A \in C : x \in A \,\} \qquad \text{and} \qquad \bigcup C = \{\, x \mid \exists A \in C : x \in A \,\}.$$

Usually, it is more natural, however, to view the set with its indexing structure.

If the index set is empty (not a likely occurrence in practice), we arbitrarily define $\bigcap C$ to be whatever universal set applies to the given situation, and we define $\bigcup C$ to be the empty set. The situation is analogous to taking products of indexed sets of numbers (Exercise 29 in Section 2.2) and sums of indexed sets of numbers (Example 9 in Section 2.2), respectively. There it is customary to define an empty product to be 1 and an empty sum to be 0.

EXAMPLE 2.   In Example 10 of Section 2.2 we looked at the collection of intervals $\{\, [x, \infty) \mid x \in \mathbf{R} \,\}$. The union of this collection includes all real numbers that are in any of these intervals; since in particular $x \in [x, \infty)$ for each real number x, we see that the union is all of $\mathbf{R}$. On the other hand, no real number is in *all* of these intervals, since $x \notin [x + 1, \infty)$; therefore the intersection is the empty set. In symbols, we have

$$\bigcup_{x \in \mathbf{R}} [x, \infty) = \mathbf{R} \quad \text{and} \quad \bigcap_{x \in \mathbf{R}} [x, \infty) = \emptyset.$$

Let L be the set of courses you are taking this semester. For each $c \in L$, let S_c be the set of students in course c. This gives us an indexed collection of sets: $S = \{\, S_c \mid c \in L \,\}$. Then $\bigcap_{c \in L} S_c$ is the set of students each of whom is taking all the courses you are taking (and perhaps some other courses as well). This set is not empty, since it includes at least you. Unless you are taking very few courses or are in a highly structured program, it is not too likely that $\left| \bigcap_{c \in L} S_c \right|$ will be more than 1 or 2. On the other hand, $\bigcup_{c \in L} S_c$ is the set of all the students you see in class this semester. Note that if you are taking only one class (discrete mathematics, no doubt), then we can write $L = \{\text{discrete mathematics}\}$ and $S = \{ S_{\text{discrete mathematics}} \}$. In this case $\bigcap S = \bigcup S = S_{\text{discrete mathematics}}$, the set of you and your classmates in discrete mathematics. $\blacklozenge$

If a collection of sets is finite (in other words, the index set is finite—we are not restricting the cardinality of the sets in the collection), then we can write the intersection and union of the collection in other forms as well. For example, if A_1, A_2, A_3, and A_4 are sets, then we can write the intersection and union of the collection $\{A_1, A_2, A_3, A_4\}$ as

$$\bigcap_{i=1}^{4} A_i = A_1 \cap A_2 \cap A_3 \cap A_4$$

and

$$\bigcup_{i=1}^{4} A_i = A_1 \cup A_2 \cup A_3 \cup A_4,$$

respectively. We will soon see that these operations are associative, so there is really no ambiguity here. For example, $A_1 \cup A_2 \cup A_3 \cup A_4$, $((A_1 \cup A_2) \cup A_3) \cup A_4$, $(A_1 \cup A_2) \cup (A_3 \cup A_4)$, and $A_1 \cup ((A_2 \cup A_3) \cup A_4)$ all represent the same set.

If $C = \{ A_i \mid i \in I \}$ is a collection of sets, then there may be no element common to all of the sets in the collection (i.e., $\bigcap C$ may be the empty set). There is an even stronger sense in which the sets in the collection may be disjoint, however: There may be no element common to even two of the sets in the collection. We say that the sets in the collection are **pairwise disjoint** if $A_i \cap A_j = \emptyset$ whenever $i \neq j$.

EXAMPLE 3. In Example 10 of Section 2.2 we defined the collection $\{ F_n \mid n \in \mathbf{N} \wedge n > 0 \}$, where F_n is the set of prime factors of the positive integer n. The sets in this collection are not pairwise disjoint, since, for instance, $F_{12} \cap F_{700} = \{2, 3\} \cap \{2, 5, 7\} = \{2\} \neq \emptyset$. The intersection of the entire collection is empty, however, since there is no prime number that is a factor of *every* n; in symbols, $\bigcap F_n = \emptyset$.

On the other hand, the sets $\{1, 3, 5, 7\}$, $\{2, 4\}$, and $\{10, 11\}$ are pairwise disjoint. ◆

COMPLEMENTS

We saw above that the operation of intersection on sets corresponds to the logical operation "and" on propositions, and the operation of union on sets corresponds to the logical operation "or" on propositions. The other basic logical operation, negation, also has a set-theoretic analogue.

To define this operation, we need to fix a universal set (a domain under discussion). Call the universal set U. Thus all the sets of interest in the given situation will be subsets of U. If A is a set, then the **complement** of A, written $\overline{A}$, is the set of all elements of the universal set U that are not in A. In symbols, $\overline{A} = \{ x \in U \mid x \notin A \}$.

EXAMPLE 4. Suppose that we are talking about natural numbers, so $U = \mathbf{N}$. The complement of the set of even natural numbers $\{0, 2, 4, 6, \ldots\}$ is the set of odd natural numbers $\{1, 3, 5, 7, \ldots\}$. If P is the set of prime numbers, then $\overline{P}$ is the set of natural numbers that are not prime: the composite numbers along with 0 and 1. The complement of $\mathbf{N}$ itself is $\emptyset$, and $\overline{\emptyset} = \mathbf{N}$. ◆

Complementation is a *unary* operation, since it acts on—deals with—just one set, rather than a *binary* operation, such as the simple case of intersection or union, which acts on two sets. We next define two binary operations that are related to complementation. Again let A and B be two sets. The **difference** of A and B, written $A - B$, is $A \cap \overline{B}$; and the **symmetric difference** of A and B, written $A \oplus B$, is $(A - B) \cup (B - A)$.

If we go back to the definitions of the earlier operations, we can write down explicit formulas for difference and symmetric difference:

$$A - B = \{ x \mid x \in A \wedge x \notin B \}$$

and

$$A \oplus B = \{\, x \mid (x \in A \wedge x \notin B) \;\vee\; (x \in B \wedge x \notin A) \,\}.$$

In words, the difference $A - B$ is the set of elements of A that are not in B, and the symmetric difference of A and B is the set of elements that are in exactly one of A and B.

   Note that "complement" is really a special case of "difference," since $\overline{A} = U - A$, where U is the universal set. Also, we leave it to the reader (Exercise 15) to show that the symmetric difference of two sets is the difference between their union and their intersection: $A \oplus B = (A \cup B) - (A \cap B)$.

EXAMPLE 5. Let $U = \mathbf{R}$, and consider the following sets, which are all intervals of real numbers (we adopt the usual convention that a parenthesis indicates that an endpoint is not included in the interval and a bracket indicates that it is):

$$A = [1,3) = \{\, x \mid 1 \le x < 3 \,\}$$
$$B = (2,4] = \{\, x \mid 2 < x \le 4 \,\}$$
$$C = (-\infty, 5) = \{\, x \mid x < 5 \,\}.$$

Succinctly express the following sets as intervals or unions of disjoint intervals:
(a) $\overline{C}$,   (b) $A - B$,   (c) $A - C$,   (d) $C - A$,   (e) $A \oplus B$.

Solution. The reader should draw pictures of all eight of these sets on the real number line.

   (a) Since $\overline{C}$ is the set of real numbers not in C, we have $\overline{C} = \{\, x \mid x \ge 5 \,\}$, which is the interval $[5, \infty)$.

   (b) $A - B$ is the set of numbers in A but not in B. Thus $A - B = \{\, x \mid 1 \le x < 3 \wedge \overline{2 < x \le 4} \,\} = \{\, x \mid 1 \le x \le 2 \,\} = [1,2]$.

   (c) Every number in A is also in C. Therefore, there are no numbers that are in A but not in C. In other words, $A - C$ is the empty set (the union of no intervals).

   (d) If we remove the elements in A from C, we are left with the numbers less than 1, together with the numbers greater than or equal to 3 but less than 5. Therefore, $C - A = (-\infty, 1) \cup [3, 5)$.

   (e) We saw in part (b) that $A - B = [1,2]$. Similarly, $B - A = [3,4]$. Therefore, $A \oplus B = [1,2] \cup [3,4]$.   ◆

VENN DIAGRAMS

So far we have been talking about sets in a rather formal, symbolic way. As is usually the case in mathematics, we will understand things better if we can draw a picture of what is going on. In set theory this is often easy to do, using **Venn diagrams**. These diagrams allow us to transform complicated formulas into visual images.

The universal set is represented in a Venn diagram by a large rectangle. Intuitively, some or all of the points inside the rectangle represent the elements of the universal set. Sets of interest are represented by circles or other simple closed shapes inside the rectangle, the interpretation being that the points inside the circle represent the elements in the set. We will often shade or otherwise indicate the regions of interest. Sometimes we even label specific points inside or outside a region. In Figure 2.1 the universe is the set of integers. The set E of even integers is represented by the lightly shaded circle labeled E, and two specific integers are shown in their correct relationship to E.

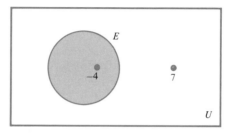

FIGURE 2.1 Venn diagram
representing the set of even integers.

The relationships of being a subset and of being disjoint, and the operations of intersection, union, complementation, difference, and symmetric difference come alive when we draw the Venn diagrams representing them. Figure 2.2 shows two sets A and B such that $A \subseteq B$. Note that every point inside the A region is also inside the B region.

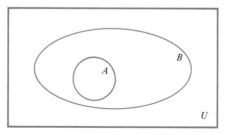

FIGURE 2.2 Venn diagram
representing $A \subseteq B$.

The intersection of two sets shows itself in a Venn diagram as the region where the regions representing the two sets overlap. Figure 2.3 shows the intersection of two sets, the double-hatched lens-shaped region.

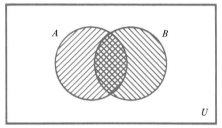

FIGURE 2.3   Venn diagram
representing $A \cap B$ (double-hatched).

If two sets are known to be disjoint, we can draw their Venn diagram with nonoverlapping circles. Figure 2.4 shows that the intersection of A and B is empty.

On the other hand, a Venn diagram showing two circles as overlapping is *not* meant to imply that the intersection is nonempty. Similarly, if the region for A is not totally contained in the region for B, we must not infer that A is not a subset of B.

The Venn diagrams in Figure 2.5 represent (a) union, (b) complement, (c) difference, and (d) symmetric difference. In each case the region of interest is shaded.

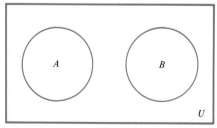

FIGURE 2.4   Venn diagram
representing $A \cap B = \emptyset$.

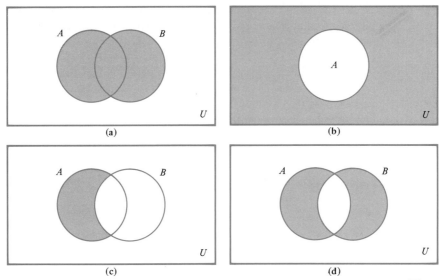

FIGURE 2.5 Venn diagrams whose shaded areas represent (a) $A \cup B$, (b) $\overline{A}$, (c) $A - B$, and (d) $A \oplus B$.

EXAMPLE 6. Draw a Venn diagram representing $A \cup (B \oplus C)$.

Solution. Since we are given no information about the sets in this problem, we must draw the diagram in the most general way possible, allowing all possible intersections. A little thought will show that Figure 2.6 (ignore the shading for a moment) gives this most general arrangement. Now we need to shade the diagram to include all points in A, together with all points that are in exactly one of B and C. We have shaded the former in one direction, the latter in the other direction. Therefore, the set of all points shaded at all represents $A \cup (B \oplus C)$. Incidentally, the double-hatched area represents $A \cap (B \oplus C)$. ◆

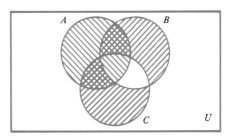

FIGURE 2.6 Venn diagram representing $A \cup (B \oplus C)$.

EXAMPLE 7. Suppose that we have three sets, A, B, and C, which are known to have empty common intersection. Draw a Venn diagram in which the pairwise intersections are shaded.

Solution. The region we want to shade is $(A \cap B) \cup (B \cap C) \cup (C \cap A)$. Figure 2.7 gives the desired picture. Note that we drew the regions (rectangles in this case) in such a way no points were common to all three. ◆

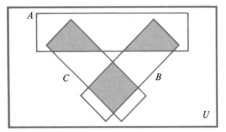

FIGURE 2.7 Venn diagram representing pairwise intersections of three sets with no common intersection.

IDENTITIES (*OPTIONAL*)

We end this section by considering, as advertised at the beginning, the identities that hold among the operations of set theory. Let us begin with a typical identity, one of the distributive laws, and indicate several ways in which we might prove that it indeed holds for all sets.

We want to show that the operation of union distributes over the operation of intersection, that is,

$$A \cup (B \cap C) = (A \cup B) \cap (A \cup C)$$

for all sets A, B, and C. One way to prove this identity is to use the definitions of the set-theoretic operations and invoke a logical identity given in part (c) of Theorem 1 of Section 1.1 (the distributivity of $\vee$ over $\wedge$). We argue as follows.

$A \cup (B \cap C)$

$$= \{\, x \mid x \in A \ \lor \ x \in B \cap C \,\} \qquad \text{(definition of } \cup \text{)}$$

$$= \{\, x \mid x \in A \ \lor \ (x \in B \land x \in C) \,\} \qquad \text{(definition of } \cap \text{)}$$

$$= \{\, x \mid (x \in A \ \lor \ x \in B) \ \land \ (x \in A \ \lor \ x \in C) \,\} \qquad \text{(logical identity)}$$

$$= \{\, x \mid (x \in A \ \lor \ x \in B) \,\} \ \cap \ \{\, x \mid (x \in A \ \lor \ x \in C) \,\} \qquad \text{(definition of } \cap \text{)}$$

$$= (A \cup B) \cap (A \cup C) \qquad \text{(definition of } \cup \text{)}$$

An alternative proof, based on the same idea, but relying on our intuition about logic (the meaning of "and" and "or"), rather than the identities from Chapter 1, proceeds as follows. Recall from Section 2.1 that if we want to show that two sets are equal, then we need to show that each is a subset of the other. In other words, we must show that whenever we have an element x that is in the set on the left-hand side of the equation, then x must also be in the set on the right-hand side of the equation; and conversely, whenever x is in the set on the right-hand side of the equation, then x must also be in the set on the left-hand side of the equation. In this case we would argue along the following lines.

First we show that $A \cup (B \cap C) \subseteq (A \cup B) \cap (A \cup C)$. To this end, let $x \in A \cup (B \cap C)$. By definition, this means that either $x \in A$ or $x \in B \cap C$. In the first case, since $x \in A$, certainly x is an element of any superset of A; in particular, $x \in A \cup B$ and $x \in A \cup C$. Therefore, by definition of intersection, $x \in (A \cup B) \cap (A \cup C)$, as desired. In the second case ($x \in B \cap C$), we know that $x \in B$ and $x \in C$. Therefore, $x \in A \cup B$ (which is a superset of B) and $x \in A \cup C$ (a superset of C). Thus again x is in their intersection, as desired. This completes half of our proof; we have shown that every element of $A \cup (B \cap C)$ is also an element of $(A \cup B) \cap (A \cup C)$.

Conversely, suppose that $x \in (A \cup B) \cap (A \cup C)$, the right-hand side of our equation; we need to show that x is also an element of the left-hand side. Now our hypothesis implies that $x \in A \cup B$ and $x \in A \cup C$. We consider two cases: either $x \in A$ or $x \notin A$. If $x \in A$, then certainly $x \in A \cup (B \cap C)$, a superset of A, which is what we were after. On the other hand, suppose that $x \notin A$. Then the only way that $x \in A \cup B$ can hold (which, we already assumed, it does) is for x to be an element of B. Similarly, x must also be in C, since $x \in A \cup C$. Therefore, $x \in B \cap C$, and thus x is also in the superset $A \cup (B \cap C)$, again what we wanted. This completes the second half of the proof showing that $(A \cup B) \cap (A \cup C) \subseteq A \cup (B \cap C)$, and so we have demonstrated that the equation $A \cup (B \cap C) = (A \cup B) \cap (A \cup C)$ is an identity.

We can also prove this identity by "considering all possible cases" as to where an element might live, relative to the three sets A, B, and C. A properly drawn Venn diagram enables us to do just that.

Consider the Venn diagrams shown in Figure 2.8. In the diagram on the left, we have shaded A and we have shaded $B \cap C$. (The overlap, $A \cap B \cap C$, having been shaded twice, is even darker.) The entire shaded area, therefore, constitutes $A \cup (B \cap C)$. In the diagram on the right, we have marked $A \cup B$ with lines slanted in one direction and $A \cup C$ with lines slanted in the other direction. Therefore, the double-hatched area represents $(A \cup B) \cap (A \cup C)$. Now we compare the two diagrams and see that, as expected, the indicated regions (the region shaded at all in Figure 2.8a and the region with double-hatching in Figure 2.8b) are the same. Therefore, the two sets are equal: $A \cup (B \cap C) = (A \cup B) \cap (A \cup C)$.

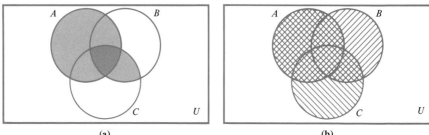

(a)   (b)

FIGURE 2.8   Proof by Venn diagram that $A \cup (B \cap C) = (A \cup B) \cap (A \cup C)$.

An approach essentially equivalent to the use of Venn diagrams (in that we consider all possible cases in a systematic way) is to use the set-theoretic analogue of truth tables. We make a table indicating the eight possibilities for an item x: either "being an element of" or "not being an element of" each of the sets A, B, and C. Let us use a 1 to indicate membership and 0 to indicate nonmembership. See the first three columns of the following table.

A	B	C	$B \cap C$	$A \cup (B \cap C)$	$A \cup B$	$A \cup C$	$(A \cup B) \cap (A \cup C)$
1	1	1	1	1	1	1	1
1	1	0	0	1	1	1	1
1	0	1	0	1	1	1	1
1	0	0	0	1	1	1	1
0	1	1	1	1	1	1	1
0	1	0	0	0	1	0	0
0	0	1	0	0	0	1	0
0	0	0	0	0	0	0	0

Now we can determine, in each case, whether $x \in A \cup (B \cap C)$ and whether $x \in (A \cup B) \cap (A \cup C)$. Again we use a 1 to indicate membership and a 0 to indicate nonmembership, and we construct the table by working on each expression from the inside out. Since the fifth column ($A \cup (B \cap C)$) has the same sequence of 0's and 1's as the eighth column $((A \cup B) \cap (A \cup C))$, we conclude that $x \in A \cup (B \cap C)$ precisely in the same cases that $x \in (A \cup B) \cap (A \cup C)$; therefore, the two sets are equal. This is exactly the same process we went through in Section 1.1 to prove logical identities by truth tables, so we will not pursue it further here.

The reader should practice proving identities by all three methods, since all three provide some valuable insights. The first method highlights the correspondence between set identities and identities for propositional logic. The second method gives the student a better understanding of how to write a mathematical proof in prose paragraph form. The third method, since it uses a picture (the Venn diagram), offers a good way for our human minds to appreciate these identities. (The "truth table" method is really just a symbolic translation of the Venn diagram method.)

Our first theorem of this chapter is a compilation of some important identities. As a comparison with Theorem 1 in Section 1.1 seems to indicate, whatever is true about propositions is also true about sets, under the following translation.

Proposition	corresponds to	set.
$\wedge$ (and)	corresponds to	$\cap$ (intersection).
$\vee$ (or)	corresponds to	$\cup$ (union).
$^{-}$ (negation)	corresponds to	$^{-}$ (complementation).
$\Longleftrightarrow$ (logical equivalence)	corresponds to	$=$ (equality).
T (true)	corresponds to	U (the universal set).
F (false)	corresponds to	$\emptyset$ (the empty set).

The proof of the following theorem, then, is left to the reader (Exercises 20–22).

THEOREM 1. *The following identities hold for sets. Here A, B, and C stand for any sets; U is the universal set, and $\emptyset$ is the empty set.*

(a) *Associative laws*

$$A \cap (B \cap C) = (A \cap B) \cap C$$
$$A \cup (B \cup C) = (A \cup B) \cup C$$

(b) *Commutative laws*

$$A \cap B = B \cap A$$
$$A \cup B = B \cup A$$

(c) *Distributive laws*

$$A \cap (B \cup C) = (A \cap B) \cup (A \cap C)$$
$$A \cup (B \cap C) = (A \cup B) \cap (A \cup C)$$

(d) *Identity laws*

$$A \cap U = A$$
$$A \cup \emptyset = A$$

(e) *Dominance laws* $\qquad$ $A \cap \emptyset = \emptyset$

$\qquad\qquad\qquad\qquad\qquad$ $A \cup U = U$

(f) *Idempotent laws* $\qquad$ $A \cap A = A$

$\qquad\qquad\qquad\qquad\qquad$ $A \cup A = A$

(g) *Complement laws* $\qquad$ $A \cap \overline{A} = \emptyset$

$\qquad\qquad\qquad\qquad\qquad$ $A \cup \overline{A} = U$

(h) *Double complement law* $\qquad$ $\overline{\overline{A}} = A$

(i) *DeMorgan's laws* $\qquad$ $\overline{A \cap B} = \overline{A} \cup \overline{B}$

$\qquad\qquad\qquad\qquad\qquad$ $\overline{A \cup B} = \overline{A} \cap \overline{B}$

Some of these laws can be generalized to more than two sets, since we have defined intersections and unions on arbitrary collections of sets. In particular, we have the following generalization of DeMorgan's laws, that the complement of the intersection is the union of the complements, and the complement of the union is the intersection of the complements.

THEOREM 2.   *Let $\{A_i\}$ be a collection of sets indexed by a set I. Then*

$$\overline{\bigcap_{i \in I} A_i} = \bigcup_{i \in I} \overline{A_i} \quad \text{and} \quad \overline{\bigcup_{i \in I} A_i} = \bigcap_{i \in I} \overline{A_i}.$$

Proof.   We prove the first equality and leave the second for the reader (Exercise 11). To show that these two sets are equal, we must show that every element of the set on the left is an element of the set on the right, and that every element of the set on the right is an element of the set on the left. For the first of these, let $x \in \overline{\bigcap_{i \in I} A_i}$. This means that $x \notin \bigcap_{i \in I} A_i$. By definition of intersection, this means that x fails to be in at least one of the sets A_i, in other words, that x is in at least one of the sets $\overline{A_i}$. But by definition of union, this means that $x \in \bigcup_{i \in I} \overline{A_i}$, as desired.

Conversely, suppose that $x \in \bigcup_{i \in I} \overline{A_i}$. By definition of union, this means that $x \in \overline{A_i}$ for at least one $i \in I$, which is the same as saying that $x \notin A_i$ for this i. But then by definition of intersection, we conclude that $x \notin \bigcap_{i \in I} A_i$, whence $x \in \overline{\bigcap_{i \in I} A_i}$, and our proof of the first of the generalized DeMorgan's laws is complete.   ■

SUMMARY OF DEFINITIONS

complement of the set A: the set of elements of the universal set that are not in A; written $\overline{A}$ (e.g., if $U = \{1, 2, 3, 4, 5\}$, then $\overline{\{1, 3\}} = \{2, 4, 5\}$).

difference of sets A and B (in that order): the set of elements in A that are not in B, written $A - B$ (e.g., $\{1, 2, \{3\}\} - \{1, 3, 5\} = \{2, \{3\}\}$).

disjoint sets: sets whose intersection is empty.

intersection of a collection of sets: the set of elements common to all sets in the collection (e.g., if $S_i = \{0, 1, 2, \ldots, i\}$, then $\bigcap_{i \in \mathbf{N}} S_i = \{0\}$).

intersection of sets A and B: the set of elements common to both A and B, written $A \cap B$ (e.g., $\{1, 2, \{3\}\} \cap \{1, 3, 5\} = \{1\}$).

pairwise disjoint collection of sets: a collection of sets, every two of which are disjoint.

symmetric difference of sets A and B: the set of elements that are in exactly one of A and B, written $A \oplus B$ (e.g., $\{1, 2, \{3\}\} \oplus \{1, 3, 5\} = \{2, 3, 5, \{3\}\}$).

union of a collection of sets: the set of elements in at least one of the sets in the collection (e.g., if $S_i = \{0, 1, 2, \ldots, i\}$, then $\bigcup_{i \in \mathbf{N}} S_i = \mathbf{N}$).

union of sets A and B: the set of elements in one or both of A and B, written $A \cup B$ (e.g., $\{1, 2, \{3\}\} \cup \{1, 3, 5\} = \{1, 2, 3, 5, \{3\}\}$).

Venn diagram: a picture for representing sets, in which the universal set is represented by points in the interior of a large rectangle, and individual sets are represented by points in the interiors of various simple regions (such as circles) inside the rectangle; Venn diagrams allow a human being to visualize notions of subset, intersection, union, and complement.

EXERCISES

1. Let $A = \{1, 3, 5, 7, 9\}$, $B = \{2, 4, 6, 8, 10\}$, and $C = \{3, 6, 9\}$. Assuming that we are working in the context of the universal set $U = \{1, 2, \ldots, 10\}$, determine the following sets explicitly.
   (a) $A \cap B$.
   (b) $A \cap C$.
   (c) $A \cup B$.
   (d) $B \cup C$.
   (e) $\overline{C}$.
   (f) $A - C$.
   (g) $B \oplus C$.

2. Let P be the set of prime numbers, let T be the set of positive even integers, and let H be the set of positive multiples of 3. The universal set for this problem is the set of positive integers. Determine the following sets: list all of their elements if they have 10 or fewer elements; list their 10 smallest elements if they have more than 10.
   (a) $P \cap H$.
   (b) $(H \cup T) - P$.
   (c) $\overline{H \cap T}$.
   (d) $H \oplus T$.
   (e) $P \cup (T - H)$.

3. Let the universal set for this problem be the set of all people. Let M be the set of all males, let C be the set of all children, and let A be the set of all Americans. Express each of the following collections as expressions involving these sets.
   (a) Boys.
   (b) Girls.
   (c) Adult women.
   (d) American adult men.
   (e) Non-American girls.

4. Draw a Venn diagram showing the correct relationship among the sets P, T, and H of Exercise 2, and indicate by a labeled point one number in each region. Regions should overlap if and only if the sets they represent are not disjoint.

5. Show that $\forall A: \forall B: A \cap B \subseteq A \cup B$.

6. Find $\{1, 2, \{1, 2\}\} - \{1, 2\}$.

7. Give an example of two nonempty disjoint sets of natural numbers
   (a) Whose symmetric difference is $\{1, 2, 3\}$.
   (b) The complement of whose union is $\{1, 2, 3\}$.
   (c) Such that the union of one with the complement of the other is $\{1, 2, 3\}$.

8. Prove (or disprove with a counterexample or Venn diagram, if not true) the following propositions about sets. Universal quantifiers are implied.
   (a) If A and B are disjoint and B and C are disjoint, then A and C are disjoint.
   (b) If A and B are disjoint and C and D are disjoint, then $A \cap C$ and $B \cap D$ are disjoint.
   (c) If A and B are disjoint and C and D are disjoint, then $A \cup C$ and $B \cup D$ are disjoint.
   (d) $A - \overline{B} = A \cap B$.
   (e) $A \subseteq B \leftrightarrow A - B = \emptyset$.
   (f) $A - B = \overline{B - A}$.
   (g) $(A \cup B) \cap C = A \cup (B \cap C)$.

9. Under what conditions is the symmetric difference of two sets equal to their union?

10. Find the following intersections and unions of intervals. The variable a ranges over the real numbers. Recall the convention that a bracket indicates that the endpoint is included in the interval, and a parenthesis indicates that the endpoint is not included in the interval: for example, $(x, y] = \{z \mid x < z \le y\}$.
    (a) $\bigcup_{0 \le a \le 1} [a, 2]$.
    (b) $\bigcap_{0 \le a \le 1} [a, 2]$.
    (c) $\bigcap_{0 \le a < 1} [a, 2]$.

(d) $\bigcap_{0 \leq a \leq 1} (a, 2]$.

(e) $\bigcap_{0 \leq a < 1} (a, 2]$.

(f) $\bigcup_{x \in \mathbf{Q}} (x - \epsilon, x + \epsilon)$, where $\mathbf{Q}$ is the set of rational numbers and ϵ is a fixed positive real number. [*Hint*: The answer to this part and the next part do not depend on ϵ.]

(g) $\bigcap_{x \in \mathbf{Q}} (x - \epsilon, x + \epsilon)$, where $\mathbf{Q}$ is the set of rational numbers and ϵ is a fixed positive real number.

11. Prove the second half of Theorem 2.

12. Suppose that $P(x)$ represents the set of biological (genetic) parents of person x. Consider the indexed collection of sets $R = \{ P(x) \mid x$ is in the United States right now $\}$.
    (a) What are the possible cardinalities of the pairwise intersections of the sets in R?
    (b) Describe in words the union of all the sets in R.
    (c) Prove that $\bigcap R = \emptyset$.

13. Give an example of an infinite pairwise disjoint collection of natural numbers whose union is all of $\mathbf{N}$.

14. Give an example of an infinite collection of subsets of $\mathbf{N}$, every two of which have nonempty intersection, such that no natural number is in *all* of the subsets.

15. Prove that $A \oplus B = (A \cup B) - (A \cap B)$.

16. Let E be the set of even integers, O the set of odd integers, F the set of multiples of 4, and H the set of multiples of 3. Draw Venn diagrams showing the correct relationships for the following collections. Take the universal set to be the set of integers.
    (a) E, F, and H.
    (b) O, F, and H.

17. Draw a Venn diagram representing each of the following sets. Be sure to state how the viewer is to interpret the diagram. (Is the answer all the shaded region? the double-hatched region? some other region?)
    (a) $A \cap (B \oplus C)$.
    (b) $(A - B) \cup (C - A)$.
    (c) $A \oplus (B \oplus C)$.

18. Neatly draw a Venn diagram to represent the most general relationship among four sets, A, B, C, and D. Then shade the region $(A \oplus B) \oplus (C \oplus D)$. [*Hint*: There need to be 16 regions.]

19. Write down expressions for each of the following shaded regions. (There are several correct answers, since there is more than one way to express the shaded set.)

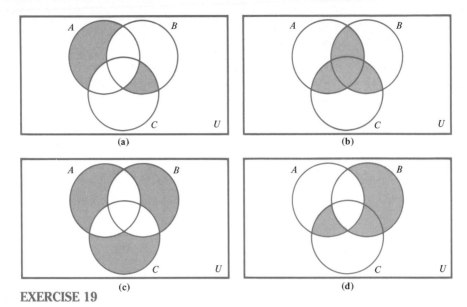

(a)

(b)

(c)

(d)

EXERCISE 19

20. Prove the following identities for sets, using the first of the methods discussed in the final subsection of this section.
    (a)   The first associative law.
    (b)   The first distributive law.
    (c)   The second identity law.
    (d)   The first idempotent law.
    (e)   The second complement law.
    (f)   The first of DeMorgan's laws.

21. Prove the following identities for sets, using the second of the methods discussed in the final subsection of this section.
    (a)   The second commutative law.
    (b)   The first distributive law.
    (c)   The first identity law.
    (d)   The second idempotent law.
    (e)   The first complement law.
    (f)   The double complement law.
    (g)   The second of DeMorgan's laws.

22. Prove the following identities for sets, using Venn diagrams.
    (a)   The second associative law.
    (b)   The first distributive law.
    (c)   The first complement law.
    (d)   The second complement law.

(e) The first of DeMorgan's laws.

(f) The second of DeMorgan's laws.

23. Determine whether the following operations on sets have the indicated properties. Either prove that they do or give a counterexample to show otherwise.

(a) Is difference commutative?

(b) Is symmetric difference associative?

(c) Does union distribute over symmetric difference?

(d) Does intersection distribute over symmetric difference?

(e) Does Cartesian product distribute over intersection?

(f) Does Cartesian product distribute over union?

24. Prove or disprove each of the following propositions about sets. (Universal quantifiers are understood.)

(a) A is a subset of B if and only if A and $\overline{B}$ are disjoint.

(b) $A \subseteq B$ if and only if $\overline{B} \subseteq \overline{A}$.

(c) $A \subseteq B$ if and only if $A \cap B = A$.

(d) $A \cap B = A \cup B$ if and only if $A = B$.

(e) $\mathcal{P}(A \cap B) = \mathcal{P}(A) \cap \mathcal{P}(B)$.

(f) $\mathcal{P}(A \cup B) = \mathcal{P}(A) \cup \mathcal{P}(A)$.

(g) $(A \cup B) \cap A = B \longleftrightarrow A = B$.

(h) $A \cap B \subseteq C \longrightarrow (A \subseteq C \wedge B \subseteq C)$.

25. Explain in what sense the difference operation on sets corresponds to "negation of implication" for propositions.

26. What logical operation corresponds to symmetric difference for sets? Justify your answer.

27. The equality relationship between sets is analogous to the logical equivalence relationship between propositions.

(a) What relationship between sets is analogous to the logical implication relationship between propositions?

(b) State and prove the set-theoretic analogue to Theorem 2, parts (b) and (c), of Section 1.1.

28. Suppose that there is a set J of jobs that need to be done in a certain company. The jobs are difficult, and people need special training to be able to do the jobs. For each $j \in J$, let E_j be the set of employees who have been trained to do job j. Let E be the set of employees in the company. Translate into set-theoretic symbols the following propositions.

(a) All the jobs can be done.

(b) Jack can do all the jobs.

(c) No employee is indispensable (even if one employee leaves, the jobs can still all be done).

(d) There are at least three different people trained to do each job.

(e) J_p is the set of jobs that person p can do.

29. Consider the collection I of infinite subsets of $\mathbf{N}$, together with the empty set. In other words, $I = \{ A \subseteq \mathbf{N} \mid A = \emptyset \ \lor \ |A| = \aleph_0 \}$. Let F be the collection of finite subsets of $\mathbf{N}$. Note that $I \cup F = \mathcal{P}(\mathbf{N})$ and $I \cap F = \{ \emptyset \}$. Determine, with justification, whether each of the following statements is true or false.
    (a) The intersection of any two elements of F is again an element of F.
    (b) The intersection of any two elements of I is again an element of I.
    (c) The union of any two elements of F is again an element of F.
    (d) The union of any two elements of I is again an element of I.

Challenging Exercises

30. Suppose that A is a given subset of a universal set U. Characterize in terms of A the solutions to the following equations in an unknown set X (some of the equations may have many solutions).
    (a) $X - A = \emptyset$.
    (b) $X \oplus A = \emptyset$.
    (c) $X - A = A - X$.

31. Find the following intersections and unions of intervals. Here $\mathbf{Q}$ is the set of rational numbers and the variable ϵ ranges over (positive) real numbers. [*Hint*: Work from the inside out.]
    (a) $\bigcup_{x \in \mathbf{Q}} \left(\bigcap_{\epsilon > 0} (x - \epsilon, x + \epsilon) \right)$.
    (b) $\bigcap_{x \in \mathbf{Q}} \left(\bigcup_{\epsilon > 0} (x - \epsilon, x + \epsilon) \right)$.
    (c) $\bigcap_{\epsilon > 0} \left(\bigcup_{x \in \mathbf{Q}} (x - \epsilon, x + \epsilon) \right)$.
    (d) $\bigcup_{\epsilon > 0} \left(\bigcap_{x \in \mathbf{Q}} (x - \epsilon, x + \epsilon) \right)$.
    (e) $\bigcap_{\epsilon > 0} \left(\bigcap_{x \in \mathbf{Q}} (x - \epsilon, x + \epsilon) \right)$.
    (f) $\bigcup_{\epsilon > 0} \left(\bigcup_{x \in \mathbf{Q}} (x - \epsilon, x + \epsilon) \right)$.

32. Suppose that $C(P, r)$ is the set of points in the plane that are exactly a distance r from the point P. In other words, $C(P, r)$ is the circle of radius r, centered at P. The variable P ranges over points in the plane, and the variable r ranges over real numbers. Describe succinctly the following sets.
    (a) $\bigcup_{0 \le r \le 1} C(P, r)$, where P is a fixed point in the plane.
    (b) $\bigcup_{0 < r < 1} C(P, r)$, where P is a fixed point in the plane.
    (c) $\bigcap_{0 \le r \le 1} C(P, r)$, where P is a fixed point in the plane.
    (d) $\bigcup_{P \in L} C(P, 1)$, where L is a fixed line in the plane.
    (e) $\bigcup_{3 \le r \le 5} C(P, r)$, where P is a fixed point in the plane.
    (f) $\bigcup_{P \in C_0} C(P, 1)$, where C_0 is a fixed circle in the plane.

33. Give an example of an infinite collection of sets such that the intersection of every two distinct sets in the collection is nonempty, but the intersection of every three distinct sets in the collection is empty; or show that no such collection exists.

34. Show that
    (a)   The union of two countable sets is countable.
    (b)   The union of a countable collection of countable sets is countable.

35. Consider $C = \{ A \subseteq \mathbf{N} \mid A \text{ is finite} \vee \overline{A} \text{ is finite} \}$; the universal set here is $\mathbf{N}$. Suppose that A and B are elements of C. Show that each of the following sets must also be an element of C.
    (a) $\overline{A}$.
    (b) $A \cap B$.
    (c) $A \cup B$.
    (d) $A - B$.

36. Determine whether the following sets are countable.
    (a) F in Exercise 29.
    (b) I in Exercise 29.
    (c) C in Exercise 35.

Exploratory Exercises

37. Consider the collection of sets $C = \{ S_n \mid n \in \mathbf{Z} \wedge |n| < 10 \}$, where $S_n = \{ x \mid x^2 = n^2 \}$.
    (a)   Describe by listing their elements S_0, S_7, and S_{-3}.
    (b)   What are the cardinalities of the elements of C?
    (c)   What is the cardinality of the index set?
    (d)   What is the cardinality of C?
    (e)   Explain in what sense the sets in C are pairwise disjoint and in what sense they are not.

38. Find out about ways to draw Venn diagrams with many regions intersecting arbitrarily by consulting Fisher, Koh, and Grünbaum [83] and Grünbaum [130]. Before doing so, you might want to try your hand at drawing a neat and useful Venn diagram for five sets (there will be 32 regions determined by intersections of these sets and their complements).

3

FUNCTIONS
AND RELATIONS

Mathematics deals with more than just static objects such as sets, numbers, and so on. It also deals with the interrelationships among them. We discover in this chapter that we can use the notion of set (together with the language of logic) to define rigorously and talk substantively about such relationships.

Functions provide certain kinds of relationships between mathematical objects. In Sections 3.1 and 3.2 we study functions, not from the perspective of a calculus student who uses them to model growth and change, but as they can be applied to discrete mathematical problems and as abstract objects in their own right. We will see that some of the more complicated objects we defined in Chapter 2 (such as strings) can be viewed profitably as functions. Then in Sections 3.3 and 3.4 we look at generalizations of functions, called relations. They provide us with the framework for discussing relationships between objects in many contexts, from something as simple as numerical order among natural numbers to something as complex as the interconnections among people in humankind's family tree.

The textbook by Fletcher and Patty [85] contains material relevant to this chapter.

SECTION 3.1
FUNCTIONS

*To every thing there is a season, and a
time to every purpose under the heaven.*

—Ecclesiastes 3 : 1

Anyone who has studied algebra or calculus has worked extensively with important mathematical functions of real numbers, such as polynomials, exponential
functions, and trigonometric functions. Calculus students have also encountered
functions *of functions*; differentiation, for example, can be viewed as a function
that takes functions to their derivatives. We can even view the process of solving
equations as a function—taking an equation to its solution. In elementary school
we study functions of integers and rational numbers, such as sum, product, and
quotient. Throughout mathematics, including discrete mathematics, we deal with
functions. In this section we confront the question of exactly what a function *is*,
and we look at several functions useful in discrete mathematics. Before we give
a formal answer to this question, let us approach the issue in a more familiar, less
formal way.

One usually learns before completing the second or third high school algebra
course that *a function f from A to B is a rule that assigns to each element x in the
set A a well-defined element f(x) in the set B*. There is nothing wrong with this
definition in the context of algebra or calculus, and we will continue to think of it,
informally, as telling us what a function is: A function is a rule.

EXAMPLE 1.   The monus function defined on the set of natural numbers, which is useful
in theoretical computer science, is given by the rule

$$m(n) = \begin{cases} n - 1 & \text{if } n > 0 \\ 0 & \text{if } n = 0. \end{cases}$$

This rule tells us exactly what number to assign to each natural number. It says
that we assign 78 to 79, for example, and that we assign 0 to both 1 and 0. In
symbols, $m(79) = 78$ and $m(1) = m(0) = 0$. Note that the variable n is a bound
(dummy) variable in this definition; we would have exactly the same function if we
had replaced n by y, say, throughout the display above. In particular, we should
refer to this *function* as m, not $m(n)$, since $m(n)$ is the *value* of the function at n.
In other words, $m(n)$ is a *number*, not a function. Numbers and functions are
two different kinds of objects: Functions are one level of abstraction higher than
numbers.   ◆

127

EXAMPLE 2.   Consider the following English sentence, which seems to describe a rule and thus determine a function. Given a natural number n, find the first occurrence of the decimal representation of that number in the decimal expansion of π, and take $f(n)$ to be the position in the expansion at which this representation begins. Since

$$\pi = 3.14159\ 26535\ 89793\ 23846\ 26433\ldots,$$

we see, for example, that $f(5) = 4$ (the first 5 occurs in the fourth decimal place), $f(43) = 23$ (the first 43 starts at the twenty-third decimal place), and $f(3) = f(31) = f(314) = 0$ (these numerals start at the "zeroth decimal place"). If we had carried out our expansion a bit further, we would have discovered that $f(0) = 32$.

Actually, no one knows whether this "rule" defines a function, because no one knows whether, in the decimal expansion of π, every finite string of digits occurs. (The expansion has been computed to over 400 million digits.) For instance, does the string 4830007375637777 occur in the expansion? If so, then $f(4830007375637777)$ is defined by this rule as the starting location of the first place the string occurs; if not, then the rule gives us no number to associate with 4830007375637777, so $f(4830007375637777)$ would be undefined.   ◆

Apparently, as we have seen in this example, we need to be quite careful if we want to use the "rule" definition of the word "function," because it is not at all clear whether a given rule is well defined. Indeed, it is not even clear what a rule *is*. In some sense, saying that a function is a rule, while it gives us an intuitive feeling that works in most cases, really begs the question and just transforms the question "What is a function?" into the question "What is a rule?" (These foundational questions were not even asked until the late nineteenth century. The practitioners of mathematics before then—and calculus, after all, had been in use since the seventeenth century—simply relied on their informal understanding of a function as a rule.)

One way to try to extricate ourselves and give a rigorous, precise definition of the concept of "function" is to focus on the process of computation itself. In Chapter 4 we will discuss at length the notion of an "algorithm" in mathematics. An algorithm is a sequence of instructions for computing a function—a well-defined *process* for taking each element of some set (such as a number) as input and producing an object (such as, again, a number) as the output assigned to that input. Informally, we say that the algorithm, and the function it computes, "takes the input to the output." Unfortunately, if we formalize a way to describe algorithms, we are forced to disqualify some things we would normally think of as functions, since it turns out—just as we saw in Section 1.3 that some true facts cannot be proved—that some functions cannot be computed. (We delve into this curious state of affairs in Section 4.4.)

Instead, we will define a function to be a certain *set*. Since there is presumably no doubt about what is and what is not a set, there will be no ambiguity. The sets that serve as functions contain all the information that we usually think of a function as providing. In essence a function for us will be *an encoding of the assignments that the function makes, all at once*. We present the formal definition first, and then discuss its interpretation. Try to make sense out of the definition before reading the discussion that follows it (this exercise will probably take quite a few minutes).

Let A and B be sets. A **function from** A **to** B is a subset f of $A \times B$ satisfying the following uniqueness condition.

$$\{\forall a \in A : \exists b \in B : (a, b) \in f\}$$
$$\wedge \ \{\forall a \in A : \forall b_1 \in B : \forall b_2 \in B : ([(a, b_1) \in f \wedge (a, b_2) \in f] \rightarrow b_1 = b_2)\}.$$

We call A the **domain** of the function f, and B its **codomain**, and we say that f is a function *on* A. Furthermore, for each $a \in A$ we write $f(a)$, read "f of a" or "f at a," for the unique b—guaranteed by the condition above—for which $(a, b) \in f$, and we call $f(a)$ the **value** of f at a or the **image** of a under f. Finally, we write $f : A \rightarrow B$ to express the fact that f is a function from A to B.

This definition is consistent with our previous discussion: The element of B "assigned to" a by the function f is that one and only one value b such that $(a, b) \in f$; the condition displayed above is what guarantees that there is exactly one such b. The proposition within the first pair of braces says that $f(a)$ has to exist for every $a \in A$; there can be no a for which $f(a)$ is not defined. The proposition within the second pair of braces says that $f(a)$ cannot be two different things at the same time; there can be no ambiguity in the definition. Note that the condition does *not* require every element in the codomain to be used. Nor does it prevent two different a's from getting assigned the same b; it is allowable for $(a_1, b) \in f \wedge (a_2, b) \in f$ even if $a_1 \neq a_2$. It does require every element in the domain to be used, though, and does prevent two different b's from getting assigned to the same a; it is *not* allowable for $(a, b_1) \in f \wedge (a, b_2) \in f$ if $b_1 \neq b_2$.

In practice, one usually defines a particular function f by giving—in words or formulas—the rule for determining, given a, the b for which $(a, b) \in f$, that is, the rule for determining $f(a)$.

EXAMPLE 3. The squaring function $s : \mathbf{R} \rightarrow \mathbf{R}$ is defined by $s(x) = x^2$ for all $x \in \mathbf{R}$. In set-theoretic terms,

$$s = \{(x, x^2) \mid x \in \mathbf{R}\}.$$

Thus s contains pairs such as $(5, 25)$, $(-5, 25)$, and (π, π^2) (i.e., $s(5) = s(-5) = 25$ and $s(\pi) = \pi^2$), but not pairs such as $(36, 6)$ or $(2, -4)$. ◆

EXAMPLE 4. Let P, B, and G be the set of all people, males, and females, respectively, that have ever lived. Note that $P = B \cup G$. Ignoring perplexing questions of human evolution, let us assume that every person has a unique father and a unique mother. Thus there are functions $f\colon P \to B$ and $m\colon P \to G$, the "father" function and the "mother" function, respectively, such that for each person x, $f(x)$ is x's father and $m(x)$ is x's mother. In particular, m consists of all ordered pairs (x, y) for which y is the mother of x. There is more than one pair (x, y) in m for which $y =$ Queen Victoria, but there is no such pair for which $y =$ Elizabeth I or $y =$ Henry VIII. On the other hand, for every person x_0 there is exactly one pair (x, y) in m for which $x = x_0$. $\blacklozenge$

In Section 3.2 we describe a visual aid for thinking about functions that is roughly analogous to Venn diagrams as a visual aid for thinking about sets.

We should be careful to distinguish between the name of a function and the name of the value of a function at a certain element of its domain. Using the notation in the definition, the function is f, not $f(a)$. Indeed, a is a bound variable in the definition. Often functions are named with single letters, typically f, although virtually all lowercase and uppercase letters, English and Greek, are fair game for function names. In some contexts, special symbols are used for function names, such as a pair of enclosing vertical bars for the absolute value function. In computer science, especially, words, rather than just single letters, are used as function names. For example, the largest value in a nonempty finite list A of numbers (integers, say) might be denoted $\max(A)$: for instance, $\max(34, -100, 34, 3) = 34$. Thus "max" is a function from the set of nonempty finite lists of integers to $\mathbf{Z}$. The function "min" is defined analogously. Similarly, "length" can be viewed as a function from V^* (the set of strings over some alphabet V) to $\mathbf{N}$; this function takes each string to its length. Thus we would write $length(discrete) = 8$ for the formal statement $(discrete, 8) \in length$.

The domain and codomain of a function come along as part of the function, so we could say that a function consists of a triple (a 3-tuple): a set A to be the domain, a set B to be the codomain, and a set of ordered pairs from $A \times B$ satisfying the uniqueness condition given in the definition. Two functions are equal if and only if their domains are the same, their codomains are the same, and they contain exactly the same ordered pairs. Usually, we assume that the domain and codomain of a function are implicitly understood if not stated explicitly and just refer to the function as the set of ordered pairs. Thus two functions f and g with the same domain A and the same codomain B are equal if and only if $\forall x \in A\colon f(x) = g(x)$.

Some authors call our definition the "graph" of the function, reserving the word "function" for the "rule" itself in the abstract. Except for the difficulties we alluded to earlier, this approach is quite reasonable, and it is consistent with the high school approach: The graph of a function is usually thought of as the set of points (x, y) in the plane (i.e., ordered pairs in $\mathbf{R} \times \mathbf{R}$) such that $y = f(x)$. We are defining this set of ordered pairs to *be* the function. Thus if f is a function with domain A, we can write

$$f = \{\, (a, f(a)) \mid a \in A \,\}.$$

EXAMPLE 5. The monus function discussed in Example 1 *is*, under our definition, the set $m = \{(0,0), (1,0), (2,1), (3,2), (4,3), \ldots\} = \{\, (x,y) \in \mathbf{N} \times \mathbf{N} \mid x = y = 0 \vee x = y + 1 \,\}$. Assuming that we consider the domain and codomain both to be the set of natural numbers, we would write $m : \mathbf{N} \to \mathbf{N}$. ◆

EXAMPLE 6. Let $X = \{2, 3, 4, 5, 6\}$ and $Y = \{1, 2, 3, 7\}$. Determine which of the following sets are functions from X to Y. For those that are, find the image of 3.

    (a) $A = \{(2,2), (4,2), (3,2), (5,1), (6,2)\}$.
    (b) $B = \{(4,2), (3,2), (5,1)\}$.
    (c) $C = \{(2,2), (4,2), (3,2), (5,1), (6,2), (3,7)\}$.
    (d) $D = \{(2,2), (4,2), (3,2), (5,1), (6,2), (3,2)\}$.
    (e) $E = \{(2,2), (4,2), (3,4), (5,1), (6,2)\}$.

Solution. The set A is a function from X to Y (i.e., $A \colon X \to Y$), since to each of the five elements in the domain X we see paired exactly one element of the codomain Y. The image of 3 is 2, since the pair $(3, 2)$ is in this set; in symbols, $A(3) = 2$. Note that $2 \in Y$ is paired with four different elements of X, and $3 \in Y$ and $7 \in Y$ are not paired with any elements of X.

    The set B is not a function because it fails to satisfy the first part of the uniqueness condition in the definition of a function: For some values of $x \in X$, namely $x = 2$ and $x = 6$, there is no $y \in Y$ such that (x, y) is in B. For a set to be a function, each element of the domain must appear as a first coordinate in exactly one of the ordered pairs. In other words, $B(2)$ and $B(6)$ are undefined.

    The set C is not a function because it fails to satisfy the second part of the uniqueness condition in the definition of a function: For some value of $x \in X$ there is more than one $y \in Y$ such that (x, y) is in C. Specifically, for $x = 3$, both $y = 2$ and $y = 7$ satisfy $(x, y) \in C$. In other words, $C(3)$ has been defined ambiguously.

    The set D *equals* the set A and hence *is* the same function. Again the image of 3 is 2, that is, $D(3) = 2$. The fact that our listing of the set mentioned this assignment twice does not invalidate it.

    Finally, E is not a function from X to Y, since E is not a subset of $X \times Y$. Specifically, $(3, 4) \notin X \times Y$, since $4 \notin Y$. This would be a function if we redefined the codomain to include 4. ◆

EXAMPLE 7. How many functions are there with the following domains and codomains?

    (a) Domain $= \{1, 2\}$, codomain $= \{3, 4\}$.
    (b) Domain $= \{1, 2\}$, codomain $= \emptyset$.
    (c) Domain $= \emptyset$, codomain $= \{3, 4\}$.
    (d) Domain $= \emptyset$, codomain $= \emptyset$.

Solution. We will develop methods for counting things like these in Chapter 6. For now we will answer these questions by brute force and ad hoc reasoning. This question is one level of abstraction above a question about a function—it asks for the cardinality of a *set of functions*.

To solve (a), we will write down all the possible functions. Each such function is a set with two elements (an ordered pair starting with 1 and an ordered pair starting with 2); we want to know *how many such sets* are possible. A little thought will show that the following list of four functions is complete: $\{(1,3),(2,3)\}$, $\{(1,3),(2,4)\}$, $\{(1,4),(2,3)\}$, $\{(1,4),(2,4)\}$. Therefore, the answer to the question is "four": There are four different functions from $\{1,2\}$ to $\{3,4\}$.

In part (b) we are looking for functions from $\{1,2\}$ to the empty set. A function of this sort would need to contain an ordered pair $(1,y_1)$ and an ordered pair $(2,y_2)$, where $y_1 \in \emptyset$ and $y_2 \in \emptyset$. Since the empty set has no elements, we cannot find any such pairs, and hence there are no such functions. Therefore, the answer to the question of how many functions there are from $\{1,2\}$ to $\emptyset$ is "zero."

In part (c), however, we need to find a set that contains exactly one ordered pair (x,y) for each $x \in \emptyset$ and no other ordered pairs. Since there are no such x's to worry about, the empty set itself satisfies this condition (vacuously), but no other set does. Therefore, the answer to the question of how many functions there are from $\emptyset$ to $\{3,4\}$ is "one."

Part (d) is handled in the same way as part (c). Therefore, there is one function (namely the empty set) from $\emptyset$ to $\emptyset$.

Mathematicians and computer scientists usually write B^A for the set of all functions from a set A to a set B. Thus in this problem we have found the cardinalities of $\{3,4\}^{\{1,2\}}$, $\emptyset^{\{1,2\}}$, $\{3,4\}^\emptyset$, and $\emptyset^\emptyset$. It turns out (see Exercise 7 in Section 6.1) that $|B^A| = |B|^{|A|}$, as long as we understand 0^0 to be 1. Indeed, the fact that there is one function from $\emptyset$ to $\emptyset$ is a good reason for *defining* 0^0 to be 1 in the context of discrete mathematics (in calculus, 0^0, like $0/0$, is best left undefined). ◆

FUNCTIONS IN DISGUISE

We put no restriction on the kinds of sets that can serve as domains and codomains for functions. With proper choices of these sets, we can use functions to deal with sequences (finite or infinite), strings, matrices, and operations.

In Section 2.2 we defined a sequence to be an ordered set—a set indexed by all the positive integers, or by the integers from 1 to n. Actually a sequence— or any indexed set—can profitably be viewed as *a function whose domain is the index set*. In symbols, we can say that a sequence $\{a_i\}_{i=1}^\infty$ from a set A is just a function $a : (\mathbf{N} - \{0\}) \to A$, from the set of positive integers to the set A. To describe a term of the sequence, we can use either the functional notation $a(i)$ or the equivalent sequence notation a_i, whichever suits our needs better in a given context. Computers, for example, will usually insist on being addressed with the notation $a(i)$ (or something similar, such as $a[i]$), simply because typing and reading subscripts is not as convenient as typing and reading symbols all in a row. As human beings, we might prefer a_i unless the subscript became particularly messy.

Similarly, the finite sequence $\{a_i\}_{i=1}^n$ from a set A is just a function $a : \{1, 2, \ldots, n\} \to A$, from the set of integers from 1 to n, to the set A.

Strings are just sequences, so they, too, can be viewed as functions. We can consider the string

happy birthday

as a function

$$g : \{1, 2, \ldots, 14\} \to (\text{the English alphabet} \cup \{space\}).$$

For example, $g(1) = h$, $g(4) = p$, and $g(6) = space$.

Until now we have been discussing functions of *one* variable. Our functions f operated on one argument a at a time, producing the value $f(a)$. If we allow the domain of a function to be itself a Cartesian product—a set of ordered pairs—then, in effect, there are two arguments. Let us define this terminology more rigorously, generalized to n arguments.

Suppose that A_1, A_2, $\ldots$, A_n and B are sets. A function

$$f : A_1 \times A_2 \times \cdots \times A_n \to B$$

from $A_1 \times A_2 \times \cdots \times A_n$ to B is called a **function of n variables** or arguments. The value of f at $(a_1, a_2, \ldots, a_n)$ is usually written $f(a_1, a_2, \ldots, a_n)$, rather than $f((a_1, a_2, \ldots, a_n))$.

Now we can view a matrix as a function. The domain of the function is essentially the "template" for the matrix—the set of positions in the matrix. The codomain is whatever set the entries in the matrix come from. For example, the matrix in Example 8 of Section 2.2 can be viewed as a function A from $\{1, 2, 3\} \times \{1, 2, 3, 4, 5\}$ to the set of real numbers. Then we have $A(1, 1) = 10$, $A(1, 2) = 3$, and so on. Again we can use the two notations interchangeably: $A(i, j) = a_{ij}$.

EXAMPLE 8. The ordinary operations of arithmetic defined in elementary school are functions. Addition and multiplication on the natural numbers are functions from $\mathbf{N} \times \mathbf{N}$ to $\mathbf{N}$. For instance, the addition function consists of pairs such as $((3, 5), 8)$; the value of this function at the pair of arguments $(3, 5)$ is the number 8. Rather than write, say, $\text{sum}(x, y)$ for the value of the addition function at the pair (x, y), we usually write, of course, $x + y$. Subtraction is not a function from $\mathbf{N} \times \mathbf{N}$ to $\mathbf{N}$, since, for instance, $3 - 8$ is not in the codomain; it is a function from $\mathbf{Z} \times \mathbf{Z}$ to $\mathbf{Z}$, however. The division operation is a little more complicated; it turns out to be quite important for discrete mathematics, and we will discuss it at length below (and again in Chapter 4). ◆

We say that a function from A^n to A is an **n-ary operation** on the set A. If $n = 2$ (the most common case), we call it a **binary** operation; if $n = 1$, we call it a **unary** operation. Addition is a binary operation on numbers, and complementation is a unary operation on sets (in the context of a given universal set).

If θ is an operation on A, and $B \subseteq A$, then θ is perforce defined on n-tuples from B as well. It may happen that $\theta(b_1, b_2, \ldots, b_n)$ is always in B when $(b_1, b_2, \ldots, b_n) \in B^n$. If so, we say that the set B is **closed** under the operation θ. For example, the set of even positive integers is closed under addition, since the sum of two even positive integers is an even positive integer, but this set is not closed under subtraction, since, for instance, $2 - 6$ is not an even positive integer.

Domains of functions can be collections of sets. Consider, for instance, the cardinality function applied to sets of natural numbers. This is a function from $\mathcal{P}(\mathbf{N})$ to $\mathbf{N} \cup \{\aleph_0\}$, where $\aleph_0$ ("aleph nought") is the symbol for the cardinality of infinite sets of natural numbers; the value of this function at the set X is $|X|$.

EXAMPLE 9.   Suppose that a universal set U is fixed, and let M be a fixed subset of U. Let B be the set of all subsets of U that are supersets of M. In symbols, $B = \{ X \mid M \subseteq X \subseteq U \}$. Consider the binary operation $\cap$ (intersection) defined on subsets of U. Show that B is closed under $\cap$.

Solution.   We need to show that if X and Y are sets that have M as a subset, then $M \subseteq X \cap Y$ as well. By definition, this means that we must show that every element of M is also an element of $X \cap Y$. To this end, let $z \in M$. Since we are given that $M \subseteq X$ and $M \subseteq Y$, we can conclude by the definition of subset that $z \in X$ and $z \in Y$. Therefore, by the definition of intersection, $z \in X \cap Y$, as desired. $\blacklozenge$

Finally, domains of functions can be sets of functions. The differentiation operation D in calculus is a function from the set of all differentiable functions of real numbers to the set of all functions of real numbers, defined by

$$D(f)(x) = \lim_{h \to 0} \frac{f(x + h) - f(x)}{h}$$

and usually written with the notation $D(f) = f'$.

SOME IMPORTANT NUMERICAL FUNCTIONS

Calculus is concerned primarily with functions whose domain is the set of real numbers (or subsets of the set of real numbers). In discrete mathematics we find much more use for functions whose domain (or codomain) is the set of integers or some other discrete set. We will introduce some of them here. We will have more to say about them in Chapter 4, where the concern will be not only with how they are defined but also with how they can be computed efficiently.

There is one function from the realm of continuous mathematics that plays a surprisingly important role in discrete mathematics, and we need to review it before turning to functions of a more discrete nature. Let b be a positive real number different from 1. The **logarithmic** function base b, written $\log_b$, can be defined rigorously using integral calculus; less formally, if x is a positive real number, then $\log_b x$ is the unique real number y such that $b^y = x$. We assume that the reader is familiar with the logarithmic function and its fundamental properties. (Recall that for positive real numbers x and y, and arbitrary real numbers z, we have $\log_b(xy) = \log_b x + \log_b y$ and $\log_b x^z = z \log_b x$; that for every positive real number x, $x = b^{\log_b x}$; and that for every real number x, $\log_b(b^x) = x$.) The domain of the logarithmic function is the set of positive real numbers, and its codomain is the set of real numbers.

In calculus, the most important value of b, the base for the logarithmic function, is, by far, Euler's constant $e = 2.718281828459\ldots$. Before the advent of electronic calculators, logarithms with $b = 10$ were invaluable as an aid in calculation. Neither of these values of b is as important in discrete mathematics as $b = 2$. Thus *we will abbreviate* $\log_2 x$ *as* $\log x$. Recall from algebra or calculus that changing the value of b does not cause much change in the function. Indeed, for any positive real numbers b and c not equal to 1, $\log_b x$ is a constant times $\log_c x$, independent of x:

$$\log_b x = \left(\frac{1}{\log_c b}\right)\log_c x = (\log_b c)\log_c x.$$

We will be noticing in later chapters that just as the exponential function 2^x grows very fast as x increases, so the logarithmic function $\log x$ grows very slowly with increasing x. For example, $\log(1000000)$ is less than 20. Figure 3.1 shows the graph of $y = \log x$.

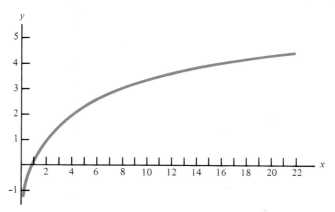

FIGURE 3.1   Graph of the logarithmic function, base 2, $y = \log x$.

Next we need to discuss functions that take real numbers to nearby integers. We might try to define a "rounding" function r by declaring $r(x)$ to be the integer closest to the real number x. This definition is not quite satisfactory, since a number such as 3.5 has two closest integers, 3 and 4. We could arbitrarily say that such ties are broken by always rounding down (or up), or by always rounding toward 0 (or away from 0), or by always rounding to the even integer (or to the odd integer)—all six methods might be acceptable. Actually, the rounding function is less important than two other related functions from $\mathbf{R}$ to $\mathbf{Z}$, partly because of this lack of consensus about ties. One of these functions rounds downward, no matter how large the fractional part, and the other always rounds upward.

Specifically, we define the **greatest integer** function, or **floor** function, from $\mathbf{R}$ to $\mathbf{Z}$ to be the function whose value at the real number x is the largest integer that is less than or equal to x. We denote the value of this function at x by $\lfloor x \rfloor$. Similarly, the **ceiling** function from $\mathbf{R}$ to $\mathbf{Z}$ is the function whose value at the real number x is the smallest integer that is greater than or equal to x. We denote the value of this function at x by $\lceil x \rceil$.

For example, $\lfloor \pi \rfloor = 3$, since 3 is the largest integer that does not exceed 3.14.... Similarly, $\lceil \pi \rceil = 4$, $\lfloor -10.4 \rfloor = -11$ and $\lceil -10.4 \rceil = -10$. If x is already an integer, then $\lfloor x \rfloor = \lceil x \rceil = x$. It is easy to show (Exercise 27) that in every case

$$x - 1 < \lfloor x \rfloor \leq x$$

and

$$x \leq \lceil x \rceil < x + 1.$$

Figure 3.2 shows the graphs of these functions. In drawing these graphs in the usual coordinate plane, we are implicitly thinking of their codomains as being $\mathbf{R}$, rather than $\mathbf{Z}$, but the point of these functions is that they take real numbers to *integers*.

The floor and ceiling functions come up repeatedly in computer science, and the reader should become familiar with them. These or similar functions are built into many calculators and computer languages, often going under the name of *TRUNC* or *INT*.

Next we turn to two important arithmetic functions of integers. The set of integers is closed under addition, subtraction, and multiplication. It is not, however, closed under the operation of division. If x and y are integers, then x/y is not necessarily an integer. Indeed, if $y = 0$, then x/y is not defined at all; and in general if $y \neq 0$, then x/y is a rational number. There are two important *integers* we can associate with a division problem of integers, however.

Let us restrict ourselves to problems in which the divisor (the denominator) is positive. This is no real restriction: If the divisor were negative, we could simply change the signs of both numerator and denominator without changing the answer; and we always prohibit division by 0. The **quotient** q of x divided by y is

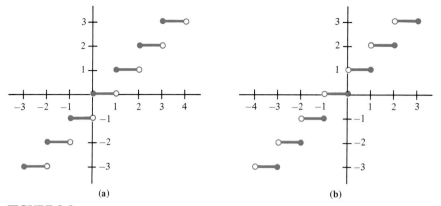

FIGURE 3.2   Graphs of (a) the floor function and (b) the ceiling function.

defined to be $\lfloor x/y \rfloor$, and the **remainder** r is defined to be $x - yq$. Note that with these definitions the remainder is always nonnegative. We denote the remainder by $x \bmod y$. These definitions are consistent with those taught in elementary school if x is also positive. For example, the quotient of 100 divided by 7 is $\lfloor 100/7 \rfloor = 14$, and the remainder is $100 - 7 \cdot 14 = 2$. On the other hand, the quotient of -100 divided by 7 is $\lfloor -100/7 \rfloor = -15$ (note that it is not $-\lfloor 100/7 \rfloor = -14$), and the remainder is $-100 - 7 \cdot (-15) = 5$. We write $100 \bmod 7 = 2$ and $(-100) \bmod 7 = 5$.

For any given x and y, the quotient q and the remainder r satisfy

$$x = qy + r, \qquad 0 \le r < y,$$

and in fact they are the only pair of integers satisfying these conditions. It is easy to see that y is a divisor of x (or, equivalently, x is a multiple of y) if and only if the remainder $r = x \bmod y$ is 0.

It turns out that the remainder is the more important of these two quantities to number theorists and, recently, to computer scientists. In Section 4.5 we will apply some number theory to problems of computation. To lay the groundwork here, let us change the setting slightly and fix a positive integer, called the **modulus**, as the divisor. The remainder $x \bmod m$ is sometimes called the **reduction** of x, modulo m. If $x \bmod m = y \bmod m$, we say that x and y are **congruent modulo m**, and write $x \equiv y \pmod{m}$. For example, since we saw that $100 \bmod 7 = 2$ and $(-100) \bmod 7 = 5$, we conclude that $100 \not\equiv -100 \pmod{7}$. On the other hand, $100 \equiv 2 \equiv -12 \pmod{7}$, since in each case the remainder is 2.

We can also look at congruence modulo m in a slightly different light. We just *defined* two numbers x and y to be congruent mod m if *they have the same remainder when divided by m*. In this case $x = q_1 m + r$ and $y = q_2 m + r$ for some integers q_1 and q_2. Subtracting, we see that $x - y = (q_1 - q_2)m$, so that *the difference*

$x - y$ is a multiple of m; solving for x, we see that $x = y + (q_1 - q_2)m$, so that x is equal to y plus a multiple of m. We now have three equivalent formulations of the statement $x \equiv y \pmod{m}$. By the last of these formulations, we see that the set of all integers congruent to x modulo m can be written as $\{\, x + km \mid k \in \mathbf{Z} \,\}$. We will have more to say about this set, called the **equivalence class** of x modulo m, in Section 3.4.

EXAMPLE 10. Find the equivalence class of 64 modulo 11.

Solution. By the discussion above, the set we want is $\{\, 64 + 11k \mid k \in \mathbf{Z} \,\}$. This is the set of all integers congruent to 64 modulo 11. We can describe this set as $\{\ldots, -13, -2, 9, 20, 31, 42, 53, 64, 75, 86, \ldots\}$. ◆

The reader is asked to show in Exercise 32 that the arithmetic operations behave nicely with respect to the operation of reduction modulo m. Specifically, if $a \equiv b \pmod{m}$ and $c \equiv d \pmod{m}$, then $a + c \equiv b + d \pmod{m}$, and $ac \equiv bd \pmod{m}$. Because of this, it is legitimate to do arithmetic entirely with the remainders. The set of remainders modulo m is $\{0, 1, 2, \ldots, m-1\}$. We define addition, subtraction, and multiplication on this set of remainders as follows. If x and y are in this set, their sum is $(x + y) \bmod m$, their difference is $(x - y) \bmod m$, and their product is $xy \bmod m$. The resulting system of arithmetic is called **modular arithmetic**, and it has many of the properties of ordinary arithmetic on the integers. The addition and multiplication tables for mod 7 arithmetic are shown in Figure 3.3.

+	0	1	2	3	4	5	6
0	0	1	2	3	4	5	6
1	1	2	3	4	5	6	0
2	2	3	4	5	6	0	1
3	3	4	5	6	0	1	2
4	4	5	6	0	1	2	3
5	5	6	0	1	2	3	4
6	6	0	1	2	3	4	5

(a)

×	0	1	2	3	4	5	6
0	0	0	0	0	0	0	0
1	0	1	2	3	4	5	6
2	0	2	4	6	1	3	5
3	0	3	6	2	5	1	4
4	0	4	1	5	2	6	3
5	0	5	3	1	6	4	2
6	0	6	5	4	3	2	1

(b)

FIGURE 3.3 Addition (a) and multiplication (b) tables for arithmetic mod 7.

The reader is already familiar with addition modulo 12, since that is the system usually used for telling time in the United States. For example, if it is currently 9:00 and you must work for 8 hours, then the hour at which your work will end is $9 + 8 \mod 12$; since $17 \mod 12 = 5$, you quit at 5:00. The movement of the hour hand around the traditional clock face gives a graphic model of addition modulo 12 (with the "12" at the top thought of as "0").

One nice property of modular arithmetic when the modulus is a prime number is that multiplicative inverses exist. If m is prime and $x \not\equiv 0 \pmod{m}$, then there exists a y such that $xy \equiv 1 \pmod{m}$. Figure 3.3b illustrates this for $m = 7$, since there is a 1 in every row of the multiplication table except for the row representing multiplication by 0. We see that the multiplicative inverse of 3, for example, is 5, since $3 \cdot 5 = 15 \equiv 1 \pmod{7}$. Because of this property, and other nice properties of modular arithmetic, these systems find wide and deep applications in such areas as cryptography and random number generation.

Another important function in discrete mathematics, especially useful in problems involving counting, is the **factorial function**. We denote the factorial of n by $n!$. The domain of this function is the set of natural numbers, and we define $n!$ to be the product of all the integers from 1 to n, that is,

$$n! = 1 \cdot 2 \cdot 3 \cdots (n - 1) \cdot n.$$

For example, we have $1! = 1$ and $4! = 1 \cdot 2 \cdot 3 \cdot 4 = 24$. Using the product notation given in Exercise 29 in Section 2.2, we can write $n! = \prod_{i=1}^{n} i$. We also understand this definition to say that $0! = 1$. In other words, the "empty product" is 1, just as the "empty sum" is 0.

SUMMARY OF DEFINITIONS

binary operation on the set A: a function from $A \times A$ to A (e.g., the difference operation on sets).

ceiling function: the function from real numbers to integers whose value at x is the smallest integer greater than or equal to x, denoted $\lceil x \rceil$ (e.g., $\lceil 4.1 \rceil = 5$).

closed set under an operation: a set for which the result of the operation is again an element of the set (e.g., the odd integers under multiplication, but not the natural numbers under the taking of square roots).

codomain of the function $f: A \to B$: the set B.

congruence of two integers modulo m: x and y are congruent modulo m, written $x \equiv y \pmod{m}$, if and only if x and y differ by a multiple of m; equivalently $x \equiv y \pmod{m}$ if and only if x and y give the same remainder when divided by m (e.g., $35 \equiv 13 \pmod{11}$).

domain of the function $f: A \to B$: the set A.

equivalence class of the integer x modulo m: the set of all integers congruent to x modulo m, namely $\{ x + km \mid k \in \mathbf{Z} \}$ (e.g., the set of even integers when $x = 14$ and $m = 2$).

factorial function: the function from **N** to **N** whose value at n is $n(n-1)(n-2)\cdots 3\cdot 2\cdot 1$; written $n!$; by definition $0! = 1$ (e.g., $10! = 3628800$).

floor function: the function from real numbers to integers whose value at x is the largest integer less than or equal to x, denoted $\lfloor x \rfloor$ (e.g., $\lfloor -4.1 \rfloor = -5$).

function from the set A **to** the set B: a subset f of $A \times B$ such that for every $a \in A$ there exists a unique $b \in B$ such that $(a,b) \in f$; written $f : A \rightarrow B$ (e.g., the set of all pairs $(x, 3x^2+1)$ is a function from the set of real numbers to the set of positive real numbers).

function of n variables: a function whose domain is a Cartesian product of n sets (e.g., the function whose value at the pair (x,y), where x is a man and y is a woman, is the number of children that x and y have together; this is a function of two variables—from the Cartesian product of the set of men and the set of women to the set of natural numbers).

greatest integer function: the floor function.

image of the element a in the domain of the function $f : A \rightarrow B$: the unique element $b \in B$ for which $(a,b) \in f$; denoted $f(a)$ (e.g., for the "father" function in Example 4, f(Elizabeth I) is Henry VIII).

logarithmic function base b, where b is a positive real number different from 1: For each positive real number x, $y = \log_b x$ if and only if $b^y = x$; when $b = 2$, the subscript is omitted, so that $\log x$ means $\log_2 x$ (e.g., $\log_{10} 3162 \approx 3.5$ and $\log 256 = 8$).

modular arithmetic: arithmetic performed on the remainders modulo some fixed modulus m, namely $\{0, 1, 2, \ldots, m-1\}$; the result of an addition, subtraction, or multiplication is reduced modulo m to end up back in this set (e.g., modulo 11 we have $7 \cdot 5 = 2$).

modulus: a fixed positive integer m in the context of modular arithmetic.

n-ary operation on the set A: a function from A^n to A—a binary operation if $n = 2$, a unary operation if $n = 1$ (e.g., for each $n \geq 2$, "sum" can be viewed as an n-ary operation on the set of integers, so that, for instance, $\text{sum}(8, 25, -100) = -67$).

quotient of the integer x divided by the positive integer y (in the context of integer arithmetic): $\lfloor x/y \rfloor$ (see "remainder").

reduction modulo m: the result of taking the remainder after division by m (e.g., $35 \bmod 11 = 2$).

remainder of the integer x divided by the positive integer y: $x - y\lfloor x/y \rfloor$; written $x \bmod y$ (e.g., if $x = 3$ and $y = 4$, then the quotient is 0 and the remainder is 3; if $x = -13$ and $y = 4$, then the quotient is -4 and the remainder is 3).

unary operation on the set A: a function from A to A (e.g., the negation operation on propositions).

value of the function $f : A \rightarrow B$ at the element $a \in A$: the image of a under f (e.g., the value of the square root function at 9 is 3).

EXERCISES

1. Determine which of the following are functions from $\{1, 2, 3\}$ to $\{1, 2, 3, 4\}$.
   (a) $(2, 3)$.
   (b) $\{(2, 3)\}$.
   (c) $\{(1, 3), (3, 1), (2, 1)\}$.
   (d) $\{(1, 3), (2, 1)\}$.
   (e) $\{(1, 3), (3, 1), (2, 1), (1, 4)\}$.
   (f) $\{(1, 3), (3, 1), (2, 5)\}$.

2. Give a simple description of the following functions from **N** to **N** by stating a rule that specifies $f(x)$ in terms of x for each $x \in \mathbf{N}$. The obvious patterns are intended.
   (a) $\{(0, 0), (1, 2), (2, 4), (3, 6), \ldots\}$.
   (b) $\{(0, 1), (1, 2), (2, 4), (3, 8), \ldots\}$.
   (c) $\{(0, 0), (1, 1), (2, 2), (3, 3), \ldots\}$.
   (d) $\{(0, 4), (1, 4), (2, 4), (3, 4), \ldots\}$.

3. List as a set of ordered pairs the functions from $\{1, 2, 3, 4\}$ to **N** given by the following rules.
   (a) $f(x) = x + 1$.
   (b) $f(x) = 8$.
   (c) $f(t) = t + 1$ if t is odd, and $f(t) = t$ if t is even.

4. Compute the following numbers exactly. Recall that log means the base 2 logarithm.
   (a) $\log 32$.
   (b) $\log \sqrt{2}$.
   (c) $2^{\log 7}$.
   (d) $\log 2^{59}$.
   (e) $\log\left(\log 2^{2^{43}}\right)$.

5. Use a calculator to compute these quantities to five decimal places. Recall that log means the base 2 logarithm and that $\log_b x = (\log_c x)/(\log_c b)$.
   (a) $\log 5$.
   (b) $\log 10$.
   (c) $\log 10^{50}$.

6. Compute the following quantities.
   (a) The quotient and remainder when 14 is divided by 7.
   (b) The quotient and remainder when 3 is divided by 7.
   (c) The quotient and remainder when 0 is divided by 7.
   (d) The quotient and remainder when -56 is divided by 7.
   (e) The quotient and remainder when -57 is divided by 7.
   (f) The quotient and remainder when -2 is divided by 7.
   (g) 10 mod 3.

    (h)   10 mod 2.

    (i) (-10) mod 3.

7. Perform the following arithmetic mod 7.
   - (a) $6 + 5$.
   - (b) $1 + 1$.
   - (c) 6×4.
   - (d) 0×2.
   - (e) 5^6.

8. Find the equivalence class of 1 modulo 7. Describe it in more than one way.

9. Write down the addition and multiplication tables for mod 5 arithmetic.

10. Compute the following quantities.
    - (a) $6!$.
    - (b) $(3 + 4)!$.
    - (c) $\lfloor 4.1 \rfloor$.
    - (d) $\lceil 4.1 \rceil$.
    - (e) $\lfloor 4.7 - 6.2 \rfloor$.
    - (f) $\lceil 4.7 - 6.2 \rceil$.

11. Let f be the father function defined in Example 4, and let the variables in this problem range over people.
    - (a)   Give an example to show that $\{ p \mid f(p) = q \}$ can be empty.
    - (b)   What can you say about $|\{ q \mid f(p) = q \}|$ for each p?
    - (c)   Can $f(p) = q$ and $f(p) = q'$ if $q \neq q'$?
    - (d)   Can $f(p) = q$ and $f(p') = q$ if $p \neq p'$?

12. For each integer $x > 1$ let $f(x)$ be the sum of all the prime numbers that exactly divide x; f is a function from the set of integers greater than 1 to itself. For example, $f(12) = 2 + 3 = 5$, since the only prime divisors of 12 are 2 and 3. Find the following numbers.
    - (a) $f(100)$.
    - (b) $f(f(30))$.
    - (c) $f(3 + 5)$.
    - (d) $f(3) + f(5)$.
    - (e) $f(3 \cdot 5)$.
    - (f) $f(3) \cdot f(5)$.

13. Rewrite in a much simpler form the uniqueness condition for the definition of a function, by using the notation $\exists! v$ for "there exists a unique v."

14. Explain why a subset of $\mathbf{R} \times \mathbf{R}$ drawn in the usual coordinate plane is (the graph of) a function f from A to $\mathbf{R}$ (where $A \subseteq \mathbf{R}$) if and only if every vertical line with equation $x = a$ intersects the subset in exactly one point if $a \in A$ and in no points if $a \notin A$. (This is called the "vertical line test" in high school algebra.)

15. Write out explicitly (as a set of ordered pairs) the function that sends each element of $\mathcal{P}(\{1, 2, 3\}) - \{\emptyset\}$ to its smallest element.

16. Explain how multiplication of real numbers is a function. Be sure to identify the domain and codomain.

17. Repeat Exercise 16 for division of real numbers.

18. Write down the set of all functions with the following domains and codomains.
    (a) From $\{1, 2\}$ to $\{3\}$.
    (b) From $\{1\}$ to $\{2, 3\}$.
    (c) From $\{a, b\}$ to $\{a, b, c\}$.
    (d) From $\{a, b, c\}$ to $\{a, b\}$.

19. Write down in symbols of logic and algebra (no words) an explicit definition of the binary operation $\odot$ on integers in which $x \odot y$ is the larger of x and y. Note that this is a function of two variables, so the answer here will be a set of things of the form $((a, b), c)$. To answer this question you need to find a proposition involving a, b, and c such that the proposition is true if and only if $((a, b), c)$ is in the function $\odot$.

20. Explain how definite integration is a function from $F \times \mathbf{R} \times \mathbf{R}$ to $\mathbf{R}$, where F is the set of continuous functions on $\mathbf{R}$.

21. Determine whether the following sets are closed under the given operations.
    (a) The set of natural numbers under subtraction.
    (b) The set of finite sets of real numbers under union (of two sets).
    (c) The set of rational numbers under addition.
    (d) The set of positive real numbers under the taking of square roots.
    (e) The set of odd integers under multiplication.
    (f) The set of odd integers under addition.

22. How many digits are there in the decimal numeral for the prime number $2^{127} - 1$?

23. Using the floor (or ceiling) function and a logarithmic function, write down an expression involving n for the number of digits in the decimal numeral for n, where n is a positive integer. For example, your expression will have the value 4 if $n = 3792$.

24. Using the floor (or ceiling) function and the remainder function (mod), write down an expression in θ for the quadrant in which the point P lies if P is the point on the unit circle obtained by starting at $(1, 0)$ and traveling θ radians counterclockwise. (To avoid ambiguity, we will say that the positive x-axis lies in the first quadrant, the positive y-axis lies in the second quadrant, the negative x-axis lies in the third quadrant, and the negative y-axis lies in the fourth quadrant.)

25. Suppose that $\log x = y$. Simplify the following expressions (i.e., write as simpler expressions involving x or y).
    (a) $\log 2x$.
    (b) $\log x^2$.
    (c) $\log_4 x$.
    (d) 2^y.
    (e) 4^y.
    (f) $x^{1/\log x}$ (if $\log x \neq 0$).

26. Draw the graphs of the following functions.
    (a) $f(x) = \lfloor 2x \rfloor$.
    (b) $g(x) = \lceil x/2 \rceil$.
    (c) $h(x) = \lfloor -x \rfloor$.

27. Explain why the following inequalities are true.
    (a) $x - 1 < \lfloor x \rfloor \leq x$.
    (b) $x \leq \lceil x \rceil < x + 1$.

28. Prove the following facts about the floor and ceiling functions. Assume that x is an arbitrary real number and n is an arbitrary integer.
    (a) $\lfloor x + n \rfloor = \lfloor x \rfloor + n$.
    (b) $\lfloor -x \rfloor = -\lceil x \rceil$.

29. Using the floor function, find a rule for the rounding function; in other words, write down an expression for the nearest integer to the real number x. You may assume that this function always rounds down for x exactly halfway between two integers.

30. Describe the following sets using congruences.
    (a) The set of odd numbers.
    (b) The set of multiples of 123.
    (c) The set of positive integers whose last digit is a 7.

31. Show that every prime number greater than 3 must be congruent either to -1 or to 1 modulo 6. (Thus every prime number greater than 3 can be written as $6k \pm 1$ for some integer k.)

32. Show that if $a \equiv b \pmod{m}$ and $c \equiv d \pmod{m}$, then
    (a) $a + c \equiv b + d \pmod{m}$.
    (b) $ac \equiv bd \pmod{m}$.

33. Prove or disprove that if $a \equiv b \pmod{m}$ and $c \equiv d \pmod{m}$, then $a^c \equiv b^d \pmod{m}$.

34. Show that every remainder mod 5 other than 0 has a multiplicative inverse mod 5.

35. Show that the statement in Exercise 34 is not true if we replace 5 by 6 throughout.

36. Show that for all positive integers m, every natural number less than m has an additive inverse mod m.

37. Let $\odot$ be a binary operation on a set A. We say that this operation is **associative** if $a \odot (b \odot c) = (a \odot b) \odot c$ for all $a, b, c \in A$. Determine whether the following operations are associative.
    (a) Multiplication mod m on $\{0, 1, 2, \ldots, m - 1\}$.
    (b) $x \odot y = x + 2y$ on the set of real numbers.
    (c) $x \odot y = x$ on the set of real numbers.

38. Let $\odot$ be a binary operation on a set A. We say that this operation is **commutative** if $a \odot b = b \odot a$ for all $a, b \in A$. Determine whether the following operations are commutative.
    (a) Multiplication mod m on $\{0, 1, 2, \ldots, m - 1\}$.
    (b) $x \odot y = x + 2y$ on the set of real numbers.
    (c) $x \odot y = x$ on the set of real numbers.

Challenging Exercises

39. Give a symbolic description of the function that takes every nonempty set of natural numbers to its smallest element.

40. Let us define a function d from the set of integers greater than 1 to the power set of the set of positive integers as follows. If $p > 1$, then $n \in d(p)$ if and only if there is an arrangement of p distinct points in the plane such that the number of different lines—not line segments—determined by pairs of these points is n. For example, $d(3) = \{1, 3\}$, since we can arrange the three points in a row, so that they determine only one line, or we can arrange the three points in a triangle so that they determine three lines (the lines containing the sides of the triangle).
    (a) Determine, with proof, $d(4)$ and $d(5)$.
    (b) Compute $d(6)$; be sure not to omit any possible configurations.
    (c) Show that $p \in d(p)$ for all $p > 2$.
    (d) Show that $2 \notin d(p)$ for all $p > 1$.

Exploratory Exercises

41. Determine what relationships exists between the following pairs of quantities, where x and y are arbitrary positive real numbers.
    (a) $\lfloor x + y \rfloor$ and $\lfloor x \rfloor + \lfloor y \rfloor$.
    (b) $\lfloor xy \rfloor$ and $\lfloor x \rfloor \lfloor y \rfloor$.
    (c) $\lfloor x - y \rfloor$ and $\lfloor x \rfloor - \lfloor y \rfloor$.

42. Why is indefinite integration not a function from the set of continuous functions on $\mathbf{R}$ to the set of continuous functions on $\mathbf{R}$, in the same way that differentiation is a function on the set of differentiable functions? What can be done in order that indefinite integration be a function? [*Hint*: The fact that

there are no closed-form answers for integrals such as $\int e^{t^2} dt$ has nothing to do with this exercise.]

43. In Example 2 we discussed the decimal expansion of π.
   (a) How might we amend the definition given there so that it will always be valid, even if certain strings of digits never appear in the decimal expansion of π?
   (b) Find out more about computing the decimal expansion of π by consulting Borwein and Borwein [28] and Castellanos [35].

44. Find out more about number theory, its applications, and its relations to computer science by consulting Hardy and Wright [145], Kirch [171], Ore [227], Richards [247], Rosen [259], Schroeder [266,267], Shapiro [271], and D. D. Spencer [287].

SECTION 3.2
FUNCTIONS IN THE ABSTRACT

Nothing will come of nothing.

—William Shakespeare, *King Lear*

In this section we return to discussing functions in general. We need some further terminology and notation for talking about functions; we need some words to describe properties that functions may or may not have; and we want to prove some general theorems about functions.

INJECTIVITY AND SURJECTIVITY

All the elements in the domain of a function must appear as first coordinates in ordered pairs in the function. All the elements of the codomain need *not* appear as second coordinates. The set of those elements of the codomain that *do* appear as values of the function is called the **range** of the function. In symbols, if $f: A \to B$, then

$$\text{range}(f) = \{ b \in B \mid \exists a \in A: (a, b) \in f \}$$
$$= \{ f(a) \mid a \in A \}.$$

Thus the range of a function is a subset of the codomain of that function. If the range equals the codomain, we say that the function is **surjective** or **onto**. (The mathematical practice of using the preposition "onto" as an adjective to describe functions is well entrenched in this context.) Equivalently,

$$f \text{ is surjective if and only if } \forall b \in B: \exists a \in A: f(a) = b.$$

EXAMPLE 1. The function $s: \mathbf{R} \to \mathbf{R}$ given by the rule $s(x) = x^2$ for all $x \in \mathbf{R}$ is not surjective. Indeed, its range is the set of nonnegative real numbers, $\{ y \in \mathbf{R} \mid y \geq 0 \} = \mathbf{R}^+ \cup \{0\}$ (recall that $\mathbf{R}^+$ represents the set of positive real numbers). If we change the codomain and instead consider the function $s_1: \mathbf{R} \to \mathbf{R}^+ \cup \{0\}$ given by the rule $s_1(x) = s(x)$ for all $x \in \mathbf{R}$, then s_1 is surjective.

   The base b logarithm function from $\mathbf{R}^+$ to $\mathbf{R}$ is surjective, where b is a positive real number other than 1. Given any real number y, we can find a positive real number x whose base b logarithm is y, namely $x = b^y$.

   The father function from people to males given by $f(x) =$ the father of x (defined in Example 4 of Section 3.1) is not onto, since there is at least one male who is not the father of anyone. ◆

   There is another way in which the domain and codomain are not on an equal footing in the definition of function. There can be only one ordered pair (a, b) in a function f with a given a as its first coordinate. On the other hand, there is no prohibition against there being several pairs (a, b) each with the same b as its second coordinate. We give a special name to functions that contain only one pair (a, b) for each b in the range: A function $f: A \to B$ is said to be **injective** or **one-to-one** if and only if

$$\forall a_1 \in A: \forall a_2 \in A: \forall b \in B: (((a_1, b) \in f \land (a_2, b) \in f) \to a_1 = a_2).$$

Equivalently, f is injective if and only if

$$\forall a_1 \in A: \forall a_2 \in A: (f(a_1) = f(a_2) \to a_1 = a_2).$$

Sometimes it is more useful to think of this definition in its contrapositive form: A function f is one-to-one if and only if for all a_1 and a_2 in its domain with $a_1 \neq a_2$, we have $f(a_1) \neq f(a_2)$. Symbolically,

$$\forall a_1 \in A: \forall a_2 \in A: (a_1 \neq a_2 \to f(a_1) \neq f(a_2)).$$

In other words, a one-to-one function takes distinct elements of the domain to distinct elements of the codomain.

EXAMPLE 2. The function $s: \mathbf{R} \to \mathbf{R}$ given by the rule $s(x) = x^2$ for all $x \in \mathbf{R}$ is not injective. Indeed, $s(7) = s(-7) = 49$, even though $7 \neq -7$. On the other hand, if we restrict the domain of s and consider instead the function $s_2: \mathbf{R}^+ \to \mathbf{R}$ from the set of *positive* real numbers to the set of real numbers, given by the rule $s_2(x) = s(x)$ for all $x \in \mathbf{R}^+$, then we find that s_2 is one-to-one. There is at most one *positive* solution to the equation $x^2 = y$, no matter what value y has.

   The father function (Example 4 of Section 3.1) is not injective, since there are certainly cases of two people having the same father. ◆

To help our human brains understand the concepts of function, injectivity, and surjectivity, we can represent a function with a diagram consisting of two ovals or other geometric shapes, one to represent the domain and one to represent the codomain. Usually, we draw the domain on the left and the codomain on the right. We depict individual elements of these two sets by points (perhaps labeled) within the ovals. Since we can view a function as "taking elements of the domain to elements of the range" (f takes a to $f(a)$), we can represent the function as a set of lines with arrowheads on them, pointing from elements in the domain to elements in the range. Figure 3.4 shows a function from $\{a, b, c, d\}$ to $\{1, 2, 3\}$, namely the function $\{(a, 2), (b, 2), (c, 1), (d, 2)\}$. If this function is named f, then we have $f(a) = f(b) = f(d) = 2$ and $f(c) = 1$; f takes a, b, and d to 2, and it takes c to 1.

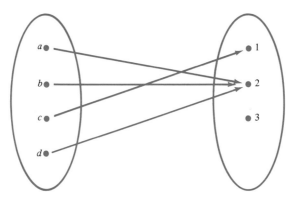

FIGURE 3.4   Schematic representation of a function.

On the other hand, Figure 3.5 shows something that is not a function. For a diagram to represent a function, there must be exactly one line leading from each element of the domain. In Figure 3.5 there are two lines leading from c and no lines leading from d. (Figure 3.5 represents something more general than a function, namely a relation, the topic of discussion in Section 3.3.)

A surjective function, in this representation, will have at least one line leading to each element of the codomain since every element of the codomain is in the range of the function. The function shown in Figure 3.4 is not surjective, since 3 is not in the range. The function shown in Figure 3.6, however, is surjective.

An injective function, in this representation, will have at most one line leading to each element of the codomain since no element of the codomain is the image of more than one element of the domain. The function shown in Figure 3.4 is not injective, since 2 is the image of more than one element of the domain. The function shown in Figure 3.7, however, is injective.

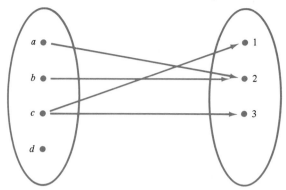

FIGURE 3.5   Schematic representation of a nonfunction.

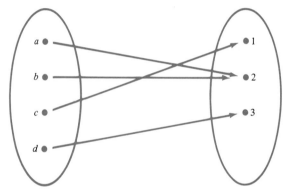

FIGURE 3.6   Schematic representation of a surjective function.

The function shown in Figure 3.8 is both injective and surjective. Functions that are both injective and surjective are extremely important in mathematics, because they set up an exact **one-to-one correspondence** between two sets. We say that a function $f: A \rightarrow B$ is **bijective** if and only if f is both injective and surjective (one-to-one and onto). Equivalently, we can express this definition with the ∃! quantifier:

$$f: A \rightarrow B \text{ is bijective if and only if } \forall b \in B: \exists! a \in A: f(a) = b.$$

Note the distinction between a one-to-one *function* and a one-to-one *correspondence*. The latter is a one-to-one function that is also surjective; thus each element in the codomain corresponds to exactly one element in the domain.

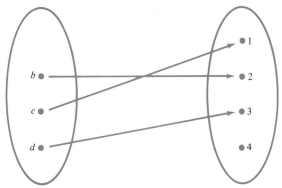

FIGURE 3.7 Schematic representation of an injective function.

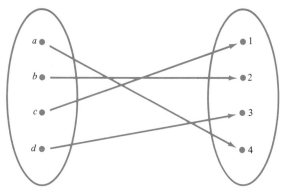

FIGURE 3.8 Schematic representation of a bijective function.

EXAMPLE 3. Let the function $s_3: \mathbf{R}^+ \to \mathbf{R}^+$ from the positive real numbers to the positive real numbers be given by the rule $s_3(x) = x^2$ for all $x \in \mathbf{R}^+$. By thus restricting the domain and codomain of the squaring function we have created a bijective function. For each positive real number y there is exactly one positive real number whose square is y, namely $\sqrt{y}$.

On the other hand, the function $f: \mathbf{R}^+ \to \mathbf{R}^+$ defined by $f(x) = x^2 + 1$ is injective but not surjective. There is no element in the domain whose image is $\frac{1}{2}$, for instance. Thus f is not bijective.

Note how s_3 "stretches" the half-line of positive real numbers and lays it back upon itself to establish a one-to-one correspondence. The number 4, for instance, now lies on top of the number 16 in the original half-line. On the other hand, f "slides" the half-line one unit to the right after stretching and so does not establish a one-to-one correspondence. There is a gap at the end of the original half-line, from 0 to 1, not covered by the stretched and slid half-line. ◆

In Section 2.1 we used the idea of a bijective function to *define* the notion of cardinality for infinite sets: For any sets A and B, $|A| = |B|$ if and only if there is a bijective function from A to B.

IMAGES AND INVERSE IMAGES (*OPTIONAL*)

Functions take elements of sets to elements of sets. Sometimes we are interested in what a given function does to a collection of elements.

Let $f: A \to B$ be a function, let $S \subseteq A$, and let $T \subseteq B$. The **image** of the set S under f, written $f(S)$, is the set of images of the elements of S under f; that is,

$$f(S) = \{ f(x) \mid x \in S \}.$$

The **inverse image** of the set T under f, written $f^{-1}(T)$, is the set of elements of S whose images under f are elements of T; that is,

$$f^{-1}(T) = \{ x \in A \mid f(x) \in T \}.$$

Let us see what is going on schematically. In Figure 3.9 we show a subset S of the domain, its image $T = f(S)$ under the function, and the inverse image $R = f^{-1}(T)$ of T under this function. We can see from the picture that $S \subseteq R$; that is, $S \subseteq f^{-1}(f(S))$. On the other hand, we also see that there may be elements in $R - S$. That is, there may be elements in the domain, not in S, that get sent to the same element that an element of S gets sent to.

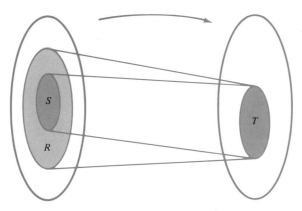

FIGURE 3.9 $T = f(S)$ and $R = f^{-1}(T)$.

EXAMPLE 4.   Suppose that g is the function shown in Figure 3.4. Let $S = \{b, d\}$. Then $g(S) = \{2\}$. Furthermore $g^{-1}(g(S)) = \{a, b, d\}$. Since 3 is not in the range of g, we have $g^{-1}(\{3\}) = \emptyset$. ◆

EXAMPLE 5.   Let m be the mother function (Example 4, of Section 3.1). Suppose that Suzanne is a certain woman. Then $m^{-1}(\{\text{Suzanne}\})$ is the set of people whose mother is Suzanne, in other words, the set of Suzanne's children. On the other hand, $m(\{\text{Suzanne}\})$ is a set consisting of one element, Suzanne's mother. The set $m^{-1}(m(\{\text{Suzanne}\}))$ consists of all the children of Suzanne's mother, that is, Suzanne and all of her brothers and sisters (and half-brothers and half-sisters on her mother's side). ◆

EXAMPLE 6.   Consider the "absolute value of the floor" function from **R** to **Z**. Let us call it F. Thus $F(x) = \big|\lfloor x \rfloor\big|$ for all $x \in \mathbf{R}$. The image of the interval $[\pi, 7)$ under this function is $\{3, 4, 5, 6\}$, since these are the possible values of $\big|\lfloor x \rfloor\big|$ if $\pi \le x < 7$. In symbols, $F([\pi, 7)) = \{3, 4, 5, 6\}$.

The inverse image of $\{-2, 0, 3\}$ under F is the set of all solutions to the equation $\big|\lfloor x \rfloor\big| = -2$, 0, or 3. There are no solutions with -2 on the right-hand side, since the absolute value of any quantity cannot be negative. Also, it is not hard to see that $\big|\lfloor x \rfloor\big| = 0$ for precisely those values of x in $[0, 1) = \{x \mid 0 \le x < 1\}$. A little thought will convince the reader that the set of values of x that satisfy $\big|\lfloor x \rfloor\big| = 3$ is precisely $[-3, -2) \cup [3, 4) = \{x \mid -3 \le x < -2 \;\vee\; 3 \le x < 4\}$. Therefore, we have

$$F^{-1}(\{-2, 0, 3\}) = [-3, -2) \cup [0, 1) \cup [3, 4).$$ ◆

Let us use this notation to see the relationship between images (and inverse images) and the operations of union and intersection. Consider the squaring function $s: \mathbf{R} \to \mathbf{R}$, given by $s(x) = x^2$. We look at the following intervals, which are subsets of the domain of s: $S_1 = [0, 5] = \{x \mid 0 \le x \le 5\}$ and $S_2 = [-3, 2] = \{x \mid -3 \le x \le 2\}$. What is the relationship between $s(S_1 \cap S_2)$ and $s(S_1) \cap s(S_2)$? The former is $s([0, 2]) = [0, 4]$. The latter is $[0, 25] \cap [0, 9] = [0, 9]$. Thus the former is a subset of the latter in this example, but the two sets are not equal. In words, the image of the intersection is not necessarily the same as the intersection of the images. As the next theorem shows, however, it is true that the image of the union of two subsets of the domain equals the union of the images of these subsets. Furthermore, the inverse image of the union of two subsets of the codomain equals the union of the inverse images of these subsets; and the inverse image of the intersection equals the intersection of the inverse images.

THEOREM 1.   *Suppose that $f: A \to B$. Let R and S be subsets of A, and let T and U be subsets of B. Then*

(a) $f(R \cup S) = f(R) \cup f(S)$.

(b) $f^{-1}(T \cup U) = f^{-1}(T) \cup f^{-1}(U)$.

(c) $f(R \cap S) \subseteq f(R) \cap f(S)$.

(d) $f^{-1}(T \cap U) = f^{-1}(T) \cap f^{-1}(U)$.

Proof. We prove parts (a) and (b), and leave the other two parts to the reader (Exercise 32). To show that two sets are equal, we show that every element of the first is an element of the second and that every element of the second is an element of the first.

(a) Suppose that $y \in f(R \cup S)$. By definition, then, $y = f(x)$ for some $x \in R \cup S$. This x is either in R or in S. There is complete symmetry here between R and S, so without loss of generality we may assume that $x \in R$. Since $x \in R$, we have $y \in f(R)$, and hence certainly y is in the superset $f(R) \cup f(S)$. Thus we have shown that $f(R \cup S) \subseteq f(R) \cup f(S)$.

Conversely, suppose that $y \in f(R) \cup f(S)$. Without loss of generality, say that $y \in f(R)$. Thus $y = f(x)$ for some x in R. But since this x is perforce in $R \cup S$, we can conclude that $y \in f(R \cup S)$, completing the proof of (a).

(b) Suppose that $x \in f^{-1}(T \cup U)$. By definition, then, $f(x) \in T \cup U$. Without loss of generality, assume that $f(x) \in T$. Then by definition $x \in f^{-1}(T)$. Hence certainly x is in the superset $f^{-1}(T) \cup f^{-1}(U)$. Thus we have shown that $f^{-1}(T \cup U) \subseteq f^{-1}(T) \cup f^{-1}(U)$.

Finally, suppose that $x \in f^{-1}(T) \cup f^{-1}(U)$. Without loss of generality, we can assume that $x \in f^{-1}(T)$. This means that $f(x) \in T$, whence certainly $f(x) \in T \cup U$, so $x \in f^{-1}(T \cup U)$, as desired. ∎

OPERATIONS ON FUNCTIONS

We close this section by raising the level of abstraction once more. There are ways of combining functions to obtain other functions. In other words, we can define operations (i.e., functions) on functions.

The simplest way to combine functions occurs when an operation already is defined on the codomain of those functions. The reader has probably encountered this situation in algebra or calculus, where numerical functions are added and multiplied. Suppose that f and g are functions from a set A to the set of real numbers. Then the (**elementwise**) **sum** of f and g, written $f + g$, is the function whose value at $x \in A$ is the sum of the values of f at x and g at x:

$$\forall x \in A \colon (f + g)(x) = f(x) + g(x).$$

Similarly, the (**elementwise**) **product** of f and g, written $f \cdot g$, is the function whose value at $x \in A$ is the product of the values of f at x and g at x:

$$\forall x \in A \colon (f \cdot g)(x) = f(x) \cdot g(x).$$

EXAMPLE 7. The rules for differentiation from calculus tell us how the differentiation operator interacts with these operations on functions. In particular, $(f+g)' = f'+g'$ and $(f \cdot g)' = f \cdot g' + f' \cdot g$. A similar rule holds for the elementwise quotient of two functions (if the divisor is constrained never to be 0). ◆

Elementwise sums and products of functions are only defined when sums and products are defined in the codomain. Next we turn to a different, more general and more important, way of combining functions to obtain new functions. This method, composition, will again probably be familiar from algebra or calculus.

Let $f: A \to B$ and $g: B \to C$ be functions with the indicated domains and codomains. Then their **composition**, written $g \circ f$ and read "g composed with f," is the function from A to C given by

$$\forall x \in A: (g \circ f)(x) = g(f(x)).$$

The schematic in Figure 3.10 shows how function composition works.

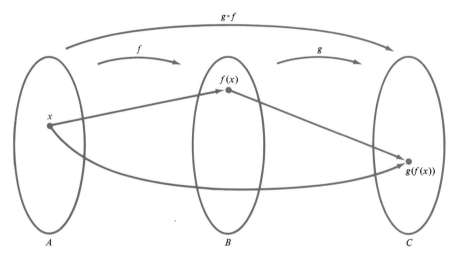

FIGURE 3.10   Schematic representation of the composition g composed with f.

The composite function $g \circ f$ has as its domain the domain of f (namely, A) and as its codomain the codomain of g (namely, C). The domain of g (namely, B) serves as an intermediary set, through which elements pass "on their way" from A to C. First f takes x to $f(x)$, and then g takes $f(x)$ to $g(f(x))$; thus $g \circ f$ takes x to $g(f(x))$. There is nothing preventing two or more of the sets A, B, and C from being the same set, although the order in which the composition is performed does make a difference.

We formally defined a function to be a set of ordered pairs. From this point of view we can write $g \circ f = \{ (a, c) \mid \exists b \in B : (a, b) \in f \wedge (b, c) \in g \}$. The b in this definition is $f(a)$ (there is always one and only one such b), and $c = g(b) = g(f(a))$. Although this form is harder to understand than the definition given above, it is the form that generalizes to relations in the next section.

EXAMPLE 8. Suppose that f is the function shown in Figure 3.4: $f(a) = f(b) = f(d) = 2$ and $f(c) = 1$. Suppose that h is the function from $\{1, 2, 3\}$ to $\{a, b, c, d\}$ given by $h(1) = a$ and $h(2) = h(3) = b$. Then we may form two compositions. First, $h \circ f$ is the function from $\{a, b, c, d\}$ to itself given by $(h \circ f)(a) = (h \circ f)(b) = (h \circ f)(d) = b$ and $(h \circ f)(c) = a$. On the other hand, $f \circ h$ is the constant function from $\{1, 2, 3\}$ to itself whose value at every element of the domain is 2. In formal terms, we have

$$f = \{(a, 2), (b, 2), (c, 1), (d, 2)\}$$

$$h = \{(1, a), (2, b), (3, b)\}$$

$$h \circ f = \{(a, b), (b, b), (c, a), (d, b)\}$$

$$f \circ h = \{(1, 2), (2, 2), (3, 2)\}. \qquad \blacklozenge$$

We can define a composition of functions under slightly more lenient assumptions about the domains and codomains. Suppose that $f : A \to B$ and $g : C \to D$, where $B \subseteq C$. Then $g(f(x))$ is well-defined for all $x \in A$, since $f(x)$, being an element of B, is necessarily an element of C.

EXAMPLE 9. Consider the mother and father functions $m : P \to G$ and $f : P \to B$ defined in Example 4 of Section 3.1; P is the set of all people, G is the set of all females, and B is the set of all males. Since $G \subseteq P$ and $B \subseteq P$, we can define such compositions as $m \circ f$, $f \circ m$, $m \circ m$, and $f \circ f$. It is easy to see that these four functions give a person's paternal grandmother, maternal grandfather, maternal grandmother, and paternal grandfather, respectively. $\blacklozenge$

EXAMPLE 10. Suppose that $f : \mathbf{R}^+ \to \mathbf{R}$ is given by $f(x) = \sqrt{2x}$, and $g : \mathbf{R} \to \mathbf{Z}$ is given by $g(x) = \lfloor x \rfloor + 5$. Then we may compose g and f to obtain the function $g \circ f : \mathbf{R}^+ \to \mathbf{Z}$, given by

$$(g \circ f)(x) = \left\lfloor \sqrt{2x} \right\rfloor + 5.$$

For example, f takes $\frac{8}{9}$ to $\frac{4}{3}$, and g takes $\frac{4}{3}$ to 6, so $g \circ f$ takes $\frac{8}{9}$ to 6. In this example, we cannot form $f \circ g$, since $\mathbf{Z}$, the codomain of g, is not equal to (or even a subset of) $\mathbf{R}^+$, the domain of f. If we tried to write down the expression for $(f \circ g)(x)$, namely $\sqrt{2(\lfloor x \rfloor + 5)}$, then for some values of x in $\mathbf{R}$ (the domain of g), the expression would be meaningless. For example, letting $x = -12.1$ would lead to $\sqrt{-16}$, which is not a real number.   ◆

EXAMPLE 11.   The chain rule in calculus tells us how function composition and differentiation interact: The derivative of the composition is determined by

$$(g \circ f)'(x) = g'(f(x)) \cdot f'(x).$$

In terms of the functions themselves, this says that $(g \circ f)' = (g' \circ f) \cdot f'$, so that we have three operations on functions appearing in the same formula—composition, differentiation, and elementwise product.   ◆

To get a better feel for working with functions in general, let us see how function composition relates to injectivity and surjectivity.

THEOREM 2.   *Let $f: A \to B$ and $g: B \to C$ be functions with the indicated domains and codomains. Then*

(a)   *$g \circ f$ is one-to-one if both f and g are one-to-one.*
(b)   *$g \circ f$ is onto if both f and g are onto.*

Proof.   The reader should draw pictures similar to Figure 3.10 as an aid in understanding this proof. The elements of the various sets that are being considered should be drawn as points in the pictures.

First let us show that if f and g are both injective, then their composition $g \circ f$ is injective. By definition, then, we need to show that if we have two elements x_1 and x_2 in A (the domain of $g \circ f$), such that $(g \circ f)(x_1) = (g \circ f)(x_2)$, then x_1 has to equal x_2. The given condition says that $g(f(x_1)) = g(f(x_2))$. Thus we have two elements of B, namely $f(x_1)$ and $f(x_2)$, whose images under g are equal. Since we are given that g is one-to-one, we can conclude (from the definition of "one-to-one") that these two elements are equal, that is, $f(x_1) = f(x_2)$. But if we now apply the definition once more, noting that f is one-to-one by our hypothesis, we can conclude that $x_1 = x_2$, exactly what we wished to show. Note that we used both parts of the hypothesis in proving that the composition was injective.

Next we show that the composition of two surjective functions is surjective. Assume that both f and g are onto. We want to show that $g \circ f$ is onto. By definition, this means that we have to show that every element z in C (the codomain of $g \circ f$) is in the range of $g \circ f$, that is, that there exists an element $x \in A$ such that $z = g(f(x))$. Again, we will use the hypotheses one at a time. First, since g is surjective, we know that z is in the range of g, so there is an element $y \in B$ such that $z = g(y)$. Next, since f is surjective, we know that y has to be in the range of f, so there is an element $x \in A$ such that $y = f(x)$. Putting these together, we have $z = g(y) = g(f(x)) = (g \circ f)(x)$, as desired.   ■

The converse of Theorem 2 is not entirely true; the reader is asked to investigate this situation in Exercises 16 and 17.

Next we look at a unary operation on a certain set of functions. This operation, too, will probably be familiar to students who have studied algebra and calculus, and it is important in many branches of mathematics. Suppose that $f: A \to B$ is a bijective function. Then the **inverse** of f, written f^{-1}, is the function from B to A given by the rule

$$f^{-1}(y) = x \quad \text{if and only if} \quad f(x) = y.$$

This rule actually determines a function precisely because f is assumed to be both one-to-one and onto. The fact that f is onto guarantees that for each $y \in B$ there is *at least one* $x \in A$ such that $f(x) = y$; and the fact that f is one-to-one guarantees that for each $y \in B$ there is *at most one* $x \in A$ such that $f(x) = y$. Thus for each $y \in B$ there is exactly one $x \in A$ such that $f(x) = y$, and that x is the value of f^{-1} at y. In terms of sets we can write $f^{-1} = \{ (b, a) \mid (a, b) \in f \}$.

EXAMPLE 12.   The function shown in Figure 3.8 has as inverse the function from $\{1, 2, 3, 4\}$ to $\{a, b, c, d\}$ which takes 1 to c, 2 to b, 3 to d, and 4 to a. The schematic for the inverse is obtained from the schematic for the original function by "turning all the arrows around." We thus obtain the schematic shown in Figure 3.11. Note that the inverse of a bijective function is always again a bijective function.   ◆

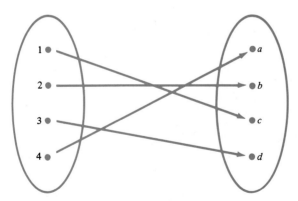

FIGURE 3.11   Schematic representation of the inverse of the function depicted in Figure 3.8.

EXAMPLE 13.   If $T(x) = 2^x$, then $T^{-1}(x) = \log x$. Here we are assuming that T is being considered as a function from $\mathbf{R}$ to $\mathbf{R}^+$; thus the inverse function (logarithm base 2) is a function from $\mathbf{R}^+$ to $\mathbf{R}$. Note that the inverse is not the same as the reciprocal; $T^{-1}(x) \neq 1/T(x)$, since $\log x \neq 1/2^x$.   ◆

Finally, let us relate the notions of inverse and composition. The **identity function** i_A on a set A is the function from A to itself that takes every element of A to itself; that is,

$$\forall x \in A: i_A(x) = x.$$

Now suppose that $f: A \to B$ is a bijective function, so that it has an inverse $f^{-1}: B \to A$. Since domains and codomains correspond, we can form two compositions, $f \circ f^{-1}: B \to B$ and $f^{-1} \circ f: A \to A$. Each of these compositions must be the appropriate identity function: $f \circ f^{-1} = i_B$ and $f^{-1} \circ f = i_A$. (To see that $f^{-1}(f(x)) = x$ for all $x \in A$, simply note that by definition $f^{-1}(y) = x \longleftrightarrow f(x) = y$—let $y = f(x)$. The other half is proved in a similar fashion.) For instance, the crucial facts about the relationship between the exponential function and the logarithmic function mentioned in Section 3.1, namely that

$$\forall x \in \mathbf{R}: \log_b(b^x) = x \quad \text{and} \quad \forall x \in \mathbf{R}^+: x = b^{\log_b x},$$

are just the statement that the composition of the logarithm function and the exponential function, in either order, is the identity function (on the set of real numbers in one case, and on the set of positive real numbers in the other).

As an everyday analogy of this fact about functions, we might notice that if, barefoot, we perform the "function" of putting on our shoes and follow it by the "inverse function" of taking them off, then we return to the shoeless state in which we began; if, wearing shoes, we perform the inverse function of taking off our shoes and follow it by the function of putting them on, then we return to the shod state in which we began.

Clothing our feet also provides a good analogy for the fact that $(g \circ f)^{-1} = f^{-1} \circ g^{-1}$, assuming that f and g are bijective functions, with the codomain of f equaling the domain of g, so that these inverses and compositions are well-defined (see Figure 3.10). Let f be the "function" of putting on socks, and g, the "function" of putting on shoes. Suppose that we begin barefoot. The composition $g \circ f$—f followed by g—consists of putting on socks, followed by putting on shoes. To reverse this process and return to the barefoot state, we need to perform the inverse function, $(g \circ f)^{-1}$. As everyone knows, we accomplish this by first taking off our shoes (performing g^{-1}) and then taking off our socks (performing f^{-1})—in other words, by performing $f^{-1} \circ g^{-1}$. We cannot take off our socks first: $(g \circ f)^{-1} \neq g^{-1} \circ f^{-1}$, in general.

We leave to the reader (Exercise 26) an actual proof that the inverse of a composition is the composition of the inverses in the opposite order.

The following theorem summarizes our discussion about inverses and compositions.

THEOREM 3. *Suppose that $f: A \to B$ and $g: B \to C$ are bijective functions. Then*

(a) $f^{-1} \circ f = i_A$ *and* $f \circ f^{-1} = i_B$.
(b) $(g \circ f)^{-1} = f^{-1} \circ g^{-1}$.

SUMMARY OF DEFINITIONS

bijective function: a function that is both injective and surjective (e.g., the function $f(x) = 2x$ from **R** to **R**).

composition of the function $g: B \to C$ and the function $f: A \to B$: the function, written $g \circ f$, with domain A and codomain C given by the rule $(g \circ f)(x) = g(f(x))$ for all $x \in A$ (e.g., the composition of the squaring function and the cubing function is the function given by $h(x) = (x^3)^2 = x^6$).

(elementwise) product of two functions f and g with the same domain, whose codomain consists of numbers: the function defined by $(f \cdot g)(x) = f(x) \cdot g(x)$ for all x in the common domain (e.g., the product of the squaring function and the cubing function is the function given by $h(x) = x^2 \cdot x^3 = x^5$).

(elementwise) sum of two functions f and g with the same domain, whose codomain consists of numbers: the function defined by $(f+g)(x) = f(x)+g(x)$ for all x in the common domain (e.g., the sum of the squaring function and the cubing function is the function given by $h(x) = x^2 + x^3$).

identity function on the set A: the function $i_A: A \to A$ given by $i_A(x) = x$ for all $x \in A$.

image of the subset S of the domain A of the function $f: A \to B$: the set of images of the elements of S, written $f(S)$; $\{ f(a) \mid a \in S \}$ (e.g., the multiples of 4 if $f(x) = 2x$ and S is the set of even integers).

injective function: a function that sends different elements of the domain to different elements of the codomain; if $f: A \to B$, then f is injective if and only if $\forall a_1 \in A: \forall a_2 \in A: (a_1 \neq a_2 \to f(a_1) \neq f(a_2))$ (e.g., the function $f(x) = 2^x$ from **R** to **R**).

inverse of the bijective function $f: A \to B$: the function $f^{-1}: B \to A$ defined by $f^{-1}(y) = x$ if and only if $f(x) = y$ (e.g., the square root function is the inverse of the squaring function from positive real numbers to positive real numbers).

inverse image of the subset T of the codomain B of the function $f: A \to B$: the set of elements of the domain whose images are in T, written $f^{-1}(T)$; $\{ a \in A \mid f(a) \in T \}$ (e.g., the set $\{1, 2, 3\}$ if the domain of f is **N**, $f(x) = 3x$ and T is the set of positive integers less than 10).

one-to-one correspondence: a bijective function.

one-to-one function: an injective function.

onto function: a surjective function.

range of the function $f: A \to B$: the set of images of the elements of the domain; $\{ f(a) \mid a \in A \}$ (e.g., the interval $[3, \infty)$ for the function $f(x) = x^2 + 3$ with domain **R**).

surjective function: a function whose range is the entire codomain; if $f: A \to B$, then f is surjective if and only if $\forall b \in B: \exists a \in A: f(a) = b$ (e.g., the function "min" from $\mathcal{P}(\mathbf{N}) - \{\emptyset\}$ to $\mathbf{N}$ that sends a nonempty subset of natural numbers to its smallest element).

EXERCISES

1. Draw a schematic representation of a function from $\{1, 2, 3\}$ to $\{a, b, c, d, e\}$ that is
   (a) Injective.
   (b) Not injective.

2. Draw a schematic representation of a function from $\{a, b, c, d, e\}$ to $\{1, 2, 3\}$ that is
   (a) Surjective.
   (b) Not surjective.

3. Give an example of a function from $\mathbf{N}$ to itself that is
   (a) One-to-one and onto.
   (b) One-to-one but not onto.
   (c) Onto but not one-to-one.
   (d) Neither one-to-one nor onto.

4. Explain the difference between *codomain* and *range*.

5. Give an example of a bijective function f from $\mathbf{N}$ to itself that differs from the identity function at exactly two elements of the domain (in other words, $f(x) \neq x$ for exactly two numbers x).

6. Determine whether each of the following functions is one-to-one and/or onto. If the function is both one-to-one and onto, find a rule for its inverse.
   (a) $f: \mathbf{R} \to \mathbf{R}$ given by $f(x) = 2x + 3$.
   (b) $f: \mathbf{Z} \to \mathbf{Z}$ given by $f(x) = 2x + 3$.

7. Determine whether the following functions from $\mathbf{R}$ to $\mathbf{R}$ are one-to-one and/or onto. [*Hint*: Look at their graphs.]
   (a) $f(x) = 2x + 1$.
   (b) $f(x) = 2x^2 + 1$.
   (c) $f(x) = 2x^3 + 1$.
   (d) $f(x) = \arctan(x)$ (the inverse tangent function).
   (e) $f(x) = x(x - 1)(x + 1)$.

8. Suppose that $f: A \to B$ is a function, R and S are subsets of A, and T and U are subsets of B. Prove the following facts.
   (a) If $R \subseteq S$, then $f(R) \subseteq f(S)$.
   (b) If $T \subseteq U$, then $f^{-1}(T) \subseteq f^{-1}(U)$.

9. What is the relationship between injectivity and surjectivity for functions from a finite set to itself?

10. Give an example of functions with the following domains and ranges.
    (a) Domain $= \{1, 2, 3, 4, 5, 6\}$, range $= \{1, 2, 4\}$.
    (b) Domain $= \mathbf{R}$, range $= \{\, x \in \mathbf{R} \mid x > 5 \,\}$.
    (c) Domain $= \mathcal{P}(\mathbf{N})$, range $= \mathbf{N}$.

11. Using the mother and father functions (Example 4 of Section 3.1), express the following symbolically.
    (a) x is the grandfather of y.
    (b) x is the grandchild of y.
    (c) x is the aunt or uncle (not by marriage) of y.
    (d) The set of x's nephews and nieces.

12. Determine whether each of the following functions is injective, surjective, neither, or both. Find the inverse if the function is bijective.
    (a) The floor function from $\mathbf{R}$ to $\mathbf{Z}$.
    (b) The maximum function from the set of finite nonempty subsets of $\mathbf{R}$ to $\mathbf{R}$.
    (c) The function $\{\, (n, 2n) \mid n \in \mathbf{N} \,\}$ from $\mathbf{N}$ to itself.
    (d) The function from $\mathbf{R} \times \mathbf{R}$ to $\mathbf{R}$ that sends the pair (x, y) to e^{x+y}.
    (e) The complementation function from $\mathcal{P}(U)$ to itself, where U is a fixed universal set.

13. Suppose that f and g are functions defined on the set of natural numbers by the rules that $f(x) = 1$ if x is odd and $f(x) = 2$ if x is even; and $g(x) = 1$ if x is even and $g(x) = 2$ if x is odd. Write down rules for the following functions.
    (a) $f + g$.
    (b) fg.
    (c) $f - g$ (whose definition is analogous to the definition of $f + g$).
    (d) $f \circ g$.
    (e) $f \circ f$.
    (f) $g \circ f$.
    (g) $g \circ g$.

14. Give an example of a function f from $\mathbf{N}$ to itself such that $\forall x : f(x) \neq x$ but $f = f^{-1}$. [*Hint*: Solve the problem for a two-element set first.]

15. Suppose that f is a function from A to B. Express each of the following sentences symbolically (like the definition of "injective" given in this section).
    (a) range$(f) = C$.
    (b) f is a two-to-one correspondence (i.e., f sends exactly two different elements of A to every element of B).

16. Let $f: A \to B$ and $g: B \to C$, and suppose that $g \circ f: A \to C$ is onto. Prove or disprove each of the following.
    (a) f must be onto.
    (b) g must be onto.

17. Repeat Exercise 16, with "one-to-one" substituted for "onto" throughout.

18. Prove that the inverse of the inverse of a bijective function is the function itself.

19. Explain why the graph of a function drawn in the usual coordinate plane $\mathbf{R} \times \mathbf{R}$ is (the graph of) an injective function on its domain if and only if every horizontal line intersects the graph in at most one point. (This is called the "horizontal line test" in high school algebra.)

20. Explain why the graph of a function from some subset $A \subseteq \mathbf{R}$ to $\mathbf{R}$ drawn in the usual coordinate plane $\mathbf{R} \times \mathbf{R}$ is (the graph of) a surjective function if and only if every horizontal line intersects the graph in at least one point.

21. Explain how to determine geometrically the range of a function whose graph is drawn in the usual coordinate plane $\mathbf{R} \times \mathbf{R}$.

22. The domain and the codomain of each of the following functions is the closed interval $[0, 1]$. Determine which of these functions are injective, which are surjective, and which are bijective.

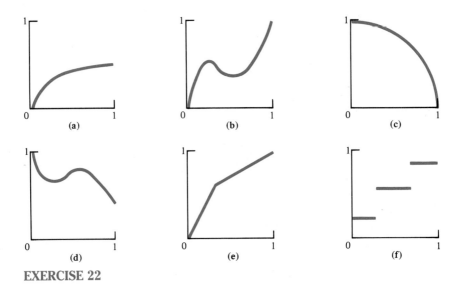

EXERCISE 22

23. Write out the formal definition of the following functions, viewing functions as sets of ordered pairs.
    (a) $f + g$.
    (b) i_A.

24. Give examples of functions f and g from a finite set A to itself satisfying each of the following conditions.
    (a) $f \circ g \neq g \circ f$.
    (b) $f \circ g = g \circ f$, but $f \neq g$ and neither f nor g is i_A.

25. Let f be a function from A to B. Prove that $f \circ i_A = f$ and $i_B \circ f = f$.

26. Prove part (b) of Theorem 3.

27. Under what conditions can the range of a function be the empty set?

28. Let $A = \{a, b, \{a, b\}\}$, and let f be the function that sends the three elements of A (in the order listed here) to 1, 2, and 3, respectively. Explain why $f(\{a, b\})$ is an ambiguous notation in this case. (This anomaly really does not come up in real life.)

29. Let $f: A \to B$. Explain why each of the following equalities holds.
    (a) $f(\emptyset) = \emptyset$. (See the quotation at the beginning of this section.)
    (b) $f^{-1}(\emptyset) = \emptyset$.
    (c) $f^{-1}(B) = A$.
    (d) $f(A) = \text{range}\,(f)$.

30. Consider the function f from **N** to itself given by the rule $f(x) = 3x$. Let A be the set of even natural numbers.
    (a) Find $f(A)$.
    (b) Find $f^{-1}(A)$.
    (c) Find the range of f.
    (d) Find $f^{-1}(\{12\})$.
    (e) Find $f^{-1}(\{5\})$.

31. Find the following images and inverse images, if r is the function that sends $x \in \mathbf{Z}$ to $x \bmod 7$.
    (a) $r^{-1}(\{0\})$.
    (b) $r^{-1}(\{1\})$.
    (c) $r^{-1}(\{8\})$.
    (d) $r^{-1}(\{0, 1, 8\})$.
    (e) $r(\text{the set of prime numbers})$.
    (f) $r(\{8\})$.
    (g) $r(\{0, 1, 8\})$.

32. Prove parts (c) and (d) of Theorem 1.

33. Give an example using the function depicted in Figure 3.4 to show that $f(R \cap S)$ need not equal $f(R) \cap f(S)$, where R and S are subsets of the domain of the function f.

34. Prove or disprove the following cancellation properties for functions.
(a)   If $f \circ g = f \circ h$, then $g = h$.
(b)   If $f \circ g = h \circ g$, then $f = h$.

35. Give an example of a pair of functions $f : A \to B$ and $g : B \to A$ such that $f \circ g = i_B$ but $g \neq f^{-1}$.

36. Consider the function from the set of points on the surface of the earth other than the north and south poles to $\mathbf{R} \times \mathbf{R}$ that sends a point P to its navigational coordinates (longitude(P), latitude(P)), where we think of longitude as a number in the interval $[-180, 180)$ giving the number of degrees west of Greenwich, England, and we think of latitude as a number in the interval $(-90, 90)$ giving the number of degrees north of the equator.
(a)   Determine whether this function is injective.
(b)   Determine the range of this function.

37. Determine the number of onto functions from $\{1, 2, 3, 4\}$ to $\{a, b\}$. [*Hint:* Do not look for anything fancy—just use brute force.]

38. Consider the function c on $\mathbf{N}$ defined as follows. If n is even, then $c(n) = n/2$; if n is odd, then $c(n) = 3n + 1$. Find a nice expression for $\left| c^{-1}(\{x\}) \right|$ in terms of x. (See also Exercise 30 in Section 5.2.)

39. Determine whether the following sets are closed under the given operations.
(a)   The set of continuous functions from $\mathbf{R}$ to $\mathbf{R}$ under addition.
(b)   The set of bijective functions from A to A under composition.
(c)   The set of bijective functions from A to B under the operation of forming the inverse.

Challenging Exercises

40. Find a bijective function from $\mathbf{N} \times \mathbf{N}$ to $\mathbf{N}$, expressing your function with an algebraic formula.

41. Find an injective function from the set of rational numbers to the set of integers.

42. Find a function whose domain is $\mathcal{P}(\mathbf{N})$ and whose range is the interval $[0, 1]$ of real numbers.

Exploratory Exercises

43. Explain in what sense $\sin^{-1}$ (the inverse sine function) is not, and in what sense it is, the inverse function of the sine function.

44. Discuss the ambiguity in the notation f^{-1}.

45. Find out how to prove the Schröder–Bernstein theorem (that if there is a one-to-one function $f: A \to B$ and a one-to-one function $g: B \to A$, then there is a one-to-one and onto function $h: A \to B$) by consulting Halmos [137] and Kamke [167].

SECTION 3.3
RELATIONS

"Hush!" said Eeyore in a terrible voice to all of Rabbit's friends-and-relations, and "Hush!" they said hastily to each other all down the line, until it got to the last one of all.

—A. A. Milne, *Winnie the Pooh*

In all of mathematics we study relationships: relationships among numbers, functions, geometric objects, and so on. Much of this study can be formalized by looking at what we will call relations between sets. In this section we discuss relations in general. Then in Section 3.4 we look at two types of relations of particular importance in mathematics and computer science: order relations and equivalence relations.

As we see from the following definition, the notion of relation generalizes the notion of function. Let A and B be sets. A **relation** R from A to B is a subset of $A \times B$ (i.e., $R \subseteq A \times B$). If $(a, b) \in R$, then we write aRb and say that a **is related to** b (under R), or that the relation R **holds** between a and b (in that order). If a is not related to b (i.e., $(a, b) \notin R$), we can write $a\not\!Rb$. We call A the **domain** of the relation and B the **codomain**. If $A = B$, as is very often the situation, we say that R is a relation **on** A.

Every function from A to B is a relation from A to B, but relations in general need not satisfy the uniqueness condition that functions do. In an arbitrary relation, there may be elements of A that are related to no elements of B, and there may be elements of A that are related to more than one element of B; but in a function, each element of A is related to exactly one element of B.

As we see from the following examples, much of familiar mathematics and other areas of human thought can be looked at in terms of relations.

EXAMPLE 1.   "Is less than" is a relation on the set **R** of real numbers (as well as on any of its subsets, such as **N**). We usually write "$<$" as the name of this relation, and of course we write $x < y$ rather than $(x, y) \in <$ to indicate that x is less than y. For instance, $3 < \pi$ and $\pi \not< 3$. Similarly, "is greater than" is a relation on **R**, as is "is less than or equal to." Note that none of these relations is a function, since in each case, for each $x \in \mathbf{R}$ there is more than one real number y related to x (an infinite number of them, in fact). ◆

EXAMPLE 2.   Let B be the set of male living persons and G be the set of living females. Then there is the "is married to" relation H from B to G, given by xHy if and only if man x is married to the woman y; we can also express this as "is the husband of." (The elements of this relation change over time, of course.) We would have (Robert)H(Nancy) if Robert and Nancy were married to each other, but (Nancy)$\not\!H$(Robert). ◆

EXAMPLE 3.   Let $A = \{1, 2, 3\}$, and let $B = \{2, 4, 5, 6\}$. There are $3 \cdot 4 = 12$ elements of $A \times B$. Therefore, by the observation made in Example 6b of Section 2.1 (that $|\mathcal{P}(X)| = 2^{|X|}$ for any finite set X), there are $2^{12} = 4096$ subsets of $A \times B$. Hence there are 4096 different relations from A to B. These include the empty set (in which no elements are related), the set $A \times B$ itself (in which every element of A is related to every element of B), and everything in between, such as $S = \{(2, 2), (2, 4), (2, 6), (3, 2)\}$. In this last case, we have, for example, $2S4$ but $1\not\!S6$ and $4\not\!S2$. ◆

EXAMPLE 4.   If M is the set of mathematicians, then we can consider the "joint authorship" relation J on M, defined by xJy if and only if $x \neq y$ and there is a published mathematical research paper coauthored by x and y (with possibly other coauthors as well). ◆

EXAMPLE 5.   The "**divides**" relation on the set of positive integers, usually written with a vertical bar ("|"), is the relation defined by $x \mid y$ if and only if y is a multiple of x. For example, we have $3 \mid 3$ and $4 \mid 192$, but $8 \not\mid 12$ and $100 \not\mid 5$. This relation should not be confused with the operation "divided by," which is often denoted with a slanted bar ("/"), as in $192/4 = 48$; "divides" is a verb that either holds or fails to hold between two numbers (4 divides 192); "divided by" is an operation that produces an answer (192 divided by 4 equals 48). ◆

We have already met another important example of a relation, "is congruent to modulo m" (which we might denote $\equiv_m$, if we need to make the value of m explicit in the notation). Thus $(x, y) \in \equiv_m$ if and only if $x \equiv y \pmod{m}$. We will study this relation further in Section 3.4.

Relations as we have defined them so far are *binary* relations, because they deal with two sets. It is also possible to define n-ary relations for $n > 2$. The definition generalizes the definition of binary relation: An **n-ary relation** among sets A_1, A_2, ..., A_n is a subset of $A_1 \times A_2 \times \cdots \times A_n$. Data bases, in which the data of government and industry are stored, are often organized as sets of interconnected n-ary relations.

EXAMPLE 6.   Let S be the set of students at your school, let C be the set of courses, and let P be the set of teachers. Consider two binary relations: the enrolled relation E from

S to C, whereby sEc if and only if student s is enrolled in course c, and the teaching relation T from P to C, whereby pTc if and only if professor p is teaching course c. Note that each student typically takes more than one course, each course typically has more than one student, each professor teaches more than one course, and there can be different professors teaching (different sections of) the same course. Thus we cannot use functions to model these relationships. We can combine E and T into one 3-ary relation R among S, C, and P, whereby $(s, c, p) \in R$ if and only if student s is being taught by professor p in course c. For instance, the triples (Mary Jones, MATH 245, Prof. Wilson) and (Sam Smith, MATH 245, Prof. Grant) might be elements of R. Symbolically, $R = \{ (s, c, p) \mid (s, c) \in E \wedge (p, c) \in T \}$. ◆

We can draw pictures to represent binary relations, just as we drew pictures in Section 3.2 to represent functions. A relation R from A to B would typically be pictured with two ovals, representing the domain and codomain, and arrows joining $a \in A$ to $b \in B$ whenever $(a, b) \in R$ (in other words, whenever aRb). For example, the relation S in Example 3 would be depicted as shown in Figure 3.12.

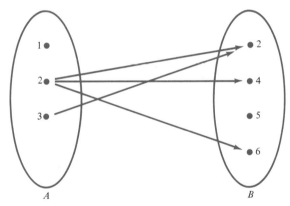

FIGURE 3.12   A picture (digraph) of a relation from A to B.

In the case of a relation *on* a set, we usually represent the domain (which is also the codomain) as a set of points, with arrows connecting pairs of points that are in the relation.

EXAMPLE 7.   Figure 3.13 shows the relation R on the set $\{0, 1, 2, 3, 4, 5\}$ defined by xRy if and only if $x + 2y$ is a multiple of 5.

Note that since $0 + 2 \cdot 0$ and $5 + 2 \cdot 5$ are multiples of 5, there is an arrow from each of 0 and 5 to itself (called a **loop**). Also note that the direction of the arrow

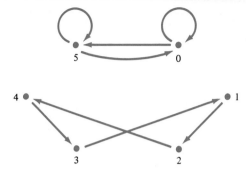

FIGURE 3.13   Picture (digraph) of a relation on the set $\{0, 1, 2, 3, 4, 5\}$.

is important. There is an arrow from 3 to 1, since $3 + 2 \cdot 1$ is a multiple of 5, but there is no arrow from 1 to 3, since $1 + 2 \cdot 3$ is not a multiple of 5. Since both $0 + 2 \cdot 5$ and $5 + 2 \cdot 0$ are multiples of 5, there are arrows from 0 to 5 and from 5 to 0.   ◆

In Chapter 8 we will study digraphs (a contraction of "directed graphs"), which are really neither more nor less than relations on a set. Abusing the terminology somewhat, we will call the kind of picture we are discussing here the **digraph of the relation**. If the domain of a relation is too large (or infinite), we cannot actually draw the digraph of the relation, but we can still imagine such a picture existing in the abstract.

EXAMPLE 8.   Consider the relation U on the set P of points in the plane given by $p_1 U p_2$ if and only if the distance between p_1 and p_2 is 1 inch. The digraph of this relation would consist of every point in the plane, with an arrow in each direction joining every pair of points 1 inch apart.   ◆

MATRIX REPRESENTATION OF RELATIONS

In Section 2.2 we saw that a matrix is a rectangular array. Let us see how we can use matrices to represent specific relations in a way that both will enable human beings to understand them better and will let computers store and work with them.

Suppose that we have a relation R from a finite set A to a finite set B. Let us list the elements of the domain and codomain in some fixed order, so that we will write $A = \{a_1, a_2, \ldots, a_m\}$ and $B = \{b_1, b_2, \ldots, b_n\}$, where $|A| = m$ and $|B| = n$. For example, if the elements of A and B are numbers, we can use their natural order as the order in which to list them; if they are letters, we can use alphabetical order. We will represent the relation R with an m by n matrix M_R (or just M if the context is clear) in which the element m_{ij} in the ith row and jth column is 1 if the ith element of A is related to the jth element of B, and is 0 if those two elements are not related. In symbols, $M_R = (m_{ij})$, where

$$m_{ij} = \begin{cases} 1 & \text{if } a_i R b_j \\ 0 & \text{if } a_i \not\mathrel{R} b_j. \end{cases}$$

Thus M_R is a matrix consisting entirely of 0's and 1's, called a **Boolean matrix**. Sometimes for clarity we will label the rows and columns of the matrix representing R by the elements of A and B, respectively.

EXAMPLE 9.   The relation S given in Example 3 is represented by the following matrix if we list the elements in A and in B in their numerical orders. (The left-hand border to this matrix shows the elements of A, and the border along the top shows the elements of B.)

$$M_S = \begin{array}{c} \\ 1 \\ 2 \\ 3 \end{array} \begin{array}{cccc} 2 & 4 & 5 & 6 \\ \left[\begin{array}{cccc} 0 & 0 & 0 & 0 \\ 1 & 1 & 0 & 1 \\ 1 & 0 & 0 & 0 \end{array}\right]. \end{array}$$

The entry 1 in the second row, second column, of the matrix proper, for instance, means that $(2, 4) \in S$; the entry 0 in the first row, first column means that $1 \not\mathrel{S} 2$. ◆

If we are dealing with a relation *on* a finite set, we insist that the elements be listed in the same order both horizontally and vertically.

EXAMPLE 10.   The relation given in Example 7 (and shown in Figure 3.13) has the following matrix representation if we use the numerical order of the elements in the domain ($\{0, 1, 2, 3, 4, 5\}$).

$$M = \begin{array}{c} \\ 0 \\ 1 \\ 2 \\ 3 \\ 4 \\ 5 \end{array} \begin{array}{cccccc} 0 & 1 & 2 & 3 & 4 & 5 \\ \left[\begin{array}{cccccc} 1 & 0 & 0 & 0 & 0 & 1 \\ 0 & 0 & 1 & 0 & 0 & 0 \\ 0 & 0 & 0 & 0 & 1 & 0 \\ 0 & 1 & 0 & 0 & 0 & 0 \\ 0 & 0 & 0 & 1 & 0 & 0 \\ 1 & 0 & 0 & 0 & 0 & 1 \end{array}\right]. \end{array}$$

Note that order matters. For example, the matrix shows that $(1, 2)$ is in the relation, but $(2, 1)$ is not. ◆

EXAMPLE 11.   Suppose that we have several models of computers, which we list in some order (alphabetical by manufacturer and model name, perhaps), and a list of features that the computers may or may not have, such as IBM compatibility, mouse, serial port, expansion slots, and so on. We may be interested in the relation from the set of computers to the set of features in which a computer is related to a feature if and only if the computer has the feature. Then a matrix in which the (i, j)th entry is 1 if the ith computer has the jth feature, and 0 if it does not, provides a compact way to represent all the relevant information. If we want to find out which features a given computer has, we need only look across the row representing that computer and pick out the columns containing 1's. If we want to find out which computers have a particular feature, we need only look down the column representing that feature and pick out the rows containing 1's. ◆

We will see later in this section that various properties of relations are reflected in their matrices. We will also see that operations on matrices enable us to find out properties of the relations they represent.

Let us look at one more example of a useful relation.

EXAMPLE 12. The **identity** or **diagonal** relation on a set A is the relation $\Delta_A = \{ (a, a) \mid a \in A \}$. In other words, $x\Delta_A y$ if and only if $x = y$, where x and y are elements of A. If $|A| = n$, then the matrix representing Δ_A is the n by n **identity matrix**, denoted I_n, which has 1's down the main diagonal (from upper left to lower right) and 0's elsewhere.

$$
I_n = \begin{bmatrix}
1 & 0 & 0 & \cdots & 0 \\
0 & 1 & 0 & \cdots & 0 \\
0 & 0 & 1 & \cdots & 0 \\
\vdots & \vdots & \vdots & \ddots & \vdots \\
0 & 0 & 0 & \cdots & 1
\end{bmatrix}.
$$

We will encounter the identity matrix several more times in this book. ◆

OPERATIONS ON RELATIONS (*OPTIONAL*)

There are several ways to form new relations from old ones, just as there are several ways to form new functions from old ones. For one thing, since relations from A to B are just subsets of $A \times B$, we can apply set operations to relations to form new relations.

If R is a relation, then the relation $\overline{R}$, called the **complementary** relation to R, is just the set $(A \times B) - R$. In other words, it is the relation that holds precisely when R does not:

$$\forall x \in A \colon \forall y \in B \colon (x\overline{R}y \leftrightarrow x\mathcal{R}y).$$

For example, the "is less than or equal to" relation is the complement of the "is greater than" relation on the set of real numbers, since $x \leq y$ if and only if

$x \not> y$. It follows directly from the definition of complement that the matrix for the complement of a relation is obtained from the matrix for the relation by changing each 0 to a 1 and each 1 to a 0. (In the terminology of Section 1.4, we form $M_{\overline{R}}$ by taking the Boolean complements of all the entries in M_R.) The digraph of $\overline{R}$ is obtained from the digraph of R by erasing every arrow in the picture and drawing an arrow from a to b whenever there was no arrow from a to b in the digraph of R.

If R and S are relations from A to B, then $R \cup S$ and $R \cap S$ are also relations from A to B. By the definition of union and intersection, we see that $R \cup S$ is the relation that holds between two elements x and y when at least one of the relations R and S holds between x and y, and $R \cap S$ is the relation that holds between x and y when both of the relations R and S hold between x and y. For example, the "is less than or equal to" relation on the set of real numbers is the union of the two relations "is less than" and "equals." The intersection of these two relations is the empty set, however, since it is impossible to have $x < y$ and $x = y$ simultaneously. Similarly, we can define the intersection and union of arbitrary collections of relations; see Exercise 18.

It follows directly from the definition that the digraph of the union of two relations on a set is obtained by drawing an arrow from a to b whenever there is an arrow from a to b in at least one of the digraphs of the two relations; and the digraph of the intersection of two relations on a set is obtained by drawing an arrow from a to b whenever there is an arrow from a to b in both of the digraphs of the two relations. An entry of the matrix for $R \cup S$ is 1 if and only if the corresponding entry in at least one of the matrices for R and S is a 1; thus we obtain $M_{R \cup S}$ by "adding" M_R and M_S entry by entry, according to the rules $0 + 0 = 0$, $0 + 1 = 1 + 0 = 1 + 1 = 1$ (this is what we called Boolean addition in Section 1.4). Similarly, an entry of the matrix for $R \cap S$ is 1 if and only if the corresponding entry in both of the matrices for R and S is a 1; thus we obtain $M_{R \cap S}$ by "multiplying" M_R and M_S entry by entry, according to the rules $0 \cdot 0 = 0 \cdot 1 = 1 \cdot 0 = 0$, $1 \cdot 1 = 1$ (this is what we called Boolean multiplication in Section 1.4).

If R is a relation from A to B, then we can form the **inverse relation** R^{-1} from B to A, defined by

$$\forall x \in A : \forall y \in B : (y R^{-1} x \leftrightarrow x R y).$$

For example, the "is less than" relation on **R** is the inverse of the relation "is greater than." As another example, consider the relations W from the set of all women to the set of all men given by $x W y$ if and only if x is the wife of y, and H from the set of all men to the set of all women given by $x H y$ if and only if x is the husband of y. Then, clearly, $H = W^{-1}$ and $W = H^{-1}$.

It follows immediately from the definition that the digraph of the inverse of a relation is obtained from the digraph of the relation by "turning all the arrows around"—replacing each arrow from a to b by an arrow from b to a. The matrix representing the inverse of a relation is obtained by interchanging the roles of the rows and the columns; if $M_R = (m_{ij})$, then $M_{R^{-1}} = (m'_{ij})$ if we set $m'_{ij} = m_{ji}$. We call $M_{R^{-1}}$ the **transpose** of M_R. Physically, we obtain the transpose by "flipping" the matrix across its diagonal from upper left to lower right.

EXAMPLE 13. In Example 7 we looked at the relation defined by xRy if and only if $5 \mid x + 2y$. Its inverse is the relation defined by $xR^{-1}y$ if and only if $5 \mid y + 2x$. Its digraph is obtained from Figure 3.13 by reversing the arrows (see Figure 3.14). Its matrix is the transpose of the matrix given in Example 10:

$$
M_{R^{-1}} = \begin{array}{c} \\ 0 \\ 1 \\ 2 \\ 3 \\ 4 \\ 5 \end{array}
\begin{array}{c}
\begin{array}{cccccc} 0 & 1 & 2 & 3 & 4 & 5 \end{array} \\
\begin{bmatrix}
1 & 0 & 0 & 0 & 0 & 1 \\
0 & 0 & 0 & 1 & 0 & 0 \\
0 & 1 & 0 & 0 & 0 & 0 \\
0 & 0 & 0 & 0 & 1 & 0 \\
0 & 0 & 1 & 0 & 0 & 0 \\
1 & 0 & 0 & 0 & 0 & 1
\end{bmatrix}
\end{array}.
$$

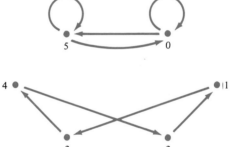

FIGURE 3.14 Digraph of the inverse of the relation depicted in Figure 3.13.

Finally, we can compose relations, just as we composed functions. Suppose that R is a relation from A to B and S is a relation from B to C. Their **composition**, written $S \circ R$ and read "S composed with R," is the relation from A to C given by

$$\forall x \in A : \forall z \in C : (x(S \circ R)z \leftrightarrow \exists y \in B : (xRy \wedge ySz)).$$

Compositions of a relation R on a set A with itself are written with exponents. Thus $R^2 = R \circ R$, $R^3 = R^2 \circ R$, $R^4 = R^3 \circ R$, and so on. (For completeness, we set $R^1 = R$, and $R^0 = \Delta_A$.)

It is not hard to see that this definition is consistent with our previous definition of the composition of functions in case the relations R and S are actually functions.

(Indeed, that is why we have written the composition of relations in what some authors consider the "wrong" order; they write "$R \circ S$" for what we have called $S \circ R$.) The schematic that we used to visualize compositions of functions (Figure 3.10) applies to relations as well, with f and g replaced by R and S, respectively. We need to think of $f(x)$ in that picture as being any element y of B related to x, and to think of $g(f(x))$ as being any element of C related to y.

EXAMPLE 14. Suppose that R is the "is a parent of" relation on the set of all people (living or dead): xRy if and only if x is a (genetic) parent of y. Note that for every y there are exactly two distinct values of x such that xRy, namely y's father and y's mother. The relation $R \circ R$ is again a relation on the set of all people, and by definition $x(R \circ R)z$ if and only if there is some y such that x is a parent of y and y is a parent of z. Clearly this condition is equivalent to the statement that x is a grandparent of z, so $R \circ R$ is the "is a grandparent of" relation.

Furthermore, R^{-1} is the "is a child of" relation. Let us see what $R \circ R^{-1}$ and $R^{-1} \circ R$ are. Suppose that $x(R \circ R^{-1})z$. Then by the definition of composition, there is some y such that $xR^{-1}y$ and yRz. In this case that means that x is a child of y, who is a parent of z. Unless $x = z$, we conclude that x and z are siblings—or at least half-siblings—since they have the common parent y. Thus $R \circ R^{-1}$ is the "is equal to or is a sibling or half-sibling of" relation. On the other hand, suppose that $x(R^{-1} \circ R)z$. Then by the definition of composition, there is some y such that xRy and $yR^{-1}z$. In this case that means that x is a parent of y, who is a child of z. Thus x and z are either the same person, or else they are the two parents of some person. We can phrase this by saying that $R^{-1} \circ R$ is the relation "is equal to or has had a child with." ◆

Interpreting composition in terms of digraphs or matrices is more involved, and we will only sketch the procedures. Suppose that we have relations R and S on a set A represented as digraphs on the same set of points (representing the elements of A). Imagine the arrows for R colored red and the arrows for S colored silver. To draw the digraph of $S \circ R$, we need to look for all paths consisting of a red arrow—say from x to y—followed by a silver arrow—from y to, say, z. For each such path we draw a black arrow from x to z. The resulting black picture is the digraph of $S \circ R$. Next suppose that we have matrices $M_R = (u_{ij})$ and $M_S = (v_{jk})$ representing the relations R from A to B and S from B to C. Then the matrix representing $S \circ R$ is given by $M_{S \circ R} = (w_{ik})$, the **Boolean matrix product** of M_R and M_S, whose (i, k)th entry w_{ik} is defined to be 1 if u_{ij} and v_{jk} are both 1 for some j, and 0 otherwise. (This operation is related to matrix multiplication, which we discuss in Section 4.5.)

Operations on relations, particularly union, intersection, and composition, as well as other operations on n-ary relations that we have not considered, play a central role in computer handling of large data bases.

DESCRIBING RELATIONS ON A SET

More often than not, the relations that we encounter have identical domain and codomain. There are certain general properties that such relations may or may not possess, and because of their importance and applicability, we need to discuss these properties in some detail.

Let R be a relation on a set A. We say that R is **reflexive** if and only if every element of A is related to itself:

$$R \text{ is reflexive} \quad \longleftrightarrow \quad \forall x \colon xRx.$$

We say that R is **symmetric** if and only if, whenever x is related to y, it also happens that y is related to x:

$$R \text{ is symmetric} \quad \longleftrightarrow \quad \forall x \colon \forall y \colon (xRy \to yRx).$$

We say that R is **antisymmetric** if and only if, whenever x is related to a *different* element y, it never happens that y is also related to x:

$$R \text{ is antisymmetric} \quad \longleftrightarrow \quad \forall x \colon \forall y \colon ((xRy \wedge x \neq y) \to y \not{R} x).$$

Equivalently, R is antisymmetric if and only if, whenever x is related to y and also y is related to x, it follows that x and y are the same, that is, $\forall x \colon \forall y \colon ((xRy \wedge yRx) \to x = y)$. We say that R is **transitive** if and only if, whenever x is related to y and y is related to z, it also happens that x is related to z:

$$R \text{ is transitive} \quad \longleftrightarrow \quad \forall x \colon \forall y \colon \forall z \colon ((xRy \wedge yRz) \to xRz).$$

We can interpret these definitions pictorially. A reflexive relation is one in whose digraph there is a loop at *every* point in the domain. In the digraph of a symmetric relation, between every pair of distinct points in the domain, either there are arrows going both ways or else there is no arrow in either direction. In the digraph of an antisymmetric relation, between every pair of distinct points in the domain, there is either no arrow in either direction or only one arrow; there cannot be arrows in both directions. Finally, in the digraph of a transitive relation, whenever we can get from a point x to a point z by following an arrow from x to some y and then another arrow from that y to z, there is already an arrow from x to z. The reader is asked to interpret the notions of reflexivity, symmetry, and antisymmetry in terms of matrices representing relations (Exercise 29).

Before we consider several examples, we should point out a common source of error. There is nothing in the definition of transitivity preventing x and z from being the same element. Thus if R is transitive and xRy and yRx, it follows that xRx (and yRy as well).

EXAMPLE 15. Determine which of the properties reflexivity, symmetry, antisymmetry, and transitivity apply to each of the following relations.

    (a) The less than relation on **R**.
    (b) The less than or equal to relation on **R**.
    (c) The relation $\{(1,1),(1,2),(2,1)\}$ on the set $\{1,2,3\}$.
    (d) The empty relation (i.e., $\emptyset$) on any nonempty set A.
    (e) The relation $A \times A$ on any nonempty set A.
    (f) The relation U on the set P given in Example 8.
    (g) The relation depicted in Figure 3.13.
    (h) The "is married to" relation on the set of all people.

Solution. We analyze these relations one at a time. Since the definitions of all these properties have universal quantifiers, in order to verify that a relation has one of these properties we need to argue that a certain condition holds *for all* elements in the set on which the function is defined; but to argue that a relation lacks one of these properties, it is enough to find one instance in which the condition fails.

    (a) The less than relation on **R** is not reflexive, since it is not true that $x < x$ for all $x \in \mathbf{R}$. For instance, $25.3 \not< 25.3$. Even more, it is the case that $\forall x \in \mathbf{R}\colon x \not< x$. A relation that is not reflexive in this very strong sense is called **irreflexive**. The less than relation is not symmetric either, since $2 < 3$ does not imply that $3 < 2$. It is antisymmetric, however, because if $x < y$, then necessarily, $y \not< x$ (the additional hypothesis that $x \neq y$ is not even needed). Finally, the fact that this relation is transitive is familiar to anyone who has studied algebra: If $x < y$ and $y < z$, then $x < z$.

    (b) The less than or equal to relation on **R** is reflexive, since it is certainly true that $x \leq x$ for all $x \in \mathbf{R}$. Again it is not symmetric: If the relation holds between x and y, there is no guarantee that it holds between y and x (in fact, it usually will not). This relation is a good example of one that is antisymmetric: If $x \leq y$ and $y \leq x$, we can conclude that $x = y$. Finally, $\leq$ is transitive just as $<$ is transitive.

    (c) This relation consists of three ordered pairs in $\{1,2,3\} \times \{1,2,3\}$: 1 is related to 1, 1 is related to 2, and 2 is related to 1. Since the pairs $(2,2)$ and $(3,3)$ are missing, the relation is not reflexive (it is not irreflexive either, since $(1,1)$ is in the relation). This relation is symmetric, the key point being that since $(2,1)$ is in the relation, so is $(1,2)$, and vice versa. The pair $(1,3)$, for example, is not in the relation, but that is acceptable since $(3,1)$ is not, either. This relation is not antisymmetric, however, because 1 is related to 2 and 2 is related to 1, even though $1 \neq 2$. To see that the relation is not transitive, let $x = 2$, $y = 1$, and $z = 2$. Then x is related to y, and y is related to z, but x is not related to z.

    (d) The empty relation on a nonempty set is not reflexive, since by assumption the domain A has at least one element a, and (a,a) is not in the empty set. The other three properties all hold, however—vacuously. Each of the other properties is a universally quantified implication; since the hypothesis of the implication never holds for the empty relation, the implication is always true. Thus the empty set, viewed as a relation on any set A, is symmetric, antisymmetric, and transitive. (If $A = \emptyset$ then reflexivity holds, vacuously, as well.)

(e) At the other extreme is the relation that always holds on the nonempty set A. This relation is clearly reflexive, symmetric, and transitive. On the other hand, it is not antisymmetric unless A has only one element (in which case the relation is vacuously antisymmetric).

(f) The relation between points in the plane that holds between points 1 inch apart is not reflexive, since no point is 1 inch away from itself. The symmetry of the defining statement of this relation makes it clear that the relation is symmetric. This relation is not antisymmetric, since there are certainly pairs of distinct points 1 inch apart. Finally, the relation is not transitive either, since if we take, for instance, the points $(0,0)$, $(1,0)$, and $(1,1)$ (in the usual Cartesian coordinate system), the first is 1 unit from the second, the second is 1 unit from the third, but the first is not 1 unit from the third.

(g) The relation depicted in Figure 3.13 is not reflexive, since, for instance, 3 is not related to 3. It is not symmetric, since, for instance, 3 is related to 1, but 1 is not related to 3. It is not antisymmetric either, since 0 is related to 5 and 5 is related to 0, even though $0 \neq 5$. Finally, this relation is not transitive, since, for instance, it contains the pair $(2,4)$ and the pair $(4,3)$, but not the pair $(2,3)$.

(h) No one is married to himself, so the "is married to" relation is not reflexive; it is, in fact, irreflexive. If x is married to y, then y is married to x, as well, so the relation is symmetric. The relation is not antisymmetric, since there are (billions of) cases of pairs of distinct people married to each other. Finally, the relation is not transitive, since there are, again, billions of cases in which x is married to y, y is married to z, but x is not married to z. ◆

CLOSURES OF RELATIONS (*OPTIONAL*)

Sometimes we would like to turn a relation that does not have a property (such as transitivity) into one that does. Obviously, we need to modify the relation to do so, and the usual procedure is to add just enough new pairs to the relation to achieve the desired result. Our next theorem says that we can accomplish this, and the discussion afterward indicates a method for doing so.

THEOREM 1. *Let R be a relation on a nonempty set A. The intersection of all reflexive relations on A that have R as a subset is a reflexive relation having R as a subset, called the* **reflexive closure** *of R. The analogous statement holds if the word "reflexive" is replaced by "symmetric," by "transitive," or by any conjunction of two or all three of these properties.*

Proof. We observed in Example 15e that the relation $A \times A$ has all three of the properties in question. Thus there is at least one reflexive, symmetric, and transitive relation on A containing R as a subset. Clearly, the intersection of any collection of sets having

R as a subset has R as a subset. Thus we need only prove that the intersection of reflexive relations is reflexive, the intersection of symmetric relations is symmetric, and the intersection of transitive relations is transitive.

All three parts are similar, so we will do only the last of these, leaving the other two parts for the reader (Exercise 19). Suppose that $\{R_i\}_{i \in I}$ is any collection of transitive relations on A (I is the index set for this collection). We need to show that $R' = \bigcap_{i \in I} R_i$ is transitive. To do so, suppose that $xR'y$ and $yR'z$. We need to show that $xR'z$. Since R' is the intersection of all the relations R_i, we know that xR_iy and yR_iz for all $i \in I$. By assumption, each of the relations R_i is transitive, so it follows that xR_iz for all $i \in I$. Therefore, by definition, $xR'z$, as desired. ∎

The closure of R (with respect to a property P) is the *smallest* relation with R as a subset that has property P. The theorem tells us what "smallest" means in this context and guarantees that the closure actually exists if P is any combination of reflexivity, symmetry, and transitivity. The proof of the theorem gives no clue, however, as to how we can efficiently form closures in practice. It does not tell us what pairs to add to achieve reflexivity, symmetry, or transitivity. The theorem is a "top-down" characterization of closures. If we want to compute a closure, we need a "bottom-up" approach. (We will encounter this important theme of top-down versus bottom-up again at other points in this book.) The following procedures effectively fill that role.

To form the reflexive closure of a relation R on a set A, we certainly need to add to R all pairs of the form (x, x) for $x \in A$. On the other hand, if we do so, we clearly have a reflexive relation with R as a subset. Thus the set $R \cup \Delta_A$ is the smallest reflexive relation with R as a subset and so must be the reflexive closure of R.

To form the symmetric closure of a relation R on a set A, we certainly need to add to R all pairs of the form (y, x) whenever $(x, y) \in R$. In other words, we need to include at least $R \cup R^{-1}$. On the other hand, if we do so, we clearly have a symmetric relation with R as a subset. Thus this set is the smallest symmetric relation with R as a subset and so must be the symmetric closure of R.

To form the transitive closure is more difficult. We certainly need to add to R all pairs of the form (x, z) whenever xRy and yRz for some y. Since such pairs are precisely the elements of $R \circ R = R^2$, we need to include at least $R \cup R^2$. After doing so, however, we may not have a transitive relation, since the added pairs may themselves create new instances of $xRy \wedge yRz$. Thus we need to repeat the process again and again, adding elements of R^3, R^4, and so on. The union of all of these relations must be included. On the other hand, this union is transitive. Therefore, the transitive closure of R is $\bigcup_{i=1}^{\infty} R^i$.

Finally, we need to be a little careful about the order in which we form closures by these procedures if we want a relation that is closed in more than one respect. The transitive closure of a symmetric relation must still be symmetric, but the symmetric closure of a transitive relation need not be transitive (see Exercise 28).

Thus if we want to form the symmetric, transitive closure of R, we must first form the symmetric closure of R (by the procedure outlined above) and then take the transitive closure of that. The relation we end up with is $\bigcup_{i=1}^{\infty}(R \cup R^{-1})^i$. It turns out not to matter at what point we throw in the diagonal elements (x, x) if we want the closure to be reflexive as well as symmetric and/or transitive.

EXAMPLE 16. Let R be the relation defined on the set $\mathbf{Z}$ of integers by xRy if and only if $y - x = 1$. In other words, each integer x is related to its successor ($y = x + 1$) but to no other integers. Compute all the closures of R with respect to nonempty subsets of $\{\text{reflexivity}, \text{symmetry}, \text{transitivity}\}$.

Solution. Clearly, R is neither reflexive, symmetric, nor transitive. To form the reflexive closure of R, we need to make each integer related to itself, as well as to its successor. The reflexive closure of R is thus the relation that holds between x and y when either $y - x = 1$ or $x = y$. To form the symmetric closure, we need to make each integer related to its predecessor as well as to its successor; thus the symmetric closure of R is the relation that holds between x and y when $|y - x| = 1$. The reflexive, symmetric closure of R is the relation that holds between x and y when $|y - x| \le 1$.

To compute the transitive closure of R, we reason as follows. First R^2 is the relation that holds between x and z when $z - x = 2$ (because xR^2z means that xRy and yRz for some y, whence we have $y - x = 1 \wedge z - y = 1$ and by adding obtain the characterization stated). Similarly, for each positive integer i, R^i is the relation that holds between x and y when $y - x = i$. Therefore, the transitive closure of R, which we saw is given by $\bigcup_{i=1}^{\infty} R^i$, is the relation that holds between x and y whenever y exceeds x by some positive integer value. Since the domain is $\mathbf{Z}$, this is the same as saying that x is related to y precisely when $x < y$. In other words, the transitive closure of R is just the "is less than" relation.

The reflexive, transitive closure of R is clearly the "is less than or equal to" relation, since we obtain it from the transitive closure by including all the pairs (x, x). A little thought will convince one that the symmetric, transitive closure and the reflexive, symmetric, transitive closure of R are both the relation that always holds. ◆

Closures can be visualized in terms of the pictures (digraphs) of relations. To obtain the reflexive closure, we need to place a loop at each point in the domain (where there is not already a loop). To obtain the symmetric closure, we need to make every "one-way" arrow "go both ways": Whenever there is an arrow from x to y, we need to add an arrow from y to x (if there is not one already there). To obtain the transitive closure, we need to add an arrow from x to y whenever it is possible to move from x to y by traveling along the arrows (in the indicated directions), no matter how long a trip is required. Two closures of the relation depicted in Figure 3.13 are shown in Figure 3.15: the symmetric closure on the left and the transitive closure on the right.

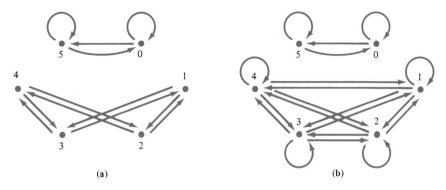

FIGURE 3.15 Symmetric closure (a) and transitive closure (b) of the relation depicted in Figure 3.13.

The reader should apply the rules given above to the digraph of the relation in Example 16 (or, rather, a portion of this infinite digraph) to see how the various closures discussed there can be obtained visually (see Exercise 30). Exercise 32 explores the question of how to obtain closures from matrix representations of relations (see also Exercise 33).

SUMMARY OF DEFINITIONS

antisymmetric relation on the set A: a relation R such that $\forall x \in A\colon \forall y \in A\colon [(xRy \wedge yRx) \to x = y]$ (e.g., the relation "divides" on $\mathbf{N}$).

Boolean matrix: a matrix all of whose entries are 0 or 1 (e.g., $\begin{bmatrix} 0 & 1 \\ 0 & 0 \end{bmatrix}$).

Boolean matrix product of matrices $M = (m_{ij})$ and $N = (n_{jk})$: the matrix whose (i, k)th entry is 1 if m_{ij} and n_{jk} are both 1 for some j, and 0 otherwise (e.g., the Boolean matrix product of $\begin{bmatrix} 0 & 1 \\ 1 & 1 \end{bmatrix}$ and $\begin{bmatrix} 0 & 1 \\ 0 & 1 \end{bmatrix}$ is $\begin{bmatrix} 0 & 1 \\ 0 & 1 \end{bmatrix}$).

closure of the relation R on the set A with respect to the property P: the intersection of all relations on A that have property P and contain R as a subset (e.g., if R is $<$ on $\mathbf{N}$, then the reflexive closure of R (the closure of R with respect to reflexivity) is $\leq$; the symmetric closure of R (the closure of R with respect to symmetry) is $\neq$; and the reflexive, symmetric closure of R (the closure of R with respect to both reflexivity and symmetry) is the relation that always holds between elements of $\mathbf{N}$).

codomain of the relation R from the set A to the set B: the set B.

complement of the relation R from A to B: the relation $\overline{R} = \{ (a, b) \in A \times B \mid a \not\!R b \}$ (e.g., the "is not a child of" relation is the complement of the "is a child of" relation).

composition of the relation R from A to B and the relation S from B to C: the relation $S \circ R$ from A to C in which a is related to c if and only if there is some b in B for which aRb and bSc (e.g., if R and S are each the "is acquainted with" relation on the set of all people, then $S \circ R$ (also written R^2 in this case) is the "have a common acquaintance" relation).

diagonal relation on the set A: the relation $\Delta_A = \{(a, a) \mid a \in A\}$.

digraph of the relation R on the set A: a picture consisting of points representing elements of A, together with a directed line or curve (arrow) from x to y whenever xRy (e.g., the picture consisting of points a and b and an arrow from each to the other is the digraph of the relation $\{(a, b), (b, a)\}$ on the set $\{a, b\}$).

divides: The nonzero integer x divides the integer y if and only if $y = qx$ for some integer q; equivalently, if and only if the remainder when y is divided by x is 0; equivalently, if and only if the real number y/x is an integer; written $x \mid y$ (e.g., $9 \mid 123456789$, since $123456789 = 13717421 \cdot 9$).

domain of the relation R from the set A to the set B: the set A.

hold: A relation R holds between x and y if and only if xRy.

identity matrix I_n: the n by n matrix whose (i, j)th entry is 1 if $i = j$ and 0 if $i \neq j$ (e.g., $I_2 = \begin{bmatrix} 1 & 0 \\ 0 & 1 \end{bmatrix}$).

identity relation on the set A: the diagonal relation on A.

inverse of the relation R from A to B: the relation R^{-1} from B to A consisting of $\{(b, a) \mid aRb\}$ (e.g., the "is a parent of" relation is the inverse of the "is a child of" relation, since b is a parent of a if and only if a is a child of b).

irreflexive relation on the set A: a relation R such that $\forall x \in A : x \not{R} x$ (e.g., the relation "is taller than" on the set of all people).

loop in the digraph of the relation R: an arrow from a point a to itself, indicating that $(a, a) \in R$.

n-ary relation among sets $A_1,\ A_2,\ \dots,\ A_n$: a subset of $A_1 \times A_2 \times \cdots \times A_n$ (e.g., the 4-ary relation among points on a fixed circle (a, b, c, d) that holds if and only if, as you go around the circle clockwise starting at a, you encounter the points in the order a, b, c, d).

reflexive relation on the set A: a relation R such that $\forall x \in A : xRx$ (e.g., the relation "is at least as tall as" on the set of all people).

related under the relation R: x and y are related under R if and only if xRy (e.g., $\{1, 2\}$ is related to $\{1, 2, 3\}$ under the relation "is a subset of").

(binary) relation R from the set A to the set B: a subset of $A \times B$; we write aRb for $(a, b) \in R$ and $a \not{R} b$ for $(a, b) \notin R$ (e.g., the relation "is an owner of" from the set of all people to the set of all houses).

(binary) relation on the set A: a relation from A to A (e.g., the relation "is a subset of" on $\mathcal{P}(U)$ for some given universal set U).

symmetric relation on the set A: a relation R such that $\forall x \in A : \forall y \in A : (xRy \to yRx)$ (e.g., the relation "is married to" on the set of all people).

transitive relation on the set A: a relation R such that $\forall x \in A : \forall y \in A : \forall z \in A :$ $[(xRy \wedge yRz) \rightarrow xRz]$ (e.g., the relation "is taller than" on the set of all people).

transpose of the matrix $M = (m_{ij})$: the matrix whose (i,j)th entry is m_{ji} (e.g., the transpose of $\begin{bmatrix} 0 & 1 & 0 \\ 0 & 0 & 1 \end{bmatrix}$ is $\begin{bmatrix} 0 & 0 \\ 1 & 0 \\ 0 & 1 \end{bmatrix}$).

EXERCISES

1. List explicitly the following relations from $\{1,2,3\}$ to $\{1,2\}$ (i.e., write down R as a set of ordered pairs).
   (a) xRy if and only if $x \le y$.
   (b) xRy if and only if $x \mid y$.
   (c) xRy if and only if $y \mid x$.
   (d) xRy if and only if $x = y$.
   (e) xRy if and only if $x = y + 3$.
   (f) xRy if and only if $y \le 2$.

2. Write down the matrix representing each relation in Exercise 1.

3. For the following relations on the set $\mathbf{N}$ of natural numbers, list three pairs in the relation and two pairs not in the relation (or explain why there are not that many).
   (a) xRy if and only if $x^2 + y^2 > 100$.
   (b) xRy if and only if $x = 3$.
   (c) xRy if and only if $xy < 0$.

4. Describe these relations on $\{1,2,3,4\}$ with rules like those in Exercises 1 and 3.
   (a) $R = \{(1,2),(1,3),(1,4),(2,3),(2,4),(3,4)\}$.
   (b) $S = \{(1,2),(2,3),(3,4)\}$.
   (c) $T = \{(2,3)\}$.
   (d) $U = \emptyset$.
   (e) $V = \{1,2,3,4\} \times \{1,2,3,4\}$.

5. Draw the digraphs of the following relations on $\{1,2,3,4\}$.
   (a) $<$.
   (b) $\le$.
   (c) $\Delta_{\{1,2,3,4\}}$.

6. Give the matrix and the digraph for $\overline{R}$, if R is the relation on $\{1,2,3,4\}$ defined by xRy if and only if $5 \mid x + 2y$ (see Example 7 and Figure 3.13, where the domain is a little larger).

7. Suppose that C is the relation "is a child of" on the set of people (living or dead).

(a) Determine whether C is a function.

(b) Explain why the digraph for C has no loops.

(c) Express $C \circ C$ in English.

8. How many different relations are there on the set $\{1, 2, 3\}$?

9. Extend the domain and codomain of the relation given in Example 7 to the set $\{0, 1, 2, 3, 4, 5, 6, 7\}$.

(a) List the pairs in the relation.

(b) Draw the digraph of the relation.

(c) Write down the matrix representing the relation.

10. Suppose that M is the relation on the set of people (living or dead) defined by xMy if and only if x and y have ever been married to each other. Explain why $\forall x\colon \forall y\colon [(xMy \wedge yMz) \to x\not My z]$.

11. Let R be the relation from $\{1, 2, 3, 4\}$ to $\{a, b, c\}$ consisting of the pairs $\{(1, a), (1, c), (3, a), (3, b), (4, c)\}$. Let S be the relation from $\{a, b, c\}$ to $\{1, 2, 3, 4\}$ consisting of the pairs $\{(a, 3), (b, 1), (b, 2), (b, 4), (c, 2), (c, 4)\}$. Find the following relations, giving the answer as a set of ordered pairs.

(a) $R \circ S$.

(b) $S \circ R$.

(c) $\overline{R}$.

(d) R^{-1}.

(e) $R^{-1} \circ S^{-1}$.

(f) $S^{-1} \circ R^{-1}$.

(g) $(R \circ S)^{-1}$.

12. Write out the elements of the 3-ary relation on $\{1, 2, 3, 4, 5\}$ that holds for a 3-tuple (x, y, z) if and only if $x + y < z$.

13. Prove or find a counterexample to each of the following statements, where R and R' are relations from A to B, and S is a relation from B to C (and A, B, and C are arbitrary sets).

(a) $R \circ \Delta_A = R$.

(b) $\Delta_B \circ R = R$.

(c) $R \circ R^{-1} = \Delta_A$.

(d) $\overline{S \circ R} = \overline{S} \circ \overline{R}$.

(e) $S \circ (R \cup R') = (S \circ R) \cup (S \circ R')$.

(f) $S \circ (R \cap R') = (S \circ R) \cap (S \circ R')$.

14. Give two interesting examples of 5-ary relations, one of which has the property of complete symmetry (by which we mean that whether or not the relation holds on a 5-tuple does not depend on the order of the elements in the 5-tuple) and one of which does not have this property.

15. What should a 1-ary relation on a set A be?

16. Show that the two definitions given for antisymmetry are actually equivalent, by showing that the proposition $(xRy \wedge x \neq y) \rightarrow y\not\!Rx$ is logically equivalent to the proposition $(xRy \wedge yRx) \rightarrow x = y$.

17. Let R be a relation on a nonempty set A. Prove or find a counterexample to each of the following statements.
  (a)   If R is reflexive, then $\overline{R}$ is not reflexive.
  (b)   If R is symmetric, then $\overline{R}$ is symmetric.
  (c)   If R is antisymmetric, then $\overline{R}$ is antisymmetric.
  (d)   If R is transitive, then $\overline{R}$ is not transitive.

18. Give a logical definition (in the form "xRy if and only if ...") for the following relations, where $\{R_i\}_{i \in I}$ is a collection of relations from A to B.
  (a) $R = \bigcup_{i \in I} R_i$.
  (b) $R = \bigcap_{i \in I} R_i$.

19. Prove (as is required in the proof of Theorem 1) that
  (a)   The intersection of any collection of reflexive relations on a set A is reflexive.
  (b)   The intersection of any collection of symmetric relations on a set A is symmetric.

20. Suppose that R and S are relations on a set A. Prove or give a counterexample to each of the following assertions.
  (a)   If R and S are symmetric, then $R \cup S$ is symmetric.
  (b)   If R and S are transitive, then $R \cup S$ is transitive.
  (c)   If $R \cup S$ is reflexive, then either R is reflexive or S is reflexive.
  (d)   If $R \cap S$ is reflexive, then both R and S are reflexive.
  (e)   If R and S are antisymmetric, then $R \cap S$ is antisymmetric.

21. Determine whether each of the following relations is reflexive, symmetric, antisymmetric, transitive.
  (a)   The relation J in Example 4.
  (b)   The relation $\{(0, 1)\}$ on the set $\{0, 1\}$.
  (c)   The relation on children defined by xRy if and only if x's father is taller than y's father and x's mother is taller than y's mother.
  (d)   The relation on children defined by xRy if and only if x's father is at least as tall as y's father and x's mother is at least as tall as y's mother.
  (e)   The relation on children defined by xRy if and only if x's father is taller than y's father or x's mother is taller than y's mother (recall that "or" is inclusive).
  (f)   The relation $\not\subseteq$ on the subsets of $\mathbf{N}$.
  (g)   The relation R on the subsets of $\mathbf{N}$ given by ARB if and only if $A = \overline{B}$.
  (h)   The relation R on $\mathbf{N}$ given by xRy if and only if $y > x + 1$.
  (i)   The relation R on $\mathbf{N}$ given by xRy if and only if $y < x + 1$.

22. Find an example of a relation on the set $\{a, b, c\}$ satisfying each of the following conditions. Express your answer by drawing the digraph.
    (a) Reflexive, symmetric, and transitive.
    (b) Reflexive and symmetric, but not transitive.
    (c) Symmetric and transitive, but not reflexive.
    (d) Reflexive and transitive, but not symmetric.
    (e) Reflexive, but not symmetric or transitive.
    (f) Symmetric, but not reflexive or transitive.
    (g) Transitive, but not reflexive or symmetric.
    (h) Neither reflexive, symmetric, nor transitive.

23. Sketch enough of the digraph of the following relation to give a clear picture of the pattern. The domain and codomain of the relation are both $\mathbf{Z} \times \mathbf{Z}$ (often called the set of **lattice points** in the plane). The relation holds between (a, b) and (c, d) if and only if $(a = c \wedge d = b + 1) \vee (b = d \wedge c = a + 1)$.

24. Find an example of a relation on the set $\{a, b, c\}$ satisfying each of the following conditions. Express your answer by drawing the digraph.
    (a) Symmetric and antisymmetric.
    (b) Symmetric but not antisymmetric.
    (c) Antisymmetric but not symmetric.
    (d) Neither symmetric nor antisymmetric.

25. Suppose that a relation R on a set A is both symmetric and antisymmetric.
    (a) Prove that R is transitive.
    (b) Give a complete and simple characterization of what R can be.

26. Find the indicated closures of the "is less than" relation on $\mathbf{R}$.
    (a) The reflexive closure.
    (b) The symmetric closure.
    (c) The transitive closure.
    (d) The symmetric, transitive closure.

27. Find the indicated closures, expressing your answer as succinctly as possible.
    (a) The transitive closure of the relation J in Example 4.
    (b) The transitive closure of the relation U in Example 8.
    (c) The transitive closure of the "is a parent of" relation on (living or dead) people.
    (d) The symmetric, transitive closure of the "is a parent of" relation on (living or dead) people.

28. Prove the following statements about closures.
    (a) The transitive closure of a symmetric relation is symmetric.
    (b) The symmetric closure of a transitive relation need not be transitive.

29. Explain how to determine whether a relation has each of the following properties by looking at its matrix.

(a)   Reflexivity.
(b)   Irreflexivity.
(c)   Symmetry.
(d)   Antisymmetry.

30. Sketch the portion of the digraph of the relation given in Example 16 that includes the integers from 1 to 5, inclusive, and all the arrows involving pairs of these points. Then sketch the corresponding portions of digraphs of the following closures of this relation.
    (a)   The reflexive closure.
    (b)   The symmetric closure.
    (c)   The transitive closure.
    (d)   The reflexive, symmetric closure.
    (e)   The reflexive, transitive closure.
    (f)   The symmetric, transitive closure.

Challenging Exercises

31. Prove the following characterizations of relation properties.
    (a)   A relation R on a set A is symmetric if and only if $R = R^{-1}$.
    (b)   A relation R on a set A is transitive if and only if $R^2 \subseteq R$.
    (c)   A relation R on a set A is reflexive if and only if $\Delta_A \subseteq R$.
    (d)   A relation R on a set A is antisymmetric if and only if $R \cap R^{-1} \subseteq \Delta_A$.

32. Explain how to find the matrix of the following closures of a relation, given the matrix for the relation.
    (a)   Reflexive closure.
    (b)   Symmetric closure.
    (c)   Transitive closure. [*Hint*: Show that $\bigcup_{i=1}^{\infty} R^i = \bigcup_{i=1}^{n} R^i$ if R is a relation on a finite set with cardinality n.]

Exploratory Exercises

33. Find out about computational procedures for finding transitive closures of relations (such as Warshall's algorithm) by consulting H. F. Smith [278].

34. Find out about relational data bases (an important application of relations to computer science) by consulting Date [55].

35. Find out about Paul Erdős and Erdős numbers by consulting Albers and Alexanderson [4] and P. Hoffman [151]. Explain what Erdős numbers have to do with this section.

SECTION 3.4
ORDER RELATIONS AND
EQUIVALENCE RELATIONS

With silver bells and cockleshells,
And pretty maids all in a row.

—Mother Goose

In the preceding section we studied various properties that a relation may or may not possess. Relations satisfying certain combinations of those properties arise repeatedly in mathematics. In this section we study the two most important such combinations.

PARTIAL ORDERS

Many relations in mathematics can be viewed as the imposition of an ordering on a set—either totally or partially. The real numbers, for example, are ordered by the usual $<$ relation. The words in a dictionary are ordered alphabetically. Sets are partially ordered by the $\subseteq$ relation. We now study order relations in some detail.

A relation R on a set A is called a **partial order** on A if and only if R is reflexive, transitive, and antisymmetric; in this case we say that the pair (A, R) (i.e., the set A together with the partial order R) is a **partially ordered set**, or **poset**. If R has the additional property that every two elements of A are related under R in one direction or the other (i.e., $\forall x \colon \forall y \colon (xRy \lor yRx)$), then we say that R is a **total order** or **linear order** on A, and that (A, R) is a **totally ordered set**.

The "is less than or equal to" relation on the set of real numbers is the prototype for a partial order. We saw in Example 15b in Section 3.3 that $\leq$ is reflexive, transitive, and antisymmetric, so $\leq$ is a partial order and the real numbers are a poset under the "is less than or equal to" relation. Furthermore, since every two real numbers are **comparable** under $\leq$ (i.e., for any two real numbers x and y, either $x \leq y$ or $y \leq x$), we conclude that $\leq$ is a total order, and the set of real numbers is totally ordered.

We often denote a partial order with the suggestive symbol $\preceq$ (which can be read "**precedes** or is equal to"). Furthermore, if $x \preceq y$ and $x \neq y$, we write $x \prec y$ (read "x precedes y" or "y **succeeds** x"). Thus $x \preceq y \leftrightarrow (x \prec y \lor x = y)$.

EXAMPLE 1. The following are all partially ordered sets.

(a) The power set of a universal set U under the relation "is a subset of" is a poset. It follows immediately from the definition that $\subseteq$ is reflexive; the reader proved that $\subseteq$ is transitive in Exercise 17 of Section 2.1; and the observation that $\subseteq$ is antisymmetric was made just before Theorem 1 in Section 2.1. This poset is not a totally ordered set, however, since two sets need not be comparable. For instance, if $U = \{1, 2, 3, 4, 5\}$, then the sets $\{1, 2\}$ and $\{2, 3, 4\}$ are incomparable, since $\{1, 2\} \nsubseteq \{2, 3, 4\}$ and $\{2, 3, 4\} \nsubseteq \{1, 2\}$.

(b) The set of integers under the greater than or equal to relation is a poset (in fact, a totally ordered set) just as it is under the less than or equal to relation. Note that neither $<$ nor $>$ are partial orders, since they are not reflexive.

(c) The set of functions from **R** to **R** is a partially ordered set under the relation defined by $f \leq g$ if and only if $\forall x \in \mathbf{R} \colon f(x) \leq g(x)$. Reflexivity and transitivity are straightforward. Antisymmetry follows from the antisymmetry of $\leq$ on real numbers together with the observation that two functions are equal if they have the same values at all points in the domain. Note that this relation is not a total order, since, for instance, the functions given by the rules $f(x) = x + 10$ and $g(x) = x^2$ are incomparable: Each exceeds the other at some point in the domain ($f(2) > g(2)$, but $g(7) > f(7)$).

(d) The equality relation gives a partial order on any set, albeit a rather trivial one. In this case, every pair of distinct elements are incomparable—the opposite extreme from the situation with a total order.

(e) Suppose that a project consists of a number of subprojects. For example, preparing dinner consists of such subprojects as roasting the meat, peeling the potatoes, and so on. Some of these subprojects cannot be started before others are completed; for example, one cannot carve the turkey before it is roasted. The set of all subprojects with the relation "is equal to or is a prerequisite for" forms a poset, as long as we assume that the project can, in fact, be completed. Reflexivity is automatic. If this relation were not antisymmetric—if each of x and y were a prerequisite for the other—then the project could never be completed, since neither x nor y could ever be started. Transitivity is clear: If z must wait for y, and y must wait for x, then certainly x is prerequisite for z. We will have more to say about this poset in Section 10.1, when we look at project scheduling. ◆

Just as we represent sets, functions, and relations with diagrams to make them more comprehensible to human beings, we can represent finite partially ordered sets conveniently and informatively with a picture called a **Hasse diagram**. Suppose that $(A, \preceq)$ is a poset with A finite. We draw a Hasse diagram for this poset by representing each element of A by a distinct point so that whenever $x \prec y$, the point representing y is situated higher than the point representing x. Furthermore, if $x \prec y$ but there is no $z \in A$ strictly between x and y (i.e., no z such that $x \prec z \prec y$), then we connect the points representing x and y by a straight line segment; in this case we say that y is an **immediate successor** of x (and that x is an **immediate predecessor** of y).

EXAMPLE 2.   Figure 3.16 shows a Hasse diagram for the "divides" relation ("$|$") on the set $\{1, 2, 3, 4, 5, 6, 7, 8, 9, 10, 11, 12\}$ (see Example 5 in Section 3.3). The reader can check (Exercise 14) that $|$ is indeed a partial order. (Actually, we can (and will) say "the" Hasse diagram, because any two Hasse diagrams for this poset have exactly the same pairs of points joined by line segments; only the placement of the points in the picture can differ. In the language of Section 2.2, any two Hasse diagrams for the same poset are isomorphic.) Since $3 \mid 12$, for instance, we have put 12 above 3. On the other hand, 12 is not an immediate successor of 3, since $3 \mid 6$ and $6 \mid 12$, so we have not drawn a line between 3 and 12. There is a line between 4 and 12, however, since there is no positive integer z other than 4 and 12 such that $4 \mid z$ and $z \mid 12$. The numbers 9 and 11 are incomparable; the fact that 9 happens to lie above 11 is immaterial (note that the definition did not specify that y was to lie above x if *and only if* $x \prec y$). ◆

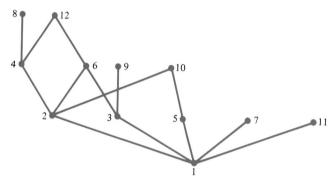

FIGURE 3.16 Hasse diagram for the poset $(\{1, 2, 3, 4, 5, 6, 7, 8, 9, 10, 11, 12\}, \mid)$.

It is easy to determine from the Hasse diagram of a poset $(A, \preceq)$, given distinct elements x and y in A, whether or not $x \prec y$: Because of the transitivity of a partial order, $x \prec y$ if and only if we can find a rising path in the picture from x to y. In Figure 3.16, for instance, the rising path from 2 to 6 to 12 tells us that $2 \prec 12$ (which here means that $2 \mid 12$). There is no rising path from 12 to 2 or from 4 to 6, so $12 \nprec 2$ and $4 \nprec 6$.

Often an order on a set is determined by an order on another set. In particular, a total order on a set A **induces** total orders on the Cartesian product $A \times A$ (or more generally on the n-fold Cartesian product A^n), and on the set A^* of strings over A. The definitions are a little tricky to state.

Let $(A, \preceq)$ be a totally ordered set. The **lexicographic** order on $A \times A$ induced by $\preceq$ is the total order (which we will also denote by $\preceq$) given by $(a_1, a_2) \prec (b_1, b_2)$ if and only if $a_1 \prec b_1$ or both $a_1 = b_1$ and $a_2 \prec b_2$. Similarly, the lexicographic order on A^n is given by

$$(a_1, a_2, \ldots, a_n) \prec (b_1, b_2, \ldots, b_n) \longleftrightarrow \exists i \geq 1 \colon ((\forall k < i \colon a_k = b_k) \wedge a_i \prec b_i).$$

EXAMPLE 3. The lexicographic order on $\{1,2\}^3$ induced by $\leq$ is $(1,1,1) \prec (1,1,2) \prec (1,2,1) \prec (1,2,2) \prec (2,1,1) \prec (2,1,2) \prec (2,2,1) \prec (2,2,2)$. ◆

Lexicographic order on A^* is similar, but the definition is complicated by the fact that strings need not have the same length. This is the order usually used in a dictionary, where the underlying total order on A (the set of letters in the alphabet) is the usual one; thus the alphabetical order on A induces an alphabetical order on A^*. To be precise, the **lexicographic order** on the set A^* of strings over A is the total order in which the string $a_1 a_2 \ldots a_n$ precedes the string $b_1 b_2 \ldots b_m$ if and only if

$$(n < m \wedge \forall i \leq n \colon a_i = b_i) \vee \exists i \geq 1 \colon ((\forall k < i \colon a_k = b_k) \wedge a_i \prec b_i).$$

This definition says that one string precedes a second string in lexicographic order either if the first is a proper initial substring of the second, or if at the first position (reading from the left) at which they differ, the symbol in the first string precedes the symbol in the second.

EXAMPLE 4. The lexicographic order on the set of strings over the set $\{a, b, c\}$ (in alphabetical order) cannot be conveniently written down in the same way as we were able to write down the lexicographic order on the Cartesian product in Example 3, since $\{a, b, c\}^*$ is infinite, and the order does not look at all similar to the usual order on the natural numbers. However, we do have, for example, the following inequalities: λ (the empty string) $\prec a \prec aa \prec aaaabac \prec aaaabba \prec aaaac \prec bc \prec cab \prec cc$. (Lexicographic order on words in a dictionary is complicated somewhat by the fact that there are at least three symbols to contend with besides the 26 letters of the alphabet—the apostrophe, the hyphen, and the blank in compound entries.) ◆

The preceding example shows that total orders can be quite complex. Entire books and research journals are devoted to orders.

Often, one needs to find extreme elements in partially ordered sets. Suppose that $(A, \preceq)$ is a poset. We say that $m \in A$ is the **maximum** element in the set if $\forall x \in A \colon x \preceq m$. In this case we also say that m **dominates** every element in A. Posets may or may not have a maximum element. For example, there is no maximum element in the totally ordered set $(\mathbf{N}, \leq)$, since each n fails to dominate $n + 1$. Similarly, there is no maximum element in the partially ordered set $(\{1, 2, 3\}, =)$, since no element of this set dominates all the others. On the other hand, 24 is the maximum element when $\{2, 4, 6, 8, 12, 24\}$ is ordered by $\leq$ and also when it is ordered by $|$. We define the **minimum** element in a poset in the analogous way (see Exercise 9).

A Hasse diagram provides a good way to recognize maximum and minimum elements in a finite poset. The maximum element (if it exists) is the point "at the top," in the sense that there is a falling path from this point down to every other point in the picture. There is no maximum element in the poset depicted in Figure 3.16. Similarly, the minimum element (if it exists) is the point "at the bottom," in the sense that there is a rising path from this point up to every other point in the picture. The number 1 is the minimum element in the poset depicted in Figure 3.16; there is a rising path from 1 to every other numbered point.

We have been referring to *the* maximum element in a poset (if one exists). Let us justify this grammatical indulgence.

THEOREM 1.   *There is at most one maximum element in a poset.*

Proof.   Let $(A, \preceq)$ be a given poset, and suppose that m_1 and m_2 are maximum elements. We need to show that $m_1 = m_2$. Since m_1 is a maximum element, it must dominate m_2, so we have $m_2 \preceq m_1$. Similarly, since m_2 is a maximum element, it must dominate m_1, so we have $m_1 \preceq m_2$. It then follows by antisymmetry that $m_1 = m_2$, as desired.   ■

There is a second notion of extreme element in a partially ordered set, which differs from the notions we have just defined if the set is not totally ordered. Let $(A, \preceq)$ be a poset. An element $m \in A$ is called **maximal** if and only if no other element dominates it, that is, $\forall x \in A \colon (m \preceq x \rightarrow x = m)$. Similarly, a **minimal** element is one that dominates no other element. In terms of the Hasse diagram, a maximal element is one from which there are no lines leading upward, and a minimal element is one from which there are no lines leading downward. For example, 7, 8, 9, 10, 11, and 12 are the maximal elements of the poset shown in Figure 3.16, and 1 is the only minimal element.

A poset need not have any maximal or minimal elements at all; for example, the set of integers under $\leq$ has none. On the other hand, a poset may have several maximal or minimal elements.

EXAMPLE 5.   Consider the set A consisting of all proper subsets of $\{1, 2, 3, 4, 5\}$, that is, $A = \mathcal{P}(\{1, 2, 3, 4, 5\}) - \{\{1, 2, 3, 4, 5\}\}$, and consider the relation "is a subset of" on A. There is exactly one minimal element of $(A, \subseteq)$, namely the empty set. The empty set is not only minimal, it is the minimum element of $(A, \subseteq)$, since $\emptyset \subseteq X$ for all subsets X of A. On the other hand, there are five different maximal elements: $\{1, 2, 3, 4\}$, $\{1, 2, 3, 5\}$, $\{1, 2, 4, 5\}$, $\{1, 3, 4, 5\}$, and $\{2, 3, 4, 5\}$. None of them is dominated by another *proper* subset of $\{1, 2, 3, 4, 5\}$. This poset has no maximum element.   ◆

Much of applied mathematics is the search for maxima (the plural of "maximum") and minima. The reader has no doubt encountered such problems in calculus. The field of operations research, a branch of mathematics with both a discrete and a continuous side, is concerned primarily with finding maxima and minima. Later in this book we will want to find such things as maximal connected subgraphs, shortest paths, minimum cost spanning trees, maximum flows, and most efficient algorithms—all examples of extreme points (minimum, minimal, maximum or maximal elements) in some ordered set.

We noted above that some posets have more than one maximal element, and that some posets have no maximal elements. Finite posets, however, are guaranteed to have at least one maximal element.

THEOREM 2.   *Every nonempty partially ordered finite set has at least one maximal element and at least one minimal element.*

Proof.   Let $(A, \preceq)$ be a poset with A finite and nonempty. We will show how to find a maximal element of A. (The proof for a minimal element is analogous; see Exercise 21.) First we will give an informal proof based on the Hasse diagram; then we will see how this proof can be made more rigorous. Let x be any element of A. If x is a maximal element, we are finished. Otherwise, by definition, there is some element $x' \in A$ that dominates x. In the Hasse diagram we think of moving upward from x to x'. If x' is a maximal element, we are finished. Otherwise, there is some $x'' \in A$ that dominates x', and we follow the line in the Hasse diagram upward to x''. We continue "climbing" up the Hasse diagram, following the lines, until we can no longer do so. At this point we have found a maximal element, one dominated by no other element of A. The fact that A is finite guarantees that we cannot keep climbing forever.

The main difficulty with this informal proof is that it relied on the Hasse diagram rather than on the formal properties of a partial order. How do we know, for instance, that our sequence x, x', x'',..., will not come back on itself and continue in a circular loop forever? We would need to exploit the transitivity and antisymmetry of $\preceq$ to show that this cannot happen, and the proof would then get rather messy. The proof in the next paragraph, by contrast, is completely rigorous, if a little dry. It has the spirit of a proof by mathematical induction, a technique that we will study in Section 5.3.

For each element $x \in A$, let $d(x)$ be the number of elements of A that dominate x; in symbols,

$$d(x) = \big|\{\, y \in A \mid x \preceq y \,\}\big|.$$

Since A is finite, the range of d, that is, the collection $D = \{\, d(x) \mid x \in A \,\}$, is a finite collection of well-defined natural numbers. Thus D has a smallest element d_0; suppose that $d_0 = d(x_0)$. We claim that x_0 is a maximal element of A (and hence that $d(x_0) = 1$). If it were not, then by definition there would be another element $x_1 \in A$ such that $x_0 \prec x_1$. Now by transitivity, any element that dominates x_1 also dominates x_0. In addition, x_0 dominates itself but not x_1 (by antisymmetry and reflexivity). Therefore, $d(x_0) > d(x_1)$. This contradicts the choice of $d(x_0)$. Thus x_0 is maximal. ∎

EXAMPLE 6. Consider the divides relation on various sets of positive integers. If we are looking at the set of all integers greater than 1, then the minimal elements are the numbers that have no proper divisors—precisely the prime numbers. There are no maximal elements in this case, since any number x divides, for instance, the number $2x$. If we look instead at the set of *all* positive integers, then there is a unique minimal element, which is also the minimum element, namely 1, since for all positive integers x we have $1 \mid x$. Given any *finite* set A of positive integers whatsoever, Theorem 2 guarantees that we can find an element $x_0 \in A$ that divides no other elements of A. In this case, the largest element of A provides one such element, but there may be others, depending on exactly which numbers are in A. For instance, in $(\{2, 3, 4, 5, 6, 7, 8\}, \mid)$, the number 7 is both a maximal element and a minimal element. ◆

EQUIVALENCE RELATIONS AND PARTITIONS

Mathematics is a search for patterns. In order to see a general pattern, one needs to overlook some differences. In other words, we need to be able to view two objects as being somehow "the same" even when they are not equal. A special type of relation, the equivalence relation, allows us to formalize this notion of considering two objects equivalent. For example, in plane geometry we consider two triangles equivalent ("the same") if they have the same size and shape, regardless of where in the plane they happen to be situated. Thus we overlook position and worry only about size and shape. In modern parlance we call two triangles that are the same in this way "congruent," but Euclid (and older geometry books) actually called them "equal."

We begin with the definition. A relation R on a set A is an **equivalence relation** if and only if R is reflexive, symmetric, and transitive. In the context of an equivalence relation R, we say that a **is equivalent to** b if and only if aRb.

EXAMPLE 7. The following are all examples of equivalence relations.

(a) Let m be a fixed positive integer. Recall from Section 3.1 the "congruent modulo m" relation defined on the set of integers: $x \equiv y \pmod{m}$ if and only if $x \bmod m = y \bmod m$, (i.e., the remainder when x is divided by m equals the remainder when y is divided by m). A portion of the digraph for this relation when $m = 4$ is shown in Figure 3.17 (here the domain is restricted to the set $\{-2, -1, 0, 1, 2, 3, 4, 5, 6\}$).

Let us show that this relation, denoted $\equiv_m$, is an equivalence relation. Certainly, the remainder when x is divided by m equals the remainder when x is divided by m (anything is equal to itself), so $\equiv_m$ is reflexive. Equally clearly, if the remainder when x is divided by m equals the remainder when y is divided by m, then the remainder when y is divided by m equals the remainder when x is divided by m (whether two things are equal does not depend on the order in which we mention them); thus $\equiv_m$ is symmetric. Finally, if the remainder when x is divided by m equals the remainder when y is divided by m, and the remainder

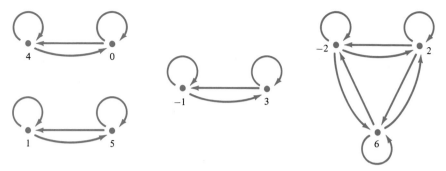

FIGURE 3.17 Digraph of the equivalence relation $\equiv_4$ on the set $\{-2, -1, 0, 1, 2, 3, 4, 5, 6\}$.

when y is divided by m equals the remainder when z is divided by m, then the remainder when x is divided by m equals the remainder when z is divided by m (as old Euclidean geometry books expressed it, "If two things are each equal to a third, then they are equal to each other"); thus $\equiv_m$ is transitive.

(b) Let R be the relation on the set of positive real numbers defined by xRy if and only if $x - y \in \mathbf{Z}$. For example, we have $4.25\,R\,9.25$. Now R is reflexive (since $x - x = 0 \in \mathbf{Z}$), symmetric (since if $x - y \in \mathbf{Z}$, then $y - x = -(x-y) \in \mathbf{Z}$), and transitive (since if $x - y \in \mathbf{Z}$ and $y - z \in \mathbf{Z}$, then their sum $x - z \in \mathbf{Z}$). Therefore, this is an equivalence relation on the set of positive real numbers.

(c) The equality relation on any set is an equivalence relation, in some sense the prototypical equivalence relation. Here two elements are being considered equivalent precisely when they are equal.

(d) The relation that always holds on a set A (i.e., the relation $A \times A$) is an equivalence relation, albeit a rather trivial one. Here every two elements are being considered equivalent. ◆

Many equivalence relations arise in the following very common situation, which generalizes what we saw in Example 7a.

THEOREM 3. *Suppose that A is a set and f is a function from A onto some set B. Then the relation R on A given by xRy if and only if $f(x) = f(y)$ is an equivalence relation, called the equivalence relation* **induced** *on A by f.*

Proof. We need to prove three things. All of them follow from the corresponding properties of the equality relation on B. First, R is reflexive, since $f(x) = f(x)$ for all $x \in A$. Second, R is symmetric, since if $f(x) = f(y)$, then certainly $f(y) = f(x)$. Finally, R is transitive, since if $f(x) = f(y)$ and $f(y) = f(z)$, then $f(x) = f(z)$. ∎

EXAMPLE 8. Let m be the mother function, so that $m(x)$ is the (biological) mother of person x. This function induces the equivalence relation in which two people are related if and only if they have the same mother (in other words, if and only if they are equal or siblings or half-siblings on their mother's side).

The equivalence relation given in Example 7b fits into the model of Theorem 3, since there two numbers are related when the "fractional part" function gives equal values.

The relation on strings in which two strings are related if and only if they have the same length is an equivalence relation by Theorem 3. Here the relation is induced by the length function. ◆

The rest of this section will be devoted to exploring the sense in which Theorem 3 has a converse—that every equivalence relation on A really is induced by some function from A onto a particular set.

Suppose that R is an equivalence relation on a nonempty set A. For each $a \in A$, the **equivalence class** of a, written $[a]_R$ (or usually just $[a]$ if the relation is clear from context), is the set of elements equivalent to a under R. In other words,

$$[a] = \{\, x \in A \mid xRa \,\}.$$

The set of equivalence classes of A under R is denoted A/R:

$$A/R = \{\, [a] \mid a \in A \,\}.$$

(It seems not inappropriate to use this symbolism, "A divided by R," because, as we will prove shortly, the set A is divided up—partitioned—into equivalence classes by the relation R. In Figure 3.17, for example, we see these equivalence classes as the connected pieces of the digraph: -2, 2, and 6 are in one class; -1 and 3 are in a second class; 0 and 4 are in a third class; and 1 and 5 are in a fourth class.) The function from A to A/R that sends a to $[a]$ is called the **canonical projection**.

EXAMPLE 9. Find the equivalence classes for the equivalence relation $\equiv_4$ on $\mathbf{Z}$.

Solution. The integers congruent to 0 modulo 4 are $\{\ldots, -8, -4, 0, 4, 8, \ldots\}$. Thus this set is the equivalence class of 0 under this relation, and we denote it by $[0]$. Similarly,

$$[1] = \{\ldots, -7, -3, 1, 5, 9, \ldots\}$$
$$[2] = \{\ldots, -6, -2, 2, 6, 10, \ldots\}$$
$$[3] = \{\ldots, -5, -1, 3, 7, 11, \ldots\}.$$

Now $[4]$ is the same set as $[0]$; $[5]$ is the same as $[1]$; and so on. Thus there are precisely four different equivalence classes; in symbols, $\mathbf{Z}/\equiv_4 = \{\, [a] \mid a \in \mathbf{Z} \,\} = \{\ldots, [-2], [-1], [0], [1], [2], [3], \ldots\} = \{[0], [1], [2], [3]\}$, and $\left|\mathbf{Z}/\equiv_4\right| = 4$.

Note that the equivalence classes are nonempty, that they are pairwise disjoint (different equivalence classes have no elements in common), and that together they cover all of $\mathbf{Z}$ (in other words, $\bigcup(\mathbf{Z}/\equiv_4) = \mathbf{Z}$). Note also that two integers are in the same equivalence class if and only if they are equivalent under the relation $\equiv_4$; thus the name "equivalence class" is an appropriate one. Finally, note that because of this last observation, the canonical projection from $\mathbf{Z}$ to $\mathbf{Z}/\equiv_4$ induces the relation $\equiv_4$ (see Theorem 3). ◆

In order to study A/R in its own right, we need another definition. A **partition** π of a set A is a collection of pairwise disjoint, nonempty subsets of A whose union is A.

EXAMPLE 10.   The set $\pi = \{\{1,3,5\}, \{2\}, \{6\}, \{4,7\}\}$ is a partition of the set $A = \{1, 2, 3, 4, 5, 6, 7\}$. We picture this partition in Figure 3.18 with the elements of the partition represented by regions in the oval representing A.

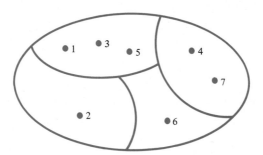

FIGURE 3.18   A partition of the set $\{1, 2, 3, 4, 5, 6, 7\}$.

Note that we can define an equivalence relation on A by declaring two elements of A to be related if and only if they are in the same set of the partition (the same region in Figure 3.18). The digraph for this relation would have arrows joining every pair of points in the same region (and loops at each point), but no arrows joining points in different regions. We will verify that this relation is reflexive, symmetric, and transitive in our proof of part (b) of Theorem 4.

There are many other partitions of $\{1, 2, 3, 4, 5, 6, 7\}$, including the extremes $\{\{1\}, \{2\}, \{3\}, \{4\}, \{5\}, \{6\}, \{7\}\}$ and $\{\{1, 2, 3, 4, 5, 6, 7\}\}$. In fact, the collection of all partitions of a set (note the level of abstraction here—we are talking about a set of sets of sets) forms a partial order under an appropriately defined relation (see Exercise 45). ◆

EXAMPLE 11. The set $\mathbf{Z}/\equiv_4$ discussed in Example 9 is a partition, illustrated in Figure 3.19 (obviously only some of the elements in each equivalence class can be shown in this finite picture). ◆

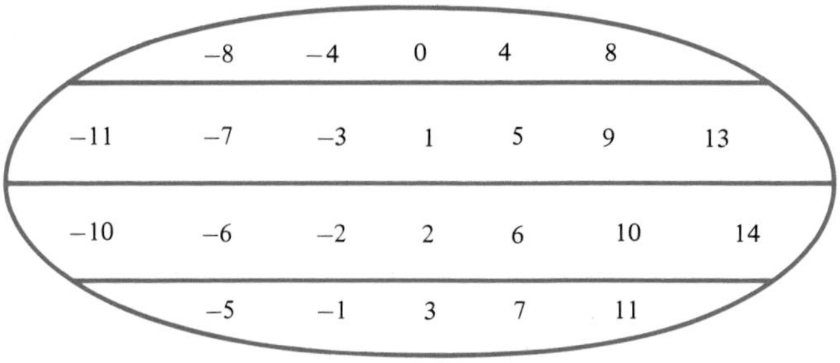

FIGURE 3.19 Partition induced by congruence modulo 4.

Now we can formalize the idea that equivalence relations and partitions are really two ways to look at the same thing.

THEOREM 4. *The relationship between equivalence relations and partitions is given by the following statements.*

(a) *If R is an equivalence relation on a set A, then A/R is a partition of A, called the partition **induced** by R, and the canonical projection from A to A/R induces R.*

(b) *If π is a partition of a set A, then the relation R on A given by*

$$x R y \longleftrightarrow \exists B \in \pi : (x \in B \wedge y \in B)$$

*is an equivalence relation, called the equivalence relation **induced** by π.*

(c) *The transformation from equivalence relations to partitions given in (a) and the transformation from partitions to equivalence relations given in (b) are inverses of each other.*

Proof. For part (a) we need to show first that the equivalence classes are nonempty, are pairwise disjoint, and cover A.

Each equivalence class $[a]$ is nonempty, since $a \in [a]$ by reflexivity.

To prove that the equivalence classes are pairwise disjoint, we will show that if two equivalence classes $[a]$ and $[b]$ have an element in common, they must be equal. So suppose that $c \in [a]$ and $c \in [b]$. Then by definition, cRa and cRb. By the symmetry of R, we have also aRc, and hence by the transitivity of R we conclude that aRb. Now we will show that any element of $[a]$ must also be in $[b]$. Let $x \in [a]$, so that xRa. Then by transitivity, since we just proved that aRb, we know that xRb, whence $x \in [b]$. This shows that $[a] \subseteq [b]$. By a similar argument, $[b] \subseteq [a]$. Thus $[a] = [b]$, as desired.

To prove that the equivalence classes cover all of A it is enough to note again that $\forall x \in A\colon a \in [a]$. (Note that we have used all three properties of an equivalence relation in this proof that A/R is a partition of A.)

Finally, to prove that the canonical projection from A to A/R induces R, we simply note that the arguments above show that aRb if and only if $[a] = [b]$. Thus we see that the converse to Theorem 3 is also true, in the sense that every equivalence relation on A is induced by a function, namely the canonical projection from A to A/R.

For part (b), we need to show that the relation induced by the partition π is reflexive, symmetric, and transitive. Reflexivity is clear, since a and a are in the same set in the partition (we use here the fact that the union of all the sets in the partition must be all of A). Symmetry is clear from the commutativity of the logical "and" connective. Finally, to prove transitivity, we need to make use of the pairwise disjointness of π. Suppose that xRy and yRz. By definition there are sets B and C in π such that x and y are in B, and y and z are in C. Since y is in both and yet the sets in the partition are pairwise disjoint, we must have $B = C$. Hence x and z are in the same set of the partition, so xRz, as desired.

Part (c) should be clear. If R is an equivalence relation on a set A, then two elements of A are related under R if and only if there is an element of A/R in which they both lie, namely their common equivalence class. Conversely, given a partition, if we define R as in part (b), then the equivalence classes for R are precisely the sets in the given partition. ∎

We saw in Example 9 how an equivalence relation induces a partition. Our final example shows how a partition induces an equivalence relation.

EXAMPLE 12. The registered voters in the United States are partitioned, by their permanent residences, into 435 congressional districts for the purpose of electing representatives to Congress. These districts are nonempty, pairwise disjoint (no one is allowed to vote for two different representatives), and cover all voters. Thus the collection of districts forms a partition. The equivalence relation on the voters induced by this partition is the relation under which two people are related if and only if they live in the same congressional district. ◆

In Section 7.2 we will look at the interesting problem of determining *how many* partitions a set with n elements has.

SUMMARY OF DEFINITIONS

canonical projection: the function from A to A/R sending each $a \in A$ to $[a] \in A/R$, where R is an equivalence relation on the set A.

comparable elements in a partially ordered set: elements x and y such that $x \preceq y \lor y \preceq x$, where $\preceq$ is the partial order (e.g., the numbers 6 and 48 are comparable, but 6 and 8 are incomparable, in the poset consisting of the positive integers under "divides").

dominate in the poset $(A, \preceq)$: x dominates y if $y \preceq x$, where x and y are elements of A (e.g., 12 dominates 2 but not 5 in the poset consisting of the positive integers under "divides").

equivalence class of the element $a \in A$ under the equivalence relation R on A: the set $\{ x \mid xRa \}$, denoted $[x]_R$ or just $[x]$ (e.g., the set of people residing at 1600 Pennsylvania Avenue, Washington, D.C., is the equivalence class of the President under the equivalence relation "lives at the same address as").

equivalence relation on the set A: a relation on A that is reflexive, symmetric, and transitive (e.g., the relation $\equiv_7$ ("is congruent to modulo 7") on the set of integers).

equivalent: related under a given equivalence relation.

Hasse diagram for the finite poset $(A, \preceq)$: a picture in which the elements of A are represented by points such that if $x \prec y$ then y lies above x, and in which there is a line joining x and y if y is an immediate successor of x.

immediate predecessor: x is an immediate predecessor of y if and only if y is an immediate successor of x.

immediate successor: If $(A, \preceq)$ is a poset, and x and y are elements of A, then y is an immediate successor of x if and only if $x \prec y \land \neg \exists z \in A: (x \prec z \land z \prec y)$—in other words, x precedes y but there is no element strictly between x and y in the partial order (e.g., 36 and 60 are both immediate successors of 12 in the poset of the positive integers under "divides").

induce a partition: An equivalence relation R on a set A induces the partition of A whose elements are all the equivalence classes $[x]_R = \{ y \mid yRx \}$ for $x \in A$ (e.g., the equivalence relation "lives at the same address as" on the set of residents of the United States induces the partition of people into households).

induce an equivalence relation (said of a function): A function f whose domain is a set A induces the equivalence relation R on A defined by xRy if and only if $f(x) = f(y)$ (e.g., the function $f(x) =$ the address of person x induces the equivalence relation "lives at the same address as").

induce an equivalence relation (said of a partition): A partition π of a set A induces the equivalence relation R on A defined by xRy if and only if $\exists S \in \pi: (x \in S \land y \in S)$ (e.g., the partition of residents of the United States into households induces the equivalence relation "lives at the same address as").

induce an order on A^n or A^*: Lexicographic order on A^n or A^* is said to be induced by a given total order on A (e.g., alphabetical order on English words without hyphens or apostrophes is induced by the order $a \prec b \prec c \prec \cdots \prec z$ on the alphabet).

lexicographic order on A^n induced by the total order $\preceq$: the order in which $(a_1, a_2, \ldots, a_n) \prec (b_1, b_2, \ldots, b_n)$ if and only if $\exists i \geq 1: ((\forall k < i: a_k =$

$b_k) \wedge a_i \prec b_i)$ (e.g., $(1, 2, 3) \prec (1, 3, 2)$ in $\{1, 2, 3, 4\}^3$ if the underlying order on $\{1, 2, 3, 4\}$ is $\leq$).

lexicographic order on A^* induced by the total order $\preceq$: the order in which $a_1 a_2 \ldots a_n \prec b_1 b_2 \ldots b_m$ if and only if $n < m \wedge \forall i \leq n : a_i = b_i$ (the first string is a proper initial substring of the second) or $\exists i \geq 1 : ((\forall k < i : a_k = b_k) \wedge a_i \prec b_i)$ (the symbol in the first string precedes the symbol in the second string at the first position at which the symbols differ) (e.g., *ant* $\prec$ *anteater* and *annex* $\prec$ *ant* among strings over the English alphabet under its usual order).

linear order: a total order.

maximal element in a poset: an element that is dominated by no other element in the poset (e.g., $\{1, 3, 4\}$ in the poset consisting of the proper subsets of $\{1, 2, 3, 4\}$ under $\subseteq$).

maximum element in a poset: an element that dominates every element in the poset (e.g., A in $(\mathcal{P}(A), \subseteq)$ where A is any set).

minimal element in a poset: an element that dominates no other element in the poset (e.g., $\{1\}$ in the poset consisting of the nonempty subsets of $\{1, 2, 3, 4\}$ under $\subseteq$).

minimum element in a poset: an element that is dominated by every element in the poset (e.g., $\emptyset$ in $(\mathcal{P}(A), \subseteq)$ where A is any set).

partial order on the set A: a relation on A that is reflexive, transitive, and antisymmetric (e.g., the relation "divides" on the set of positive integers).

partially ordered set: a set together with a partial order on the set (e.g., $(\mathcal{P}(A), \subseteq)$ where A is any set).

partition of a set A: a collection of pairwise disjoint, nonempty subsets of A whose union is A (e.g., {freshmen, sophomores, juniors, seniors} is a partition of the set of undergraduates at a university).

poset: a partially ordered set.

precede: is related to but not equal to in a given partial order; often written $\prec$ (e.g., 12 precedes 7 in the poset $(\mathbf{N}, \geq)$).

succeed: x succeeds y if and only if y precedes x (e.g., 12 succeeds 7 in the poset $(\mathbf{N}, \leq)$).

total order on the set A: a partial order on A with the property that every pair of elements of A are comparable (e.g., alphabetical order on English words).

totally ordered set: a set together with a total order on the set (e.g., $(\mathbf{R}, \leq)$).

EXERCISES

1. Consider the Hasse diagram on p. 200.
   (a) Write down the entire partial order that this diagram depicts.
   (b) Redraw the diagram and add a dashed line joining every pair of distinct points that are comparable under this partial order.

2. Draw the Hasse diagrams for the following posets.

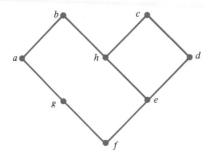

EXERCISE 1

(a) The set $\{2, 4, 6, 8, 12, 24\}$ under the relation $\leq$.

(b) The set $\{2, 4, 6, 8, 12, 24\}$ under the "divides" relation.

(c) $(\{2, 4, 6, 8, 12, 24\}, \geq)$.

(d) $(\mathcal{P}(\{1, 2, 3\}), \subseteq)$.

(e) The set of all subsets of $\{1, 2, 3, 4\}$ that contain an odd number of elements, under the relation $\supseteq$ (where $A \supseteq B$ means $B \subseteq A$).

3. List the elements of $\{a, b, c\} \times \{a, b, c\}$ in lexicographic order.

4. List the strings of length less than or equal to 3 over the alphabet $\{a, b\}$, in lexicographic order.

5. Consider the relation $\{(1, 1), (1, 5), (5, 1), (5, 5), (2, 2), (2, 6), (6, 2), (6, 6), (3, 3), (4, 4)\}$.

(a) Draw the digraph of the relation.

(b) Verify that this is an equivalence relation.

(c) Write down the partition induced by this relation.

(d) Draw a diagram to illustrate the partition.

(e) List explicitly the equivalence classes [1], [2], [3], [4], [5], and [6].

6. Write down the equivalence classes for the relation "congruent modulo 3" on the set of integers, and draw a picture to illustrate the partition induced by this relation.

7. Consider the partition $\{\{1, 2, 3\}, \{4, 5\}, \{6\}\}$.

(a) Write down the equivalence relation induced by this partition.

(b) Draw the digraph of the equivalence relation.

8. Consider the relation $\{(a, a), (a, d), (a, f), (b, b), (b, e), (b, f), (b, g), (c, c), (c, e), (c, g), (d, d), (d, f), (e, e), (e, g), (f, f), (g, g), (h, h), (i, i), (i, j), (j, j)\}$ on the set $\{a, b, c, d, e, f, g, h, i, j\}$.

(a) Verify that this relation is a partial order.

(b) Draw the Hasse diagram for this relation.

9. Write in symbols the definitions of the following concepts.

    (a)   The minimum element in the poset $(A, \preceq)$.

    (b)   A minimal element in the poset $(A, \preceq)$.

10. Prove or disprove: If R is a partial order on a set A, then R^{-1} is also a partial order on A.

11. Give examples of three relations, each of which just fails to be a partial order by failing to satisfy a different one of the three defining properties (but satisfying the other two properties in each case).

12. Determine whether each of the following relations is a partial order and, if it is, whether it is a total order.

    (a) xRy if and only if $x < y - 2$ on the set of integers.

    (b) xRy if and only if $x > y - 2$ on the set of integers.

    (c) ARB if and only if $|A| \leq |B|$ on the power set of $\mathbf{N}$.

    (d) xRy if and only if $x = y$ or x beat y, on the set of teams entered in a round-robin tournament. (Assume that each team plays every other team once, and games cannot end in a tie. Make no other assumptions about the outcomes of the games.)

    (e)   The relation R on the set $\mathbf{N} \times \mathbf{N}$ given by $(a, b)R(c, d)$ if and only if $a \leq c \wedge b \leq d$.

    (f)   The relation R on the set $\mathbf{N} \times \mathbf{N}$ given by $(a, b)R(c, d)$ if and only if $a \leq c \vee b \leq d$.

13. Model with a poset the following set of tasks that must be performed in order to prepare dinner. The partial order is the relation "is equal to or must be performed prior to." Draw the Hasse diagram for this poset. The tasks are carving the turkey, preheating the oven, putting the potatoes in to boil, putting the turkey in the oven to roast, boiling a pot of water for the beans, putting the beans in to boil, peeling the potatoes for boiling, setting the table, boiling a pot of water for the potatoes, draining the potatoes, draining the cooked beans, stuffing the turkey, and combining the cooked potatoes and beans in a casserole.

14. Verify that the "divides" relation on any set of positive integers is a partial order.

15. Consider the "divides" relation on the set of positive integers.

    (a)   Give a satisfying explicit characterization of when x is an immediate predecessor of y under this relation.

    (b)   Find a rule for the number of immediate predecessors a given y has.

    (c)   Determine how many immediate successors each x has.

16. Draw the Hasse diagrams for the following posets.

    (a) $(\mathcal{P}(\{a, b, c, d\}), \subseteq)$.

    (b) $(\{1, 2, 3, 4, 5\}, =)$.

(c)   The set of integers from 1 to 11, under the relation in which each even integer $2x$ dominates $2x \pm 1$ (no other pairs are related, except as dictated by reflexivity).

(d) $(\{1, 2, 3\} \times \{1, 2, 3\}, R)$, where $(a, b)R(c, d)$ if and only if $a \leq c \wedge b \leq d$.

17. Prove or provide a counterexample for each of the following statements.

(a)   The intersection of two partial orders on the same set is a partial order.

(b)   The union of two partial orders on the same set is a partial order.

(c)   The complement of a partial order is a partial order.

18. Define what it should mean for two posets to be isomorphic (see the last part of Example 1 in Section 2.2). Then determine which pairs among $(\mathbf{N}, \leq)$, $(\mathbf{N}, \geq)$, $(\mathbf{Z}, \leq)$, and $(\mathbf{Z}, \geq)$ are isomorphic.

19. Characterize the minimum element, the maximum element, the minimal elements, and the maximal elements of the following posets. (Some of these may not exist.)

(a) $\mathcal{P}(\mathbf{N})$ under the $\subseteq$ relation.

(b) $\{1, 2, 3\} \times \{1, 2, 3\}$ under the relation R given by $(a, b)R(c, d)$ if and only if $a \leq c \wedge b \leq d$.

(c)   The set $\{a, b, c, d, e\}$ under the relation $\{(a, a), (a, e), (b, b), (b, c), (b, d), (b, e), (c, c), (c, e), (d, d), (d, e), (e, e)\}$.

(d)   The poset given in Exercise 8.

(e)   The poset whose Hasse diagram is given in Exercise 1.

(f)   The poset $\{(1, 1)\}$.

(g)   The poset $\{(1, 1), (2, 2)\}$.

20. Prove or disprove each of the following assertions.

(a)   If a poset has a unique maximal element, then that element is a maximum element.

(b)   If a poset has a maximum element, then that element is the only maximal element.

21. Prove the part of Theorem 2 dealing with minimal elements.

22. Describe in English what the digraph of an equivalence relation has to look like.

23. Give examples of three relations, each of which just fails to be an equivalence relation by failing to satisfy a different one of the three defining properties (but satisfying the other two properties in each case).

24. Give examples of three collections of subsets of a given set, each of which just fails to be a partition by failing to satisfy a different one of the three defining properties (nonemptyness, pairwise disjointness, and having the correct union).

25. Determine whether the following relations are equivalence relations.

(a)   The relation R on $\mathbf{R}$ given by xRy if and only if $\lfloor x \rfloor = \lfloor y \rfloor$.

    (b)   The relation R on **R** given by xRy if and only if $|x - y| \leq 1$.

    (c)   The "is similar to" relation on triangles in the plane.

    (d)   The relation R on the set of positive integers given by xRy if and only if $\forall p : (p \mid x \leftrightarrow p \mid y)$, where the variable p ranges over all prime numbers.

    (e)   The relation on people in which two people are related if and only if they have the same mother and father.

    (f)   The relation on people in which two people are related if and only if they have the same mother or the same father.

26. For those relations in Exercise 25 that are equivalence relations, describe the equivalence classes (in other words, describe the partition induced by the relation).

27. Suppose that R is the equivalence relation induced by a function $f: A \to B$, according to Theorem 3. What are the inverse images of elements of B called?

28. Define a relation R on **Z** by xRy if and only if $x^2 - y^2$ is divisible by 3.

    (a)   Show that R is an equivalence relation.

    (b)   Determine the partition induced by R.

29. One way to define the relation $\equiv_m$ is to declare $x \equiv y \pmod{m}$ to mean "$x - y$ is a multiple of m." Show directly from this definition that $\equiv_m$ is an equivalence relation.

30. Write down in symbols a definition of "the set π is a partition of the set A."

31. Determine how many partitions the following sets have. [*Hint*: List them all explicitly. Be careful!]

    (a) $\emptyset$.

    (b) $\{a\}$.

    (c) $\{a, b\}$.

32. List all the partitions of the set $\{a, b, c\}$.

33. Let S be the set $\{A, B, C, D\}$.

    (a)   How many partitions are there of the set S?

    (b)   How many equivalence relations are there on the set S?

34. In each case find the number of partitions of $\{a, b, c, d\}$ satisfying the given property.

    (a)   The cardinality of the partition is 2 (in other words, there are two parts).

    (b)   The cardinality of the partition is 3.

    (c)   No set in the partition has cardinality more than 2.

    (d)   No set in the partition has cardinality less than 2.

35. Prove or provide a counterexample for each of the following statements.

    (a)   The intersection of two equivalence relations on the same set is an equivalence relation.

    (b)   The union of two equivalence relations on the same set is an equivalence relation.

    (c)   The inverse of an equivalence relation is itself.

    (d)   The complement of an equivalence relation is an equivalence relation.

36. Consider the function from the set of integers greater than 1 to the set of prime numbers given by $f(n) =$ the smallest prime factor of n. Describe the equivalence classes of the equivalence relation induced by f by listing several elements of several of the equivalence classes.

37. List the partitions induced by the following equivalence relations on the set of positive integers. In each case there is an infinite number of sets in the partition, and each one is infinite, so your answer will have lots of dots to indicate "and so on in this pattern."

    (a) x is related to y if and only if the highest power of 2 that is a divisor of x equals the highest power of 2 that is a divisor of y.

    (b) x is related to y if and only if the quotient of x divided by the highest power of 2 that is a divisor of x equals the quotient of y divided by the highest power of 2 that is a divisor of y.

38. Consider the relation R on the unit square S in which each point in the interior of the square is related only to itself; each point on the interior of an edge of the square is related only to itself and the corresponding point straight across on the opposite edge; and all four corners of the square are related to each other (and to themselves). Explain why R is an equivalence relation on S, and give a geometric interpretation of S/R. [*Hint*: Imagine the square as a sheet of rubber, and make corresponding points coincide by bending the sheet and gluing appropriately.]

39. Let R be a relation on a set A. Explain how to find the smallest equivalence relation containing R.

40. Suppose that R is the relation on $\mathbf{N} \times \mathbf{N}$ given by $(a, b)R(c, d)$ if and only if $c = a + 1$ and $d = b + 1$.

    (a)   Draw the digraph for R. [*Hint*: Represent the points of $\mathbf{N} \times \mathbf{N}$ in their usual position in the coordinate plane.]

    (b)   Explain how your picture shows the partition of $\mathbf{N} \times \mathbf{N}$ induced by the smallest equivalence relation containing R.

41. Define a relation R on $\mathcal{P}(\mathbf{N})$ by ARB if and only if there exists a bijective function from A to B.

    (a)   Show that R is an equivalence relation.

    (b)   Fully describe the equivalence classes of R.

Challenging Exercises

42. Recall from Section 3.3 that a relation on a finite set can be represented by a matrix.
    (a) Describe the set of matrices that represent equivalence relations.
    (b) Describe the set of matrices that represent total orders.

43. A partial order $\preceq$ on a set A is called **dense** if and only if, given any two elements x and y in A such that $x \prec y$, there exists an element z such that $x \prec z \wedge z \prec y$.
    (a) Give an example of a dense total order.
    (b) Give an example of a dense partial order that is not a total order.
    (c) Give an example of a total order on an infinite set that is not dense.
    (d) Show that if R is a dense partial order on A, and A has at least two comparable elements, then A is infinite.

44. A nonempty set A is said to be **well ordered** by a total order R if every nonempty subset of A contains a minimum element.
    (a) Explain why **N** is well ordered by $\leq$.
    (b) Explain why **Z** is not well ordered by $\leq$.
    (c) Determine whether the set of nonnegative real numbers is well ordered by $\leq$.
    (d) Suppose that A is well ordered by R. Is $A \times A$ well ordered by the lexicographic order induced by R?
    (e) Suppose that A is well ordered by R. Is A^* well ordered by the lexicographic order induced by R?

45. Consider the relation $\preceq$ on the set of all partitions on a set A given by $\pi_1 \preceq \pi_2$ if and only if $\forall X \in \pi_1 : \exists Y \in \pi_2 : X \subseteq Y$. This relation is called the refinement relation; if $\pi_1 \preceq \pi_2$, we say that π_1 is a **refinement** of π_2.
    (a) Find two partitions π_1 and π_2 of the set $\{1, 2, 3, 4, 5\}$, neither of which is $\{\{1, 2, 3, 4, 5\}\}$ or $\{\{1\}, \{2\}, \{3\}, \{4\}, \{5\}\}$, such that $\pi_1 \prec \pi_2$.
    (b) Find two partitions of $\{1, 2, 3, 4, 5\}$ that are incomparable under the refinement relation.
    (c) Show that the refinement relation is a partial order.
    (d) Find the maximum element of this poset.
    (e) Find the minimum element of this poset.

Exploratory Exercises

46. Explain how the set of rational numbers can be interpreted as the set of equivalence classes of pairs of integers under a suitable equivalence relation. Your discussion should include exactly what the rational numbers are and how the operations of addition and multiplication and the relation "$<$" are defined for them. You may not assume anything about the rational or real numbers ahead of time; all you are given are the integers with their addition and multiplication and "$<$" relation.

47. Find out more about well-ordered sets (see Exercise 44) by consulting Halmos [137]. In particular, find out how uncountable sets can be well ordered.

48. Find out more about partial orders and equivalence relations by consulting Bogart [23].

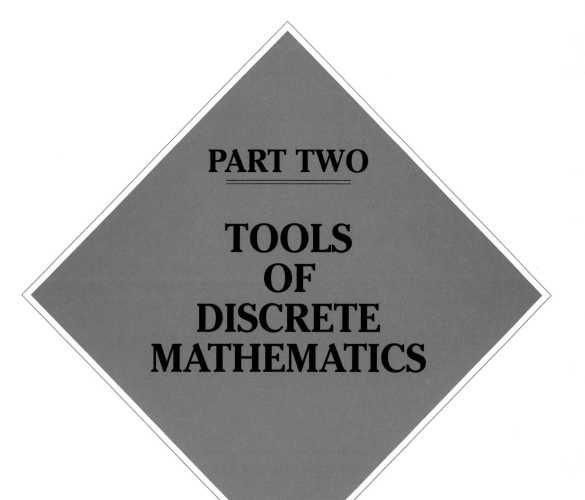

PART TWO

TOOLS
OF
DISCRETE
MATHEMATICS

4

ALGORITHMS

One aspect of mathematics, though by no means the only one, is computation. In this chapter we study computational procedures—algorithms—as a mathematical subject. We give a working definition of what an algorithm is and look at some of its characteristics (Section 4.1). We discuss ways to formulate algorithms precisely so that human beings and computers can understand them. In particular, we lay out rules for a language for describing algorithms—a "pseudocode" similar to the programming language Pascal—in Section 4.2. We judge various algorithms on their efficiency (after developing a theory for doing so) in Section 4.3, and we look briefly at the theoretical limitations of algorithms (Section 4.4). Finally, in Section 4.5 we survey some of the algorithms for a number of important arithmetical and algebraic computations; indeed, a secondary goal of this chapter is to expose the reader to some of the rudimentary number theory and algebra that underlie much of the mathematical computation that goes on daily in computers throughout the world.

There are several good books on algorithms in general and "combinatorial" algorithms (those dealing with discrete mathematics) in particular. See, for example, Aho, Hopcroft, and Ullman [2,3], Baase [12], Brassard and Bratley [30], three volumes of Knuth [175–177], Kronsjö [189], Sedgewick [268], and Wilf [320]. In addition, the reader is urged to look at Hofstadter [153], which, among other things, has an excellent discussion of algorithms.

SECTION 4.1
THE IDEA OF AN ALGORITHM

Go directly to jail. Do not pass GO.
Do not collect $200.

—Monopoly "chance" card

Essentially, an **algorithm** is an unambiguous set of instructions for carrying out a specific task. The task might be to calculate something, such as the decimal numeral for $2^n - 1$, given n. It might be to find something, such as a word in a dictionary. We might need to rearrange something—perhaps to sort a list of words into alphabetical order. Often in discrete mathematics we need to optimize something—perhaps to find the shortest route from one point to another along the interstate highways. In all cases we can think of an algorithm as being *a clearly specified method for obtaining a solution to a problem.* Note that we have not formally *defined* the word "algorithm" in the same sense that we defined, for example, the word "subset." To do so would require us to specify a model of computation and get into very messy technical details (see the references listed in Exercise 22 in Section 4.4). The informal "definitions" given in this paragraph, together with the discussion of the characteristics of algorithms toward the end of this section, should suffice for our purposes.

We begin our discussion with a case study, which will reappear briefly in Sections 4.2, 5.2, and 7.2. We take an extensive look at a specific algorithm, one of the oldest and most elegant algorithms in mathematics: the Euclidean algorithm for finding the greatest common divisor of two natural numbers. As the name suggests, this algorithm dates back to ancient Greece.

Throughout this discussion, we work with the natural numbers, so the word "number" will mean an element of **N** unless otherwise specified. In particular, we will have no need in this section for negative integers. Recall that a nonzero number d is said to divide (or to be a divisor of) a number x (written symbolically as $d \mid x$) if x is a multiple of d (i.e., $x = dm$ for some number m). For example, 13 is a divisor of 468, written $13 \mid 468$, since $468 = 13 \cdot 36$. By this definition every nonzero number is a divisor of 0; but each nonzero number x has only a finite number of divisors, since clearly any divisor of x must be no larger than x. Sometimes we need to find a **common divisor** of two numbers, that is, a number that is a divisor of both. For example, 3 is a common divisor of 468 and 54, since $468 = 3 \cdot 156$ and $54 = 3 \cdot 18$.

Furthermore, it is usually desirable in such cases to find the **greatest common divisor**—the largest number among the set of common divisors. Such a need arises, for example, when we try to simplify fractions. The example mentioned here allows us to see that $54/468$ can be simplified to $18/156$; knowing the greatest common

210

divisor would allow us to simplify a fraction immediately to its *lowest* terms. We denote the greatest common divisor of natural numbers x and y by $\gcd(x, y)$.

EXAMPLE 1. We have $\gcd(468, 54) = 18$, because 18 is the largest number in the intersection of the set of positive divisors of 468 (namely $\{1, 2, 3, 4, 6, 9, 12, 13, 18, 26, 36, 39, 52, 78, 117, 156, 234, 468\}$) and the set of positive divisors of 54 (namely $\{1, 2, 3, 6, 9, 18, 27, 54\}$). Similarly,

$$\gcd(468, 270)$$
$$= \max(\{1, 2, 3, 4, 6, 9, 12, 13, 18, 26, 36, 39, 52, 78, 117, 156, 234, 468\}$$
$$\cap \{1, 2, 3, 5, 6, 9, 10, 15, 18, 27, 30, 45, 54, 90, 135, 270\})$$
$$= \max(\{1, 2, 3, 6, 9, 18\}) = 18,$$

and $\gcd(14, 0) = \max(\{1, 2, 7, 14\}) = 14$.   ◆

Note that $\gcd(x, y)$ always exists if at least one of x and y is not zero, since we have defined it as the largest element in a nonempty finite set of numbers: 1 is always a common divisor, and no common divisor can exceed the larger of x and y, that is, $\gcd(x, y) \leq \max(x, y)$. Note that $\gcd(x, y) = \max(x, y)$ if exactly one of x and y is 0. If both x and y are nonzero, then no common divisor can exceed the smaller of x and y, so $\gcd(x, y) \leq \min(x, y)$. We will not define $\gcd(0, 0)$, since the set of common divisors of 0 and 0 is infinite and has no largest element. We take note of one useful property of the greatest common divisor, whose proof is left to the reader (Exercise 11): *Every* common divisor of a pair of numbers divides their greatest common divisor. For example, since $\gcd(468, 270) = 18$, we know that the only other common divisors of 468 and 270 are just the proper divisors of 18, namely 1, 2, 3, 6, and 9.

THEOREM 1.   *Let x and y be two natural numbers, not both zero. If d is a common divisor of x and y, then* $d \mid \gcd(x, y)$.

Now the act of having *defined* the greatest common divisor, stated some of its properties, and shown that it always exists does not mean that we know of any way to compute it, let alone a good way to compute it. In other words, we have not yet indicated explicitly an unambiguous set of instructions by which a person or computer can determine the value of $\gcd(x, y)$, given the values of x and y. What we need, then, is an algorithm for computing $\gcd(x, y)$.

Our algorithms for computing greatest common divisors will need to use algorithms for doing basic arithmetic. Let us assume for now that we already know algorithms for the four arithmetic operations on natural numbers: addition, subtraction, multiplication, and division. This is certainly a valid assumption, since such algorithms are taught in the first five or six grades of elementary school. (A more complete discussion of algorithms for the arithmetic operations appears in Section 4.5.) Specifically, we assume that we have a procedure for finding, given natural numbers x and y, the sum $x + y$; the difference $x - y$ (if $x \geq y$); the product $x \cdot y$; and, if $y \neq 0$, the quotient and remainder when x is divided by y.

This last operation needs a bit of elaboration, because we will need to exploit it heavily in what follows. Recall from Section 3.1 that the quotient q and the remainder r when x is divided by y are the unique numbers satisfying

$$x = qy + r, \qquad 0 \leq r < y;$$

we have $q = \lfloor x/y \rfloor$ and $r = x - qy$. The remainder is denoted $x \bmod y$. For example, when 36 is divided by 13, we obtain $q = \lfloor 36/13 \rfloor = 2$ and $r = 36 \bmod 13 = 10$; when 36 is divided by 3, $q = \lfloor 36/3 \rfloor = 12$ and $r = 36 \bmod 3 = 0$. In particular, carrying out this division operation allows us to determine whether $y \mid x$, since y *divides* x *if and only if* $x \bmod y = 0$. It is easy to determine whether $y \mid x$ on a calculator: We simply perform the division x/y and note whether the answer is a natural number (i.e., has no fractional part).

Now let us consider possible algorithms for computing $\gcd(x, y)$, where x and y are not both 0. Here is a very naive algorithm based on the definition and our observation that $\gcd(x, y)$ is no greater than $\min(x, y)$, unless x or y is 0. If x or y is 0, then $\gcd(x, y)$ is the nonzero argument. Otherwise, we look at the numbers from 1 to $\min(x, y)$, one by one, and check for each such number d whether $(d \mid x) \wedge (d \mid y)$, that is, whether $x \bmod d = 0 = y \bmod d$. Among all the values of d we find that satisfy this compound condition, we take the largest as our answer.

We could make this algorithm more efficient by starting at $d = \min(x, y)$ and working down toward 1, rather than the other way around, since then we could stop as soon as we found the first common divisor. Still, this algorithm seems to involve a lot of arithmetic. For example, to determine that $\gcd(468, 54) = 18$ by this method, we would need to divide each of 468 and 54 by all the numbers 54, 53, 52, ..., 18, not a pleasant prospect.

The method usually taught in school for computing the greatest common divisor of two positive integers is first to obtain the prime factorization of each number and then to write down the prime factorization of the answer in the following way: For each prime number p appearing in both factorizations, the power to which p appears in the prime factorization of $\gcd(x, y)$ is the smaller of the powers to which p appears as a factor in x and in y. In our example, we would first obtain $468 = 2^2 \cdot 3^2 \cdot 13$ and $54 = 2 \cdot 3^3$, and then deduce that the prime 2 can occur only to the first power, the prime 3 only to the second power, and the prime 13 not at all in $\gcd(468, 54)$, whence $\gcd(468, 54) = 2 \cdot 3^2 = 18$. The problem with this algorithm is that it relies on obtaining the prime factorizations of the two arguments. It is not too difficult to think of an algorithm for finding the prime factorization of a number m; the reader is asked to supply one in Exercise 20. What schoolteachers fail to point out, however, is that obtaining prime factorizations is very time consuming for large numbers. For example, the reader is challenged to try to find gcd(16273346, 12954622) by this method. (We will have a little more to say about the difficulty of factoring large numbers in Section 4.4.)

So far we have indicated two algorithms for finding greatest common divisors: trying all possible divisors and working from the prime factorization. Neither of them is very good in the sense of giving us the correct answer without too much effort. We now turn to a discussion of a venerable and wonderful algorithm for solving this problem.

THE EUCLIDEAN ALGORITHM

Ancient Greek mathematicians often conceived of arithmetic mainly as it applied to geometry. They might have looked at the problem of finding gcd(x, y) in terms of commensurability of the sides of a rectangle. Suppose that you have a rectangle x millimeters by y millimeters (x and y are positive integers), and you wish to divide it into a square grid, using as large a square as possible. A little thought will convince you that the side of the largest square you can use is precisely gcd(x, y). In Figure 4.1, for example, is a picture of a rectangle 152 mm by 57 mm, drawn to scale. Since gcd(152, 57) = 19, the rectangle can be covered with a grid of 19 mm by 19 mm squares, as shown in Figure 4.1. Thus the two sides of the rectangle are commensurate, using 19 mm as the common measure, and in terms of this measure, the rectangle is 8 units by 3 units.

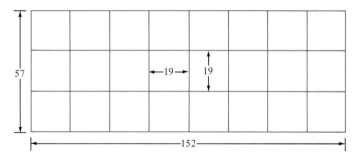

FIGURE 4.1   A 152 by 57 rectangle, covered with 19 by 19 squares.

The key idea behind the algorithm the Greeks devised for finding greatest common divisors is the following theorem.

THEOREM 2. *Let x be a natural number and let y be a positive integer. Let $r = x \bmod y$ be the remainder when x is divided by y. Then*

$$\gcd(x, y) = \gcd(y, r).$$

Proof. By the definition of remainder, we can write

$$x = qy + r \qquad \text{or} \qquad r = x - qy.$$

The first of these equations shows that any divisor of both y and r is also a divisor of x; the second shows that any divisor of both x and y is also a divisor of r. Thus the set of common divisors of x and y is the same as the set of common divisors of y and r. It follows that the greatest common divisors are equal. ∎

To find $\gcd(x, y)$ we can proceed as follows. If $y = 0$, there is nothing to do, for the answer is x (recall that x cannot be 0 if y is 0). Otherwise, we divide y into x, obtaining a quotient q_1 (which we ignore) and a remainder r_1. By Theorem 2 the answer we desire is the same as $\gcd(y, r_1)$, and we have transformed our original problem into a simpler one, since the second argument is now r_1 rather than y, and $r_1 < y$. Now we repeat the process. If $r_1 = 0$, then the answer is $\gcd(y, 0) = y$. Otherwise, we divide y by r_1 and obtain a quotient q_2 and a remainder r_2, with $0 \le r_2 < r_1$. Invoking Theorem 2 again, we see that the answer we desire is equal to $\gcd(r_1, r_2)$, and the second argument (now r_2) is even smaller than it was at the last step. If $r_2 = 0$, we are done (and the answer is r_1); if not, we repeat the process. Since the remainders are constantly getting smaller, eventually the remainder must be 0; at that point we are done. In practice, the number of repetitions (**iterations**) required is usually quite small. This procedure is known as the **Euclidean algorithm**.

EXAMPLE 2. Find $\gcd(468, 54)$ and $\gcd(16273346, 12954622)$ by the Euclidean algorithm.

Solution. In the first problem, we divide 468 by 54 and obtain a quotient of 8 and a remainder of 36. Thus the answer we want is equal to $\gcd(54, 36)$, the greatest common divisor of the number we were dividing by and the remainder. Repeating the process (dividing 54 by 36), we get a quotient of 1 and a remainder of 18; thus our answer is the same as $\gcd(36, 18)$. One final iteration gives us $\gcd(18, 0)$, which is equal to 18. Thus our answer is $\gcd(468, 54) = 18$.

The second problem requires nine iterations:

$$\gcd(16273346, 12954622) = \gcd(12954622, 3318724) = \gcd(3318724, 2998450)$$

$$= \gcd(2998450, 320274) = \gcd(320274, 115984)$$

$$= \gcd(115984, 88306) = \gcd(88306, 27678)$$

$$= \gcd(27678, 5272) = \gcd(5272, 1318)$$

$$= \gcd(1318, 0) = 1318 \, .$$

The prime factorizations turn out to be $16273346 = 2 \cdot 659 \cdot 12347$ and $12954622 = 2 \cdot 659 \cdot 9829$, but it would have taken a lot of work to find the common prime factor 659 by trial and error.

Incidentally, there is an easy way to find remainders on most calculators. Suppose that you wish to find the remainder when 468 is divided by 54. Press "468 ÷ 54 =" and the display will read 8.666666667. Thus the quotient is 8. Now press "− 8 =" to get the remainder as a decimal fraction: 0.666666667. Finally, press "× 54 =" (and round a bit if necessary) to get the remainder as a natural number rather than as a decimal fraction: 36.00000002, or just 36. Thus $468 \bmod 54 = 36$. ◆

DESCRIPTIONS OF ALGORITHMS

We described the Euclidean algorithm above in a paragraph of English prose. We explained the sequence of steps that were required to go from the given values of x and y to the desired value, $\gcd(x, y)$. As long as our set of instructions is unambiguous and complete (in the sense of always specifying what to do next), there is nothing wrong with this form of algorithm description.

We need to look at some other ways to describe algorithms, since the prose paragraph style is not always appropriate. For example, given the present state of technology, a computer could make little sense out of such a verbal description. (Although the subject of artificial intelligence is growing rapidly, we are still far away from machine understanding of natural language, especially natural language as complex as that in the description of an algorithm.) Also, if an algorithm is complicated, involving many steps and substeps, with difficult logical dependencies, then even a competent human being would be hard pressed to make sense of a verbal description. Finally, verbal descriptions of algorithms often run too great a risk of being ambiguous and unclear.

We will discuss a range of methods for describing algorithms, from the verbal to the highly symbolic, using the Euclidean algorithm as our example. In all of the discussion that follows, the arguments or data on which the algorithm operates (x and y in our example) are called the **input**; they are available at the start of the algorithm. The answer or result of the computation ($\gcd(x, y)$ in our example) is called the **output**; it is produced before the algorithm stops.

One way to make a prose paragraph description somewhat more mechanical is to use numbered steps, explaining exactly what piece of computation is to be performed at each step, and what step is to be performed next. Unless explicit directions are given to the contrary, the algorithm proceeds by starting at step 1 and performing the steps in numerical order. Here is the Euclidean algorithm in this style.

Euclidean algorithm for computing $\gcd(x, y)$, where the inputs x and y are natural numbers, not both zero.

1. If $y = 0$, then output x and stop. Otherwise, continue with steps 2, 3, and 4.
2. Divide x by y, obtaining a quotient q and a remainder r.
3. Replace x by y.
4. Replace y by r, return to step 1, and continue from there.

The reader should check that this is really the same algorithm as that given above in paragraph form. Since this form of algorithm description allows the same numbered step to be repeated many times, we are able to reuse variable names, as we have done here. What we called r_1, $r_2, \ldots$ in our prose description—the successive remainders—are all called r here.

Many parts of the IRS-recommended algorithm for filling out some schedules of one's federal income tax return are given in this format, although rarely with instructions to return to a previous step. The tax booklet for Form 1040 also contains algorithms written in the next form we present, the **flowchart**.

In a flowchart, the numbered steps are replaced by boxes and arrows, with transitions from step to step indicated by the arrows. The statements within the boxes are usually written symbolically. In particular, we will use the **left-pointing arrow** ($\leftarrow$) as the symbol that means "is set equal to." Thus

$$x \leftarrow \text{an expression}$$

means to compute the value of the expression and assign its value to the variable x. Rectangular boxes can be used for simple computational steps (such as $r \leftarrow x \bmod y$), diamond-shaped boxes for decision steps (e.g., "Is $y = 0$?"), with the possible answers indicated on the arrows leaving the box, and ovals for "Start" and "Stop." Figure 4.2 gives a flowchart description of the Euclidean algorithm. We assume again that x and y are the inputs, with the usual restriction that they are not both zero.

For a "computer" to implement an algorithm, it is necessary for the algorithm to be given in, or translated into, a language the "computer" understands. We put quotation marks around "computer" in the last sentence, because the computer can be viewed on many different levels. At the highest level, the computer can be thought of as understanding languages such as BASIC or Pascal, and thus we can

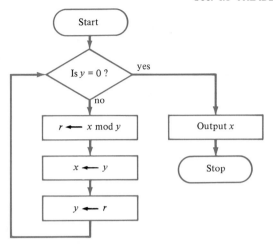

FIGURE 4.2 Flowchart for the Euclidean algorithm.

describe algorithms by giving them in such a **high-level programming language**. Figures 4.3 and 4.4 give descriptions of the Euclidean algorithm in BASIC and Pascal, respectively. When we get down to this level of detail, we have to worry about how the input is to be put in, and how the output is to be put out; in our examples here we use commands "INPUT" and "PRINT" in BASIC, and "readln" and "writeln" in Pascal. The reader who is familiar with these programming languages (or ones like them) should be able to see that these programs do describe the Euclidean algorithm. [For the benefit of the reader who does not know these languages, the assignment statement in BASIC uses an equal sign (=) and the (optional) word "LET," while the assignment statement in Pascal uses ":="; we used ← in the flowchart in Figure 4.2. Also, "INT" in BASIC means the natural number part of a real number, so that "INT (X / Y)" gives the quotient when X is divided by Y, and hence R is the desired remainder.]

```
10  REMARK. THE EUCLIDEAN ALGORITHM
20  INPUT X,Y : REMARK. THEY CANNOT BOTH BE 0
30  IF Y = 0 THEN 80
40  LET R = X - Y * INT ( X / Y )
50  LET X = Y
60  LET Y = R
70  GOTO 30
80  PRINT "THE GCD IS "; X
90  END
```

FIGURE 4.3 A BASIC program for the Euclidean algorithm.

```
program EUCLID;
    var X, Y, R : integer;
    begin
        readln (X,Y); {cannot both be zero}
        while (Y <> 0) do begin
            R := X mod Y;
            X := Y;
            Y := R
            end;
        writeln ('The gcd is ', X)
    end.
```

FIGURE 4.4. Pascal program for the Euclidean algorithm.

Finally, we could consider a computer as operating only at its most fundamental level. With this view we would need to describe an algorithm by giving a program in the **assembly language** or **machine language** of a particular (real or fictitious) computer. A reader familiar with such a language is invited to program the Euclidean algorithm in it, but we will not do so. For our purposes, a high-level language description will be a sufficient level of detail in giving algorithms.

In this book we present most of the algorithms in the form suggested by the Pascal program in Figure 4.4. However, we want to maintain as much flexibility as possible in our descriptions, and we do not wish to get bogged down with the particular syntactical rules of Pascal. Furthermore, we do not want to worry too much about the **data structures** (ways to organize and store the data that the algorithm needs to operate on) necessary for actually carrying out the algorithms. Therefore, we will use a so-called **pseudocode** to describe our algorithms. It will look very much like structured programming languages (such as Pascal, PL/1, or C), and it should usually be quite straightforward for a human being to tell, from the pseudocode, what the steps of the algorithm are. Section 4.2 describes our pseudocode in detail.

CHARACTERISTICS OF MOST ALGORITHMS

We have already mentioned in passing some of the properties that algorithms can be expected to have. Let us now take a somewhat more formal approach and make a list of the properties. Just as most rules have exceptions, most of these properties have exceptions, as our discussion will indicate.

1. An algorithm usually has output. If there were no output, then the answer to the computation would never be known. For example, the Euclidean algorithm has $\gcd(x, y)$ as its output. On the other hand, suppose that we wanted to determine whether the existentially quantified proposition $\exists x: f(x) = 0$ were true (assume that the quantifier ranges over the positive integers). We could design an algorithm that computed $f(1)$, $f(2)$, $\ldots$, halting when the value 0 was obtained. If the computation ever halted,

then by the very fact that it halted we would know that the proposition was true, and we would not need any output to tell us so. (Since this algorithm could never let us conclude that the proposition was false, it is probably not a very good algorithm.) Almost all the algorithms mentioned in this book will have output.

2. An algorithm usually has input and can operate on a large set of possible values of that input. It would not be very useful to have an algorithm that only computed the greatest common divisor of 468 and 54, for example ("Output 18" would be the simplest such algorithm). Instead, a greatest common divisor algorithm should be able to compute $\gcd(x, y)$ for *any* allowable pair of values x and y, as the Euclidean algorithm does. Another way to say this is that *an algorithm should be general.*

3. An algorithm usually is finite, in the sense that no matter what the input values are, the computation will stop after a finite length of time. The length of time can depend on the input; it might well happen that the larger the input, the longer the algorithm takes before it halts. In our example for property 1, however, the algorithm was not finite if 0 was not in the range of f. Unless specified otherwise, all of our algorithms will be finite.

4. An algorithm usually is **deterministic**, with no elements of chance entering into it. The result of each step of the computation is a well-determined, nonrandom value. For example, an algorithm usually would not have a step of the form "Flip a coin, and if the result is heads, then set x equal to 1; otherwise, set x equal to 0." However, there is a whole class of algorithms of great importance to computer scientists that are called **probabilistic** algorithms, and they do have precisely such (controlled) randomness built into them. We will say a little more about probabilistic algorithms in Sections 4.3 and 4.5.

5. The instructions in an algorithm should be unambiguous. Ambiguity would arise, for instance, if an algorithm used the value of a variable that had never been assigned a value. Also, a statement such as "Select an element from list L" is ambiguous, for it does not specify which element is to be picked.

6. Finally, each step in an algorithm must be capable of being performed. For example, an algorithm cannot have a step such as "Assign to x the value of the largest prime number," since there is no such value—see Exercise 8a in Section 1.3. Similarly, we cannot have an algorithm that "begs the question" by containing a step that says essentially "Write down the answer," without telling how this is to be done. For example, an algorithm to compute $\gcd(x, y)$ cannot say simply, "Write down the largest integer that divides both x and y." We will not always reduce all the steps of the algorithms in this book to basic arithmetic, but in all cases it should be clear that each step *can* be done effectively.

Let us conclude with examples of two more algorithms, both a lot simpler than the Euclidean algorithm. As a matter of fact, these algorithms are almost trivial, but they illustrate the *idea* of an algorithm. We present them here in prose form; the reader is asked to give them in other forms (Exercise 14).

EXAMPLE 3. The following algorithm takes as input a list of words $w_1, w_2, \ldots, w_n$ (i.e., a list of strings over the alphabet $\{a, b, \ldots, z\}$), and gives as output the length of a longest word in the list (one whose length is at least as long as the length of every other word in the list), as well as this word itself. Thus the output will be $\max(\text{length}(w_1), \text{length}(w_2), \ldots, \text{length}(w_n))$, together with one of the words in the list whose length is this value. We assume that $n \geq 1$.

> First set *maxword* equal to w_1 and set *maxlength* equal to the length of w_1. Now successively for each i from 2 to n, inclusive, compare the length of w_i to *maxlength*; if $\text{length}(w_i) > maxlength$, then set *maxword* equal to w_i and set *maxlength* equal to the length of w_i. The values of *maxlength* and *maxword* at the end are the desired answer. ◆

EXAMPLE 4. The following algorithm takes as input a list of numbers $s_1, s_2, \ldots, s_n$, and gives as output the sum of the numbers in the list $(s_1 + s_2 + \cdots + s_n)$, that is, $\sum_{i=1}^{n} s_i$. We assume that $n \geq 0$ (recall that the sum is defined to be 0 if the list has no numbers in it, i.e., if $n = 0$).

> First set *sum* equal to 0. Now successively for each i from 1 to n, inclusive, add s_i to the current value of *sum*, and make that the new value of *sum*. The value of *sum* at the end is the desired answer. ◆

Much of mathematics and computer science involves the creative process of constructing algorithms to solve problems. You are called upon to construct some algorithms in the exercises that follow, as well as in subsequent exercise sets. *There is no algorithm for constructing algorithms.* You need to figure out how you would go about solving a given problem (try watching yourself solve a few instances of the given problem), and then you need to explain your process completely and unambiguously. In the next section we present a kind of computer language with which to encode algorithms, and you will have an opportunity to write algorithms in this language; for now, concentrate on the algorithms themselves, and write them clearly in the form requested.

SUMMARY OF DEFINITIONS

algorithm: an unambiguous set of instructions for carrying out a task; a clearly specified method for obtaining a solution to a problem.

assembly language: a language with well-defined syntax that can be used to communicate an algorithm to a computer essentially at the level of detail at which the computer will actually work in executing it.

common divisor of two integers: a natural number that divides them both (e.g., 3 is a common divisor of 123456789 and 987654321).

data structure: a construct for organizing the data that an algorithm needs to operate on (e.g., a stack, which is a list to which new elements can be added or current elements deleted, but additions and deletions can occur only at the top of the stack).

determinism: the property that most algorithms have of producing the same output every time they are executed; a deterministic algorithm has no elements of chance in it.

Euclidean algorithm: a particular algorithm for finding the greatest common divisor of two nonnegative integers.

flowchart: a pictorial representation of an algorithm, in which the steps are contained in boxes and the order in which the steps are to be executed is indicated by directed lines between boxes.

greatest common divisor of two natural numbers x and y, not both of which are 0: the largest integer that is a common divisor of x and y, denoted $\gcd(x, y)$ (e.g., 9 is the greatest common divisor of 123456789 and 987654321, as can be verified by the Euclidean algorithm).

high-level programming language: a language with well-defined syntax that can be used to communicate an algorithm to a computer without the need to specify the steps at the level of detail at which the computer will actually carry them out (e.g., Pascal, LISP, Prolog, FORTRAN).

input to an algorithm: the data on which an algorithm acts (e.g., the natural numbers x and y, not both 0, of which the Euclidean algorithm computes the greatest common divisor).

iteration: repetition (e.g., the Euclidean algorithm consists of iterating the step of replacing the pair (x, y) by the pair $(y, x \bmod y)$ until the second element of the pair is 0).

left-pointing arrow: is set equal to (e.g., the instruction "$x \leftarrow x + 1$" means to set the variable x equal to its old value plus 1).

machine language: a language with well-defined syntax in which the steps of an algorithm can be specified to a computer exactly as the computer will execute them.

output of an algorithm: the answer that the algorithm produces for the given input (e.g., the output of the Euclidean algorithm is $\gcd(x, y)$ when the input is x and y).

probabilistic algorithm: an algorithm that contains one or more steps involving a random choice (e.g., an algorithm to estimate the area of an irregularly shaped region within the unit square may work essentially by repeatedly simulating the process of throwing a dart at the square and keeping track of the proportion of times that the dart lands inside the region).

pseudocode: a method for describing algorithms, similar to a high-level programming language but with flexibility in describing individual steps (for the pseudocode used in this book, see Section 4.2).

EXERCISES

1. Describe an algorithm that takes as input three distinct natural numbers and gives as output the input value that is neither the largest nor the smallest of these three numbers.

2. Describe, in a prose paragraph, an algorithm for adding positive fractions. Be precise. Assume only the four basic arithmetic operations on natural numbers. The answer need not be in lowest terms. (The input will be four positive integers a, b, c, and d, and the output will be the numerator and denominator of $(a/b) + (c/d)$.)

3. Describe, in a prose paragraph, an algorithm for the following problem: Given a positive integer n, find the largest perfect square less than or equal to n. (For example, if the input is 11, then the output is 9, since $3^2 = 9 \leq 11$ but $4^2 = 16 > 11$.) Be precise. Assume only the four basic arithmetic operations on natural numbers.

4. Describe your algorithm in Exercise 3 using numbered steps.

5. Describe your algorithm in Exercise 3 using a flowchart.

6. Find the greatest common divisor of 50 and 60 by explicitly listing the divisors of each.

7. Find the following greatest common divisors by factoring the arguments into their prime factorizations.
   (a)   gcd(900, 1750).
   (b)   gcd(46, 27).
   (c)   gcd(2323, 9191).

8. Find the greatest common divisors in Exercise 7 using the Euclidean algorithm.

9. Express $d = \gcd(x, y)$ totally in logical and arithmetical symbols. Assume that not both x and y are 0.

10. Find the following greatest common divisors using the Euclidean algorithm.
    (a)   gcd(2584, 1597).
    (b)   gcd(19287341, 11111).
    (c)   gcd(5787250564, 3382740674).

11. Prove Theorem 1. [*Hint*:   For each prime number p, look at the power to which p occurs in the prime factorizations of d and $\gcd(x, y)$.]

12. Describe, in a prose paragraph, a naive algorithm for finding the **least common multiple** of two positive integers. The least common multiple of x and y, written lcm(x, y), is the smallest positive integer that is a multiple of both x and y. For example, lcm$(23, 5) = 115$ and lcm$(8, 12) = 24$ (see Exercise 13 for a deeper look at least common multiples).

13. The least common multiple of two positive integers x and y is intimately related to their greatest common divisor.
    (a) Express $m = \text{lcm}(x, y)$ totally in logical and arithmetical symbols (see Exercise 12).
    (b) Find an algorithm for computing $\text{lcm}(x, y)$ based on the prime factorizations of x and y.
    (c) State and prove an analogue of Theorem 1 for least common multiples.
    (d) Prove that $\text{lcm}(x, y) \cdot \gcd(x, y) = x \cdot y$. [*Hint*: See part (b).]
    (e) Use part (d) to devise a good algorithm for computing least common multiples.
    (f) Apply your algorithm in part (e) to find $\text{lcm}(50059, 98789)$.

14. Describe the following algorithms in the following forms.
    (a) The algorithm given in Example 3 in numbered steps.
    (b) The algorithm given in Example 3 as a flowchart.
    (c) The algorithm given in Example 4 in numbered steps.
    (d) The algorithm given in Example 4 as a flowchart.

15. Prove the converse of Theorem 1.

16. Show that $\gcd(x, x + 1) = 1$ for every positive integer x.

17. Show that $\gcd(x^2 - 1, x) = 1$ for every positive integer x.

18. Describe an algorithm, which uses the Euclidean algorithm, for reducing a positive fraction to lowest terms. Be specific about what the input and output are.

19. Reduce the following fractions to lowest terms. [*Hint*: See Exercise 18.]
    (a) $\dfrac{238133}{341936}$.

    (b) $\dfrac{238134}{341936}$.

    (c) $\dfrac{238135}{341936}$.

20. Describe an algorithm for obtaining the prime factorization of a positive integer $m \geq 2$.

21. Describe an algorithm for finding, given a positive integer N, four integers whose squares add up to N.

22. Describe an algorithm for determining, given a natural number N, whether or not N can be written as the sum of the squares of two natural numbers.

23. Describe an algorithm for finding the second largest element of a set of distinct numbers $\{s_1, s_2, \ldots, s_n\}$, assuming that $n \geq 2$.

24. Describe an algorithm that takes as input a list of words $(w_1, w_2, \ldots, w_n)$ together with a word w and gives as output "yes" or "no," depending on whether or not w is in the list.

25. Describe an algorithm that takes as input a list of words $(w_1, w_2, \ldots, w_n)$ together with a word w and gives as output the subscript of the first occurrence of w in the list (if w is not in the list, the output is to be 0).

26. Describe an algorithm that takes as input a list of words $(w_1, w_2, \ldots, w_n)$ together with a word w and gives as output the subscript of the last occurrence of w in the list (if w is not in the list, the output is to be 0).

27. Suppose that the input to an algorithm is two lists of natural numbers, each in strictly increasing numerical order. The input can be viewed as being two sets of natural numbers. Describe an algorithm for producing each of the following.

    (a)   A list, in strictly increasing numerical order, of the numbers in the intersection of the two lists.

    (b)   A list, in strictly increasing numerical order, of the numbers in the union of the two lists.

    (c)   A list, in strictly increasing numerical order, of the numbers in the difference of the two lists (i.e., the numbers that are in the first list but not in the second).

Challenging Exercises

28. Describe an algorithm for determining, given a positive integer n, whether the decimal expansion for $1/n$ terminates.

29. Find a totally geometric implementation of the Euclidean algorithm for finding the length of the greatest common measure between the two sides of a rectangle.

30. One property of the greatest common divisor is as follows. For any positive integers x and y there exist integers m and n such that $xm + yn = \gcd(x, y)$. Show how to use the ideas in the Euclidean algorithm to find m and n, given x and y.

31. Describe an algorithm for computing $\sqrt{2}$ to within 10^{-n}, for n a positive integer.

32. Describe an algorithm for the following problem. The input is a set of pairs of integers, each integer lying between 1 and n, inclusive. The output is to be an ordering of the integers from 1 to n such that for all x and y, if (x, y) is in the input set, then x precedes y in the ordering. Assume that the input is such that an ordering satisfying this condition is possible. For example, if $n = 5$ and the input is $\{(1, 2), (1, 4), (5, 2), (5, 3)\}$, then one possible output is $(1, 5, 4, 2, 3)$. (This problem is known as **topological sorting**; it can be formulated as finding a total order that is a superset of a given partial order.)

33. Describe an algorithm for determining, given a subset R of $\{1, 2, \ldots, n\} \times \{1, 2, \ldots, n\}$ whether
    (a) R is an equivalence relation on $\{1, 2, \ldots, n\}$.
    (b) R is a partial order on $\{1, 2, \ldots, n\}$.

34. Describe an algorithm for producing the transitive closure of a relation, which is given as input as a list of ordered pairs.

Exploratory Exercises

35. Write computer programs to implement some of the algorithms given in the examples and exercises in this section and run them on a computer.

36. It is a theorem of mathematics that the probability that two integers chosen "at random" are **relatively prime** (i.e., have a greatest common divisor of 1) is $6/\pi^2$. Conduct an experiment with a computer to see if this statement seems plausible.

37. Consider the following two-player game related to the Euclidean algorithm. Two unequal positive integers are given. The players (alternately) replace the larger number by itself minus some positive multiple of the smaller number, without forming a negative number. The winner is the player who is able to replace a number by 0. For example, the numbers might be 12 and 5. The first player may choose to replace 12 by $12 - 2 \cdot 5 = 2$; this would leave the numbers 5 and 2. Suppose that the second player then replaces the 5 by $5 - 2 = 3$, leaving 5 and 3. The first player's next move is forced to be replacing 3 by $3 - 2 = 1$; the numbers are now 2 and 1. The second player now wins by replacing 2 by $2 - 2 \cdot 1 = 0$.
    (a)   Play the game several times with some friends.
    (b)   Determine the optimal strategies, and decide which player (first or second) will win the game if both players use their best strategies. (Obviously, the answer will depend on the starting numbers.)
    (c)   Consult Spitznagel [290] for more details.

38. Discuss in a two- or three-page essay what it means to "solve" a mathematical problem. (Find out more by consulting P. J. Davis [61].)

39. We have taken basic arithmetic steps as our smallest unit in describing algorithms. It is possible to go to an even more primitive level, using only adding or subtracting 1 as the basic operation. See Guy [132] for an interesting example of what can be done in this mode.

40. See Maurer et al. [209] for a lively discussion of the importance—or lack thereof—of the algorithmic way of thinking about mathematics.

41. Designing algorithms is one aspect of solving problems. For some good general heuristics on problem solving, consult Pólya [236].

42. Find out more about the mathematician and master problem solver George Pólya by consulting Albers and Alexanderson [4].

43. Find out more about algorithms by consulting Knuth [174] and Wilf [323].

SECTION 4.2
PSEUDOCODE DESCRIPTION
OF ALGORITHMS

A mathematician's algorithm for boiling water:
if the water is already boiling **then** quit
  else. . .

—Anonymous

In this section we explain the pseudocode we will use for describing algorithms in this book. In some sense, this section is a quick course in computer programming.

The basic structure of an algorithm given in pseudocode is as follows:

procedure *procedure_name*(arguments)
  {comments}
    statement
    statement
    ⋮
    statement

The statements are the steps of the algorithm and may be of several types, discussed below. They are executed in order until the algorithm encounters a **return** statement (which we will explain presently), at which time it stops. The end of each statement is usually indicated by the end of the line on which it is written, but if there is not enough room, a statement may continue onto subsequent lines; this will usually be clear from indentation and context. The words in boldface type in an algorithm written in pseudocode are special words in the pseudocode and appear precisely as written. Our pseudocode description of an algorithm will always begin with the word "**procedure**," followed by the name of the procedure, which can be any name, usually chosen to be descriptive of what the procedure does. (We will tend to use the words "algorithm" and "procedure" interchangeably in our discussions.) The **arguments**—the inputs to the procedure—follow, enclosed in parentheses and separated by commas if there are two or more of them. In the unlikely event that there are no arguments, the argument list and the parentheses are omitted. We will usually also indicate what sort of objects the arguments are expected to be—integers, strings, sequences, graphs, and so on. These labels are given after each argument or group of similar arguments, separated from their arguments by colons. Comments, enclosed in braces, may be placed anywhere in the pseudocode description; these are not part of the procedure itself but help explain it to a human reader. There should be a comment at the beginning to explain what the algorithm is supposed to do and what the important variables stand for. Note that the only input to the procedure is through its argument list; there are no "read" statements for a "user" to supply information. Similarly, in most cases we will have no "write" statements; instead, the output is indicated by a **return** statement.

A procedure does its work by assigning values to **variables**. Each variable has a name, usually chosen to be descriptive of what it represents in the algorithm. Variable names can be anything from a single letter to a descriptive phrase constructed from letters, digits, the prime symbol ($'$), and the underscore character (_). For example, a tree (an object we discuss in Chapter 9) might be called T, T'', T_1, or *new_tree*. We do not want to get bogged down with keeping track of data types, so we will be rather freewheeling when it comes to what variables can represent. Variables can have values that are numbers (integers, real numbers, etc.), sets, lists, matrices, graphs, or more complicated data structures. They can also be logical (sometimes called **Boolean**) variables that can take only the values **true** and **false**. We will not "declare" the variables at the beginning of the procedure; their types should be clear from context and comments. Any reasonable operations are allowed on variables, such as addition of numbers or the extraction of the first element of a list. Algebraic and logical expressions will be used freely, as will constructions in English when their meaning is clear.

A typical statement might be an **assignment** statement of the form

variable $\leftarrow$ expression

which indicates that the expression is to be evaluated and its value assigned to the variable, replacing any previous value the variable might have had. The variable itself might appear in the expression. For example, the assignment statement $x \leftarrow x + 1$ causes the current value of x to be incremented by 1. The left-hand side of an assignment statement may even be something more complicated than a variable, like an ordered pair of variables; the expression on the right-hand side must represent a value of the same type, of course. The expression might be an algebraic expression (such as $\sqrt{3x/5}$) or it might be given wholly or partially in English, as long as the statement is unambiguous and capable of being carried out effectively. For example, in Algorithm 1 of Section 8.2 we will find the following statement:

$x \leftarrow$ lowest numbered vertex adjacent to w in H

Given the definitions of that section and the context in which the statement occurs, this statement does meet these criteria.

To indicate the output of an algorithm written in pseudocode, we use a statement of the form

return(*output*)

where *output* is a list of the output. When this statement is executed, the algorithm stops. We think of the "person" performing the algorithm as "returning" from work, answers in hand. Usually there is only one output (one number, one matrix, or whatever), but occasionally there is more than one. In the unlikely event that an algorithm provides no output, the syntax of the statement is just "**return**." In Example 8 we will construct a procedure that prints something, rather than returning any values. A statement of the form "**print**(...)" indicates that the items enclosed in parentheses are to be written down, in the given order.

A statement can be a block of statements, indicated by

begin
   statement
   statement
   $\vdots$
   statement
end

which is simply a sequence of statements to be executed in the order written. We will need blocks of statements in order to construct algorithms with looping and branching, which we discuss in the next two subsections.

EXAMPLE 1.   Here is the pseudocode description of an algorithm to compute two different means of two positive real numbers.

procedure *means*$(x, y :$ positive real numbers$)$
   {this procedure computes the arithmetic and geometric means of the two input
     values}
     arithmetic_mean $\leftarrow (x + y)/2$
     geometric_mean $\leftarrow \sqrt{xy}$
     return(*arithmetic_mean*, *geometric_mean*)

This algorithm produces two outputs, which are returned to the "person" calling the procedure.   ◆

LOOPS

With the syntax explained so far, only so-called "straight-line" programs can be described, programs in which simple statements are executed in order; there is no returning to statements executed previously. Most algorithms are not so simple as to be describable in a predetermined small finite number of steps. Instead, they usually involve some repetition of the same steps, a process called iteration. We have two constructs for iteration in our pseudocode: the **while** loop and the **for** loop.

The **while** loop is a statement of the form

> **while** condition **do**
>   statement

where "statement" may actually be a block of statements enclosed between the words **begin** and **end**. The "condition" is any logical expression (i.e., a proposition), usually with one or more variables. A **while** loop is executed as follows. First, the condition is evaluated. If it is false, nothing more is done; execution continues with the statement *following* the statement in the **while** loop. If it is true, "statement" is executed, after which "condition" is evaluated again. If the condition is now false, nothing more is done. If it is still true, "statement" is executed again, after which "condition" is evaluated again. This process continues until "condition" finally becomes false; then execution continues with the statement following the statement in the **while** loop. (There is one exception: The procedure may stop somewhere inside "statement" because of a **return** statement. In that case execution immediately returns to whoever called this procedure. It is sometimes useful to have "condition" be simply **true**, in which case presumably the procedure terminates with a **return** statement somewhere inside the loop.)

EXAMPLE 2. We can now write the Euclidean algorithm in pseudocode. The reason this procedure works correctly is implicit in our discussions in Section 4.1, where we wrote the same algorithm in other forms.

> **procedure** *euclid*$(x, y :$ natural numbers$)$
>   {this procedure calculates gcd(x, y) using the Euclidean algorithm; it assumes that x and y are not both 0}
>   **while** $y \neq 0$ **do**
>   **begin**
> $r \leftarrow x \bmod y$ {r is the remainder when x is divided by y}
> $x \leftarrow y$
> $y \leftarrow r$
>   **end**
>   **return**(x)
>   {at the end, x is the greatest common divisor of the original inputs}

In real programming, it would probably be poor practice to use the same variable (x) for both input and output of this procedure, but we allow such simplifications in our pseudocode. ◆

We adopt the convention that logical "and" operations are evaluated from left to right, with only as much of the expression evaluated as is needed to determine its truth value (as soon as one false subexpression is encountered, the expression is known to be false). For example, evaluation of the expression

$$i \leq n \, \wedge \, x_i > 0$$

when $i = 5$ and $n = 4$ would yield the value **false**, *without x_5 being looked at*. This is useful in **while** loops, since we often have compound conditions, parts of which might not make sense under certain conditions (there may be no x_5 in the case just cited).

EXAMPLE 3. The following algorithm computes the arithmetic mean of a sequence of numbers. Note that the argument A is a sequence, whose terms are called a_1, a_2, ..., a_n, according to the comment. We also assume that the length of the sequence, n, is available to the procedure.

> **procedure** *sequence_mean(A)*
> $\{A = (a_1, a_2, \ldots, a_n)$ is a sequence of one or more numbers; *mean* will be their mean$\}$
>   *mean* $\leftarrow 0$
> $i \leftarrow 1$
>   **while** $i \leq n$ **do**
>   **begin**
>   *mean* $\leftarrow$ *mean* $+ a_i$
> $i \leftarrow i + 1$
>   **end**
>   *mean* $\leftarrow$ *mean*$/n$
>   **return**(*mean*)

The reader should "play computer" with this procedure, stepping through it for, say, $n = 3$, to see exactly how it works and to be convinced that it is correct.   ◆

Many iterations are of the form seen in this example: The statement in the **while** loop is executed once for each value of an index variable in a certain sequence. In Example 3 the index variable is i and the sequence of values that i takes on is $(1, 2, \ldots, n)$. We have a second, simpler loop construction in our pseudocode for such cases, written as follows:

> **for** iteration **do**
>   statement

Here "iteration" is an unambiguous description of some finite sequence, usually in terms of a variable that assumes successively the values in the sequence. In most cases "iteration" will be something like

$$i \leftarrow 1 \textbf{ to } n$$

which indicates that the sequence in question is $(1, 2, \ldots, n)$. If we had wanted the iteration to take place in the reverse order, we could have written it as follows:

$$i \leftarrow n \textbf{ down to } 1$$

If we want only the even numbers from 10 to 100, inclusive, then we can write the iteration with the step size indicated:

$$i \leftarrow 10 \textbf{ to } 100 \textbf{ by } 2$$

Some iterations indicate the empty set. A loop that begins

for $k \leftarrow 1 \textbf{ to } n \textbf{ do}$

for example, would have its statement executed no times if $n \leq 0$. Occasionally the iteration can be more complicated; in Algorithm 1 in Section 10.3 we find "**for** each unlabeled vertex w in G **do**"; in the context of that algorithm, this iteration is well defined. Note that, as with **while** loops, a **for** loop may be interrupted by a **return** within "statement."

EXAMPLE 4.   The following algorithm counts the number of 1's in a bit string s. Recall from Example 6 in Section 2.2 that a bit string is a finite sequence of 0's and 1's; for instance, 110100, 111, and λ are bit strings of lengths 6, 3, and 0, respectively.

procedure *count_ones*(s : bit string)
  {$s = s_1 s_2 \ldots s_n$ is a bit string of length $n \geq 0$; the number of 1's in s is
   recorded in *count*}
    count $\leftarrow 0$
    for $i \leftarrow 1 \textbf{ to } n \textbf{ do}$
      count $\leftarrow$ *count* $+ s_i$
    return(*count*)

Dealing somewhat loosely with data types, we exploited the fact that the bits could be viewed as numbers to be added to *count*. Each s_i is either 0 or 1. Each time $s_i = 1$ (as i progresses from 1 to n), *count* is incremented by 1; when $s_i = 0$, *count* remains the same. Note that the correct answer is obtained if s is the empty string, since in that case the **for** loop is executed no times.   ◆

It is certainly possible for loops to be nested one inside another. In other words, the "statement" in the body of a **while** or **for** loop can be another loop or can be a block of statements that contains one or more loops. Indeed, many procedures will contain several cases of such nesting.

IF...THEN STATEMENTS

Most algorithms require decisions to be made at one or more points in their execution. In finding the larger of two numbers x and y, we need to give one answer if x is larger than y and another answer if y is larger than x. In pseudocode we indicate such decision points with the **if...then** statement, which, in its simplest form, has the following syntax.

if condition **then** statement

When this statement is performed, first the logical expression "condition" is evaluated; if it is true, "statement" is executed; otherwise, "statement" is not executed. In either case execution then continues with the statement following the **if...then** statement, as long as there was no **return** within "statement."

EXAMPLE 5.   The procedure for determining the larger of two inputs could be written as follows:

procedure *maximum*$(x, y :$ numbers$)$
    {*max* is to be the larger of the two inputs}
      max $\leftarrow x$
      if $y > x$ **then** *max* $\leftarrow y$
      return(*max*)   ◆

Other times we want the algorithm to take one action if a condition is satisfied and another action if it is not. We use the following form of the **if...then** statement for such situations:

if condition **then** statement **else** statement

Again, the "condition" is evaluated first. If the condition is true, then the first "statement" is executed, but not the second, whereas if the condition is false, then the second "statement" is executed, but not the first.

EXAMPLE 6.   We can now rewrite the procedure *maximum* in Example 5 more simply as follows:

procedure *simpler_maximum*$(x, y :$ numbers$)$
    {*max* is to be the larger of the two inputs}
      if $y > x$ **then** **return**(y) **else** **return**(x) ◆

Conditions in **while** loops and **if...then** statements usually involve relations such as $<$ or $=$. It is important not to confuse the *equality relation*, which we denote in the usual way by =, with the assignment operator, which we have denoted with a left-pointing arrow ←. In our pseudocode one cannot write a conditional statement as "**if** $x \leftarrow y$ **then** ..." or an assignment statement as "$x = x + 1$." The assignment operator ← is used as an imperative verb, whereas the relation symbol = is used as a declarative (or, more to the point, an interrogative) verb.

If...**then** statements can be nested, since there is nothing preventing the statements *within* an **if**...**then** statement from themselves being **if**...**then** statements. There is a potential for ambiguity, however, because of the **else** clauses. Unless we specify a rule, it will not be clear which "**if**" an **else** action is to attach to. For example, is the action D in

> **if** A **then**
> **if** B **then** C
> **else** D

to be executed in lieu of the action C (in the case in which A is true and B is false), or is it to be executed in lieu of the second **if** statement (in the case in which A is false)? We adopt the convention that the former interpretation is intended. In particular, *an* **else** *clause attaches itself to the closest* **if** *clause that it can*.

ARRAYS

In many of the algorithms that we will discuss in this book, it is necessary to keep track of many things at once. Often we do not even know ahead of time how many things there are. For example, suppose that we need to compute and calculate with the distances of n cities from Washington, D.C. It would not make sense to have separate variable names for each of these distances. Instead, we can think of the distances as being collected into a data structure called a list or one-dimensional **array**. Thus we might have an array called *distance* for keeping track of the n distances in our example. Arrays are indispensable in our pseudocode descriptions of algorithms. We usually refer to the elements in the list with normal functional notation, but sometimes we may use the equivalent subscript notation. In our example, we would refer to the distance between Washington and the ith city with the expression *distance*(i). Note that conceptually an array is just an indexed set (see Section 2.2). We do not concern ourselves with how the values in the array might be physically stored in a real computer.

EXAMPLE 7. Suppose that we wish to design an algorithm to record the frequency with which each of the digits 0 through 9 appears in a string of digits. We can describe such an algorithm in pseudocode as follows:

procedure *tabulate*(s : string of decimal digits)
{this procedure records in the array *count* the frequencies with which the digits 0 through 9 occur in the string $s = s_1 s_2 \ldots s_n$; *count* is indexed from 0 to 9, and *count*(j) is the number of occurrences of the digit j in the string s, for $j = 0, 1, \ldots, 9$}
  for $j \leftarrow 0$ **to** 9 **do**
    count(j) $\leftarrow 0$ {initialize all counts to 0}
  for $i \leftarrow 1$ **to** n **do**
    count(s_i) $\leftarrow$ *count*(s_i) $+ 1$
  return(*count*)

Note that the entire array is returned. If later we wished to find the digit that occurred most frequently, the array *count* contains all the necessary information (see Exercise 15).   ◆

EXAMPLE 8. Suppose that we are given a list of names of teams in a sports league, and we want to print out a list of all possible pairings of two distinct teams. Let us structure the input as an array *team*, containing the names of the teams (thus these are strings) and indexed by $\{1, 2, \ldots, n\}$. The following procedure will print the desired list. The iterations for i and j here are a little tricky, since we want to print each pair only once. The reader should walk through this procedure for $n = 4$ to see that it performs correctly.

procedure *season*(*team* : array of length n of strings)
  {this algorithm prints all pairings of distinct teams}
    for $i \leftarrow 1$ **to** $n - 1$ **do**
      for $j \leftarrow i + 1$ **to** n **do**
        print(*team*(i) "—" *team*(j))
    return

Note that this particular procedure does not return any output; instead, it prints the desired information as it executes.   ◆

We will also have need for matrices or two-dimensional arrays, indexed by an ordered pair of variables, and there is nothing preventing us from using arrays with more than two dimensions.

EXAMPLE 9. Design an algorithm that takes as input an n by n array D and gives as output the largest value appearing in D. (We can imagine an application in which the array D gives the distances between pairs of cities served by an airline company; our algorithm will compute the maximum distance between two cities served by that airline.)

Solution. The following algorithm searches through the array D to find the desired maximum. We need to have a loop within a loop in order to access all the entries.

procedure *max_distance*($D : n$ by n matrix)
   {this algorithm finds the maximum value in D}
     $max \leftarrow -\infty$
     for $i \leftarrow 1$ **to** n **do**
       for $j \leftarrow 1$ **to** n **do**
         if $D(i, j) > max$ **then** $max \leftarrow D(i, j)$
     return(max)

Note that we initialized *max* to be $-\infty$, which from a practical point of view need merely be a large negative number. Alternatively, we could have initialized it to be $D(1, 1)$ or any other value in D.   ◆

We will have more to say about matrices and algorithms on matrices in Section 4.5.

SUBPROCEDURES

We must deal finally, in our description of pseudocode, with how a procedure can invoke other procedures; we call the latter **subprocedures** in this context. This is an important aspect of algorithm construction, for two reasons. First, it enables us to build up complex algorithms piece by piece. We can decompose a task into a series of smaller tasks, develop algorithms for the smaller tasks, and then put them together as an algorithm for the entire task. Modern computer programming theory and practice is founded on this principle. Second, permitting procedures to call other procedures allows us, in Section 5.2, to discuss recursive algorithms, which call upon themselves. Recursiveness, as we will see in Chapter 5, is perhaps the most important concept in all of discrete mathematics.

Suppose that we want to compute the greatest common divisor of three positive integers x, y, and z. A little thought will show that

$$\gcd(x, y, z) = \gcd(\gcd(x, y), z).$$

Thus we can compute $\gcd(x, y, z)$ by invoking the Euclidean algorithm twice. If we want to write a procedure for computing $\gcd(x, y, z)$, it would be wasteful to write out the steps of the Euclidean algorithm twice. Instead, we write the Euclidean algorithm as a subprocedure that returns the greatest common divisor of its two arguments, and then write a procedure that **calls** upon this subprocedure.

When a procedure calls a subprocedure, execution of the calling procedure is temporarily suspended, and control of execution passes to the subprocedure. When the subprocedure finishes, control of execution passes back to the suspended calling procedure, which continues where it left off. A good metaphor to keep in mind here is that of a contractor (building a house, say), who hires subcontractors to do specific tasks associated with the project (such as putting in the electrical wiring). The contractor is the main procedure; the subcontractors are the subprocedures, invoked by the contractor. While the subcontractors are doing their work, the contractor can be taking a break. Note that the contractor need not be concerned with *how* the subcontractors do their jobs; he or she only cares that the jobs are completed to specification.

A subprocedure has exactly the same form as a procedure. One or more of the statements in a subprocedure will be the **return** statement. Once the **return** statement is executed, the subprocedure is finished; no further statements in the subprocedure are executed, and control of the computation passes back to the calling procedure.

A procedure can call a subprocedure in one of two ways, depending on whether the subprocedure returns any output. If the subprocedure *sub_proc* returns output, the calling procedure invokes it by using *sub_proc*(arguments) in an expression. The arguments enclosed in parentheses when this call is made are the inputs that the subprocedure will work with. The value returned by the subprocedure is incorporated into whatever calculation the main procedure is engaged in. For example, if the subprocedure $power(b, n)$ computes b^n, we could write the statement

$$z \leftarrow \sqrt{power(x, 2) + power(y, 2)}$$

in a procedure. The effect is that z is assigned the value $\sqrt{x^2 + y^2}$. Note that the subprocedure acted on x in the first case and y in the second, and that it computed the second power both times, since it took its inputs (x and 2 the first time, y and 2 the second time) from the calling expression. As another example, suppose that we have a subprocedure *sub_proc* that takes one input and returns a pair of outputs. We could reference this subprocedure in a main procedure by means of the following statement:

$$(n, B) \leftarrow sub\_proc(G)$$

This statement has the effect of assigning the first of the outputs to n and the second to B.

If the subprocedure *sub_proc* does not return output, the calling procedure invokes it with the following statement:

call *sub_proc*(arguments)

Again, the arguments in the calling statement provide the actual inputs to the subprocedure.

Before giving some substantive examples to illustrate the use of subprocedures, we must discuss the way in which arguments are passed to subprocedures. Although this is a rather technical point, we need to discuss it so that our pseudocode will be unambiguous. (On the other hand, this discussion will be brief, since this section is not intended to be a computer programming manual.) Our convention is that *all arguments are passed to subprocedures by value and not by reference*. This means that the variables that appear in the argument list of a subprocedure are local to the subprocedure, and no changes made to these variables in the subprocedure affect any variables in the calling procedure. At the point in the execution of the algorithm at which the subprocedure is called, these local variables are assigned their initial values, as dictated by the values of the arguments in the argument list provided by the calling procedure.

For example, suppose that there is a call to $sub\_proc(x, A, 3)$ in the main procedure, and suppose that the subprocedure sub_proc begins with the following heading:

procedure $sub\_proc(x, C, n)$

When the execution of the subprocedure begins, its local variables x, C, and n are assigned the value of x in the calling procedure, the value of A in the calling procedure, and 3, respectively. After the subprocedure has finished executing and control is returned to the main procedure, the values of x and A in the main procedure are exactly as they were when the subprocedure was called, even if within the subprocedure there might have been some new values assigned to x or C.

Furthermore, all the variables used in a subprocedure are local to the subprocedure. For example, if a variable j is used as a loop index in both a subprocedure and the calling main procedure, then the value of j in the suspended main procedure is not affected by anything that might have happened to j in the subprocedure. In other words, we are permitting no global variables. All information that a subprocedure needs must be passed to it explicitly through the argument list.

Finally, we will allow arbitrarily deep nesting of procedure calls (procedures can call procedures that call procedures that call procedures that ...), and we place no restrictions on procedures' calling themselves.

Let us return to the problem with which we began this subsection.

EXAMPLE 10. Write a simple procedure to calculate the greatest common divisor of a list of positive integers, using the procedure *euclid* from Example 2 as a subprocedure.

Solution. The following procedure accomplishes the task by making use of the observation we made at the beginning of this subsection. We start by finding the greatest common divisor of the first two numbers in the list, and then we repeatedly find the greatest common divisor of our previous gcd and the next number in the list. The final gcd is then the greatest common divisor of the entire list.

procedure $gcd\_of\_list(A)$
  {$A = (a_1, a_2, \ldots, a_n)$ is a list of $n \geq 2$ positive integers; the greatest common
    divisor of these integers is calculated}
      $d \leftarrow euclid(a_1, a_2)$
      for $i \leftarrow 3$ **to** n **do** {note that this loop is skipped if $n = 2$}
        $d \leftarrow euclid(d, a_i)$
      return(d)

Note that the subprocedure *euclid* is invoked at the beginning of the procedure
gcd_of_list, and again in the iterated step. The first time, the local variables x and
y in *euclid* are given the values of a_1 and a_2, respectively. Thereafter, the values
of x and y in *euclid* are given the values of d and a_i, respectively. Each time
euclid is called, it operates with a clean slate, oblivious to the fact that it might
have been called before. (When we discuss recursive procedures in Section 5.2,
we will see that this notion of a subprocedure's beginning with a clean slate when
it is called is crucial; in that section we will have subprocedures called anew even
when they are in the middle of their own operation.) Furthermore, the fact that x
and y get "destroyed" in the subprocedure *euclid* has no effect on the variables a_i
or d, which get passed to the subprocedure, since only the *values* of the variables
are passed. ◆

We close this section with one final example, which illustrates many of the
features of our pseudocode, as well as some of the problem-solving techniques that
go into constructing algorithms.

EXAMPLE 11.   Describe in pseudocode an algorithm for finding, given a positive integer N,
the smallest pair of twin primes greater than N. (A pair of twin primes is a pair
of prime numbers whose difference is 2, such as 41 and 43.) Thus the output, if
it exists, will be a pair of prime numbers whose difference is 2, both of which are
greater than N. For instance, if the input is 90, then the output is $(101, 103)$. No
one knows whether there is any input for which the output does not exist; in other
words, no one knows whether there is a largest pair of twin primes, beyond which
there are no twin primes. Twin primes have been found as high as anyone has
looked—well above 10^{700}—so if there is a largest pair of twin primes, it must be
very large.

Solution.   Suppose that we denote the pair of primes we are looking for by p and $p + 2$.
We can assume that p is odd and $p \geq 3$. We want to find the smallest such pair for
which $p > N$. Clearly, it will be helpful to have as part of our algorithm a method
for determining whether a positive integer is prime. Let us suppose for the time
being that we *already have* a subprocedure for performing a test for primeness;
suppose that $is\_prime(x)$ returns **true** if x is prime and **false** if x is not prime,
where x is an odd number greater than or equal to 3. Then our main procedure
can be written as follows:

procedure *next_twin_primes*(N : positive integer)
  {this procedure finds the next pair of twin primes greater than N; if no such
  pair exists, then the procedure never halts}
    if N is even **then** $p \leftarrow N + 1$
      else $p \leftarrow N + 2$ {start at first odd number $> N$}
    while true do
      begin
        if *is_prime*(p) **then**
            if *is_prime*($p + 2$) **then return**($p, p + 2$)
              else $p \leftarrow p + 4$
          else $p \leftarrow p + 2$
      end

The logic is a little complex here, so let us comment on how this procedure works. Our search for the twin primes starts at the first odd number greater than N; the first statement in the procedure sets p to this value. Then the main **while** loop executes, apparently forever, since the condition (**true**) is, tautologically, always true. If there is no pair of twin primes greater than N, then the procedure will, indeed, run forever. Let us see why the procedure may eventually halt by encountering the **return** statement within the loop; we hope that it will do so as soon as it has found a pair of twin primes.

The loop begins with a test to see whether p is prime, using our yet-to-be-written subprocedure *is_prime*. Note that we invoke *is_prime* simply by referring to it; since *is_prime* returns a logical value (**true** or **false**), we can write *is_prime*(p) as the condition of an **if** clause. If p is not prime, we need to continue the search, so p is incremented by 2 (this is the final **else** clause). On the other hand, if p is prime, then possibly p and $p + 2$ are the desired twin primes. To see whether this is so, the procedure calls *is_prime* on $p + 2$. If $p + 2$ is prime, we are done, so the procedure returns the answer, namely the pair $(p, p + 2)$. Otherwise, the search needs to continue, and since we have just found that $p + 2$ is *not* prime, we may as well continue our search from $p + 4$.

To finish this exercise, we need to write the pseudocode description of *is_prime*. We will give a rather naive algorithm for testing primeness, based on the following observation, which follows fairly easily from the definition of a prime number: An odd positive integer $x \geq 3$ is not prime if and only if x has an odd divisor greater than or equal to 3 and less than or equal to $\sqrt{x}$. Thus our procedure will simply search for such a divisor, returning **true** if it finds one, **false** if it does not. (We will say a little more about algorithms for testing for primeness in Section 4.5.)

procedure *is_prime*(x : odd positive integer ≥ 3)
  {naive algorithm for determining whether x is prime}
    for $q \leftarrow 3$ **to** $\lfloor \sqrt{x} \rfloor$ **by** 2 **do**
      if x mod $q = 0$ **then return**(**false**)
    return(**true**)

The iteration has the index variable q taking on successively the values 3, 5, 7, ..., as long as those values remain less than or equal to $\lfloor \sqrt{x} \rfloor$. (Since q takes on integer values, being less than or equal to $\sqrt{x}$ is equivalent to being less than or equal to $\lfloor \sqrt{x} \rfloor$.) If during this iteration it is found that q is a divisor of x (which we test for using the remainder operation "mod"), we conclude that x is not prime and return the value **false**. (The subprocedure would be finished executing at this point, so the "**return(true)**" statement would never be executed.) If the entire iteration is completed without finding such a divisor, then x must be prime, so the value **true** is returned. ◆

The exercises that follow ask you to design algorithms for solving various problems and write pseudocode descriptions of your algorithms. Try a two-step approach, first analyzing the problem and designing the algorithm in a general way, as we have done in Example 11, and then figuring out how to express it in pseudocode. As we said at the end of Section 4.1, there is no algorithm for you to follow in designing algorithms to solve problems; that is one of the reasons that mathematics is a creative process and not just mindless calculations. Bring all your insight, cleverness, and ingenuity to bear and you will be pleased with what you can create.

SUMMARY OF DEFINITIONS

arguments of a procedure in pseudocode: the values passed to the procedure as input; listed in parentheses, separated by commas, and identified as to type (using a colon), after the name of the procedure (e.g., in **procedure** $square(x :$ real number), the procedure $square$ has one argument, a real number).

array in pseudocode: a one-dimensional or multidimensional list (an indexed set) of variables (e.g., if $labeled$ is a one-dimensional array, indexed by the set $\{1, 2, \ldots, n\}$, we would refer to the third variable in the list as $labeled(3)$; if $distance$ is a two-dimensional array (a matrix) indexed by $\{1, 2, \ldots, n\} \times \{1, 2, \ldots, n\}$, we would write $distance(4, 5)$ to refer to the element in the fourth row, fifth column).

assignment statement: a statement of the form "variable ← expression," which causes the value of the expression to be calculated and then assigned to the variable.

begin...**end**: delimiters of a block of statements; the statements within the block are executed sequentially; the entire block plays the role of one statement.

Boolean variable in pseudocode: a variable that takes on the value **true** or **false**.

call a subprocedure: to suspend operation of the current procedure and invoke the subprocedure; when the subprocedure has finished execution, execution of the current procedure is resumed.

for loop: the construction "**for** iteration **do** statement"; the statement is executed once for each value specified in the iteration condition, unless interrupted by a **return** statement within "statement" (e.g., in the execution of "**for** $i \leftarrow 10$

to 20 **do** $x \leftarrow x + 1$," the value of x is incremented 11 times, once for each value of i from 10 to 20, inclusive).

if...then statement: a pseudocode statement of the form "**if** condition **then** statement"; the condition is evaluated—if it is true, the statement is performed, whereas if it is false, the statement is not performed.

if...then...else statement: a pseudocode statement of the form "**if** condition **then** statement_1 **else** statement_2"; the condition is evaluated—if it is true, statement_1 is performed but statement_2 is not, whereas if it is false, statement_2 is performed but statement_1 is not.

procedure: an algorithm written in pseudocode to perform a certain task or subtask.

return statement: a statement in pseudocode terminating the execution of the current procedure; the items listed in parentheses following the word **return**, if any, are the output of the procedure (e.g., **return**$(x_1, \ldots, x_n)$ returns the list $x_1, x_2, \ldots, x_n$).

subprocedure: a procedure that is called by (and returns to) another procedure in the execution of an algorithm.

variable in pseudocode: a named object that is assigned values as a procedure executes; the values can be numbers, strings, arrays, and so on.

while loop: the construction "**while** condition **do** statement"; repeatedly, the condition is checked and, if it is true, the statement is executed; the **while** loop terminates either when it encounters a **return** statement within "statement" or when condition becomes **false**.

EXERCISES

1. Write in pseudocode an algorithm that takes as input a list of integers and gives as output the sum of all those elements of the list that are less than 100.

2. Write in pseudocode the algorithm for computing the absolute value of a real number, based on the definition of absolute value (see the proof of Theorem 7 in Section 1.3).

3. Write a pseudocode description of algorithms for the following problems.
   (a) Exercise 1 in Section 4.1.
   (b) Exercise 2 in Section 4.1 (there are two outputs—numerator and denominator).
   (c) Exercise 3 in Section 4.1.
   (d) Example 3 in Section 4.1.

4. Write in pseudocode an algorithm that takes as input a list of numbers and gives as output the first element in the list that is greater than 100. If no such element exists, the algorithm should **print** an error message to that effect and return the value 0.

5. Rewrite the following constructions using a **while** statement.

    (a)   **for** $i \leftarrow m$ **to** n **do** statement.

    (b)   **for** $i \leftarrow n$ **down to** 1 **do** statement.

    (c)   **for** $i \leftarrow m$ **to** n **by** k **do** statement (assume that $k > 0$).

6. Suppose that we want to add a **repeat** loop to our pseudocode, where a repeat loop has the form

> **repeat**
>   statement
>   statement
> $\vdots$
>   statement
> **until** condition

   and indicates that the statements within the loop are to be executed over and over again, in order, until the condition is satisfied. The sequence of statements is always executed at least once, since the test is made at the end.

    (a)   Explain how to achieve the same result in the pseudocode as we have defined it. [*Hint*: Use a **while** loop.]

    (b)   Rewrite the procedure *euclid* using a **repeat** loop (see Example 2).

7. Prove that $\gcd(x, y, z) = \gcd(\gcd(x, y), z)$ for positive integers x, y, and z.

8. Prove that if x is an odd integer greater than or equal to 3, then x is composite (i.e., not prime) if and only if x has an odd divisor d such that $3 \leq d \leq \sqrt{x}$.

9. Write a pseudocode description of the most improved naive algorithm for computing $\gcd(x, y)$ discussed in the fourth and fifth paragraphs following Theorem 1 in Section 4.1. Assume that x and y are both positive.

10. Write in pseudocode your algorithms for the following problems.

    (a)   Exercise 21 in Section 4.1.

    (b)   Exercise 23 in Section 4.1.

    (c)   Exercise 24 in Section 4.1.

    (d)   Exercise 27a in Section 4.1.

    (e)   Exercise 27b in Section 4.1.

11. Write in pseudocode an algorithm for finding the smallest prime factor of an integer greater than 1.

12. Write in pseudocode an algorithm for finding the largest prime factor of an integer greater than 1.

13. Write in pseudocode an algorithm for finding the prime factorization of an integer greater than 1. (You will need to decide how to report the output.)

14. Write in pseudocode an algorithm for finding the largest power of x that is a divisor of y, where x and y are integers greater than 1.

15. Using procedure *tabulate* (Example 7) as a subprocedure, write a procedure to find the most frequently occurring digit in a string of one or more decimal digits. You may assume that all the frequencies will turn out to be distinct, so that the answer will be unique.

16. Write in pseudocode an algorithm that takes as input a list (a, b, c) of three distinct numbers and gives as output a list of the same three numbers put into increasing order.

17. Repeat Exercise 16 for lists of four numbers.

18. Write in pseudocode an algorithm that produces a list of all 3-tuples of integers between 1 and n, inclusive, meeting the following conditions on the entries in the 3-tuples. Assume that n is a positive integer.
    (a)   There are no restrictions on the entries.
    (b)   The entries in each 3-tuple must be distinct and appear in increasing order.
    (c)   The entries in each 3-tuple must appear in nondecreasing order (but need not be distinct).

19. Write in pseudocode an algorithm that will compute, given a list of nonzero real numbers, the difference between the average of the positive numbers in the list and the absolute value of the average of the negative numbers in the list. You may assume that there will be at least one positive number and at least one negative number in the list.

20. Write in pseudocode an algorithm that takes as input a string of digits and gives as output a 10 by 10 array whose (i, j)th entry is the number of times that the digit j immediately follows the digit i in the string. Assume that the array is indexed from 0 to 9 in each dimension.

21. Write in pseudocode an algorithm that takes as input a one-dimensional array $(a_1, a_2, \ldots, a_n)$ of length n and gives as output the n by n array shown here (called a **circulant** matrix), in which each successive row is obtained by placing the first element in the previous row at the end.

$$\begin{bmatrix} a_1 & a_2 & a_3 & \cdots & a_{n-1} & a_n \\ a_2 & a_3 & a_4 & \cdots & a_n & a_1 \\ a_3 & a_4 & a_5 & \cdots & a_1 & a_2 \\ \vdots & \vdots & \vdots & \ddots & \vdots & \vdots \\ a_n & a_1 & a_2 & \cdots & a_{n-2} & a_{n-1} \end{bmatrix}.$$

22. Write in pseudocode an algorithm that takes as input an m by n array of real numbers and gives as output two one-dimensional arrays, one giving the row sums and one giving the column sums of the input matrix.

23. Let n be a positive integer. Then $\phi(n)$ is defined to be the number of positive integers less than or equal to n that are relatively prime to n (two numbers are

called relatively prime if their greatest common divisor is 1). The function ϕ is called the **Euler phi function**. For example, $\phi(12) = 4$, because the only numbers from 1 to 12 that are relatively prime to 12 are the four numbers 1, 5, 7, and 11. Write a procedure in pseudocode that computes and returns $\phi(n)$ (it will probably call the subprocedure *euclid*).

24. We explained in Example 1 of Section 1.1 that Goldbach conjectured that every even number greater than 4 is the sum of two (necessarily odd) primes. For example, $6 = 3 + 3$, $8 = 5 + 3$, and so on. The conjecture has never been proved or disproved, so no one knows if it is true or false. Write in pseudocode an algorithm that will search for a counterexample to Goldbach's conjecture. Your algorithm must have the property that if Goldbach's conjecture is false, then it will find a counterexample, whereas if Goldbach's conjecture is true, then it will never halt. (This is an example of the rare case in which an algorithm has no input.)

Challenging Exercises

25. Repeat Exercise 15 without the assumption that there is only one most frequently occurring digit. Your algorithm must find all the digits that occur most frequently.

26. Write in pseudocode an algorithm that computes the relative frequency of the word lengths in a passage of text. The input is a string consisting of letters, punctuation marks, and blanks. A word is a maximal substring (consecutive subsequence) made up entirely of letters. The output is an array f, where $f(l)$ is the relative frequency of words of length l in the input (i.e., the number of words of length l divided by the total number of words). Assume that there are no words of length greater than 20.

27. Write in pseudocode an algorithm that takes two strings, s and t, as input, and determines whether s is a substring of t, giving **true** or **false** as output (see also Exercise 36 in Section 2.2).

28. Write in pseudocode an algorithm that will find a counterexample to Fermat's last theorem if there is one (see Exercise 31 in Section 1.2).

29. Write in pseudocode an algorithm that, given an integer $n > 1$, will compute the length of the repeat in the decimal expansion of $1/n$. For example, the length of the repeat for $n = 3$ is 1, since $\frac{1}{3} = 0.\overline{3}$, and the length of the repeat for $n = 28$ is 6, since $\frac{1}{28} = 0.03\overline{571428}$. (Define the length of the repeat to be 0 for fractions that terminate, such as $\frac{1}{40} = 0.025$.)

Exploratory Exercises

30. Write computer programs to implement some of the algorithms given in the examples and exercises in this section and run them on a computer.

31. Discuss the ways in which our pseudocode differs from real structured, procedural programming languages, such as Pascal or others with which you may be familiar.

32. Find out about Don Knuth, one of the most prominent present-day computer scientists, by consulting Albers and Alexanderson [4] and Albers and Steen [5].

33. Find out about various opinions on the relationships among mathematics, computer science, and computer programming by consulting Dijkstra [69] and Knuth [178,179].

34. Find out more about computer programming languages by consulting Feldman [80], Tesler [303], and A. B. Tucker [307].

35. Find out about representing data structures in a computer and using data structures in programming by consulting Naps and Singh [220], H. F. Smith [278], Standish [292], Stubbs and Webre [295], and Wirth [329–331].

SECTION 4.3
EFFICIENCY OF ALGORITHMS

Though I am always in haste,
I am never in a hurry.

—John Wesley

Usually, there are several different algorithms for solving the same problem. As we saw in Section 4.1, for example, we can compute $\gcd(x, y)$ in a variety of totally different ways: (1) by trying all the numbers that could conceivably divide both x and y and choosing the largest one that actually does; (2) by obtaining the prime factorizations of x and y and taking the product of the smaller powers of all the primes that occur in both; or (3) by the Euclidean algorithm. We noted that the first two of these methods seem to be painfully inefficient if x and y are large or have large common prime factors. The Euclidean algorithm, on the other hand, seems to work remarkably fast, even for very large arguments.

In this section we want to explore the idea of the efficiency of algorithms in more detail. As a secondary consideration, we also look at the issue of *correctness* of algorithms: How do we know that an algorithm does what we want it to do? For analyzing efficiency, we will develop a useful way to measure algorithm performance, and we will find that some algorithms are much more efficient than others.

Two things are important in algorithm performance: time and space. If an algorithm for solving a problem will take too long or require too much computer memory, it may not be feasible. For example, if the payroll department for a large corporation with thousands of employees uses an algorithm for processing its monthly payroll that takes 800 hours on the only computer available, then there will soon be a lot of disgruntled employees, since there are fewer than 800 hours in a month. Even if feasibility is not an issue, money is: If a good algorithm works thousands of times faster than, or with a tenth of the storage requirements of, a poor one, the difference in cost of the necessary computer equipment can be substantial. We will concentrate primarily on the time efficiency of algorithms in this book.

COUNTING STEPS

It would not make sense to measure the time it takes an algorithm to complete its task in real temporal units like seconds, since the actual time depends on the machine used to implement the algorithm and the details of the encoding of the algorithm into the language of the machine. Instead, we will be concerned with the number of "basic steps" an algorithm requires to solve a problem. What we will consider to be a basic step will vary from problem to problem. After a little practice, the reader will get a feeling for what should be considered a basic step. Essentially, a basic step is one that takes only a small *constant* amount of time, in the context of a given problem. Furthermore, we will usually only count the *important* basic steps in an algorithm, and ignore things that might be classified as the overhead or bookkeeping costs of implementing an algorithm. The examples below should help to make this clear.

Usually the number of steps depends on the input. For example, consider the usual grade-school algorithm for multiplication of positive integers. Here a basic step is a digit-by-digit multiplication. Clearly it takes many more basic steps to multiply 4377265 by 9820361743 than it does to multiply 23 by 7. Thus we will view execution time (we call it "time" even though it really is "number of steps") as a *function of input*. In general, this function is extremely complicated, so we will make two important simplifications in our study of algorithm efficiency.

First, rather than thinking of execution time as a function of the input, we will think of it as a function of the *size* of the input, as measured by some simple parameters. This notion of size is best illustrated by means of some examples. (It is possible to write down a rigorous definition of "size," but we will not do so; the definition is technical, not too enlightening, and not important to our discussion.) The execution time for an algorithm for putting elements of a list into alphabetical order (the **sorting problem**) will be judged as a function of the length of the list. The execution time for an algorithm for multiplying an m-digit number by an n-digit number will be judged as a function of m and n. If our problem is to find an element in a sorted list (such as the problem of looking up a word in a dictionary), we will judge algorithm efficiency as a function of the length of the list. Or consider this more abstract, mathematical example: An algorithm to determine whether a compound proposition in logic is a tautology will be judged as a function of the length of the proposition when it is written out in symbols. In our discussions we will often let the variable n stand for the size of the input to an algorithm.

There is an obvious difficulty with saying that execution time is a function of input size: It is not a well-defined function. In particular, two different inputs of the same size might require drastically different numbers of steps for the algorithm to work on them. For example, an algorithm to sort a list may work much faster if the input list is already sorted. We get around this difficulty by talking about the *worst-case performance* of an algorithm. (It is also possible to talk about the *average-case performance* of an algorithm, and we will occasionally do so informally, but a proper discussion of average-case performance requires rather complicated probability calculations.)

Let us define exactly what we mean. Let A be an algorithm. Then the **time complexity** of A, written T_A, or simply T, is the function whose value at n is the maximum, taken over all inputs I of size n, of the number of basic steps required for algorithm A to complete its task on input I. Sometimes we refer to the time complexity informally as the **running time** or the **efficiency** of the algorithm.

It is time to look at some examples that will indicate how to count steps and compute $T(n)$.

EXAMPLE 1. Consider the problem of searching a list to find a particular object that may or may not be in the list. For example, as part of an algorithm to translate English to French, we would certainly need to consult frequently a list of English words with their French equivalents, containing entries like "cat" ("le chat"), "red" ("rouge"), and "eat" ("manger"). When we come across the word "red" in a text we are trying to translate, we need to find "red" in our list in order to know that its French translation is "rouge." Let us analyze a naive (and not very efficient) algorithm for performing such a search. For specificity we will assume that our list is a list of words (i.e., strings) over some alphabet, and that the elements in the list are distinct. Our algorithm, called **linear search**, is shown as Algorithm 1. Its input is the list to be searched and the item being searched for. It returns as output the place in the list at which the item occurs if the item is an element of the list (this will be a positive integer), and it returns 0 if the item is not in the list.

procedure *linear_search*(A, x)
    $\{A = (a_1, a_2, \ldots, a_n)$ is a list of $n \geq 0$ distinct words; x is a word; the algorithm returns the index at which x appears in the list, if it does (i.e., the value of i such that $x = a_i$); and returns 0 if x is not in the list$\}$
      for $i \leftarrow 1$ **to** n **do**
        if $x = a_i$ **then return**(i)
      return(0)

Algorithm 1. Linear search.

Before we look at the number of basic steps this algorithm takes to complete its task, let us ask two even more fundamental questions about it: *How* does it work, and how do we know that it *does* work? The elements of the list are compared to x one at a time; if and when a match is found, the subscript for that element is returned, as desired, terminating the execution of this procedure. If the algorithm gets all the way through the **for** loop, having compared x to all n of the elements in the list without having found a match, then the value 0 is returned, as desired. Since this algorithm is so simple, there can be no doubt that it works. Many of the algorithms we will see in this book are quite complex, and often we will give proofs that they produce the intended answers, especially when the fact that they actually work is far from obvious.

To compute $T_L(n)$—we use the subscript L to stand for *linear* search—we first need to agree on what to count as a basic step. The key component of this algorithm is the repeated comparison of x to an element of the list, occurring in the **if** statement. Let us agree, therefore, to count such comparisons as our measure of efficiency for this algorithm. The number of comparisons will depend on how lucky we are. At best, $x = a_1$, so after just one comparison the algorithm has finished. At worst, we will need to make n comparisons, since the loop can be iterated n times (this will be the case either if x is the last element in the list or if x is not in the list). Since we agreed to look at the worst-case performance, $T_L(n) = n$. (To do an average-case analysis, we would need to worry about the relative frequency with which the words were searched for, compared to the placement of those words in the list. If words frequently sought after were at the beginning of the list, the algorithm would run faster, on the average, than if they were near the end. We will not pursue this analysis here.)

What about all the other things that need to be done when this algorithm is executed? In order for the algorithm to execute the **for** loop, it needs to compare i to n once during each pass through the loop (to see whether the iteration is finished) and a final time to get out of the loop. Thus there may really be as many as $n + 1$ additional comparisons (in the worst case—that in which x is not in the list). Perhaps we should take these into account. To start the loop, i must be assigned the value 1, requiring, in effect, one step (execution of the statement $i \leftarrow 1$). Furthermore, i has to be incremented each time through the loop (in effect, execution of the implied statement $i \leftarrow i + 1$). There could be as many as n such operations. Moreover, in any real implementation of this algorithm, a computer would have to keep track of which instruction it was at, and it would take some time to jump from instruction to instruction. Finally, we might want to consider the overhead involved in starting the algorithm running and returning the answer when it finishes. Our attitude is that all of these other costs do not really matter. The two or three extra steps at the beginning and end pale in comparison to the n steps we have already considered. The overhead for the loop (the bookkeeping involved in the iteration) would perhaps add another $3n + 2$ steps to the n that we already counted, for a total of $4n + 2$, but the difference is not significant: Each pass through the loop requires some small constant amount of time, and there are n passes (in the worst case). Thus $T_L(n) = n$ is a fair expression of the time complexity of this algorithm. (The second simplifying assumption about algorithm analysis that we make below will justify this "handwaving" in a more formal way.) ◆

The linear search algorithm in this example requires n comparisons in the worst case: $T_L(n) = n$ for linear search. It is natural to ask whether we can do better. A typical real English–French dictionary, for example, might contain several thousand words, say $n = 10,000$. Does it really take a person 10,000 steps to look up the translation of "red"? Of course it does not, because a real person does not use linear search in a real dictionary. We do not look up "red" by comparing it successively to "a," "aardvark," "aback," "abacus," "abandon," and so on. Instead, we take advantage of the fact that the words in the dictionary are arranged in alphabetical order. Our next example gives a much more efficient searching algorithm for a case like this in which the list to be searched is ordered.

EXAMPLE 2. The **binary search** algorithm for finding an element x in an ordered list of words is presented as Algorithm 2. We use the notation $x \prec y$ to indicate that x precedes y in lexicographic order (see Section 3.4). We represent our list with an array (writing $A(i)$ rather than a_i), both for variety and for typographical considerations.

procedure *binary_search*(A, x)
    {A is an array of $n \geq 1$ distinct words, ordered so that
    $A(1) \prec A(2) \prec \cdots \prec A(n)$; x is a number; if x is in the array, then the
    procedure returns the index at which x appears in the array (i.e., the value of
    i such that $x = A(i)$); the procedure returns 0 if x is not in the array}
    low ← 1
    high ← n
    {the search will always be restricted to the portion of the array from
      $A(low)$ to $A(high)$}
    while *low* $<$ *high* **do** {there is still searching to do}
      begin
        middle ← $\lfloor(low + high)/2\rfloor$
        if $x \preceq A(middle)$ **then**
            high ← *middle* {x was not in the upper half}
          else *low* ← *middle* + 1 {x was not in the lower half}
      end
    if $A(high) = x$ **then** **return**$(high)$
    else **return**(0)

Algorithm 2. Binary search.

This time it is less clear how the algorithm works, or that it indeed always gives the right answer. [Before proceeding, in order to get a feeling for what the algorithm does, the reader should try the algorithm on a small example or two, say looking for $x = more$ and $x = eating$ in the array $(a, eating, little, miss, muffet, on, sat, tuffet)$.] Entire books have been written on techniques for proving algorithms correct. We will tend to keep our proofs fairly informal, and we will often omit them entirely, since this is not a book on techniques for proving the correctness of algorithms.

To show that binary search always works, we first claim that the position of x in the array if x *is* in the array (i.e., the index i such that $x = A(i)$) is always between *low* and *high*, inclusive. After a little thought, it should be clear that the **if**...**then**...**else** statement inside the loop guarantees this. A little more thought will convince us that when the **while** loop has finished, *low* and *high* must be equal (see Exercise 7). Thus the final **if**...**then** statement will return the correct answer. Note that we might have "found" x earlier, in the sense that x might have equaled $A(middle)$; the algorithm could be modified to use this observation, but we prefer to keep it as simple as possible.

So far we have argued that the algorithm always gives the correct answer— assuming that it gives an answer at all. Is there any guarantee that the loop will not iterate forever? We answer this question and analyze the time complexity $T_B(n)$—B standing for *binary* search—at the same time. Again, let us agree to count comparisons of x to elements in the list as our basic steps. In this procedure, those comparisons occur in the statement "**if** $x \preceq A(middle)$ **then**..." within the **while** loop, and in the final statement "**if** $A(high) = x$ **then**...". In other words, essentially we are counting the number of passes through the **while** loop. This makes sense, since during each pass through the loop only a small finite number of things happen, and outside the loop also only a small finite number of things happen.

The loop is repeated as long as $low < high$, in other words, as long as the integer $(high - low)$ is greater than 0. Therefore, we need to see what happens to $(high - low)$ on one pass through the loop. There are really four cases to consider, depending on whether $(high - low)$ is even or odd, and whether $x \preceq A(middle)$; it is not hard to show that in each case the value of $(high - low)$ at least decreases by a factor of 2 (actually, by slightly more than a factor of 2 if $(high - low)$ is odd or $A(middle) \prec x$); see Exercise 7. For example, suppose that $n = 101$, so that $(high - low)$ starts out equal to 100. Then $middle = \lfloor (1 + 101)/2 \rfloor = 51$, so the search narrows either to the interval $[1, 51]$ or to the interval $[52, 101]$. Thus after the first pass through the loop, $(high - low)$ will be reduced to at most 50. The iteration is over when $(high - low)$ becomes less than 1, since, being an integer, it necessarily becomes 0 in that case.

To get a handle on these successive divisions of $(high - low)$ by 2, it will be useful to write $n - 1$, which is the initial value of $(high - low)$, as a power of 2, say as 2^k for some real number k. (If $n = 1$, then no iterations are required, so we assume that $n \geq 2$.) We can certainly do this, just by letting $k = \log(n - 1)$, recalling our convention from Section 3.1 that "log" with no subscripts means the logarithm to the base 2. Now if $(high - low)$ starts out as 2^k and is at least halved each time through the loop, then after p passes through the loop its value will be at most $2^k/2^p = 2^{k-p}$. Thus after at most $p = k + 1$ iterations we must have $(high - low) \leq 2^{-1} < 1$, which forces $(high - low)$ to be 0. Therefore, the algorithm requires at most $k + 1$ iterations through the **while** loop, and hence at most $k + 2$ comparisons in all. In summary, then, $T_B(n) \leq \log(n - 1) + 2$. If, in fact, $n - 1$ is an integral power of 2 and x precedes all the elements in the list, then this bound is the best we can do; and a little more messy analysis will show that actually $T_B(n) = \lfloor \log(n - 1) \rfloor + 2$ for all $n \geq 2$.   ◆

Now let us compare the time complexity for linear search, found in Example 1 to be $T_L(n) = n$, to the complexity of binary search, found in Example 2 to be $T_B(n) = \lfloor \log(n-1) \rfloor + 2$. For $n = 5$, the difference is not all that significant. Linear search requires $T_L(5) = 5$ comparisons, while binary search requires $T_B(5) = 4$, and the difference between these could easily be overshadowed by higher overhead costs for binary search. But let us see what happens if we let n grow. For $n = 200$ we find that $T_L(200) = 200$ and $T_B(200) = 9$, a difference that is hard to ignore if the time required for one basic step is at all comparable between the two algorithms. For $n = 10,000$ the results are devastating: Linear search may require as many as 10,000 steps, but binary search requires at most $\lfloor \log(9999) \rfloor + 2 = 15$ steps. Thus linear search is probably a poor choice if n is large and the searching must be done repeatedly (as would be the case in an English-to-French translation program), but binary search is quite efficient.

This study of searching algorithms should convince us that the rate of growth of $T(n)$ as n gets large can make or break an algorithm. If $T(n)$ becomes too large too quickly as n grows, the algorithm will be limited in the size of the problems it can handle in practice. Table 4.1 below shows this dramatically for some typical complexity functions. Let us suppose that a computer can perform 1 million (10^6) basic steps per second. The table shows how long it will take such a computer to perform an algorithm with various time complexities for various values of n. (It is a routine exercise in arithmetic to figure out the entries in the table: Plug the given

Table 4.1   Time required to perform $T(n)$ steps, each taking 10^{-6} second.

| | \multicolumn{5}{c}{$T(n) =$} |
	$\log n$	n	n^2	n^5	2^n
$n = 10$	3×10^{-6} sec	10^{-5} sec	10^{-4} sec	0.1 sec	10^{-3} sec
$n = 20$	4×10^{-6} sec	2×10^{-5} sec	4×10^{-4} sec	3 sec	1 sec
$n = 50$	6×10^{-6} sec	5×10^{-5} sec	2×10^{-3} sec	5 min	40 years
$n = 100$	7×10^{-6} sec	10^{-4} sec	10^{-2} sec	3 hours	4×10^{16} years
$n = 1000$	1×10^{-5} sec	10^{-3} sec	1 sec	30 years	$> 10^{100}$ years
$n = 100,000$	2×10^{-5} sec	0.1 sec	20 min	3×10^{11} years	$> 10^{100}$ years

value of n into the given expression involving n, multiply by 10^{-6} second, and convert to other temporal units by appropriate divisions, dividing by 60 to convert to minutes, etc.) All numbers in Table 4.1 are rounded to one significant digit.

Suppose that we were to obtain a faster computer, say one able to do 1 billion (10^9) operations per second rather than only 1 million. Then all the times in the table would improve (i.e., decrease) by a factor of 1000. That might be significant for reducing a 3-hour job to 11 seconds, but it makes no difference to an entry such as 3×10^{11} years: 3×10^8 years is still too long to wait. No conceivable increase in computer speed could ever make feasible an algorithm that took 10^{100} steps. The only way to solve the problem that such an algorithm solves is to find a better algorithm.

We now turn to a mathematical formalism for implementing our desire that the complexity functions be thought of only in very approximate terms. This is the second important simplification in our study of algorithm efficiency, promised several pages ago.

BIG-OH NOTATION

In our analysis of the linear search algorithm, we explained that we really did not want to be bothered with the difference between a function such as $T(n) = 4n + 2$ and the simpler function $T(n) = n$. We will now make this sloppiness precise. In essence we want a "fuzzy ruler" with which to measure algorithm performance, or equivalently, a fuzzy ruler to measure the rate of growth of functions, a ruler that does not distinguish between minor differences. On the other hand, we *do* want our ruler to distinguish between major differences, such as the difference between $T(n) = n$ and $T(n) = \log n$; as we saw in Table 4.1, the former grows significantly faster than the latter as n increases. In brief, our attitude will be that *functions will be considered essentially the same if they differ at most by a constant multiple*. Thus we will "throw away" from algebraic expressions giving the rules for functions (1) constant factors and (2) additive pieces that grow more slowly than pieces that we retain.

Let us formalize these ideas. In what follows, we will be dealing with functions whose domain and codomain are intentionally left vague. For simplicity *we will assume that their domains are the set of positive integers, and their codomains are the set of nonnegative real numbers*. Also, traditionally the argument of the function (usually called n) is included in the notation; thus we will sometimes abuse terminology a little and refer to "the function $f(n)$" rather than "the function f." For example, we will speak of "the function $\log n$."

In this setting, then, let g be a function. We define the set of functions $O(g)$, read "**big-oh** of g," as follows: $f \in O(g)$ if there exist constants C and k such that $f(n) \leq Cg(n)$ for all $n \geq k$. (If we were allowing negative function values, we would have put absolute value signs around $f(n)$ and $g(n)$.) Certainly, $g \in O(g)$, since we can take $C = 1$ and $k = 1$. Any multiple of g is in $O(g)$, and more generally, $O(g)$ contains every function that is less than some constant multiple of

g for all sufficiently large values of the argument. We can think of $O(g)$ as a swath surrounding g, including g, all functions that grow (as n increases) at more or less the same rate that g does (up to a multiplicative constant), and all functions that grow more slowly than g does.

EXAMPLE 3. In Example 1 we saw that the function $f(n) = 4n + 2$ more accurately describes the number of basic steps in linear search than does the function $g(n) = n$. The function $f(n)$ is in big-oh of the function $g(n)$, since $4n + 2 \leq 4n + n = 5n$ for all $n \geq 2$. In the notation of the definition, we can take $C = 5$ and $k = 2$. Informally, the reason that $4n + 2$ is in $O(n)$ is that we can "throw away" the 2 (an additive term that is smaller than the term $4n$) and then "throw away" the constant multiplier 4, to obtain n. (Even something apparently bigger, such as $20n + 1000$, is in $O(n)$: Eventually, for large enough n, the term $20n$ dominates the term 1000, so we ignore the 1000, and then $20n$ is just a constant times n.) Also, $g(n) \in O(f(n))$ in this example (take $C = 1$ and $k = 1$, for instance). Thus in terms of our fuzzy ruler, these two functions are "the same," growing in essentially the same fashion.

On the other hand, $g(n) = n$ is not in big-oh of $h(n) = \sqrt{n}$, since there is no constant C for which $n \leq C\sqrt{n}$ for all large n. (To see this rigorously, simply note that if $n \leq C\sqrt{n}$, then $\sqrt{n} \leq C$, which is obviously impossible for the *constant* C to satisfy for *all* large n.) In other words, $\sqrt{n}$ is a more slowly growing function than n. ◆

For a more complicated example, note that

$$\frac{n(n-1)}{2} + n \log n < n^2 + n^2 = 2n^2 < 2n^3,$$

(since $\log n < n$), so we can conclude that

$$\frac{n(n-1)}{2} + n \log n \in O(n^3).$$

In other words, the function $h(n) = \big(n(n-1)/2\big) + n \log n$ grows no faster than a constant multiple times the function n^3. However, this last statement is not the "best" statement that can be made about the rate of growth of the function $h(n) = \big(n(n-1)/2\big) + n \log n$; it grows no faster than n^2, either, so a stronger (better) statement is that $h(n) \in O(n^2)$.

Suppose that we have an algorithm that requires $\big(n(n-1)/2\big) + n \log n$ basic steps in the worst case for inputs of size n. Rather than saying that its complexity function is $\big(n(n-1)/2\big) + n \log n$, we will make the simpler statement that its complexity function is in $O(n^2)$. Even more informally, we might say that this algorithm uses $O(n^2)$ steps in the worst case; thus we will sometimes write "$O(g)$"

when referring to *an element of* $O(g)$. (Some authors even write $f(n) = O(g(n))$ for $f(n) \in O(g(n))$, but we see no reason to abuse the notation this much.)

A useful way to check whether $f \in O(g)$ is to look at the limit

$$\lim_{n \to \infty} \frac{f(n)}{g(n)}.$$

In other words, we look at the *asymptotic* behavior of f and g. If this limit exists (in practice it usually does) and is a finite number (possibly 0), we can conclude that $f \in O(g)$. If the limit is infinity, then $f \notin O(g)$. For example, $7n^3 + 100n - 3 \in O(n^3)$ because the limit of the ratio of these functions, as $n \to \infty$, is the finite number 7. In fact, if the limit is a finite *nonzero* number, then $O(f) = O(g)$, as is the case with this example. L'Hôpital's rule will often aid in evaluating this limit.

We need to issue a word of warning about big-oh. Sloppiness, even formalized sloppiness, is not without its price. Sometimes in the act of simplifying we lose important information. Suppose that you are presented with a problem to solve, for which you have a choice between two algorithms, one with time complexity $O(n \log n)$ and the other with time complexity $O(n^2)$. Which should you choose? Obviously, there is a lot of information omitted from this question, such as the space requirements of the two algorithms, comparative potential for programming errors, the round-off errors, if any, in the algorithms, the amount of time required for the computer to perform one basic step, and so on. Suppose that all of these are similar for the two algorithms. Since $n \log n$ is a more slowly growing function than n^2, it seems that you would save execution time by choosing the former algorithm over the latter; and in most situations this will be correct. You have to be careful, however. First of all, the big-oh terminology provides only an *upper bound*. It might be that the second algorithm takes only $3n$ basic steps. Since $3n \in O(n^2)$, the statement that the second algorithm has time complexity $O(n^2)$ would be correct, but not useful for comparing the two algorithms. We will get around this difficulty by trying to give the best possible big-oh estimate for functions that we encounter. So let us suppose that the second algorithm has a time complexity function for which $O(n^2)$ is the best that can be said. It may still be that for the values of n for which you need to apply the algorithm, the second algorithm works faster than the first. For example, suppose that the first algorithm requires approximately $5n \log n + 20n$ steps, while the second requires $n(n-1)/2$ steps. For large n, the latter function takes on substantially larger values than the former, such as 12,497,500 as opposed to 407,193 when $n = 5000$. But for small and moderate values of n, the second function takes on smaller values, such as 45 as opposed to 366 when $n = 10$. Thus big-oh comparison of time complexity is only one—albeit an important one—of the considerations in the design and analysis of algorithms.

Our intent is to use *simple* functions inside the big-oh. Furthermore, as we have seen, whenever we say that $f \in O(g)$, we would like to be making the best statement possible, in other words choosing the most slowly growing simple g we can that yields a true statement. Listed in Table 4.2 is a set of reference functions from which we will usually pick g. The list is in increasing order: Each function in the list is in big-oh of all the functions after it, but not in big-oh of any of the functions before it. The functions that have verbal names are particularly important. The letter k is meant to stand for a constant.

Table 4.2 Reference functions for big-oh comparisons, in increasing order.

Function	Verbal Description
1	Constant
$\log n$	Logarithmic
$\sqrt{n}$	
$n/\log n$	
n	Linear (polynomial of degree 1)
$n \log n$	
$n^{3/2}$	
n^2	Quadratic (polynomial of degree 2)
n^k $(k > 2)$	Polynomial of degree k (when k is an integer)
k^n $(k > 1)$	Exponential
$n!$	Factorial
n^n	

These, then, are some of the markings on our fuzzy ruler. Given a function f, such as one representing the time complexity of an algorithm, we want to see where on this ruler f falls. In other words, we want to find the smallest simple function g (usually one on this list) for which we can say $f \in O(g)$. It will usually be the case that $g \in O(f)$ as well.

EXAMPLE 4. Linear search (Example 1) has linear time complexity, since $T_L \in O(n)$. Thus we know that in the worst case the number of steps needed to find the desired item (or discover that it is not in the list) is at most a constant times the size of the list. Binary search (Example 2) has logarithmic time complexity, since $T_B \in O(\log n)$. Thus we know that in the worst case the number of steps needed to find the desired item (or discover that it is not in the list) is at most a constant times the logarithm of the size of the list. ◆

This fuzzy ruler metaphor can be made more formal by considering the equivalence relation among functions defined by declaring that f is big-oh equivalent to

g if and only if $O(f) = O(g)$. In this case we say that g is a **big-oh equivalent** for f. Thus the markings on the ruler are really the equivalence classes of this equivalence relation, and we usually want to describe an equivalence class by the simplest function in it. The following theorem shows another way of looking at this equivalence relation.

THEOREM 1.   *Suppose that f and g are positive-valued functions of positive integers. Then $f \in O(g)$ and $g \in O(f)$ if and only if $O(f) = O(g)$.*

Proof.   The "if" part is trivial, since $f \in O(f)$ and $g \in O(g)$. Let us consider the "only if" part. By symmetry, it is enough to show only one-half of the equality, say that $O(f) \subseteq O(g)$. To this end, let $h \in O(f)$. Then there exist constants C_1 and k_1 such that $h(n) \leq C_1 f(n)$ for all $n \geq k_1$. Now since $f \in O(g)$ there exist constants C_2 and k_2 such that $f(n) \leq C_2 g(n)$ for all $n \geq k_2$. Let $k = \max(k_1, k_2)$ and let $C = C_1 C_2$. Then for all $n \geq k$ we have $h(n) \leq Cg(n)$; hence $h \in O(g)$. ∎

We end this section by applying the big-oh time complexity analysis to one of the most important tasks in data processing: sorting. For specificity, let us assume that we are working with numbers. The problem is as follows: Given a list of numbers $A = (a_1, a_2, \ldots, a_n)$, produce another list $B = (b_1, b_2, \ldots, b_n)$, with the same elements as A but arranged in nondecreasing order, so that $b_1 \leq b_2 \leq \cdots \leq b_n$. The algorithm is thought of as operating on A by interchanging pairs of elements of A until A is in nondecreasing order.

Algorithm 3 gives the pseudocode description of one straightforward sorting algorithm, called **bubble sort**. The name comes from the fact that the smaller elements in the list seem to move gradually toward the front of the list just as bubbles of air move toward the top of a column of water.

procedure *bubble_sort*(A)
   $\{A = (a_1, a_2, \ldots, a_n)$ is a list of numbers; at the completion of the procedure, the list will have been rearranged into nondecreasing order$\}$
     for $i \leftarrow 1$ **to** $n - 1$ **do**
       for $j \leftarrow 1$ **to** $n - i$ **do**
         if $a_j > a_{j+1}$ **then** interchange a_j and a_{j+1}
     return(A) $\{$which is now sorted into nondecreasing order$\}$

Algorithm 3.   Bubble sort.

Let us first follow the operation of the algorithm on a simple example, then argue that it always works, and finally analyze its time complexity. Suppose that

A is initially $(6, 1, 5, 3, 8, 2)$, so that $n = 6$. The algorithm consists of five phases (as the index variable i goes from 1 to 5). During the ith phase, adjacent elements in the list are compared, and if they are out of order, then they are interchanged; this happens $n - i$ times—to the first and second elements of the list (when $j = 1$), then to the second and third (when $j = 2$), and so on until finally to the $(n - i)$th and $(n - i + 1)$th elements of the list (when $j = n - i$). In our example, during the first phase, we compare $a_1 = 6$ to $a_2 = 1$ and find that they are out of order, since $6 > 1$. Therefore, these two elements are interchanged, and the list becomes $(1, 6, 5, 3, 8, 2)$. Next the second and third elements of the list as it now stands are compared, and we see that $a_2 = 6$ and $a_3 = 5$ are out of order; thus the list changes to $(1, 5, 6, 3, 8, 2)$. The next pass through the j-loop results in another interchange: $(1, 5, 3, 6, 8, 2)$. Next (i is still 1, and j is now 4) we compare 6 to 8, and since they are in the correct order ($6 < 8$), no change is made by the **if...then** statement. Finally, we compare 8 to 2 and perform the required interchange. Thus at the end of the first phase, the list has become $(1, 5, 3, 6, 2, 8)$. Note that the last element in the list is correctly placed. The second phase requires four comparisons, and the list becomes $(1, 3, 5, 2, 6, 8)$ at the end of this phase. Note that the *two* largest elements are correctly placed after the second phase. Similarly, the third phase ends with $A = (1, 3, 2, 5, 6, 8)$, the fourth with $A = (1, 2, 3, 5, 6, 8)$ (which, as it happens by accident, is already correctly sorted), and the fifth and final phase ends again with $A = (1, 2, 3, 5, 6, 8)$, as desired.

The key to seeing that the algorithm must work correctly is the observation that after the ith pass the i largest elements are correctly placed in the last i locations in the list. A formal proof of this fact requires a technique called mathematical induction, which we will discuss in Section 5.3, so we postpone the proof until then.

Finally, we analyze the time complexity of bubble sort. There is only one operative statement in the procedure: the **if...then** statement. It requires one comparison and sometimes one interchange of a pair of elements in the list (a_j and a_{j+1}). The interchange can be done with three quick assignment statements: *temp* $\leftarrow a_j$ followed by $a_j \leftarrow a_{j+1}$ followed by $a_{j+1} \leftarrow$ *temp*. Thus to compute $T(n)$ for bubble sort, we simply need to count the number of times the procedure executes this **if...then** statement. The procedure consists of a loop within a loop, so we have to see how many times the inner loop is executed for each pass through the outer loop. For $i = 1$ the inner loop is iterated $n - 1$ times; for $i = 2$ the inner loop is iterated $n - 2$ times; ...; for $i = n - 2$ the inner loop is iterated two times; and for $i = n - 1$ the inner loop is iterated one time. Thus the total number of times the **if...then** statement is executed is $(n - 1) + (n - 2) + \cdots + 2 + 1$, which as the reader should recall from algebra (and will prove by mathematical induction in Exercise 2 of Section 5.3) is equal to $n(n - 1)/2$. Thus for bubble sort, $T(n) = n(n - 1)/2 \in O(n^2)$.

We will look at some other sorting algorithms in Exercises 19 and 20, and presently we will discuss (in Section 5.2) and analyze (in Section 7.3) another, more efficient sorting algorithm called "merge sort." It is worth pointing out that bubble sort, or any other $O(n^2)$ sorting algorithm, is unsatisfactory for real applications with large values of n. For example, suppose that there are $n = 30,000$ students at a large university whose addresses are stored by student number, and we want to print out a telephone directory. To sort these records into alphabetical order using bubble sort would require $30000 \cdot 29999/2$, or about half a billion comparisons and thus tie up the computer facilities for quite awhile. Efficient sorting, it turns out, has time complexity in $O(n \log n)$. Suppose that such an algorithm actually used $5n \log n$ basic steps of the same size as the basic steps in bubble sort. Since $5 \cdot 30000 \cdot \log 30000$ is only about 2 million, we would realize over a 100-fold increase in efficiency by using the more efficient algorithm.

SUMMARY OF DEFINITIONS

big-oh of a nonnegative-valued function of positive integers: $O(g) = \{ f \mid \exists C \colon \exists k \colon \forall n \geq k \colon f(n) \leq Cg(n) \}$, where f ranges over nonnegative-valued functions of positive integers (e.g., $10n^2 + 45n + 10 \in O(n^2)$).

big-oh equivalent of the function f: a function g such that $O(f) = O(g)$, which is the same as saying that $f \in O(g)$ and $g \in O(f)$; one tries to find simple big-oh equivalents in describing complexity of algorithms (e.g., a simple big-oh equivalent of $n(n-1)/2$ is n^2).

binary search: an algorithm for locating an object in a sorted list that works by repeatedly bisecting the "live" portion of the list and restricting the search to one-half of the live portion, until a portion of length 1 remains (at which point the sought-after object is either the unique object in that portion, or not in the original list).

bubble sort: an $O(n^2)$ sorting algorithm that sorts a list by successively changing the positions of adjacent elements that are out of order.

efficiency of an algorithm: the time complexity of the algorithm.

linear search: an algorithm for locating an object in a list that works by comparing the elements of the list, one by one, with the object being sought, until a match is found (if no match is found, the object is not in the list).

running time of an algorithm: the time complexity of the algorithm.

sorting problem: the problem of taking a list of n elements from a set with a total order and arranging them in order (e.g., sorting a list of numbers into numerically nondecreasing order, or sorting a list of strings into lexicographical order).

time complexity of algorithm A: the number of basic steps required for algorithm A to run to completion in the worst case among all inputs of a given size, expressed as a function $T_A(n)$ of the size n, or expressed as a big-oh class of such a function (e.g., binary search has time complexity $O(\log n)$, where n is the length of the list being searched).

EXERCISES

1. Carry out the following searches in the list $(a, curds, eating, her, little, miss, muffet, on, sat, tuffet)$.
   (a) Linear search for the entry *eating* (Algorithm 1).
   (b) Linear search for the entry *more* (Algorithm 1).
   (c) Binary search for the entry *eating* (Algorithm 2).
   (d) Binary search for the entry *more* (Algorithm 2).

2. Determine the running time for each of the following calculations. Assume that 10^6 steps can be performed per second.
   (a) A calculation with 2^{30} steps—give the answer in minutes.
   (b) A calculation with 2^{40} steps—give the answer in days.
   (c) A calculation with 2^{200} steps—give the answer in powers of 10 years.

3. Find simple big-oh equivalents, chosen from Table 4.2 if possible, for the following functions. You may use intuition, rather than providing a formal proof from the definition.
   (a) $10n^3 + 4n$.
   (b) $n^2 + n \log n$.
   (c) $n^{1.5}/(n + 1)$.
   (d) $2^n + n^2$.

4. Show that if $f \in O(g)$ and $g \in O(h)$, then $f \in O(h)$.

5. Determine, in simple big-oh terms as a function of n, the number of basic steps required in the following algorithms. This is essentially equivalent to finding the value x that the procedure returns.
   (a) **procedure** $a(n :$ positive integer$)$
   ```
        x ← 0
        i ← n
        while i ≥ 1 do
          begin
            i ← i − 2
            x ← x + 1
          end
        return(x)
   ```
   (b) **procedure** $b(n :$ positive integer$)$
   ```
        x ← 0
        i ← n
        while i ≥ 1 do
          begin
            i ← i/3
            x ← x + 1
          end
        return(x)
   ```

(c)   **procedure** $c(n :$ positive integer)

      $x \leftarrow 0$

      $i \leftarrow n$

      while $i \geq 1$ **do**

        begin

          $i \leftarrow 0$

          $x \leftarrow x + 1$

        end

      return(x)

6. Sort the following list using bubble sort: $(12, 4, 5, 8, 6, 10, 2)$.

7. In this exercise we check the details of the proof given in Example 2 that Algorithm 2 gives the correct output.
    (a) Write down, as a function of the current value of $(high - low)$, exactly what values $(high - low)$ can have the next time through the loop.
    (b) Using part (a), show that $high = low$ when Algorithm 2 ends.

8. Find simple big-oh equivalents, chosen from Table 4.2 if possible, for the following functions.
    (a) $\dfrac{n^3 + 5n^2 + 4}{n + 3}.$
    (b) $\sqrt{n^2 + 3n + 2}.$
    (c) $\dfrac{n^2 \log n}{1 + n}.$
    (d) $\left(\dfrac{n + 1}{n}\right)^3.$

9. Show that $\log_a n \in O(\log_b n)$ for all real numbers a and b greater than 1. In other words, whether we take logarithms to the base 2 or to the base 10 (or to the base e or to any other base) is of no importance in big-oh analysis.

10. Show that $n^5 \in O(2^n)$ but $2^n \notin O(n^5)$. [*Hint:* L'Hôpital's rule can be used here.]

11. Verify the following big-oh statements, where k is a constant bigger than 1.
    (a) $k^n \in O(n!)$.
    (b) $n! \in O(n^n)$.
    (c) $n! \notin O(k^n)$.
    (d) $n^n \notin O(n!)$.

12. Suppose that f and g are functions such that $f \in O(g)$, but $g \notin O(f)$. Let us say that a function h is "strictly between" f and g if $f \in O(h)$ and $h \in O(g)$, but $h \notin O(f)$ and $g \notin O(h)$. Find functions that are strictly between the following pairs of functions.
    (a) n^2 and n^3.
    (b) n and $n \log n$.
    (c) 1 and $\log n$.

13. Suppose that $f_1 \in O(g)$ and $f_2 \in O(g)$. Show that $f_1 + f_2 \in O(g)$.

14. Suppose that $f_1 \in O(g_1)$ and $f_2 \in O(g_2)$. Show that $f_1 \cdot f_2 \in O(g_1 \cdot g_2)$.

15. Prove that the following statements are true. These are the justifications for our heuristic, informal way of dealing with big-oh: "Throw away smaller pieces, and ignore multiplicative constants."
    (a)   If $f(n) \leq g(n)$ for all large n, then $f(n) + g(n) \in O(g(n))$.
    (b)   If m is a positive constant, then $m f(n) \in O(f(n))$.

16. Find a function that grows faster than n^n (on the big-oh scale).

17. Determine, in simple big-oh terms as a function of n, the number of basic steps required in the following algorithms. This is essentially equivalent to finding the value x that the procedure returns.
    (a)   **procedure** $a(n : \text{positive integer})$
    $$x \leftarrow 0$$
    $$i \leftarrow n$$
    while $i \geq 1$ **do**
       begin
          $i \leftarrow i - 1$
          for $j \leftarrow 1$ **to** n **do**
             $x \leftarrow x + 1$
       end
       return(x)
    (b)   **procedure** $b(n : \text{positive integer})$
    $$x \leftarrow 0$$
    $$i \leftarrow n$$
    while $i \geq 1$ **do**
       begin
          $i \leftarrow i - 1$
          for $j \leftarrow 1$ **to** i **do**
             $x \leftarrow x + 1$
       end
       return(x)
    (c)   **procedure** $c(n : \text{positive integer})$
    $$x \leftarrow 0$$
    $$i \leftarrow n$$
    while $i \geq 1$ **do**
       begin
          $i \leftarrow i/2$
          for $j \leftarrow 1$ **to** n **do**
             $x \leftarrow x + 1$
       end
       return(x)

(d)   **procedure** $d(n :$ positive integer)

$x \leftarrow 0$

$i \leftarrow n$

while $i \geq 1$ **do**

   begin

      $i \leftarrow i/2$

      for $j \leftarrow 1$ **to** i **do**

         $x \leftarrow x + 1$

   end

return(x)

18. Analyze the efficiency of the procedure *is_prime*, given in Example 11 of Section 4.2, by counting the number of iterations of the loop,
    (a)   As a function of the input x.
    (b)   As a function of the length (number of digits) of the input x.

19. The following algorithm is called **selection sort**.

    procedure *selection_sort*(A)
       $\{A = (a_1, a_2, \ldots, a_n)$ is a list of real numbers; at the completion of the procedure, the list will have been rearranged into nondecreasing order$\}$
       for *end* $\leftarrow n$ **down to** 2 **do**
         begin
            $k \leftarrow 1$ $\{k$ will be index of largest element not yet placed$\}$
            for $j \leftarrow 2$ **to** *end* **do**
               if $a_j \geq a_k$ **then** $k \leftarrow j$
            interchange a_k and a_{end}
         end $\{$now a_{end} is correct$\}$
       return(A)

    (a)   Sort the list $(12, 4, 5, 8, 6, 10, 2)$ using selection sort.
    (b)   Explain why the procedure works.
    (c)   Analyze the efficiency of selection sort.

20. The following algorithm is called **radix sort**. It operates on natural numbers with no more than k digits (it could also be easily modified to operate on strings of a fixed length). We count digits from the right. Thus, for example, the first digit of 03417 is 7, and its fifth digit is 0. A **queue** is a list to which elements are added at the end.

    procedure *radix_sort*(A)
       $\{A = (a_1, a_2, \ldots, a_n)$ is a list of natural numbers, each with k digits (padded with leading 0's if necessary); at the completion of the procedure, the list will have been rearranged into nondecreasing order; there are 10 queues, labeled 0 to 9, for holding numbers temporarily during the sorting process$\}$

```
for p ← 1 to k do
   begin
      empty all the queues
      for i ← 1 to n do {with the current list A}
         put a_i at the end of queue d if its pth digit is d
      A ← contents of queue 0
      for d ← 1 to 9 do
         A ← A followed by contents of queue d
   end
return(A)
```

For example, to sort the list $(473, 465, 123, 358, 248, 163, 258)$, there are three passes. After the first pass, the last digit is in nondecreasing order: $(473, 123, 163, 465, 358, 248, 258)$. After the second pass, the list has become $(123, 248, 358, 258, 163, 465, 473)$; the list is now sorted by the last two digits. The final pass (when $p = 3$) puts the list into correct order.

(a)   Sort the list $(142, 844, 145, 338, 126, 810, 882)$ using radix sort.

(b)   Explain informally why the algorithm works.

(c)   Show that the time complexity of radix sort is $O(kn)$. We can assume that the number of digits in the numbers we wish to sort is proportional to the logarithm of the number of numbers. Conclude that radix sort has time complexity $O(n \log n)$ and hence is much more efficient than bubble sort for large n.

Challenging Exercises

21. We saw that Algorithm 1 uses $2n + 1$ comparisons if we count the bookkeeping comparisons. Modify the algorithm so that it uses no more than $n + 1$ comparisons of any kind.

22. Find two functions, neither of which is in big-oh of the other.

23. Show that if $f \in O(g)$ but $g \notin O(f)$, then there is a function strictly between f and g (see Exercise 12).

24. Find, with justification, a simple big-oh equivalent for the function $\left(\dfrac{n+1}{n}\right)^n$.

25. In this exercise we compare the logarithmic function to small power functions.

(a)   Show that n^a grows faster than $\log n$ (in other words, that $\log n \in O(n^a)$ but $n^a \notin O(\log n)$), for every constant $a > 0$.

(b)   Consider the functions $n^{0.01}$ and $\log n$. At approximately what value does the former become and remain larger than the latter?

26. Determine all the big-oh relationships among $\log(n!)$, $\log(n^n)$, and $\log(a^n)$, where a is a constant bigger than 1. (Justification that your list of relationships is complete involves showing that certain big-oh relationships do hold and that certain others do not.)

27. Find a function that is bigger than all polynomial functions (n^k) but smaller than all increasing exponential functions (k^n, for $k > 1$) on the big-oh ruler.

28. Determine, in simple big-oh terms as a function of n, the number of basic steps required in the following algorithm. This is essentially equivalent to finding the value x that the procedure returns.

    procedure *fast*(n : positive integer)
       $x \leftarrow 0$
       $i \leftarrow 2$
       while $i \leq n$ **do**
         begin
           $i \leftarrow i^2$
           $x \leftarrow x + 1$
         end
       return(x)

Exploratory Exercises

29. Look up the word *discrete* in a regular dictionary, and write down every word you looked at until you found it. Describe your algorithm with as much precision as possible, in a well-written English paragraph or two.

30. Look up the nonword *reflantiguous* in a regular dictionary, and write down every word you looked at until you could conclude that it was not in the dictionary. Describe your algorithm with as much precision as possible, in a well-written English paragraph or two.

31. Suppose that a program to translate from English to French is being constructed. In the repeated searches for words in the dictionary, the program will end up spending most of its time searching for a relatively few words (such as "the" or "is"). Discuss a good search method for handling this situation.

32. Make a set of 60 file cards, with the ith card (for $1 \leq i \leq 60$) containing the last name of the first person on page $\lfloor ni/61 \rfloor$ of your local telephone book, where n is the number of pages in the book (white pages only). Shuffle the cards well each time before sorting them.
    (a) Sort them into alphabetical order by the method that first occurs to you to use. Comment on the algorithm you employed.
    (b) Experiment with other sorting methods, looking for an efficient one. Which method works best for you?
    (c) Ask two or three friends to sort them, and comment on their algorithms.

33. Program bubble sort on a microcomputer and investigate how large a list can be sorted in 10 seconds, in 2 minutes, in an hour, or in a week (extrapolate for this last one, bearing in mind the complexity analysis of this section).

34. Program linear search and binary search for a real computer. Run them with some lists of various sizes and compare their efficiency in practice.

35. Find out more about algorithm efficiency by consulting Lewis and Papadimitriou [196].

36. Find out more about sorting and searching algorithms by consulting Baase [12], Kenner [169], and Knuth [177].

SECTION 4.4
INTRACTABLE AND UNSOLVABLE PROBLEMS

The situation is hopeless, but not serious.

—old Austrian saying

In this short but intellectually demanding section we continue our discussion, begun in the preceding section, of the efficiency of algorithms. Here we are concerned with more theoretical issues. The basic goal of this section is to separate all problems for which we might wish to find algorithmic solutions into three classes: those for which there are fairly efficient algorithms, those for which there appear to be only very inefficient algorithms, and those for which there are no algorithms at all.

As we saw in Table 4.1, exponential functions grow so rapidly that an algorithm requiring an exponential number of steps is infeasible if n, the size of the input, gets at all large. (In this section the size of the input will almost always roughly correspond to the number of symbols needed to write down the input.) On the other hand, if we have a fast computer, then algorithms with polynomial time complexity $T(n) \in O(n^k)$, especially if k is very small, are reasonably feasible for moderately large values of n. Let us call an algorithm **good** if its worst-case time complexity is in $O(n^k)$ for some constant k. In other words, a good algorithm is one with polynomial time complexity. An algorithm that is not good is called **bad**. For example, if an algorithm requires 2^n steps to finish its computation on an input of size n, then it is bad. Roughly, then, "good" means (reasonably) fast, and "bad" means (painfully) slow.

EXAMPLE 1. Let S be a compound proposition consisting of propositional variables P_1, P_2, ..., P_k, joined in legal ways by negation, conjunction, and disjunction symbols. For example, S might be $(P_1 \wedge P_2) \vee \overline{P}_2$. We will consider the problem of determining whether S is *not* a tautology. (The reason that we look at the problem in this apparently negative way, instead of asking whether the given proposition *is* a tautology, will become clear presently.) The size of this problem is the length of (i.e., the number of symbols in) the compound proposition. A naive algorithm for solving the problem is to construct a truth table and in effect look at all 2^k possible assignments of truth values (T or F) to the variables $P_1, P_2, \ldots, P_k$. If we find an assignment of truth values that makes the proposition false, then we know that it is not a tautology; if *all* truth assignments make the proposition true, then we know that it is a tautology. In our example above, we would try all $2^2 = 4$ truth assignments and discover that for P_1 false and P_2 true the proposition S is false; hence we conclude that S is not a tautology. Clearly, this algorithm requires at least 2^k steps in the worst case (the case in which the proposition *is* a tautology). Now the lengths n of the propositions we are interested in (the size for this problem) may very well be linear in k (say, $n = Ck$ for some constant C). In this case the time complexity of this naive algorithm is an exponential function of the length of the input, because $2^k = 2^{n/C} = (2^{1/C})^n$. This exponential function grows faster than any polynomial, so this is a bad algorithm. ◆

EASY PROBLEMS AND HARD PROBLEMS

We now back up one level, from talking about algorithms for solving problems to talking about the problems themselves. Computer scientists, in their search for optimal algorithms for solving problems, have classified the problems themselves on the basis of how good the algorithms for solving them *could possibly be*. Let us restrict ourselves, in this discussion, to problems that have yes/no answers. Thus a **problem** consists of a description of the input (i.e., a set from which input will come) and a question about the input that can be answered either yes or no. In Example 1 the input was a compound proposition S (an element of the set of compound propositions) and the question was, "Is S not a tautology?" Another example of a problem is this:

> Input: an integer $N > 1$.
> Question: Is N a composite number?

Here the input set is the set of integers greater than 1. The size n of this problem is to be thought of as the number of digits in the input N; thus $n \approx \log_{10} N$.

An algorithm A is said to **solve** problem M if for every input in the input set of M, the algorithm A provides the correct answer to the question of M. In Example 1 we outlined an algorithm for solving the nontautology problem. The problem of determining whether a number N is composite can be solved by the naive algorithm of trying all possible factors from 2 to $N - 1$. Both of these seem to be rather inefficient algorithms, however. We saw in Example 1 that the naive algorithm for the nontautology problem is bad (in the technical sense in which we have defined the word "bad"), and it is not hard to see that the naive algorithm for solving the composite number problem is also bad (the number of possible factors to try is essentially $N \approx 10^n$, an exponential function of the size of the input).

Are there better algorithms? In particular, are there "good" (in the technical sense) algorithms for solving these two problems?

Computer scientists formalize these questions with the following definition, in essence a distinction between "easy" problems and "hard" problems from an algorithmic point of view. A problem M is said to be in class P (the letter P stands for "polynomial") if there is some good algorithm for solving M. A problem not in class P is said to be **intractable**.

It can be proved that there *do* exist intractable problems. For many of the problems that arise time and again in applications and *appear* to be hard, however, no one has been able to prove that no good algorithms for their solution exists. Just because no one has *found* a good algorithm for a problem does not mean that none exists.

Let us explore this topic further, since it is generally considered to be *the* major open research area in theoretical computer science today. It often happens that a positive answer to a question can be given in polynomial time if someone gives you the right hint. Suppose, for example, that the question is, "Is 2714757533 a composite number?" It would take a lot of computation to answer this question (the reader is invited to try to do so). But suppose that a genie came and whispered in your ear, "Compute 62753 × 43261." Taking this hint, you dutifully perform the multiplication (which takes only a few steps) and, behold, you find that the product is 2714757533. Therefore, you know that the answer to the question is yes. Generalizing from this example, we make the following definition. A problem M is said to be in class NP (the letters NP stand for "nondeterministic polynomial") if there is some good algorithm for verifying that the answer to M is yes whenever the answer to M *is* yes, under the assumption that the algorithm is provided with an appropriate hint (the hint will usually depend on the input).

Of course, an algorithm that requires a hint is no algorithm at all, since it is not deterministic (being deterministic is one of the characteristics of algorithms we listed in Section 4.1). The algorithm has no way to figure out what hint to use—one just comes out of the blue. If the genie in our example above had suggested that you compute 46721 × 58393 (which equals 2728179353), you would have been no better off than you were before—you still would not know whether 2714757533 was composite. Also note that you are not allowed to *rely* on anything the genie says; you must verify for yourself, after hearing the hint, that the input is composite. It would do you no good, for example, for the genie to whisper only, "2714757533 is composite."

EXAMPLE 2. Let us show that the nontautology problem is in the class NP. If a given proposition is not a tautology, then its truth value is F for at least one assignment of truth values to the variables in the proposition. If a genie tells you what truth assignment to try, you can quickly verify that the proposition is false for that set of truth values and hence that the proposition is not a tautology. On the other hand, if the proposition *is* a tautology, it is not clear how any hints would be helpful. ◆

Most problems that seem to be hard, when appropriately formulated, turn out to be in the class NP. We will encounter some in our discussion of graph theory in Chapters 8, 9, and 10. Here are a couple more.

EXAMPLE 3. Surprisingly, the problem of determining whether a number is prime is also in class NP. That this is so is not at all obvious—and requires some advanced ideas from number theory. Given a prime integer N *and the appropriate hint* (the complete prime factorization of $N - 1$ actually turns out to be a useful hint), we can quickly (in time polynomial in the number of digits of N) demonstrate, using theorems of number theory, that N is prime. We have already seen that with the appropriate hint (a factorization of N) we can demonstrate quickly that N is composite if it is composite. Thus both of these problems—determining primeness and determining compositeness—are in the class NP. It is not known, however, whether these problems are in P, that is, whether *without any hints* we can determine whether N is prime in time bounded by a polynomial function of the number of digits of N. Most researchers in this branch of theoretical computer science believe that these problems *are* in P. Furthermore, there are probabilistic algorithms (algorithms requiring random numbers to work) that will answer these questions in polynomial time with probability arbitrarily close to certainty that the answers are correct. We will have a little more to say about these problems in Section 4.5.

Although testing an integer for primeness is in practice fairly fast, the related problem of factoring numbers into their prime factors may well be intractable. This intractability is behind some of the encoding schemes used in applications where secrecy is important. To illustrate this idea, consider the following 199-digit number.

5,388,903,257,091,073,500,275,961,426,493,432,403,238,064,589,620,604,

809,638,831,230,890,717,370,569,110,101,934,865,874,780,426,924,589,

858,592,299,020,936,560,708,242,199,722,590,878,787,886,552,536,051,

932,487,611,848,981,598,053,453,558,834,177,377,613,539,237.

It turns out to be easy, with the number theory discussed in Section 4.5 (Theorem 3), to show that this number is composite; a few seconds of computation on a good mainframe computer would be more than adequate. Finding the factors is another matter. The author will pay the list price of this book to the first person who finds them. ◆

EXAMPLE 4. Consider the following **setsum problem**, a simple case of the problem known as the knapsack problem. The input is a finite set S of positive integers and a positive integer g. The question is whether there exists a subset of S whose sum is g. The size of the input is taken to be $n = |S|$. This problem is in class NP.

For example, is there a subset of $\{3, 5, 10, 15, 19, 25, 29, 32\}$ that sums to 49? If a genie suggests that $49 = 5 + 15 + 29$, then you can see immediately that the answer is yes. Knapsack problems turn out to have applications in finding the most efficient use of resources such as computer memory. ◆

Now clearly the class P is a subset of the class NP: If we have a good (polynomial time) algorithm for solving a problem, then we have, *a fortiori*, a good algorithm for verifying that the answer to a problem whose answer is yes is yes. What about the converse: Are the apparently hard problems that are in NP actually intractable (i.e., can be solved only by bad algorithms), or is it simply the case that no one has found the good algorithms for solving them? In symbols, the issue becomes: *Does $P = NP$?* No one knows the answer. Most computer scientists think that $P \neq NP$, if for no other reason than the fact that hundreds of very intelligent people all over the world have been working without success for nearly two decades to find good algorithms for various notorious problems in NP. Furthermore, it has been proved that most of the apparently hard problems in NP all stand or fall together. For example, if there is a good algorithm for solving the nontautology problem, then there is also a good algorithm for solving the setsum problem, and conversely. Indeed, it has been proved that if $P \neq NP$, then the nontautology problem and the setsum problem are in $NP - P$. Equivalently, if someone were to find a good algorithm for either of these problems, then $P = NP$, and hence all problems in NP would have polynomial time algorithms. Problems such as nontautology and setsum are called NP-**complete** problems, and there are literally hundreds of NP-complete problems that have surfaced over the past two decades, in such diverse areas as set theory, graph theory, computer operations, number theory, logic, games, and programming.

Our study of algorithm efficiency has taken us, then, to the frontiers of research in computer science. We end this section by going one step further: from problems that are *hard* to solve algorithmically to problems that *cannot* be solved algorithmically.

UNDECIDABLE PROBLEMS

If there were algorithmic solutions to all problems that could be formulated mathematically, then in some sense mathematicians would be out of business. Suppose, for example, that there were an algorithm for determining whether a given arithmetic equation had an integral solution (i.e., a solution in which all the variables took on integer values). Thus the input to this algorithm would be an equation, and the output would be yes or no. Consider Fermat's last theorem, the conjecture that there do not exist positive integers a, b, and c, and an integer $n > 2$ such that $a^n + b^n = c^n$ (see Exercise 31 in Section 1.2). No one knows whether the conjecture is true or false, and much of the progress in number theory and related fields over the past couple of centuries has been motivated by a search for a proof of this conjecture. (It has been proved that there is no solution with $n < 125{,}000$ or with

$c < 10^{1,800,000}$, however.) If there were an algorithm for answering the question of whether integral solutions to such equations existed, the algorithm could be applied to this particular conjecture and, sooner or later, we would know whether it was true or false. In theory, there would be no need for creative mathematicians to come up with a clever proof.

Now the fact that the existence of an algorithm such as this would in some sense trivialize the life work of many men and women of genius over the course of civilization does not mean that such an algorithm does not exist. If we want to know for sure that no such algorithm exists, then someone must *prove* this mathematically. In 1970 the Soviet mathematician Y. Matiyasevič did just that. We cannot go into the details of Matiyasevič's work here. Instead, we will discuss another problem that has no algorithmic solution, one that is even more fundamental. Let us first formalize these ideas a bit.

A problem is called **unsolvable**, or **undecidable**, if there is no algorithm that will give the correct answer to the question for all possible inputs. Thus an unsolvable problem is even harder than any of the intractable problems we discussed in the preceding subsection: Not only is there no *good* algorithm for solving it, there is not even a bad algorithm for solving it. Keep in mind, however, that saying that a problem is unsolvable does not mean that it has no answer. For each *specific* input to the problem, the answer is definitely either yes or no; it is just that there is no one algorithm (or finite number of algorithms) that will find this answer for all possible inputs.

Our example of an unsolvable problem is the classical one, which was first shown to be unsolvable by the father of theoretical computer science, A. Turing of Great Britain, in 1936. It is known as the **halting problem**. The problem is easy enough to state: *The input is an algorithm (computer procedure) A that takes one input, together with an input I for A; and the question is whether A will ever halt on input I.* (Recall that a procedure halts when it encounters a **return** statement during its execution.) For example, if the procedure A is

procedure $A(n :$ positive integer)
   $k \leftarrow 1$
   while $k \neq n$ **do**
     $k \leftarrow k + 2$
   return(n)

then the procedure will halt if the input I (the value for n) is 7 (or any odd positive integer) but will run on forever if the input is 24 (or any even positive integer). In this case, it was easy enough to determine, given the procedure and the input, whether or not the procedure would halt. We want to know whether there is an *algorithm* for making this determination. In a sense, such an algorithm would be a detector of "infinite loops" in computer programs. One cannot detect an infinite loop simply by letting the program run, since if the program has not stopped after running for awhile, you would not know whether its failure to stop during that time was caused by an infinite loop or merely by the fact that the time period was too short.

Having such an algorithm would obviously be a boon to programmers. Unfortunately, no such algorithm exists. On a higher philosophical plane, the fact that there is no algorithm for answering questions like this says that *creativity is not algorithmic*. Students often want to reduce problem solving in mathematics to algorithms. The existence of unsolvable problems tells us that sometimes it cannot be done. Thus in some sense our claim at the end of Section 4.1 that there is no algorithm for constructing algorithms is a mathematical theorem.

THEOREM 1. *The halting problem is unsolvable.*

Proof. We cannot give a completely formal, rigorous proof, because we have not sufficiently formalized the notions of computations and procedures. Nevertheless, we can give a pretty good idea of what is involved in the unsolvability of the halting problem. The proof is rather subtle, and its key idea is self-reference. The proof is related in spirit to the proof that the cardinality of any set is strictly less than the cardinality of its power set (Example 6b in Section 2.1). Our proof proceeds by contradiction.

Suppose that there is an algorithm H for solving the halting problem. Let us explicitly show the arguments of H by writing $H(A, I)$ to mean that H looks at the procedure A and input I. The output of H is either yes or no, depending on whether or not A will ever halt when fed input I. (Note that one of the inputs to H is a procedure, which can be thought of simply as a long string. The dual role of a procedure both as something that acts and as something that is acted upon is central to this proof.) Now consider the following algorithm, which we will call Q. Algorithm H will be a subprocedure of Q. Algorithm Q takes one argument—let us call it X—which is itself a procedure. This input X will play the dual role of both procedure and data. Algorithm Q first calls $H(X, X)$. In other words, Q invokes the hypothetical halting-problem-solver H and applies it to the *procedure* X and the *input* X. If H returns the answer yes (indicating that yes, the procedure X will halt when fed itself as input), then Q intentionally goes into an infinite loop (such as by executing the statement "**while true do** $z \leftarrow z$"). If H returns the answer no (indicating that no, the procedure X will never halt when fed itself as input), then Q halts immediately.

Now comes the self-referential trick. Since Q is an algorithm, it can be encoded as a procedure, and hence it can serve as its own input. Let us see what would happen if Q is executed with Q as input. There are two possibilities: Either it halts or it does not halt. If it halts, then by the way we constructed Q, we know that $H(Q, Q)$ must have answered no. In other words, H must have said that Q does not halt when fed Q as input. This is a contradiction, since we just said that Q *did* halt in the case under consideration. The other possibility is that Q does not halt when executed with Q as input. By the way we constructed Q, this will happen if and only if $H(Q, Q)$ answered yes, that is, H said that Q *does* halt when fed Q as input. Again this is a contradiction. The only way out of the dilemma is to reject our hypothesis that H exists. Hence there is no algorithm for solving the halting problem. ∎

Thus we have just *proved* that there are certain problems that a computer cannot solve.

SUMMARY OF DEFINITIONS

bad algorithm: an algorithm that is not good (e.g., an algorithm that determines whether a compound proposition is a tautology by constructing its truth table).

good algorithm: an algorithm whose time complexity is in $O(n^k)$ for some constant k, where n is the size of the input (e.g., bubble sort).

halting problem: the problem whose input is a procedure and an input for the procedure, and whose question is whether the procedure will halt when given the input.

intractable problem: a problem not in the class P.

the class *NP* (**nondeterministic polynomial** problems): the set of all problems for which there is a good algorithm to verify that the answer to the problem is yes when it is yes, provided that the algorithm is given an appropriate hint (e.g., the problem whose input is a natural number $N > 1$ (the number of digits of N is the size n for the problem) and whose question is whether N is composite; the hint is a pair of factors of N, both greater than 1—the good algorithm works by multiplying the two factors together, using $O(n^2)$ steps, and verifying that their product is N).

NP-**complete** problem: a problem in *NP* that is not in P unless $P = NP$ (e.g., the nontautology problem).

the class P (**polynomial** problems): the set of all problems that can be solved by a good algorithm.

problem (in complexity theory): a set and a yes/no question about an arbitrary element of the set (e.g., the problem of determining whether an element of the set of natural numbers is prime).

setsum problem: the problem whose input is a set of n positive integers and a positive integer g, and whose question is whether there is a subset whose sum is g).

solve a problem (in complexity theory): Algorithm A solves problem X if A always gives the correct answer to the question of X for the given input.

undecidable problem: a problem that no algorithm solves (e.g., the halting problem).

unsolvable problem: an undecidable problem.

EXERCISES

1. Determine whether these compound propositions are nontautologies; write out a proof that is complete but as short as possible. Compare the lengths of the proofs in the two cases.
   (a) $\overline{P} \lor (Q \land R)$.
   (b) $((P \land Q) \lor R) \lor \overline{P} \lor \overline{Q}$.

2. We can solve the composite number problem by dividing the input N by all possible factors from 2 to $\sqrt{N}$, not from 2 to $N - 1$. Show that even with this vast improvement (by about a factor of $\sqrt{N}$), this is still a bad algorithm.

3. Explain why each of the following problems is in the class NP. The size of the input is considered to be the number of digits in N, which is assumed to be a natural number.
   (a)   Input: N. Question: Is N a multiple of 2?
   (b)   Input: N. Question: Is N a multiple of 3?
   (c)   Input: N. Question: Is N a perfect square?
   (d)   Input: N. Question: Can N be written as the sum of the squares of four natural numbers?

4. Explain why the problems in Exercise 3a and 3b are in the class P.

5. Prove that $2^{67} - 1$ is not prime. [*Indispensable hint:* Compute 193707721 × 761838257287.]

6. Explain why the problems in Exercise 3c and 3d are in the class P.

7. For each of these problems, explain why the problem is in the class NP. (These are all NP-complete problems; thus no one knows whether they are in P, but the educated guess of computer scientists is that they are not.) The size of the input in each case is the number of symbols needed to write down the input.
   (a)   Input: a collection of m finite sets of natural numbers, and a positive integer $k \leq m$. Question: Are there k pairwise disjoint sets in the collection?
   (b)   Input: a finite set of natural numbers. Question: Can the set be partitioned into two parts so that the sum of the numbers in the first part is equal to the sum of the numbers in the second part?
   (c)   Input: a finite set S of pairs of natural numbers. Question: Does there exist a finite set C of natural numbers such that every pair in S contains one element in C and one element not in C?

8. Show that the following problem is in the class NP. Input: positive integers a, b, and c. Question: Are there any *positive integer* solutions (x, y) to the equation $ax^2 + by = c$? (There is something subtle that needs checking here: Why must the algorithm for verifying a yes answer to the question be a good algorithm, that is, polynomial in the size of the input, which is taken to be the

total number of digits in the coefficients? What if the plus sign were a minus sign?)

9. Describe an algorithm for the setsum problem, and show that it is not a good algorithm, in the technical sense.

10. Show that the following problems are in the class P by describing good algorithms for solving them.
    (a) Input: a list of words. Question: Does the list have any duplicates? (The size is the number of words in the list.)
    (b) Input: three positive integers x, y, and z. Question: Is xy greater than z? (The size is the total number of digits in the input.)

11. Show that the following special case of the setsum problem is in P. Input: a set S of powers of 3, together with a number x. Question: Is x the sum of some subset of S? [*Hint*: Any power of 3 is greater than the sum of all smaller powers of 3.]

12. What is wrong with each of the following arguments, which purport to show that the halting problem is solvable?
    (a) Given an algorithm A and input I for that algorithm, either A halts on I or it doesn't. Answer yes if it does, and answer no if it doesn't.
    (b) Given an algorithm A and input I for that algorithm, to determine whether A will halt on I, just run the algorithm A on the input I. If it stops, then answer yes, and if it doesn't stop, then answer no.

Challenging Exercises

13. Show how an algorithm for the halting problem would give an algorithm for determining whether a polynomial equation in one or more variables has an integral solution.

14. Prove that there is an algorithm for solving the following problem. Input: a natural number N. Question: Is there a pair of twin primes greater than N? [*Hint*: The problem asks you to prove that such an algorithm exists, not to produce one.]

Exploratory Exercises

15. Make up a setsum problem (or some other problem in *NP* but apparently not in *P*) for which you know that the answer is yes, but that you think no one will be able to solve, even with the help of a computer. Give it to your smartest friends and offer them $20 (or some other amount of money that it will pain you to lose) if they can prove, within one week, that the answer is yes. At the end of the week, either pay up or show them your proof that the answer is yes.

16. The apparent ease of determining whether a number is prime and apparent difficulty of finding its prime factors if it is composite lie behind some recent

important discoveries in cryptography—the making and breaking of secret codes. Find out about how number theory and the topics of this section are involved in cryptography by consulting Deneen [65], Denning [66], Gardner [103], Hellman [146], Luciano and Prichett [201], Meyer and Matyas [211], Simmons [272], and Sloane [275].

17. Find out how the difficulty in factoring can be exploited to permit such communication feats as playing real poker over the telephone, by consulting Shamir, Rivest, and Adleman [270].

18. Find out more about the *P* versus *NP* question by consulting Garey and Johnson [112] and Stockmeyer and Chandra [294].

19. Find out more about unsolvable problems by consulting Charlesworth [38], M. Davis [58], and Davis and Hersh [59].

20. Find out more about Alan Turing by consulting Hodges [149].

21. Find out more about algorithms for factoring by consulting Blair, Lacampagne, and Selfridge [21], Dixon [70], Riesel [248,249], and H. C. Williams [324].

22. Find out about the theory of computation, especially different assumptions or "models" of what a computer does and what it can compute under those assumptions, by consulting Cohen [47], Hopcroft [157], Hopcroft and Ullman [159], and Lewis and Papadimitriou [197].

23. Find out more about the relationship between algorithms and creativity by consulting Hofstadter [153].

SECTION 4.5
ALGORITHMS FOR
ARITHMETIC AND ALGEBRA

"Reeling and Writhing, of course, to begin with," the
Mock Turtle replied; "and then the different branches of
Arithmetic—Ambition, Distraction, Uglification and Derision."

—Lewis Carroll, *Alice's Adventures in Wonderland*

We now turn to some specific important algorithms of discrete mathematics. In each case we present the algorithm and analyze its time complexity. We cover algorithms for natural number arithmetic, polynomial manipulation, and matrix operations. In addition to the applications presented in this section, we will see applications in future chapters.

First let us discuss algorithms for the arithmetic of natural numbers. If we only cared about performing arithmetic on fairly small numbers, say those that fit inside one word of computer memory on the typical computer (generally, about eight decimal digits long), we would not have much to say. The algorithmic questions would be ones for the designers of the electronic circuits of the computer to worry about, and we would simply think of each arithmetic operation as requiring one step. Many applications, however, require large natural numbers, with an unlimited number of digits. We will discuss algorithms for working with numbers of arbitrary length; indeed, *the lengths of the numbers will be the size parameters for our efficiency analyses.*

Our approach will be fairly naive. Most of the algorithms we discuss are the ones taught in elementary school. Thus we can think of ourselves as elementary school teachers, instructing bright shiny young computers how to do arithmetic on natural numbers. Before discussing the algorithms themselves, we need to review briefly, but formally, the numeration system.

We usually work in the base 10 numeration system. This means that any natural number x is represented as a string of **digits**,

$$x = d_{n-1}d_{n-2}\ldots d_2 d_1 d_0,$$

where each digit d_i is an element of the set $\{0,1,2,3,4,5,6,7,8,9\}$. We say that the digit d_i is in the ith **column**; in other words, we are numbering the columns from right to left, starting with 0. This string is called the (**base 10** or **decimal**) **numeral** for x, and we call x an **n-digit number**. The number that this string represents is given by

$$\sum_{i=0}^{n-1} d_i \cdot 10^i = d_0 + 10d_1 + 10^2 d_2 + \cdots + 10^{n-1}d_{n-1}.$$

In general we will allow leading zeros, so that, for example, 00473 can be considered a five-digit number, even though it is equal to the three-digit number 473.

There is nothing special about the number 10 other than the fact that we human beings have 10 fingers. Any other base $\beta \geq 2$ could be chosen to represent natural numbers. For example, if we let $\beta = 4$, then the numeral 1302 represents the number $2 + 4 \cdot 0 + 4^2 \cdot 3 + 4^3 \cdot 1 = 114$. There are some straightforward algorithms, discussed in Exercises 9–12, for converting from one base to another.

ARITHMETIC OPERATIONS

Let us begin with the operation of adding two natural numbers. With the usual grade-school algorithm, we add digit by digit, from right to left, mindful of a possible carry from column to column. In pseudocode, we get Algorithm 1.

We were convinced in elementary school that Algorithm 1 works. The time analysis is easy, since the algorithm consists of one **for** loop and little else. Indeed,

procedure $sum(x, y :$ natural numbers)
    $\{x$ and y are represented by the decimal numerals $a_{n-1} \ldots a_0$ and $b_{m-1} \ldots b_0$,
    respectively; the answer $z = x + y$ will appear as the decimal numeral
    $c_k \ldots c_0$, where $k = \max(n, m)$; we assume that a_i is defined to be 0 for $i \geq n$
    and that b_i is defined to be 0 for $i \geq m\}$
    $k \leftarrow \max(n, m)$
    $carry \leftarrow 0$
    for $i \leftarrow 0$ **to** $k - 1$ **do**
      begin
        $c_i \leftarrow a_i + b_i + carry$
        if $c_i > 9$ **then**
          begin
            $c_i \leftarrow c_i - 10$ {the 10 gets carried to the next column}
            $carry \leftarrow 1$
          end
        else $carry \leftarrow 0$
      end
    $c_k \leftarrow carry$
    return(z) {represented as $c_k \ldots c_0\}$

Algorithm 1. Addition of two natural numbers.

the time required for adding is in $O(\max(n, m))$, which is equivalent to $O(n + m)$. There is no way to improve this performance, at least as far as big-oh approximation is concerned, since any algorithm for adding two numbers must at least *look at* the two numbers, and that already requires $n + m$ steps.

Subtraction is similar to addition and is left for the reader (Exercise 13).

Multiplication is more complicated. The elementary school algorithm requires the computation of partial products (products of the multiplicand with each digit of the multiplier), which are then summed to obtain the answer. We can streamline the procedure somewhat by thinking of each digit-by-digit multiplication as contributing to the final answer, shifted an appropriate number of places to the left. The main messiness comes in managing the carries, which we handle each time we do a digit-by-digit multiplication. See Algorithm 2.

To see that this algorithm is correct, we need to note that when we multiply the digits b_i and a_j, we are really multiplying the parts of y and x that these digits represent; that is, we are really multiplying $b_i 10^i$ by $a_j 10^j$, obtaining $b_i a_j 10^{i+j}$. Hence the correct column for this part of the answer is the $(i + j)$th column from the right. The rest of the inner **for** loop takes care of carrying from the column that we have just incremented to the next column to the left.

To analyze the time complexity of this algorithm, we note that except for the initialization (which requires $O(m + n)$ steps), the entire body of the procedure is a **for** loop, iterated m times. Within this loop is another **for** loop, iterated n times.

procedure *product*$(x, y :$ natural numbers)
$\{x$ and y are represented by the decimal numerals $a_{n-1} \dots a_0$ and $b_{m-1} \dots b_0$,
respectively; the answer z will appear as the decimal numeral $c_{m+n-1} \dots c_0\}$
   for $i \leftarrow 0$ **to** $m + n - 1$ **do**
      $c_i \leftarrow 0$ {initialize the answer to be 0}
   for $i \leftarrow 0$ **to** $m - 1$ **do** {multiply ith digit of y}
      for $j \leftarrow 0$ **to** $n - 1$ **do** {by jth digit of x}
         begin
            $k \leftarrow i + j$ {correct column for this product}
            $c_k \leftarrow c_k + b_i \cdot a_j$
            carry $\leftarrow \lfloor c_k / 10 \rfloor$
            $c_k \leftarrow c_k - 10 \cdot$ *carry*
            $c_{k+1} \leftarrow c_{k+1} +$ *carry*
         end
   return(z) {represented as $c_{m+n-1} \dots c_0\}$

Algorithm 2. Multiplication of two natural numbers.

Within the inner loop, only a small finite number of calculations occur. Thus the nested loops require $O(mn)$ steps, so that the total running time is $O(m + n + mn)$, which is clearly $O(mn)$.

In discussing arithmetic on large numbers, we usually assume that the numbers involved are roughly the same size. Thus we can summarize this analysis in the following way.

THEOREM 1. *The grade-school algorithms compute the sum of two n-digit numbers in* $O(n)$ *steps and the product of two n-digit numbers in* $O(n^2)$ *steps.*

It turns out that the grade-school algorithm is not the most efficient algorithm for multiplication of two natural numbers. More sophisticated algorithms for the multiplication of two n-digit numbers, such as the Schönhage–Strassen algorithm, can accomplish the task in time $O(n^{\log 3}) \approx O(n^{1.58})$ or even less. We will not pursue the subject in this book (see Exercise 34).

Natural number (long) division is usually not taught in school until the fifth grade or so, perhaps because it is messier than multiplication. The ambitious reader can grapple with the details (Exercise 24); we will content ourselves here with observing the following fact about efficiency, which should at least be plausible at this point.

THEOREM 2. *The grade-school algorithm computes the natural number quotient and remainder of two n-digit numbers in* $O(n^2)$ *steps.*

Because of the applications, we discuss the exponentiation operation not for the arithmetic of natural numbers but rather for modular arithmetic.

MODULAR ARITHMETIC (*OPTIONAL*)

In Section 3.1 we defined x mod m to be the remainder when x is divided by m. We also pointed out that we can define addition and multiplication on the set of possible remainders $\{0, 1, \ldots, m-1\}$. In particular, the mod m sum (or product) of two numbers x and y from this set was defined to be their ordinary sum (or product) mod m. We will assume that in addition to algorithms for performing addition and multiplication of natural numbers we also have an algorithm for reduction mod m, which is the same thing as obtaining the remainder of a natural number division. In our efficiency analyses we will count each addition and multiplication mod m as one basic step, even though in fact the time they require depends on the number of digits in the numbers. In particular, we will not count reductions mod m at all. The philosophy here is that m is fixed (although possibly as large as several hundred digits in practice), so all the numbers we are working with are no larger than m, hence bounded in size by a constant. We will discuss two interesting applications of modular arithmetic: primeness testing and random number generation.

First, we will find an efficient way to implement exponentiation. This operation has application to primeness testing (as we will see) and to cryptography (which we will not pursue). The problem is to compute a^n mod m, where a is a natural number less than m, and n is any natural number. The naive algorithm, utilizing the definition of exponentiation as repeated multiplication, is shown as Algorithm 3.

procedure *modular_power*$(a, n, m :$ integers$)$
    $\{$this procedure computes a^n mod m, where $m \geq 2$, $n \geq 0$, and $0 \leq a < m$;
    the answer is accumulated in $x\}$
      $x \leftarrow 1$
      for $i \leftarrow 1$ **to** n **do**
        $x \leftarrow x \cdot a$ mod m
      return(x)

Algorithm 3. Modular exponentiation.

Clearly, the time efficiency is in $O(n)$: the procedure uses n multiplications. In many applications n can be extremely large (say, around 10^{100}); for such values of n, this algorithm is out of the question. Instead, we will develop an algorithm for exponentiation that uses successive squaring. Suppose that we want to compute 3^{16}. We can square 3, obtaining $3^2 = 9$, then square 9, obtaining $3^4 = 81$, then square 81, obtaining $3^8 = 6561$, and finally, square 6561 to obtain $3^{16} = 43,046,721$.

Rather than 16 multiplications, we use only four. The calculation works nicely since 16 is a power of 2, but with a little more work we can do any exponentiation in this manner. Suppose that we want to obtain a^{100}. Successive squarings will give us a, a^2, a^4, a^8, a^{16}, a^{32}, and a^{64}. Since $100 = 64 + 32 + 4$, we can compute a^{100} as the product $a^{64}a^{32}a^4$. Rather than 100 multiplications, this calculation uses only nine. Algorithm 4 is just a fancy version of this idea, done mod m.

procedure *fast_modular_power*$(a, n, m$: integers)
    {this procedure computes a^n mod m, where $m \geq 2$, $n \geq 0$, and $0 \leq a < m$;
    successive squares are stored in s; the answer is accumulated in x}
    $s \leftarrow a$
    $x \leftarrow 1$
    while $n > 0$ **do**
      begin
        if n is odd **then** $x \leftarrow x \cdot s$ mod m
        $s \leftarrow s \cdot s$ mod m
        $n \leftarrow \lfloor n/2 \rfloor$
      end
    return(x)

Algorithm 4. Fast modular exponentiation.

For example, if $n = 100$, then during the third, sixth, and seventh (last) pass through the **while** loop, x is multiplied by a^4, a^{32}, and a^{64}, yielding its correct final value of a^{100} (all calculations mod m).

The analysis is similar to the analysis of binary search in Section 4.3. The complexity is the number of passes through the loop, and since n is halved each time through the loop, the number of passes is roughly $\log n$. Thus our algorithm has time complexity in $O(\log n)$.

As an application, we cite the following theorem, first discovered by Fermat. In contrast to his famous last theorem (Exercise 31 in Section 1.2), which no one has been able to prove, this theorem can be proved using the tools of number theory and abstract algebra (although the proof is beyond the scope of this book and we omit it).

THEOREM 3. *If p is an odd prime number, then $2^{p-1} \equiv 1 \pmod{p}$.*

To see the significance of this theorem, suppose that we have an odd number N and we want to discover whether N is prime. Suppose that we compute 2^{N-1} mod N, which we can do efficiently by Algorithm 4. If the answer is *not* equal to 1, then we know by the contrapositive of Theorem 3 that N is not prime. In this case we will have no idea what the proper factors of N are—we only know that it has some. For example, since 2^{20} mod $21 = 4 \neq 1$, we know that 21 is not prime.

Unfortunately, the converse of this theorem is not true, although the exceptions to the converse are few and far between. The theorem says that *if* N is an odd prime, *then* the congruence holds. It is not always true that *if* the congruence holds for an odd number N, *then* N must be prime. But most of the time, if we find 2^{N-1} mod N to be 1, then we are looking at a prime number N. For example, 2^{22} mod 23 = 1 and 23 is indeed prime. On the other hand, 2^{340} mod 341 = 1, but 341 = 11 × 31 is not prime. Numbers such as 341 are called **pseudoprimes**.

Using some refinements of these ideas, as well as other ideas from number theory, mathematicians have constructed elaborate and effective tests for primeness, some probabilistic in nature, some totally deterministic. In one version of a probabilistic primeness test, one is able, in polynomial time, either to demonstrate with certainty that the input is not prime, or else to demonstrate with *any* degree of certainty desired, short of absolute certainty, that the input is prime. Unfortunately, we cannot go into the details of this fascinating story (see Exercise 38).

Our second application of modular arithmetic is to the problem of generating "random" numbers. For the purposes of this discussion, we will say that a process generates a sequence of **random numbers** r_1, r_2, r_3, ..., if each r_i is a real number between 0 (inclusive) and 1 (exclusive), each r_i is equally likely to fall anywhere within the interval $[0, 1)$, and the value of r_i is totally independent of the values of r_1, r_2, ..., r_{i-1}. (A full definition of these terms would require a digression into probability theory that we do not wish to make.) As we will prove in Section 6.4, it is impossible to generate such a sequence by computer. Nevertheless, the following approach, generating a sequence of **pseudorandom numbers**, often works reasonably well for many applications requiring randomness, such as simulations or probabilistic algorithms. It is called a **linear congruential pseudorandom number generator**.

First we pick a large natural number m as the base of our generator, and we pick two more large constants, A and C, both less than m. We will generate a sequence of integers a_1, a_2, ..., such that $0 \le a_i < m$ for all i. Our sequence of pseudorandom numbers is then a_1/m, a_2/m, ... Clearly, each of these numbers is in the desired interval $[0, 1)$. We generate the sequence of a_i's iteratively. We choose an arbitrary natural number a_0 less than m, called the **seed** of the sequence. Each subsequent a_i comes from the previous term in the sequence, a_{i-1}, by the formula

$$a_i = (A \cdot a_{i-1} + C) \bmod m.$$

If m, A, and C are chosen carefully (and there have been both theoretical and empirical studies to help with these choices), the sequence that is generated appears to have the properties of a random sequence. One popular version of Pascal for microcomputers uses $m = 2^{31} - 1 = 2,147,483,647$, $A = 7^5 = 16,807$, and $C = 0$.

EXAMPLE 1.   Let us see what happens for $m = 100$, $A = 31$, and $C = 72$. Choosing the seed to be 1, our sequence starts out

$$3, \ 65, \ 87, \ 69, \ 11, \ 13, \ 75, \ 97, \ 79, \ 21, \ \ldots.$$

There seems to be no particular pattern, and the numbers seem to hop around rather randomly. The corresponding sequence of real numbers is just the quotients of these numbers divided by m, namely 0.03, 0.65, 0.87, 0.69, 0.11, and so on. In practice, m is taken to be fairly large, say around 10^{10}. ◆

POLYNOMIALS (*OPTIONAL*)

Many applications of mathematics involve computer calculations with functions, particularly with polynomial functions. A **polynomial function** of degree n is a function of the form

$$f(x) = a_0 + a_1 x + a_2 x^2 + \cdots + a_n x^n,$$

where each coefficient a_i is a real number, and $a_n \neq 0$. (The zero function, that is, the function f for which $f(x) = 0$ for all x, is considered to be a polynomial of degree $-\infty$.) A polynomial function has as its domain the entire set of real numbers. In many discrete mathematics applications, the coefficients and argument will be integers. We will discuss three aspects of polynomial arithmetic: addition, multiplication, and evaluation. Since the coefficients completely determine the polynomial, we can view a polynomial as its list of coefficients $(a_0, a_1, \ldots, a_n)$.

Adding polynomials, as beginning algebra students are taught, is simply a matter of collecting like terms: The coefficient of x^i in the sum of f and g is the sum of the coefficient of x^i in f and the coefficient of x^i in g. Thus Algorithm 5 for adding polynomials is almost trivial. It can be viewed as a simplification of the algorithm for adding natural numbers (the coefficient of x^i is like the digit in the ith column)—simpler because no carrying is involved.

procedure *polynomial_sum*(f, g : polynomials)
    {assume that $f(x) = a_0 + a_1 x + a_2 x^2 + \cdots + a_n x^n$ and $g(x) = b_0 + b_1 x + b_2 x^2 + \cdots + b_m x^m$; the sum will be $(f + g)(x) = c_0 + c_1 x + c_2 x^2 + \cdots + c_l x^l$, where $l = \max(n, m)$; assume that a_i is defined to be 0 for $i > n$ and b_i is defined to be 0 for $i > m$}
    $l \leftarrow \max(n, m)$
    for $i \leftarrow 0$ **to** l **do**
        $c_i \leftarrow a_i + b_i$
    return($f + g$) {i.e., all the coefficients c_i}

Algorithm 5. Addition of two polynomials.

Obviously, the time complexity is in $O(n + m)$, if we consider the arithmetic operations as the basic steps.

Multiplication is somewhat more complicated. By definition from elementary algebra (just multiplying term by term and regrouping), the product of

$$f(x) = \sum_{i=0}^{n} a_i x^i \quad \text{and} \quad g(x) = \sum_{j=0}^{m} b_j x^j \quad \text{is} \quad (fg)(x) = \sum_{k=0}^{m+n} c_k x^k,$$

where the coefficients c_k of the product fg are given by

$$c_k = \sum_{i=0}^{k} a_i b_{k-i}.$$

Turning this definition into a procedure, we obtain Algorithm 6 for multiplying polynomials. Again, it is really just a simplification of the corresponding algorithm for multiplying natural numbers—no carrying is needed.

---

procedure *polynomial_product*(f, g : polynomials)
  {assume that $f(x) = a_0 + a_1 x + a_2 x^2 + \cdots + a_n x^n$, and
  $g(x) = b_0 + b_1 x + b_2 x^2 + \cdots + b_m x^m$; the product will be
  $(fg)(x) = c_0 + c_1 x + c_2 x^2 + \cdots + c_{n+m} x^{n+m}$}
  $l \leftarrow n + m$
  for $k \leftarrow 0$ **to** l **do**
    $c_k \leftarrow 0$
  for $i \leftarrow 0$ **to** n **do**
    for $j \leftarrow 0$ **to** m **do**
      $c_{i+j} \leftarrow c_{i+j} + a_i \cdot b_j$
  return(fg) {i.e., all the coefficients c_k}

Algorithm 6. Multiplication of two polynomials.

To analyze the complexity, let us consider each arithmetic operation as a basic step. Then the major time consumed by Algorithm 6 comes from the nested **for** loops; the assignment statement within them must be executed $O(nm)$ times. Hence the time complexity for this method of polynomial multiplication is in $O(nm)$, which is $O(n^2)$ when the polynomials are of the same degree. Computations equivalent to polynomial multiplication arise in many scientific and engineering applications, and it is fortunate that a faster algorithm is available. We cannot go into the details of this fast algorithm, known as the fast Fourier transform (see Exercise 35); it uses fairly advanced data structures and mathematics. Remarkably, it cuts the number of arithmetic steps from $O(n^2)$ to $O(n \log n)$.

Finally, we look at the important problem of evaluating (computing the value of) a polynomial $f(x) = a_0 + a_1 x + a_2 x^2 + \cdots + a_n x^n$ at a particular value of the argument x. Apparently, we have to compute x, x^2, x^3, ..., x^n, multiply each by its appropriate coefficient, and add. If we do this in the most naive way, it will take $O(n^2)$ steps, since the computation of *each* term in the sum will require $O(n)$ multiplications in order to form the appropriate power of x. A faster exponentiation algorithm, along the lines of that presented in Algorithm 4 for modular exponentiation, would reduce the overall time complexity to $O(n \log n)$. That would be the wrong approach, however. In order to have an efficient algorithm for polynomial evaluation, it is enough to note that the value of x^{i+1} can be computed from the value of x^i with one multiplication; thus there is no need to "start over" in the computation of the power of x for each term in the polynomial. Developing this idea into an algorithm, we have the straightforward procedure shown as Algorithm 7.

procedure *polynomial_value*(f : polynomial, x : real number)
    {assume that $f(x) = a_0 + a_1x + a_2x^2 + \cdots + a_nx^n$}
      power_of_x $\leftarrow 1$
      value $\leftarrow 0$
      for $i \leftarrow 0$ **to** n **do**
        begin
          value $\leftarrow$ *value* $+ a_i \cdot$ *power_of_x*
          power_of_x $\leftarrow x \cdot$ *power_of_x*
        end
      return(*value*)

Algorithm 7.   Evaluation of a polynomial.

Clearly the time complexity is now linear; approximately $2n$ multiplications and n additions of real numbers are required. We cannot improve this performance in big-oh terms, since it takes at least n steps just to look at the coefficients. We can however, get by with half as many multiplications as in Algorithm 7, using Algorithm 8, known as **Horner's method**.

procedure *horner*(f : polynomial, x : real number)
    {assume that $f(x) = a_0 + a_1x + a_2x^2 + \cdots + a_nx^n$}
      value $\leftarrow a_n$
      for $i \leftarrow n - 1$ **down to** 0 **do**
        value $\leftarrow x \cdot$ *value* $+ a_i$
      return(*value*)

Algorithm 8.   Horner's method for the evaluation of a polynomial.

This algorithm works because of the algebraic identity

$$a_0 + a_1 x + a_2 x^2 + \cdots + a_n x^n = a_0 + x(a_1 + x(a_2 + \cdots + x(a_{n-1} + x a_n) \cdots)).$$

The algorithm proceeds from right to left on the right-hand side of this equation.

EXAMPLE 2. Evaluate $f(x) = 3 + 5x - 6x^2 + x^5 - x^6$ at $x = 3$, using Horner's method.

Solution. In the notation of the procedure, we have $n = 6$, with coefficients $a_0 = 3$, $a_1 = 5$, $a_2 = -6$, $a_3 = 0$, $a_4 = 0$, $a_5 = 1$, and $a_6 = -1$. Thus *value* begins at -1. It is then multiplied by 3 and incremented by 1 to give -2. During the next pass through the loop, *value* becomes $3 \cdot (-2) + 0 = -6$; then $3 \cdot (-6) + 0 = -18$; then $3 \cdot (-18) - 6 = -60$. Two more passes yield the final answer, -522.

In summary, as a calculation from left to right, we have computed

$$((((((-1) \cdot 3 + 1) \cdot 3 + 0) \cdot 3 + 0) \cdot 3 + (-6)) \cdot 3 + 5) \cdot 3 + 3.$$

It is interesting to note that implementation of Horner's method is easier on an inexpensive calculator than it is on a sophisticated one. Because the latter follows the usual algebraic hierarchy that multiplication takes precedence over addition, you need to press the equals button to complete every addition operation when performing Horner's method on a sophisticated calculator. Thus the calculation above would be effected by pushing the following sequence of buttons:

$$-1 \times 3 + 1 = \times 3 + 0 = \times 3 + 0 = \times 3 - 6 = \times 3 + 5 = \times 3 + 3 =$$

On the \$5 models, there is no need to press the equals button until the end, because with the pressing of each operation button, the pending calculation is automatically completed. Thus on a cheap calculator this calculation is implemented with the following keystrokes:

$$-1 \times 3 + 1 \times 3 + 0 \times 3 + 0 \times 3 - 6 \times 3 + 5 \times 3 + 3 =$$

◆

MATRIX OPERATIONS

A matrix, as we saw in Section 2.2, can be thought of as a rectangular array of entries. We will assume that the entries are numbers, but applications in which the entries are polynomials or other more complex objects also arise. We will need to add and multiply matrices in Chapter 8, so we end this section with a definition of these operations and a discussion of algorithms for their implementation.

Suppose that an m by n matrix A has a_{ij} as the entry in its ith row, jth column. Thus the matrix looks like

$$\begin{bmatrix} a_{11} & a_{12} & \cdots & a_{1n} \\ a_{21} & a_{22} & \cdots & a_{2n} \\ \vdots & \vdots & \ddots & \vdots \\ a_{m1} & a_{m2} & \cdots & a_{mn} \end{bmatrix}.$$

Recall that we abbreviate this by writing $A = (a_{ij})$. If $B = (b_{ij})$ is another m by n matrix, we define the **sum** $A + B$ to be the matrix whose (i, j)th entry is the sum of the (i, j)th entries of A and B, namely $a_{ij} + b_{ij}$. Thus the sum is again an m by n matrix. We do not define addition for matrices whose sizes (number of rows and number of columns) are not equal.

EXAMPLE 3.   Let

$$A = \begin{bmatrix} 4 & 1 & -1 \\ -2 & 1 & 2 \\ 0 & 1 & 1 \\ 2 & 3 & 1 \end{bmatrix} \quad \text{and} \quad B = \begin{bmatrix} 2 & 6 & 1 \\ 0 & 0 & 2 \\ 5 & -1 & 1 \\ 2 & 3 & 1 \end{bmatrix}.$$

Then the sum is

$$A + B = \begin{bmatrix} 6 & 7 & 0 \\ -2 & 1 & 4 \\ 5 & 0 & 2 \\ 4 & 6 & 2 \end{bmatrix}.$$

For example, the entry -2 in the second row, first column of $A + B$, is the sum of the corresponding entries, -2 and 0, in A and B.   ◆

The multiplication operation on matrices, whose definition is not so obvious, plays an important role in discrete mathematics, as we will see in Section 8.2. It also plays a key role in linear algebra. Suppose that $A = (a_{ik})$ is an m by n matrix, and $B = (b_{kj})$ is an n by p matrix. Then the **product** $A \times B$ is the m by p matrix $C = (c_{ij})$ whose entries are given by the formula

$$c_{ij} = \sum_{k=1}^{n} a_{ik}b_{kj}.$$

In particular, matrix multiplication is not defined "entrywise," as matrix addition was defined. In order for $A \times B$ to be defined, the number of columns of A must equal the number of rows of B. A mental heuristic for multiplying A by B is to imagine picking up each row of A one at a time and lining it up successively with the columns of B. When the ith row of A is lined up with the jth column of B, we obtain the (i, j)th entry of $A \times B$ by taking the sum of the products of corresponding entries.

EXAMPLE 4.   Let

$$
A = \begin{bmatrix} 4 & 1 & -1 \\ -2 & 1 & 2 \\ 0 & 1 & 1 \\ 2 & 3 & 1 \end{bmatrix} \quad \text{and} \quad B = \begin{bmatrix} 2 & -1 \\ 1 & 0 \\ 0 & 0 \end{bmatrix}.
$$

Then the product $A \times B$, obtained from the formula above, is

$$
A \times B = \begin{bmatrix} 9 & -4 \\ -3 & 2 \\ 1 & 0 \\ 7 & -2 \end{bmatrix}.
$$

For example, the entry -3 in the second row, first column of $A \times B$ is the sum of the second row of A, $[-2 \ \ 1 \ \ 2]$, multiplied term by term with the first column of B, $[2 \ \ 1 \ \ 0]$, giving $(-2) \cdot 2 + 1 \cdot 1 + 2 \cdot 0 = -3$. Note that the order of matrix multiplication matters, unlike the order of matrix addition or real number multiplication. The latter two operations are commutative, but matrix multiplication definitely is not. Indeed, in this example, $B \times A$ is not even *defined*, and even in those cases in which $A \times B$ and $B \times A$ are both defined, they will usually not be equal (or even necessarily of the same size). ◆

The algorithms for matrix addition and multiplication are straightforward; writing them down is left to the reader (Exercise 19). In analyzing their efficiency, we will assume that the additions and multiplications of numbers are the basic steps. Clearly, the algorithm for matrix addition has time complexity in $O(mn)$, since each of the mn entries in the sum can be computed with one arithmetic operation. If we transcribe the definition of matrix multiplication into the obvious algorithm, it is clear that it requires $O(n)$ steps to compute each of the mp entries in the product of an m by n matrix with an n by p matrix. This gives total time complexity in $O(mnp)$. In many applications, all the matrices are square, so $m = n = p$ and we see that straightforward matrix multiplication has time complexity in $O(n^3)$. We summarize this discussion for future reference.

THEOREM 4.   *The sum of two n by n matrices can be computed in $O(n^2)$ steps. The straightforward algorithm for computing the product of two n by n matrices requires $O(n^3)$ steps.*

Just as with numerical multiplication, this turns out not to be the best possible result. A more complicated algorithm for matrix multiplication, which is not at all obvious from the definition and which we will discuss briefly in Section 7.2, requires only $O(n^{\log 7}) \approx O(n^{2.8})$ steps. If n is very large, this improvement can be important.

We have just scratched the surface of arithmetic and algebraic algorithms. The reader with access to sophisticated software for performing exact multidigit arithmetic and algebra on a computer—systems such as MACSYMA for mainframe computers, or Maple, Mathematica, or muMath for smaller machines—should explore the features of such systems.

SUMMARY OF DEFINITIONS

base 10 numeral: a decimal numeral.

ith **column** in a decimal numeral: the position in which the digit d represents $d \cdot 10^i$; thus the rightmost position is the 0th column (or ones' column), the position to its left is the first column (or tens' column), the position to its left is the second column (or hundreds' column), and so on.

decimal numeral: a string $d_{n-1}d_{n-2}\ldots d_2 d_1 d_0$ where each d_i is a digit, which represents the natural number $\sum_{i=0}^{n-1} d_i \cdot 10^i$.

digit: an element of the set $\{0, 1, 2, 3, 4, 5, 6, 7, 8, 9\}$.

Horner's method for evaluating the polynomial $a_0 + a_1 x + a_2 x^2 + \cdots + a_n x^n$ at x: the calculation $a_0 + x(a_1 + x(a_2 + \cdots + x(a_{n-1} + xa_n)\cdots))$, performed from right to left.

linear congruential pseudorandom number generator: a method of generating pseudorandom numbers by letting $r_i = a_i/m$, where the seed a_0 is picked arbitrarily and a_i for $i > 0$ is computed from a_{i-1} by the formula $a_i = (A \cdot a_{i-1} + C) \bmod m$; here m, A, and C are carefully chosen constants.

n**-digit number**: a natural number that can be represented by a decimal numeral of length n.

polynomial function of degree n: a function of the form $f(x) = a_0 + a_1 x + a_2 x^2 + \cdots + a_n x^n$, where each coefficient a_i is a real number, and $a_n \neq 0$; the zero polynomial $f(x) = 0$, has degree $-\infty$ (e.g., $f(x) = 3x^4 - \pi x^3 + x$ is a polynomial of degree 4).

product of two matrices: If $A = (a_{ik})$ is an m by n matrix and $B = (b_{kj})$ is an n by p matrix, the product $A \times B$ is the m by p matrix whose (i, j)th entry is $\sum_{k=1}^{n} a_{ik}b_{kj}$ (e.g., $\begin{bmatrix} 1 & 2 & 3 \\ 4 & 5 & 6 \end{bmatrix} \times \begin{bmatrix} 7 & 8 \\ 9 & 10 \\ 11 & 12 \end{bmatrix} = \begin{bmatrix} 58 & 64 \\ 139 & 154 \end{bmatrix}$).

pseudoprime: an integer $n > 1$ that is not prime but satisfies the congruence $2^{n-1} \equiv 1 \pmod{n}$ (which all prime numbers satisfy); there are far fewer pseudoprimes than primes, so if a randomly chosen large number satisfies the congruence, it is a safe bet that it is prime (e.g., 645 is a pseudoprime).

pseudorandom numbers: a sequence of numbers generated by a deterministic (nonrandom) process that appear to be random (e.g., a sequence generated by a linear congruential generator, with good choices of C and m).

random numbers: a sequence of numbers r_1, r_2, r_3, $\ldots$, generated by some chance process, where each r_i is a real number between 0 (inclusive) and 1 (exclusive), each r_i is equally likely to fall anywhere within the interval $[0, 1)$, and the value of r_i is totally independent of the values of $r_1, r_2, \ldots, r_{i-1}$.

seed of a pseudorandom number generator: an arbitrarily chosen number used to start the sequence.

sum of two matrices of the same size: the matrix whose (i, j)th entry is the sum of the (i, j)th entries of the two matrices (e.g., $\begin{bmatrix} 1 & 2 \\ 3 & 4 \end{bmatrix} + \begin{bmatrix} 5 & 6 \\ 7 & 8 \end{bmatrix} = \begin{bmatrix} 6 & 8 \\ 10 & 12 \end{bmatrix}$).

EXERCISES

1. Carry out Algorithm 1 by hand for the following inputs.
   (a) $x = 375$ and $y = 259$.
   (b) $x = 999$ and $y = 99$.

2. Carry out Algorithm 2 by hand for the following inputs.
   (a) $x = 375$ and $y = 259$.
   (b) $x = 999$ and $y = 99$.

3. Compute the following numbers mod 100, using pencil and paper (and calculator if desired). [*Hint*: Use Algorithm 4 and/or clever common sense.]
   (a) 2^{99}.
   (b) 98^{10}.
   (c) 52^{52}.

4. Consider the linear congruential pseudorandom number generator with $m = 100$, $A = 33$, and $C = 73$.
   (a) Using the seed 4, generate the first 15 numbers in the sequence.
   (b) Comment generally on the apparent randomness or lack thereof in the sequence obtained in part (a).
   (c) Explain why this method cannot generate a truly random sequence.

5. Find the sum and the product of the following polynomials, using Algorithms 5 and 6, respectively.
   (a) $3x^2 + 4x - 2$ and $x + 5$.
   (b) $3x^2$ and $-3x^2$.

6. Evaluate the following polynomials using Horner's method.
   (a) $f(x) = x^6 - 10x^5 + 3x^3 - 4x + 8$ at $x = 2$.
   (b) $q(x) = x + 2x^2 + 3x^3 + 4x^4 + 5x^5$ at $x = 1.5$.

7. Find the sum $A + B$ of the following matrices.

$$A = \begin{bmatrix} 3 & 8 & -2 & 3 \\ 0 & 0 & 2 & -1 \end{bmatrix} \quad \text{and} \quad B = \begin{bmatrix} 5 & -8 & -4 & 6 \\ 4 & 0 & -2 & 11 \end{bmatrix}.$$

8. In each case, compute the products $A \times B$ and $B \times A$ if possible, explaining why they do not exist if they do not. In case both products exist, note whether they are equal.

(a) $A = \begin{bmatrix} 3 & 8 & -2 \\ 0 & 2 & -1 \end{bmatrix}$ and $B = \begin{bmatrix} 5 & -8 & -4 & 6 \\ 4 & 0 & -2 & 11 \\ 0 & 1 & -1 & 3 \end{bmatrix}$.

(b) $A = \begin{bmatrix} 4 & 1 & -2 \\ 0 & 2 & 3 \\ 1 & 4 & -1 \end{bmatrix}$ and $B = \begin{bmatrix} 2 & 3 & -2 \\ 1 & 0 & 1 \\ 1 & 2 & 3 \end{bmatrix}$.

(c) $A = \begin{bmatrix} 2 & 3 & 5 & 7 & 11 \end{bmatrix}$ and $B = \begin{bmatrix} 1 \\ 2 \\ 3 \\ 4 \\ 5 \end{bmatrix}$.

9. Determine what numbers the following numerals in base 6 represent. Base 6 is just like base 10, except that the set of digits is $\{0, 1, 2, 3, 4, 5\}$, and the number 10 is replaced by the number 6 every place it occurs in the definition at the beginning of this section.
   (a)   24.
   (b)   301.
   (c)   2.
   (d)   10000.

10. Find the base 6 representations (see Exercise 9) of the following numbers, given here in their usual base 10 form.
    (a)   10.
    (b)   3.
    (c)   47.
    (d)   12345.

11. Write in pseudocode an algorithm for evaluating numbers, given their base $\beta \geq 2$ representation. Specifically, the input is an integer $\beta \geq 2$ and a string of digits $d_{n-1}d_{n-2}\ldots d_1 d_0$, where each $d_i \in \{0, 1, \ldots, \beta - 1\}$, and the output is the number x that the string represents. Assume that x is small enough to fit inside one computer word. For example, if $\beta = 6$ and the input is the string 123, then the output is the number 51. [*Hint*: Use Horner's method.]

12. Develop an algorithm for finding base β numerals. Specifically, the input is an integer $\beta \geq 2$ and a natural number x (assume that x is small enough to fit inside one computer word), and the output is the string of digits $d_{n-1}d_{n-2}\ldots d_1 d_0$ (where each $d_i \in \{0, 1, \ldots, \beta - 1\}$) that represents x in base β. For example, if $\beta = 6$ and the input is the number 51, then the output is the string 123. [*Hint*: Divide repeatedly by β.]

13. Develop and write in pseudocode an algorithm similar to Algorithm 1 for subtraction of two natural numbers. You may assume that the subtrahend is no larger than the minuend. Analyze the time complexity of your algorithm.

14. Compute $76^{99^{99}}$ mod 100.

15. Verify that 341 is a pseudoprime. (A calculator will make the arithmetic easier.)

16. Figure out what is wrong with the linear congruential pseudorandom number generator with $m = 100$, $A = 30$, and $C = 73$.

17. Describe an algorithm for producing pseudorandom integers between 1 and N, inclusive. Your algorithm can call upon a pseudorandom number generator as a subprocedure.

18. Suppose that $f(x)$ is a polynomial of degree n and $g(x)$ is a polynomial of degree m. (Recall that the zero polynomial has degree $-\infty$.) What can be said about the degree of
    (a)   The sum of $f(x)$ and $g(x)$?
    (b)   The product of $f(x)$ and $g(x)$?

19. Write out in pseudocode the obvious algorithms (based directly on the definitions) for the following operations.
    (a)   Matrix addition.
    (b)   Matrix multiplication.

20. Matrix multiplication is associative. That is, if A is a p by q matrix, B is a q by r matrix, and C is an r by s matrix, then both of the products $(A \times B) \times C$ and $A \times (B \times C)$ are defined, both are p by s matrices, and they are equal.
    (a)   Verify that this statement is true for the following matrices.

$$A = \begin{bmatrix} 3 & -2 \\ 0 & 1 \\ 5 & 1 \end{bmatrix} \quad B = \begin{bmatrix} 2 & 5 & -2 & 1 \\ 1 & 0 & 3 & -2 \end{bmatrix} \quad C = \begin{bmatrix} 1 & 7 \\ 0 & 3 \\ 4 & 0 \\ 1 & -1 \end{bmatrix}.$$

    (b)   Determine the number of numerical multiplications (as a function of p, q, r, and s) required to compute the product by each of the associations, and show that in general they are not the same number.
    (c)   Determine the more efficient way to multiply a 10 by 7 matrix by a 7 by 9 matrix by a 9 by 3 matrix.

21. Find a matrix O that is the identity for addition of m by n matrices (i.e., such that $A + O = O + A = A$ for all m by n matrices A).

22. Show that the n by n identity matrix I_n (Example 12 in Section 3.3) is the identity for matrix multiplication of n by n matrices (i.e., satisfies $A \times I_n = I_n \times A = A$ for all n by n matrices A).

23. Use Theorem 3 to test the number 25 for primeness.

Challenging Exercises

24. Develop an algorithm for obtaining the natural number quotient and remainder when a natural number is divided by a positive integer. Use the basic idea of long division as it is taught in elementary school. The inputs are assumed to be given as strings of digits (base 10), and the outputs are also to appear in that form. Describe the algorithm in pseudocode and analyze its efficiency.

25. Continuing with the notation of Exercise 20, determine after a little algebraic manipulation, a simple rule for determining when it is faster to multiply $A \times B$ first and when it is faster to multiply $B \times C$ first. (Each of the four variables p, q, r, and s should appear only once in your rule.)

26. The largest prime number known in 1988 was $2^{216091} - 1$. It was proved to be prime on a supercomputer with several hours of computation.
  (a) How many decimal digits does this number have?
  (b) What are its last three (rightmost) decimal digits?
  (c) What are its first three (leftmost) decimal digits?

27. Suppose that a_1, a_2, ..., a_n is a sequence of real numbers, and you wish to find

$$\sum_{1 \le i < j \le n} a_i a_j = a_1 a_2 + a_1 a_3 + \cdots + a_1 a_n + a_2 a_3 + a_2 a_4 +$$

$$\cdots + a_2 a_n + \cdots + a_{n-1} a_n.$$

Assume that real number arithmetic is available.
  (a) Describe in pseudocode and analyze the efficiency of the obvious algorithm for obtaining this answer.
  (b) Find and analyze an algorithm that works significantly faster.

Exploratory Exercises

28. Have a computer produce the decimal numeral for the prime number $2^{216091} - 1$. [*Hint*: See Exercise 26.]

29. Explore an arithmetic and algebraic manipulation system such as MACSYMA, Maple, Mathematica, or muMath. Perhaps use it to find your own private large number that is probably prime (it passes the pseudoprime test).

30. Implement some of the algorithms discussed in this section on a computer.

31. Can a negative integer be used as the base of a number system? Try to develop a number system, base -3 or -4. For more details, consult Gilbert [115].

32. Find out about the hardware implementation of computer arithmetic by consulting Hwang [161].

33. **Sparse polynomials**, those with few nonzero coefficients, can be handled more efficiently than we have indicated in this section. Find out more about algorithms for sparse polynomials by consulting Sedgewick [268].

34. Find out about fast algorithms for multiplying natural numbers by consulting Kronsjö [189].

35. Find out about the fast Fourier transform by consulting Sedgewick [268] and Kronsjö [189].

36. Find out more about pseudorandom number generators by consulting Knuth [176] and Park and Miller [233].

37. Find out how computers and calculators compute functions that are not defined in terms of basic arithmetic—things like exponential, logarithmic, and trigonometric functions—by consulting Fike [81] and Kropa [190].

38. Find out more about computer methods for determining whether a number is prime by consulting Pomerance [238,239], Slowinski [276], and Wagon [313].

INDUCTION AND RECURSION

If there is one unifying theme in discrete mathematics, it is the idea of recursion. Suppose that you have a problem P, which you need to solve. Suppose that you can reduce P to another problem P', which you already know how to solve. Then in essence you have a solution to P. In this chapter, the reduced problem, P', will be *a simpler instance of P itself*. Although it may seem that such an approach is circular and can lead nowhere, in fact this recursive method is extremely powerful and extremely effective. It is also conceptually satisfying and allows us to concentrate on key ideas without getting bogged down in details of implementation. We will see recursion in three settings: It can be used to make definitions that would otherwise be nearly impossible to make or to work with (Section 5.1); it can be used to write elegant and conceptually simple algorithms for solving problems (Section 5.2); and it can be used to prove theorems that defy proof in any other way (Section 5.3).

One general reference for this chapter is E. S. Roberts [253].

SECTION 5.1
RECURSIVE DEFINITIONS

*And there was evening and
there was morning, a fifth day.*

—Genesis 1 : 23

In this section we look at recursive or inductive (we use the terms interchangeably) definitions of sequences, sets, and functions that arise in discrete mathematics.

Suppose that we wish to define an object X that contains many "pieces." For example, we might wish to define a sequence or a set; the pieces are the terms in the sequence or the elements in the set. Often we can define X in one fell swoop, perhaps by giving a characterization of all the pieces or a general formula for each piece. For example, we might define a sequence $\{a_n\}$ by the formula $a_n = 2^n - 1$, or we might define a set S of bit strings as $\{\, s \in \{0,1\}^* \mid$ first symbol of $s =$ last symbol of $s \,\}$. Many times, however, we cannot give such an explicit definition. We may not know (or there may not be) a nice formula or general characterization of the thing X we want to define, even though it really is clear what X is. In such cases, a recursive definition can often be used.

A **recursive** (or **inductive**) **definition** of an object X consists of two parts. The first part describes a few of the pieces of X, usually one, sometimes two or three, occasionally more, and on rare occasions, none. This part is called the **base case** of the definition. The second part of a recursive definition describes how new pieces are determined by other pieces already defined; this part is called the **inductive** (or **recursive**) **part** of the definition. It is understood that X consists precisely of all pieces given in the base case, together with all additional pieces that can be generated successively by the inductive part of the definition. Before continuing the general discussion, we illustrate these ideas with a classical example.

EXAMPLE 1. A sequence of natural numbers that arises again and again in discrete mathematics (and, apparently, in nature as well) was defined (recursively) by an Italian mathematician in the thirteenth century. This sequence is usually denoted f_0, f_1, f_2, ..., and is known as the **Fibonacci sequence**. We do not define the sequence by giving an explicit formula for the nth term of the sequence (although we will find an explicit formula—annoyingly complicated, unfortunately—in Section 7.3). Instead, we define $\{f_n\}$ by stating explicitly what the first two terms in the sequence are (the base case of the definition), and then giving a formula which shows how each of the remaining terms is determined by terms that appear earlier in the sequence (the recursive part of the definition). Specifically, we set

Base case: $f_0 = 1, \quad f_1 = 1.$
Recursive part: $f_n = f_{n-1} + f_{n-2}, \quad$ for all $n \geq 2.$

295

Note how this definition in fact determines the entire sequence in a definite and unambiguous way. The base case tells us what f_0 and f_1 are. What about f_2? The inductive part of the definition tells us that $f_2 = f_1 + f_0$. Since we already know f_1 and f_0, we can compute f_2, namely $f_2 = 1 + 1 = 2$. What about f_3? By definition, $f_3 = f_2 + f_1$; but since we know f_2 because of the calculation we just finished, and since we know f_1 from the base case, we can compute that $f_3 = 2 + 1 = 3$. Obviously, we can continue in this way as long as we wish, finding successively that $f_4 = 3 + 2 = 5$, $f_5 = 5 + 3 = 8$, $f_6 = 13$, $f_7 = 21$, $f_8 = 34$, and so on. ◆

Let us stand back and look at what was going on in this example from an iterative or "bottom-up" point of view. We see from the preceding discussion that the definition allows us to pull ourselves up by our bootstraps, starting at the beginning and generating more and more terms in the sequence, in an iterative manner. We will think of creating the object we are building *one day at a time*. Initially, on what we might call day 0, only the terms given by the base case "exist." We can think of more and more terms of the sequence as "coming into existence" each day by virtue of the recursive part of the definition. Thus on day 1 we can apply the inductive step of the definition to any terms that existed on day 0 to create new terms. In this case, f_2 comes into existence on day 1, since its calculation uses only f_0 and f_1, both of which were already in existence on day 0. Similarly, f_3 comes into existence on day 2, f_4 on day 3, and so on. Each term in the sequence eventually comes into existence; by the "end of time" the entire sequence is there. This **days of creation model** is quite useful, both as an aid in understanding recursive definitions and as a vehicle for computing elements in the sequence or set being defined. It will also enable us (in Section 5.3) to prove things about recursively defined sequences and sets.

Next we turn to the question of actually computing the terms of the Fibonacci sequence with a procedure in our pseudocode. The following algorithm computes *and stores*, using an array, all the terms in the sequence up to and including f_n. It is not hard to modify the algorithm so that it keeps only the most recent two terms in the sequence. We will look at this improvement in Section 5.2.

procedure *iterative_fibonacci* (n : natural number)
  {this algorithm computes and stores as $f_0, f_1, \ldots, f_n$ the first $n + 1$ terms of
  the Fibonacci sequence}
    $f_0 \leftarrow 1$
    $f_1 \leftarrow 1$
    for $i \leftarrow 2$ **to** n **do**
      $f_i \leftarrow f_{i-1} + f_{i-2}$
    return(f_n)

Algorithm 1. Generating the Fibonacci sequence.

Note that the assignment statement within the **for** loop is valid, since when it is being executed the terms on the right-hand side have already been determined. As for efficiency, if we assume that each addition takes one step, then this algorithm takes $O(n)$ steps to compute f_n. It also requires $O(n)$ space.

EXAMPLE 2. J. Conway, a contemporary British mathematician, has invented a unified mathematical approach to analyzing two-person games. Suppose that the players are called Lefty and Righty; they play alternately until the person whose turn it is has no legal moves, at which point that player loses. Many games, such as chess or tic-tac-toe, can be formulated in this way. It is also the natural setting for mathematical games such as nim (which is defined in Exercise 1 of Section 9.5). We will take another, more extensive, look at games in Section 9.5.

Conway begins with a strange abstract recursive definition of a **game**: *A game is an ordered pair* (L, R) *such that* L *and* R *are themselves* sets *of games*. Intuitively, a game is thought of as the position of the game board as a player is about to move; L is the set of game board positions that Lefty can choose by making his move (if it is his turn), and R is the set of game board positions that Righty can choose by making her move (if it is her turn). The game, in essence, *is* its set of rules.

Let us find some games from this definition. First note that the base case seems to be missing—it is vacuous and provides us with no games! In other words, no games exist on day 0. Fortunately, that does not present an obstacle to there ever being any games, because although there are no *games* in existence yet, there is a *set* of games, namely the empty set. Thus we can apply the recursive definition to create exactly one game on day 1, namely $(\emptyset, \emptyset)$. Conway calls this game 0. Note that whoever gets to move first when playing game 0 loses, since no legal moves are available (the set of choices for each player is the empty set).

Now (we are at the end of day 1) two sets of games exist, namely $\emptyset$ and $\{0\}$. Thus on day 2 we have a bit of flexibility for the members of an ordered pair of sets of games. In particular, there are these four possibilities:

$$(\emptyset, \emptyset) \quad (\emptyset, \{0\}) \quad (\{0\}, \emptyset) \quad (\{0\}, \{0\}).$$

The game $(\emptyset, \emptyset)$ already existed on day 1. The other three are new, and Conway calls them -1, 1, and $*$, respectively (these names turn out to have profound significance). At the end of day 2, then, there are four games. The game 1, for example, namely $(\{0\}, \emptyset)$, is a game that Lefty always wins, no matter who goes first: If Righty goes first, then she has no moves and so loses immediately, whereas if Lefty goes first, then he makes the only move he has available, to game 0, and then Righty, faced with game 0, loses.

On day 3, many new games come into existence, such as $(\{0\}, \{1\})$, $(\{1\}, \emptyset)$ and $(\{*, 1\}, \{-1\})$. This process can be repeated forever, resulting in more and more games. (In fact, this recursive definition, unlike most of the recursive definitions that we will encounter in this book, can be extended "beyond forever," since there is nothing preventing us from putting infinite sets of games into the ordered pairs.)   ◆

Now we introduce an alternative way to look at recursive definitions. Even more important conceptually than the iterative (days of creation) approach to recursion, and a little harder to appreciate, is the truly recursive or "top-down" approach. We should be able to think of a recursive definition not only in the forward-looking direction, as telling us how to take pieces already defined and make new pieces out of them, but also in the backward-looking direction, as telling us how we can *reduce* the problem of finding new pieces, to a problem involving *simpler* pieces. Under this point of view, for example, we would see the definition of the Fibonacci sequence as a means of reducing the problem of determining f_n to the problems of determining f_{n-1} and f_{n-2}. These two values have smaller subscripts and therefore are simpler than f_n. Suppose that we want to know what f_{20} is. The definition tells us that $f_{20} = f_{19} + f_{18}$. Since 19 and 18 are both less than 20, we have essentially solved our problem, by reducing it to the simpler problems of computing f_{19} and f_{18}. This point of view will be extremely valuable in Sections 5.2 and 5.3, where we will have much more to say about it.

Finally we look at two examples of recursively defined functions. Recall that functions are really sets (sets of ordered pairs), so this is a special case of the recursive definition of a set. Furthermore, a function f whose domain is **N** is really nothing more than a sequence (the sequence $f(0)$, $f(1)$, $f(2)$, ...), so in that sense this is a special case of the recursive definition of a sequence.

EXAMPLE 3.   We gave an informal definition of the factorial function in Section 3.1. A rigorous definition requires recursion, namely

$$0! = 1, \quad \text{and} \quad n! = n \cdot (n-1)! \quad \text{for } n \geq 1.$$

The base case tells us what $0!$ is; the recursive part gives $n!$ in terms of $(n-1)!$. Thus, for example,

$$1! = 1 \cdot 0! = 1 \cdot 1 = 1$$
$$2! = 2 \cdot 1! = 2 \cdot 1 = 2$$
$$3! = 3 \cdot 2! = 3 \cdot 2 = 6$$
$$4! = 4 \cdot 3! = 4 \cdot 6 = 24$$
$$5! = 5 \cdot 4! = 5 \cdot 24 = 120$$

and so on. The informal definition we have been using until now,

$$n! = n(n-1)(n-2)\cdots 3 \cdot 2 \cdot 1,$$

is in effect the iterative interpretation of the recursive definition.

We defined matrix multiplication in Section 4.5. We can also define natural number powers of matrices recursively as follows (A is a p by p matrix of numbers):

$$A^0 = I_p$$
$$A^n = A^{n-1} \times A \quad \text{for } n \geq 1,$$

where I_p is the p by p identity matrix, defined in Example 12 of Section 3.3. Note that we have just given a recursive definition of a function from $\mathbf{N} \times \mathcal{M}_p$ to $\mathcal{M}_p$, where $\mathcal{M}_p$ is the set of all p by p matrices whose entries are numbers. ◆

The exercises contain many more examples of recursively defined functions. We will return to recursively defined sequences in Sections 7.2 and 7.3, where we will use them to model and solve counting problems.

WELL-FORMED FORMULAS

An important use of recursive definitions in discrete mathematics and computer science is for defining sets of strings. In a programming language, for example, certain strings of symbols (letters, digits, punctuation marks, etc.) are valid variable names, expressions, statements, or programs, while others are not. Rules of syntax determine which strings are allowed. In most cases, rules of syntax can be described recursively. It is satisfying conceptually to give recursive definitions in this context. Furthermore, such definitions make it easier to write compilers to recognize strings that are valid programs in high-level programming languages and translate them into machine-language programs.

We look first at some examples that occur in most high-level programming languages.

EXAMPLE 4. Suppose that a variable name is allowed to be any string of one or more characters, each of which is either a letter or a digit, the first of which must be a letter. We can describe the set V of all variable names as follows.

Base case: If x is a letter, then x is a variable name.
Recursive part: If α is a variable name and x is a letter or a digit, then αx is also a variable name.

The first statement is the base case, and from this we get all the valid variable names of length 1, such as W or M. The second statement is the inductive part of the definition. It tells us how to construct valid variable names from other valid variable names. Specifically, it tells us that we may take any valid variable name and concatenate onto the end of it any letter or digit. For example, since W is a valid variable name, so are $W8$ and WE. Then since $W8$ is a valid variable name, so is $W8R$. A little thought with our "days of creation" heuristic will show that all variable names of length n come into existence on day $n - 1$. ◆

We should reiterate the assumption that we have been making all along: Our recursive definition tells us not only that certain elements are in the set we are defining, but also that *the only* elements in the set are the ones that are forced to be there by the definition, in other words, the objects that can be built up according to the rules given in the definition.

EXAMPLE 5.   Algebraic expressions in a high-level programming language are rather complicated strings involving variable names, numerals, operator symbols, and parentheses. We can describe the set of valid algebraic expressions with a recursive definition. For simplicity, we will insist on more parentheses than are usually required. Also, we assume that the notions of variable name (see Example 4) and unsigned integer (a string of digits, such as 3817009, representing a natural number—see Exercise 10) have already been defined.

A **fully parenthesized algebraic expression**, or **FPAE**, is given by the following rules:

> Base case: If x is a variable name or an unsigned integer, then x is a FPAE.
>
> Recursive part: If α and β are FPAE's, then so are $(-\alpha)$, $(\alpha + \beta)$, $(\alpha - \beta)$, $(\alpha * \beta)$, and (α/β). Notice that the parentheses are part of the expressions.

Let us see, for example, why the string $(((-1) - s) + ((-s) * (3 + 1)))$ is a FPAE. First, s is a FPAE by the base case of the definition, since it is a variable name. Therefore, using the first statement in the recursive part of the definition, $(-s)$ is a FPAE. (Note, however, that $-s$ is not a FPAE; it is not a variable name or an unsigned integer, and it cannot possibly arise through an application of the recursive part of the definition, since it contains no parentheses.) Similarly, 3 is a FPAE (being an unsigned integer), as is 1, so $(3 + 1)$ is a FPAE. Applying the recursive part of the definition once more, we see that $((-s)*(3+1))$ is a FPAE. By similar reasoning we can show that $((-1) - s)$ is a FPAE, and one final application of the definition shows that the given string $(((-1) - s) + ((-s) * (3 + 1)))$ is a FPAE. The reader can verify that this string comes into existence on day 3. ◆

Sets of strings given by recursive definitions are often called **well-formed formulas**, or **wff**'s for short. The exercises contain some other examples of wff's in various contexts.

One advantage of a recursive definition for the syntax of objects we wish to study is that such a definition often enables us to give a recursive definition to the *meaning* or semantics of the objects. In Example 5 we specified which strings were to be considered valid expressions, but we gave no indication of what they meant. A computer needs to do more than realize that $(1+1)$ is a valid FPAE; it also needs to know that it is meant to be evaluated as the sum of 1 and 1—in other words, that it has the value 2. The next example shows how we can define—recursively, of course—an evaluation function on FPAEs.

EXAMPLE 6. Suppose that an evaluation function $EVAL: N \to \mathbf{R}$, from the set N of all variable names to the set $\mathbf{R}$ of real numbers, is given. We can think of such a function as existing at each point in the execution of a computer program, where $EVAL(w)$ is the value currently assigned to variable w. Then we can extend this function to have as its domain the set of all FPAEs. There is one technical point we cannot avoid here: We need to expand the codomain as well, to recognize the fact that division by zero is not defined. Thus our function $EVAL$ will have as codomain the set $\mathbf{R} \cup \{undefined\}$; and arithmetic involving this new element of the codomain is defined by saying that anything divided by 0 is *undefined*, and the result of any arithmetic operation on *undefined* is *undefined*.

The recursive definition of $EVAL$ reads as follows.

Base case: If x is an unsigned integer, then $EVAL(x) = x$.
   If x is a variable name, then $EVAL(x)$ is already defined.
Recursive part: If x has the form $(-\alpha)$, then $EVAL(x) = -EVAL(\alpha)$.
   If x has the form $(\alpha + \beta)$, then
$$EVAL(x) = EVAL(\alpha) + EVAL(\beta).$$
   If x has the form $(\alpha - \beta)$, then
$$EVAL(x) = EVAL(\alpha) - EVAL(\beta).$$
   If x has the form $(\alpha * \beta)$, then
$$EVAL(x) = EVAL(\alpha) \cdot EVAL(\beta).$$
   If x has the form (α/β), then
$$EVAL(x) = EVAL(\alpha)/EVAL(\beta).$$

Now x can satisfy only one line of this definition; in other words, we can determine by looking at x whether it is an unsigned integer, a variable name, a string of the form $(-\alpha)$, a string of the form $(\alpha+\beta)$, and so on. Therefore, there is no ambiguity in this definition. Note that the arithmetic operator symbols are being used in two distinctly different ways in our recursive definition. Used to describe x they are mere symbols; used on the right-hand side of these formulas, they mean the honest-to-goodness arithmetic operations. For example, consider $EVAL((4+9))$. The argument is a *string of five symbols*, namely "$(4 + 9)$"; the outer parentheses in the expression $EVAL((4+9))$ are part of the usual functional argument notation. The value of this function at this argument is a number, namely the sum of $EVAL(4)$ and $EVAL(9)$; this follows from the recursive part of our definition. Since by the base case of our definition these two numbers are 4 and 9, respectively, we have $EVAL((4 + 9)) = 4 + 9 = 13.$ ◆

We give one more example of how a function on wff's can be defined rigorously, once a recursive definition of the domain is given.

EXAMPLE 7. We defined a string over an alphabet V nonrecursively in Section 2.2, and we commented in Section 3.1 that a string can be viewed formally as a function. The concept can also be defined recursively as follows. Base case: The empty string λ is a string over V. Recursive part: If α is a string over V, and $x \in V$, then αx is a string over V. With this definition, all the strings of length n appear on day n. Now we can define the **reverse** of a string recursively. We will write w^R for the reverse of the string w.

$$\lambda^R = \lambda$$
$$(\alpha x)^R = x(\alpha^R) \text{ whenever } x \in V \text{ and } \alpha \text{ is a string over } V.$$

We have stopped explicitly indicating the base case and the recursive part; the reader should have no difficulty determining which is which. With this definition, we learn on day n what the reverses of strings of length n are. (Note that the parentheses in this definition are not symbols in the strings; rather, we need to use them to indicate exactly which part of the strings we mean to reverse. We could avoid using them by writing the recursive part in words: The reverse of the string αx, where $x \in V$ and α is a string over V, is the string formed by following x with the reverse of α.)

Let us see how the definition tells us recursively what the reverse of *stop* is. (Presumably, the alphabet V in this case includes at least the four letters in this word.) The reason that *stop* is a string is that it is the string *sto* followed by the letter p. Therefore, by the recursive part of the definition, the reverse of *stop* is p followed by the reverse of *sto*. The latter item will take more calculation. To see what the reverse of *sto* is, we need to note that *sto* is the string consisting of the string *st* followed by the symbol o. Therefore, $(sto)^R$ is $o((st)^R)$, but $(st)^R$ requires further calculation. That penultimate calculation shows that $(st)^R = t(s^R)$. Finally, since $s = \lambda s$, we have $s^R = s(\lambda^R) = s\lambda = s$ (invoking the base case at last). Summarizing, we have performed the following calculation:

$$(stop)^R = p((sto)^R) = po((st)^R) = pot(s^R) = pots(\lambda^R) = pots\lambda = pots. \quad \blacklozenge$$

THE ACKERMANN FUNCTION (*OPTIONAL*)

Recursive definitions can become much more complicated than we have indicated so far. To end this section we discuss a function that uses a kind of "double" recursion. This function on positive integers embodies the usual arithmetic operations of addition, multiplication, and exponentiation, as well as higher order analogues of these operations.

We begin by showing that the familiar arithmetic operations can be given precise recursive definitions, once we know how to count. As a way to formalize the process of counting, let the **successor function** from the set of natural numbers $\{0, 1, 2, 3, \ldots\}$ to itself be the function that takes each natural number to the next larger natural number. We write x^+ for the successor of x. The successor function is simply what enables us to count. Thus, for example, $7^+ = 8$, $739^+ = 740$, and $0^+ = 1$. Every natural number other than 0 is the successor of a natural number, namely of the number that precedes it in the ordered list of natural numbers; in particular, every natural number other than 0 can be written as n^+ for some n.

Now let us see how addition can be defined recursively in terms of the successor function. First, adding 0 to a number is easy enough: $x + 0 = x$ for all natural numbers x. This will be the base case of our definition. Next, $x + 1 = x^+$, the successor of x. Since $1 = 0^+$, we can write this as $x + (0^+) = (x + 0)^+$. What about $x + 2$? Since $x + 2 = x + (1^+) = x + (1 + 1) = (x + 1) + 1 = (x + 1)^+$, we can compute $x + 2$ for any x once we know how to compute $x + 1$ for any x. More generally, suppose that we want to compute $x + (n^+)$. Since $x + (n^+) = x + (n + 1) = (x + n) + 1 = (x + n)^+$, we can reduce the problem of computing $x + (n^+)$ to the problem of computing $x + n$. This is exactly the idea behind recursion. Thus we make the following recursive definition of the operation of addition on natural numbers. For all x,

$$x + 0 = x$$
$$x + (n^+) = (x + n)^+ \qquad \text{for all } n \geq 0.$$

The recursive part of the definition takes care of all addition problems that the base case does not cover, since every natural number other than 0 is a successor.

EXAMPLE 8. Compute $5 + 3$ from the recursive definition of addition.

Solution. To add 3 to a number, we have to invoke the recursive definition four times: three times to reduce the number we are adding from 3 to 2 to 1 to 0 (using the recursive part of the definition), and a final time to know what $5 + 0$ is. The calculation proceeds as follows:

$$5 + 3 = (5 + 2)^+ = ((5 + 1)^+)^+ = (((5 + 0)^+)^+)^+$$
$$= ((5^+)^+)^+ = (6^+)^+ = 7^+ = 8.$$

Computing $3 + 5$ from the definition would have involved a different set of calculations, passing through $((((3^+)^+)^+)^+)^+$. ◆

Next we will define multiplication in the same manner, drawing on addition, which we have just defined. In essence, multiplication arises out of addition just as addition arises out of counting, since $x \cdot (n^+) = x \cdot (n+1) = (x \cdot n) + x$. Thus for all x we set

$$x \cdot 0 = 0$$

$$x \cdot (n^+) = (x \cdot n) + x \qquad \text{for all } n \geq 0.$$

EXAMPLE 9. Compute $5 \cdot 3$ from the recursive definition of multiplication.

Solution. To multiply a number by 3, we have to invoke the recursive definition four times: three times to reduce the number we are multiplying by from 3 to 2 to 1 to 0 (using the recursive part of the definition), and a final time to know what $5 \cdot 0$ is. The calculation proceeds as follows:

$$5 \cdot 3 = (5 \cdot 2) + 5 = ((5 \cdot 1) + 5) + 5 = (((5 \cdot 0) + 5) + 5) + 5$$

$$= ((0 + 5) + 5) + 5 = (5 + 5) + 5 = 10 + 5 = 15.$$

Note that if we wanted to reduce everything all the way down to applications of the successor function, we would need six applications of the definition of addition to compute $0 + 5$, six more to compute $5 + 5$, and six more to compute $10 + 5$. Thus the total number of steps needed to compute $5 \cdot 3$ from the definitions is 22. In general, it would take $O(xy)$ steps to compute $x \cdot y$ directly from the definitions using only the successor function. Since the definition gets us to the answer by counting, we need xy applications of the successor function, together with some bookkeeping. This compares very poorly with the efficiency of the algorithms in Section 4.5, where we took advantage of the base 10 representations of x and y. There it took only $O((\log x)(\log y))$ steps to find the product. ◆

The reader can try defining exponentiation in a similar way (Exercise 13a). Just as addition is really repeated applications of the successor function, and multiplication is repeated addition, so exponentiation is repeated multiplication. Can we go further? What about repeated exponentiation, or repeated applications of that, or then repeated applications of that? Can we define one function that somehow embodies all of these operations at once?

The answer is a resounding yes, by using recursion. Our development here follows fairly closely the discussion given in the 1920s by the German mathematician D. Hilbert and his student W. Ackermann. A more modern version of this function is discussed in Exercise 17. All versions of this function are usually named after Ackermann.

We will define a function of three variables, $\varphi(x, n, i)$, which, for each i, gives an arithmetic function of x and n. Rather than starting with the successor function, we will start with addition (for $i = 0$), followed by multiplication (repeated addition) for $i = 1$, followed by exponentiation (repeated multiplication) for $i = 2$, followed by repeated exponentiation, and so on. Thus we want the function φ to have the following properties:

$$\varphi(x, n, 0) = x + n$$

$$\varphi(x, n, 1) = x \cdot n$$

$$\varphi(x, n, 2) = x^n,$$

with some suitable extension for $i = 3$, 4, and so on.

Our definition uses double recursion: Both the third argument (i) and the second argument (n) of our function φ enter the definition in a recursive manner. It turns out that we need four base cases to handle $i = 0$ or $n = 0$. Two of these cases correspond to the facts that $x \cdot 0 = 0$ and $x^0 = 1$ for all x. The recursive part of the definition will show how the calculation of $\varphi(x, n, i)$ reduces to calculations with a smaller value of the third argument and a smaller value of the second argument.

The **Ackermann function** is defined as follows:

$$\varphi(x, n, 0) = x + n$$

$$\varphi(x, 0, 1) = 0$$

$$\varphi(x, 0, 2) = 1$$

$$\varphi(x, 0, i) = x \quad \text{for } i \geq 3$$

$$\varphi(x, n, i) = \varphi(x, \varphi(x, n - 1, i), i - 1) \quad \text{for } i \geq 1 \text{ and } n \geq 1.$$

Calculations from this definition are tedious, to say the least: The existence of this function has theoretical, not practical significance. Nevertheless, it is a very useful exercise in the use of recursion to make one or two such calculations. The effort will be rewarded when we come to Sections 5.2 and 5.3.

EXAMPLE 10. Calculate $\varphi(2, 3, 2)$ from the definition of the Ackermann function.

Solution. By the recursive part of the definition, we find that the desired value is equal to $\varphi(2, \varphi(2, 2, 2), 1)$. Thus we need to begin by computing the value of $\varphi(2, 2, 2)$, after which we can proceed with the calculation. But by the definition, $\varphi(2, 2, 2) = \varphi(2, \varphi(2, 1, 2), 1)$, so we need to find $\varphi(2, 1, 2)$ first. Again by the recursive part of the definition, $\varphi(2, 1, 2) = \varphi(2, \varphi(2, 0, 2), 1)$. Now the third line of the base case applies: Since $\varphi(2, 0, 2) = 1$, we now know that $\varphi(2, 1, 2) = \varphi(2, 1, 1)$. To calculate this value, we use the recursive part of the definition again, obtaining $\varphi(2, \varphi(2, 0, 1), 0)$, and then parts of the base case to show that this equals $\varphi(2, 0, 0) = 2 + 0 = 2$. Hence we have shown that $\varphi(2, 1, 2) = 2$, and therefore $\varphi(2, 2, 2) = \varphi(2, 2, 1)$. Now we need to perform the following calculation:

$$\varphi(2, 2, 1) = \varphi(2, \varphi(2, 1, 1), 0)$$

$$= \varphi(2, 2, 0) \quad \text{(by part of our earlier calculation)}$$

$$= 2 + 2 = 4.$$

Therefore, $\varphi(2, 3, 2) = \varphi(2, 4, 1)$. Several more steps give us the final answer:

$$\varphi(2, 4, 1) = \varphi(2, \varphi(2, 3, 1), 0)$$

$$= \varphi(2, \varphi(2, \varphi(2, 2, 1), 0), 0)$$

$$= \varphi(2, \varphi(2, 4, 0), 0) \quad \text{(by the calculation above)}$$

$$= \varphi(2, 2 + 4, 0)$$

$$= \varphi(2, 6, 0) = 2 + 6 = 8.$$

Thus $\varphi(2, 3, 2) = 8$. ◆

We will see in Section 5.3 that φ has the properties we wanted: $\varphi(x, n, 1) = x \cdot n$ and $\varphi(x, n, 2) = x^n$ (our calculation in Example 10 showed that this last statement is true for $x = 2$ and $n = 3$). Furthermore, we will also see that

$$\varphi(x, n, 3) = x^{x^{x^{\cdot^{\cdot^{\cdot^{x}}}}}}$$

with n x's in the exponent, a kind of repeated exponentiation (this symbolism is meant to require evaluation from the top to the bottom—for instance, $3^{3^3} = 3^{(3^3)} = 3^{27} \approx 8 \times 10^{12}$, not $\left(3^3\right)^3 = 27^3 \approx 2 \times 10^5$). As i increases, $\varphi(x, n, i)$ becomes a faster- and faster-growing function of x and n. For example,

$$\varphi(3, 3, 0) = 3 + 3 = 6$$

$$\varphi(3, 3, 1) = 3 \cdot 3 = 9$$

$$\varphi(3, 3, 2) = 3^3 = 27$$

$$\varphi(3, 3, 3) = 3^{3^3} \approx 3^{8 \times 10^{12}},$$

which is a number with more than 3 trillion digits. The number $\varphi(3, 3, 4)$ is too big to comprehend; there is essentially no way to write it or describe it or discuss it that is any more meaningful than simply to say that it is exactly what the recursive definition says it is. What about a number like $\varphi(100, 100, 1000)$? It is surely a positive integer, but the mind boggles at its enormity.

In summary, we have, with a recursive definition, incorporated *all* these faster- and faster-growing arithmetic functions into one function, but the only operation actually appearing in the definition was lowly addition.

SUMMARY OF DEFINITIONS

Ackermann function (recursively defined): $\varphi(x, n, 0) = x + n$, $\varphi(x, 0, 1) = 0$, $\varphi(x, 0, 2) = 1$, $\varphi(x, 0, i) = x$ for $i \geq 3$, $\varphi(x, n, i) = \varphi(x, \varphi(x, n - 1, i), i - 1)$ for $i \geq 1$ and $n \geq 1$ (see also Exercise 17 for a simpler version).

base case of a recursive definition: the explicit specification of one or more elements in the set or sequence being defined.

days of creation model: a way to imagine a set or sequence being defined by a recursive definition; the base case specifies what elements exist on day 0, and the elements that come into existence on day $n + 1$ are those that make use of the recursive part of the definition utilizing only elements that were already in existence by day n (e.g., the term f_3 of the Fibonacci sequence comes into existence on day 2, since it is the sum of f_1 (which existed on day 0, being part of the base case) and f_2 (which came into existence, through the recursive part, on day 1)).

Fibonacci sequence: the sequence defined recursively by $f_0 = f_1 = 1$ (the base case) and $f_n = f_{n-1} + f_{n-2}$ for $n \geq 2$ (the recursive part); thus $f_2 = 2$, $f_3 = 3$, $f_4 = 5$, $f_5 = 8$, and so on.

FPAE: a fully parenthesized algebraic expression.

fully parenthesized algebraic expression (recursively defined): Variable names and unsigned integers are fully parenthesized algebraic expressions, and if α and β are fully parenthesized algebraic expressions, then so are $(-\alpha)$, $(\alpha + \beta)$, $(\alpha - \beta)$, $(\alpha * \beta)$, and (α/β) (e.g., $(x + (2 - y))$).

game (Conway's definition): recursively defined as an ordered pair of sets of games; no base case is necessary in this definition (e.g., the game $(\{(\emptyset, \emptyset), (\{(\emptyset, \emptyset)\}, \{(\emptyset, \emptyset)\})\}, \{(\{(\emptyset, \emptyset)\}, \emptyset)\})$, which appears on day 3).

inductive definition: a recursive definition.

inductive part of a recursive definition: the recursive part of the definition.

recursive definition: a definition of a set or sequence given by specifying a few elements of the set or sequence explicitly (the base case) and then indicating how subsequent elements are determined by elements already in the set or sequence (the recursive part) (e.g., the definition of the Fibonacci sequence).

recursive part of a recursive definition: the specification of how elements in the set or sequence being defined determine other elements.

reverse of the string w: the string w^R, which is w written backwards; this can be defined recursively by $\lambda^R = \lambda$ and $(\alpha x)^R = x(\alpha^R)$ for all strings α and symbols x (e.g., $(sleep)^R = peels$).

successor function: the function from $\mathbf{N}$ to $\mathbf{N}$ whose value at n is the natural number following n (in the usual ordering of the natural numbers); the successor of n is written n^+ (e.g., $425^+ = 426$).

well-formed formula: an element of a given set of strings, usually defined recursively (e.g., a fully parenthesized algebraic expression).

wff: a well-formed formula.

EXERCISES

1. Find the term f_{11} in the Fibonacci sequence.

2. Consider the sequence defined recursively by $a_0 = 0$, and $a_n = 2a_{n-1} + 1$ for $n \geq 1$.
   (a) Compute $a_1, a_2, \ldots, a_7$.
   (b) Conjecture a nonrecursive definition (explicit formula) for this sequence.

3. Consider the sequence defined recursively by $a_1 = a_2 = a_3 = 1$, and $a_n = a_{n-1} + a_{n-2} + a_{n-3}$ for $n \geq 4$. Find a_{10}.

4. In Example 3 we saw that $5! = 120$. Use this fact to find $6!$, $7!$, and $8!$.

5. Let A be the 2 by 2 matrix

$$A = \begin{bmatrix} 1 & -1 \\ 3 & 2 \end{bmatrix}.$$

   Compute the following powers of A.
   (a) A^2.
   (b) A^0.
   (c) A^3.
   (d) A^1.

6. Determine if the following strings are valid variable names according to the definition given in Example 4.
   (a) *HELLO*.
   (b) *HELLO_THERE*.
   (c) *2BY4*.
   (d) *W4UNH*.

7. Determine whether the following strings are valid FPAEs according to the definition given in Example 5.
   (a) 46.
   (b) (46).
   (c) $(4 \cdot 6)$.
   (d) $4 + 6$.
   (e) $(3 + x2)$.
   (f) $(3 + x * 2)$.

8. Consider the sequence defined by $T(0) = 1$, and $T(n) = 2T(\lfloor n/2 \rfloor)$ for $n \geq 1$.
   (a) Compute the first 16 terms of the sequence after $T(0)$.
   (b) Conjecture a nonrecursive definition (explicit formula) for this sequence, or at least explain in English what seems to be going on.

9. Show that the FPAE $(((-1) - s) + ((-s) * (3 + 1)))$ comes into existence on day 3, using the definition given in Example 5.

10. Give recursive definitions of the following sets of strings of digits.
    (a) Unsigned integers, with leading 0's not allowed if the string does not represent the number 0 (e.g., 38217 and 0 and 902 are included, but 0902 is not).
    (b) Unsigned integers, with leading 0's allowed (e.g., 38217 and 000902 are included).

11. Consider the sequence P defined by $P(1) = 1$, and $P(n) = P(n - P(n-1)) + 1$ for $n \geq 2$.
    (a) Compute the first 16 terms of the sequence.
    (b) Explain why this might *not* be a valid definition. [*Hint*: What would need to be true for it to be valid?] In fact it is valid.
    (c) Make a conjecture as to what seems to be going on with this sequence.

12. Give a recursive definition of a palindrome, which is a string that reads the same forward as backward. Assume a fixed alphabet V.

13. Give a recursive definition of the following arithmetic functions.
    (a) Exponentiation, using multiplication. (In other words, define x^n for natural numbers x and n. Recall that in discrete mathematics we define 0^0 to be 1.)
    (b) Iterated exponentiation, using exponentiation. (In other words, define $x^{x^{x^{\cdot^{\cdot^{\cdot^x}}}}}$, with n x's in the exponent, for natural numbers x and n. You might notice what happens for $x = 0$.)

14. In Section 1.1 we represented propositions with formulas made up of propositional variables, the constants T and F, and the logical connectives for conjunction ($\wedge$), disjunction ($\vee$), and negation ($^{-}$) (as well as some other connectives that we ignore here). Let us insist for this exercise that our formulas contain no propositional variables—only constants—and that the formulas be fully parenthesized in the sense that disjunctions and conjunctions must be enclosed in a pair of parentheses. For example, $((T \wedge \overline{\overline{F}}) \vee \overline{(T \vee F)})$ is one such formula.
    (a) Give a recursive definition of this set of wff's. (Note that these are not exactly strings, since the negation symbols sit above other groups of symbols. We could, alternatively, use the in-line negation operator symbol "$\neg$".)
    (b) Show that the example given above is in the set as you have defined it, by determining on which day of creation it finds its way into the set.
    (c) Define recursively a truth evaluation function on the set, which gives the truth value of every element in the set.

15. Compute the following values of the Ackermann function, using only the definition. Check your answers with the facts stated below Example 10.
    (a) $\varphi(3, 3, 1)$.
    (b) $\varphi(3, 2, 2)$.

16. Assume, as we will prove in Section 5.3, that $\varphi(x, n, 3) = x^{x^{x^{\cdot^{\cdot^{\cdot^{x}}}}}}$, with n x's in the exponent. Show that there are exactly 58 pairs (x, n) with $x \geq 2$ and $n \geq 1$ such that $\varphi(x, n, 3) < 10^{100}$.

17. A simpler version of the Ackermann function was defined by the American mathematician R. Robinson in 1948. It is a function with two variables, defined as follows:

$$A(n, 0) = n + 1$$

$$A(0, i) = A(1, i - 1) \qquad \text{for } i > 0$$

$$A(n, i) = A(A(n - 1, i), i - 1) \qquad \text{for } n > 0 \text{ and } i > 0.$$

    (a)   Compute $A(1, 2)$ from the definition.
    (b)   Explain why this is a valid recursive definition.

18. Suppose that the set of prime numbers is known. Give a recursive definition of the set of composite numbers.

19. Give a recursive definition of the set of Boolean expressions (Section 1.4).

20. Consider the sequence of real numbers defined recursively by $a_1 = 1$, and $a_n = \sqrt{a_{n-1} + 1}$ for $n \geq 2$.
    (a)   Find the first 16 terms of the sequence on a calculator. Be sure to use the calculator efficiently, in the sense that you must not reenter a_{n-1} by hand in order to compute a_n. Write your answers to seven decimal places.
    (b)   Make a conjecture as to what is happening. Be as specific as possible.

21. Repeat Exercise 20 for the sequence of real numbers defined by $a_1 = 1$, and $a_n = (a_{n-1} + 1)^{-1}$ for $n \geq 2$.

22. What is wrong with the following purported recursive definitions of sequences of real numbers?
    (a) $a_1 = 100$, and $a_n = (\sin a_{n+1}) - a_{n+1}^2 + 2^{a_{n+1}}$ for $n \geq 2$.
    (b) $a_1 = 100$, and $a_n = \sqrt{a_n^2 - 1}$ for $n \geq 2$.
    (c) $a_1 = 2$, and $a_n = 1/(a_{n-1} - 1)$ for $n \geq 2$.

23. A generalization of dominoes (without the spots on them) can be defined recursively. We define n-**nominoes** to be physical objects constructed from 2-inch by 2-inch by $\frac{1}{4}$-inch plain pieces of wood (called "squares") as follows: There is only one 1-nomino, namely a square; if you have an $(n - 1)$-nomino,

then you obtain an n-nomino by gluing the edge of a new square onto an exposed edge of a square in the $(n-1)$-nomino. (An edge is one of the 2-inch by $\frac{1}{4}$-inch faces.) Assume that the glue is not part of the object; in other words, two n-nominoes are considered the same if the arrangement of squares looks the same, even if the pattern in which they happen to be glued together is different. Note also that these objects can be rotated and turned over. The following picture shows five 4-nominoes.

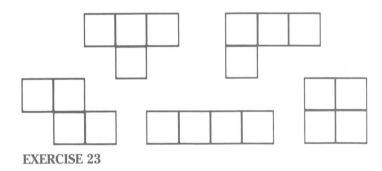

EXERCISE 23

(a) Explain why the set of 4-nominoes shown here is a complete set of all 4-nominoes. [*Hint*: Concentrate on the longest string of squares in a row. Your proof will thus have lots of cases.]

(b) Find the complete set of all 5-nominoes.

24. Let A be a set, and let $\bullet$ be a binary operator defined on A whose codomain is not necessarily a subset of A. (In other words, $x \bullet y$ may fail to be in A, even though x and y are in A; the examples below will make this clear.) The **closure** of A under this operator, written cl(A), is defined recursively as follows: $a \in$ cl(A) for each $a \in A$; and if x and y are elements of cl(A), then so is $x \bullet y$. Find the closures of the following sets under the specified operators.

(a) The natural numbers under subtraction.

(b) The positive integers under real number division by nonzero quantities (as opposed to the natural number quotient).

(c) The natural numbers under averaging (in other words, $x \bullet y = (x+y)/2$).

25. One way to define a **list** in computer science is recursively as follows. We will simultaneously define two things recursively: *list* and *inside*. Our intent is for an inside to be a sequence of things separated by commas, where the things are either numbers or lists; and a list is an inside enclosed in parentheses. The alphabet consists of left and right parentheses, numbers, and a comma.

Every number is an inside.
( ) is a list.
If α is a list, then α is an inside.
If α and β are insides, then α, β is an inside.
If α is an inside, then (α) is a list.

(a)   Determine everything that exists on day 0, day 1, and day 2, identified as list or inside (or both). For this part of the exercise, assume that 5 is the only number.

(b)   Show that $(3, (4), (7, 2), (( ), ( )))$ is a list and determine on which day it appears.

26. Give recursive definitions of the following sets, and then figure out explicit definitions for them by constructing the first several days' production.

(a)   The set of positive amounts of postage that can be put on an (arbitrarily large) envelope using only 3-cent stamps and 4-cent stamps.

(b)   The set of numbers that a college football team can score in a game (a team scores 6, 7, or 8 points for a touchdown and possible point(s) after, 3 points for a field goal, and 2 points for a safety).

(c)   The set of positive amounts of postage that can be put on an (arbitrarily large) envelope using only 5-cent stamps and 9-cent stamps.

(d)   The set of positive amounts of postage that can be put on an (arbitrarily large) envelope using only 4-cent stamps and 6-cent stamps.

27. Define a subset E of the set of integers recursively as follows: $5 \in E, 7 \in E$, and $\forall x: \forall y: [(x \in E \wedge y \in E) \to (x + y \in E \wedge x - 1 \in E)]$.

(a)   Determine which elements of E come into existence on days 0, 1, and 2.

(b)   Prove that $E = \mathbf{Z}$.

28. One definition genealogists use for cousinhood is illustrated by the following example: If A and A' together have two children B and C, if B has child D and C has child E, and if D and E have children F and G, respectively, then D and E are first cousins, F and G are second cousins, and D and G are first cousins, once removed. For this exercise assume that there are no intermarriages of cousins or other relatives.

(a)   Give a recursive definition of nth cousin. Extending this backward, what would 0th cousins be?

(b)   Give a recursive definition of nth cousin, m times removed; the variable for the induction is m, and the base case is $m = 0$, which is handled in part (a).

Challenging Exercises

29. Give a recursive definition of signed or unsigned decimals, which are strings of digits containing exactly one decimal point and an optional plus sign or minus sign at the beginning; leading 0's are allowed. For example, the set includes 46.0, $+.6$, $3.$, and -003.57.

30. Study the following sequence of geometric figures, called the **approximations to the snowflake curve**. Figure out how each one is built from the one before it.

    (a) Write down a precise recursive definition of the sequence S_n suggested by these first three terms.

    (b) These figures "converge" to a limiting figure, which has the perverse properties that (1) it has only a finite area and yet (2) it has infinite perimeter. Show that statements (1) and (2) are true.

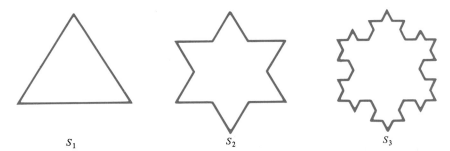

S_1 S_2 S_3

EXERCISE 30

31. The Cantor middle third set is the intersection of the following sequence of sets, defined recursively. $C_0 = [0, 1]$ is the closed interval of real numbers from 0 to 1 (it includes the endpoints). For $n \geq 1$, we obtain C_n by removing the open middle third from *each* closed interval comprising C_{n-1}. For example, $C_1 = [0, \frac{1}{3}] \cup [\frac{2}{3}, 1]$, obtained when the open middle third $(\frac{1}{3}, \frac{2}{3})$ is removed from $[0, 1]$. To obtain C_2, we need to remove $(\frac{1}{9}, \frac{2}{9})$ and $(\frac{7}{9}, \frac{8}{9})$ from C_1.

    (a) Draw a picture showing C_0, C_1, C_2, and C_3.

    (b) Find the total length of all the segments that have been removed.

    (c) Show that the Cantor set is not empty; in fact, show that it has an infinite number of points in it.

    (d) Show that the Cantor set has an uncountable number of points in it.

32. For most human and computer purposes, algebraic expressions do not have to be fully parenthesized. An expression such as $4 * (3 - 2 * x)$ is considered valid (certain parentheses are implied by precedence rules for operators), as is an expression such as $(((1 + 1)))$ with superfluous parentheses.

    (a) Rewrite the definition of FPAE (see Example 5) to conform to this convention.

    (b) Explain what the difficulty is with defining the *EVAL* function (see Example 6) under this definition.

33. There are four possibilities for the outcomes of games (in Conway's sense—see Example 2) if both players play optimally: Either Lefty will win (regardless of who goes first), or Righty will win (regardless of who goes first), or the first player to move will win, or the second player to move will win.

    (a) Give a simultaneous recursive definition of the four notions **winning for Lefty**, **winning for Righty**, **winning for the first player to move**, and **winning for the second player to move** as applied to games. There is no need for a base case. [*Hint*: A game (L, R) is winning for Lefty if and only if there exists a game in L that is either winning for Lefty or winning for the second player, and every game in R is either winning for Lefty or winning for the first player.]

    (b) Classify the seven games mentioned in Example 2 as to who wins.

Exploratory Exercises

34. Consider the sequence P defined by $P(1) = P(2) = 1$, and $P(n) = P(n - P(n-1)) + P(n - P(n-2))$ for $n \geq 3$. (This sequence and those in the next exercise were defined by D. Hofstadter.)

    (a) Compute the first 16 terms of this sequence.

    (b) Explain why this might *not* be a valid definition. (In fact, no one knows whether it is valid.)

    (c) Compute several thousand terms of the sequence, and see how it behaves. Make some conjectures supported by the data you obtain. One thing to look at is the values of $P(3 \cdot 2^n)$.

35. Investigate the following recursively defined functions (sequences) by hand or with a computer.

    (a) $P(1) = P(2) = 1$, and $P(n) = P(n - P(n-1)) + P(P(n-1))$ for $n \geq 3$.

    (b) $P(0) = 0$, and $P(n) = n - P(P(n-1))$ for $n \geq 1$.

    (c) $P(0) = 0$, and $P(n) = n - P(P(P(n-1)))$ for $n \geq 1$.

    (d) In this part we define simultaneously a pair of functions recursively; each enters into the definition of the other as well as itself. The base case is that $F(0) = 1$ and $M(0) = 0$. The recursive part is that $F(n) = n - M(F(n-1))$ and $M(n) = n - F(M(n-1))$ for $n \geq 1$.

36. Investigate the sequence of real numbers defined by $a_1 = 1$, $a_2 = x$, and $a_n = (a_{n-1} + 1)^{-1} a_{n-2}$ for $n \geq 3$, for various real numbers x. Concentrate on whether the sequence (or a subsequence of it) converges.

37. Make up your own recursively defined sequences and investigate them.

38. Recursively define a sequence of integers e_n as follows: $e_1 = 1$, and for each $n \geq 2$ we choose e_n to be the smallest integer greater than e_{n-1} that will not form a three-term arithmetic progression with any two of the numbers $e_1, e_2, \ldots, e_{n-1}$ (i.e., so that for all i and j with $1 \leq i < j < n$ it is not the case that $e_n - e_j = e_j - e_i$).

(a) Find the first 16 terms of this sequence.

(b) Discuss the pattern that you find. Come up with a conjecture as to what is going on, and test your conjecture by calculating more terms in the sequence.

39. The limit of the ratio of successive terms of the Fibonacci sequence,

$$\lim_{n \to \infty} \frac{f_{n+1}}{f_n},$$

exists.

(a) Compute the limit from this definition to what appears to be four decimal-place accuracy.

(b) Derive the fact that the limit equals $(1 + \sqrt{5})/2$, assuming that it exists. Your proof should be independent of the base case, so that similar sequences with different starting values (such as the **Lucas sequence** 1, 3, 4, 7, 11, 18, ...) will yield the same limit.

(c) Find out more about this limit, called the **golden ratio**, by consulting Gardner [110].

40. Find out more about games by consulting Berlekamp, Conway, and Guy [17], Conway [49–51], and Guy [134].

41. Certain recursively defined sequences of real numbers lead to very bizarre objects called **strange attractors**. Find out about them by consulting Hofstadter [154] and Ruelle [261].

42. Find out more about the Fibonacci sequence by consulting Gardner [97], Jean [162], and Wall [314].

43. Find out about John H. Conway (who defined games recursively) by consulting Albers and Alexanderson [4] and Guy [133].

44. Find out more about very large numbers, such as those produced by Ackermann's function, by consulting Knuth [180,181] and Smoryński [280].

45. A variation of the real number line (of which the real numbers are only a very small part) can be defined recursively in the same manner that games are defined (Example 2). Find out about these so-called **surreal numbers** by consulting Conway [51] and Knuth [182].

46. Find out more about n-nominoes by consulting Gardner [109] and Lunnon [202].

SECTION 5.2
RECURSIVE ALGORITHMS

Q: How do you live to be 100?
A: Get to 99 and be very careful.

—Anonymous

We now consider a second aspect of recursion in discrete mathematics: **recursive algorithms**.

Recall from our development of pseudocode descriptions of algorithms that we allow a procedure to invoke other procedures (subprocedures) to perform subtasks that the first (main) procedure requires. Control of the computer passes temporarily from the main procedure to the subprocedure, and everything that was going on in the main procedure is temporarily put "on hold." It is as if the computer takes out a clean sheet of paper, places it directly on top of the sheet of paper it was currently writing on, and begins to work on the subprocedure. When this call to the subprocedure is made, the main procedure passes arguments to the subprocedure, which assigns their values to its own variables. When the subprocedure has finished its work, it may pass some information back to the procedure that called it (through the **return** statement), but in any case the subprocedure ceases its work. It is as if the computer throws away that new sheet of paper, revealing the former sheet of paper, exactly as it was when execution of the main procedure was suspended. In particular, *values of variables have not changed*. Information returned from the subprocedure (if any) will be used by the main procedure in evaluating the expression that contained the subprocedure call. The main procedure can be interrupted like this as often as necessary.

Subprocedures can in turn call on other subprocedures to help with their tasks. If a subprocedure A calls another subprocedure B, then execution in A is temporarily suspended while B executes. Any variables in A retain their values, again as if a clean sheet of paper on which to work on B were placed directly on top of the paper on which the work on A was proceeding (which in turn was covering the sheet of paper on which work for the main procedure was in progress). Subprocedure A can pass values to subprocedure B through the arguments, and receive information back through the **return** statement. When B has finished its work, work on A continues. This process—procedures calling other procedures— can be nested as deeply as desired, and the stack of papers indicating work in progress can get arbitrarily high.

Now we come to the key idea in recursive algorithms. There is no reason that the subprocedure doing the calling cannot be the same as the subprocedure being called. In other words, *a procedure can call itself*, perhaps better viewed as a copy of itself. The important thing to remember is that *each invocation of a procedure has an existence of its own, totally unrelated to anything going* on in either suspended or finished invocations of that procedure. If there is a local variable n in the procedure, then n may equal 7 in one invocation of the procedure that is currently suspended, 3 in an invocation of the procedure that already terminated, 8 in an invocation of the procedure that was suspended prior to the first one, and 10 in the invocation currently in operation. Suppose that the current invocation terminates, and control of the execution returns to the most recently suspended invocation of the procedure. There n still has the value 7 that it had when the procedure was suspended, and that invocation knows nothing about the n that just had the value 10, nor about the n that has the value 8 in the previously suspended and still pending invocation. Similarly, when that invocation has finished its work and control passes back to the earlier invocation, n there is still 8; that invocation is totally oblivious to anything that might have happened in other invocations.

In brief, the only way that two procedures communicate with each other, even if the two are different invocations of the same procedure, is through the arguments and returned values. We will illustrate these ideas with a simple example.

EXAMPLE 1.   The recursive definition of $n!$ can be turned directly into a **recursive procedure** for computing $n!$ shown as Algorithm 1. This procedure is very simple, but we will trace through the calculation of 2! step by step to illustrate the mechanics of recursive calls.

> **procedure** *recursive_factorial*(n : natural number)
>   {this procedure returns $n!$}
>   **if** $n = 0$ **then return**(1) {base case}
>   **else return**($n \cdot$ *recursive_factorial*($n - 1$)) {recursive part}

Algorithm 1.   Recursive calculation of factorial.

A recursive procedure consists of a **base case** and a **recursive part**. The base case tells what to do for some small inputs (in the case of factorial, for input $n = 0$). The rest of the procedure, the recursive part, tells what to do the rest of the time, and it involves invoking the same procedure with an input smaller than the input with which it is currently being invoked. The last line of Algorithm 1 shows that this procedure calls upon itself with argument $n - 1$, which is one less than the argument with which the current invocation was called.

Let us trace through the execution of this procedure when it is requested to compute 2!. We start with a clean sheet of paper on the table; let us label it "invocation 1" and write $n = 2$ on it. The algorithm compares n (which is 2) to 0 in the first step of the procedure and sees that equality does not hold. Hence the **else** clause must be executed. The **else** clause is a **return** statement, so the value to be returned must be calculated. That value is a product of two numbers. The first of the numbers is n, which is 2, but the second is *recursive_factorial*$(n - 1)$. This requires a call to the subprocedure *recursive_factorial*, with $n - 1 = 2 - 1 = 1$ as argument.

Therefore, at this point execution of invocation 1 is suspended, a new clean sheet of paper is placed on top of the current one, and invocation 2 of *recursive_factorial* begins. In addition to labeling this sheet with its invocation number, we write $n = 1$, since 1 is the value of the argument that was passed by the calling procedure. Invocation 2 proceeds in the expected manner. Since $n \neq 0$ (i.e., $1 \neq 0$), the **else** clause must be executed. To execute this **return** statement, the value to be returned must be calculated, but before the multiplication can be performed, the second factor, *recursive_factorial*$(n - 1)$, has to be computed. Since $n = 1$, this requires a call to the subprocedure with argument $1 - 1 = 0$. Hence execution of invocation 2 is temporarily suspended, and a clean sheet of paper is placed on the stack, labeled "invocation 3." (Invocation 1 is buried two levels down at this point, but it will be resurrected at the appropriate time.)

For invocation 3, the argument is 0. Hence we write $n = 0$ on this sheet of paper. The **if** statement is executed, and, since the condition is satisfied, the **then** clause is executed. That clause is simply **return**(1), so invocation 3 has finished its work, and it passes the value 1 back to the procedure that called it. Therefore we throw this top sheet of paper into the shredder, remembering only that value 1.

Now invocation 2 (the top sheet of paper in our stack) resumes its operation, armed with the knowledge that the value it requested, *recursive_factorial*(0), is 1. Therefore, it can finish its calculation—multiplying n (which, as we wrote at the top of this sheet, is 1) by 1, obtaining the answer 1. Invocation 2's work is now finished, since the statement said to **return** this value, 1, to the calling procedure. Therefore, we commit this sheet of paper to the shredder and resume execution of invocation 1, knowing that the answer being returned is 1.

Finally, invocation 1 can complete its multiplication, obtaining $2 \cdot 1 = 2$, since $n = 2$ for this invocation. Therefore, invocation 1 returns the value 2, and our work is complete. We have found that $2! = 2$. ◆

The recursive call in *recursive_factorial* occurred only once, at essentially the end of the procedure. In cases such as this, the stack of sheets of paper representing the suspended invocations simply grows steadily to its maximum height, then shrinks to nothing. Often the recursive structure is more complicated (as we will see in the next subsection), with more than one recursive call. In such cases the stack grows and shrinks repeatedly as the number of suspended invocations of the procedure varies wildly.

It is important to have gone through this calculation once, to see exactly what it means for a computer to execute a recursive procedure. Really, though, these are details of implementation. Much more important is the approach to recursive procedures that we will discuss next. Human nature tends to balk at this other approach and wants to retreat to the "let's see what's happening as we go deeper and deeper into the recursion" approach that we have just finished. The reader is cautioned to consciously suppress these desires, and *not* allow him/herself to do the recursing. Our moral will be: *Let the computer do the recursing; human beings need only understand the recursive step.*

How shall we view the *recursive-factorial* procedure, then, from the proper human perspective? By definition, in order to compute $n!$, we need only compute $(n-1)!$ and multiply by n. The computation of $(n-1)!$ is a *simpler problem* than the computation of $n!$, because it is *closer to the base case* of $n = 0$. Therefore, if we write a procedure that handles the base case correctly, *and* correctly reduces the problem at hand to the simpler problem, then we know *in one fell swoop* that the procedure performs correctly. Algorithm 1 does both of these things: The base case is explicitly handled in the first line, and in the second line the problem of computing $n!$ is correctly reduced to the problem of computing $(n-1)!$ and multiplying it by n. What we make sure *not* to worry about is *how* the computer will go about computing $(n-1)!$ when we ask it to; it will do so correctly *because we are writing a correct procedure* to compute factorials.

If we can borrow from a famous novel, *recursion means not having to say "and so on."* A nonrecursive algorithm for computing $n!$ would probably be something like "multiply n times $(n-1)$ times $(n-2)$ *and so on* down to 2." Recursion gives us a way to conceptualize this iterative process in just one step. Looked at symbolically, recursion can eliminate the ellipses (...) that pervade informal discussions of processes whose duration depends on the input. Our perspective will become clearer as we look at more examples.

THE TOWERS OF HANOI PUZZLE

We now consider an old puzzle called the towers of Hanoi. Stripped of its mystical trappings, the problem is as follows. You (the person working the puzzle) are presented with three tall pegs sticking up from a solid base. On one of the pegs stands a tower of n solid disks with holes in their centers, all of different diameters. No disk sits on a disk of smaller diameter, so the stack of disks on the peg looks like a cone, wide at the bottom and narrow at the top. Figure 5.1 shows this initial position when $n = 5$. We label the pegs A, B, and C, as shown, and we label the disks 1 to n, from smallest to largest.

Your task is to move the disks so that the entire stack, which began on peg A, ends up on peg B. Two rules must be followed. First, you may move only one disk at a time, removing it from the top of the stack on its current peg and placing it on the top of the stack on some other peg. Second, a disk may never be placed on top of a smaller disk.

For example, you may, as your first move, transfer disk 1 from peg A to peg C. As your second move you will probably want to move disk 2, but it may not be moved to peg C (where it would rest on the smaller disk 1). Suppose that you move it to peg B. Then, if you wish, you may move disk 1 from peg C to peg B, resulting in the position shown in Figure 5.2.

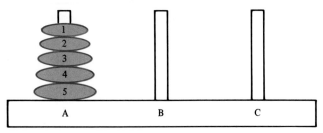

FIGURE 5.1   Initial position for the towers of Hanoi puzzle.

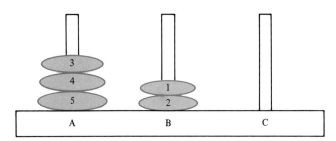

FIGURE 5.2   A position for the towers of Hanoi puzzle after three moves.

The problem is to solve the puzzle: Find a sequence of moves that will result in the entire stack of n disks standing on peg B. As a secondary problem, we might ask how many moves the most efficient algorithm will take to perform this task.

This problem is ideal for illustrating the recursive approach. All we have to do is to reduce the given problem to a simpler problem. The obvious choice of a simpler problem here is the same problem with fewer disks. After all, even a 3-year-old child could solve the problem if $n = 1$, and we seemed to have stumbled on the solution for $n = 2$ as we explained the rules (see Figure 5.2). Thus the base case is done, and we will have solved the problem completely if we can show how to incorporate a solution to the towers of Hanoi puzzle for $n - 1$ disks into a solution for n disks. A few moments' thought will yield the following insight: In order to accomplish our goal and abide by the rules, we should, about halfway through the operation, transfer disk n from peg A to peg B, and this can only be done if all the other disks are sitting on peg C at that time. What is more, after we have moved disk n, we can forget about it, leaving it forever at the bottom of peg B; that is where we want it to end up, and since it is the largest disk, its being on the bottom will not interfere with any other transfers. Therefore, a solution to the problem will consist of three stages, performed in order:

1. Move all but the largest disk from peg A to peg C, using peg B as the "scratch" peg.
2. Move disk n from peg A to peg B.
3. Move all the disks on peg C to peg B, using peg A as the "scratch" peg.

Unfortunately, this does not quite reduce our problem to a simpler instance of *the same* problem, but it almost does. The only way in which steps 1 and 3 differ from the original problem for $n - 1$ disks is that the roles of the pegs have been switched around. We circumvent this difficulty by generalizing our problem somewhat. Conceive of the problem not as "move the stack of n disks from peg A to peg B" but rather as "move the stack of n disks from peg X to peg Y, using peg Z as a scratch peg," where X, Y, and Z are arguments to the algorithm (just as n is an argument to the algorithm).

We seem to be converging on a solution. Let $hanoi(X, Y, Z, n)$ be the algorithm which, we hope, solves the problem just stated. We will assume that a solution to the problem consists of printing a sequence of statements of the form "Move disk i from peg p to peg q." If, for example, we call $hanoi(A, B, C, 2)$, then we hope to see the output:

> Move disk 1 from peg A to peg C.
> Move disk 2 from peg A to peg B.
> Move disk 1 from peg C to peg B.

To write this algorithm recursively, we need only do two things. First we need to handle the base case, $n = 1$; that much will be trivial. Then we need to handle the recursive or inductive step, and we figured that out above in our analysis of the three stages. *What we do not have to do is to worry about how the computer will plunge deeper and deeper into the various levels of recursion, as the procedure we write calls itself again and again.* We human beings are done as soon as we have written the procedure itself.

Almost as an anticlimax, then, Algorithm 2 gives the details.

Thus to solve the problem presented originally and illustrated in Figure 5.1, we simply

> **call** $hanoi(A, B, C, 5)$,

and the computer takes care of the rest, bobbing up and down through the levels of recursion (sometimes as much as five levels deep), printing out the necessary moves. Although we have warned the reader *not* to feel compelled to see what the computer is doing, it would probably not do any harm, just this one last time, to play computer for $n = 3$, to see exactly how the recursive algorithm does what it is supposed to do (see Exercise 1).

Our recursive solution has a by-product that might be unexpected. We asked not only for a sequence of moves that accomplished the given task; we also wanted to know how many moves would be required by *the most efficient* algorithm.

procedure *hanoi*(X, Y, Z : peg names, n : positive integer)
  {this procedure prints out in order the moves needed to transfer n disks from
    peg X to peg Y, following the rules of the towers of Hanoi puzzle; the peg
    names X, Y, and Z must be A, B, and C, in some order}
  if $n = 1$ **then print**("Move disk 1 from peg " X " to Peg " Y ".")
    else
      begin
        call *hanoi*($X, Z, Y, n-1$)
        print("Move disk " n " from peg " X " to peg " Y ".")
        call *hanoi*($Z, Y, X, n-1$)
      end
  return

Algorithm 2. Recursive solution to the towers of Hanoi puzzle.

THEOREM 1.   *Procedure hanoi solves the towers of Hanoi puzzle with the fewest possible moves.*

    Exercise 43 in Section 5.3 will request a formal proof of this theorem; the proof needs to have the same recursive (inductive) flavor as the algorithm. Intuitively, though, it seems right; after all, how else could we have efficiently transferred the stack of n disks from peg X to peg Y if not by efficiently transferring the top $n-1$ disks to peg Z, then moving the nth disk from peg X to peg Y, and finally efficiently transferring the top $n-1$ disks from peg Z back to peg Y? There was no way to get to move that nth disk until the others were all stacked out of the way on peg Z.

    Thus the number of moves (disk transfers) required by the most efficient algorithm is the number of moves required by Algorithm 2, since Algorithm 2 is the most efficient algorithm. We can figure out how many moves Algorithm 2 uses by giving a recursive description of the function that counts the number of moves. Suppose that $T(n)$ is the number of moves used by Algorithm 2 to solve the problem for n disks. Because of the way the algorithm handles the base case, clearly $T(1) = 1$. For $n \geq 2$ the algorithm requires $T(n-1)$ moves to transfer all but the largest disk to peg Z, one move to transfer disk n to peg Y, and $T(n-1)$ more moves to transfer all but the largest disk from peg Z to peg Y: a total of $2\,T(n-1) + 1$ moves. Thus the recursive part of our definition of T is

$$T(n) = 2\,T(n-1) + 1.$$

    In Section 7.3 we will develop specific ways to take equations like this and find explicit formulas for the functions they define. At this point, however, we should not really be dissatisfied with this answer as it stands. After all, it *does* define the function that we wanted to find. We can compute, for example, that $T(2) = 3$, $T(3) = 7$, $T(4) = 15$, $T(5) = 31$, and so on. And from the pattern that seems to be followed here, we can guess that $T(n) = 2^n - 1$. (In Section 5.3 we will see how to prove this formula.) Assuming that this formula is correct, we see that the towers of Hanoi problem is feasible only for fairly small values of n. For $n = 20$, for example, over 1 million moves are required, since $2^{20} - 1 = 1,048,575$.

RECURSION VERSUS ITERATION

In Section 5.1 we discussed recursive definitions of functions (or sequences, which are really just functions) and gave many examples. It is usually straightforward to take a recursive definition of a function and turn it into a recursive procedure for computing the function. An **if**...**then**...**else** statement is used to determine if we are in the base case (in which case the procedure computes the appropriate values, usually very simply) or the recursive case (in which case a more complicated calculation is performed, including a recursive call to the procedure itself, with a smaller argument). This was certainly the paradigm we encountered in Algorithm 1 for computing $n!$.

In this subsection we look at a few more examples and see that although a recursive algorithm is conceptually simple and satisfying, it is not always necessarily efficient. We begin with the classical example of the Fibonacci sequence.

EXAMPLE 2.   Recall the recursive definition of the Fibonacci sequence, which we write in functional notation to suit our needs in this example:

$$f(0) = f(1) = 1, \qquad \text{and} \qquad f(n) = f(n-1) + f(n-2) \text{ for } n \geq 2.$$

The recursive procedure for computing f (Algorithm 3) is short and simple.

procedure $f(n : \text{natural number})$
   {this procedure computes the value of $f(n)$ in the Fibonacci sequence,
     recursively from the definition}
     if $n < 2$ **then return**(1) **else return**($f(n-1) + f(n-2)$)

Algorithm 3.   Recursive computation of the Fibonacci sequence.

The difficulty with this procedure is that it is quite inefficient. Let us see how the recursive definition as implemented here would compute $f(5)$. We will write down the sequence of equalities that represent the recursive calculation.

$$f(5) = f(4) + f(3) = (f(3) + f(2)) + f(3)$$
$$= ((f(2) + f(1)) + f(2)) + f(3)$$
$$= (((f(1) + f(0)) + f(1)) + f(2)) + f(3)$$
$$= (((1 + f(0)) + f(1)) + f(2)) + f(3)$$
$$= (((1 + 1) + f(1)) + f(2)) + f(3)$$
$$= ((2 + f(1)) + f(2)) + f(3) = ((2 + 1) + f(2)) + f(3)$$
$$= (3 + f(2)) + f(3) = (3 + (f(1) + f(0))) + f(3)$$
$$= (3 + (1 + f(0))) + f(3) = (3 + (1 + 1)) + f(3)$$
$$= (3 + 2) + f(3) = 5 + f(3) = 5 + (f(2) + f(1))$$
$$= 5 + ((f(1) + f(0)) + f(1)) = 5 + ((1 + f(0)) + f(1))$$
$$= 5 + ((1 + 1) + f(1)) = 5 + (2 + f(1)) = 5 + (2 + 1) = 5 + 3 = 8.$$

In fact, the number of steps required to compute $f(n)$ is in $O(f(n))$, and these numbers tend to grow quite rapidly. (We will see in Section 5.3 that $f(n)$ grows exponentially.) Furthermore, the amount of space needed to perform this calculation turns out to be in $O(n)$, since space is needed to store the "pieces of paper" currently on the stack of suspended invocations of the procedure, and that stack can get as high as n—see the expression following the fourth equals sign in the calculation above. In other words, the recursion can get n levels deep when this procedure executes.

By comparison, Algorithm 1 in Section 5.1 uses only $O(n)$ steps to compute $f(n)$. That algorithm computes and stores *all* of the function values $f(0)$ through $f(n)$, so it also requires $O(n)$ storage space. A minor modification, however, in which we throw away all but the last two terms in the sequence after we calculate them, reduces the space requirement to $O(1)$. See Algorithm 4.

The moral of this example is that sometimes an iterative algorithm is much more efficient than a recursive one. Even when an iterative and a recursive algorithm are more or less equally efficient (as is the case, for example, in computing $n!$, where either approach takes $O(n)$ multiplications), the iterative one will probably run faster on a computer, since the overhead of recursion is not needed (storage of the stack of suspended invocations, for example). On the other hand, the conceptual elegance and simplicity of recursive algorithms make them ideal for *studying* processes that lend themselves to this approach. ◆

In Section 5.1 we discussed a rapidly increasing function φ called the Ackermann function. Although no one would really want to write an algorithm to compute values of the Ackermann function (since it is merely addition, multiplication, or exponentiation for small values of the third argument, and generates values too

procedure *iterative_f* (*n* : natural number)
  {this procedure computes the value of $f(n)$ in the Fibonacci sequence,
    iteratively, using $O(1)$ space}
    if $n < 2$ **then return**(1)
      else
        begin
          $y \leftarrow 1$ {the latest number in the sequence}
          $x \leftarrow 1$ {the number in the sequence before y}
          for $i \leftarrow 2$ **to** n **do**
            begin
              $z \leftarrow x + y$ {the next number in the sequence}
              $x \leftarrow y$
              $y \leftarrow z$
            end
          return(z)
        end

Algorithm 4. Iterative computation of the Fibonacci sequence.

big even to think about after that), the only really simple way to do so is through
a recursive algorithm. Again all we have to do is to copy down the definition.

EXAMPLE 3. Algorithm 5 computes the Ackermann function φ.

procedure *phi* (*x*, *n*, *i* : natural numbers)
  {this procedure computes the value of the Ackermann function $\varphi(x, n, i)$,
    recursively from the definition}
    if $i = 0$ **then return**($x + n$) {first base case}
      else if $n = 0$ **then**
        if $i < 3$ **then return**($i - 1$) {next two base cases}
          else return(x) {fourth base case}
        else return(phi (*x*, phi (*x*, $n - 1$, i), $i - 1$)) {recursive part}

Algorithm 5. Recursive computation of the Ackermann function.

It is possible to write nonrecursive procedures for performing this calculation,
but they are much harder to understand (see Exercise 33).   ◆

In support of our assertion that recursive procedures are the natural embodiment of some algorithms, we rewrite the Euclidean algorithm recursively. Our iterative version appears in Example 2 of Section 4.2. The amount of raw computation done by the two versions is the same; they really differ only in the way they handle the bookkeeping. In the iterative version, the human programmer and reader of the procedure have to deal with the bookkeeping, playing with **while** loops and assignment statements; it is too easy to make mistakes, and too hard to see the big picture for all the details. In the recursive version shown as Algorithm 6, the computer handles all these details. Human beings can see at a glance that the recursive algorithm is correct, because it is just a restatement of Theorem 2 in Section 4.1.

procedure *recursive_euclid*(x, y : natural numbers)
{this procedure calculates gcd(x, y) recursively, using the Euclidean algorithm; it assumes that x and y are not both 0}
    if $y = 0$ **then return**(x)
      else return(*recursive_euclid*(y, x mod y))

Algorithm 6. Recursive computation of the greatest common divisor by the Euclidean algorithm.

We give one more example of how a recursive version of an algorithm is conceptually simpler than a nonrecursive version.

EXAMPLE 4. Rewrite the fast modular exponentiation algorithm (Algorithm 4 in Section 4.5) using recursion.

Solution. We want to compute a^n mod m for large n. To speed up execution, we must choose the correct way to reduce this problem to a simpler problem. The poor way is to note that $a^n = a \cdot a^{n-1}$. That was the idea behind the slow modular exponentiation procedure given as Algorithm 3 in Section 4.5, with running time in $O(n)$. The fast way is to note that $a^n = (a^{n/2})^2$ if n is even, and $a^n = (a^{\lfloor n/2 \rfloor})^2 \cdot a$ if n is odd. Rather than the annoying details of the iterative version shown in Section 4.5, we get the conceptually simple recursive version (Algorithm 7).

procedure *exp*(a, n, m : integers)
{this procedure computes a^n mod m, where $m \geq 2$, $n \geq 0$, and $0 \leq a < m$}
    if $n = 0$ **then return**(1)
      else if n is even **then return**(($exp(a, n/2, m))^2$ mod m)
        else return(($exp(a, \lfloor n/2 \rfloor, m))^2 \cdot a$ mod m)

Algorithm 7. Recursive computation of modular powers.

The procedure works—the recursion eventually stops—because if $n \geq 1$, then $n/2$ or $\lfloor n/2 \rfloor$ is smaller than n; thus eventually the base case is invoked. ◆

As our last two examples have shown, a recursive procedure is often the algorithm of choice in discrete mathematics, especially when the goal is human understanding. A recursive algorithm is usually the most mathematically elegant way to solve a problem.

MERGE SORT

We end this section with a beautiful application of recursion to the problem of sorting. Recall that in Section 4.3 we discussed a straightforward sorting algorithm, bubble sort, that is rather inefficient: To sort a list of n items, bubble sort requires $O(n^2)$ time. The **merge sort** algorithm we present now requires only $O(n \log n)$ steps, although we will not derive this result until Section 7.3. The merge sort is useful, for example, in trying to sort about 100 physical items by hand; the author regularly uses a variant of it to sort students' tests into alphabetical order.

The key to making merge sort work is the fact that two sorted lists can be merged into one sorted list efficiently, taking advantage of the fact that the two lists being merged are already sorted. The algorithm is given in pseudocode below, but let us describe it in English first, in a rather anthropomorphic vein.

Given sorted lists $A = (a_1, a_2, \ldots, a_n)$ and $B = (b_1, b_2, \ldots, b_m)$, which we may as well imagine to be lists of numbers with $a_1 \leq a_2 \leq \cdots \leq a_n$ and $b_1 \leq b_2 \leq \cdots \leq b_m$, we wish to produce a list $C = (c_1, c_2, \ldots, c_{n+m})$ which contains all the elements of the two lists, totally sorted so that $c_1 \leq c_2 \leq \cdots \leq c_{n+m}$. We will find the elements of C one at a time, first c_1, then c_2, and so on. We begin with our left index finger pointing to a_1 and our right index finger pointing to b_1. We compare the two numbers we are pointing at and find the smaller. We put the smaller number into C and advance the index finger that was pointing to that number. For example, if $a_1 > b_1$, then we set c_1 equal to b_1, and advance our right index finger to point to b_2. We repeat this process—compare numbers being pointed at, put the smaller as the next element of C, and advance that finger—until we have filled C.

It is convenient to put artificial infinite numbers at the ends of lists A and B, so that when we come to the end of one list we are pointing at ∞ and are thereby forced to finish putting the elements in the other list into C. Also, ties are broken arbitrarily, say in favor of B. The reader should try this method on, for example, the lists $A = (3, 5, 9, 20)$ and $B = (1, 6, 6, 9, 24, 31, 49)$.

In pseudocode, we have Algorithm 8.

Note that we did not make the merge algorithm recursive, although we could have done so; it is conceptually more of an iterative process than a recursive one. It is interesting to note that some computer languages, such as LISP and Logo,

procedure *merge*$(A, B :$ sorted lists of numbers$)$
$\{A = (a_1, a_2, \ldots, a_n),$ with $a_1 \leq a_2 \leq \cdots \leq a_n;$ $B = (b_1, b_2, \ldots, b_m),$ with $b_1 \leq b_2 \leq \cdots \leq b_m;$ this procedure returns the two lists merged into one sorted list $C = (c_1, c_2, \ldots, c_{n+m}),$ with $c_1 \leq c_2 \leq \cdots \leq c_{n+m}\}$
$\quad j \leftarrow 1$ {points to element currently being scanned in A}
$\quad k \leftarrow 1$ {points to element currently being scanned in B}
$\quad a_{n+1} \leftarrow \infty$ {set artificial wall at end of A}
$\quad b_{m+1} \leftarrow \infty$ {set artificial wall at end of B}
$\quad$**for** $i \leftarrow 1$ **to** $n + m$ **do** {compare elements and put smaller into C}
$\quad\quad$**if** $a_j < b_k$ **then**
$\quad\quad\quad$**begin**
$\quad\quad\quad\quad c_i \leftarrow a_j$ {element of A goes into C}
$\quad\quad\quad\quad j \leftarrow j + 1$ {advance to scan next element of A}
$\quad\quad\quad$**end**
$\quad\quad$**else**
$\quad\quad\quad$**begin**
$\quad\quad\quad\quad c_i \leftarrow b_k$ {element of B goes into C}
$\quad\quad\quad\quad k \leftarrow k + 1$ {advance to scan next element of B}
$\quad\quad\quad$**end**
$\quad$**return**(C)

Algorithm 8. Merging two sorted lists.

naturally make almost all nontrivial procedures recursive; an implementation of *merge* in one of these languages would undoubtedly be recursive.

Now we come to the algorithm for sorting a list. Using *merge* we can solve our problem of sorting a list L if we can arrive at a point where we have two sorted lists to merge. But this we can do recursively. Thus our algorithm is to split the original list in half, sort the first half, sort the second half, and then merge the two halves. Since half a list is shorter than the whole list, the recursive call is to a simpler version of the sorting problem, and the recursion is valid. The base case is a list of length 1, for which there is nothing to do: A list of length 1 is already sorted.

Translating this discussion into pseudocode, we arrive at Algorithm 9.

The technique illustrated by merge sort is sometimes called "divide and conquer," and it is an extremely powerful tool. We divide a problem into smaller problems and thereby conquer it. The smaller problems are smaller instances of the original problem.

With the recursive approach, we do not have to go through the computation to see how or why it works. We are able to stay one level above the fray and simply bask in the beauty of it all.

procedure *merge_sort*(L : nonempty list of numbers)
  $\{L = (l_1, l_2, \ldots, l_n)$; the procedure returns a list of the same numbers
  rearranged into nondecreasing order; the technique is to sort two halves
  recursively and then merge$\}$
    if $n = 1$ **then return**(L) $\{$base case—nothing to do$\}$
      else
        begin
          $m \leftarrow \lfloor n/2 \rfloor$ $\{$middle of list $L\}$
          $L_1 \leftarrow (l_1, l_2, \ldots, l_m)$ $\{$first half of $L\}$
          $L_2 \leftarrow (l_{m+1}, l_{m+2}, \ldots, l_n)$ $\{$second half of $L\}$
          return(*merge*(*merge_sort*(L_1), *merge_sort*(L_2)))
        end

Algorithm 9.   Recursive merge sort.

SUMMARY OF DEFINITIONS

base case of a recursive procedure: the part of the procedure that handles a few
  small values of the argument without a recursive call (e.g., the calculation of
  $f(0)$ and $f(1)$ in Algorithm 3).
merge sort: an algorithm for sorting a list of n objects that works by splitting the
  list into two roughly equal parts, sorting each half by merge sort (recursively),
  and then merging the two sorted lists into one; its time complexity is in
  $O(n \log n)$.
recursive algorithm: a recursive procedure.
recursive part of a recursive procedure: the part of the procedure that handles
  the general case by a call to the same procedure with a smaller argument (e.g.,
  the calculation of $f(n)$ for $n \geq 2$ in Algorithm 3).
recursive procedure: a procedure that calls itself as a subprocedure (e.g., procedure f in Example 2).

EXERCISES

1. Write down the output from executing Algorithm 2 with a call to *hanoi*(A, B, C, 3).

2. A legend behind the towers of Hanoi puzzle states that the world will end
when priests in an Asian temple succeed in transferring all the disks from
peg A to peg B, following the rules; there are 64 disks. Assume that it takes
1 second to move a disk from one peg to another.
  (a)   Approximately how deep will the recursion get in this case? In other
      words, what is the largest number of suspended invocations of the pro-
      cedure pending at any one time?

(b)   How many moves will be required in this case? (Assume that the conjecture we made about $T(n)$ for this problem is correct—it is.)

(c)   When will the world end?

3. Write in pseudocode a recursive algorithm for finding the reverse of a string (see Example 7 in Section 5.1).

4. Write in pseudocode a recursive algorithm for computing Robinson's version of the Ackermann function, given in Exercise 17 of Section 5.1.

5. Carry out the merge procedure on the lists $A = (3, 5, 9, 20)$ and $B = (1, 6, 6, 9, 24, 31, 49)$.

6. Carry out merge sort on the list $(5, 7, 24, 3, 1, 6, 9, 15, 12, 2)$.

7. Rewrite Algorithm 2 so that the base case is $n = 0$; this simplifies the procedure slightly.

8. Write in pseudocode a recursive algorithm for finding the sum of a list of numbers.

9. Write in pseudocode a recursive algorithm for finding
   (a)   The largest element in a nonempty list of numbers.
   (b)   The index of the largest element in a nonempty list of distinct numbers.

10. Write in pseudocode a recursive algorithm for finding the largest and smallest elements in a nonempty list of numbers, simultaneously. (Your procedure must return two values—think of it as returning an ordered pair.)

11. Consider the sequence $\{a_n\}$ given in Exercise 3 of Section 5.1.
    (a)   Write in pseudocode a recursive algorithm for calculating a_n.
    (b)   Write in pseudocode an iterative algorithm using only $O(1)$ storage space for calculating a_n.
    (c)   Analyze the efficiency of the iterative algorithm.

12. Consider the sequence $\{T(n)\}$ given in Exercise 8 of Section 5.1. Do not use any insights you may have gained about a possible explicit formula for T.
    (a)   Write in pseudocode a recursive algorithm for calculating $T(n)$.
    (b)   Analyze the efficiency of the recursive algorithm.
    (c)   Write in pseudocode an iterative algorithm for calculating $T(n)$.
    (d)   Analyze the efficiency of the iterative algorithm.

13. Write in pseudocode a recursive algorithm for recognizing valid variable names, as defined in Example 4 of Section 5.1. The input is a string of one or more symbols, and the output is a logical value—either **true** or **false**. Your procedure can have statements that begin, for example, "**if** c is a digit **then**...."

14. Write in pseudocode a recursive algorithm whose input is a string of one or more digits and whose output is the number represented by that string in base 10.

15. Write in pseudocode a pair of recursive algorithms for computing the functions F and M defined in Exercise 35d of Section 5.1.

16. Write in pseudocode a recursive algorithm that recognizes palindromes (strings, such as "madam" and "noon," that read the same forward or backward). Input is a string of one or more symbols, and output is **true** or **false**.

17. Write in pseudocode a recursive version of linear search (see Algorithm 1 in Section 4.3). [*Hint*: Work from the back.]

18. Write in pseudocode a recursive version of binary search (see Algorithm 2 in Section 4.3). [*Hint*: Pass *high* and *low* as arguments.]

19. Analyze the efficiency of procedure *merge* (Algorithm 8), in terms of m and n.

20. Often merge sort is combined with ad hoc procedures for very small values of n, since using merge sort to sort, say, a list of two numbers seems to be a case of overkill.
    (a)   Write in pseudocode an efficient algorithm for sorting lists of numbers with three or fewer items. The input is a short list, and the output is a list of the same length, sorted.
    (b)   Modify Algorithm 9 to call on your procedure in case $n \leq 3$.

21. One way to handle sets (of numbers, say) algorithmically is to represent a set by a list. The representation is not unique: Lists can contain repeated elements and can present elements in different orders, so two different lists might represent the same set. Write algorithms in pseudocode with the following specifications. All but the third and fourth algorithms should be recursive, with base cases that involve the empty list.
    (a)   *element_of*(x, B) takes as input a number x and a set B (represented as a list of numbers) and returns the truth of the statement $x \in B$.
    (b)   *subset_of*(A, B) takes as input two sets, A and B (represented as lists of numbers), and returns the truth of the statement $A \subseteq B$.
    (c)   *equals*(A, B) takes as input two sets, A and B (represented as lists of numbers), and returns the truth of the statement $A = B$ (as sets).
    (d)   *adjoin*(x, B) takes as input a number x and a set B (represented as a list of numbers) and returns the list consisting of B with x adjoined to its end.
    (e)   *reduce*(A) takes as input a set A (represented as a list of numbers) and returns a list containing all the numbers in A but with no number occurring more than once in the list.

22. Give recursive algorithms (in English) for the following problems.
    (a) Walking n paces.
    (b) Finding all of a person's nth cousins (no times removed) (see Exercise 28 in Section 5.1.)

Challenging Exercises

23. Write in pseudocode a recursive procedure to print all the subsets of the set $\{1, 2, \ldots, n\}$, one subset per line. (For example, if $n = 2$, then the first line will be blank, the second will read "1," the third will read "2," and the fourth will read "1 2.")

24. Write in pseudocode a recursive algorithm for recognizing valid FPAEs, as defined in Example 5 of Section 5.1. The input is a list of items (the items making up the FPAE, from left to right), each of which is either an unsigned integer, a variable name, one of the four arithmetic operation symbols, or a left or right parenthesis; assume that you can recognize each, so, for example, a statement beginning "**if** c is a variable name **then**..." is allowed in your algorithm. You may ignore unary minus. [*Hint*: The hard part is to find the outermost operator.]

25. Write in pseudocode a recursive algorithm for evaluating a FPAE (see Example 6 of Section 5.1). Assume that *EVAL_atom* is already defined on variable names and numbers (see Exercise 24). You may ignore unary minus and the problem of division by 0.

26. Consider the following recursive procedure.

    procedure $close(x, y, \epsilon : \text{positive real numbers})$
        if $|x^2 - y| < \epsilon$ **then return**(x)
            else return$(close((x + (y/x))/2, y, \epsilon))$

    (a) Explain what the procedure does if asked to compute $close(1, 10, 10^{-4})$. What is the output?
    (b) Rewrite the procedure as an iterative procedure.

27. Consider the following iterative procedure.

    procedure $gauss(x, y, \epsilon : \text{positive real numbers, with } x \le y)$
        while $y - x > \epsilon$ **do**
          begin
            $geom \leftarrow \sqrt{xy}$
            $arith \leftarrow (x + y)/2$
            $x \leftarrow geom$
            $y \leftarrow arith$
          end
        return(x)

(a) Explain what the procedure does if asked to compute $gauss(1, 10, 10^{-4})$. What is the output?

(b) Rewrite the procedure as a recursive procedure.

28. Write in pseudocode a recursive algorithm for finding the second largest element in a list of distinct numbers. Assume that the list has length at least 2.

Exploratory Exercises

29. Merge sort a shuffled deck of 60 file cards with names on them (see Exercise 32 in Section 4.3 for instructions on constructing such a set), except use ad hoc methods for sets with four or fewer cards. Also experiment with merge sorting the deck by merging three subdecks at a time, instead of just two.

30. Consider the following algorithm.

procedure $collatz(n$: positive integer)
    if $n = 1$ **then return**(1)
      else if n is even **then return**$(collatz(n/2))$
        else return$(collatz(3n + 1))$

(a) Compute $collatz$ of the numbers 1 through 17.

(b) What possible values can $collatz(n)$ have? In other words, what is the range of this function?

(c) Discuss the problem of whether this is a valid finite algorithm. (No one knows whether it is valid—you are to explain how there could be any question about it.)

(d) Find out more about this function by consulting Lagarias [191] and Wagon [311].

31. Find out more about the recursive "logic" programming language Prolog by consulting Malpas [204] and Rogers [258].

32. Find out more about the recursive programming language LISP by consulting Winston and Horn [328].

33. Find out about an iterative method for computing the Ackermann function by consulting Grossman and Zeitman [129].

34. Find out more about the towers of Hanoi problem by consulting Cull and Ecklund [53].

35. Find out about other recursive puzzles by consulting Dewdney [68].

SECTION 5.3
PROOF BY MATHEMATICAL INDUCTION

*Induction is the process of discovering general laws
by the observation and combination of particular
instances. It is used in all sciences, even in mathematics.
Mathematical induction is used in mathematics alone
to prove theorems of a certain kind. It is rather
unfortunate that the names are connected because there
is very little logical connection between the two
processes. There is, however, some practical connection;
we often use both methods together.*

—**George Pólya,** *How to Solve It*

In this section we consider one of the most important techniques in discrete mathematics. We have studied definitions and algorithms from a recursive point of view. We now complete the trilogy and study *proofs* from a recursive point of view.

We have seen that the idea behind an inductive (or recursive) definition is that we can define an object in terms of smaller instances of itself, as long as we also define the smallest instance(s) explicitly. The idea behind a recursive algorithm is that we can solve a problem by reducing it to smaller instances of the same problem, as long as we know how to solve the smallest instance(s) explicitly. In this section we extend this pattern to recursive proofs of propositions. Let us first state the **principle of mathematical induction** verbally:

We can establish the truth of a proposition if we can show that it follows from smaller instances of the same proposition, as long as we can establish the truth of the smallest instance (or instances) explicitly.

We need to formulate this principle more precisely, and give a compelling argument as to why it is true, but first we will use it to prove the first half of the fundamental theorem of arithmetic, a classical and beautiful truth about the structure of the positive integers.

THEOREM 1. The Fundamental Theorem of Arithmetic.   *Every integer greater than 1 is the product of one or more prime numbers. Every integer greater than 1 can be written as the product of prime numbers in only one way (ignoring order).*

Here by a "product" of one number we simply mean that number itself.

Before we proceed with the proof, let us see what the theorem is saying and appreciate its profoundness. The number 66687192, for example, is not prime, because it can be factored as $9144 \cdot 7293$. These two numbers are not prime, however. The theorem says that 66687192 can be factored into prime factors, and the factorization is unique (except for the order in which the factors are written); it happens to be $2 \cdot 2 \cdot 2 \cdot 3 \cdot 3 \cdot 3 \cdot 11 \cdot 13 \cdot 17 \cdot 127$. The fundamental theorem of arithmetic describes the multiplicative structure of the set of positive integers: Every number greater than 1 is built up of basic multiplicative building blocks (the prime numbers), which are the numbers that cannot be broken down into smaller multiplicative pieces, and the decomposition of a number into its building blocks is unique. Something like this certainly does not hold, for example, for the rational numbers: There are no indecomposable rational numbers whose products give all rational numbers uniquely.

We will prove the half of the theorem that states that every number greater than 1 *has* a prime factorization (the other half is messier and we omit it).

Proof of the existence of prime factorizations. Let $k > 1$ be the number we want to factor into prime factors. If k is itself prime, we are done: It is the product of one prime number, namely itself. If k is not prime, then, *by definition*, k can be factored into $k = a \cdot b$, with $1 < a < k$ and $1 < b < k$. (Recall that we can define a prime number as simply "an integer greater than 1 that *cannot* be written as the product of two smaller positive integers.") The problem is that a and b may not be prime, so we are not finished with our proof at this point.

However, suppose that we already knew that the theorem was true for numbers smaller than k. Since a and b are smaller than k, this means that each of them is the product of primes. In other words, we can write $a = p_1 p_2 \cdots p_r$ and $b = p_{r+1} p_{r+2} \cdots p_s$, where all the p_i's are (not necessarily distinct) prime numbers. Then $k = ab = p_1 p_2 \cdots p_s$, which is exactly what we wanted to show: k can be written as a product of primes.

Thus we have reduced the truth of the statement "k can be written as the product of primes" to smaller instances of the same statement: "a can be written as the product of primes" and "b can be written as the product of primes." Furthermore, the smallest instance is certainly true, since 2, the smallest number of interest here, is prime. The principle of mathematical induction says that this reduction is all we need to do to demonstrate the truth of the proposition that every integer greater than 1 is the product of primes; therefore, the proof is complete. ■

Note that this proof has the same flavor as the recursive algorithms we encountered in Section 5.2.

THE TOP-DOWN APPROACH

We want to formalize this discussion, and give a template, or model, for constructing proofs by mathematical induction. Just as with recursive definitions and recursive algorithms, we will break things down into a **base case** and an **inductive**

step. It is important to remember that we are talking about the truth of *propositions*. The objects we are dealing with are propositions, not numbers or algebraic expressions. To set the notation, let us suppose that $S(n)$ is a proposition with a free variable n that usually is assumed to range over the set of positive integers. (Sometimes n ranges over the set of natural numbers, or, as in the theorem above, over some other set of integers, and simple modifications handle such cases.) In other words, we have a sequence of propositions $S(1)$, $S(2)$, $S(3)$, ...

For example, $S(n)$ might be the equation

$$1 + 2 + 3 + \cdots + n = \frac{n(n+1)}{2}.$$

Then $S(1)$ is the true statement that $1 = 1 \cdot (1+1)/2$ or $1 = 1$; $S(2)$ is the true statement that $1 + 2 = 2 \cdot (2+1)/2$ or $3 = 3$; $S(3)$ is the true statement that $1 + 2 + 3 = 3 \cdot (3+1)/2$ or $6 = 6$; and so on. Perhaps $S(n)$ is true for *all* positive integers n. If not, then we should be able to produce a counterexample—a value of n for which $S(n)$ is false. But if $\forall n: S(n)$ is true, how can we prove it? (In fact, $S(n)$ is true for all n, as the reader is asked to prove in Exercise 2.)

Let us translate the verbal statement of the principle of mathematical induction given above into the symbolism we have introduced. To establish $\forall n: S(n)$ we only have to show, for all numbers k, that the truth of $S(n)$ for all smaller values of the free variable (i.e., $n = i$ for all i less than k) implies the truth of $S(n)$ for $n = k$; and establish that S holds for the smallest value (i.e., that $S(n)$ is true for $n = 1$). The following theorem gives the precise statement of the principle of mathematical induction.

THEOREM 2. The Principle of Mathematical Induction (Top-Down Form). *Suppose that $S(n)$ is a proposition with a free variable n that ranges over the positive integers. Then if*

(base case)   *$S(1)$ is true,*

and if

(inductive step)   *for every $k > 1$ the implication*
                   $\big(\forall i{<}k: S(i)\big) \rightarrow S(k)$ is true,

then $S(n)$ is true for all n.

In symbols, the principle of mathematical induction states that

$$\{S(1) \wedge [\forall k{>}1: ((\forall i < k: S(i)) \rightarrow S(k))]\} \longrightarrow \forall n: S(n).$$

We call this the "top-down form" of the principle of mathematical induction to distinguish it from the "bottom-up form," which we will discuss later. Theorem 2 is also called the "strong" form of mathematical induction. The idea here is that we begin at the top (the k that we wish to prove the inductive step for) and descend down to lower values ($i < k$) as needed. We have stated the theorem with the base case of $n = 1$. Sometimes we need to start at 0 or other values of n, and the obvious modifications can be made.

At the end of this subsection, we will argue that the principle of mathematical induction is true—in other words, that the set of positive integers is so constructed that this method of "proof by mathematical induction" is valid. First we look at the question of just how one goes about constructing a proof by mathematical induction and consider two examples.

As we see from the statement above, we can reach the desired conclusion ($\forall n: S(n)$) if we can do two things, labeled "base case" and "inductive step." The base case is usually pretty easy, often quite trivial: All we have to do is show that $S(1)$ is true. The inductive step is the heart of the matter. We have to let $k > 1$ be arbitrary and prove the implication ($\forall i < k: S(i)) \rightarrow S(k)$. We usually do this with a direct proof. As we saw in Section 1.3, this means that we must prove the conclusion $S(k)$, *assuming the hypothesis* $\forall i < k: S(i)$, which is called the **inductive hypothesis**. In other words, we need to reduce the case we are working on, $S(k)$, to smaller cases $S(i)$ for one or more values of i less than k. Note that we do not have to prove $S(k)$ outright, only to show that the truth of $S(k)$ *follows from* the truth of $S(i)$ for all $i < k$.

As a practical matter, every proof by mathematical induction will have essentially the following form.

0. The first step is to write down explicitly exactly what proposition (statement) about n it is we are trying to prove for all n. (Remember that propositions must have verbs.) Also, we identify the exact set of integers over which n ranges—usually, but not always, $n \geq 1$ or $n \geq 0$.

1. Then we check the base case (and label it as such); in other words, we perform the (usually trivial) verification that the statement under consideration holds for $n = 1$ (or $n = 0$, or some slightly more complicated set of values, as the case may be).

2. Next we do the inductive step, which is the hard part. We should always begin this step by (1) labeling it as such, (2) explicitly stating that we are *assuming* the inductive hypothesis (and explicitly stating what that hypothesis is), and (3) explicitly stating exactly what it is we are trying to derive under this assumption. (In our statement of the principle of mathematical induction we used k as the name of the variable in this step, rather than n. We could just as easily have called it n, but the use of a different letter helps us remember that we are only proving an implication within the inductive framework, not proving $S(n)$ outright.

At the beginning of our encounter with proofs by mathematical induction we will mostly use k during the inductive step, but gradually we will revert to n, which is the common practice among mathematicians and computer scientists.)

3. Once the inductive step is complete (i.e., once we have derived $S(k)$, using the inductive hypothesis $\forall i < k: S(i)$), our proof is finished. Good style demands a statement to that effect, since proofs by induction usually seem to end abruptly.

EXAMPLE 1. An interesting discrete geometry problem is to determine the set of values of b for which the following construction can be carried out. You are given a b by b checkerboard, with one square deleted, together with a supply of L-shaped pieces consisting of three checkerboard-sized squares, as shown in Figure 5.3. The problem is to cover the entire board, except for the deleted square, with L-shaped pieces, which must lie on the board, squares covering squares, without overlapping. Clearly, we must use exactly $(b^2 - 1)/3$ such pieces to cover the $b^2 - 1$ squares on the board.

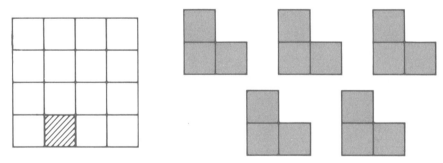

FIGURE 5.3 Problem: Cover the checkerboard with one square deleted, using L-shaped pieces.

It is easy to find some values of b for which the covering cannot be done, and some for which it can. If $b = 6$, then $(b^2 - 1)/3$ is not a natural number, so no covering is possible; whereas if $b = 4$, then Figure 5.4 shows one way in which the covering can be accomplished.

We will not solve the total problem here (see Exercise 47), but we will show that the covering is possible when $b \geq 2$ is a power of 2. In other words, the covering can be done for $b = 2, 4, 8, 16, 32, \ldots$. Let us write $b = 2^n$, where n ranges over the positive integers. Our proof will be by mathematical induction; indeed, it is hard to see how any other proof technique could work.

Let us follow the steps outlined above.

0. We will prove the following statement for all $n \geq 1$:

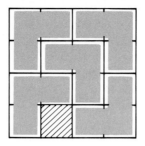

FIGURE 5.4 Covering
of a 4 by 4 checkerboard
with one square deleted.

$S(n)$: given a 2^n by 2^n checkerboard with one square deleted, there
exists a covering of this board by L-shaped pieces.

1. Base case ($n = 1$). If $n = 1$, then we are trying to cover a 2 by 2 board, with
one square deleted. There are four cases, depending on which square is deleted,
but in each case it is clear that by properly rotating the L-shaped piece, we can
make it exactly cover the remaining three squares. As often happens, the base case
here is utterly trivial.

2. Inductive step. Let k be an arbitrary number greater than 1. Assume the
inductive hypothesis, namely that for all positive integers i less than k, it *is* possible
to cover any square-deleted 2^i by 2^i board with L-shaped pieces. We want to show
that, under this assumption, it is possible to cover any square-deleted 2^k by 2^k
board. Now comes the hard part: We have to figure out just how to reduce the
case of the 2^k by 2^k board to the case of a smaller board. The insight we need
to make is that by bisecting the given board horizontally and vertically, we obtain
four 2^{k-1} by 2^{k-1} boards; the deleted square in the given board lies in one of them
(see Figure 5.5.)

We position one L-shaped piece in the middle of the board, as shown in
Figure 5.5, so that each of the four 2^{k-1} by 2^{k-1} boards is missing exactly one
square. Three of them are missing a square occupied by the L-shaped piece in the
center, and the fourth is missing the originally missing square. (The reader should
not be deceived here: There is no reason that anyone should come up with this
ingenious trick right away. It is not unreasonable to expect that weeks of playing
with this problem would precede this idea's suddenly flashing into someone's mind.
Once the trick is suggested, however, the proof itself follows the induction model.)

Now *by the inductive hypothesis* it is possible to cover each quarter of the
board, since each is a 2^{k-1} by 2^{k-1} board with one square missing, and $k - 1$ is a
positive integer less than k. These four coverings, together with the one L-shaped
piece that we placed in the center, give us the desired covering of the 2^k by 2^k
board. (Note that the covering shown in Figure 5.4 is of this form.)

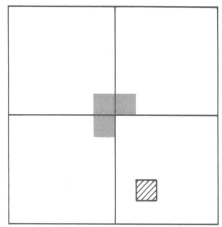

FIGURE 5.5 Reducing the 2^k by 2^k problem to four 2^{k-1} by 2^{k-1} problems.

3. Since we have shown that $S(k)$ is true, using the assumption that $S(i)$ is true for all $i < k$ (actually we only used the inductive hypothesis with $i = k - 1$), we have completed our proof by mathematical induction that $S(n)$ is true for all n. ◆

The example is very typical. To construct a proof by mathematical induction, it is necessary to verify the base case, as we did, explicitly, in step 1; and then to reduce the general case ($n = k$) to one or more smaller cases ($n = i$ for $i < k$), as we did in step 2.

EXAMPLE 2. Show that the sum of the squares of the first n positive integers is equal to $n(n + 1)(2n + 1)/6$, for all $n \geq 1$. In other words, prove that

$$(*) \qquad \sum_{j=1}^{n} j^2 = 1^2 + 2^2 + \cdots + n^2 = \frac{n(n + 1)(2n + 1)}{6}.$$

Solution. Before we solve this problem, let us comment on where this formula comes from. In Chapter 7 we will develop a technique for obtaining formulas like this, but suppose that we did not have the tools discussed there. If locked in a room for 6 hours with a pencil and a pad of paper, how would you figure out such a formula? The answer is: *by treating mathematics like the experimental science that it sometimes is*. Whenever you are first presented with a problem, it is good practice to "get your hands dirty" by playing around with it. Look at special cases. Here, for instance, see what the sums are for $n = 1$, $n = 2$, ..., $n = 10$. This is an increasing sequence of 10 positive integers; perhaps try factoring these numbers. See if there is any analogous problem that might be easier to solve, and look at it first (here, for instance, look at the problem of finding just the sum of the first n positive integers, rather than the sum of their squares). Look for patterns in the data, make and test conjectures. Let your mind wander. Eventually, you might come up with an answer that you can then try to prove correct. This all might be called the inductive approach to solving problems, as opposed to the deductive process of presenting a logical proof as to why the result is correct. (Proof by mathematical induction, like any other proof technique, is a deductive method, not an inductive one.) It is rarely the case that you can come up with an interesting and true proposition, let alone a valid deductive argument for it, without a lot of inductive experimentation.

Now let us prove the formula by mathematical induction. Let $S(n)$ be the proposition $(*)$ displayed above. Note that $S(n)$ is this entire *equation*; it is not the algebraic expression on the right-hand side.

Base case. If $n = 1$, then the left-hand side of the equation is the sum from $j = 1$ to $j = 1$ of j^2, in other words, just 1^2, or 1. The right-hand side is $1 \cdot (1 + 1) \cdot (2 \cdot 1 + 1)/6 = 6/6 = 1$. Therefore, equality does, indeed, hold, and our verification of the base case is complete.

Inductive step. Let $k > 1$, and assume that equation $(*)$ is valid for all values less than k. In other words, assume that

$$\sum_{j=1}^{i} j^2 = \frac{i(i + 1)(2i + 1)}{6},$$

if $1 \leq i < k$. This is the inductive hypothesis. We need to prove, under this assumption, that

$$\sum_{j=1}^{k} j^2 = \frac{k(k + 1)(2k + 1)}{6}.$$

This time the reduction is quite straightforward. The sum on the left from 1 to k has, within it, the sum from 1 to $k - 1$. Explicitly,

$$\sum_{j=1}^{k} j^2 = \left(\sum_{j=1}^{k-1} j^2 \right) + k^2.$$

By the inductive hypothesis, since $k - 1 < k$, we *already know* what the sum from 1 to $k - 1$ is, namely the right-hand side of $(*)$ with $k - 1$ plugged in for n. We put this all together and do some algebra, to obtain the desired equation:

$$\sum_{j=1}^{k} j^2 = \left(\sum_{j=1}^{k-1} j^2\right) + k^2$$

$$= \frac{(k-1)((k-1)+1)(2(k-1)+1)}{6} + k^2 \quad \text{(by the inductive hypothesis)}$$

$$= \frac{(k-1)k(2k-1) + 6k^2}{6}$$

$$= \frac{k[(k-1)(2k-1) + 6k]}{6}$$

$$= \frac{k(2k^2 + 3k + 1)}{6}$$

$$= \frac{k(k+1)(2k+1)}{6}.$$

Thus our proof by induction is complete. Note that we showed the point at which we invoked the inductive hypothesis; this point should always be made explicit. ◆

Similar formulas can be obtained for other sums. We have already mentioned

$$\sum_{j=1}^{n} j = 1 + 2 + \cdots + n = \frac{n(n+1)}{2},$$

which appears in Exercise 2. In Exercise 4 the reader is asked to prove that

$$\sum_{j=1}^{n} j^3 = 1^3 + 2^3 + \cdots + n^3 = \frac{n^2(n+1)^2}{4}.$$

Let us now argue that the principle of mathematical induction, as stated above (Theorem 2), is valid.

Proof of Theorem 2. We want to show that if the base case and inductive step have been proved, then $S(n)$ must be true for all n. We will argue by contradiction. If $\forall n: S(n)$ is not true, there must be one or more counterexamples, positive integers n that make $S(n)$ false. Let n_0 be the *smallest* counterexample. Certainly $n_0 \neq 1$, since the proof of the base case explicitly showed that $S(1)$ *is* true. On the other hand, if $n_0 > 1$, then look at the inductive step, with $k = n_0$; it was proved that

$$(\forall i < n_0: S(i)) \longrightarrow S(n_0).$$

Since n_0 is the *smallest* counterexample, the inductive hypothesis $\forall i < n_0 \colon S(i)$ is true. Therefore, the conclusion of the inductive step must also be true, namely $S(n_0)$. But this contradicts our assumption that n_0 was a counterexample to $\forall n \colon S(n)$. Therefore, $\forall n \colon S(n)$ has no counterexamples, and our argument is complete. ▨

The only property of the positive integers that the preceding argument relied on was the fact that every nonempty set of positive integers has a smallest element; here we looked at the set of purported counterexamples to the proposition proved by mathematical induction. As long as we believe this fundamental fact about the positive integers (and who can doubt it?), we must believe in the method of proof by mathematical induction.

Proof by mathematical induction is difficult for nearly all students. You may find it helpful to apply this "minimum counterexample" argument for the validity of the principle of mathematical induction to specific propositions. For example, we can recast our proof of half of the fundamental theorem of arithmetic (that all integers greater than 1 can be written as the product of one or more primes) along the following lines. If the theorem were not true, there would be a smallest integer, $k > 1$, which was not the product of one or more primes. Clearly, k could not be prime, so we could write $k = ab$ for some integers a and b greater than 1 and smaller than k. But since k was the *smallest* counterexample, each of a and b *is* the product of primes, so their product (namely k) is also the product of primes. This contradiction shows that no such counterexample k exists.

THE BOTTOM-UP APPROACH

In Examples 1 and 2, the proof of $S(k)$ in the inductive step in our proof by mathematical induction relied on only one "smaller instance," namely $S(k-1)$. (This contrasts with our proof of half of Theorem 1, in which we needed the full force of the inductive hypothesis, namely not only $S(k-1)$ but also $S(k-2)$, $S(k-3), \ldots, S(3)$, and $S(2)$, since we did not know where in the range from 2 to $k-1$ we would find a and b.) Many proofs by mathematical induction have this property, that the weaker inductive hypothesis $S(k-1)$ is sufficient for proving $S(k)$. We state this simpler form as a theorem. Also, we change the notation slightly (without affecting the meaning) by replacing the variable k by $k+1$; this makes the statement slightly more appealing (see Exercise 19).

THEOREM 3. *The Principle of Mathematical Induction (Bottom-Up Form).*
Suppose that $S(n)$ is a proposition with a free variable n that ranges over the positive integers.

Then if

        (base case)   *S(1) is true,*

and if

        (inductive step)   *for every $k \geq 1$*
                                 the implication $S(k) \rightarrow S(k+1)$ is true,

then S(n) is true for all n.

The point of view in this formulation of mathematical induction differs slightly from the viewpoint of Theorem 2. Rather than thinking of the inductive step of the proof as reducing $S(k+1)$ to $S(k)$, we often think of it instead as starting with $S(k)$ and producing $S(k+1)$. In other words, it is a bottom-up, iterative approach rather than a top-down approach. There is no real mathematical or logical difference, just a change of emphasis. This form is sometimes called the "weak form" of mathematical induction. Although it appears to be less forceful than Theorem 2, in that you need to prove more in the inductive step in order to conclude $\forall n: S(n)$, in fact one can prove Theorem 2 from Theorem 3 (see Exercise 44).

Looking at it in this way, we can view the justification for the method of mathematical induction in terms of a row of dominoes. Suppose that we think of each of the statements $S(1)$, $S(2)$, and so on, as represented by a domino. When a domino falls over, we know the truth of the corresponding proposition. Line up this infinite collection of dominoes in a row: the first domino standing at the front of the row, the second domino standing behind it, the third domino standing behind the second, and so on. The inductive step says that whenever one domino falls down, it knocks the next one down as well. For example, if and when domino 326 falls, it will knock over domino 327. The base case says that we can push over the first domino with our finger. Knocking over the first domino sets in motion a chain reaction that causes the entire row to topple: The first knocks over the second, which in turn knocks over the third, and so on. In terms of our propositions, the truth of $S(1)$ (the base case) forces the truth of $S(2)$ by one application of the inductive step. Then the truth of $S(2)$ forces the truth of $S(3)$ by one more application of the inductive step. This process continues forever. For every n, eventually the truth of $S(n)$ is forced, after a sufficient number of applications of the inductive step. In other words, $S(n)$ must be true for all n.

Alternatively, the justification for the validity of proof by mathematical induction given earlier can be carried over almost word for word to justify Theorem 3.

EXAMPLE 3.   The Ackermann function φ was defined inductively in Section 5.1. Show that $\varphi(x, n, 1) = x \cdot n$ for all $x \geq 0$ and all $n \geq 0$.

Solution.   The first step is to identify exactly what statement we are trying to prove. It is

$$S(n): \qquad \forall x \geq 0 : \varphi(x, n, 1) = x \cdot n.$$

Note that the domain of discussion is the set of natural numbers here, not the set of positive integers, so we need to start our proof at $n = 0$ rather than at $n = 1$.

Base case. A base case of $n = 0$ is handled just like a base case of $n = 1$: We must verify $S(0)$. In this case, that says $\varphi(x, 0, 1) = x \cdot 0 = 0$, which is indeed true by the second line in the definition of φ given in Section 5.1.

Inductive step. Let $k \geq 0$ be given (it is "$k \geq 0$" rather than "$k \geq 1$" because our base case only took care of $n = 0$). Suppose that $S(k)$ is true, in other words, that $\varphi(x, k, 1) = x \cdot k$ for all x. We want to show that $S(k+1)$ is true, that is, that $\varphi(x, k+1, 1) = x \cdot (k+1)$ for all x. The proof is straightforward:

$$\varphi(x, k+1, 1) = \varphi(x, \varphi(x, k, 1), 0) \quad \text{(by the definition of } \varphi)$$

$$= \varphi(x, x \cdot k, 0) \quad \text{(by the inductive hypothesis)}$$

$$= x + x \cdot k \quad \text{(by the definition of } \varphi)$$

$$= x \cdot (k+1) \quad \text{(by algebra)}.$$

Note that we have put the quantifier on x inside $S(n)$. This proof would also work if we pulled it outside and viewed x as a constant throughout the whole proof. In other cases it is often crucial that variables such as this be inside, however. ◆

EXAMPLE 4. In Section 5.2 we conjectured, in connection with the towers of Hanoi problem, that the sequence determined by the recursive definition

$$T(1) = 1, \qquad T(n) = 2\,T(n-1) + 1 \qquad \text{for } n > 1$$

is given explicitly by $T(n) = 2^n - 1$. Prove this conjecture.

Solution. The statement we wish to prove for all $n \geq 1$ is

$$S(n): \qquad T(n) = 2^n - 1.$$

The base case ($n = 1$) is clear, since $2^1 - 1 = 2 - 1 = 1$, and $T(1) = 1$ by definition. So $S(1)$ is true. For the inductive step, we assume that $S(k)$ is true, namely that $T(k) = 2^k - 1$; and we want to prove that $T(k+1) = 2^{k+1} - 1$. By definition, $T(k+1) = 2\,T(k) + 1$; but we are assuming that $T(k) = 2^k - 1$. Thus $T(k+1) = 2 \cdot (2^k - 1) + 1 = 2 \cdot 2^k - 2 + 1 = 2^{k+1} - 1$, exactly as desired. ◆

VARIATIONS IN PROOF BY MATHEMATICAL INDUCTION

Proofs by mathematical induction pervade the mathematical literature. Theorems about sets, numbers, graphs, and practically every other mathematical entity are proved by induction. Computer scientists prove that their programs are correct by induction. Strategies for games are justified by induction. In this last subsection we look at some more examples and encounter a few of the variations that can arise in the general method we have outlined.

As we have formulated the principle of mathematical induction, we always need to have an integer variable n "on which to induct." It is not always clear what n is. Sometimes the base case needs to be expanded a bit. We will investigate these variations and see what sort of theorems we can prove with mathematical induction.

EXAMPLE 5. Prove that the number of left parentheses is equal to the number of right parentheses in every FPAE, as defined in Section 5.1. In other words, if we let $LP(x)$ and $RP(x)$ be the number of left and right parentheses in the FPAE x, respectively, then $\forall x: LP(x) = RP(x)$.

Solution. On the surface there is no variable on which to induct. The universally quantified variable here is x, and it takes on values that are strings, not integers. In fact there are two candidates for a variable to induct on, lurking behind the scenes in the inductive definition. Either one will serve our needs. One approach is to let n be the day on which a FPAE is born, in the "days of creation" model for inductive definitions. Since every FPAE is born on some day, all we have to do is to prove that for all n, every FPAE x born on day n satisfies $LP(x) = RP(x)$. The other approach is to let n be the length of a FPAE. Then all we have to do is to prove that for all n, every FPAE x of length n satisfies $LP(x) = RP(x)$. We will use the latter approach and leave the former as Exercise 25.

Let the length of a FPAE be defined as the number of symbols occuring in x, where we count each variable name, each unsigned integer, each arithmetic operator, and each parenthesis as one symbol each time it appears. For example, the FPAE $(((-789) + x) * newton)$ has length 12. We will prove by induction (using Theorem 2) that a FPAE with length n has an equal number of left and right parentheses.

The base case is $n = 1$. The only way for a FPAE x to have length 1, in view of the definition, is for it to be either an unsigned integer or a variable name; all other FPAEs have length at least 4. In both of these cases, x has no parentheses of either type, and certainly $0 - 0$.

For the inductive step we fix $k > 1$ and assume the inductive hypothesis, that for every FPAE x of length *less than* k, $LP(x) = RP(x)$. We need to prove that for every FPAE x of length *equal to* k, $LP(x) = RP(x)$. To this end, let x be a FPAE of length k. Since $k > 1$, x must have arisen through one of the five clauses in the recursive part of the definition. They are all similar, so almost without loss of generality we will assume that x is $(\alpha + \beta)$. (The only clause that is a little different in syntax is the one that allows x to be $(-\alpha)$, but there, too, the ideas are the same.) Now clearly $LP(x) = LP(\alpha) + LP(\beta) + 1$ and $RP(x) = RP(\alpha) + RP(\beta) + 1$. But by the inductive hypothesis, $LP(\alpha) = RP(\alpha)$ and $LP(\beta) = RP(\beta)$, since α

and β have lengths less than k, the length of x (note the use of the strong inductive hypothesis). Adding these equations, and then adding 1 to both sides, we obtain the desired equality. This completes the proof.   ◆

EXAMPLE 6. Mathematical induction can be used to prove inequalities. We claimed in Section 5.2 that the terms of the Fibonacci sequence grow exponentially. We will now show that in fact $f_n > \left(\frac{3}{2}\right)^n$ for $n \geq 5$. Two base case values of n need checking here, $n = 5$ and $n = 6$. We compute that $\left(\frac{3}{2}\right)^5 = 7.59375 < 8 = f_5$; and $\left(\frac{3}{2}\right)^6 = 11.390625 < 13 = f_6$. Now let $k > 6$, and assume the inductive hypothesis that $f_i > \left(\frac{3}{2}\right)^i$ for all $i < k$ (we will need it only for $i = k - 1$ and $i = k - 2$). Then we have

$$f_k = f_{k-1} + f_{k-2}$$

$$> \left(\frac{3}{2}\right)^{k-1} + \left(\frac{3}{2}\right)^{k-2} \qquad \text{(by the inductive hypothesis)}$$

$$= \left(\frac{3}{2}\right)^{k-2} \left(\frac{3}{2} + 1\right) \qquad \text{(factoring)}$$

$$= \left(\frac{3}{2}\right)^{k-2} \left(\frac{5}{2}\right)$$

$$> \left(\frac{3}{2}\right)^{k-2} \left(\frac{9}{4}\right) \qquad \left(\text{since } \frac{5}{2} > \frac{9}{4}\right)$$

$$= \left(\frac{3}{2}\right)^{k-2} \left(\frac{3}{2}\right)^2$$

$$= \left(\frac{3}{2}\right)^k, \text{ as desired.}$$

The variations here were that we needed to check more than one value of n in the base case (otherwise, we could not have backed up by two in the inductive step and claimed that $f_{k-2} > \left(\frac{3}{2}\right)^{k-2}$), and that the induction began at $n = 5$ rather than $n = 0$ or 1 (in fact, the statement $f_n > \left(\frac{3}{2}\right)^n$ is not true for $n < 5$).   ◆

EXAMPLE 7. Prove that if S is a finite set with cardinality n, then $|\mathcal{P}(S)| = 2^n$. Prove also that $n < 2^n$; it then follows that the power set of any finite set is bigger than the set itself.

Solution. First we prove the statement about the cardinality of the power set of S, using induction on $n = |S|$. If $n = 0$, then $S = \emptyset$, and $\mathcal{P}(\emptyset) = \{\emptyset\}$. Therefore,

$|\mathcal{P}(S)| = 1 = 2^0$, as desired. Next assume the inductive hypothesis, that $|\mathcal{P}(T)| = 2^n$ for all sets T of cardinality n. We need to prove that $|\mathcal{P}(S)| = 2^{n+1}$ if S is a set with cardinality $n + 1$. Note that we are using n in place of k as the variable for the inductive step in our proof. The variable k is a dummy variable in the statement of Theorem 3 (it is bound by the universal quantifier "for every"), so we can call it anything we wish. It is common practice to do what we are doing here, that is, to use the same variable name in the inductive step as is used in the statement we are trying to prove.

Let S be any set with cardinality $n + 1$, and let x be a fixed element of S. Let $T = S - \{x\}$; thus T is the set of all the elements of S other than x, so clearly, $|T| = (n + 1) - 1 = n$. Now the subsets of S fall into two categories, those containing x and those not containing x. The subsets of S not containing x are precisely the same as the subsets of T, and by the inductive hypothesis there are 2^n of these. On the other hand, each subset A of S that does contain x can be written as $A = \{x\} \cup A'$, where A' does not contain x; this gives us a one-to-one correspondence between such subsets A and the corresponding $A' = A - \{x\}$. But A' ranges over all the subsets of T, so by the inductive hypothesis there are 2^n of these as well. Therefore, the number of subsets of S altogether is $2^n + 2^n = 2^{n+1}$, as desired. (In other words, we have partitioned $\mathcal{P}(S)$ into two sets (those subsets of S that do not contain x and those that do), and found a bijective function between them, given by $f(A') = A' \cup \{x\}$ for each $A' \subseteq T$. The inductive hypothesis tells us that each of these parts has cardinality 2^n, so $\mathcal{P}(S)$ has cardinality $2 \cdot 2^n = 2^{n+1}$.)

Next we show that $n < 2^n$ for all natural numbers n. The base case is trivial: $0 < 1 = 2^0$. Assume the inductive hypothesis that $n < 2^n$, for a fixed $n \geq 0$. We must show that $n + 1 < 2^{n+1}$. If we start with the inductive hypothesis and add 1 to each side, we obtain $n + 1 < 2^n + 1$. But 1 is certainly less than or equal to 2^n, so we have $n + 1 < 2^n + 1 \leq 2^n + 2^n = 2^{n+1}$, as desired. ◆

EXAMPLE 8. Let a be a real number, and let r be a real number different from 1. Prove that the sum of the finite geometric series starting at a with ratio r is given by

$$a + ar + ar^2 + ar^3 + \cdots + ar^{n-1} = a \cdot \frac{1 - r^n}{1 - r}$$

for all $n \geq 1$.

Solution. We prove this by induction on n. The base case $n = 1$ is the identity

$$a = a \cdot \frac{1 - r}{1 - r}.$$

Next we assume the inductive hypothesis

$$a + ar + ar^2 + ar^3 + \cdots + ar^{n-1} = a \cdot \frac{1 - r^n}{1 - r}$$

and want to derive the same formula for $n + 1$, namely,

$$a + ar + ar^2 + ar^3 + \cdots + ar^{n-1} + ar^n = a \cdot \frac{1 - r^{n+1}}{1 - r}.$$

The left-hand side of this last display is the same as the left-hand side of the preceding one, plus ar^n. Thus it equals, by the inductive hypothesis,

$$a \cdot \frac{1 - r^n}{1 - r} + ar^n.$$

A little straightforward algebraic manipulation with this expression easily transforms it into the right-hand side we are after. ◆

EXAMPLE 9. Prove that bubble sort (Algorithm 3 in Section 4.3) works correctly.

Solution. The hard part here (and in most proofs of program correctness) is to choose the correct statement to prove by induction. Let $S(i)$ be the statement that "after the ith pass through the outer loop of the procedure, the list still contains the original numbers, and the largest i numbers are correctly placed in the last i locations in the list." If we can prove this statement for all relevant values of i (namely, $0 \le i \le n - 1$), then we will be done, because for $i = n - 1$ the statement is equivalent to the statement that the algorithm correctly sorted the entire list (if the last $n - 1$ numbers are in their correct places, the first number must be in its correct place, as well). Thus we will prove the statement in quotation marks, using induction on i.

For $i = 0$, the statement is vacuous, so there is nothing to prove in the base case. Thus let us assume the inductive hypothesis, that after the $(i - 1)$th pass through the outer loop of the procedure, the list still contains the same numbers, and the largest $i - 1$ numbers are correctly placed in the last $i - 1$ locations in the list; in other words, $a_{n-i+2}, a_{n-i+3}, \ldots, a_n$ are the largest $i - 1$ numbers in the list, and they occupy their correct positions at this point (i.e., $a_{n-i+2} \le a_{n-i+3} \le \cdots \le a_n$). First, since the inner loop does not touch any number in the list beyond a_{n-i+1}, the largest $i - 1$ numbers remain in their correct positions. Second, the overall contents of the list are unaltered in the loop, since we only tamper with the list by interchanging pairs of numbers. Thus the only thing we have to show is that the ith largest number in the list, which is somewhere in $\{a_1, a_2, \ldots, a_{n-i+1}\}$, ends up as a_{n-i+1} after this pass is completed. That this is true is not hard to see by looking at the inner loop. For $j = 1, 2, \ldots, n - i$, after the jth pass, the largest number among $(a_1, a_2, \ldots, a_{j+1})$ is forced to be a_{j+1} by the **if**...**then** statement. Therefore, after the inner loop is completed, a_{n-i+1} is indeed the ith largest number in the list, and our proof is complete. ◆

SUMMARY OF DEFINITIONS

base case in a proof of $\forall n\colon S(n)$ by mathematical induction: the proof that $S(n)$ is true for the smallest value (or values) of n (usually, $n = 0$ and/or $n = 1$).

inductive hypothesis in a proof of $\forall n\colon S(n)$ by mathematical induction: the assumption, made in carrying out the inductive step (deriving $S(k)$ from this assumption), that $S(i)$ is true for all $i < k$ (strong form) or that $S(k - 1)$ is true (weak form).

inductive step in a proof of $\forall n\colon S(n)$ by mathematical induction: the proof, for an arbitrary k greater than the values covered in the base case, that $S(k)$ follows from $\forall i < k\colon S(i)$ (strong form) or from $S(k - 1)$ (weak form).

principle of mathematical induction: the technique of proving a proposition $\forall n\colon S(n)$ (where n ranges over the natural numbers or some similar set) by establishing the truth of $S(0)$ (or other base cases) explicitly and then showing that for all larger k, $S(k)$ follows from $S(i)$ for $i < k$ (the inductive step); see Theorems 2 and 3 for precise statements.

EXERCISES

1. Draw a covering of an 8 by 8 checkerboard with the lower left-hand corner deleted, using L-shaped pieces. [*Hint*: Use the proof given in Example 1.]

2. Use mathematical induction to prove that for all $n \geq 1$,

$$1 + 2 + 3 + \cdots + n = \frac{n(n + 1)}{2}.$$

3. Use mathematical induction to prove that for all $n \geq 2$,

$$1 \cdot 2 + 2 \cdot 3 + 3 \cdot 4 + \cdots + (n - 1) \cdot n = \frac{(n - 1)n(n + 1)}{3}.$$

4. Use mathematical induction to prove that for all $n \geq 1$,

$$1^3 + 2^3 + 3^3 + \cdots + n^3 = \left[\frac{n(n + 1)}{2}\right]^2.$$

5. Write the bottom-up form of the principle of mathematical induction totally in logical and mathematical symbols.

6. Prove that $n! > n^2$ for $n \geq 4$.

7. Prove that the following inequality holds for all $n \geq 1$.

$$\frac{1}{2} \cdot \frac{3}{4} \cdot \frac{5}{6} \cdots \frac{2n - 1}{2n} \geq \frac{1}{2n}$$

8. Determine with proof the value of $1+3+5+\cdots+999$. Do not perform a lot of arithmetic or use a computer. [*Hint*: Prove a general statement involving n by mathematical induction; then plug in a particular value for n.]

9. Use mathematical induction to prove that for all $n \geq 1$,

$$\sum_{j=1}^{n} j \cdot j! = (n+1)! - 1.$$

10. Define a sequence recursively by $a_1 = 7$, and $a_n = \sqrt{2 + a_{n-1}}$ for $n > 1$. Thus the sequence begins $7, 3, \sqrt{5}, \ldots$ [*Hint*: Use mathematical induction.]
    (a) Show that $a_n \leq 7$ for all n.
    (b) Show that $a_n \geq 2$ for all n.
    (c) Show that $a_n < 2.1$ for all $n \geq 4$.
    (d) Show that the sequence is decreasing (i.e., that $a_{n+1} < a_n$ for all n).

11. Prove that Algorithm 3 in Section 5.2 requires $f(n) - 1$ addition operations to compute $f(n)$.

12. Consider the sequence defined recursively by $a_n = a_{n-1} + 2a_{n-2}$ for $n \geq 2$, with each of the starting conditions given below. In each case compute enough terms of the sequence to enable you to guess an explicit formula for a_n in terms of n, and then prove, using mathematical induction, that your formula is correct.
    (a) $a_0 = 1$, $a_1 = 2$.
    (b) $a_0 = 2$, $a_1 = 1$.

13. Determine with proof the exact value of the following rational number.

$$\frac{1}{1 \cdot 3} + \frac{1}{3 \cdot 5} + \frac{1}{5 \cdot 7} + \cdots + \frac{1}{(99999)(100001)}.$$

14. Prove that $n^3 < 3^n$ for all $n \geq 4$. [*Hint*: Write $(k+1)^3 = [k^3] + [3k^2] + [3k+1]$, and note that each of the terms in brackets is at most k^3.]

15. Determine, with proof, which natural numbers n satisfy $2^n > n^2$.

16. Prove that for all natural numbers n, the following quantities are divisible by 7.
    (a) $8^n - 1$. [*Hint*: $8^{k+1} - 1 = 8^{k+1} - 8^k + 8^k - 1$.]
    (b) $11^n - 4^n$.

17. Determine what is wrong with each of the following arguments, which purport to prove false propositions by mathematical induction. Be specific.
    (a) Proposition: $1 + 2 + 3 + \cdots + n = (n^2 + n + 4)/2$ for all $n \geq 1$.
        Proof: by mathematical induction. Assume the statement for k. Then for $k+1$ we have $1+2+3+\cdots+(k+1) = [1+2+3+\cdots+k]+(k+1) =$

$[(k^2 + k + 4)/2] + (k + 1)$ (by the inductive hypothesis), which equals (via a little algebra) $((k + 1)^2 + (k + 1) + 4)/2$, as desired.

(b)   Proposition: All odd numbers greater than or equal to 3 are prime.

Proof: by induction. The base case is 3, and 3 is certainly prime. For the inductive step, we want to show that if 3 is prime, then 5, the next odd number, is also prime. Here both the hypothesis and the conclusion are true, so by the truth table definition of "implies," this implication is correct. Therefore, by mathematical induction we have shown that all odd numbers greater than or equal to 3 are prime.

(c)   Proposition: If A is any nonempty finite set of positive integers, then either every element of A is even, or every element of A is odd. (Note that this is not claiming the trivial tautology that every element of A is either even or odd; it is saying that all the elements of A have the *same* parity.)

Proof: by induction on the cardinality n of the set. If $n = 1$, then A has just one element, which is either even or odd, so the statement is trivially true. Assume the statement for all sets of cardinality $k - 1$. Let A be a set of cardinality k, say $A = \{a_1, a_2, \ldots, a_k\}$. Look first at the subset of A obtained by discarding a_k; this set is $B = \{a_1, a_2, \ldots, a_{k-1}\}$. It has $k - 1$ elements, so by the inductive hypothesis either all its elements are even or all are odd. Without loss of generality, suppose that all of them are even. Next look at the subset of A obtained by discarding a_1; this set is $C = \{a_2, a_3, \ldots, a_k\}$. Again it has $k - 1$ elements, and so again by the inductive hypothesis, either all are even or all are odd. But we already assumed that all the elements of B were even, so in particular a_2 is even. Therefore, since $a_2 \in C$, all the elements of C must be even, as well. Hence we have shown that *all* the elements $a_1, a_2, \ldots, a_n$ are even, and our proof by induction is complete.

(d)   Proposition: Let $a > 0$ be a real number and n a natural number; then $a^n = 1$.

Proof: by induction on n. The base case ($a^0 = 1$) is true by definition of exponentiation. So assume the inductive hypothesis that $a^i = 1$ for all $i < k$. Now by the rules of exponents we have $a^k = a^{2(k-1)-(k-2)} = a^{k-1}a^{k-1}/a^{k-2}$. But by the inductive hypothesis, each of the three terms in the last expression is equal to 1, so $a^k = 1 \cdot 1/1 = 1$, and our proof is complete.

(e)   Proposition: For every positive integer n, $n^2 + n$ is odd.

Proof: by induction on n. The base case is true since $n = 1$ is odd. Assume the inductive hypothesis that $k^2 + k$ is odd. We want to show that $(k + 1)^2 + (k + 1)$ is odd. By algebra we have $(k + 1)^2 + (k + 1) = k^2 + 2k + 1 + k + 1 = (k^2 + k) + 2(k + 1)$. Now $k^2 + k$ is odd by the inductive hypothesis, and $2(k + 1)$ is even by definition, so their sum is odd (see Exercise 2c in Section 1.3). This completes the proof.

18. Restate the top-down form of the principle of mathematical induction (Theorem 2) using $k + 1$ as the new value of n under consideration in the inductive step rather than k.

19. Restate the bottom-up form of the principle of mathematical induction (Theorem 3) using k as the new value of n under consideration in the inductive step rather than $k + 1$.

20. Restate both forms of the principle of mathematical induction (Theorems 2 and 3) using n in place of k throughout. (Many authors of books and articles in mathematics and computer science use this notation.)

21. State a top-down form of the principle of mathematical induction that applies to a situation in which special arguments are needed to establish the truth of the statement for $n = 0, 1, 2$ (i.e., the base case consists of these three values of n).

22. Explain how the base case in the top-down form of the principle of mathematical induction is really just a special case of the inductive step with the restriction $k > 1$ removed. Thus it would be equally valid (perhaps more elegant, but not as informative or user friendly) to state Theorem 2 with just one clause.

23. Show that the 1000th term in the Fibonacci sequence has more than 170 digits (in its base 10 numeral). [*Hint*: Use the result of Example 6 and the base 10 logarithm.]

24. Let $\{f_n\}$ be the Fibonacci sequence.
    (a)   Show that $f_n < \left(\frac{5}{3}\right)^n$ for every $n \geq 1$.
    (b)   Find the smallest number n_0 for which $f_{n_0} > \left(\frac{8}{5}\right)^{n_0}$, and then prove that $f_n > \left(\frac{8}{5}\right)^n$ for all $n \geq n_0$. [*Hint*: A computer will come in handy here; n_0 is not too small.]

25. Give a slightly different proof of the claim in Example 5, using day of birth, rather than length, as the variable to induct on.

26. Prove that the number of variable names and unsigned integers in a FPAE (see Example 5) is always less than or equal to one more than the number of operator symbols. (Each object is counted as many times as it occurs in the FPAE.)

27. Give an argument to justify Theorem 3, similar to the argument used to justify Theorem 2.

28. Prove the following facts about the Ackermann function (defined in Section 5.1).
    (a) $\varphi(x, n, 2) = x^n$ for all x and n.
    (b) $\varphi(x, n, 3) = x^{x^{\cdot^{\cdot^{\cdot^{x}}}}}$ (with n x's in the exponent), for all x and n.

29. Let A be Robinson's version of the Ackermann function, defined in Exercise 17 of Section 5.1. Find nice algebraic expressions for each of the following, and prove your answers correct (by induction).
    (a) $A(n, 1)$.
    (b) $A(n, 2)$.
    (c) $A(n, 3)$.

30. Consider the checkerboard problem discussed in Example 1.
    (a) Show that there is no solution for $b = 3$, no matter where the missing square is.
    (b) Generalize from part (a) and find an infinite collection of values of b for which there is no solution.
    (c) Find a solution when $b = 5$ and the missing square is in a corner.
    (d) Prove that there is a solution when $b = 80$.
    (e) Discuss the truth or falsity of this statement: Our discussion in Example 1 demonstrated that if a $2b$ by $2b$ checkerboard with one square deleted can be covered with L-shaped pieces, then so can a b by b board with one square deleted.

31. Discover and prove a theorem that explicitly gives the parity (oddness or evenness) of f_n as a function of n, where f_n is the nth term in the Fibonacci sequence.

32. Find and prove nice formulas for the following expressions involving the terms of the Fibonacci sequence, in terms of other terms in the sequence.
    (a) $f_0 + f_1 + f_2 + \cdots + f_n$.
    (b) $f_1 + f_3 + f_5 + \cdots + f_{2n-1}$.

33. Prove the following identities for the terms of the Fibonacci sequence.
    (a) $f_{n+2}^2 - f_{n+1}^2 = f_n f_{n+3}$.
    (b) $f_{n+1}^2 = f_n f_{n+2} + (-1)^{n+1}$.

34. In calculus, the derivative of a function f is defined by

$$\frac{d}{dx} f(x) = \lim_{h \to 0} \frac{f(x+h) - f(x)}{h}.$$

One of the early theorems (which is proved without too much difficulty, using the definition and some theorems about limits) is the product rule, which states that the derivative of the product of differentiable functions f and g is given by

$$\frac{d}{dx}(f(x) \cdot g(x)) = f(x) \cdot \left(\frac{d}{dx} g(x) \right) + g(x) \cdot \left(\frac{d}{dx} f(x) \right).$$

Give a rigorous proof that the derivative of x^n is nx^{n-1}, using mathematical induction, as follows.

(a) Prove the base cases $n = 0$ and $n = 1$ directly from the definition.

(b) Prove the inductive step using the product rule.

35. Prove the following statements about the parity of a sum. [*Hint*: Prove these by induction on the number of numbers in the sum, using the results of Exercise 2 in Section 1.3.]

(a) The sum of any number of even numbers is even.

(b) The sum of an even number of odd numbers is even.

(c) The sum of an odd number of odd numbers is odd.

36. Explain what is going on in the inductive step in the proof that $|\mathcal{P}(S)| = 2^{|S|}$ (Example 7), using the example $S = \{a, b, c\}$ and $x = a$.

37. Prove that $n^3 - 4n + 6$ is divisible by 3 for all natural numbers n

(a) Using mathematical induction.

(b) Without using mathematical induction.

38. Prove that for every natural number n, $\dfrac{n^3}{3} + \dfrac{n^5}{5} + \dfrac{7n}{15}$ is an integer.

Challenging Exercises

39. A game is played in the following way. Initially, one pile with n stones lies on the table between the two players. The players take turns until no further moves are possible. At his or her turn to play, a player must divide a pile containing at least two stones into exactly two piles, each with at least one stone in it. If a player cannot play, because the table contains only piles with 1 stone each, then that player loses.

(a) Prove by induction that the number of plays is equal to $n-1$, regardless of how the players choose to divide the piles. [*Hint*: Be sure to formulate precisely the statement S that you are trying to prove. It is not sufficient to have S deal only with the beginning of the game; it must take into account how many piles there are, as well.]

(b) Give a noninductive proof of the result given in part (a), based on viewing the stones as arranged in a row, with a play consisting of placing a stick between two adjacent stones.

(c) Use the result of part (a) to determine who will win the game (first player to play, or second player), as a function of n. (The analysis in this problem is typical of the analyses done in one type of game theory.)

40. A game is played in the following way. Initially, one pile with n stones lies on the table between the two players. The players take turns until no further moves are possible. At his or her turn to play, a player must remove one, two, or three stones from the pile. If a player cannot play, because the pile contains no more stones, then that player loses. Analyze this game, figuring out the winning strategies, and deciding who will win, first player or second (assuming that both players adopt their best strategies), as a function of n.

41. The reverse of a string was defined recursively in Example 7 of Section 5.1. Give an inductive proof that for any strings u and v, $(uv)^R = (v^R)(u^R)$. [*Hint*: Induct on the length of v.]

42. Let A be Robinson's version of the Ackermann function, defined in Exercise 17 in Section 5.1. Find a nice algebraic expression for $A(n, 4)$ and prove your answer correct.

43. Give a rigorous proof of Theorem 1 of Section 5.2. Be careful to state very explicitly exactly what the statement $S(n)$ is that you are proving by induction.

44. Derive Theorem 2 from Theorem 3. [*Hint*: Let $S'(n) = S(1) \wedge S(2) \wedge \cdots \wedge S(n)$.]

Exploratory Exercises

45. Use the result of Exercise 9 to show how to define a numeration system based on factorials, as opposed to powers of a fixed base.

46. A sequence of strings of positive integers is defined as follows. Let s_1 be 1. Given s_n, we obtain s_{n+1} in the following manner. We read s_n (aloud) from left to right in the manner "k_1 copies of d_1, followed by k_2 copies of $d_2, \ldots$," where $d_1 \neq d_2$, $d_2 \neq d_3$, and so on. Then we let s_{n+1} be the string $k_1 d_1 k_2 d_2 \ldots$. Thus since s_1 contains "one 1," we set $s_2 = 11$. Then since s_2 contains "two 1's," we set $s_3 = 21$. Next, since s_3 contains "one 2, followed by one 1," we set $s_4 = 1211$.
    (a) Write down the first 10 terms of the sequence.
    (b) Prove that the length of s_n is always even, for $n > 1$.
    (c) Prove that the only numbers ever appearing in s_n are 1, 2, and 3.
    (d) Try to discover and prove other facts about the sequence.

47. Find out more about covering checkerboards by consulting Chu and Johnsonbaugh [45] and Singmaster [273].

48. We have found formulas in terms of n for $\sum_{i=1}^{n} i^k$ for $k = 1, 2, 3$. Find out more about the general problem of expressing this sum as a function of n for arbitrary natural numbers k by consulting Acu [1] and Kelly [168].

49. Consult Guy [135] for a fascinating collection of patterns that seem to appear in sequences of natural numbers defined in myriad ways. Some of these patterns lead to theorems, and some are mere coincidences.

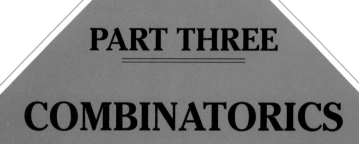

PART THREE

COMBINATORICS

6

ELEMENTARY
COUNTING
TECHNIQUES

Everyone learned early in life to count, that is, to determine the size (cardinality) of a finite set by enumerating its elements one by one. Sometimes it is necessary to determine the size of a set without explicitly counting its elements. For example, to determine the number of cents in your bank account after you write a check for $34.97 when you had a balance of $213.48, you do not physically count pennies, but rather perform some arithmetical calculations based on the data in the problem. In this chapter we look at the problem of determining the size, and in some cases also the structure, of various sets that arise in such diverse applications as computer program analysis, games, probability, and mathematics itself. This branch of discrete mathematics is known as combinatorics, and the problems and techniques we discuss are called combinatorial.

We discuss the basic principles in Section 6.1; we will see that these straightforward ideas (and some clever reasoning) will allow us to solve very complicated combinatorial problems. In Section 6.2 we introduce the notions of permutations and combinations, which bring some simplification and unification to the solutions of these problems. Some harder counting problems and techniques for solving them are discussed in Section 6.3. Finally, in Section 6.4 we show how an utterly obvious idea called the pigeonhole principle can lead to some very interesting theorems.

Good general references on combinatorics for this chapter and the next are Bose and Manvel [29], Brualdi [31], Even [78], Hall [136], Hu

[160], Liu [198], Lovász [199], Niven [223], Pólya, Tarjan, and Woods [237], Reingold, Nievergelt, and Deo [243], and Riordan [251], F. S. Roberts [255], Tomescu [304], and A. Tucker [306].

SECTION 6.1
FUNDAMENTAL PRINCIPLES OF COUNTING

One, two—buckle your shoe,
Three, four—knock on the door.

—Mother Goose

To get a feeling for the type of problem we will want to solve, consider the following brain-teaser: Find three integers between 10 and 30, inclusive, whose sum is 70 and whose squares together contain all the digits from 1 to 9, inclusive. Let us suppose that we want to try to find a solution by trial and error (perhaps with the aid of a short computer program). We could plan to generate all triples of integers in the required range (call them x, y, and z) whose sum is 70 (i.e., so that $x + y + z = 70$), and for each such triple (x, y, z) compute the squares $(x^2, y^2,$ and $z^2)$ and check to see if every digit from 1 to 9 appeared exactly once among these three three-digit squares. Now before attempting this (either by hand or by computer), we would do well to figure out *how many* such triples we would need to consider; that is, we would want to know the cardinality of the set

$$\{ (x, y, z) \mid 10 \leq x, y, z \leq 30 \ \wedge \ x + y + z = 70 \}.$$

If, for example, it turns out that there are 2500 such triples, it would be out of the question to attempt to test them all by hand, but perhaps quite feasible by computer. If there were 10^{100} such triples, then even the fastest modern-day computer could not possibly try them all. Furthermore, some preliminary analysis would probably allow us to cut down on the number of triples we need to test. For example, we might realize that the search for a solution could be restricted to triples (x, y, z) for which $x < y < z$; this cuts the checking time by a factor of more than 6. The moral of this example (which appears as Exercise 26 in Section 6.3) is that rather than plowing straight into an exhaustive search for a solution, we should do a little preliminary mathematical analysis to focus our search and estimate its duration. Even once we had decided to go ahead with the search for a solution, we would still need to figure out how to list the set of triples we desire in a systematic manner.

Combinatorics is the study of such counting and listing problems. In this section we present the most fundamental tools for counting.

THE MULTIPLICATION PRINCIPLE
AND THE ADDITION PRINCIPLE

The answer to the question "How many elements are in a given finite set?" can in many cases be obtained by applying one or both of two very simple principles: the **multiplication principle** and the **addition principle**. We state these two basic principles first in terms of sets, in which setting their truth should be clear. Then we restate them in terms of "tasks," the setting in which we will usually apply them to counting problems. In addition, we introduce one other principle, the **overcounting principle**, which must often be combined with the two basic principles to obtain the correct answers to counting problems.

Recall in what follows that $|X|$ denotes the cardinality of a set X (i.e., the number of elements in X).

THEOREM 1. The Multiplication Principle—Set Formulation.   *If A and B are finite sets, then*

$$|A \times B| = |A| \cdot |B|.$$

In other words, the cardinality of the Cartesian product of two sets is the product of their cardinalities.

The validity of this principle should be clear. The Cartesian product of A and B consists of all ordered pairs (x, y) such that $x \in A$ and $y \in B$. There are $|A|$ possible values that the first coordinate, x, can have, and *for each of these*, there are $|B|$ pairs (x, y) in $A \times B$, one for each y in B. Indeed, if one were asked to *define* multiplication of natural numbers in terms of more fundamental concepts (such as sets), this principle could be taken as a definition.

EXAMPLE 1.   Let $A = \{2, 4\}$ and $B = \{1, 2, 5\}$. Then $|A| = 2$ and $|B| = 3$. Furthermore, $A \times B$ consists of the pairs

$$(2, 1), (2, 2), (2, 5)$$

$$(4, 1), (4, 2), (4, 5)$$

and has cardinality $2 \cdot 3 = 6$.   ◆

THEOREM 2. The Addition Principle—Set Formulation.   *If A and B are disjoint finite sets, then*

$$|A \cup B| = |A| + |B|.$$

In other words, the cardinality of the union of two *disjoint* sets is the sum of their cardinalities. Note that although the two sets were not required to be disjoint for the multiplication principle, they must be disjoint in order that the addition principle be correct. We will use the overcounting principle (and, more generally, the inclusion–exclusion principle in Section 7.4) to handle the cases, which occur more often than not, in which the sets of interest are not disjoint.

Again, this principle is really nothing more than a definition of addition, as we learned it as a young child ("If Debbie has three apples and Billy gives her two more apples, how many apples does Debbie have in all?").

EXAMPLE 2. Let A be the set of prime numbers between 10 and 30, and let B be the set of multiples of 7 in the same interval. It is easy to check that $|A| = 6$ and $|B| = 3$. Since no multiple of 7 greater than 7 is prime, A and B are disjoint sets. Thus the cardinality of $A \cup B$ (the set consisting of numbers between 10 and 30 that are either prime or multiples of 7) is $6 + 3 = 9$. ◆

Both of these principles have obvious extensions to more than two sets, the details of which we leave to the reader (Exercises 5 and 10). We will use this more general setting in our second, and more applicable, formulation of the two principles, to which we now turn.

In most of our combinatorial problems, the set whose cardinality we wish to find will be *a set of ways to perform a certain task*. For example, suppose that we want to know how many possible license plates there are, consisting of either three letters followed by three digits, or three digits followed by three letters (a question of interest to a state's motor vehicle department). Then we need to find the cardinality of the set of ways to choose six symbols satisfying the stated conditions. In other words, we are asking: In how many ways can we perform the task of choosing symbols for a license plate?

We begin with a reformulation of the multiplication principle.

THEOREM 1′. The Multiplication Principle—Task Formulation. *Suppose that a task consists of a sequence of steps or subtasks, say subtask 1, subtask 2, ..., subtask n. In other words, to perform the task one must first perform subtask 1, then perform subtask 2, ..., then perform subtask n. Further suppose that subtask 1 can be performed in t_1 ways; that subtask 2 can be performed in t_2 ways after subtask 1 has been performed; that subtask 3 can be performed in t_3 ways after subtask 1 and subtask 2 have been performed; and in general that for each i from 1 to n, subtask i can be performed in t_i ways after the subtasks numbered 1, 2, ..., i − 1 have been performed. Then the number of ways in which the entire task can be performed is the product $t_1 \cdot t_2 \cdots t_n$.*

Although this theorem is a bit messy to state, it is easily understood by means of an example. Its truth should become evident as we look at the example, and we will not attempt to give a formal proof.

EXAMPLE 3. Consider the task performed by a man getting dressed in the morning and having to decide what to wear. Suppose that he owns a green shirt, a red shirt, and a yellow shirt; brown trousers and charcoal trousers; and a green tie, a red tie, and a yellow tie. He wants to don a shirt, trousers, and a tie, and his sense of aesthetics dictates that the tie have a color different from that of the shirt. In how many ways can he get dressed; that is, how many different outfits can he construct subject to these constraints?

Solution. The task here can be viewed as a sequence of three subtasks. Subtask 1 is choosing (and putting on) a shirt, and since there are three shirts, this subtask can be performed in three ways: $t_1 = 3$. Subtask 2 consists of choosing trousers. After subtask 1 has been performed, there are two trousers available (since there are no color restrictions between shirt and trousers), so $t_2 = 2$. Finally, subtask 3 consists of choosing a tie. Although there are three ties, t_3 is not equal to 3, since subtask 3 cannot be performed in three ways *after* the first two subtasks have been performed. In fact, since the choice of shirt always rules out one possible tie (because of the man's self-imposed aesthetic rule), the number of ways of choosing a tie after a shirt and trousers have already been chosen is only 2; thus $t_3 = 2$.

Therefore, by the multiplication principle, the answer to the problem (i.e., the number of outfits that the man can wear) is $3 \cdot 2 \cdot 2 = 12$. In effect the problem was to construct a 3-tuple (s, p, t), where s is a shirt color, p is a trousers color, and t is a tie color, subject to the stated conditions. There are three choices for the first coordinate s (green, red, and yellow). For *each* choice for s there are two choices for the second coordinate p (brown and charcoal). Thus there are $3 \cdot 2 = 6$ choices for the first two coordinates of the 3-tuple. Finally, for each of these six choices, there are two choices for t (two of green, red, and yellow, depending on what s is). Thus there are $6 \cdot 2 = 12$ triples in all.

These 12 choices can be listed explicitly as follows:

	Shirt	Trousers	Tie
1.	Green	Brown	Red
2.	Green	Brown	Yellow
3.	Green	Charcoal	Red
4.	Green	Charcoal	Yellow
5.	Red	Brown	Green
6.	Red	Brown	Yellow
7.	Red	Charcoal	Green
8.	Red	Charcoal	Yellow
9.	Yellow	Brown	Green
10.	Yellow	Brown	Red
11.	Yellow	Charcoal	Green
12.	Yellow	Charcoal	Red

The order in which the 12 choices have been listed is the one that naturally follows from the analysis above, using the order of colors stated in the problem. Finally, let us look at our solution in one more way: as a **tree diagram** (see Figure 6.1).

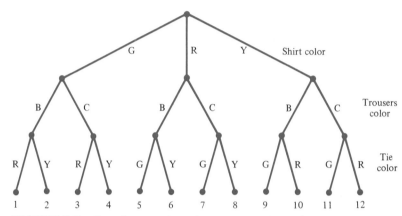

FIGURE 6.1 Tree diagram for solving the man's dressing problem.

As is customary in discrete mathematics, we draw our trees with their "roots" at the top and their "leaves" at the bottom. (In Chapter 9 we will formally define, and have a lot more to say about, trees.) Each branching of the tree corresponds to a choice made in one of the subtasks, and each path down the tree, from root to leaf, corresponds to one way to perform the entire task. For example, the leaf labeled 8 corresponds to choosing a red shirt, charcoal trousers, and a yellow tie. Since there is a unique path from the root to each of the leaves, the number of ways to perform the entire task is just the number of leaves. In this case, that number is 12. ◆

We now reformulate the second major elementary counting principle.

THEOREM 2′. The Addition Principle —Task Formulation. *Suppose that a task consists of performing exactly one subtask from among a collection of disjoint (mutually exclusive) subtasks, say subtask 1, subtask 2, . . . , subtask n. In other words, to perform the task one must either perform subtask 1, or perform subtask 2, . . . , or perform subtask n, and it is impossible to perform more than one of the subtasks simultaneously. Further suppose that subtask 1 can be performed in t_1 ways; that subtask 2 can be performed in t_2 ways; that subtask 3 can be performed in t_3 ways; and in general that for each i from 1 to n, subtask i can be performed in t_i ways. Then the number of ways in which the task can be performed is the sum $t_1 + t_2 + \cdots + t_n$.*

A simple example should clarify what this theorem is saying. Again we will not give a formal proof.

EXAMPLE 4. A woman needs to decide what to wear. She can wear a skirted suit with a white blouse, and she owns three skirted suits. Alternatively, she can wear one of her six dresses, or she can choose any of four slacks along with her only sweater. In how many different ways can she get dressed?

Solution. The task at hand (choosing an outfit to wear) consists of performing exactly one of three mutually exclusive subtasks (choosing a suit, choosing a dress, choosing a pair of slacks). There are three ways to perform the first subtask (she has three suits to choose from, $t_1 = 3$), six ways to perform the second (six dresses, $t_2 = 6$), and four ways to perform the third ($t_3 = 4$). Therefore, by the addition principle there are $3 + 6 + 4 = 13$ ways to perform the task (i.e., there are 13 different ways to get dressed). ◆

A convenient rule of thumb will usually help the reader to decide whether to apply the multiplication principle or the addition principle in a given situation. If the problem consists of performing *all* of several subtasks (subtask 1 *and* subtask 2 *and*...), then the multiplication principle is invoked. If the problem consists of performing only *one* of several mutually exclusive subtasks (subtask 1 *or* subtask 2 *or*...), then the addition principle is invoked. In brief,

"and" means multiply,
"or" means add.

In many situations, both principles need to be used, often more than once. It is usually best to organize your work from the top down, decomposing a problem into simpler problems.

EXAMPLE 5. Suppose that the woman in Example 4 buys a new sweater to wear with her slacks. In how many different ways can she dress now?

Solution. We draw a tree diagram to organize our work (see Figure 6.2). We first decompose the problem into three problems, using the addition principle. The number of ways she can get dressed is the number of ways that she can wear a suit, plus the number of ways that she can wear a dress, plus the number of ways that she can wear a sweater and slacks. In terms of our diagram, the number of leaves in the tree is the sum of the numbers of leaves in the three branches of the tree. There are three ways to dress if she decides to wear a suit, represented by the three leaves in the leftmost branch. Similarly, there are six ways to dress if she decides to wear a dress, shown in the middle branch. If she decides on sweater and slacks, she can first choose the sweater (two ways), and then for each of these choices (i.e., in each subbranch on the right), there are four ways for her to choose the slacks. By the multiplication principle, there are $2 \cdot 4 = 8$ ways to choose sweater and slacks, as shown by the eight leaves in the rightmost of the three branches of our tree. Thus the final answer is $3 + 6 + 8 = 17$, corresponding to the fact that there are 17 leaves in the tree; each path from the root (top) of the tree to a leaf uniquely specifies a way to perform the task of getting dressed. ◆

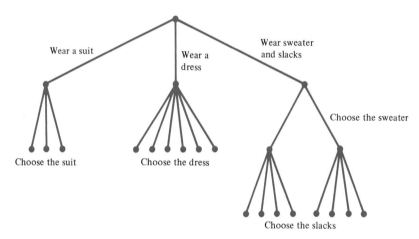

FIGURE 6.2   Tree diagram for counting the woman's outfits.

Our next example illustrates a seemingly universal rule for combinatorial problems: *Every combinatorial problem can be solved in at least two distinct ways*.

EXAMPLE 6.   Michigan license plates consist of either three letters followed by three digits (integers from 0 to 9), or three digits followed by three letters. Ignoring the fact that certain letter combinations are not actually used (lest they arouse the prurient interest of tailgating motorists), determine how many license plates are possible.

Solution 1.   The task of constructing a license plate consists of either the subtask of constructing a plate of the form *letter-letter-letter-digit-digit-digit* or the subtask of constructing a plate of the form *digit-digit-digit-letter-letter-letter*. Thus by the addition principle, we need to compute the number of license plates of the first form and add to that the number of license plates of the second form.

The problem of counting the number of license plates of the form *letter-letter-letter-digit-digit-digit* is easily solved by the multiplication principle. There are six subtasks, each of which must be performed. Let subtask i denote the subtask of deciding what symbol to put into the ith position on the license plate, for $i = 1, 2, 3, 4, 5, 6$. Since there are 26 possible letters to choose from and 10 possible digits, and since there are no restrictions that limit the choices of subsequent symbols after some symbols have been chosen, it is easy to see that the number of ways t_i to perform subtask i is given by $t_1 = t_2 = t_3 = 26$ and $t_4 = t_5 = t_6 = 10$. Therefore by the multiplication principle there are $26 \cdot 26 \cdot 26 \cdot 10 \cdot 10 \cdot 10 = 26^3 \cdot 10^3 = 17{,}576{,}000$ different license plates of the form *letter-letter-letter-digit-digit-digit*.

An exactly analogous argument shows that there are $10 \cdot 10 \cdot 10 \cdot 26 \cdot 26 \cdot 26 = 10^3 \cdot 26^3 = 17{,}576{,}000$ different license plates of the form *digit-digit-digit-letter-letter-letter*. Finally, by the addition principle, there are $17{,}576{,}000 + 17{,}576{,}000 = 35{,}152{,}000$ license plates possible altogether.

Solution 2. Here is a different approach to the same problem, one that uses only the multiplication principle. Let us view the construction of a license plate as a sequence of seven subtasks, performed one after the other. The first three subtasks are the choices of the three letters to appear on the plate, in order of appearance (left to right). The next three subtasks are the choices of the digits to appear on the plate, in order of appearance. The seventh subtask is the choice as to whether the letters are to precede the numbers or vice versa. Since $t_1 = t_2 = t_3 = 26$, $t_4 = t_5 = t_6 = 10$, and $t_7 = 2$, the number of ways to perform the entire task of constructing a license plate is $26 \cdot 26 \cdot 26 \cdot 10 \cdot 10 \cdot 10 \cdot 2 = 26^3 \cdot 10^3 \cdot 2 = 35{,}152{,}000$. ◆

Note that in the last example we obtained the same answer working the problem in two different ways; this lends quite a bit of believability to our answer. When solving combinatorial problems, the reader should try to obtain the same answer by two different methods: If the answers agree, they are probably correct; if they disagree, then clearly at least one of them is wrong, and further cogitation is in order.

Let us look at one more example, this one much subtler than before. In this example, as in the example of the man's dressing, subsequent choices are restricted by earlier ones; the rub here is that unless the solver is clever, the nature of the restriction is not clear.

EXAMPLE 7. Determine the number of odd integers between 1000 and 10,000 in whose base 10 representation no digit appears more than once. Thus we want to find the cardinality of a set which includes, for example, 8715, but does not include 873 (outside the range), 7347 (the digit 7 appears more than once), or 8346 (not odd). [We are going to make our solution unnecessarily long by including two false starts that practically everyone would be likely to make. It is just as worthwhile to learn from making and watching mistakes as it is to learn from seeing slick, correct solutions presented out of the blue. More important, we hope that the reader will realize (1) that the solution of mathematical problems, and combinatorial problems in particular, often begins with approaches that do not work, and (2) that solving combinatorial problems is much more an art (the art of being clever) than a science. In particular, *there is no algorithm for solving combinatorial problems in general*.]

First false start toward a solution.   The set we are counting consists of certain four-digit numbers. Let us see what choices are involved in specifying such a number. It appears that we need to apply the multiplication principle, where the four subtasks that must be performed one after the other are the choices of the four digits. Let subtask 1 be the choice of the leftmost (thousands') digit; subtask 2, the choice of the hundreds' digit; subtask 3, the choice of the tens' digit; and subtask 4, the choice of the ones' digit (the rightmost digit in the numeral). Since the numeral cannot begin with 0 (that would make it less than 1000), the number of ways to perform subtask 1 is $t_1 = 9$, there being nine nonzero digits. Now after subtask 1 has been performed, there are nine digits remaining as candidates for the hundreds' place (because of the prohibition against digits' appearing more than once); thus $t_2 = 9$. Similarly, after both of the first two subtasks have been performed, there are $10 - 2 = 8$ digits left for the tens' place, so $t_3 = 8$. ["We seem to be on a roll here—one more step and we'll be done; then we just multiply these four numbers to obtain the answer."] Now how many digits are available for the ones' place? Recalling that the number was required to be odd, we note that only five digits are possible: 1, 3, 5, 7, and 9. However, the earlier subtasks may have used some of these digits, reducing the number available and making t_4 less than 5. The problem is—and it is a fatal flaw in this solution—that we have no way of knowing how many of the first three digits were odd, and so we do not know whether t_4 should be 5 or something less than 5. In other words, we cannot write down t_4 at all, and so this attack on the problem has failed. ["Oh, we were so close!"]

Second false start toward a solution.   The difficulty with our first analysis was that when we came to the ones' place, we did not know which odd digits were left. Maybe the trick is to do the subtasks in the opposite order. Let subtask 1 be the choice of a digit for the ones' place; subtask 2, a choice of a digit for the tens' place; subtask 3, a choice of a digit for the hundreds' place; and finally subtask 4, a choice of a digit for the thousands' place. We see that the number of ways to perform subtask 1 is $t_1 = 5$, since there are five possible ending digits for an odd number. Then $t_2 = 9$ and $t_3 = 8$, since there are $10 - 1 = 9$ and $10 - 2 = 8$ digits left available for those positions, respectively. Unfortunately, we run into the same difficulty as before, this time with the thousands' place. Since our number has to be bigger than 1000, the leading digit cannot be a 0. Depending on whether or not a 0 was selected as one of the other three digits, there may be $9 - 2 = 7$ or $9 - 3 = 6$ digits left for the thousands' place. Thus we cannot determine whether $t_4 = 7$ or $t_4 = 6$. ["This is getting discouraging."]

Solution.   Suppose that we try the following order of subtasks: first choose the ones' digit, then the thousands', then the hundreds', then the tens'. As in our last attempt, $t_1 = 5$. Now t_2, the number of ways to choose the thousands' digit, must be $9 - 1 = 8$, since one of the nine nonzero digits was definitely used in the ones' place. Having passed this hurdle, we are practically home. Clearly, $t_3 = 8$, since two of the 10 available digits have been preempted; and similarly, $t_4 = 7$. Thus the solution to our problem, by the multiplication principle, is $5 \cdot 8 \cdot 8 \cdot 7 = 2240$. ◆

The reader should keep in mind exactly what we have done by solving the problem posed in Example 7. We were presented with a short description of a very specific, fairly large finite set. *Without actually writing down even one element of the set*, we were able to tell exactly how many elements it had. Someone with no knowledge of (or intuition to figure out) the counting principles we have used here would be hard pressed to come up with an answer, short of actually generating the set. These are powerful methods.

THE OVERCOUNTING PRINCIPLE

Many times when trying to apply the multiplication principle or addition principle to a counting problem, we count more than we should. We might include in our count some elements not in the set we are counting, or we might end up counting some elements more than once. Obviously, then, the answer we obtain will be wrong.

This apparent disadvantage can, however, be turned into an advantage. If our answer is too large by an amount k because we counted k items that we should not have counted, then we can simply subtract k to obtain the correct answer. If our answer is too large because we counted k of the elements twice when we wanted to count them only once, then again we can simply subtract k to obtain the correct answer. If our answer is too large by a factor m because we counted each element we wanted to count, not once, but m times, then we can simply divide by m to obtain the correct answer. A cynic might well call our next principle the "fudge rule": If the answer isn't right, make it right. Actually, for many problems we *purposely* overcount, and then correct the overcount, as the neatest way to obtain the answer. We formalize this discussion as a theorem.

THEOREM 3. The Overcounting Principle. *Let A be a finite set whose cardinality is to be computed.*

(a) *If $A \subseteq B$ and $|B| = n$ and $|B - A| = k$, then $|A| = n - k$.*
(b) *If n is the answer obtained when counting the number of elements in a set A, except that in this count exactly k of the elements were counted twice, then $|A| = n - k$.*
(c) *If n is the answer obtained when counting the number of elements in a set A, except that in this count each element was counted exactly m times, then $|A| = n/m$.*

By applying part (b), we obtain the following rule for finding the cardinality of the union of two sets that are not necessarily disjoint. It is a generalization of Theorem 1, and it in turn will be generalized to more than two sets in Section 7.4.

THEOREM 4. If $A = B \cup C$, then $|A| = |B| + |C| - |B \cap C|$.

Proof. To count the number of elements of A, we count the elements of B (they are all in A), and we count the elements of C (they are all in A), but then we note that we have counted the elements that were in both B and C (in other words, in $B \cap C$) twice. Since we only wanted to count each element once, we must subtract the overcount, that is, the number of elements in $B \cap C$. ∎

EXAMPLE 8. How many even integers are there between 1000 and 10,000 in whose base 10 representation no digit appears more than once?

Solution 1. First we solve this problem using a direct approach along the lines of Example 7, not using the overcounting principle. We need to consider two cases: A numeral of the desired type either ends with a 0 or it ends with an even digit other than 0. If the numeral ends with a 0, then by choosing the digits in the order *ones'-thousands'-hundreds'-tens'*, we find (by the multiplication principle) that there are $1 \cdot 9 \cdot 8 \cdot 7 = 504$ such numbers. If the numeral ends with an even digit other than 0, then, with the same order of choices, we see that there are $4 \cdot 8 \cdot 8 \cdot 7 = 1792$ different possibilities. Thus by the addition principle, there are $504 + 1792 = 2296$ numbers of the desired type.

Solution 2. We use the answer obtained in Example 7, together with part (a) of Theorem 3, to compute the correct answer. First let us count *all* the numbers between 1000 and 10,000 in whose base 10 representation no digit appears more than once. Using the multiplication principle and working from left to right, we see that there are $9 \cdot 9 \cdot 8 \cdot 7 = 4536$ such numbers. Then since we found in Example 7 that 2240 of them are odd, we know that the remaining $4536 - 2240 = 2296$ must be even, in agreement with Solution 1. ◆

EXAMPLE 9. How many bit strings of length $n \geq 2$ either begin or end with a 1?

Solution. A bit string that begins with a 1 can be constructed by choosing the bits one at a time, working from left to right. Since there are two choices for each bit other than the first, the multiplication principle tells us that there are $1 \cdot 2 \cdot 2 \cdots 2 = 2^{n-1}$ such strings. Similarly, there are $2 \cdot 2 \cdots 2 \cdot 1 = 2^{n-1}$ bit strings of length n that end with a 1. These two sets of strings are not disjoint, however, since a bit string can both begin and end with a 1, so that if we were simply to add 2^{n-1} and 2^{n-1}, we would have overcounted by the number of bit strings that both begin and end with a 1. There are $1 \cdot 2 \cdot 2 \cdots 2 \cdot 2 \cdot 1 = 2^{n-2}$ such strings. Thus by part (b) of Theorem 3 (or, equivalently, Theorem 4) we obtain the answer $2^{n-1} + 2^{n-1} - 2^{n-2} = 3 \cdot 2^{n-2}$. ◆

EXAMPLE 10. Dominoes for the children's game are a set of wooden rectangles, each consisting of two adjacent squares with some number of dots painted on each square. The number of dots on each square is between zero and six, inclusive, and the two squares on a domino *are* allowed to contain the same number of dots. How many dominoes are there in a standard set containing exactly one domino for each allowable dot pattern?

Solution. We will distinguish between those dominoes that contain two different numbers of dots and those that contain the same number of dots in both of their squares. There are clearly exactly seven of the latter type (two 0's, two 1's, ..., two 6's). To count the number of dominoes that contain two different numbers of dots on their squares, we note that to specify such a domino is to specify two distinct numbers from 0 to 6, inclusive. By the multiplication principle, there are $7 \cdot 6 = 42$ ways to make those choices: seven ways to choose the first number, and then six ways to choose the second, different, number after the first has been chosen. There are *not* 42 dominoes with different numbers of dots on their squares, however: We have overcounted. Indeed, every such domino has been counted twice—once when we specified the smaller number first and the larger number second, and again when we specified the larger number first and the smaller number second. For example, the domino with three dots and five dots was counted once by the sequence of choices "3, 5" and a second time by the sequence of choices "5, 3." By part (c) of Theorem 3, there are therefore $42/2 = 21$ dominoes with different numbers of dots on their two squares. Combining this with the seven "doubles" dominoes (using the addition principle), we obtain the answer $21 + 7 = 28$. ◆

EXAMPLE 11. A pair of distinct positive integers is called an "amicable pair" if the sum of all the proper divisors of the first is equal to the second, and the sum of all the proper divisors of the second is equal to the first. For example, 220 and 284 form an amicable pair, since the sum of all the proper divisors of 220 is $1 + 2 + 4 + 5 + 10 + 11 + 20 + 22 + 44 + 55 + 110 = 284$ and the sum of all the proper divisors of 284 is $1 + 2 + 4 + 71 + 142 = 220$. A naive algorithm for finding all such pairs within a given range would proceed as follows:

```
procedure amicable(n : positive integer)
    {this algorithm finds and prints all amicable pairs (i, j) with 1 ≤ i < j ≤ n}
        for i ← 1 to n − 1 do
            for j ← i + 1 to n do
                if (sum of proper divisors of i)  =  j
                    and (sum of proper divisors of j)  =  i
                        then print(i, j)
    return
```

Suppose that we want to estimate the number of steps required by this algorithm to print all the amicable pairs (i, j) with $1 \le i < j \le n$, so that we can tell whether running this algorithm is feasible for a particular n. We will compute the number of times the condition in the **if** clause is checked.

The number of times the condition is checked is equal to the number of pairs (i, j) with $1 \leq i < j \leq n$. To count this number it will be convenient to ignore temporarily the restriction that $i < j$ and simply require that i and j be distinct. Then by the multiplication principle, since there are n choices for the value of i, followed by $n - 1$ choices for the value of j once i has been chosen, there are $n(n - 1)$ such pairs. By symmetry, in exactly half of these pairs is $i < j$ (in the remaining half, $j < i$). Thus we have overcounted what we really wanted by a factor of 2. Therefore, the number of pairs (i, j) with $1 \leq i < j \leq n$, that is, the number of times the condition is checked, is equal to $n(n - 1)/2$. (We could also obtain this answer by an analysis similar to that given for bubble sort in Section 4.3.) ◆

AMBIGUITY

We end with a word of warning about combinatorial problems. It is often very difficult to state in English exactly what set the author of a problem wants the solver to count. For example, consider the following problem: "In how many ways can five men and three women be seated around a circular table so that no two women are seated next to each other?" Each of the following is a reasonable question to ask about this problem:

- Are the eight people to be considered distinct, or do we only care about the sexes of the people, so that, in effect, we are placing symbols M and W in a circle, rather than human beings?
- Do we care about the compass directions in which the people face, or are two arrangements to be considered identical if the entire party is rotated around the table? In other words, do we only care about who sits to the immediate right and to the immediate left of each person?
- Are right and left to be distinguished, or do we only care about who the two immediate neighbors of each person are?
- Are we just interested in the final arrangement around the table (what we would see in a snapshot taken of the party), or do we care about the temporal process of seating—who sits down first, who second, and so on (what we would see in a movie taken of the seating process)?

To avoid some of the ambiguity, let us agree on the following conventions, unless a problem explicitly states otherwise:

1. Objects that can be distinguished on sight (people, for example) are to be considered distinct; objects that appear identical (green balls, for example) are to be considered indistinguishable. We will use the words "identical" and "indistinguishable" interchangeably.
2. Circular arrangements are to be considered identical if one can be obtained from the other by a rotation around the circle.

3. Right and left, front and back, top and bottom, clockwise and counter-clockwise are considered different.
4. The time order in which a process is carried out is of no concern.

Under these conventions, the answer to the problem stated above is 1440, obtained as follows. Because of the convention on circular seating (2), we do not care where the first man (see convention 1) is seated: He has to be seated somewhere. Next we look at the *men* we encounter as we proceed clockwise (see convention 3) from the first man. The first such man we encounter can be any of the four other men; the second can be any of the other three; the third can be any of the other two; and the fourth is the only man remaining at that point. Thus by the multiplication principle there are $4 \cdot 3 \cdot 2 \cdot 1 = 24$ ways in which the men can be seated around the table, relative to each other. Now we must also seat the women (by convention 4, it is legitimate to assume that all the men sit down first). Since no two women may sit next to each other, the women must choose distinct gaps between men in which to sit. There are five such gaps. The first woman may choose any of the five gaps; after she has been seated, the second woman may sit in any of the four remaining gaps; and the third woman may sit in any of the three gaps then remaining. Thus there are $5 \cdot 4 \cdot 3 = 60$ ways for the women to be seated, after the relative circular order of the men has been determined. Finally, by the multiplication principle, there are $24 \cdot 60 = 1440$ ways to seat the entire ensemble.

The exercises that follow can be solved by using some combination of the three fundamental principles, common sense, cleverness, and, occasionally, ad hoc reasoning.

SUMMARY OF DEFINITIONS

addition principle: The number of ways to perform a task that consists of exactly one of a number of subtasks, no two of which can be performed simultaneously, is the sum of the number of ways that the subtasks can be performed (e.g., the number of ways to choose one item for lunch at a certain restaurant is the sum of the number of sandwiches on the menu, the number of salads on the menu, and the number of soups on the menu; no food is both a sandwich and a soup, for instance, simultaneously).

combinatorics: the branch of mathematics dealing with enumerating and counting various finite sets, especially sets of arrangements of objects or ways to perform tasks.

multiplication principle: The number of ways to perform a task that consists of a sequence of subtasks is the following product: the number of ways to perform the first subtask, times the number of ways to perform the second subtask after the first subtask has been performed, times the number of ways to perform the third subtask after the first two subtasks have been performed, ..., times the number of ways to perform the last subtask after all the other subtasks have been performed; if these numbers depend on the actual choices made at previous stages, the principle cannot be used, and ad hoc reasoning, perhaps with the aid of a tree diagram, should be tried (e.g., the number of four-digit natural numbers, the first digit of which is odd and all of whose digits are distinct, is $5 \times 9 \times 8 \times 7 = 2520$).

overcounting principle: If an answer computed for a combinatorial problem is too large because k items were counted that should not have been counted, subtract k to obtain the correct answer; if the answer is too large because k items were counted twice, subtract k; if the answer is too large because every item was counted m times, divide by m (e.g., $|A \cup B| = |A| + |B| - |A \cap B|$).

tree diagram: a picture that aids in the application of the addition and multiplication principles; each leaf of the tree (or, equivalently, each path from root to leaf) represents a way to perform an entire task.

EXERCISES

1. Let $X = \{a, b\}$ and $Y = \{a, b, c\}$.
   (a) Without listing any of its elements, determine the cardinality of $X \times Y$.
   (b) Write out explicitly the set $X \times Y$.

2. An identifier in one version of the programming language FORTRAN consists of one to six alphanumeric characters, the first of which is a letter. (An **alphanumeric character** is a letter from A to Z or a digit from 0 to 9. Assume that only uppercase letters are used.)
   (a) How many different identifiers of length exactly 3 are there?
   (b) How many different identifiers are there?

3. A pizza parlor offers the following toppings for its pizza: sausage, anchovies, onions, olives, mushrooms. Any combination of zero or more toppings may be ordered.
   (a) How many different pizzas can be ordered?
   (b) How many pizzas with at least two toppings can be ordered? [*Hint:* Use part (a) of Theorem 3.]

4. How many rows will there be in the truth table for a compound proposition involving five propositional variables, P, Q, R, S, and U?

5. State a correct generalization of Theorem 1 that applies to n sets rather than just two.

6. Determine the number of license plates that can be made according to each of the following rules. Assume that only capital letters are used.
   (a) The plate must consist of two letters followed by four digits.
   (b) The plate must consist of two letters followed by four digits or two digits followed by four letters.

(c)   The plate must consist of two letters followed by four digits or four digits followed by two letters.

(d)   The plate must consist of three letters followed by three digits, all preceded by an optional 1. (For example, CUP–885 and 1–VIN–002 are valid plates.)

7. Let $X = \{a, b, c, d\}$ and $Y = \{a, b, c, d, e\}$. [*Hint:* You may want to review Example 7 in Section 3.1.]

(a)   Write out three elements of the set of all functions from X to Y. Recall that a function is a set of ordered pairs.

(b)   How many different functions are there from X to Y?

(c)   In general, determine how many different functions there are from a set with k elements to a set with n elements. Note what happens if $k = 0$ or $n = 0$.

8. Let $|X| = 7$ and $|Y| = 10$.

(a)   How many different one-to-one functions are there from X to Y?

(b)   How many different functions from X to Y are there that are not one-to-one?

(c)   How many different one-to-one functions are there from Y to X?

(d)   How many different functions from Y to X are there that are not one-to-one?

9. A cook must prepare dinner every day for a week (seven days, starting with Sunday). The main course each day consists of one of the following: beef, chicken, lamb, fish. How many menus for the week are possible

(a)   If there are no restrictions on the main course?

(b)   If there must be fish on Friday?

(c)   If beef must be served at least once?

(d)   If beef must be served at least twice?

(e)   If the same food may not be served two days in a row?

(f)   If no food may be served if it was served on either of the previous two days?

(g)   If there must be fish on Friday and the same food may not be served two days in a row?

(h)   If the foods served on the first four days must all be different and the foods served on the last four days must all be different?

10. State a correct generalization of Theorem 2 that applies to n sets rather than just two.

11. Let A be the set of all propositions that use only the five propositional variables P, Q, R, S, and U, together with logical connectives. For example, $(P \wedge S) \rightarrow (U \vee \overline{Q} \rightarrow P)$ is in A.

(a)   What is the cardinality of A?

(b) What is the largest possible cardinality of a subset B of A such that no two elements of B are logically equivalent? [*Hint*: How many truth tables are possible? (See Exercise 4.)]

(c) Logical equivalence is an equivalence relation on the set A. What is the cardinality of the set of equivalence classes under this equivalence relation?

12. How many strings of length 6 are there consisting of letters (A to Z)
    (a) If there are no restrictions?
    (b) If no letter may appear more than once?
    (c) If the letter A must appear at least once?
    (d) If the letter A must appear at least once and no letter may appear more than once?
    (e) If the string must begin or end with the letter A?
    (f) If the string must begin or end with the letter A and no letter may appear more than once?

13. The original version of Michigan Lotto is a lottery in which a person may purchase an entry and choose six different numbers from 1 to 40, inclusive. After all entries have been made, the state (randomly) determines a winning set of six numbers. Any entrants who picked all six winning numbers share the grand prize. All entrants who correctly picked exactly five of the six winning numbers share the second prize. How many different choices of six numbers allow a person to share the second prize?

14. How many bit strings containing exactly eight 0's and twelve 1's either have all the 0's consecutive or have all the 1's consecutive? (For example, the string 00011111111111100000 is allowed.)

15. Consider strings of length n consisting of symbols from the set $\{a, b, c\}$.
    (a) How many such strings are there?
    (b) How many such strings are there with exactly two a's?
    (c) How many such strings are there with at least two a's?
    (d) How many such strings are there with exactly two a's and at least one of each of the other two letters?

16. Suppose that a pizza can have meat toppings and/or vegetable toppings. The meat toppings can be sausage, pepperoni, and hamburger; and the vegetable toppings can be mushrooms, peppers, onions, and olives. A pizza can have from zero to all seven of these toppings.
    (a) How many different pizzas can be ordered?
    (b) How many different pizzas contain no meat?
    (c) How many different pizzas contain at most one meat topping?

17. Consider the set of all strings of length 5 over the alphabet $\{a, b, c, d, e, f, g, h\}$.
    (a) How many such strings are there?
    (b) How many of these strings contain neither of the letters a or b?

(c) How many of these strings contain the letters a or b (or both)?

(d) How many of these strings contain the letter a but not the letter b?

(e) How many of these strings do not contain both the letter a and the letter b?

(f) How many of these strings contain both the letter a and the letter b?

(g) How many of these strings begin or end with an e?

(h) In how many of these strings does the letter c occur exactly once?

18. Suppose that there are n_1 flights every morning from Detroit to Washington, n_2 flights every afternoon from Detroit to Washington, n_3 flights every morning from Washington to Detroit, and n_4 flights every afternoon from Washington to Detroit. How many choices of a pair of flights are possible if a person must leave Detroit for Washington on Tuesday, return the following day, and spend at least one complete morning or one complete afternoon in Washington?

19. Of all the numbers between 10,000 and 99,999, inclusive, how many

(a) Do not contain the digit 5?

(b) Do contain the digit 5?

(c) Are odd and contain no digit more than once?

(d) Have no two consecutive digits the same?

20. A palindrome is a string that reads the same backward as forward. (For example, the words "madam" and "noon" are palindromes.) How many palindromes are there

(a) Of length 5 using letters (A to Z)?

(b) Of length 6 using letters (A to Z)?

(c) Of length n using symbols chosen from a set of k symbols? (Each symbol may be used as many times as desired.)

21. Six men and four women need to line up at a box office to buy concert tickets.

(a) In how many ways can this be done?

(b) In how many ways can this be done if all the men precede all the women?

(c) In how many ways can this be done if no two women may stand next to each other?

(d) In how many ways can this be done if no two men may stand next to each other?

22. Six men and four women need to sit down for dinner at a circular table.

(a) In how many ways can this be done?

(b) In how many ways can this be done if no two women may sit next to each other?

(c) In how many ways can this be done if no two men may sit next to each other?

23. How many bit strings of length $n \geq 2$ are there such that

(a) No two consecutive bits are the same?

    (b)   The first and last bit are not both 0?

    (c)   There is exactly one 1?

    (d)   There are exactly two 1's?

    (e)   Assuming that n is odd, there are more 0's than 1's?

24. A 10-member mathematics department must select one person to chair the discrete mathematics committee and two other people to co-chair the calculus committee.

    (a)   In how many ways can this be done?

    (b)   In how many ways can this be done if Professor Blue refuses to chair the discrete mathematics committee?

    (c)   In how many ways can this be done if Professor Green refuses to co-chair the calculus committee?

    (d)   In how many ways can this be done if Professor Blue refuses to chair the discrete mathematics committee and Professor Green refuses to co-chair the calculus committee?

    (e)   In how many ways can this be done if Professor Black and Professor White cannot work together as co-chairs?

    (f)   In how many ways can this be done if Professor Gray must be given one of the three jobs?

25. Determine the cardinality of the following sets in terms of $|A|$, $|B|$, and $|C|$. Assume that all the sets are finite and that B and C are disjoint.

    (a) $\mathcal{P}(A)$ (the power set of A). [*Hint*: What are the subtasks involved in specifying a subset of A?]

    (b) $\mathcal{P}(B \cup C)$.

    (c) $\mathcal{P}(B) \times \mathcal{P}(C)$.

    (d) $\{\, X \mid X \subseteq A \wedge |X| > 1 \,\}$.

26. How many ways are there for 15 boys and 15 girls at a dance to pair up into 15 heterosexual dance couples?

27. How many integers between 1 and 1500, inclusive, have a factor greater than 1 in common with the number 15? [*Hint*: Look at multiples of 3 and multiples of 5.]

28. Find a formula for the number of positive integer divisors of a positive integer in terms of its prime factorization. For example, the number $300 = 2^2 \cdot 3 \cdot 5^2$ has 18 divisors, namely 1, 2, 3, 4, 5, 6, 10, 12, 15, 20, 25, 30, 50, 60, 75, 100, 150, and 300. How can we compute this answer 18 without listing all the divisors?

Challenging Exercises

29. How many bit strings of length 20 either have all the 0's consecutive or have all the 1's consecutive? (For example, the string 00011111111111100000 is allowed.)

30. How many ways are there to make change for a dollar, using quarters, dimes, and nickels? [Hint: Consider the subtasks to be deciding how many quarters and how many dimes to include. Organize the counting with a tree diagram—although you will probably not draw the entire tree. This exercise requires more than just an application of Theorem $1'$, since the number of ways to do subtask $(i + 1)$ depends on the choice made in performing subtask i.]

31. How many sequences of the numbers from 1 to 10, inclusive, are there with the property that except for the first term in the sequence, whenever a number n appears in the sequence then either the number $n-1$ or the number $n+1$ appears earlier in the sequence? For example, the sequence $3, 4, 5, 2, 6, 1, 7, 8, 9, 10$ is allowed.

32. Determine by brute force, ad hoc reasoning the number of different sets of positive integers, whose elements sum to 12. For example, $\{12\}$ and $\{3, 4, 5\}$ are to be included (but $\{3, 3, 6\}$ is not since as a set its sum is 9).

33. Suppose that we want to distribute one big marble and five small marbles among three boxes, in such a way that each box contains at least one big marble or two small ones. All the marbles must be distributed. Determine the number of ways this can be done, under each of the following assumptions.
    (a) The boxes are indistinguishable and the small marbles are indistinguishable.
    (b) The boxes are distinct but the small marbles are indistinguishable.
    (c) The boxes are indistinguishable but the small marbles are distinct.
    (d) The boxes are distinct and the small marbles are distinct.

34. What is the sum of the 4536 numbers from 1000 to 10,000 that have all their digits distinct?

35. How many ways are there for a chemistry class of 30 students to split up simultaneously into 15 pairs for laboratory work?

Exploratory Exercises

36. Combinatorics deals with more than just counting. One important area is the theory of **designs**, which are useful in statistics, among other areas. A typical problem would be to fill an n by n matrix with the natural numbers from 1 to n, inclusive, using each number n times, in such a way that each number appears once in each row of the matrix and once in each column. A design of this type is called a **Latin square**. For example, the following is a Latin square of order 3:

$$\begin{bmatrix} 1 & 2 & 3 \\ 3 & 1 & 2 \\ 2 & 3 & 1 \end{bmatrix}.$$

(a) Construct two substantially different Latin squares of order 4 (and explain how "substantially different" should be defined).

(b) Find out about combinatorial designs by consulting Bogart [23], Gardner [108], Liu [198], and Ryser [263].

37. Find out more about how combinatorial problems can usually be solved in more than one way by consulting Golomb [117].

38. Combinatorial geometry deals with combinatorial questions in geometry, such as questions of what kinds of tiles can fit together to form patterns that cover the plane. Find out more about tilings by consulting Grünbaum and Shephard [131].

SECTION 6.2
PERMUTATIONS AND COMBINATIONS

If you don't play, then you can't win.

—State lottery advertisement

In this section we begin a more systematic study of combinatorial problems. Recall that the general problem we want to solve is this:

Given a description of a finite set, determine its cardinality.

The set in question will usually be a set of ways to make some choices. In one situation that arises frequently—the one we will begin to study in this section—the question is even more specific:

How many ways are there to choose k things from a collection of n things?

Even in this form the question is still quite ambiguous, for we are not told

- Whether the n things are to be considered distinct or indistinguishable, or whether perhaps there are several identical copies of distinct objects.
- Whether the selected items are to be considered as forming a set, without any specific order to the elements, or as arranged into an *ordered* collection.
- Whether repetition is allowed (i.e., whether items in the original collection can be chosen more than once).

In this section we assume that the n items in the collection are indeed *distinct*, and that repetition is *not* allowed. (We will look at many of the other situations in Section 6.3.) We will discuss both permutations, which are *ordered arrangements* of objects from the given collection, and combinations, which are *unordered selections* from (in other words, simply subsets of) the given collection.

PERMUTATIONS

We deal first with counting ordered selections. Let S be a set with cardinality $n \geq 0$ (i.e., an n-set), and let k be an integer such that $0 \leq k \leq n$. Then a **permutation** of k objects from S, sometimes called a **k-permutation** from S, is an ordered arrangement of k of the elements of S, with repetition not allowed. If $k = n$, then we call such an ordered arrangement simply a **permutation** of S. The *number* of different k-permutations of a set with n elements is denoted by $P(n, k)$.

EXAMPLE 1. There are twenty-four 3-permutations from the set $\{a, b, c, d\}$:

abc	acb	bac	bca	cab	cba
abd	adb	bad	bda	dab	dba
acd	adc	cad	cda	dac	dca
bcd	bdc	cbd	cdb	dbc	dcb

Thus $P(4, 3) = 24$. We could have computed $P(4, 3)$ without explicitly listing all the permutations, using the theory developed in Section 6.1. Indeed, to specify a 3-permutation from a set with four elements, we must first choose one of the four elements to be the first item in the ordered arrangement (and there are clearly four ways to do this), then one of the three remaining elements of the set to be the second item in the ordered arrangement (and there are clearly three ways to do this), and finally one of the two remaining elements of the set to be the third item in the ordered arrangement (and there are two ways to do this). Thus by the multiplication principle, $P(4, 3) = 4 \cdot 3 \cdot 2 = 24$. If, instead, we had asked for the number of permutations of the set $\{a, b, c, d\}$ (i.e., the number of 4-permutations of this 4-set), we would have found that there are still only 24, since once the first three items have been selected for the permutation, there is only one way to choose the fourth item. In symbols, $P(4, 4) = 24$. ◆

The argument given in this example can be extended to justify the following formula for computing the number of permutations in general.

THEOREM 1. *The number of k-permutations from an n-set, where $0 \leq k \leq n$, is given by*

$$P(n, k) = n(n - 1)(n - 2) \cdots (n - k + 1)$$

$$= \frac{n!}{(n - k)!}.$$

In particular, the number of permutations of an n-set, $n \geq 0$, is given by $P(n, n) = n!$.

Proof. The first equality follows directly from the multiplication principle, by the argument used in Example 1. Since there are k subtasks involved, there are k factors in the product, decreasing as shown because of the prohibition on repetition. (The reader should verify that the kth term in the product really is $n - k + 1$.) The second equality follows from the definition of factorial (see Sections 3.1 and 5.1). Its numerator is the product of all the integers from 1 through n; since the desired product is missing all the factors from 1 through $n - k$, we needed to divide by $(n - k)!$. Substituting n for k in either formula, we obtain the final assertion. Note that this theorem provides another motivation for defining $0! = 1$: It makes the formula correct when $k = n$. ■

In actually computing the number of permutations, the first equality in Theorem 1 is more useful than the second: We simply write down the decreasing product, starting with n and including k terms.

EXAMPLE 2. A permutation of the set $\{1, 2, 3, \ldots, n\}$ is called "graceful" if the absolute values of the $n - 1$ differences between successive numbers in the permutation are all distinct. For example, the permutation 41532 is graceful, since the absolute differences are $|4 - 1| = 3$, $|1 - 5| = 4$, $|5 - 3| = 2$, and $|3 - 2| = 1$, all distinct; but the permutation 45132 is not graceful since $|4 - 5| = |3 - 2|$. We might want to program a computer to count the number of graceful permutations of the set $\{1, 2, 3, \ldots, n\}$ in the following way: Generate each permutation and if the defining condition is satisfied, add one to a counter. Before we did so, we would do well to ask whether the project is feasible: For what values of n might we expect a computer to be able to carry out this task in a reasonable length of time?

Since the number of permutations that need to be checked is $n!$ by Theorem 1, and since each permutation has n numbers in it, the algorithm will require at least $n \cdot n!$ steps just to generate and look at the permutations. This may be feasible on a fast computer for, say, $n \leq 10$, since $10 \cdot 10! \approx 3.6 \times 10^7$, but is certainly out of the question for, say, $n \geq 20$, since $20 \cdot 20! \approx 4.9 \times 10^{19}$. Thus if we needed to know the number of graceful permutations for $n = 20$, then we would need a more sophisticated approach than just this brute force method. ◆

EXAMPLE 3. Eight horses have entered a race. If a spectator were to pick three different horses at random to bet on for win (first), place (second), and show (third), how likely is he to be completely correct? Assume that there are no dead heats (ties).

Solution. The spectator is specifying a 3-permutation from an 8-set, because the eight horses are distinct (they have names) and the selection to be made is an ordered one (the spectator is specifying which horse will come in first, which second, and which third). By Theorem 1 there are $P(8, 3) = 8 \cdot 7 \cdot 6 = 336$ such permutations.

Only one of these is completely correct, namely the selection of the actual winner of the race as the winner, the actual place horse as the place, and the actual show horse as the show. Since the problem states that the selections are made at random, the spectator is equally likely to choose each of the 336 permutations. (This appeal to symmetry is in essence the definition of the word "random" as it is being used here.) Hence there is 1 chance in 336 that the spectator will win all three bets. ◆

COMBINATIONS

Next we turn to the problem of counting selections in which order is ignored. Again let S be a set with cardinality $n \geq 0$ (i.e., an n-set), and let k be an integer such that $0 \leq k \leq n$. Then a **combination** of k objects from S, sometimes called a **k-combination** from S, is an unordered collection of k of the elements of S, with repetition not allowed (i.e., a subset of S with cardinality k). The *number* of different k-combinations of a set with n elements is denoted by $C(n, k)$ or by $\binom{n}{k}$.

EXAMPLE 4. There are four 3-combinations from the set $\{a, b, c, d\}$:

$$abc$$
$$abd$$
$$acd$$
$$bcd$$

Thus $C(4, 3) = 4$. Similarly, there are six 2-combinations from the same set, namely, ab, ac, ad, bc, bd, and cd, so $\binom{4}{2} = 6$. ◆

Developing a formula for calculating $C(n, k)$ is a little more complicated than the straightforward argument we used for Theorem 1. Before stating and proving our next theorem, which contains the general formula, let us look at what happened in Examples 1 and 4 with the 3-permutations and the 3-combinations of the set $\{a, b, c, d\}$. Each of the four 3-combinations listed in Example 4 can be thought of as corresponding to an entire row of six 3-permutations listed in Example 1. For example, the first 3-combination listed, namely abc, can be ordered in the six ways shown in the first row of Example 1: abc, acb, bac, bca, cab, and cba. Since by the same token *each* 3-combination can be ordered in six ways, what we have done in counting the 3-permutations is to overcount the 3-combinations by a factor of 6. Therefore, by part (c) of Theorem 3 of Section 6.1, the number of 3-combinations is equal to the number of 3-permutations divided by 6, that is, $P(4, 3)/6 = 24/6 = 4$. We merely need to generalize this argument for a proof of the next theorem.

THEOREM 2.   *The number of k-combinations from an n-set, where* $0 \leq k \leq n$*, is given by*

$$C(n, k) = \frac{P(n, k)}{P(k, k)}$$

$$= \frac{n!}{k!(n - k)!}$$

$$= \frac{n(n - 1)(n - 2) \cdots (n - k + 2)(n - k + 1)}{k(k - 1)(k - 2) \cdots \quad 2 \quad \cdot \quad 1}.$$

Proof.   We will count the k-combinations of the given n-set by counting instead the k-permutations of the n-set and then dividing by the factor by which we have over-counted. Now each k-combination can be ordered in $P(k, k)$ different ways to produce a k-permutation, since we use all k of the chosen elements to form the permutation. Thus to count the k-permutations is to count each k-combination $P(k, k)$ times. Therefore, the first equality above follows from part (c) of Theorem 3 of Section 6.1 (and the definition of $P(n, k)$). The second and third lines follow from Theorem 1. ■

From a computational point of view, the last line of Theorem 2 is the most useful. You can compute $C(n, k)$ with a calculator quite easily, without using parentheses or having to write down intermediate results. Simply enter the factors in the numerator, touching the "times" key between entries; followed by a "divided by"; followed by all the factors in the denominator, touching the "divided by" key between entries; followed by "equals." If you are forced to compute by hand, it is best to write down the entire arithmetic expression (last line in Theorem 2) and then cancel common factors to get rid of the entire denominator before multiplying; thus no divisions are necessary. In either case, it is good to have in the "quick recall" area of your brain the facts that $C(n, 0) = 1$, $C(n, 1) = n$, and $C(n, 2) = n(n - 1)/2$. Finally, note that $C(n, k) = C(n, n - k)$. The easy algebraic proof of this identity is left to the reader (Exercise 6), but here is a more conceptual, combinatorial proof (of the type that we will study in Section 7.1). First $C(n, k)$ is the number of ways to select k elements from an n-set; we can think of these as the elements to include in a subset of cardinality k. On the other hand, $C(n, n - k)$ is the number of ways to select $n - k$ elements from the n-set; we can think of these as the elements to leave out of the subset. Since the subset is uniquely determined either by specifying which elements are to be included or by specifying which elements are to be left out, these two quantities must be equal. Thus, for example, it is much quicker to compute $C(8, 6)$ by computing $C(8, 2) = 8 \cdot 7/2 = 28$ than by using the five multiplications and six divisions required by applying the formula directly.

EXAMPLE 5. How many different committees with four or five members is it possible to appoint from the U.S. House of Representatives, which has 435 members?

Solution 1. By the addition principle, we can solve this problem by finding the number of four-member committees and the number of five-member committees, and adding (we want to choose a four-member committee *or* a five-member committee, and no four-member committee is at the same time a five-member committee). There are

$$C(435, 4) = \frac{435 \cdot 434 \cdot 433 \cdot 432}{4 \cdot 3 \cdot 2 \cdot 1} = 1{,}471{,}429{,}260$$

four-member committees and, by a similar calculation, there are $C(435, 5) = 126{,}837{,}202{,}212$ committees with five members, for a total of $128{,}308{,}631{,}472$ possible four- or five-member committees.

Solution 2. Let us add a phantom member to the House, making a total of 436 "members." To specify a four- or five-member committee of the House is to specify a five-member committee of this enlarged House (if the phantom is chosen for the committee, then the committee is a four-member committee). Thus the number of possible choices is

$$\binom{436}{5} = 128{,}308{,}631{,}472,$$

as before. ◆

As we saw in Example 5, Theorems 1 and 2 alone are not sufficient to solve most interesting combinatorial problems. Often we need to combine the formulas shown there with some clever analysis and one or more applications of the multiplication principle, the addition principle, or the overcounting principle. The trick to solving most combinatorial problems is to look at the problem in the right way. The reader should study the remaining examples carefully, note some of the tricks that can be used, and then try as many of the exercises as possible.

EXAMPLE 6. Solve the following combinatorial problems about strings.

   (a) How many bit strings of length 10 are there with exactly four 1's?
   (b) How many strings using symbols from $\{a, b, c\}$ are there having length 10 and using exactly four a's?
   (c) How many strings using symbols from $\{a, b, c\}$ are there having length 10 and using exactly four a's and three b's?

Solution. Trying to solve these problems by viewing them as a sequence of 10 subtasks (successively choosing symbols to put into each of the 10 positions in the string) may seem like a reasonable approach, but it will not work. The difficulty is that the number of choices available for the later positions depends on what symbols were chosen for the earlier positions. Instead, we need to look at these problems in a totally different way.

(a) Specifying a string of length 10 consisting of 0's and 1's is equivalent to specifying which of the 10 positions in the string are to contain the 1's. Once we have specified which positions contain the 1's, we know that the remaining positions contain 0's. Thus we simply need to choose a *set* of four positions from the set of 10 positions. In other words, we need to choose a 4-combination from a 10-set. There are

$$C(10,4) = \frac{10 \cdot 9 \cdot 8 \cdot 7}{4 \cdot 3 \cdot 2 \cdot 1} = 210$$

such subsets of four positions, so there are 210 bit strings of length 10 with exactly four 1's.

(b) As in part (a), we need to choose the four positions in which to put the a's, and this can be done in $C(10,4) = 210$ ways. In this problem, however, there are more choices to be made. After deciding where to put the a's, we need to decide whether to put a b or a c in each of the remaining six positions. Since each such decision can be made in two ways, there are $2 \cdot 2 \cdot 2 \cdot 2 \cdot 2 \cdot 2 = 2^6 = 64$ ways to determine the contents of the remaining six positions. Finally, by one more application of the multiplication principle, we see that the entire process can be carried out in $210 \cdot 64 = 13{,}440$ ways (i.e., there are 13,440 strings of the specified type).

(c) This time there are two subtasks that need to be performed. First we need to choose the positions for the four a's, and this can be done in $C(10,4)$ ways. Then we need to choose the positions for the three b's from the six positions not yet spoken for; this can be done in $C(6,3)$ ways. No further choices need to be made, since the remaining three positions must contain c's. (Looked at another way, we need to choose three of the remaining three positions for the c's, and this can be done in only $C(3,3) = 1$ way.) Thus by the multiplication principle, there are $C(10,4) \cdot C(6,3) = 210 \cdot 20 = 4200$ strings of the specified type.   ◆

EXAMPLE 7.   The Michigan Lotto (original version) was explained in Exercise 13 of Section 6.1. What are the chances of winning the grand prize by matching all six numbers from 1 to 40, inclusive, picked by the state? What are the chances of winning the third prize, given to any entrant whose ticket contains exactly four of the six numbers picked by the state?

Solution.   Since a ticket consists of a set of six numbers between 1 and 40, inclusive, the number of possible tickets is the number of 6-combinations from a 40-set, namely $C(40,6) = 3{,}838{,}380$. Only one of these tickets is the winner, so the probability of winning the grand prize is 1 in 3,838,380 (each ticket has an equal chance to win, since the winning 6-set is chosen totally at random).

To answer the question about the third prize, we need to compute the number of subsets of $\{1, 2, \ldots, 40\}$ of cardinality 6 that contain exactly four of the state's six numbers. To specify such a set, we need to choose first a subset of four of the state's six numbers to match; this can be done in $C(6, 4)$ ways. Next we need to choose two more numbers to complete the ticket, and these two numbers cannot be any of the state's six numbers. Since there are 34 other numbers to choose from, this subtask can be done in $C(34, 2)$ ways. Therefore, by the multiplication principle, there are $C(6, 4) \cdot C(34, 2) = 15 \cdot 561 = 8415$ possible choices for a third prize ticket. Since there are 3,838,380 possible tickets altogether, the probability of holding a third prize ticket is $8415/3,838,380 \approx 0.0022$ or about 1 chance in 456. ◆

EXAMPLE 8. How many ways are there to put k identical balls into n distinct boxes if at most one ball can be put into each box? Assume that $0 \le k \le n$.

Solution. We simply need to choose k of the boxes to be the recipients of the balls, and this can be done in $C(n, k)$ ways. We will return to this example in the next section, and in some of the exercises in this section and the next, and ask what happens if we make some or all of the following changes: let some or all of the balls be distinct, let some or all of the boxes be identical, allow more than one ball per box, require a certain minimum number of balls in some of the boxes, let some of the balls go unplaced, or worry about the time order in which the balls are put into the boxes. ◆

EXAMPLE 9. How many bit strings are there containing exactly fifteen 0's and six 1's, if every 1 must be followed immediately by a 0?

Solution 1. A "gluing" technique can be applied here. Since each 1 must be followed immediately by a 0, we can think of gluing a 0 to the right of each 1. This leaves $15 - 6 = 9$ loose 0's. Then the tokens we have to arrange are six 10's and nine 0's, a total of 15 symbols. An arrangement is determined by which positions are chosen for the 0's, say, so there are $C(15, 9) = 5005$ such strings.

Solution 2. Consider fifteen 0's in a row. To determine a string of the desired type, we need only decide which of the gaps formed by the 0's will have a 1 put into them. Since each 1 must be followed immediately by a 0, there can be at most one 1 per gap. Furthermore, the "gap" to the left of the leftmost 0 is a legal position in which to place a 1, but the "gap" to the right of the rightmost 0 is not. Thus there are 15 legal gaps altogether, and we need to choose six of them for our six 1's. Therefore, the answer is $C(15, 6) = 5005$, as in the first solution. ◆

EXAMPLE 10. Determine the number of injective (one-to-one) functions from a k-set A to an n-set B. Assume that $0 \le k \le n$.

Solution. Let us label the elements of the domain $a_1, a_2, \ldots, a_k$. Recall from Section 3.2 that an injective function $f: A \to B$ is one in which no element of the codomain is the image of two different elements of the domain. Therefore, the sequence $f(a_1)$, $f(a_2), \ldots, f(a_k)$ is an ordered arrangement of elements of B, without repetitions, that is, a k-permutation from the n-set B. Conversely, any k-permutation of B can be identified with the function that sends a_i to the ith object in that permutation, for each i. Therefore, the number of injective functions from A to B is just the number of k-permutations of B, namely $P(n, k)$, which equals $n!/(n - k)!$. Note that we needed to assume that $k \le n$; if $k > n$, there are no injective functions from A to B.

Since we saw in Exercise 7c in Section 6.1 that there are a total of n^k functions from A to B, we can conclude by the overcounting principle that there are $n^k - (n!/(n - k)!)$ functions from A to B that are *not* injective. The problem of counting surjective (onto) functions from A to B is harder; we solve that problem in Section 7.4. ◆

In our final example we consider a model that is useful in such applications as gambling and, appropriately generalized to three dimensions, the motion of gas molecules.

EXAMPLE 11. Suppose that an object is located at the origin (0) of the number line. At each unit of time, the object can either take a 1-unit step to the right, or take a 1-unit step to the left. For example, in 11 units of time, the object's "walk" might be $-1, 0, 1, 2, 3, 2, 3, 2, 1, 2, 3$. In a gambling application, a step to the right would indicate that the gambler had won one dollar, and a step to the left would indicate that he had lost one dollar. The path given here represents a gambler who ended up three dollars ahead of the game after 11 plays. If the moves are determined randomly, as they are in the gambling setting, motion like this is called a "random walk."

Determine the number of walks of length n, and determine how many of them end at 3. Suppose that a gambler wins one dollar if a tossed coin comes up heads and loses one dollar if the coin comes up tails. How likely is the gambler to end up three dollars ahead after 11 tosses?

Solution. A walk of length n can be represented simply by a string of R's and L's of length n, where an R indicates a step to the right and an L indicates a step to the left. The walk given above would be represented by $LRRRRLRLLRR$. By the multiplication principle there are 2^n walks of length n; for $n = 11$, there are $2^{11} = 2048$ walks. If we want the walk to end up at 3, then there will have to be three more R's than L's in the sequence: in other words, there will have to be $r = (n+3)/2$ R's and $l = (n-3)/2$ L's. (These numbers are obtained by solving the equations $r + l = n$ and $r - l = 3$ simultaneously.) This can happen only if n is an odd integer greater than or equal to 3. If n satisfies this condition, then by the same reasoning as in Example 6a, there are $C(n, (n + 3)/2)$ walks ending at 3. For $n = 11$, this is $C(11, 7) = 330$. Since our gambler is equally likely to win or lose at each play (independent of previous plays), all 2048 walks are equally likely. Thus his chance of ending up exactly three dollars ahead after 11 plays is $330/2048$, which is about 16%.   ◆

SUMMARY OF DEFINITIONS

combination of k objects from the set S with n elements: an unordered collection of k of the elements of S, with repetition not allowed, in other words, a k-element subset of S; the number of such subsets is denoted $C(n, k)$ or $\binom{n}{k}$ (e.g., $\{3, 4, 6, 7\}$ is a combination of 4 objects from the set $\{1, 2, 3, 4, 5, 6, 7, 8\}$; there are $C(8, 4)$ such combinations in all).

k-combination from the set S: a combination of k objects from S; a subset of S having cardinality k.

k-permutation from the set S: a permutation of k objects from S.

permutation of k objects from the set S with n elements: an ordered arrangement of k of the elements of S, with repetition not allowed; the number of such arrangements is denoted $P(n, k)$ (e.g., 6437 is a permutation of 4 objects from the set $\{1, 2, 3, 4, 5, 6, 7, 8\}$; there are $P(8, 4)$ such permutations in all).

permutation of the set S with n elements: an ordered arrangement of all of the elements of S, with repetition not allowed; the number of such arrangements is denoted $P(n, n)$ (e.g., 643512 is a permutation of the set $\{1, 2, 3, 4, 5, 6\}$; there are $P(6, 6)$ such permutations in all).

EXERCISES

Some of these exercises require fairly clever insights, and it is impossible to give examples of all the insights that might be required. You can only learn to solve combinatorial problems by working many of them.

1. Compute the following quantities.
   (a) $P(7, 3)$.
   (b) $P(7, 1)$.
   (c) $P(7, 7)$.
   (d) $P(7, 6)$.
   (e) $P(10, 8)$.

2. Compute the following quantities.
   (a) $C(7, 3)$.
   (b) $C(7, 1)$.
   (c) $C(7, 7)$.

(d) $C(10, 3)$.

(e) $C(10, 8)$.

3.   Let $A = \{1, 3, 5, 7, 9\}$.
   (a)   Without listing any of them, determine how many 2-permutations from the set A there are.
   (b)   Write down explicitly all the 2-permutations from A.
   (c)   Without listing any of them, determine how many 3-combinations from the set A there are.
   (d)   Write down explicitly all the 3-combinations from A.

4.   How many different four-member committees can be appointed from the U.S. Senate (which contains 100 members)?

5.   How many bit strings of length 10 are there with exactly
   (a)   Two 1's?
   (b)   Six 1's?
   (c)   No 1's?
   (d)   One 1?

6.   Using the formula involving factorials, show that $C(n, k) = C(n, n - k)$.

7.   The Watson family contains five members. The family has a lot of money to put in the bank, but the bank will only insure each account for $100,000. However, a separate account may be opened for each member of the family, and, in addition, a separate joint account for each distinct pair of family members may also be opened. How much insured money can the Watson family deposit?

8.   A password for a certain computer system consists of either one capital letter or two capital letters in order. Determine the number of possible passwords.

9.   A pizza parlor offers the following toppings for its pizza: sausage, hamburger, pepperoni, anchovies, onions, olives, mushrooms, peppers, and pineapple. Any combination of zero or more toppings may be ordered. How many different pizzas can be formed using a prime number of toppings?

10.   How likely is it for the gambler in Example 11 to be five dollars behind after 11 plays, if he is equally likely to win or lose one dollar on each play (independent of previous plays)?

11.   A club of 21 people is meeting for the first time. They will vote to decide whether to elect a president, secretary, and treasurer as their officers, or instead to elect an executive committee of three to run the club. Then they will immediately elect three people to fill whichever roles they decide on. How many outcomes are possible?

12.   Determine the number of strings of length five consisting of five distinct capital letters (A to Z)

   (a)   That do not contain an A.

   (b)   That contain an A.

   (c)   That contain an A, a B, and a C as their first three symbols, in that order.

   (d)   That contain an A, a B, and a C as their first three symbols, in any order.

   (e)   That contain an A, a B, and a C somewhere, in that order.

   (f)   That contain an A, a B, and a C somewhere, in any order.

13. Determine the number of subsets of cardinality 9 from a set consisting of 10 adult men, 12 adult women, and 17 children

   (a)   If there are no restrictions.

   (b)   If the subset must contain three adult men, three adult women, and three children.

   (c)   If the subset must contain only adult men or only adult women or only children.

   (d)   If the subset cannot contain both adult men and adult women.

14. The new Michigan Lotto is similar to the original version (see Exercise 13 in Section 6.1 and Example 7 in the present section), except that the choice is from numbers from 1 to 44, inclusive.

   (a)   Determine the number of different possible tickets.

   (b)   Determine the number of different tickets that can win second prize and express as a fraction with numerator 1 the approximate probability of winning second prize.

   (c)   Repeat part (b) for third prize.

15. One state lottery has the player pick seven numbers from $\{1, 2, 3, \dots, 80\}$. The state picks 11 numbers, randomly, in this range. The player wins the jackpot if his or her seven numbers are a subset of the state's numbers. Determine the probability of holding a winning ticket.

16. Let n be a natural number. Be sure to address the case $n = 0$ as well as the cases $n > 0$ in your discussion.

   (a)   Explain why $P(n, 0) = 1$, and show that the correct answer is obtained from the formula given in Theorem 1.

   (b)   Explain why $C(n, 0) = 1$, and show that the correct answer is obtained from the formula given in Theorem 2.

17. Consider strings of length 12 using symbols from $\{a, b, c, d\}$.

   (a)   How many such strings have exactly five a's?

   (b)   How many such strings have three of each letter?

   (c)   How many such strings have exactly five a's and four b's?

   (d)   How many such strings have exactly five a's or exactly four b's (remember that the word "or" is used in the inclusive sense)?

18. How many ways are there to put k distinct balls into n distinct boxes if at most one ball can be put into each box? Assume that $0 \le k \le n$.

19. How many ways are there to put k balls into n identical boxes if at most one ball can be put into each box? Assume that $0 \le k \le n$. (The answer to this exercise does not depend on whether the balls are distinct or identical.)

20. How many ways are there to put five identical red balls and eight identical blue balls into 20 distinct boxes
    (a) If at most one ball can be put into each box?
    (b) If at most one ball of each color can be put into each box?

21. How many bit strings are there with exactly eighteen 1's and exactly four 0's, such that every 0 is followed by at least two 1's? [*Hint*: Glue is helpful here.]

22. How many different ranges are possible for functions from $\{1, 2, 3, 4\}$ to $\{1, 2, 3, 4, 5, 6, 7, 8, 9, 10\}$?

23. How many different ways can 12 people be arranged around a circular table?

24. Let m and w be integers greater than 1, which represent the number of men and the number of women, respectively, who are dining together at a circular table.
    (a) How many ways are there for the people to be arranged around the table?
    (b) How many ways are there for the people to be arranged around the table if no two men may sit together? [*Hint*: There are at least two cases to consider, depending on the relative sizes of m and w.]
    (c) How many ways are there for the people to be arranged around the table if $m = w$, the people are married couples, and each person must sit next to his or her spouse?
    (d) How many ways are there for the people to be arranged around the table if $m = w$, the people are married couples, each person must sit next to his or her spouse, and no one may sit next to a person of the same sex? (See also Exercise 53.)

25. How many different ways can 12 people be arranged around
    (a) A rectangular table that seats one person at each end and five people along each long side?
    (b) A square table that seats three people along each side?

26. How many strings of capital letters (A to Z) of length 5 are there with the properties that no letter appears more than once, and if the letter Q appears, then it is followed immediately by the letter U?

27. How many strings are there of left and right parentheses, containing n left parentheses and n right parentheses?

28. Describe an algorithm for computing $C(n, k)$ that is not prone to overflow. For example, to compute that $C(150, 75) \approx 9.3 \times 10^{43}$ directly from the formula requires the intermediate answer of $P(150, 75)$, which exceeds 10^{100}; hence most programmable pocket calculators would not be able to compute $C(150, 75)$ from the formula, even though the answer is well within its range.

29. Let k and l be two positive integers whose sum is less than the integer n. How many ways are there to choose two subsets A and B from $\{1, 2, \ldots, n\}$, with $|A| = k$ and $|B| = l$, such that every element of A is less than every element of B?

30. The Hare system of voting is sometimes used for elections in which there are a large number of candidates and more than one candidate is to be elected, as, for example, to serve on a city council. Under this method, each voter is asked to rank order (no ties permitted) as many candidates as he or she finds acceptable. For example, if the candidates are Heawood, Euler, Hardy, and Gauss, a voter may choose to submit a ballot that ranks Euler first and Gauss second, and leaves Heawood and Hardy unranked (the voter likes Euler best and cannot accept Hardy or Heawood).
  (a) Determine how many different ballots are possible if there are five candidates.
  (b) Find a formula for the number of different ballots possible if there are n candidates.

31. How many different relations are there on the set $\{1, 2, \ldots, n\}$? [*Hint:* See the suggestions for Exercise 32.]

32. How many different relations are there, with each of the following properties, on the set $\{1, 2, \ldots, n\}$? [*Hint:* Remember that a relation on a set A is simply a subset of $A \times A$. You can think of determining a relation by writing either a 0 or a 1 in each position of the matrix representing the relation (see Section 3.3). For each part of this exercise, you need to figure out what restrictions to apply to the process of writing down the matrix, and count the number of ways to write down the matrix with these restrictions. Alternatively, and equivalently, you can think of specifying the digraph for the relation.]
  (a) Reflexivity.
  (b) Irreflexivity.
  (c) Symmetry.
  (d) Reflexivity and symmetry.
  (e) Antisymmetry.
  (f) Reflexivity and antisymmetry.

33. How many bit strings of length 16 have more 0's than 1's?

34. A **standard deck of cards** consists of 52 cards, with 13 of them in each of the four **suits** (spades ♠, hearts ♡, diamonds ◇, and clubs ♣) and in each suit one card of each **rank** $(2, 3, 4, 5, 6, 7, 8, 9, 10, J, Q, K, A)$. A **poker hand** consists of a subset of cardinality 5 from a standard deck of cards.

(a) How many different poker hands are there?

(b) A poker hand is called a **four-of-a-kind** if it contains four cards of one rank and one card of another rank. Determine how many different poker hands are four-of-a-kinds.

(c) A poker hand is called a **flush** if all five cards are of the same suit. How many different poker hands are flushes?

(d) A poker hand is called a **straight** if the ranks of the five cards are consecutive in the sequence $A, 2, 3, 4, 5, 6, 7, 8, 9, 10, J, Q, K, A$. Thus the ace (A) can be considered either below 2 or above king (K), but not both simultaneously; $\{K, A, 2, 3, 4\}$, for example, is not a straight. Determine how many different poker hands are straights.

(e) A poker hand is called a **straight flush** if it is both a straight and a flush. Determine how many different poker hands are straight flushes.

(f) A poker hand is called a **full house** if the hand consists of three cards of one rank and two cards of another rank. Determine how many different poker hands are full houses.

(g) A poker hand is called a **three-of-a-kind** if it contains three cards of one rank and one card of each of two other ranks. Determine how many different poker hands are three-of-a-kinds.

(h) A poker hand is called a **two-pair** if it contains two cards of one rank, two cards of another rank, and one card of a third rank. Determine how many different poker hands are two-pairs.

(i) A poker hand is called a **pair** if it contains two cards of one rank and one card of each of three other ranks. Determine how many different poker hands are pairs.

(j) How many different poker hands are neither flushes, straights, full houses, three-of-a-kinds, two-pairs, or pairs?

35. A **bridge hand** consists of a subset of cardinality 13 from a standard deck of cards. (See Exercise 34 for the definition of a "standard deck of cards.")

(a) How many different bridge hands are there?

(b) How many different bridge hands contain four spades, four hearts, three diamonds, and two clubs?

(c) How many different bridge hands contain four cards in each of two suits, three cards in a third suit, and two cards in the fourth suit?

(d) How many different bridge hands contain exactly three cards in each suit?

(e) How many different bridge hands contain no card better than a 9 (the cards better than a 9 are 10, J, Q, K, and A)? Bridge players call such a hand a **Yarborough**; it is a very poor hand to hold at bridge. How often is a bridge player likely to hold a Yarborough (state your answer in the form "once out of every N hands")?

36. The U.S. House of Representatives has 435 members, at least one from each of the 50 states, and the Senate has 100 members, two from each state.
    (a) How many different joint House–Senate committees with six members can be appointed, if there must be at least two members of each body on the committee?
    (b) How many different joint House–Senate committees with six members can be appointed, if there must be three members from each body and the senators must all come from different states?

37. In the original version of Michigan Lotto (see Exercise 13 of Section 6.1), you win nothing if your ticket matches exactly three of the state's six numbers. How likely is it that the one ticket you bought will come so frustratingly close?

38. How many ways are there to sink 15 distinct balls in six distinct pockets when playing pool,
    (a) If the order in which the balls are sunk is ignored?
    (b) If the order in which the balls are sunk is taken into consideration?

39. Suppose that k couples on dates go to a movie theater and want to sit in the front row, which has n distinct seats. (For this to be possible, we must have $n \geq 2k$.) Naturally, each couple wants to sit in adjacent seats. (For example, if $n = 11$ and $k = 3$, then one couple could occupy seats 1 and 2, another could occupy seats 6 and 7, and the third could occupy seats 8 and 9.) In how many ways can the people be seated under each of the following assumptions?
    (a) All the people are indistinguishable. [*Hint*: All that matters is which seats are occupied, but they have to be occupied in groups of 2.]
    (b) The couples are distinct, but the two people within a couple are not; in other words, we do not care, within each couple, who sits on the right and who sits on the left. [*Hint*: Modify the answer to part (a).]
    (c) All the people are distinct. [*Hint*: Modify the answer to part (b).]

Challenging Exercises

40. How many ways are there to put five identical red balls and eight identical blue balls into 20 distinct boxes if at most one ball can be put into each box, and the order of placing the balls is taken into account? In other words, how many different movies can be made of someone placing balls into boxes, under the given restrictions?

41. See Exercise 35 for the definition of a bridge hand.
    (a) How many different bridge hands contain cards from exactly two suits?
    (b) How many different bridge hands contain more red cards than black cards? (The hearts and the diamonds are red, and the spades and the clubs are black.)
    (c) How many different bridge hands contain more spades than hearts?

42. How many different four-member committees can be appointed from the U.S. Senate (which contains 100 members, two from each state) if each member of the committee must come from a different state? Give two distinct ways to solve this problem.

43. Let $\{a_i\}_{i=1}^n$ be a sequence of n distinct numbers. Suppose that $0 \le k \le n$.
    (a)   How many subsequences of length k does this sequence have?
    (b)   How many subsequences does this sequence have?
    (c)   How many consecutive subsequences of length k does this sequence have?
    (d)   How many consecutive subsequences does this sequence have?

44. Suppose that all $C(n, 2)$ diagonals are drawn in a convex polygon with n sides. Suppose that no three of these diagonals pass through the same point inside the polygon. (See the picture on the left in Exercise 50, where $n = 6$.) Determine the number of points of intersection inside the polygon.

Exploratory Exercises

45. Investigate the probability that two people in a group of n people will have the same birthday (same month and date of birth, not necessarily same year). Find a formula in terms of n, and compute the probability for various values of n. Find the smallest value of n for which the probability exceeds $\frac{1}{2}$. Be sure to make your assumptions explicit (it is acceptable to simplify the calculations by ignoring February 29, for example).

46. If you pick six numbers at random from the set of positive integers from 1 to 40, what is the probability that all six will be greater than 20? Discuss each of these statements about playing the original Michigan Lotto (see Exercise 13 in Section 6.1).
    (a)   "It is very unlikely that the winning ticket will contain only numbers greater than 20."
    (b)   "If you want to win, you should not choose only numbers greater than 20; your chance of winning will be greater if your choices are more spread out."

47. Simulate the gambler in Example 11, using a pseudorandom number generator on a computer. Have him take 2048 walks of length 11, for example, and see where they end. Compare the results of your simulation to the theoretical results you obtain by the kind of analysis done in Example 11. See what happens if you change the chances of winning each play, say from 50% to 48%.

48. Consider a picture, similar to Figure 4.1, that consists of a large rectangle divided into squares: m squares horizontally, n squares vertically. Assume that $m \ge n$.
    (a)   How many rectangles are there in such a picture?

   (b)   How many squares are there in such a picture?

   (c)   What are the answers if we generalize this to three dimensions?

   (d)   Consult Duncan and Litwiller [73] for solutions to these problems.

49. Consider the following combinatorial geometry problem. All $C(n, 2)$ diagonals are drawn in a convex polygon with n sides. Suppose that no three of these diagonals pass through the same point inside the polygon. (See the picture on the left in Exercise 50, where $n = 6$.) The problem is to determine the number of regions that the interior of the polygon is split up into. Try to solve this problem, or see Freeman [89] and Honsberger [155] for a solution.

50. Consider the following triangle-counting problems. In the first, all $C(n, 2)$ diagonals are drawn in a convex polygon with n sides. Suppose that no three of these diagonals pass through the same point inside the polygon. See the leftmost picture, with $n = 6$. In the second problem, a figure like that shown on the right is constructed, with each side of the outer triangle divided into n intervals ($n = 4$ in the figure shown). The problem in each case is to determine the number of triangles (of all sizes) in the figures, as a function of n. Try working on these rather difficult problems; consult Cormier and Eggleton [52] for a solution.

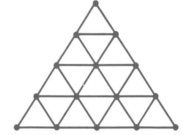

EXERCISE 50

51. Find out more about applications of mathematics to gambling and games, such as poker, by consulting Packel [230].

52. How would you go about generating permutations and combinations (all of them systematically, or one of them at random)? There are various interesting algorithms for these problems. Find out more by consulting Nijenhuis and Wilf [222] and Reingold, Nievergelt, and Deo [243].

53. Suppose that n married couples wish to sit around a circular table so that men and women alternate and so that no one sits next to his or her spouse (see also Exercise 24). Counting the number of ways that this can be done is a famous and fairly difficult problem known as the "*ménage* problem." Find out how to solve it by consulting Bogart and Doyle [24] (preferably after grappling with it yourself).

54. Find out more about random walks by consulting Jewett and Ross [164].

SECTION 6.3
COMBINATORIAL PROBLEMS
INVOLVING REPETITIONS

Tomorrow, and tomorrow, and tomorrow
Creeps in this petty pace from day to day,
To the last syllable of recorded time ...

—William Shakespeare, *Macbeth*

In this section we look at some combinatorial problems in which the items to be selected or arranged are not all distinct. We will see that some of these problems are fairly easy, and we can develop formulas for obtaining answers to them. We will also see that some problems of this type are extremely hard.

PERMUTATIONS WITH DUPLICATES

We saw in the preceding section that there are $P(n, k)$ ways to choose k objects from a set of n *distinct* objects and arrange them in order. In this section we ask the same question with the relaxed condition that some of the objects in the collection may be indistinguishable. For example, if we have 10 blue cards, seven red cards, five white cards, two yellow cards, and one green card, we might want to know how many ways there are to arrange eight of these cards in a row. The general problem would be of the following form.

> How many ordered arrangements are there of k things taken from a collection of n things if the collection of n things contains n_1 identical things of one type, n_2 identical things of a second type, ..., and n_t identical things of a tth type, where $n_1 + n_2 + \cdots + n_t = n$?

In the example mentioned above we have $k = 8$, $t = 5$, $n = 10+7+5+2+1 = 25$.

Unfortunately, the problem in this generality is rather hard. We will consider only the case in which $k = n$. Now the problem is quite manageable using the techniques from Section 6.1, and the solution is contained in the following theorem.

THEOREM 1. *Consider a collection of n objects classified into t distinct types, including n_1 identical objects of the first type, n_2 identical objects of the second type, ...,* and n_t *identical objects of the tth type (so that $n_1 + n_2 + \cdots + n_t = n$). Then the number of permutations of all n of these objects is given by*

$$\frac{n!}{n_1! \, n_2! \cdots n_t!}.$$

Proof. We temporarily pretend that the n objects are in fact all distinguishable. This could be effected by attaching labels to them, say numbering the objects of the first type with the numbers 1, 2, ..., n_1, objects of the second type with the numbers 1, 2, ..., n_2, and so on. Since there are n objects altogether, there are $P(n,n) = n!$ permutations of these distinguishable objects. Now we have overcounted by a certain factor, namely the number of ways to permute the labeled objects of each type, since permuting the objects within each type does not change the permutation of the objects as unlabeled objects. Clearly, there are $n_i!$ ways to permute the n_i objects of the ith type, for $i = 1, 2, \ldots, t$, so by the multiplication principle there are $n_1! n_2! \cdots n_t!$ ways to rearrange all the labeled objects without changing their arrangement as unlabeled objects. Thus the theorem follows from part (c) of Theorem 3 in Section 6.1. ∎

EXAMPLE 1. Suppose that we have 10 blue cards, seven red cards, five white cards, two yellow cards, and one green card. How many ways are there to arrange these 25 cards in a row?

Solution. By Theorem 1, with $n = 25$, $t = 5$, $n_1 = 10$, $n_2 = 7$, $n_3 = 5$, $n_4 = 2$, and $n_5 = 1$, we see that the answer is

$$\frac{25!}{10! 7! 5! 2! 1!} = 3{,}533{,}791{,}060{,}800. \qquad \blacklozenge$$

EXAMPLE 2. How many different ways are there to deal out a standard deck of cards to four people so that each person gets 13 cards? (By our nonambiguity conventions people are to be considered distinct, and the time order in which the operation takes place is not considered. Thus in this example all that matters is which set of 13 cards each of the four distinct card players receives.)

Solution 1. Consider the 52 cards as laid out in a row in an arbitrary order. Below each card in the row we will write a letter from the set $\{N, S, E, W\}$, which represents the four card players (traditionally, the names of the four players at a card game are the compass directions North, South, East, and West). Since the cards are to be dealt evenly, we need to write 13 of each letter. Theorem 1 applies, with 52 letters to be permuted, 13 of each type. There are therefore $52!/(13!^4) \approx 5.4 \times 10^{28}$ possible deals.

Solution 2. Despite the fact that cards are dealt out in a circular pattern, there is no reason not to assume that North chooses *all* his cards first (and he can do so in

$C(52, 13)$ ways), and then East chooses her cards from those that North did not choose ($C(39, 13)$ possible choices), after which South chooses her cards (in one of $C(26, 13)$ ways); West gets what is left over. By the multiplication principle there are

$$C(52, 13) \cdot C(39, 13) \cdot C(26, 13) = \frac{52!}{13!39!} \cdot \frac{39!}{13!26!} \cdot \frac{26!}{13!13!} = \frac{52!}{13!^4}$$

ways for this to be done, and we obtain the same answer as before.   ◆

EXAMPLE 3.   In Example 11 of Section 6.2 we looked at one-dimensional discrete walks, in which an object began at the origin of the number line (0) and took successive unit steps either right or left. In a two-dimensional discrete walk, the object starts at the origin in the plane ($(0, 0)$) and at each step either moves one unit to the right (from (x, y) to $(x + 1, y)$), one unit to the left (from (x, y) to $(x - 1, y)$), one unit up (from (x, y) to $(x, y + 1)$), or one unit down (from (x, y) to $(x, y - 1)$). For example, a walk of length 9 might proceed, as shown in Figure 6.3, from the origin to $(1, 0)$, to $(1, 1)$, to $(1, 2)$, to $(0, 2)$, to $(0, 1)$, to $(-1, 1)$, to $(-1, 0)$, to $(-1, 1)$, and end at $(0, 1)$.

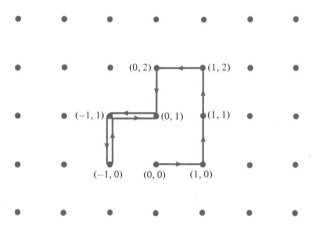

FIGURE 6.3   Walk $RUULDLDUR$ in the plane, starting at $(0, 0)$, ending at $(0, 1)$.

We can represent a walk of this type by a string from the alphabet $\{L, R, U, D\}$, where these letters stand for moves to the left, to the right, up, and down, respectively. Let us count the number of such walks of a given length that begin and end at the origin.

A walk will return to the origin if and only if the number of left steps equals the number of right steps and the number of up steps equals the number of down steps. In particular, the total number of steps must be even, so let us call it $2n$.

Let l be the number of left steps in the walk. Then there are also l right steps, and there are $n - l$ up steps and $n - l$ down steps (in order for the total number of steps to be $2n$). Thus the number of walks from $(0,0)$ to $(0,0)$ that contain exactly l left steps (where $0 \le l \le n$) is the number of strings of length $2n$ containing l L's, l R's, $n - l$ U's, and $n - l$ D's. By Theorem 1, this number is

$$\frac{(2n)!}{l!^2(n-l)!^2}.$$

By the addition principle the number of walks of length $2n$ from $(0,0)$ to $(0,0)$ is the sum of these quantities over all the relevant values of l, so the answer to our question is

$$\sum_{l=0}^{n} \frac{(2n)!}{l!^2(n-l)!^2}. \qquad \blacklozenge$$

COMBINATIONS WITH REPETITIONS ALLOWED

Questions of resource allocation often boil down to finding solutions of the equation

$$x_1 + x_2 + \cdots + x_t = k$$

subject to possible constraints on the values of the x_i's; here k is the total amount of the resource to be allocated, and x_1 through x_t are the amounts allocated to the t distinct competitors for the resource.

For example, three political candidates in different districts might be competing for $100,000 in campaign funds to be spent by their party in the next election. Here x_i represents the number of dollars that candidate i receives, for $i = 1, 2, 3$, and $k = 100,000$.

Obviously, it makes sense in such a problem to require that each amount allocated be nonnegative ($x_i \ge 0$ for each i), and, because this is a book in *discrete* mathematics, we will assume that the total amount of the resource available and the amounts allocated to each competitor are all natural numbers. (In the real world, this is still a reasonable assumption; for example, money is allocated in discrete units, namely cents.) Since we are interested in counting things in this chapter, we will ask for the *number of solutions* to a given resource allocation problem:

> How many distinct solutions are there in natural numbers to the equation
> $x_1 + x_2 + \cdots + x_t = k$?

Let us reformulate the resource allocation problem in terms of choosing objects (ignoring order) from a collection, with repetitions allowed. Suppose that we have a collection of t distinct objects, and we wish to choose k objects from the collection, with the freedom to choose each object more than once. In how many ways can we make the choices? Equivalently, imagine that we have objects of t different types, with an unlimited supply of objects of each type (actually, k objects of each type will suffice): Objects of different types are distinguishable, but objects of the same type are indistinguishable. In how many ways can we select k of these objects? We are ignoring order, so this is a problem of "combinations with repetitions allowed."

Before showing that these two formulations—counting the number of solutions of an equation and counting the number of combinations with repetitions allowed—are equivalent, and before deriving a formula that answers the question, we will look at a simple example and actually enumerate all the possibilities.

EXAMPLE 4. A child wants to buy some pet fish. The pet store sells goldfish, guppies, and angelfish. How many different selections are possible if the child wants to take home six fish?

Solution. The parameters in this problem are small enough that we can solve the problem by brute force. Let O, U, and A stand for goldfish, guppies, and angelfish, respectively. Being careful to enumerate the possibilities systematically (in alphabetical order reading down the columns), we see that the child may choose any of the following combinations.

AAAAAA	*AAAOOU*	*AAUUUU*	*OOOOOO*
AAAAAO	*AAAOUU*	*AOOOOO*	*OOOOOU*
AAAAAU	*AAAUUU*	*AOOOOU*	*OOOOUU*
AAAAOO	*AAOOOO*	*AOOOUU*	*OOOUUU*
AAAAOU	*AAOOOU*	*AOOUUU*	*OOUUUU*
AAAAUU	*AAOOUU*	*AOUUUU*	*OUUUUU*
AAAOOO	*AAOUUU*	*AUUUUU*	*UUUUUU*

Thus there are 28 possible choices. Note that we did not list, for example, *AOAUUA*, since that represents the same selection as *AAAOUU*—three angelfish, a goldfish, and two guppies. These are *combinations* of three objects (A, O, and U) with repetitions allowed, not *permutations*. ◆

Let us first show how to think of the "resource allocation problem" in terms of the "combinations with repetitions allowed" problem, and vice versa.

THEOREM 2. *Let k and t be natural numbers. Then the number of ways to choose k objects from a collection of t distinct objects with repetitions allowed is equal to the number of solutions in natural numbers to the equation*

$$x_1 + x_2 + \cdots + x_t = k.$$

Proof. We will show how to associate combinations (with repetitions allowed) to solutions (to the equation), in a one-to-one manner. To specify a selection of k objects from a collection of t distinct objects with repetitions allowed, it is necessary and sufficient to specify how many times the first object is picked, how many times the second object is picked, and so on. If we let x_i be the number of times the ith object is picked, for $i = 1, 2, \ldots, t$, then a selection of the type we desire is uniquely determined by, and uniquely determines, a t-tuple of integers $(x_1, x_2, \ldots, x_t)$ such that each $x_i \geq 0$ and $x_1 + x_2 + \cdots + x_t = k$. ◼

For example, the problem of finding the number of ways for the child in Example 4 to choose a collection of six fish from the three varieties available is equivalent to the problem of finding the number of solutions in natural numbers to $a + o + u = 6$, where a is the number of angelfish, o the number of goldfish, and u the number of guppies that the child buys. In the notation of Theorem 2, $t = 3$ and $k = 6$.

We can also view a solution of the resource allocation equation in terms of dropping balls into boxes. (This image will be quite useful later on when we look at solutions of equations of this type with further constraints.) Imagine t distinct boxes. Think of dropping k indistinguishable balls into the boxes, as many balls per box as desired. The final distribution of the balls in the boxes clearly corresponds to a solution to the equation: x_i is the number of balls in the ith box. But the number of ways to drop the k balls into the t boxes is equal to the number of ways to choose k objects from t distinct objects with repetitions allowed: The objects being chosen are the k boxes into which we drop the balls.

To obtain a formula for the number of solutions to the resource allocation problem (or the combinations with repetitions allowed problem), we need to resort to a bit of cleverness that, in retrospect, should be very satisfying.

THEOREM 3. *Let k and t be natural numbers. Then the number of solutions in natural numbers to the equation*

$$x_1 + x_2 + \cdots + x_t = k,$$

or equivalently the number of ways to choose k objects from a collection of t objects with repetitions allowed, is

$$C(t + k - 1, k).$$

Proof. Consider the set of all strings of length $t + k - 1$ containing exactly k stars and $t - 1$ bars [for example, the string $** \mid * * * \mid * \mid * \mid\mid ** \mid$, when $t = 7$ and $k = 9$]. Using either the techniques of Section 6.2 (choose the positions for the stars) or Theorem 1 of the present section, we see that the cardinality of this set is $C(t+k-1, k)$. Therefore, it will be sufficient to show how such strings correspond to solutions of the equation $x_1 + x_2 + \cdots + x_t = k$. The $t - 1$ bars in the string divide the string into t substrings of stars: the stars to the left of the first bar, the stars between the first bar and the second bar, $\ldots$, the stars between the $(t - 2)$th bar and the $(t-1)$th bar, and the stars to the right of the $(t-1)$th bar. The numbers of stars in these t substrings are the values of x_1 through x_t; since there are k stars altogether, the sum is k. [For example, the string $**\mid***\mid*\mid*\mid\mid**\mid$ corresponds to the solution $2+3+1+1+0+2+0 = 9$ when $t = 7$ and $k = 9$.] It should be clear that this correspondence between strings and solutions is a one-to-one correspondence, so our theorem is proved. ∎

In our pet fish example, we had $t = 3$ and $k = 6$. Thus the number of combinations is $C(3+6-1,6) = C(8,6) = C(8,2) = 8\cdot7/2 = 28$, as we found by explicitly enumerating them in Example 4. Obviously, if t and k are large, then enumeration would be out of the question, and the power of Theorem 3 becomes evident.

EXAMPLE 5. A computer program consists of checking a certain condition for all m-tuples $(I_1, I_2, \ldots, I_m)$ such that $1 \le I_1 \le I_2 \le \cdots \le I_m \le N$. How many such tuples must be checked?

Solution. We want to know how many ways there are to choose the m numbers I_1, I_2, $\ldots$, I_m from the set $\{1, 2, \ldots, N\}$, with repetitions allowed, as the coordinates of the m-tuple. We want an unordered selection (i.e., a combination), since the order in the tuple is fixed (namely, it has to be nondecreasing) once the numbers are chosen. Thus we have a classic example of a combination with repetitions allowed; using the notation of Theorem 3, $t = N$ and $k = m$. The answer is $C(N+m-1, m)$. ◆

EXAMPLE 6. Three children are being given a snack of milk and cookies. The cookie jar contains ten identical cookies. How many ways are there to serve this snack? (Assume that cookies may not be broken.)

Solution. As asked, the problem seems a bit unrealistic, since apparently each child can receive as few as zero and as many as 10 cookies. Note that not all the cookies have to be served. Under this interpretation, we can let x_1, x_2, and x_3 be the numbers of cookies received by the three children (recalling our convention that people are distinct) and let x_4 be the number of cookies remaining in the cookie jar. We call x_4 here a **slack variable**. Then we are asking for the number of solutions to the equation $x_1 + x_2 + x_3 + x_4 = 10$, subject only to the condition that each $x_i \ge 0$. By Theorem 3 with $t = 4$ and $k = 10$, we see that there are $C(13, 10) = C(13, 3) = 286$ possibilities.

The clever insight of introducing a slack variable here made the solution easier than it would have been without that insight, but the following method of solving the problem would also work. For each s from 0 to 10, inclusive, we can serve s cookies. The number of ways to serve s cookies is the number of solutions in natural numbers to $x_1 + x_2 + x_3 = s$, which by Theorem 3 is $C(3+s-1, s)$, or equivalently $C(s+2, 2)$. Therefore, the solution is $C(2, 2) + C(3, 2) + C(4, 2) +$

$\cdots + C(12, 2) = 1 + 3 + 6 + \cdots + 66 = 286$. The fact that this expression equals $C(13, 3)$ is reconsidered in Example 2 of Section 7.1. ◆

We can count solutions to equations with more complicated constraints if we use a bit of ingenuity. The following example shows the kind of reasoning required.

EXAMPLE 7. Consider the situation of Example 6; the children's names are Janek, Pamela, and Sasha.

(a) In how many ways can the snack be served if each child must receive at least one cookie?

(b) In how many ways can the snack be served if Pamela and Sasha must get between one and four cookies each, but Janek insists on getting between two and five cookies? (These ranges are meant to be inclusive.)

Solution. Again we want to solve the equation $x_1 + x_2 + x_3 + x_4 = 10$, but the constraints are more subtle. In the analysis that follows, we can think of the children (and the collector of undistributed cookies) as distinct boxes and the cookies as indistinguishable balls being dropped into these boxes.

(a) If each child must receive at least one cookie, then the inequalities that the variables must satisfy are no longer

$$x_1 \geq 0, \qquad x_2 \geq 0, \qquad x_3 \geq 0, \qquad x_4 \geq 0,$$

but rather

$$x_1 \geq 1, \qquad x_2 \geq 1, \qquad x_3 \geq 1, \qquad x_4 \geq 0.$$

To solve this problem, we can think of each child's receiving one cookie at the outset. No choice is involved here: There is only one way to hand out the first round of cookies. That leaves seven cookies, which can be distributed with no other constraints. (With the box metaphor, we take three of the 10 balls and drop one into each of the first three boxes before doing anything else.) Now Theorem 3 applies with $t = 4$ and $k = 7$, so there are $C(4 + 7 - 1, 7) = C(10, 7) = 120$ solutions. If we wanted to be more formal about it, we could create new variables to represent the "excess" cookies given to the children:

$$x_1' = x_1 - 1, \qquad x_2' = x_2 - 1, \qquad x_3' = x_3 - 1.$$

Thus x_i' is the number of cookies given to child i in addition to the one required cookie. Equivalently,

$$x_1 = x_1' + 1, \qquad x_2 = x_2' + 1, \qquad x_3 = x_3' + 1.$$

The original equation becomes, upon making these substitutions,

$$x_1' + x_2' + x_3' + x_4 = 7$$

with each variable only constrained to be nonnegative. Thus there are $C(4+7-1, 7)$ solutions.

(b) Again, we can distribute the required cookies before we start counting possibilities: two to Janek, one each to Pamela and Sasha. This leaves six cookies, and the constraints on gluttony imposed in the original problem are that each child must now get from zero to three *additional* cookies, while the slack variable can have any nonnegative value. Letting the variables stand for the incremental distribution, we want to count the number of solutions to

$$x_1 + x_2 + x_3 + x_4 = 6$$

subject to the constraints

$$0 \le x_1 \le 3, \qquad 0 \le x_2 \le 3, \qquad 0 \le x_3 \le 3, \qquad x_4 \ge 0.$$

If we ignore these upper bounds on x_1, x_2, and x_3, then there are $C(4+6-1, 6) = C(9, 6)$ solutions. By ignoring the constraints, however, we have overcounted. To see how many cases we have counted that we did not want to count, let us see how the problem would read if one of the constraints is violated. (In this example it would not be possible for more than one of the constraints to be violated simultaneously; if, say, x_1 and x_2 were both bigger than three, then the sum would have to be at least 8 and could not equal 6.) There are three symmetric variables at this point, so without loss of generality let us assume that $x_1 > 3$, in other words, that $x_1 \ge 4$. Letting $x_1' = x_1 - 4$, we see that the number of solutions is equal to the number of solutions to

$$x_1' + x_2 + x_3 + x_4 = 2$$

subject to the constraints

$$x_1' \ge 0, \qquad x_2 \ge 0, \qquad x_3 \ge 0, \qquad x_4 \ge 0.$$

By Theorem 3, there are $C(4 + 2 - 1, 2) = C(5, 2)$ solutions. Since there were three children, each of whom could have received more than the allowed number of cookies in a solution we wish to exclude, the total overcount is $3C(5, 2)$. Therefore, by part (a) of Theorem 3 in Section 6.1, the number of solutions to the problem as posed is $C(9, 6) - 3C(5, 2) = 54$. Although this analysis may seem a bit complex, the reader should compare it to the task of finding the answer by enumerating all the possible distributions of cookies. ◆

PARTITIONS OF POSITIVE INTEGERS

As we have seen, many combinatorial problems can be viewed in terms of putting balls into boxes. Theorem 3 deals with the problem of counting the number of ways to put k identical balls into t distinct boxes if each box can hold any number of balls. Such a distribution might be called an "ordered partition" of k into t nonnegative parts. For example, we can partition 6 into five parts as $6 = 1 + 3 + 0 + 1 + 1$. A much harder problem results if we consider the boxes as being indistinguishable as well. A distribution of k identical balls into t identical boxes might be called an "unordered partition" of k into t nonnegative parts. Thus the partition of 6 into five parts mentioned above would be the same as $6 = 3 + 1 + 1 + 1 + 0$ when viewed as an unordered partition.

We usually ignore the boxes containing no balls, and so make the following definition. Let k be a natural number. A **partition** of k is a decomposition of k into a sum of positive integers, with order ignored. The number of partitions of k will be denoted $P(k)$. Note that $P(0) = 1$, since the empty sum (with no summands) is the only partition of 0. We have $P(1) = 1$ (the only partition of 1 is the sum with just 1 in it), and $P(2) = 2$, since we can write $2 = 2$ and $2 = 1 + 1$.

This definition complements the definition of a partition of a *set* given in Section 3.4: A partition of a positive integer k is an unordered listing of the cardinalities of the disjoint subsets comprising the partition of a set with cardinality k.

EXAMPLE 8.   The partitions of 6 are 6, $5+1$, $4+2$, $4+1+1$, $3+3$, $3+2+1$, $3+1+1+1$, $2+2+2$, $2+2+1+1$, $2+1+1+1+1$, and $1+1+1+1+1+1$. (Since order was to be ignored, we have followed the convention of writing the summands in nonincreasing order.) Thus $P(6) = 11$.   ◆

In contrast to Theorem 3, there is no good formula for $P(k)$. We will have more to say about $P(k)$, however, in Section 7.2, where we will develop a recursive approach to computing it. Let us look at one final problem that we *can* solve explicitly.

EXAMPLE 9.   Determine the number of ordered partitions of the positive integer k into t positive parts, and determine the number of ordered partitions of the positive integer k into positive parts.

Solution.   The ordered partition problem is the same as the resource allocation problem covered by Theorem 3. If we let $x_1, x_2, \ldots, x_t$ be the parts into which we wish to partition k, then we are asking for the number of solutions to $x_1 + x_2 + \cdots + x_t = k$ subject to the constraints $\forall i: x_i \geq 1$. Following the same technique as we used in Example 7, we see that this problem is equivalent to counting the number of solutions to $x_1' + x_2' + \cdots + x_t' = k - t$ subject to the constraints $\forall i: x_i' \geq 0$; we are letting $x_i' = x_i - 1$. By Theorem 3 the number of solutions is $C(t + (k - t) - 1, k - t) = C(k - 1, k - t)$, which can also be written as $C(k - 1, t - 1)$, using the identity $C(n, r) = C(n, n - r)$.

An alternative way to obtain this answer directly is to imagine k balls laid out in a row. An ordered partition of k into t positive parts is obtained by placing $t - 1$ barriers between adjacent balls. The partition is then read from left to right, counting the balls between successive barriers. For example, the arrangement $* \mid * * * \mid * \mid *$ corresponds to the partition $6 = 1 + 3 + 1 + 1$. Since there are $k - 1$ pairs of adjacent balls, we simply need to choose $t - 1$ of these pairs to receive the barriers. Therefore, the answer is $C(k - 1, t - 1)$.

This second model is ideal for solving the second part of this problem, determining the number of ordered partitions of k into any number of positive parts. Again a partition corresponds to a placement of barriers between some pairs of adjacent balls. We specify the partition simply by specifying which of the $k - 1$ pairs of adjacent balls are to receive the barriers. Since there are 2^{k-1} subsets of such pairs that we can choose, the answer is 2^{k-1}. For example, the $2^{3-1} = 4$ partitions of $k = 3$ are 3, $2 + 1$, $1 + 2$, and $1 + 1 + 1$, corresponding to the pictures $* * *$, $* * \mid *$, $* \mid * *$, and $* \mid * \mid *$, respectively.   ◆

SUMMARY OF DEFINITIONS

partition of the natural number k: a decomposition of k into a sum of positive integers, order ignored; $P(k)$ is the number of partitions of k (e.g., $4+3+3+1$ is a partition of 11).

slack variable: a variable added to a resource allocation problem to represent the amount of the resource that is not distributed.

EXERCISES

1. How many permutations are there of all the letters in the following words?
    (a) *discrete.*
    (b) *mathematical.*
    (c) *noon.*

2. How many two-dimensional walks of the following lengths are there from the origin to the origin (see Example 3)?
    (a) Length 2.
    (b) Length 10.
    (c) Length 9.
    (d) Length 0.

3. John eats only frozen dinners at his evening meal. The supermarket he shops at has 10 kinds of frozen dinners to choose from. John plans his menus a week (7 days) at a time.
    (a) In how many ways can John buy a week's worth of dinners with one trip to the supermarket?
    (b) In how many ways can John plan his dinner menus for a week before he goes shopping?

4. Solve the domino counting problem (Example 10 of Section 6.1) by applying Theorem 3 of the present section.

5. A collection of at least four red pegs, four white pegs, four yellow pegs, four blue pegs, four green pegs, and four tan pegs is available.
   (a) How many ways are there to arrange four colored pegs in a row?
   (b) How many ways are there to choose four colored pegs from the collection?

6. Compute the following quantities by enumerating all the partitions.
   (a) $P(3)$.
   (b) $P(4)$.
   (c) $P(5)$.
   (d) $P(7)$.

7. A bag contains four identical white balls, three identical red balls, and three identical blue balls. Selections of balls are made from the bag without replacement.
   (a) How many ways are there to choose three balls in order (e.g., the selection *blue–red–blue* is to be considered different from the selection *blue–blue–red*)?
   (b) How many ways are there to choose three balls, ignoring order?
   (c) How many ways are there to choose 10 balls in order?
   (d) How many ways are there to choose 10 balls, ignoring order?
   (e) How many ways are there to choose five balls in order? [*Hint*: Some slightly messy ad hoc reasoning is needed for this part.]
   (f) How many ways are there to choose five balls, ignoring order?

8. Give a proof of Theorem 1 along the following lines. To specify a permutation of the $n_1 + n_2 + \cdots + n_t = n$ objects, we first need to choose the positions in the permutation for the n_1 identical objects of the first type; this can be done in $C(n, n_1)$ ways. Then we need to choose n_2 positions, from the $n - n_1$ remaining, for the objects of the second type. This process continues until finally the objects of the tth type must occupy the n_t remaining positions. Use the multiplication principle, Theorem 2 of Section 6.2, and algebra to obtain the desired formula.

9. How many two-dimensional walks (see Example 3) of the following lengths are there from the origin to the point $(2, 3)$?
   (a) Length 5.
   (b) Length 3.
   (c) Length 15.
   (d) Length 14.
   (e) Length n.

10. *List explicitly* all solutions in integers to the following allocation problems.
    (a) $x_1 + x_2 + x_3 = 5$, each $x_i \geq 0$.

(b) $x_1 + x_2 + x_3 + x_4 + x_5 = 3$, each $x_i \geq 0$.

(c) · $a + b + c = 15$, $a \geq 4$, $b \geq 2$, $1 \leq c \leq 3$.

11. Find the number of solutions in integers to the following problems.

(a) $x_1 + x_2 + x_3 + x_4 = 40$, each $x_i \geq 0$.

(b) $x_1 + x_2 + x_3 + x_4 < 40$, each $x_i \geq 0$.

(c) $x_1 + x_2 + x_3 + x_4 = 40$, $x_1 > 0$, $x_2 > 0$, $x_3 \geq 0$, $x_4 > 5$.

(d) $x_1 + x_2 + x_3 + x_4 = 40$, $x_1 \geq -2$, $x_2 \geq 0$, $x_3 \geq 0$, $x_4 \geq 2$.

12. Sylvia has $100 to distribute to 10 distinct charities this year. She always gives amounts that are multiples of $5. In how many ways can she make her donations

(a)   If she has to give away the entire $100?

(b)   If she has to give away the entire $100 and she wants to give at least $5 to each of the 10 charities?

(c)   If she need not give away the entire $100 (or any of it, for that matter)?

(d)   If she has to donate at least $50 this year?

(e)   If she need not give away the entire $100 and she wants to give at least $5 to each of the 10 charities?

13. How many ways are there to distribute 25 cookies to four people if each person must get between 2 and 10 cookies, inclusive? Assume that all the cookies must be distributed.

14. In how many ways can 30 cookies be distributed to five children under each of the following conditions?

(a)   All 30 cookies must be distributed.

(b)   All 30 cookies must be distributed and each child receives at least four cookies.

(c)   All 30 cookies must be distributed, each child receives at least four cookies, and no child receives more than eight cookies.

15. How many ways are there to distribute eight blue balls and nine red balls to 10 distinct boxes if each box can hold any number of balls?

16. How many natural numbers

(a)   Less than 100,000 have the sum of their digits equal to 15?

(b)   Less than 50,000 have the sum of their digits equal to 10?

(c)   Between 50,000 and 100,000 have the sum of their digits equal to 17?

17. While Theorem 3 gives a formula for the number of k-combinations from an n-set with repetitions allowed, no formula is given in this section for the number of k-permutations from an n-set with repetitions allowed. Write down and justify such a formula.

18. Let A be a set with cardinality k. Explain how the following problems can be viewed using the "balls in boxes" paradigm. Neither of these problems has a nice formula for its solution.

(a)   "How many partitions does *the set A* have?"

(b)   "How many partitions does *the set A* have into exactly t classes?"

19. The owner of a baseball team has a $20 million salary budget for the 25 players. All salaries are to be in multiples of $100,000, and the entire budget is to be spent. Each player, according to union rules, must receive at least $100,000, but the 15 first-string players must receive at least $200,000. The first-string shortstop, who was last year's Most Valuable Player, has asked for a cool million, and the ace first-string relief pitcher wants a million and a half; both of these demands must be met.

   (a)   How many possible salary distributions are there?

   (b)   What is the maximum possible salary a player can receive under these conditions?

   (c)   What is the maximum possible salary a second-string (i.e., not first-string) player can receive under these conditions?

   (d)   What is the maximum possible salary a second-string player can receive under these conditions if each first-string player must get at least as much as each second-string player gets?

20. A golfer shoots a 72 over 18 holes. Assuming that she makes no holes-in-one, how many different score cards are possible?

21. How many partitions are there into exactly two parts

   (a)   Of the natural number k? (The parts have to be positive.)

   (b)   Of a k-set? (The parts have to be nonempty.)

22. How many ways are there to place k identical balls into m distinct boxes, so that none of the boxes remains empty?

Challenging Exercises

23. Suppose that six distinct numbers are chosen at random from the set of integers from 1 to 40. How likely is it that at least two of them are adjacent?

24. Among all the partitions of an integer $N > 1$, which one has the maximum product? (For example, the partition $10 = 5 + 3 + 2$ has the product $5 \cdot 3 \cdot 2 = 30$, but the partition $2 + 2 + 2 + 2 + 2$ has the product 32. Is that the largest product for $N = 10$?)

25. A bowl contains balls numbered 1 through 10, together with 10 identical unnumbered balls.

   (a)   How many unordered selections of 10 balls from the bowl are possible?

   (b)   How many ordered selections of all 20 balls from the bowl are possible?

   (c)   How many ordered selections of 10 balls from the bowl are possible?

26. Recall the brain-teaser described at the beginning of Section 6.1.

   (a)   How many 3-tuples of integers (x, y, z) satisfy the conditions $x + y + z = 70$ and $10 \le x, y, z \le 30$?

(b)   How many 3-tuples of integers (x, y, z) satisfy the conditions $x+y+z = 70$ and $10 \leq x < y < z \leq 30$?

(c)   Solve the brain-teaser.

27. Under the assumption given in Exercise 5, determine how many ways there are to arrange four colored pegs

(a)   In a circle.

(b)   In a circle if we do not consider two patterns different if one read clockwise is the same as the other read counterclockwise (e.g., $RWYY$ is the same as $WRYY$).

(c)   In a row if we do not consider two patterns different if one is the same as the other read backwards (e.g., $RWYY$ is the same as $YYWR$).

Exploratory Exercises

28. By explicitly listing them, count the number of partitions of k into odd parts (each summand is an odd number), the number of partitions of k into an odd number of parts, and the number of partitions of k into distinct parts (each summand is distinct), for k from 2 to 8. Then formulate a conjecture about two or more of these quantities. (For example, for $k = 5$, the partitions into odd parts are 5, $3 + 1 + 1$, and $1 + 1 + 1 + 1 + 1$; the partitions into an odd number of parts are 5, $3 + 1 + 1$, $2 + 2 + 1$, and $1 + 1 + 1 + 1 + 1$; and the partitions into distinct parts are 5, $4 + 1$, and $3 + 2$.)

29. Find out about **Pólya's theory of enumeration**, which is a systematic method to deal with problems like 27, by consulting Liu [198] and Pólya, Tarjan, and Woods [237].

30. Find out more about partitions by consulting Andrews [8].

SECTION 6.4
THE PIGEONHOLE PRINCIPLE

At evening, casual flocks of pigeons make
Ambiguous undulations as they sink
Downward to darkness, on extended wings.

—Wallace Stevens, "Sunday Morning"

A patently obvious fact about finite sets, known as the **pigeonhole principle**, turns out to be surprisingly useful in proving theorems in discrete mathematics. In this section we take a look at the principle, its generalizations, and some applications.

ONE PIGEON PER HOLE

Fix a positive integer n. *If each pigeon in a collection of more than n pigeons occupies one of n pigeonholes, then some hole must contain more than one pigeon.* This fact is clear, since if each hole contained at most one pigeon, then there would be at most n pigeons altogether. We formalize this principle as a theorem in several equivalent ways. In each case the proof is immediate. (For part (c), recall the definition of equivalence relation and equivalence classes from Section 3.4.)

THEOREM 1. The Pigeonhole Principle.

(a)   *If $f : A \rightarrow B$ is a function from a finite set A to a finite set B, and if $|A| > |B|$, then f is not injective (i.e., f is not a one-to-one function).*

(b)   *Let k and n be positive integers with $k > n$. If k balls are put into n boxes, then some box contains at least two balls.*

(c)   *If R is an equivalence relation on a finite set A, and if the number of equivalence classes for R is less than $|A|$, then there exist two different elements x and y in A such that xRy.*

EXAMPLE 1.   At least two members of the U.S. House of Representatives have the same birthday (month and date, not year). This follows from the pigeonhole principle, since there are only 366 possible birthdays (these are the pigeonholes) and 435 members of the House (these are the pigeons). It is possible that each of the 100 U.S. Senators has a different birthday, since $100 \not> 366$. The probability that 100 randomly chosen people all have different birthdays is *extremely* small, however (see Exercise 45 in Section 6.2). ◆

EXAMPLE 2.   A deterministic computer cannot be programmed to produce a sequence of random numbers. To see why this is so, we need to make two observations about a computer. First, it contains only a finite number of bits of memory, say N, each of which can at any point in time be in any of two states. Therefore, there are only a finite number, 2^N, of possible internal states or configurations for the entire machine. Second, the configuration of the computer at any point in time is determined by its previous configuration. (We assume that state changes take place at discrete time intervals, say every 100 nanoseconds.) Now by the pigeonhole principle, after $2^N + 1$ units of time have passed, the computer must have been in the same state twice, and therefore by the second observation it will continually loop through the same sequence of configurations. Thus any sequence of supposedly random numbers that the computer tries to produce will eventually cycle and hence cannot be random forever (under any reasonable definition of randomness, which we purposely avoid making precise here). This theoretical argument, while absolutely correct, is of no consequence in practice. Since 2^N is so large, the forced cycling would not have to begin for quite a long time, so a finite sequence of numbers meeting a user's requirements for randomness could conceivably be produced. In fact, fairly good, apparently random finite sequences of numbers can be generated by fairly simple arithmetical algorithms, as we saw in Section 4.5. ◆

The next example, an application of the pigeonhole principle to a mathematical problem, is striking because the problem is so difficult to approach any other way.

EXAMPLE 3. Show that if S is any subset of $\{1, 2, \ldots, 2n\}$ with cardinality $n + 1$, then some element of S is a multiple of some other element of S.

Solution. Note that we cannot weaken the hypothesis, since the subset $\{n + 1, n + 2, \ldots, 2n\}$, which has cardinality n, has no element being a multiple of another element. To prove the proposition at hand, partition $\{1, 2, \ldots, 2n\}$ by the following equivalence relation:

$$a R b \longleftrightarrow \mathrm{odd}(a) = \mathrm{odd}(b),$$

where for any positive integer x we define the **odd part** of x, $\mathrm{odd}(x)$, to be the largest odd divisor of x. Note that we can write every number x as a power of two (possibly 2^0) times its odd part. For example, $\mathrm{odd}(140) = 35$, and $140 = 2^2 \cdot 35$. It is not hard to see that the equivalence classes for R are as follows:

$$[1] = \{\, x \leq 2n \mid \mathrm{odd}(x) = 1 \,\} = \{1, 2, 4, 8, 16, \ldots\}$$
$$[3] = \{\, x \leq 2n \mid \mathrm{odd}(x) = 3 \,\} = \{3, 6, 12, 24, \ldots\}$$
$$[5] = \{\, x \leq 2n \mid \mathrm{odd}(x) = 5 \,\} = \{5, 10, 20, 40, \ldots\}$$

$$\vdots$$

$$[2n - 1] = \{\, x \leq 2n \mid \mathrm{odd}(x) = 2n - 1 \,\} = \{2n - 1\}.$$

In particular, there are exactly n equivalence classes, since there are n odd numbers from 1 to $2n$. Since we are given that $|S| = n + 1$, the pigeonhole principle guarantees that S contains two different elements x and y in the same equivalence class, that is, such that $\mathrm{odd}(x) = \mathrm{odd}(y) = k$, say. Then we must have

$$x = 2^i k \qquad \text{and} \qquad y = 2^j k$$

for some pair of distinct natural numbers i and j. If $i < j$, then clearly y is a multiple of x, and if $i > j$, then clearly x is a multiple of y, as desired. ◆

MANY PIGEONS PER HOLE

We turn next to a generalization of the pigeonhole principle in which we can conclude, under appropriate hypotheses, that there is at least one pigeonhole with *many* pigeons in it. Precisely, *if each pigeon in a collection of more than mn pigeons occupies one of n pigeonholes, then some hole must contain more than m pigeons*. Again the proof is clear: If each hole contained at most m pigeons, then there would be at most mn pigeons altogether. We formalize this in Theorem 2, in a slightly more computationally useful form. Recall from Section 3.1 that the ceiling of a real number x, denoted $\lceil x \rceil$, is x rounded up if necessary to an integer—in other words, the smallest integer greater than or equal to x. In particular we always have $x \leq \lceil x \rceil < x + 1$.

THEOREM 2. The Generalized Pigeonhole Principle. *Let k and n be positive integers. If k balls are put into n boxes, then some box contains at least $\lceil k/n \rceil$ balls.*

Proof. If each box contained fewer than $\lceil k/n \rceil$ balls, then there would be at most $n(\lceil k/n \rceil - 1)$ balls altogether. But $n(\lceil k/n \rceil - 1) < n((k/n) + 1 - 1) = k$, a contradiction. ■

EXAMPLE 4. If 479 students are enrolled in Calculus I, and if there are seven sections of the course being offered, then some section has at least $\lceil 479/7 \rceil = \lceil 68.4 \ldots \rceil = 69$ students in it. ◆

For a substantial application of the generalized pigeonhole principle, we consider the monotone subsequence problem. A **monotone subsequence** of a sequence of distinct numbers is a subsequence that is either totally increasing (each term bigger than the one before it) or totally decreasing (each term smaller than the one before it). The problem is to determine the length of—and usually to find, as well—the longest increasing, or decreasing, subsequence of a given finite sequence. In symbols, if we let $x_1, x_2, \ldots, x_n$ be a sequence of distinct numbers, then the monotone subsequence problem is to find the largest k such that for some choice of indices $i_1, i_2, \ldots, i_k$, with $1 \leq i_1 < i_2 < \cdots < i_k \leq n$, the sequence $x_{i_1}, x_{i_2}, \ldots, x_{i_k}$ is either increasing, that is,

$$x_{i_1} < x_{i_2} < \cdots < x_{i_k},$$

or decreasing, that is,

$$x_{i_1} > x_{i_2} > \cdots > x_{i_k}.$$

EXAMPLE 5. The sequence $23, 9, 11, 6, 5, 68, 1, 62, 95$ has a longest increasing subsequence of length 4 (one such subsequence is $9, 11, 62, 95$). Its longest decreasing subsequence has length 5 (the subsequence $23, 11, 6, 5, 1$, for example). On the other hand, the sequence $7, 8, 9, 4, 5, 6, 1, 2, 3$ has no monotone subsequence (of either type) of length more than 3. ◆

The reader should think about how to devise algorithms for finding such subsequences (see Exercises 27 and 37 in this section and Exercise 24 in Section 10.1). Here we will prove that the length of the longest monotone subsequence must be at least $\sqrt{n}$.

THEOREM 3. *Let m be a natural number. Any sequence of $m^2 + 1$ distinct numbers has either an increasing subsequence of length $m + 1$ or a decreasing subsequence of length $m + 1$.*

Proof. Let the given sequence be $x_1, x_2, \ldots, x_{m^2+1}$. There is nothing to prove if $m = 0$, so assume that $m \geq 1$. We present a subtle proof by contradiction using the generalized pigeonhole principle. Suppose that there is no monotone subsequence of length $m + 1$. For each i from 1 to $m^2 + 1$, let l_i be the length of the longest increasing subsequence that starts at x_i. By our assumption, each of the $m^2 + 1$ numbers l_i is between 1 and m. Therefore, by the generalized pigeonhole principle, at least

$$\left\lceil \frac{m^2 + 1}{m} \right\rceil = m + 1$$

of these numbers are the same (the l_i's are the pigeons, and the numbers from 1 to m are the pigeonholes). Now if $i < j$ and $l_i = l_j$, then $x_i > x_j$: Otherwise x_i followed by the longest increasing subsequence starting at x_j would be an increasing subsequence of length $l_j + 1$ starting at x_i, a contradiction since $l_i = l_j$. Therefore, the (at least) $m + 1$ terms x_i in the original sequence with a common value of l_i form a decreasing subsequence of length at least $m + 1$. ∎

COROLLARY. *Let n be a positive integer. Then every sequence of n distinct numbers has a monotone subsequence of length at least $\sqrt{n}$.*

Proof. Take $m = \lceil \sqrt{n} \rceil - 1$ in Theorem 3. The details are left to the reader (Exercise 15). ∎

The result given in this corollary is the best possible. In other words, there are sequences of length m^2 for which the length of the longest monotone subsequence is m (see Exercise 16). We can also ask about monotone subsequences of infinite sequences of distinct numbers. It follows immediately from the corollary that such a sequence must have arbitrarily long monotone subsequences. Even more is true, however; such a sequence must have an infinite monotone subsequence. The proof of this last fact is rather subtle (see Exercise 32).

RAMSEY THEORY

It is possible to generalize the pigeonhole principle even further. In fact, the pigeonhole principle has spawned an abundance of interesting theorems and a whole new area of mathematical research, covering many fields within mathematics, including set theory, logic, number theory, graph theory, geometry, and functional analysis. These profound, and often subtle, results comprise what is called Ramsey theory, named after the British logician who proved some of the first results in this area around 1930. In this last subsection we look at a small corner of this fascinating subject. (The Hungarian mathematicians P. Erdős and G. Szekeres began this study in the 1930s. Many of the results mentioned here are due to them.)

We begin with the simplest classical example of a Ramsey theorem.

THEOREM 4.   *Suppose that six people attend a party. Then among the six, there are either three mutual acquaintances or three mutual strangers. In other words, either there are three party-goers X, Y, and Z, such that X and Y are acquainted with each other, and X and Z are acquainted with each other, and Y and Z are acquainted with each other; or there are three party-goers X, Y, and Z such that X and Y are not acquainted with each other, and X and Z are not acquainted with each other, and Y and Z are not acquainted with each other.*

Proof.   Consider one person at the party; call her A. Then of the five other people, either A is acquainted with at least three, or A is not acquainted with at least three (this is an application of the generalized pigeonhole principle; the five people are the pigeons and the properties "acquaintance" and "nonacquaintance" are the pigeonholes; at least one of the pigeonholes has $\lceil 5/2 \rceil = 3$ pigeons). Since acquaintance and nonacquaintance have a symmetric role in this theorem, we can assume, without loss of generality, that the former case holds, that is, that A is acquainted with at least three others; call them B, C, and D. Now if any two of these three people are acquainted with each other, then those two together with A form the desired triangle of mutual acquaintances. On the other hand, if no two of B, C, and D are acquainted with each other, then B, C, and D form the desired triangle of mutual strangers.   ■

Note that if only five people attend the party, then there need not be three mutual acquaintances or three mutual strangers present. Indeed, let the party-goers be Carol, Cheryl, Claudia, Grace, and Alex. Suppose that the following pairs are acquainted: Carol and Cheryl, Cheryl and Claudia, Claudia and Grace, Grace and Alex, and Alex and Carol. The reader can verify that there is no subset of three of these people who are all acquainted with each other or all not acquainted with each other. We express the fact that the theorem is true but that the number 6 in the theorem could not have been reduced to 5 by saying that the **Ramsey number** of 3 is 6, or, in symbols, $R(3) = 6$.

There is nothing special about the number 3 in Theorem 4. For example, it can be proved—but the proof is quite subtle and messy—that $R(4) = 18$. In other words, if 18 people attend a party, then either some four of them are mutual acquaintances (i.e., every pair among those four are acquainted) or some four of them are mutual strangers (i.e., every pair among those four are not acquainted);

whereas there is an example of a set of 17 people, no four of whom are mutual acquaintances and no four of whom are mutual strangers. Indeed, the general statement of the finite version of Ramsey's classical theorem is as follows.

THEOREM 5.   *For each pair of integers k and l greater than 1, there exists a natural number N, depending on k and l, such that if N people attend a party, then either some k of them are mutual acquaintances or some l of them are mutual strangers. The smallest such number N is denoted $R(k, l)$.*

The proof of Theorem 5, using the idea in the proof of Theorem 4 and mathematical induction, is left to the reader (Exercise 30). In fact, the proof gives us a recursively defined upper bound on how large $R(k, l)$ can be. The values of $R(k, l)$ grow exponentially with k and l, but no one knows exactly how; in fact, no one even knows the exact value of $R(5, 5)$. In the notation of this theorem, $R(k)$ is $R(k, k)$. There is also a sense in which $R(\infty) = \infty$ (see Exercise 32).

We carry the generalization one step further. Instead of considering the irreflexive, symmetric binary relation "is acquainted with" defined on the set of people, we imagine a "relation" T holding or not holding on subsets of cardinality r of some set S. For irreflexive, symmetric binary relations, $r = 2$. For example, we could consider the 3-ary relation "adds up to 10" on the set of positive integers; then, for example, the relation holds for $\{1, 2, 7\}$ but not for $\{3, 4, 8\}$. Note that we are not using the word "relation" in quite the same way that we used it in Section 3.3. We are imposing a symmetry condition by insisting here that when T holds on an r-tuple of distinct elements of S, then T also holds on every permutation of that r-tuple. In other words, we can think of T as a subset of the set of all subsets of S having cardinality r.

Here, then, is a more general form of Ramsey's theorem.

THEOREM 6.   *Suppose that p, q, and r are integers with p and q both greater than or equal to $r \geq 2$. Then there is a natural number N, depending on p, q, and r, such that if S is a set with cardinality N, and if T is a relation that either holds or does not hold for each r-element subset of S, then either there is a p-element subset of S on all of whose r-element subsets T holds, or there is a q-element subset of S on none of whose r-element subsets T holds. The smallest such number N is denoted by $R(p, q; r)$.*

Theorems 4 and 5 are special cases of Theorem 6. In Theorem 4, for instance, $p = q = 3$, $r = 2$, $N = 6$, S is the set of six people at the party, and T is the "is acquainted with" relation. Incidentally, this is still not the most general form, since we can generalize to more parameters (see Exercise 31).

EXAMPLE 6.   Suppose that we consider the relation T that holds on a set of four people if and only if those four people have played bridge together (bridge being a game that requires exactly four players). Let $r = 4$. Also let $p = 10$ and $q = 12$.

Theorem 6 guarantees that there is a number N (the smallest such number is denoted by $R(10, 12; 4)$) such that given any set of N people whatsoever, either we can find a subset of 10 of them such that *every* foursome from that subset (and there are $C(10, 4) = 210$ such foursomes) have played bridge together; or else we can find a subset of 12 of them such that *not even one* of the $C(12, 4) = 495$ foursomes from that subset have played bridge together. Theorem 6 does not tell us how big $N = R(10, 12; 4)$ is, nor how to find it, but its determination is a finite problem whose solution can be obtained in principle by brute force. In general, the numbers $R(p, q; r)$ arising in Ramsey theory are unfathomably large, like the numbers produced by the Ackermann function (defined in Section 5.1). ◆

Finally, let us apply Theorem 6 to a problem in combinatorial geometry. Suppose that we have some distinct points in the plane and we want to know under what conditions we can find n of these points that are the vertices of a convex n-gon. (A figure is **convex** if for every two points in the interior of the figure, the line segment joining the two points lies entirely in the interior of the figure. An **n-gon** is a polygon with n vertices.) For example, the six points in the right half of Figure 6.4 do contain a subset of five points that form a convex pentagon (as shown), whereas the six points in the left half do not, as the reader may verify with a little trial and error. (The need to determine whether points form convex polygons arises in computer graphics.)

FIGURE 6.4 Six points may or may not form a convex pentagon.

We will assume that the points are in what is called **general position**, that is, that no three or more are collinear. The next theorem says that *if there are enough points, then some n of them do form a convex n-gon.* We begin with two lemmas.

LEMMA 1. *If five points in the plane are in general position, then some four of them are the vertices of a convex quadrilateral.*

Proof. Look at what we will call the **convex shell** of the five points—the smallest convex polygon that encloses all the points. There are three possibilities for the shape of

  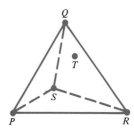

FIGURE 6.5 Possible shells for a collection of five points.

the convex shell: It can contain all five of the points, four of them, or three of them, as shown in Figure 6.5.

  If the convex shell is a pentagon, as in the picture on the left in Figure 6.5, then any four of the points are the vertices of a convex quadrilateral. If the convex shell is a quadrilateral (with one point inside), as in the picture in the middle of Figure 6.5, then the four points of this quadrilateral are the points we are looking for. Finally, suppose that the convex shell is a triangle $\triangle PQR$ (see the rightmost picture in Figure 6.5). Using one of the two points inside the triangle, say S, we can split the interior into three triangular regions, as the dashed lines in Figure 6.5 show. The fifth point T lies within one of these regions, say the one that does not have P on its boundary. Then we claim that either $PSTQ$ or $PSTR$ is a convex quadrilateral. To see why, extend the line PS. If T is on the same side of this line as Q is, then $PSTQ$ is a convex quadrilateral; otherwise, $PSTR$ is a convex quadrilateral. ■

LEMMA 2.   *If $n \geq 4$ points in the plane are in general position, and if all quadrilaterals formed by subsets of four of the points are convex, then the n points are the vertices of a convex n-gon.*

Proof.   As in the proof of Lemma 1, look at the convex shell of the n points. If this polygon contains all n of the points, then we are done. Otherwise, there is at least one point Q inside. Split the interior into triangles by drawing a line segment from one of the vertices of the convex shell, say P, to each of the others. Then, as we see in Figure 6.6, Q lies inside one of the triangles, say $\triangle PRS$. But then the points P, Q, R, and S contradict the hypothesis. ■

THEOREM 7.   *Given any integer $n \geq 3$, there is an integer N such that every set of N or more points in general position in the plane contains a subset of n points that are the vertices of a convex n-gon.*

Proof.   We let N be the Ramsey number $R(n, 5; 4)$, guaranteed by Theorem 6, and we let the 4-ary relation T on the set of points in the plane be "the four points form a

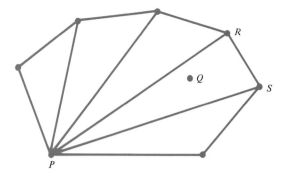

FIGURE 6.6   Situation that cannot occur in
the proof of Lemma 2.

convex quadrilateral." Let S be any set of N or more points in general position
in the plane. By Theorem 6 there is either a subset of n points of S on which T
always holds, or there is a subset of five points of S on which T never holds. If
the latter were the case, then there would be a set of five points, none of whose
four-element subsets determined a convex quadrilateral; this contradicts Lemma 1.
Thus the first possibility must hold; that is, there is a subset of n points, every four
of which determine a convex quadrilateral. This is precisely what Lemma 2 told
us would guarantee that the n points form a convex n-gon, as desired.   ◼

SUMMARY OF DEFINITIONS

convex geometric figure: a figure such that for every two points in its interior, the
line segment joining these two points also lies entirely in its interior (e.g., the
perimeter of a stop sign is convex, but the perimeter of a star on the American
flag is not).

convex shell of a set of points in the plane: the smallest convex polygon that
encloses all the points (e.g., for the points $(-1, 0)$, $(0, 0)$, $(1, 0)$, $(0, 1)$, and
$(0, 2)$ in the usual coordinate system, the convex shell is the triangle with
vertices $(-1, 0)$, $(1, 0)$, and $(0, 2)$).

general position: A collection of points is in general position if and only if
no three of them are collinear (e.g., Baltimore, Dallas, Los Angeles, and
Minneapolis are in general position).

generalized pigeonhole principle: If more than mn pigeons occupy n pigeonholes, then some hole contains more than m pigeons.

monotone subsequence of a sequence: a subsequence in which the terms are
either all increasing or all decreasing (e.g., the subsequence $3, 6, 7, 9$ is a
monotone subsequence of the sequence $3, 10, 6, 1, 2, 7, 9, 4, 5$).

n-gon: a polygon with n vertices (e.g., a triangle is a 3-gon, the perimeter of a
stop sign is an 8-gon, and the perimeter of a star on the American flag is a
10-gon).

odd part of the positive integer x: the largest odd divisor of x, denoted odd(x) (e.g., the odd part of 104 is 13).

pigeonhole principle: If more than n pigeons occupy n pigeonholes, then some hole contains more than one pigeon.

Ramsey number $R(k)$, where k is an integer greater than 1: the smallest positive integer N such that among any collection of N people there is either a subset of k people who are mutual acquaintances or a subset of k people who are mutual strangers; generalized versions of Ramsey numbers, with more parameters, are defined by Theorems 5 and 6 (e.g., $R(3) = 6$).

EXERCISES

1. Use the pigeonhole principle (or generalized pigeonhole principle) explicitly to solve these simple brain-teasers.
    (a) Your bureau drawer contains 10 pairs of loose socks, each pair a different color. How many socks must you take out of the drawer without looking to be sure of getting a matched pair?
    (b) Your bureau drawer contains 10 loose red socks and 10 loose blue socks. How many socks must you take out of the drawer without looking to be sure of getting a matched pair?

2. Show that there are more than 10,000 people in the United States who all have the same height (to the nearest half-inch) and the same weight (to the nearest pound).

3. Ten thousand dollars is available for raises this year in a company with nine employees. Show that at least one of the employees will get less than a $1200 raise.

4. Determine the value of the Ramsey number $R(2)$.

5. Explain how Theorem 5 is a special case of Theorem 6.

6. Show that the configuration that the computer in Example 2 is in after $2^N + 1$ units of time have passed is a state that it was previously in. (This is saying slightly more than was claimed in the discussion in Example 2.)

7. Show that if S is a set containing 10 positive integers less than 54, then at least two different (not necessarily disjoint) four-element subsets of S have the same sum.

8. Show that if S is a set containing 10 positive integers less than 118, then at least two different (not necessarily disjoint) subsets of S have the same sum.

9. Show that any subset of $\{1, 2, \ldots, 2n\}$ having cardinality $n + 1$ contains two distinct elements that are relatively prime (i.e., have a greatest common divisor of 1). [*Hint*: Find the right partition, as we did in Example 3.]

10. Doctor Dodge needs to see 65 patients next week; she works Monday through Saturday.
    (a) Give a lower bound to the number of patients she will see on the busiest day next week.
    (b) Give an upper bound to the number of patients she will see on the least busy day next week.

11. Show that given any five points in the interior of a square 2 inches on a side, at least two of the points are within $\sqrt{2}$ inches of each other.

12. Show that given any five points in the interior of an equilateral triangle 2 inches on a side, at least two of the points are within 1 inch of each other.

13. Prove that if each pigeon in a collection of fewer than mn pigeons occupies one of n pigeonholes, then some hole must contain fewer than m pigeons.

14. A group of people attend a convention and during the course of the convention some pairs of people shake hands. Show that at least two of these people shook hands with the same number of other people at the convention.

15. Carefully and completely fill in the details in the proof of the corollary to Theorem 3.

16. Verify the claim made after the corollary to Theorem 3 that there are sequences of distinct numbers of length m^2 for which the length of the longest monotone subsequence is m. [*Hint*: Generalize the second sequence in Example 5.]

17. As was pointed out in the subsection on Ramsey theory, it can be proved that $R(4) = 18$. Show at least that $R(4) > 6$. For a challenge, see for how large a $k \leq 17$ you can show that $R(4) > k$.

18. Determine the value of $R(2, n)$ as a function of n. Assume that $n \geq 2$.

19. Write Theorem 6 in symbols, paying special attention to the correct use of quantifiers. For this exercise, ignore the last sentence of the theorem.

20. There is a Ramsey-type theorem for arithmetic progressions in the integers, called van der Waerden's theorem. (An **arithmetic progression** of length n is a sequence $a, a + d, a + 2d, \ldots, a + (n-1)d$; for example, $113, 119, 125$ is an arithmetic progression of length 3.) The simplest version of the theorem says that for every positive integer n there is a number N such that if S is any subset of the set $U = \{1, 2, \ldots, N\}$, then either S or $U - S$ contains an arithmetic progression of length n. (It is helpful to think of the elements of S as being "colored red" and the elements of $U - S$ as being "colored blue"; then the theorem says that there is a monochromatic (one-colored) arithmetic progression of length n.) Show that for $n = 3$ the smallest such value of N is 9. In other words,

(a) Find a subset S of $\{1, 2, 3, 4, 5, 6, 7, 8\}$ such that neither S nor $\{1, 2, 3, 4, 5, 6, 7, 8\} - S$ contains an arithmetic progression of length 3.

(b) Show that if S is any subset of $\{1, 2, 3, 4, 5, 6, 7, 8, 9\}$, then either S or $\{1, 2, 3, 4, 5, 6, 7, 8, 9\} - S$ must contain an arithmetic progression of length 3. [*Hint*: The brute force approach works well here. Without loss of generality, assume that 5 is colored red and 4 is colored blue. (Why is this legitimate?) Consider the two cases for the color of 6, and see what follows in each case.]

Challenging Exercises

21. Show that there is a multiple of 127 whose base 10 numeral is comprised entirely of 0's and 7's. [*Hint*: Look at the numbers 7, 77, 777, ...]

22. Show that given any positive number $\epsilon > 0$ and any infinite collection of numbers between 0 and 1, there are two of the numbers whose absolute difference is less than ϵ. [*Hint*: State, justify, and then apply an infinite version of the pigeonhole principle.]

23. Ten workers in a certain high-tech office have purchased a fleet of voice-activated robots to help them with various tasks, such as getting coffee, delivering mail, and carrying furniture. The robots are expensive, so they bought only seven of them, assuming that no more than seven workers would be needing robots simultaneously. Each robot can serve any number of masters (though only one at a time), but in order to respond to voice commands, a time-consuming and expensive training session is needed between each robot and each master it will serve. To avoid waste, the workers want to conduct as few training sessions as possible.

(a) Show that at least 28 training sessions are necessary if it is desired that whenever any seven workers simultaneously want the services of robots there will be a way for them to find robots with the proper training to serve them.

(b) Show that 28 training sessions are sufficient if it is desired that whenever any seven workers simultaneously want the services of robots there will be a way for them to find robots with the proper training to serve them. [*Hint*: Write down a set of 28 master–robot pairs and show that this set suffices.]

(c) Generalize to n workers and k robots; assume that $k \leq n$.

24. A baseball player plays in 40 consecutive games and gets a total of 57 hits during those games, with at least one hit in each game. Show that there is a subset of one or more *consecutive* games during which he got exactly 22 hits. [*Hint*: Let h_n be the number of hits the player got in or before the nth game; thus the sequence $h_1, h_2, \ldots, h_{40}$ is strictly increasing. Consider the 80 numbers consisting of the h_i's and the numbers $h_i + 22$, and apply the pigeonhole principle.]

25. Let n be an integer greater than 2. What is the minimum number of bit strings of length n needed to guarantee that at least two of them agree in at least two positions?

26. A sequence of numbers is called **unimodal** if the terms in the sequence increase for a while (possibly not at all) and then decrease from then on. For example, the sequences $(2, 5, 6, 10, 9, 8, 7, 4, 3, 1)$ and $(1, 2, 3, 4, 5, 6, 7, 8, 9, 10)$ are unimodal.
    (a) Give the definition of unimodal sequence in symbols.
    (b) Determine the number of different unimodal sequences of length n that use each of the numbers 1 through n exactly once. (The sequences given above are two such sequences for $n = 10$.)

27. Describe an algorithm for finding the longest increasing subsequence of a given sequence of distinct numbers. Analyze the time complexity of your algorithm. [*Hint*: Using the notation in the proof of Theorem 3, have your algorithm calculate the numbers l_i, from back to front.]

28. Show that among any 10 people there are either three mutual acquaintances or four mutual strangers. [*Hint*: Use both the method of proof and the result of Theorem 4. In particular, start by observing that a person must have either four acquaintances or six nonacquaintances.]

29. Show that $R(4) \leq 20$. [*Hint*: Use both the method and the result of Exercise 28.]

30. Extend the ideas employed in the previous two exercises to prove Theorem 5. In particular,
    (a) Prove that $R(k, l) \leq R(k - 1, l) + R(k, l - 1)$ by induction on the value of $k + l$ (take $R(1, l) = R(k, 1) = 1$ as the base case).
    (b) Use part (a) and the fact, to be proved in Section 7.1, that $C(n, r) = C(n - 1, r - 1) + C(n - 1, r)$ to show that $R(k, l) \leq C(k + l - 2, k - 1)$.

31. Suppose that there are three mutually exclusive and exhaustive possibilities for the relationship between every pair of people: Either they do not know each other, or they are friends, or they are enemies. Show that at any party with 17 people you will find either three mutual strangers or three mutual friends or three mutual enemies.

32. The ideas in this section can be extended to the infinite, as well.
    (a) Show that $R(\infty) = \infty$ in the following sense. Let T be an irreflexive, symmetric binary relation on the set of positive integers; in other words, for every pair of distinct positive integers, either T holds or T does not hold. Show that either there is an infinite set S of positive integers on which T always holds (i.e., $\forall(x, y) \in S \times S : (x \neq y \rightarrow xTy)$), or there is an infinite set S of positive integers on which T never holds (i.e., $\forall(x, y) \in S \times S : (x \neq y \rightarrow x\overline{T}y)$). [*Hint*: Construct a sequence n_1,

$n_2, \ldots,$ such that for each i, either $\forall j > i : n_i T n_j$ or $\forall j > i : n_i \not\!T n_j$. The set S is a subset of the n_i's.]

(b) Give an example of an infinite sequence of distinct numbers that has arbitrarily long decreasing subsequences but no infinite decreasing subsequence.

(c) Apply either the proposition proved in part (a) or the idea of the proof to show that an infinite sequence of distinct numbers must have an infinite monotone subsequence.

33. Suppose that you are given $2n$ points in the plane in general position (i.e., no three of them are collinear). Half of them are red and half of them are blue. Show that it is possible to draw n line segments, each segment joining a red point to a blue point (so that each red point is paired with a unique blue point, and vice versa) in such a way that none of the line segments intersect.

Exploratory Exercises

34. Ramsey theorems give rise to games, of which the following is typical. Six points are evenly spaced around an imaginary circle in the plane. Two players, Red and Blue, take turns by connecting with a straight line of their own color two of the points that were not previously connected (the lines themselves may cross). Blue goes first. The winner of the game (or, in a variation, the loser) is the first player to form a triangle of his or her own color whose vertices are three of the original six points.

(a) Show that neither the original game nor the variation can end in a draw.

(b) Play the games with some friends, and try to develop some strategies for winning.

(c) Show that Blue has a winning strategy for the original game.

(d) Is there a winning strategy for the original game if there are five points rather than six?

(e) Invent some similar games based on other Ramsey theorems discussed in this section or exercise set.

(f) Find out more about such games by consulting Harary and Plochinski [143] and Shader [269].

35. We showed in Theorem 4 that there had to be at least one "triangle" of acquaintances or strangers among any group of six people. Try to prove that in fact there have to be two such triangles, or consult A. W. Goodman [120] for a proof.

36. Find out about Schur's theorem, a Ramsey-type theorem which guarantees monochromatic (one-colored) solutions of certain equations if all the integers are colored, by consulting Honsberger [156].

37. Find out about another approach to the monotone subsequence problem by consulting Schensted [265].

38. Find out more about Ramsey theory by consulting Gardner [104], Graham, Rothschild, and Spencer [126], and F. S. Roberts [254].

39. Ron Graham is one of the principal figures in Ramsey theory research. Find out about his interesting life by consulting Albers and Alexanderson [4] and Kolata [185].

ADDITIONAL TOPICS
IN COMBINATORICS

We cannot hope to cover the entire field of combinatorics in just two chapters of this book. In this chapter we discuss several topics beyond the basics covered in Chapter 6. In Section 7.1 we explore combinatorial identities. These are equations, similar to trigonometric identities, that always hold among various combinatorial quantities, notably the numbers $C(n, k)$. In Sections 7.2 and 7.3 we discuss recurrence relations. Here we will develop techniques that enable us to obtain explicit formulas for recursively defined functions. The methods are quite powerful and find important application, for example, in analyzing the efficiency of algorithms. A generalization of the overcounting principle from Section 6.1, called the inclusion–exclusion principle, is discussed in Section 7.4; it allows us to determine the cardinalities of some rather messy sets. Finally, in Section 7.5 we look at generating functions, a powerful tool for working with sequences; with generating functions we can transform combinatorial problems into algebraic ones.

The general references on combinatorics given in the introduction to Chapter 6 are appropriate for this chapter as well.

SECTION 7.1
COMBINATORIAL IDENTITIES

*"I can't explain myself, I'm afraid, Sir," said
Alice, "because I'm not myself, you see."*

—Lewis Carroll, *Alice's Adventures in Wonderland*

In solving combinatorial problems, the numbers $C(n, k)$ occur repeatedly. These numbers have some nice properties and other useful applications that we explore in this section.

The results we will obtain are called **combinatorial identities**, since they are algebraic identities (equations with variables in them that hold for *all* relevant values of the variables) involving combinatorial quantities like $C(n, k)$. There is another sense in which the identities we will study are combinatorial, and that is that we can often prove them by combinatorial arguments. The reader should compare the identities we discuss here with trigonometric identities (such as $\cos 2x = 1 - 2\sin^2 x$) that prove quite useful in calculus (e.g., in integrating $\sin^2 x$).

THE BINOMIAL THEOREM

Every beginning algebra student memorizes the fact that $(a + b)^2 = a^2 + 2ab + b^2$. Students who go further in algebra may also memorize the corresponding result for the third power of a binomial, after they have multiplied it out a few times: $(a+b)^3 = a^3 + 3a^2b + 3ab^2 + b^3$. There is a general formula for expanding $(a+b)^n$ for any positive integer n, which is known as the **binomial theorem**. To state it we need to use the numbers $C(n, k)$, which we first defined in Section 6.2. Since the numbers $C(n, k)$ arise in expanding the powers of a binomial, they are often called the **binomial coefficients**.

THEOREM 1. The Binomial Theorem.   *If n is a natural number, then*

$$(a + b)^n = \sum_{k=0}^{n} C(n, k)a^k b^{n-k}.$$

Proof.   There are several ways to prove this result. We give one proof here that relies on the meaning of $C(n, k)$; a proof by mathematical induction will be given later in this section.

When we expand $(a + b)^n = (a + b)(a + b) \cdots (a + b)$, we will get 2^n terms, each one a product of some (zero or more) a's and some (zero or more) b's, with a total of n factors in all; in other words, the general term will look like $a^k b^{n-k}$, where k is an integer between 0 and n, inclusive. These statements follow from

429

the fact that in order to obtain a term of the expansion, we must choose either the summand a or the summand b in each of the n factors $(a + b)$. Now we need to figure out how many of these terms are $a^k b^{n-k}$ for each possible value of k. The term $a^k b^{n-k}$ will arise whenever we choose, in forming the product, exactly k of the a summands and exactly $n - k$ of the b summands. The number of ways to do this is just the number of subsets of cardinality k that the set of the n factors has, since we need only specify which of the n factors we want to choose the a's from. Therefore, there are $C(n, k)$ terms $a^k b^{n-k}$ in the expansion of $(a + b)^n$; in other words, after collecting terms, $C(n, k)a^k b^{n-k}$ is part of this expansion. Since the expansion consists precisely of all these terms for all relevant values of k, the result follows. ∎

As written, the formula gives the expansion in increasing powers of a; we could just as well have given it in terms of decreasing powers of a (as we are used to doing when expanding $(a + b)^2$, for example). Written this way (with the alternative notation for the binomial coefficients for variety), the theorem tells us that

$$(a + b)^n = \sum_{k=0}^{n} \binom{n}{k} a^{n-k} b^k.$$

EXAMPLE 1.   We can write down $(x + 2y)^5$ without multiplying it out by invoking the binomial theorem:

$$(x + 2y)^5 = \sum_{k=0}^{5} C(5, k) x^k (2y)^{n-k}$$

$$= C(5, 0) x^0 (2y)^5 + C(5, 1) x^1 (2y)^4 + C(5, 2) x^2 (2y)^3$$
$$+ C(5, 3) x^3 (2y)^2 + C(5, 4) x^4 (2y)^1 + C(5, 5) x^5 (2y)^0$$
$$= 2^5 y^5 + 5 x 2^4 y^4 + 10 x^2 2^3 y^3 + 10 x^3 2^2 y^2 + 5 x^4 2y + x^5$$
$$= 32 y^5 + 80 x y^4 + 80 x^2 y^3 + 40 x^3 y^2 + 10 x^4 y + x^5.$$ ◆

By substituting various quantities for a and b in Theorem 1, we obtain some interesting corollaries of the binomial theorem.

COROLLARY 1.   *If n is a natural number, then*

$$\sum_{k=0}^{n} C(n, k) = 2^n.$$

Proof.   We simply let $a = b = 1$ in Theorem 1. ∎

For example, $C(4, 0)+C(4, 1)+C(4, 2)+C(4, 3)+C(4, 4) = 1+4+6+4+1 = 16 = 2^4$.

COROLLARY 2. *If n is a positive integer, then*

$$\sum_{k=0}^{n}(-1)^k C(n, k) = 0.$$

Proof. We simply let $a = -1$ and $b = 1$ in Theorem 1. ∎

For example, $C(4, 0)-C(4, 1)+C(4, 2)-C(4, 3)+C(4, 4) = 1-4+6-4+1 = 0$. This fact should already have been clear to us in the case of odd n, since in that case the term $(-1)^k C(n, k)$ in the sum is exactly canceled out by the term $(-1)^{n-k} C(n, n-k)$. (For example, with $n = 5$ we have $1-5+10-10+5-1 = 0$.) When n is even, however, as in our example here with $n = 4$, something more subtle seems to be going on.

Another fundamental identity involving the binomial coefficients is known as **Pascal's identity**.

THEOREM 2. Pascal's Identity. *If $1 \leq k < n$, then $C(n, k) = C(n - 1, k) + C(n - 1, k - 1)$.*

Proof. We give an algebraic proof, using the formula for the binomial coefficients from Theorem 2 in Section 6.2. Starting with the left-hand side, we have

$$C(n - 1, k) + C(n - 1, k - 1)$$

$$= \frac{(n - 1)!}{k!(n - 1 - k)!} + \frac{(n - 1)!}{(k - 1)!((n - 1) - (k - 1))!}$$

$$= \frac{(n - 1)!}{k!(n - k - 1)!} + \frac{(n - 1)!}{(k - 1)!(n - k)!}$$

$$= \frac{(n - 1)!}{k(k - 1)!(n - k - 1)!} + \frac{(n - 1)!}{(k - 1)!(n - k)(n - k - 1)!}$$

$$= \frac{(n - 1)!(n - k) + (n - 1)!k}{k(k - 1)!(n - k)(n - k - 1)!}$$

$$= \frac{(n - 1)!n}{k(k - 1)!(n - k)(n - k - 1)!}$$

$$= \frac{n!}{k!(n - k)!} = C(n, k). \qquad \blacksquare$$

For example, this theorem tells us that $C(5,3) + C(5,4) = C(6,4)$, as the calculation $10 + 5 = 15$ confirms.

Pascal's identity gives us a recursive way to calculate the binomial coefficients, since it tells us the value of $C(n,k)$ in terms of binomial coefficients with a smaller value of n. The base cases are $C(n,0) = C(n,n) = 1$ for all $n \geq 0$, since Theorem 2 only applies for $1 \leq k < n$. This recursive approach allows us to form **Pascal's triangle**, the display of the binomial coefficients shown in Figure 7.1.

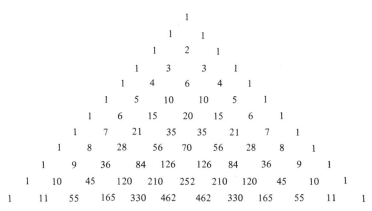

FIGURE 7.1 Pascal's triangle.

The nth row of Pascal's triangle gives the binomial coefficients $C(n,k)$, as k goes from 0 (at the left) to n (at the right); the top row, consisting of just the number 1, is for $n = 0$. The left and right borders are all 1's, reflecting the fact that $C(n,0) = C(n,n) = 1$ for all n. Each entry in the interior of the triangle is the sum of the two entries immediately above it; for example, each 15 in row 6 (remember that we are starting the count of rows at 0) is the sum of the 10 and the 5 immediately above it.

By Corollary 1 to Theorem 1 the sum of the entries in the nth row of Pascal's triangle is 2^n. By Corollary 2 the alternating sum of the entries in the nth row ($n \geq 1$) is 0 (i.e., if we alternately subtract and add the terms in any row we get 0 for the answer). The diagonals of Pascal's triangle are also interesting; they correspond to constant values of k. The left edge, consisting of all 1's, corresponds to $k = 0$, reflecting the fact that $C(n,0) = 1$. The diagonal parallel to the left edge but moved one unit to the right reads (from the top down) $1, 2, 3, 4, 5, \ldots$, reflecting the fact that $C(n,1) = n$ for $n \geq 1$. The next diagonal to the right, reading $1, 3, 6, 10, 15, \ldots$, reflects the fact that $C(n,2) = n(n-1)/2$ for $n \geq 2$; these numbers are called "triangular" numbers, and their differences increase by 1 as we move down the diagonal.

Other interesting patterns, theorems, and conjectures emerge from Pascal's triangle. The reader is urged to find some; there are a few suggestions in the exercises.

As an application of Pascal's identity, let us give an inductive proof of the binomial theorem. The base case is clear, since $(a + b)^0 = 1 = 1 \cdot a^0 b^0 = \sum_{k=0}^{0} C(0, k)a^k b^{0-k}$. Now suppose that the theorem is true for n; we want to prove it for $n + 1$. We have

$$(a + b)^{n+1} = (a + b)(a + b)^n$$

$$= (a + b) \sum_{k=0}^{n} C(n, k)a^k b^{n-k} \quad \text{(by the inductive hypothesis)}$$

$$= \sum_{k=0}^{n} C(n, k)a^{k+1} b^{n-k} + \sum_{k=0}^{n} C(n, k)a^k b^{n-k+1}$$

$$= \sum_{j=1}^{n+1} C(n, j - 1)a^j b^{n+1-j} + \sum_{j=0}^{n} C(n, j)a^j b^{n+1-j}$$

(we let $j = k + 1$ in the first sum, and $j = k$ in the second)

$$= a^{n+1} b^0 + \left(\sum_{j=1}^{n} (C(n, j - 1) + C(n, j))a^j b^{n+1-j} \right) + a^0 b^{n+1}$$

$$= a^{n+1} b^0 + \left(\sum_{j=1}^{n} C(n + 1, j)a^j b^{n+1-j} \right) + a^0 b^{n+1}$$

(by Pascal's identity)

$$= \sum_{j=0}^{n+1} C(n + 1, j)a^j b^{n+1-j},$$

which is exactly what we wanted.

COMBINATORIAL PROOFS

We turn now to a proof technique that is totally different from the algebraic or inductive proofs that we are used to. We will illustrate the technique, called combinatorial proof, by using it to prove Pascal's identity,

$$C(n, k) = C(n - 1, k) + C(n - 1, k - 1).$$

In order to show that the left-hand side equals the right-hand side, we will show that both sides count the same thing.

The left-hand side tells us how many subsets of cardinality k a set S with n elements has. We will show that the right-hand side also counts the number of subsets of cardinality k. Let us distinguish one particular element of S and call it x. Either a subset of cardinality k from S will contain x, or it will not; and these two possibilities are disjoint. If we want to choose a subset that does not contain x, then there are $C(n-1, k)$ ways to make the choice, since there are $n-1$ elements other than x in S, and we need to choose k of them. On the other hand, if we want to choose a subset that does contain x, then there are $C(n-1, k-1)$ ways to make the choice, since we need to include x (no choice involved) and $k-1$ elements chosen from the $n-1$ other elements in S. Therefore, by the addition principle there are $C(n-1, k) + C(n-1, k-1)$ ways to choose a subset of cardinality k from S, and our proof is complete.

Next let us give a combinatorial proof of Corollary 1 to the binomial theorem. We want to prove that $\sum_{k=0}^{n} C(n, k) = 2^n$. We claim that both sides represent the number of subsets of a set with n elements. Certainly the right-hand side does so by the multiplication principle, since to specify a subset we merely need to choose, for each of the n elements in the set, whether to include that element or to exclude it from the subset. The left-hand side also represents the number of subsets, indexed by cardinality; for each possible k from 0 to n, there are $C(n, k)$ subsets of cardinality k, so, by the addition principle, there are $\sum_{k=0}^{n} C(n, k)$ subsets altogether.

Our next example involves a more complicated identity.

EXAMPLE 2. Prove that $C(k, k)+C(k+1, k)+C(k+2, k)+\cdots+C(n, k) = C(n+1, k+1)$ for all natural numbers $k \le n$.

Solution. First we give a combinatorial proof. We will count bit strings of length $n+1$ that contain exactly $k+1$ 1's (and thus $n-k$ 0's). Certainly, the right-hand side of our identity is the number of such strings, since a string is determined by choosing the $k+1$ positions that are to be occupied by the 1's. Now we must show that the left-hand side counts the same quantity.

A bit string of length $n+1$ with $k+1$ 1's must have its last 1 in some position p from $k+1$ to $n+1$, inclusive. There will be 0's in all the positions beyond position p, whereas in the $p-1$ positions before the last 1 there must be exactly k 1's. Therefore, to specify such a bit string we must specify which of these $p-1$ positions are to contain the 1's, and this can be done in $C(p-1, k)$ ways. By the addition principle, therefore, the number of bit strings must be

$$\sum_{p=k+1}^{n+1} C(p-1, k) = C(k, k) + C(k+1, k) + C(k+2, k) + \cdots + C(n, k),$$

exactly the left-hand side of our identity.

Let us also give an inductive proof of this identity. We will induct on the variable n. The base case is trivial, since if $n = 0$, then $k = 0$ as well, and the equation is simply $C(0, 0) = C(1, 1)$, which is true. We assume the inductive hypothesis (the equation for n), and try to verify the statement for $n + 1$, that is,

$$C(k, k) + C(k + 1, k) + C(k + 2, k) + \cdots + C(n + 1, k) = C(n + 2, k + 1)$$

for all $k \leq n + 1$. If $k = n + 1$, then the left-hand side has only one term and the equation reduces to the true statement $C(n + 1, n + 1) = C(n + 2, n + 2)$. Now assume that $k \leq n$. The left-hand side is $[C(k, k) + C(k + 1, k) + C(k + 2, k) + \cdots + C(n, k)] + C(n + 1, k)$, which by the inductive hypothesis equals $C(n + 1, k + 1) + C(n + 1, k)$. But by Pascal's identity, this is the same as $C(n + 2, k + 1)$, as desired.

There is an interesting interpretation of this identity in sports. Suppose that two teams, A and B, are in a playoff series (such as the World Series in baseball), with the series lasting at most $n + 1$ games, where n is even ($n = 6$ for the World Series). The winner of the series is the first team to win $k + 1$ games, where $k = n/2$ (thus $k = 3$ for the World Series). How many different ways can the games of the series come out, assuming that team A wins the series? One analysis uses the left-hand side of our identity. The series will last p games, where p is the number of the game in which team A won the $(k + 1)$th game; thus p is between $k + 1$ and $n + 1$, inclusive. The number of ways to specify an outcome that lasts p games is the number of ways to choose the k games among the first $p - 1$ games that team A also wins. (The bit strings we used in our solution above thus represent records of the series, with a 1 meaning a win for team A and a 0 meaning a win for team B.) The right-hand side looks at the series in the following way: As soon as team A has won its $(k + 1)$th game, the series is over, so in effect we may as well assume that team A forfeited the remaining games of the series (making the final tally $k + 1$ wins for team A and k wins for team B). Note that it is much easier to compute the right-hand side (it involves only one binomial coefficient) than to compute the left-hand side (which involves several). In the case of the World Series, the identity is $C(3, 3) + C(4, 3) + C(5, 3) + C(6, 3) = 1 + 4 + 10 + 20 = 35 = C(7, 4)$; there are 35 different ways that team A can win the series.

Another combinatorial proof of this identity follows from the discussion in Example 6 in Section 6.3. In that case we had $k = 2$ and $n = 12$. ◆

EXAMPLE 3. Give an algebraic and a combinatorial proof of the identity

$$\frac{(2n)!}{2^n n!} = (2n - 1)(2n - 3) \cdots 3 \cdot 1.$$

Solution. For an algebraic proof, we write the left-hand side of this equation as

$$\frac{2n(2n - 1)(2n - 2)(2n - 3)(2n - 4)(2n - 5) \cdots 5 \cdot \quad 4 \cdot 3 \cdot \quad 2 \cdot 1}{2n \qquad 2(n - 1) \qquad 2(n - 2) \qquad \cdots \quad (2 \cdot 2) \quad (2 \cdot 1)}.$$

It is now clear that after canceling common factors in numerator and denominator we obtain the right-hand side.

A combinatorial proof is harder to find but more satisfying. Consider the following problem. Given $2n$ distinct points, labeled $1, 2, \ldots, 2n$, how many ways are there to draw n line segments, each segment connecting a pair of points, so that each point is paired up with one other point? See Figure 7.2 for one such pairing with $n = 3$. (For a more human setting, see Exercise 35 in Section 6.1.)

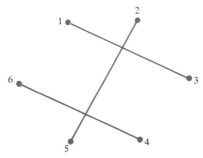

FIGURE 7.2   A pairing of six points.

One way to look at this problem is to choose a permutation of the $2n$ points (which can be done in $(2n)!$ ways), move the points into a straight line according to this permutation, and then pair the first two points in the permutation, the next two points, and so on. Clearly, we have overcounted here, since many such permutations lead to the same pairing. Both of the permutations shown in Figure 7.3 lead to the pairing shown in Figure 7.2, for instance.

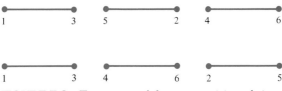

FIGURE 7.3   Two views of the same pairing of six points.

Let us analyze how many such permutations we can construct for the same pairing. There are n pairs, so there are $n!$ orders in which we can lay the pairs along the line. Furthermore, each pair can be laid with its smaller number on the left (as with the pair $(2, 5)$ in the lower line of Figure 7.3) or with its smaller number on the right (as with the pair $(5, 2)$ in the upper line of Figure 7.3). Since there are two choices for each of the n pairs, there are 2^n choices in all. Thus there are a total of $2^n n!$ permutations that lead to the same pairing, so the number of pairings is $(2n)!/(2^n n!)$, the left-hand side of our identity.

On the other hand, we can construct the pairing as follows. Take the lowest numbered point (1) and decide which other point to pair it with. There are $2n - 1$ other points, so there are $2n - 1$ ways to do this. Then take the lowest numbered point that has not been paired (it will be either 2 or 3, but we do not care what its number is), and decide which other unpaired point to pair it with. There are now $2n - 3$ other unpaired points, so there are $2n - 3$ ways to do this. Continue in this manner. At the third stage, there are $2n - 5$ other unpaired points to pair with the lowest numbered unpaired point. If we continue this process for n stages, we will have paired up all the points. By the multiplication principle, there are therefore $(2n - 1)(2n - 3)(2n - 5) \cdots 5 \cdot 3 \cdot 1$ ways to do the pairing, the right-hand side of our identity. This completes the proof, since we have shown that both sides of the identity count the same quantity. ◆

The exercises that follow ask for proofs of various identities by all the methods we have discussed in this section—combinatorial (which in most cases gives the most elegant and satisfying proof), algebraic, and inductive.

SUMMARY OF DEFINITIONS

binomial coefficient: a number of the form $C(n, k)$.

binomial theorem: $(a + b)^n = \sum_{k=0}^{n} C(n, k) a^{n-k} b^k$, where n is a natural number.

combinatorial identity: an equation between two expressions involving combinatorial functions such as $C(n, k)$ that holds for all values of the variables (e.g., Pascal's identity).

Pascal's identity: If $0 < k < n$, then $C(n, k) = C(n - 1, k) + C(n - 1, k - 1)$ (e.g., $\binom{10}{4} = \binom{9}{4} + \binom{9}{3}$, namely $210 = 126 + 84$).

Pascal's triangle: a triangular display of the binomial coefficients, with the values $C(n, k)$ in the nth row (the first row being the zeroth row), with k increasing from left to right, in such a way that $C(n, k)$ is below the space between $C(n - 1, k - 1)$ and $C(n - 1, k)$ (see Figure 7.1).

EXERCISES

1. Expand the following expressions, using the binomial theorem.
   (a) $(x + y)^6$.
   (b) $(a - 3)^4$.

    (c) $(2x + 3y)^5$.

    (d) $(x + x^{-1})^8$.

2. Find the coefficient of $x^4 y^5$ in the expansion of each of the following expressions.

    (a) $(x + y)^9$.

    (b) $(x + y)^7$.

    (c) $(4x + 5y)^9$.

    (d) $(x - 3y)^9$.

3. Fill in the next three rows of Pascal's triangle (Figure 7.1) using Pascal's identity, and verify Corollaries 1 and 2 for these rows.

4. Verify Pascal's identity explicitly for $n = 10$, $k = 6$.

5. Verify the identity in Example 2 explicitly for $n = 10$, $k = 6$.

6. Expand $(x + y + z)^3$ by thinking combinatorially, not by multiplying it out and combining like terms.

7. Find the coefficient of $xy^3 z^2$ in the expansion of $(x+y+z)^6$. [*Hint*: Understand the proof of the binomial theorem.]

8. The **multinomial theorem** states that the coefficient of $x_1^{n_1} x_2^{n_2} \cdots x_m^{n_m}$ in the expansion of $(x_1 + x_2 + \cdots + x_m)^n$, where $n_1 + n_2 + \cdots + n_m = n$, is given by

$$\frac{n!}{n_1! n_2! \cdots n_m!}.$$

It is sometimes denoted $C(n; n_1, n_2, \ldots, n_m)$.

    (a)   Show that this specializes to the binomial theorem when $m = 2$.

    (b)   Prove this result.

    (c)   Use this result to find the coefficient of $a^6 b^2 d^2$ in the expansion of $(2a + 3b - c - 2d)^{10}$.

9. Find closed forms (simple expressions involving n) for the following sums. [*Hint*: Use the binomial theorem.]

    (a) $\sum_{k=0}^{n} C(n, k) 2^k$.

    (b) $\sum_{k=0}^{n} C(n, k)(-2)^k$.

10. Note that apparently (see Figure 7.1) every entry in the pth row of Pascal's triangle, except the 1's, is divisible by p as long as p is a prime number. Prove that this statement is always true.

11. Explain why the numbers $C(n, 2)$ should be called "triangular numbers" by looking at triangular arrangements of dots.

12. Prove that $C(n, 0)+C(n+1, 1)+C(n+2, 2)+\cdots+C(2n, n) = C(2n+1, n+1)$.

13. Consider the following situation: George and Martha invite n married couples to their house for a party. They need to choose two people from among them

to help set out the food. To count the number of ways to make this choice, one could reason in one of two ways. First, there are $2n$ people, and George and Martha need to choose two of them (order does not matter). Second, there are three possibilities: They could choose two men, they could choose two women, or they could choose one man and one woman.

(a) Write down the combinatorial identity suggested by this analysis.

(b) Prove algebraically the identity suggested by this analysis.

14. Consider the following situation: n people are gathered for a game requiring two teams (such as soccer), each with k players. One team will wear blue shirts, and the other team will wear red shirts. Assume that $2k \leq n$. The $n - 2k$ people not on either team will just watch the game. We can choose the teams in either of two ways. One approach is to choose the k players to wear the red shirts, from among the n people, and then choose the k players to wear the blue shirts, from among the remaining $n - k$ people. The other approach is to choose the $2k$ players from among the n people, and then choose k of them to wear the red shirts (giving blue shirts to the other k players).

(a) Write down the combinatorial identity suggested by this analysis.

(b) Prove algebraically the identity suggested by this analysis.

15. Consider the identity $kC(n, k) = nC(n - 1, k - 1)$, where $1 \leq k \leq n$.

(a) Verify the identity for $n = 10$, $k = 5$.

(b) Give an algebraic proof of the identity.

(c) Prove the identity by induction on n, using Pascal's identity.

(d) Give a combinatorial proof of the identity.

16. Consider the identity $k(k - 1)C(n, k) = n(n - 1)C(n - 2, k - 2)$.

(a) Verify the identity for $n = 10$, $k = 5$.

(b) Give an algebraic proof of the identity.

(c) Give a combinatorial proof of the identity.

17. Give a combinatorial proof of the identity found in Exercise 9a. [*Hint*: Consider painting the elements of an n-set red, white, and blue.]

18. Consider the identity $\sum_{k=1}^{n} kC(n, k) = n2^{n-1}$.

(a) Verify the identity for $n = 6$.

(b) Prove the identity by induction, using Pascal's identity.

(c) Prove the identity by a combinatorial argument based on choosing a committee with a chairperson from a group of n people.

(d) Prove the identity by a combinatorial argument based on the sum of the cardinalities of all the subsets of an n-set.

19. Give a combinatorial proof of the identity $\sum_{k=0}^{n} C(n, k)^2 = C(2n, n)$. [*Hint*: Think of each term on the left as $C(n, k)C(n, n - k)$, and consider the number of ways to choose a set of n people from a collection of n married couples.]

20. Consider the identity

$$\binom{n}{k+l}\binom{k+l}{l} = \binom{n}{k}\binom{n-k}{l}.$$

(a) Verify the identity for $n = 10$, $k = 3$, $l = 2$.
(b) Give an algebraic proof of the identity.
(c) Give a combinatorial proof of the identity.

21. Give an algebraic proof of the identity

$$\binom{2m+2n}{m+n}\binom{m+n}{m}^2 = \binom{2m+2n}{2m}\binom{2m}{m}\binom{2n}{n}.$$

22. Give a combinatorial proof of the identity $P(n,r) - P(n-1,r) = rP(n-1, r-1)$ by counting the number of r-permutations of the set of the first n positive integers that include the number n.

23. Give a combinatorial proof that $(kn)!$ is divisible by $(k!)^n$, where k and n are positive integers.

24. By using the model of $2n$ tennis players who need to be assigned to n distinct courts, prove the following identities.

$$\binom{2}{2}\binom{4}{2}\binom{6}{2}\cdots\binom{2n-2}{2}\binom{2n}{2} = n![(2n-1)(2n-3)\cdots 5\cdot 3\cdot 1] = \frac{(2n)!}{2^n}.$$

25. A group of n boys and n girls try out for a debating team. The team will have n persons on it. The chauvinist sponsor of the team insists that a boy be the captain of the team, but otherwise there are no restrictions on who will get to be on the team. We want to find out the number of ways that the team can be picked and a captain designated.
(a) Show that the answer is $C(2n - 1, n - 1)$.
(b) Show that the answer is also $\sum_{i=1}^{n} iC(n, i)^2$. [*Hint*: Here i stands for the number of boys on the team; recall that $C(n, i) = C(n, n - i)$.]

Challenging Exercises

26. Give a combinatorial proof for Exercise 25 in Section 1.3.

27. In a human pyramid, n people kneel on their hands and knees, side by side. Then $n - 1$ people kneel on the backs of these people, each person in the second layer straddling two people on the ground. Next, $n - 2$ people kneel in a third layer, and so on, until one person kneels at the top. Construct a Pascal-like triangle for computing the weight born by each person in the pyramid. (Assume that each person weighs the same, say w pounds.) State the recursive rule carefully, and carry out the triangle to six rows.

28. Give a combinatorial proof of the identity given in Exercise 21.

29. Discover and prove a satisfying identity involving $\sum_{i=0}^{\lfloor n/2 \rfloor} C(n-i, i)$.

Exploratory Exercises

30. Look at Pascal's triangle and try to guess for what values of n all the binomial coefficients $C(n, k)$ are odd. Try to prove your conjecture, or consult Dacić [54] for a proof.

31. Construct a Pascal-like triangle in which there are 1's down the left border (as in Figure 7.1), powers of 2 down the right border, and such that each internal entry is equal to the sum of the number above it and to its right plus twice the number above it and to its left. Thus the 0th row is 1; the first row is $1, 2$; the second row is $1, 4, 4$; the third row is $1, 6, 12, 8$; and so on. Give one or more interpretations or applications of the numbers in this triangle.

32. The **Stirling numbers of the first kind** are defined as follows. For n a positive integer and $0 \leq k \leq n$, $s(n, k)$ is the coefficient of x^k in the expansion of the "falling factorial" with n factors, $x(x-1)(x-2)\cdots(x-n+1)$. For example, take $n = 3$; then since $x(x-1)(x-2) = x^3 - 3x^2 + 2x$, we have $s(3, 0) = 0$, $s(3, 1) = 2$, $s(3, 2) = -3$, and $s(3, 3) = 1$.
  (a) Find some relationships among these numbers; you probably want to arrange them in a Pascal-like triangle and determine how each row can be obtained from the row above it.
  (b) Find out more about these numbers by consulting Bogart [23], Oullette and Bennett [229], and Pólya, Tarjan, and Woods [237].

33. Investigate the identity $\sum_{k=j}^{i} C(k, j)C(i, k)(-1)^k = 0$, where i and j are natural numbers with $j < i$. Consult Flanders [84] for a nice combinatorial proof.

34. Try to find a combinatorial proof that $\sum k^3 = \left(\sum k\right)^2$. Consult Stein [293] for a solution.

35. Investigate the identity $4^n = \sum_{r=0}^{n} C(2r, r)C(2n - 2r, n - r)$. Consult Sved [299] for more information.

36. Construct Pascal's triangle "mod 2"; in other words, replace each entry in the triangle by its remainder when divided by 2. (Note that you do not have to compute binomial coefficients and then reduce them mod 2; instead, simply use Pascal's identity, working mod 2. Thus there is virtually no arithmetic involved in doing this exercise.) Extend the triangle for about 80 rows (with a computer), and comment on any interesting patterns. Find out more about patterns in Pascal's triangle by consulting Gardner [96], Hilton and Pedersen [148], K. J. Smith [279], and Usiskin [310].

37. Find out more about combinatorial identities by consulting Riordan [252] and Sved [298].

38. Find out more about Blaise Pascal by consulting Bell [15].

SECTION 7.2
MODELING COMBINATORIAL PROBLEMS
WITH RECURRENCE RELATIONS

*But Peter, who was very naughty, ran straight away
to Mr. McGregor's garden, and squeezed under the gate!*

—**Beatrix Potter,** *The Tale of Peter Rabbit*

In Section 5.1 we showed how sequences can be defined recursively. Indeed, some sequences have no simple definition other than a recursive one. In this section we look at sequences that are not *defined* recursively (they may be defined in terms of an application) but for which a recursive formula can be written down. Such a formula is called a recurrence relation for the sequence. The advantage of such a formula is twofold. First, it allows us to compute the terms in the sequence, one at a time. Second, as we will see in Section 7.3, it sometimes allows us to derive a closed-form, nonrecursive formula for the terms in the sequence.

We start by defining some terminology. Suppose that a_0, a_1, a_2, ... is a sequence of numbers. A **recurrence relation** for the sequence $\{a_n\}$ is an identity of the form $a_n = f(a_{n-1}, a_{n-2}, \ldots, a_1, a_0)$. In other words, a recurrence relation is a formula giving the nth term in the sequence in terms of the previous terms of the sequence. Throughout our discussion, the expression defining f can also involve n, even though we do not show it explicitly as an argument. For example, $a_n = na_{n-1}^2 + a_3 + 5$ is a recurrence relation. Often only the immediately preceding term of the sequence enters into the recurrence relation, so that $a_n = f(a_{n-1})$ for all $n \geq 1$; in this case we say that we have a first-order recurrence relation. For example, $a_n = n^2 + 1/a_{n-1}$ is a first order recurrence relation. More generally, if we have $a_n = f(a_{n-1}, a_{n-2}, \ldots, a_{n-k})$ for all $n \geq k$, then the recurrence relation is said to be of **order** k. Often, the restriction $n \geq k$ is not written explicitly, but it is to be understood nonetheless, since if $n < k$, then the term a_{n-k} would make no sense.

Given a sequence that satisfies a kth-order recurrence relation, together with specific values for a_0, a_1, ..., a_{k-1}, we can write down as many terms of the sequence as we wish. The specifications of the values of a_0 through a_{k-1} are called **initial conditions**. Occasionally, the sequence begins at an index other than 0. For example, we may have a kth-order recurrence relation valid for all $n > k$ with initial conditions specifying $a_1, a_2, \ldots, a_k$.

EXAMPLE 1. We saw in Section 5.1 that the Fibonacci sequence $\{f_n\}$ satisfies the second-order recurrence relation $f_n = f_{n-1} + f_{n-2}$ for all $n \geq 2$, together with the initial conditions $f_0 = 1$ and $f_1 = 1$. Indeed, in this case the recurrence relation and initial conditions form the *definition* of the sequence. Knowing the initial conditions and recurrence relation, we can compute the terms of the sequence, one by one. In this case, we find that

$$f_2 = f_1 + f_0 = 1 + 1 = 2$$
$$f_3 = f_2 + f_1 = 2 + 1 = 3$$
$$f_4 = f_3 + f_2 = 3 + 2 = 5$$

and so on. ◆

EXAMPLE 2. The sequence given by the explicit formula $a_n = n(n-1)/2$ satisfies the first-order recurrence relation $a_n = a_{n-1} + (n-1)$ for all $n \geq 1$, since we have

$$a_{n-1} + (n-1) = \frac{(n-1)(n-2)}{2} + (n-1) = (n-1) \cdot \frac{n}{2} = a_n.$$

The initial condition here is that $a_0 = 0$. ◆

EXAMPLE 3. The sequence $1, 2, 4, 8, 16, \ldots$ satisfies the recurrence relation $a_n = a_{n-1} + a_{n-2} + \cdots + a_1 + a_0 + 1$. In other words, each term in this sequence is the sum of all the previous terms, plus 1. It also satisfies other recurrence relations, such as the first order relation $a_n = 2a_{n-1}$ and the third-order relation $a_n = a_{n-1} + 4a_{n-3}$. ◆

Our main goal in this section is to set up recurrence relations for solving problems. Typically, we are given a sequence defined in concrete, not algebraic, terms; we want to write down a recurrence relation that the sequence must satisfy. To illustrate, we begin with an application to finance.

EXAMPLE 4. Some people are allowed by the tax laws to set up tax-deferred savings accounts called individual retirement accounts (IRAs). In a typical IRA the saver deposits $2000 at the beginning of every year. The bank holding the account adds interest to the account, say at a rate of 5%, on the money on deposit for the year. We can model an IRA mathematically by finding a recurrence relation and initial condition that is satisfied by the amount of money in such an account.

Let A_n be the amount in the account at the end of n years after the account begins. As an initial condition, we can take $A_0 = 0$, since there is no money in the account before we begin. Now let us see how the amount in the account after n years depends on the amount in the account after $n - 1$ years. First, $2000 is added to the amount already in the account at the beginning of the year. Next, the bank adds the interest, which will be $0.05 \cdot (A_{n-1} + 2000)$; both the previous balance and the new deposit earn the interest. Therefore the amount in the account at the end of n years will be

$$A_n = A_{n-1} + 2000 + 0.05 \cdot (A_{n-1} + 2000)$$
$$= 2100 + 1.05 A_{n-1}.$$

Let us see how the account grows. After 1 year we have $2100, since $A_1 = 2100 + 1.05 A_0 = 2100 + 1.05 \cdot 0 = 2100$. After 2 years we have $A_2 = 2100 + 1.05 \cdot 2100 = 4305$ dollars. After 3 years, there is $6620.25 in the bank. We can continue this calculation as far as we wish to find the balance after any desired length of time. For instance, we would find that after 30 years the account would contain about $140,000. ◆

Next we look at the biological application that led Fibonacci to consider the sequence named after him.

EXAMPLE 5. Suppose that a new-born pair of rabbits—a male and a female—is placed on an island. We assume that a pair of rabbits needs to mature for one full month before producing any offspring, but that after they are mature, each pair of rabbits produces one pair of rabbits (again a male and a female) once a month. We also assume that the rabbits never die. The problem is to determine how many rabbits there will be on the island after n months.

Anticipating the solution, we denote by f_n the number of pairs of rabbits on the island after n months. Thus we are told that $f_0 = 1$. Furthermore, since these are newborn rabbits, they produce no offspring the first month, so $f_1 = 1$ as well. During the second month, however, the original pair of rabbits produces a new pair, so after 2 months there are $f_2 = 2$ pairs of rabbits on the island—the original pair and a new pair. During the third month the original pair again produces a pair of rabbits, but the rabbits born in the second month are not yet mature, so after 3 months there are $f_3 = 3$ pairs of rabbits on the island—the original pair, the rabbits born in the second month, and the rabbits born in the third month. During the fourth month, both the original rabbits and the pair born in the second month produce a pair, so after 4 months there are $f_4 = 5$ pairs of rabbits (the original pair, the pair born the second month, the pair born the third month, and the two pairs born the fourth month).

We want to write down a recurrence relation for the sequence $\{f_n\}$. We ask ourselves, "How does the number of pairs of rabbits on the island after n months relate to the number of pairs of rabbits on the island earlier?" Since we are assuming that no rabbits die, certainly all the rabbits on the island after $n - 1$ months survive to the end of the nth month, so f_n is at least f_{n-1}. In addition, some new rabbits are born during the nth month. The number of pairs of rabbits born during the nth month is precisely equal to f_{n-2}, the number of pairs of rabbits on the island at the end of the $(n - 2)$th month, since these are precisely the ones (under our assumptions) that are mature enough to reproduce during the nth month. Thus the recurrence relation we seek is $f_n = f_{n-1} + f_{n-2}$.

This analysis was valid as long as $n \geq 2$. For $n = 0$ or $n = 1$ it makes no sense to talk about what was the case after $n - 2$ months, so the recurrence relation does not apply in these cases. Instead, we note the initial conditions $f_0 = 1$ (we are told that there is one pair of rabbits initially) and $f_1 = 1$ (the original rabbits are not yet mature, so no new rabbits are born the first month). In summary, then, the recurrence relation and initial conditions that model the rabbit problem are

$$f_n = f_{n-1} + f_{n-2}, \qquad n \geq 2$$
$$f_0 = f_1 = 1.$$

These are precisely the defining conditions for the Fibonacci sequence. Thus the number of pairs of rabbits on the island after n months is the nth term in the Fibonacci sequence. ◆

Let us look at another example, whose answer may turn out to be a bit of a surprise.

EXAMPLE 6. Let a_n be the number of bit strings of length n that do not contain a pair of consecutive 0's as a substring. Find a recurrence relation and initial conditions for the sequence $\{a_n\}$.

Solution. We must answer the question: How can we relate the number of bit strings of length n that do not contain a pair of consecutive 0's to the number of such sequences of smaller length? Consider a sequence of length n, not containing a pair of consecutive 0's. The sequence may begin with a 1. If so, it must be followed by a string of $n - 1$ bits, not containing a pair of consecutive 0's, and by definition there are a_{n-1} of these. Thus a_n is at least as large as a_{n-1}. On the other hand, the sequence may begin with a 0. If so, the second bit has to be a 1, since otherwise there would be a pair of consecutive 0's at the beginning of the string. Following the initial 01, however, the string can continue as any bit string of length $n - 2$ not containing a pair of consecutive 0's, and there are a_{n-2} of these. Thus there are another a_{n-2} bit strings of length n not containing a pair of consecutive 0's, namely those beginning 01. We have considered all the possibilities, and they are mutually exclusive, so by the addition principle the desired recurrence relation is $a_n = a_{n-1} + a_{n-2}$. Note that this is the same recurrence relation satisfied by the Fibonacci sequence.

The initial conditions must still be determined. The analysis above was valid for $n \geq 2$. There is exactly one string of length 0 (the empty string), and it contains no pair of consecutive 0's, so one initial condition is $a_0 = 1$. There are exactly two strings of length 1, and neither contains a pair of consecutive 0's, so the second initial condition is $a_1 = 2$. Note that these are not the same initial conditions as the Fibonacci sequence satisfied, so the sequences are not identical. They are almost the same, however. Since $a_0 = f_1$ and $a_1 = f_2$, and since the recurrence relation, which governs the rest of each sequence, is the same, we know that $a_n = f_{n+1}$ for all n.

The recurrence relation allows us to compute the rest of the sequence as far as we wish: $a_2 = a_1 + a_0 = 2 + 1 = 3$ (and indeed the three strings are 01, 10, and 11); $a_3 = a_2 + a_1 = 3 + 2 = 5$ (and indeed the five strings are 010, 011, 101, 110, and 111); and so on. We would be hard pressed to make an accurate list of, say, all the bit strings of length 8 containing no pair of consecutive 0's, but we can count them easily enough by using our recurrence relation: The sequence continues $a_4 = 8$, $a_5 = 13$, $a_6 = 21$, $a_7 = 34$, and $a_8 = 55$. ◆

If we wish to count the number of bit strings of length n that *do* contain a pair of consecutive 0's, one solution is simply to take the sequence $\{a_n\}$ obtained in Example 6 and subtract a_n from 2^n, since there are a total of 2^n bit strings of length n. For example, there are $2^8 - 55 = 201$ bit strings of length 8 containing a pair of consecutive 0's. Alternatively, we could set up a recurrence relation directly, as shown in the next example.

EXAMPLE 7. Let b_n be the number of bit strings of length n containing a pair of consecutive 0's. Find a recurrence relation and initial conditions for the sequence $\{b_n\}$.

Solution. There are three mutually exclusive ways that such a sequence might start: 1, 01, and 00. If it starts with a 1, it must continue with a bit string of length $n - 1$ containing a pair of consecutive 0's, and there are b_{n-1} of these. If it starts with 01, it must continue with a bit string of length $n - 2$ containing a pair of consecutive 0's, and there are b_{n-2} of these. Finally, if it starts 00, it can be followed by *any* bit string of length $n-2$ (since a pair of consecutive 0's is already present), and there are 2^{n-2} of these. Therefore, the desired recurrence relation is $b_n = b_{n-1} + b_{n-2} + 2^{n-2}$. Clearly, the initial conditions are $b_0 = b_1 = 0$, since no strings of length less than 2 can contain 00 as a substring.

With this recurrence relation, we can compute the terms in the sequence. We have

$$b_2 = b_1 + b_0 + 2^0 = 0 + 0 + 1 = 1$$
$$b_3 = b_2 + b_1 + 2^1 = 1 + 0 + 2 = 3$$
$$b_4 = b_3 + b_2 + 2^2 = 3 + 1 + 4 = 8$$
$$b_5 = b_4 + b_3 + 2^3 = 8 + 3 + 8 = 19$$
$$b_6 = b_5 + b_4 + 2^4 = 19 + 8 + 16 = 43$$
$$b_7 = b_6 + b_5 + 2^5 = 43 + 19 + 32 = 94$$
$$b_8 = b_7 + b_6 + 2^6 = 94 + 43 + 64 = 201$$

and so on. ◆

Unfortunately, there is no algorithm to tell us how to analyze an applied problem, such as the ones we have been considering here, to come up with a recurrence relation. A successful analysis often takes a bit of cleverness and usually involves one or more false starts. In the rest of this section we turn to problems that are somewhat more involved than the ones we have looked at so far.

RECURRENCE RELATIONS NOT IN CLOSED FORM

In Section 3.4 we discussed partitions of a set, and in Section 6.3 we introduced partitions of a natural number. In neither case is it possible to write down a nice formula for the number of partitions. We can write down recurrence relations, though, and therefore we will be able to compute the number of partitions, without having to list them all (a process that is both time consuming and error-prone).

Let $p(n)$ be the number of partitions of a set with n elements. (By Theorem 4 in Section 3.4, $p(n)$ is also the number of different equivalence relations on a set with n elements.) The numbers $p(n)$ are known as the **Bell numbers**, after the American mathematician E. T. Bell. (We use functional notation this time because the arguments are going to get too messy to look nice as subscripts.) For example, $p(3) = 5$, since the partitions of $\{1, 2, 3\}$ are $\{\{1, 2, 3\}\}$, $\{\{1, 2\}, \{3\}\}$, $\{\{1, 3\}, \{2\}\}$, $\{\{2, 3\}, \{1\}\}$, and $\{\{1\}, \{2\}, \{3\}\}$. In order to get a recurrence relation for $p(n)$, we need to see how partitions of smaller sets help to determine partitions of larger ones.

We count the partitions of $\{1, 2, \ldots, n\}$ as follows. The element n must be in one of the sets of the partition. It can be in a set by itself, or it can have one or more (possibly even all) of the other elements of $\{1, 2, \ldots, n\}$ with it. Let k be the *number* of elements other than n in the same set with n in a partition of $\{1, 2, \ldots, n\}$. For example, if $n = 3$, then the partition $\{\{2, 3\}, \{1\}\}$ has $k = 1$, since only 2 is in the same set as 3. Note that $0 \leq k \leq n - 1$. In order to specify a partition with this value of k, we can first decide which k elements are to be in the same set as n (and we can do this in $C(n - 1, k)$ ways), and then decide how to partition the remaining elements of $\{1, 2, \ldots, n\}$ (and we can do this in $p(n-k-1)$ ways, since there are $n - k - 1$ elements left to be partitioned). Therefore, by the multiplication principle there are $C(n - 1, k)p(n - k - 1)$ partitions of $\{1, 2, \ldots, n\}$ in which exactly k elements are in the same set as n. Finally, by the addition principle, the total number of partitions of $\{1, 2, \ldots, n\}$ is given by

$$p(n) = \sum_{k=0}^{n-1} C(n - 1, k)p(n - k - 1).$$

This formula is our recurrence relation; it specifies $p(n)$ in terms of the numbers $p(n - k - 1)$, all of which have arguments smaller than n (since $k \geq 0$). The only initial condition needed is $p(0) = 1$, reflecting the fact that the empty set is the only partition of the empty set.

Note that this recurrence relation is not of a fixed order, as were most of the recurrence relations we considered earlier in this section. Instead, the recurrence relation expresses $p(n)$ as a function of all the numbers $p(0)$, $p(1)$, ..., $p(n-1)$ (as well as n).

EXAMPLE 8.   Find the number of partitions of a set with five elements.

Solution.   We need to compute $p(5)$, and we can do so if we first compute $p(0)$, $p(1)$, $p(2)$, $p(3)$, and $p(4)$. The first three of these we may as well do directly, since they are so simple. We already noted that $p(0) = 1$, and it is clear that $p(1) = 1$ as well. Also, $p(2) = 2$, since we can put the two elements either in one set together or in separate sets. We already computed that $p(3) = 5$ several paragraphs above. To compute $p(4)$ we use the recurrence relation:

$$p(4) = \sum_{k=0}^{3} C(3, k)p(3 - k)$$

$$= C(3, 0)p(3) + C(3, 1)p(2) + C(3, 2)p(1) + C(3, 3)p(0)$$

$$= 1 \cdot 5 + 3 \cdot 2 + 3 \cdot 1 + 1 \cdot 1 = 15.$$

Finally, we use the recurrence relation again to find $p(5)$:

$$p(5) = \sum_{k=0}^{4} C(4, k)p(4 - k)$$

$$= C(4, 0)p(4) + C(4, 1)p(3) + C(4, 2)p(2) + C(4, 3)p(1) + C(4, 4)p(0)$$

$$= 1 \cdot 15 + 4 \cdot 5 + 6 \cdot 2 + 4 \cdot 1 + 1 \cdot 1 = 52.$$

Thus there are exactly 52 partitions of the set $\{1, 2, 3, 4, 5\}$ (or any other set with five elements). It would have been difficult to be sure of obtaining the right answer by trying to list these 52 partitions. ◆

The computation of the number of partitions of the *natural number n* requires one additional wrinkle in the use of recurrence relations. Until now we have had only one independent variable, which we usually called n. We were finding recurrence relations for functions of one (natural number) variable (i.e., sequences). This might be thought of as a one-dimensional problem. Recurrence relations also apply to functions of more than one variable. In such cases we get recurrence relations for multidimensional arrays, rather than one-dimensional sequences. (It was far from obvious that the trick of introducing an additional variable into the problem of computing the number of partitions of a number was going to be a useful thing to do; think of this idea as the last after a long series of failures by whoever first solved this problem.)

To compute the number of partitions of the natural number k, which we will denote $P(k)$ as we did in Section 6.3, we will let $P_m(k)$ denote the number of partitions of the number k into parts no bigger than m each, where m is a positive integer. Think of $P_m(k)$ as a function of two variables, k and m. For instance, $P(4) = 5$, since the partitions of 4 are 4, $3+1$, $2+2$, $2+1+1$, and $1+1+1+1$; but $P_2(4) = 3$, since three of these partitions use only numbers less than or equal to 2. We will write down a rather complicated recurrence relation for $P_m(k)$. This will solve our problem, since $P(k) = P_k(k)$: No parts bigger than k are used in any partition of k.

As usual, we must ask: How does $P_m(k)$ depend on values of P with smaller arguments? (Both k and m are being viewed as arguments.) There are a few special cases, which we can view as the initial conditions for this analysis. If $m > k$, then $P_m(k) = P_k(k)$, since no parts bigger than k are ever used in a partition of k. Also, $P_1(k) = 1$, since there is only one way to write k as the sum of numbers no bigger than 1, namely, $k = 1 + 1 + \cdots + 1$. Finally, $P_m(0) = 1$, since there is only one partition of 0, namely, the empty sum. Now suppose that $1 < m \le k$. Then a partition of k into parts no bigger than m is either a partition of k into parts no bigger than $m - 1$ (and there are $P_{m-1}(k)$ of these), or a partition of k into parts no bigger than m, at least one of which is actually m. There are $P_m(k - m)$ of the latter type, since once we use one m in the partition, there remains $k - m$ left to be partitioned. Therefore, our recurrence relation and initial conditions read as follows:

$$P_m(k) = \begin{cases} 1 & \text{if } k = 0 \\ 1 & \text{if } m = 1 \\ P_k(k) & \text{if } 0 < k < m \\ P_{m-1}(k) + P_m(k - m) & \text{if } 1 < m \le k. \end{cases}$$

EXAMPLE 9. Use the recurrence relation to find the number of partitions of 5.

Solution. We just apply the recurrence relation repeatedly to compute $P_5(5)$:

$$P_5(5) = P_4(5) + P_5(0) = P_4(5) + 1$$
$$= P_3(5) + P_4(1) + 1 = P_3(5) + P_1(1) + 1 = P_3(5) + 1 + 1 = P_3(5) + 2$$
$$= P_2(5) + P_3(2) + 2 = P_2(5) + P_2(2) + 2 = P_2(5) + P_1(2) + P_2(0) + 2$$
$$= P_2(5) + 4 = P_1(5) + P_2(3) + 4 = P_1(5) + P_1(3) + P_2(1) + 4$$
$$= P_1(5) + P_1(3) + P_1(1) + 4 = 7.$$

Therefore, $P(5) = P_5(5) = 7$; there are seven ways to write 5 as the sum of positive integers, ignoring order. ◆

The exercises show how recurrence relations can be used to find the number of surjective ("onto") functions, the binomial coefficients, and other similar quantities.

SYSTEMS OF RECURRENCE RELATIONS (*OPTIONAL*)

Sometimes a problem deals with several interrelated dependent variables at once. In such cases rather than modeling the situation with one recurrence relation, we need to find a system of recurrence relations that are satisfied simultaneously by the several variables involved. We illustrate this situation with a simplified model for the spread of a communicable disease.

EXAMPLE 10. Suppose that an incurable but relatively harmless disease is spread by sexual contact between males and females. It is reasonable to assume that during an epidemic of this disease the number of new persons acquiring the disease each month will be proportional to the number of possible contacts between infected and uninfected persons that month. To set up a model, let M and F be the total populations of sexually active males and females, respectively, and let M_n and F_n be the number of infected males and infected females among them at the end of the nth month of the epidemic. We need to write down recurrence relations for M_n and F_n.

To compute M_n, note that during the nth month there are $M - M_{n-1}$ uninfected males. The number of females who are infected is F_{n-1}, so the number of new cases of the disease will (according to our assumption) be proportional to $(M - M_{n-1})F_{n-1}$, say $\alpha(M - M_{n-1})F_{n-1}$ for some constant α. Thus the recurrence relation for the number of infected males is

$$M_n = M_{n-1} + \alpha(M - M_{n-1})F_{n-1}.$$

Similarly, the recurrence relation for the number of infected females is

$$F_n = F_{n-1} + \beta(F - F_{n-1})M_{n-1},$$

for some constant β. To make things specific, let us suppose that $M = F = 1000$, $M_0 = 100$, $F_0 = 50$, $\alpha = 1.0 \times 10^{-4}$, $\beta = 1.2 \times 10^{-4}$, and the values of M_n and F_n are actually given by the floor of the values shown in the recurrence relations.

This model then allows us to compute the sequences $\{M_n\}$ and $\{F_n\}$. Some of the values are shown in the following table.

n	M_n	F_n
0	100	50
1	104	61
2	109	72
3	115	84
4	122	96
5	130	109
10	185	184
15	269	284
20	379	406
30	635	674
40	831	866
60	972	983
74	990	992
75	990	992

Note that the higher chance of contacting the disease for females ($\beta > \alpha$) is reflected in the faster growth of $\{F_n\}$ as compared to $\{M_n\}$. Although initially the females had only half the number of cases that the males had, after 10 months they are just as infected as the males. Also note that after 74 months, the sequences have become constant: All those that are going to become infected have already been infected (and that includes all but 10 males and 8 females). ◆

ANALYSIS OF ALGORITHMS

One of the most important applications of recurrence relations is in modeling the performance of recursive algorithms. Recall that a recursive algorithm is one that calls itself. Let $T(n)$ be the number of steps used by a given recursive algorithm to complete its work on input of size n. The algorithm will call itself with an input of size less than n. Thus we should be able to write down a recurrence relation expressing $T(n)$ in terms of $T(i)$ for one or more values of i less than n.

We have already seen one example of a recurrence relation used to analyze the efficiency of an algorithm. In Section 5.2 we devised a recursive algorithm (Algorithm 2) to solve the towers of Hanoi problem (in which size is the number of disks). We found that we could model the number of disk moves required to transfer n disks with the recurrence relation

$$T(n) = 2T(n-1) + 1, \qquad n \geq 2$$

$$T(1) = 1.$$

Thus we can compute, for instance, that

$$T(2) = 2T(1) + 1 = 2 \cdot 1 + 1 = 3$$

$$T(3) = 2T(2) + 1 = 2 \cdot 3 + 1 = 7$$

$$T(4) = 2T(3) + 1 = 2 \cdot 7 + 1 = 15$$

and so on.

In Section 5.2 we also promised that we would analyze the efficiency of the merge sort (Algorithms 8 and 9). Let us develop that recurrence relation model now. The reader should review the material at the end of Section 5.2 before continuing.

Let $T(n)$ be the number of comparisons of numbers needed to sort a list of n numbers into nondecreasing order, using Algorithm 9 in Section 5.2. The first step in the algorithm tells us that $T(1) = 0$: If the list has only one number, it is already sorted and no comparisons are needed. If $n > 1$, then the only step in Algorithm 9 that will cause any comparisons to be made is the **return** statement:

$$\textbf{return}(merge(merge\_sort(L_1), merge\_sort(L_2)))$$

This statement implies that the algorithm calls itself twice, once on the first half of the list (L_1) and once on the second half (L_2). The lists L_1 and L_2 have length $\lfloor n/2 \rfloor$ and $n - \lfloor n/2 \rfloor = \lceil n/2 \rceil$, respectively, so the number of comparisons needed to sort them is $T(\lfloor n/2 \rfloor) + T(\lceil n/2 \rceil)$. After these two lists are sorted, they need to be merged by Algorithm 8 in Section 5.2 (**procedure** *merge*). Every time the comparison of a_j to b_k is made in the middle of Algorithm 8, another element is correctly placed in the merged list; therefore, the number of comparisons required to merge the two lists, whose total length is n, will be n. Thus we have $T(n) = T(\lfloor n/2 \rfloor) + T(\lceil n/2 \rceil) + n$. That, together with our initial condition $T(1) = 0$, gives the desired model. We will see how to solve this rather complicated recurrence relation in Section 7.3.

EXAMPLE 11. Find the number of comparisons needed by merge sort for a list of 10 numbers.

Solution. By the recurrence relation, we know that $T(10) = T(5) + T(5) + 10$. Thus we first need to find $T(5)$. Again from the recurrence relation we have $T(5) = T(2) + T(3) + 5$. At this point we find $T(2) = T(1) + T(1) + 2 = 0 + 0 + 2 = 2$, and then $T(3) = T(1) + T(2) + 3 = 0 + 2 + 3 = 5$. Therefore, $T(5) = 2 + 5 + 5 = 12$, and so finally, $T(10) = 12 + 12 + 10 = 34$. ◆

Not all recursive algorithms can be analyzed this easily. With merge sort we were able to express $T(n)$ in terms of specific values of T at smaller arguments. With the Euclidean algorithm for computing greatest common divisors (Algorithm 6 in Section 5.2), since the arguments for the recursive calls depend on the remainders that happen to occur as the algorithm proceeds, we cannot say which value of j

is needed to express $T(n)$ in terms of $T(j)$. We will not get into the analysis of the Euclidean algorithm in this book; suffice it to say that it turns out that the number of iterations (computations of remainder and recursive calls) needed to find $\gcd(x, y)$ is in $O(\log(x + y))$ (see Exercise 39).

We close with one more example of how to set up a recurrence relation to analyze a recursive algorithm. Matrix multiplication was discussed at the end of Section 4.5. We saw that using the definition led to an algorithm with efficiency in $O(n^3)$ for multiplying two n by n matrices. In other words, $O(n^3)$ multiplications and additions of numbers were needed in order to find all the entries in the product matrix. We mentioned at the time that there were more efficient algorithms for multiplying matrices. Here we will discuss one such algorithm, developed by V. Strassen, without going into much detail (see Exercise 40).

Suppose that we have two n by n matrices, A and B, which we wish to multiply. Our approach will be to split the matrices "in half" horizontally and vertically, so we will assume for simplicity in what follows that n is a power of 2. We can think of each matrix as broken up into four quadrants as shown:

$$A = \begin{bmatrix} A_{11} & A_{12} \\ A_{21} & A_{22} \end{bmatrix} \quad \text{and} \quad B = \begin{bmatrix} B_{11} & B_{12} \\ B_{21} & B_{22} \end{bmatrix},$$

where A_{11}, A_{12}, $\ldots$, B_{22} are each $(n/2)$ by $(n/2)$ matrices. For example, A_{12} consists of the entries in rows 1 through $n/2$, columns $(n/2) + 1$ through n, of matrix A. It can easily be shown that the product $A \times B$ can be broken up into four quadrants as well, with the submatrices in the quadrants as shown:

$$A \times B = \begin{bmatrix} A_{11} \times B_{11} + A_{12} \times B_{21} & A_{11} \times B_{12} + A_{12} \times B_{22} \\ A_{21} \times B_{11} + A_{22} \times B_{21} & A_{21} \times B_{12} + A_{22} \times B_{22} \end{bmatrix}.$$

Essentially, we just "multiply" the quadrants as if they were entries of a 2 by 2 matrix.

Let $T(n)$ be the number of operations on numbers—additions, subtractions, or multiplications—that need to be performed in order to do this calculation for n by n matrices. Recursively, we use eight multiplications of $(n/2)$ by $(n/2)$ submatrices, which will take $8T(n/2)$ operations; also we use four additions of $(n/2)$ by $(n/2)$ matrices. Adding two $(n/2)$ by $(n/2)$ matrices requires $(n/2)^2$ numerical operations—one sum for each entry in the answer. This leads to the recurrence relation

$$T(n) = 8T(n/2) + 4(n/2)^2 = 8T(n/2) + n^2.$$

The initial condition is $T(1) = 1$, since the product of two 1 by 1 matrices is just the product of the two numbers in the matrices. Thus if n is a power of 2, we can compute $T(n)$ by using this recurrence relation. We obtain

$$T(2) = 8 \cdot 1 + 4 = 12$$

$$T(4) = 8 \cdot 12 + 16 = 112$$

$$T(8) = 8 \cdot 112 + 64 = 960$$

$$T(16) = 8 \cdot 960 + 256 = 7936$$

and so on. In fact, $T(n) = 2n^3 - n^2$ if n is a power of 2 (see Exercise 2 in Section 7.3). Thus this algorithm has time complexity in $O(n^3)$, so it is similar in efficiency to the usual algorithm discussed in Section 4.5.

By reorganizing the calculation, however, it is possible to get by with seven multiplications and 18 additions/subtractions of $(n/2)$ by $(n/2)$ matrices, rather than eight multiplications and four additions. The trick is to compute several intermediate quantities and combine them in clever ways; we will not go into the details. Therefore, there is a recursive algorithm for multiplying matrices that forms these seven products recursively and then performs the 18 additions/subtractions using $18(n/2)^2 = 4.5n^2$ additions/subtractions of numbers. Under this scheme, the recurrence relation for T becomes

$$T(n) = 7T(n/2) + 4.5n^2.$$

still with initial condition $T(1) = 1$. Thus we have

$$T(2) = 7 \cdot 1 + 4.5 \cdot 2^2 = 25$$

$$T(4) = 7 \cdot 25 + 4.5 \cdot 4^2 = 247$$

$$T(8) = 7 \cdot 247 + 4.5 \cdot 8^2 = 2017$$

$$T(16) = 7 \cdot 2017 + 4.5 \cdot 16^2 = 15271$$

and so on. Although these numbers exceed the corresponding numbers obtained under the first scheme for the values of n for which we have done the computation, the numbers obtained under the second scheme are smaller than those of the first scheme for moderately large values of n; the reader is invited to find the point at which the second scheme becomes faster (Exercise 26).

SUMMARY OF DEFINITIONS

Bell numbers: The Bell number $p(n)$ is the number of partitions of a set with n elements.

initial conditions for a sequence: the specification of values for the first few terms of the sequence.

kth-order recurrence relation for the sequence $\{a_n\}$: a recurrence relation that gives a_n as a function of $a_{n-1}, a_{n-2}, \ldots, a_{n-k}$ (e.g., $a_n = 1.04a_{n-1} + a_{n-2}$ is a second-order recurrence relation).

recurrence relation for a sequence: an equation that gives the nth term in a sequence as a function of the previous terms (e.g., $a_n = a_{n-1} + \sqrt{n}$).

EXERCISES

1. A bank adds 5% interest to savings accounts at the end of every year. A saver deposits $1200 at the beginning of a certain year, but makes no further deposits to the account. Let A_n be the balance n years later.
   (a) Write down a recurrence relation for A_n, together with an initial condition.
   (b) Use the recurrence relation and initial condition to compute the balance at the end of 4 years.

2. After you graduate you accept a job that promises a starting salary of $40,000 and a raise at the end of each year equal to 5% of your current salary plus $1000. For example, your raise at the end of the first year is $3000. Let s_n be your salary after n years, so that $s_0 = 40000$.
   (a) Find a recurrence relation for $\{s_n\}$.
   (b) Determine how much you will be making after 2 years, after 5 years, and after 10 years.

3. Use the recurrence relation developed in this section to find the number of partitions of the set $\{A, B, C, D, E, F\}$.

4. Use the recurrence relation developed in this section to find the number of partitions of 6 into positive integers.

5. List explicitly the partitions of 8 into parts no larger than 3. Check your answer by computing $P_3(8)$ using the recurrence relation developed in this section.

6. Compute the number of infected males and infected females in the population described in Example 10 after 6 months.

7. Compute the number of comparisons of elements needed by the merge sort algorithm to sort lists of length n for the following values of n.
   (a) $n = 8$.
   (b) $n = 9$.
   (c) $n = 20$.
   (d) $n = 50$.

8. A bank adds 5% interest to savings accounts at the end of every year. A saver deposits $2500 at the beginning of a certain year. Then at the end of every year the saver withdraws $100 (this amount does earn interest for the year, however). Let A_n be the balance at the end of n years.
   (a) Write down a recurrence relation for A_n, together with an initial condition.

(b) Use the recurrence relation and initial condition to compute the balance at the end of 4 years.

(c) Suppose that the annual withdrawal is $200 instead of $100. Explain how the model eventually breaks down.

9. Find a recurrence relation and initial condition to model the IRA investor in Example 4 if, instead of depositing $2000 at the *beginning* of every year, he or she deposits that amount at the end of the year (so that it does not earn interest for the year).

10. Suppose that Fibonacci's rabbits in Example 5 take 2 months to mature instead of 1 month.
   (a) Write down the recurrence relation and initial conditions to model the growth of the rabbit population.
   (b) Find the number of rabbits present at the end of 6 months.

11. Find a recurrence relation and initial conditions for the sequence $\{a_n\}$ if a_n is the number of bit strings of length n that do not contain three consecutive 0's.

12. Find a recurrence relation and initial conditions for the sequence $\{a_n\}$ if a_n is the number of bit strings of length n that contain three consecutive 0's.

13. Find a recurrence relation and initial conditions for the number of sequences over the alphabet $\{a, b, c\}$ that do not contain two consecutive a's.

14. Find a recurrence relation and initial conditions for the number of sequences over the alphabet $\{a, b, c\}$ that contain two consecutive a's.

15. A child takes either big steps or little steps. The big steps cover 20 inches, and the little steps cover 10 inches. Let a_n be the number of ways there are for the child to walk $10n$ inches.
   (a) Write down a recurrence relation and initial conditions for $\{a_n\}$.
   (b) Determine the number of ways for the child to walk 10 feet.

16. Find an explicit formula for the number of partitions of the positive integer k into parts no larger than 2.

17. Let $q(n)$ be the number of partitions of a set with n elements into sets with at most four elements each.
   (a) Write down a recurrence relation for q, and give the appropriate initial conditions.
   (b) Compute $q(6)$.

18. Let $r(n)$ be the number of partitions of a set with n elements into sets with at least two elements each.
   (a) Write down a recurrence relation for r, and give the appropriate initial conditions.
   (b) Compute $r(6)$.

19. Show that $p(n) = \sum_{k=0}^{n-1} C(n-1, k)p(k)$.

20. We can give a fuller analysis of partitions of sets by paying attention to the number of subsets in each partition. Define the **Stirling numbers of the second kind** as follows. For natural numbers n and k, $S(n, k)$ is the number of partitions of an n-set into exactly k parts (recall that the parts have to be nonempty). For example, since the partitions of $\{1, 2, 3\}$ are $\{\{1, 2, 3\}\}$, $\{\{1, 2\}, \{3\}\}$, $\{\{1, 3\}, \{2\}\}$, $\{\{2, 3\}, \{1\}\}$, and $\{\{1\}, \{2\}, \{3\}\}$, we see that $S(3, 1) = 1$, $S(3, 2) = 3$, $S(3, 3) = 1$, and $S(3, k) = 0$ for all other values of k. Thus the total number of partitions of an n-set (the Bell number) is the sum of these numbers: $p(n) = \sum_{k=0}^{n} S(n, k)$. Prove the following facts about the Stirling numbers of the second kind.
    (a) $S(n, 0) = 0$ and $S(n, 1) = 1$ for $n > 0$, and $S(0, 0) = 1$.
    (b) $S(n, 2) = 2^{n-1} - 1$ for $n \geq 1$.
    (c) $S(n, n-1) = C(n, 2)$ if $n \geq 1$ (we take $C(x, y)$ to be 0 by definition if $x < y$).
    (d) $S(n, n-2) = C(n, 3) + 3C(n, 4)$ for $n \geq 2$.
    (e) $S(n, k) = S(n-1, k-1) + kS(n-1, k)$ if n and k are positive integers.
    (f) $S(n, k) = \sum_{m=0}^{n-1} C(n-1, m)S(m, k-1)$ if n and k are positive integers.

21. Find a recurrence relation and initial conditions for the numbers $L(n, k)$, if $L(n, k)$ is the number of bit strings of length n having exactly k 0's. Is there another common notation for $L(n, k)$?

22. Let $M(n, k)$ be the number of strings over the alphabet $\{a, b, c\}$ of length n having exactly k a's.
    (a) Write down a recurrence relation and initial conditions for the numbers $M(n, k)$.
    (b) Compute $M(8, 3)$.

23. Let a_n be the number of ordered partitions of the positive integer n into a sum of 1's and 2's. For example, $a_4 = 5$, since we can write 4 as $2 + 2$, $2 + 1 + 1$, $1 + 2 + 1$, $1 + 1 + 2$, or $1 + 1 + 1 + 1$.
    (a) Write down a recurrence relation and initial conditions for the sequence $\{a_n\}$.
    (b) Compute a_5.

24. Let a_n be the number of ordered partitions of an n-set into sets of cardinality 1 and 2. For example, $a_3 = 12$, since the ordered partitions of the set $\{1, 2, 3\}$ into sets of cardinality 1 and 2 are $\{\{1\}, \{2\}, \{3\}\}$, $\{\{1\}, \{3\}, \{2\}\}$, $\{\{2\}, \{1\}, \{3\}\}$, $\{\{2\}, \{3\}, \{1\}\}$, $\{\{3\}, \{1\}, \{2\}\}$, $\{\{3\}, \{2\}, \{1\}\}$, $\{\{1, 2\}, \{3\}\}$, $\{\{1, 3\}, \{2\}\}, \{\{2, 3\}, \{1\}\}$, $\{\{1\}, \{2, 3\}\}$, $\{\{2\}, \{1, 3\}\}$, and $\{\{3\}, \{1, 2\}\}$.

(a)   Write down a recurrence relation and initial conditions for the sequence $\{a_n\}$.

(b)   Compute a_4.

25. Let a and b be natural numbers. Consider polygonal paths in the coordinate plane from the origin $(0,0)$ to the point (a,b) that turn only at lattice points (points whose coordinates are both integers) and proceed only in the positive x or y direction. For example, a path might go from $(0,0)$ to $(1,0)$ to $(2,0)$ to $(2,1)$ to $(3,1)$ to $(3,2)$. Let $T(a,b)$ be the number of such paths.

(a)   Write down a recurrence relation and initial conditions for $T(a,b)$.

(b)   Analyze the problem nonrecursively and obtain an explicit formula for $T(a,b)$.

26. We saw in the discussion at the end of the section that the number of arithmetic operations for multiplying two n by n matrices (where n is a power of 2) using the "obvious" recursive algorithm satisfies the recurrence relation $T(n) = 8T(n/2)+n^2$, but that the number of operations for the reorganized calculation satisfies the recurrence relation $T(n) = 7T(n/2) + 4.5n^2$, with $T(1) = 1$ in both cases. Compute the number of operations required under each scheme for successive powers of 2 until the number of operations for the reorganized calculation becomes less than the number of operations for the "obvious" algorithm.

27. In the second recursive matrix multiplication algorithm discussed at the end of this section, we used the initial condition $T(1) = 1$, which led to $T(2) = 25$. The first recursive algorithm uses only eight multiplications and four additions of numbers, however, so if we modify our algorithm to revert to the first algorithm for $n = 2$, then we can take $T(2) = 12$ as another initial condition.

(a)   With this initial condition, find $T(4)$ and $T(8)$.

(b)   Determine at what point it is more efficient to invoke the second algorithm (with this one improvement) rather than the first algorithm. Restrict yourself to n being a power of 2.

28. Let $M = (m_{ij})$ be the p by p matrix in which each off-diagonal entry (m_{ij} with $i \neq j$) is 1, and in which each diagonal entry (m_{ii}) is 2. (Here p is a fixed integer greater than 1.) Let d_n be the value of each diagonal entry of M^n, and let a_n be the value of each off-diagonal entry of M^n.

(a)   Write down a recurrence relation and initial conditions for the interrelated sequences $\{a_n\}$ and $\{d_n\}$.

(b)   Find M^4 when $p = 5$.

(c)   Show by mathematical induction that $d_n = a_n + 1$ for all n.

29. Let $T(n)$ be the number of comparisons needed in the worst case to find an element in a list of n distinct numbers, using binary search (Algorithm 2 in Section 4.3). Note that binary search is essentially recursive.

(a)   Write down a recurrence relation and initial conditions for $T(n)$.

(b)   Use the recurrence relation to compute $T(20)$.

Challenging Exercises

30. Suppose that Fibonacci's rabbits in Example 5 die immediately after producing their second pair of offspring.
  (a)   Write down the recurrence relation and initial conditions to model the growth of the rabbit population.
  (b)   Find the number of rabbits present at the end of 8 months.

31. Let a_n be the number of bit strings of length n that do not contain 101 as a substring (i.e., there is no occurrence of a 1 immediately followed by a 0 immediately followed by a 1).
  (a)   Show that $\{a_n\}$ satisfies the recurrence relation $a_n = a_{n-1} + a_{n-3} + a_{n-4} + \cdots + a_1 + a_0 + 2$ for $n \geq 3$. Write down the initial conditions that go along with this recurrence relation.
  (b)   Find the number of such bit strings of length 10.
  (c)   Obtain a recurrence relation of order 3 for $\{a_n\}$, using the result in part (a).

32. Let $f(k, n)$ be the number of surjective functions from a set with k elements onto a set with n elements, where $1 \leq n \leq k$.
  (a)   Show that f satisfies the recurrence relation

$$f(k, n) = n^k - \sum_{i=1}^{n-1} C(n, i) f(k, i).$$

  (Note that we can think of this as a recurrence relation in the single variable n, if we think of k as a constant.)
  (b)   Write down appropriate initial conditions.
  (c)   Find $f(6, 4)$ using the recurrence relation and initial conditions. (We will see this problem again in Section 7.4.)

33. Let $f(k, n)$ be as defined in Exercise 32.
  (a)   Show that f satisfies the recurrence relation $f(k, n) = n(f(k - 1, n) + f(k - 1, n - 1))$.
  (b)   Write down appropriate initial conditions.
  (c)   Find $f(6, 4)$ using the recurrence relation and initial conditions.

34. Let $N(k)$ be the number of partitions of the natural number k into integers greater than 1.
  (a)   Compute $N(k)$ for $k = 2, 3, 4, 5, 6$ by finding all such partitions.
  (b)   Find a recurrence relation and initial conditions for $N(k)$.
  (c)   Use the answer in part (b) to find $N(7)$.

35. Consider polygonal paths as in Exercise 25, with the added restriction that the path must stay on or below the line $y = x$; in other words, it may pass only through points (x, y) with $y \leq x$. Let $T'(a, b)$ denote the number of such paths from $(0, 0)$ to (a, b).

(a)   Show by listing the paths that $T'(3, 3) = 5$.

(b)   Write down a recurrence relation and initial conditions for $T'(a, b)$.

(c)   Using part (b), compute $T'(5, 5)$.

36. Let R_n be the number of regions in the plane determined by n lines, no two of which are parallel and no three of which pass through a common point. For instance, $R_2 = 4$.

(a)   Find a recurrence relation and initial conditions for R_n.

(b)   Compute the number of regions formed by seven lines.

37. It is a remarkable fact that the number of partitions of the natural number k into at most m parts is equal to the number of partitions of k into parts none of which is larger than m.

(a)   Verify this theorem for $k = 7$ and $m = 3$ by listing all the partitions in each case.

(b)   Prove the theorem. [*Hint*: Represent a partition of k into m parts by its **Ferrers diagram**, an array of k dots neatly arranged into m rows, the ith row containing the number of dots in the ith part; for example, the Ferrers diagrams for two partitions of 7 are shown here.]

7 = 3 + 2 + 2   7 = 3 + 3 + 1

EXERCISE 37

Exploratory Exercises

38. Find out more about the Bell numbers by consulting Gardner [102,105].

39. Find out about the efficiency of the Euclidean algorithm by consulting Rosen [259].

40. Find out more about Strassen's fast matrix multiplication algorithm by consulting Kronsjö [189].

41. Implement the recursive matrix multiplication algorithm discussed in this section in Pascal or some other programming language; better yet, implement Strassen's algorithm (see Exercise 40).

42. Find out about **Catalan numbers**, another recursively defined sequence of integers that occurs repeatedly in combinatorial situations, by consulting Camp-

bell [33], Eggleton and Guy [75], and Gardner [102]. These numbers arise, for example, as $T'(n, n)$ in Exercise 35.

43. Investigate the problem of computing the number of partitions of a positive integer into a given number of unequal parts. Consult Konečný [187] for a discussion of this problem.

44. The problem posed in Exercise 36 is harder if we allow some of the lines to be parallel. Consult Wetzel [318] to see what happens in that case.

45. Find out about algorithms for generating partitions by consulting Nijenhuis and Wilf [222].

46. Find out more about Stirling numbers of the second kind (see Exercise 20), and their relationship to Stirling numbers of the first kind (see Exercise 32 in Section 7.1), by consulting Bogart [23], Oullette and Bennett [229], and Pólya, Tarjan, and Woods [237].

SECTION 7.3
SOLVING RECURRENCE RELATIONS

Alice made a short calculation,
and said "Seven years and six months."
"Wrong!" Humpty Dumpty exclaimed triumphantly.

—Lewis Carroll, *Through the Looking Glass*

In the preceding section we found out how to write down recurrence relations and initial conditions for various sequences and functions. This certainly gives us a way to compute the values we need, but it does not give us an explicit formula for those values. For instance, we know that the number of rabbits on the island in Example 5 of Section 7.2 after n months is f_n, the nth term of the Fibonacci sequence, and we can calculate this number fairly efficiently using the iterative Algorithm 4 in Section 5.2, but we do not have an algebraic formula for f_n in terms of n. We could not very easily, for example, determine the order of magnitude of f_{5000} using just the recurrence relation and initial conditions. In this section we develop some techniques that enable us to solve many, although not all, recurrence relations with initial conditions to obtain explicit formulas for the terms of the sequences they describe.

To begin we must give some definitions, so that we know exactly what we are talking about. Let $a_n = f(a_{n-1}, a_{n-2}, \ldots, a_1, a_0)$, $n \geq k$, be a recurrence relation, and let $a_0 = b_0$, $a_1 = b_1$, ..., $a_{k-1} = b_{k-1}$ be k initial conditions (the b_i's are constants). We say that the formula $a_n = g(n)$ is a **solution** or an **explicit solution** of the recurrence relation if the recurrence relation becomes a true statement (an identity that holds for all values of n) when this formula is substituted into the recurrence relation. A solution $a_n = g(n)$ may have one or more **arbitrary constants**, which can take on any values. If *every* solution of the recurrence relation is equal to $a_n = g(n)$ for some choice of the arbitrary constants, then $a_n = g(n)$ is said to be **the general solution** of the recurrence relation. The recurrence relation *with the initial conditions* uniquely determines the sequence; an explicit solution of the recurrence relation and initial conditions is called the **specific solution**.

We can determine whether a given formula $a_n = g(n)$ is a solution of a recurrence relation simply by plugging it into the relation and manipulating the resulting equation algebraically until we can tell whether we have an identity. We can determine whether it satisfies initial conditions by explicitly checking them. More important, *given a solution of a recurrence relation that contains arbitrary constants, we can often determine the values of those constants that make the solution the specific solution of the recurrence relation with initial conditions, by plugging in the initial conditions and solving for the constants.*

EXAMPLE 1. Show that $a_n = A + B \cdot 2^n$ is a solution of the recurrence relation $a_n = 3a_{n-1} - 2a_{n-2}$. Determine the values of the constants A and B that make this formula the specific solution of the given recurrence relation with the initial conditions $a_0 = 3$ and $a_1 = 7$.

Solution. First we plug the proposed solution into the recurrence relation. The left-hand side (a_n) is just $A + B \cdot 2^n$. To write down the right-hand side we note that the proposed solution gives us $a_{n-1} = A + B \cdot 2^{n-1}$ and $a_{n-2} = A + B \cdot 2^{n-2}$. Thus the right-hand side $(3a_{n-1} - 2a_{n-2})$ is $3(A + B \cdot 2^{n-1}) - 2(A + B \cdot 2^{n-2})$. Therefore, we want to verify that the equation

$$A + B \cdot 2^n = 3(A + B \cdot 2^{n-1}) - 2(A + B \cdot 2^{n-2})$$

holds for all $n \geq 2$. This is an exercise in algebra. Starting from the right-hand side we have

$$3(A + B \cdot 2^{n-1}) - 2(A + B \cdot 2^{n-2}) = 3A + 3B \cdot 2^{n-1} - 2A - 2B \cdot 2^{n-2}$$

$$= A + B \cdot 2^{n-2}(3 \cdot 2 - 2)$$

$$= A + 4B \cdot 2^{n-2} = A + B \cdot 2^n,$$

which is exactly the left-hand side.

Note that we have verified an infinite number of different solutions of the recurrence relation, each obtained by using different values for A and B. For instance, we know that $a_n = 17 - 2^n$ is a solution, as are $a_n = -4$, $a_n = 46 \cdot 2^n$, and $a_n = 0$. Each of these solutions satisfies a different set of initial conditions. It turns out in fact that $a_n = A + B \cdot 2^n$ is the general solution of this recurrence relation: All solutions are of this form. On the other hand, $a_n = 5 - 3^n$ is not a solution of the recurrence relation; if we substitute it into the relation, we obtain $5 - 3^n = 5 - 7 \cdot 3^{n-2}$ (after a little algebra), which is clearly not an identity.

Next we are asked to find the specific solution of the recurrence relation with given initial conditions. Plugging $n = 0$ into our solution, and invoking the given condition that $a_0 = 3$, we obtain the equation $3 = A + B \cdot 2^0 = A + B$. Plugging $n = 1$ into our solution, and invoking the given condition that $a_1 = 7$, we obtain the equation $7 = A + B \cdot 2^1 = A + 2B$. Thus we need to solve the system of simultaneous linear equations

$$A + B = 3$$

$$A + 2B = 7.$$

An easy exercise in elementary algebra yields $A = -1$ and $B = 4$. Therefore, the unique solution of the recurrence relation with initial conditions is $a_n = -1 + 4 \cdot 2^n$. This tells us, for instance, that $a_7 = -1 + 4 \cdot 2^7 = 511$, without having to compute the terms of the sequence one by one until we reach a_7.   ◆

It is important to keep in mind that a solution of a recurrence relation is a *function*, not a number; in particular, n is not the unknown in recurrence relations. Recurrence relations are one level of abstraction higher than algebraic equations. We can check whether a given number is a solution of an algebraic equation by plugging the number into the equation and checking that the two sides are equal numbers. Here we check whether a given function of n is a solution of a recurrence relation by plugging the function into the recurrence relation and checking that the two sides are equal as functions of n.

We usually work with recurrence relations of order k for some small k. (We had $k = 2$ in the last example.) The recurrence relation will express a_n as a function of a_{n-1} through a_{n-k}, for all $n \geq k$, and there will be k initial conditions, $a_0 = b_0$, $a_1 = b_1$, ..., $a_{k-1} = b_{k-1}$. In this section we consider two cases that sometimes lend themselves to explicit solutions.

First, we consider problems in which we have a first-order recurrence relation $a_n = f(a_{n-1})$ for $n \geq 1$ together with an initial condition $a_0 = b_0$. Such problems can often be solved by what might be called "brute force" using an iterative approach; we discuss this method in the next subsection. There are no restrictions on what form f might take, but in order for the method to be feasible, f needs to be fairly simple.

Second, we consider problems in which the function f is of a special form, given in the following definition. A recurrence relation

$$a_n = c_1 a_{n-1} + c_2 a_{n-2} + \cdots + c_k a_{n-k} + h(n),$$

where $c_1, c_2, \ldots, c_k$ are constants (independent of n) with $c_k \neq 0$, is called a kth-order **linear recurrence relation with constant coefficients**. If $h(n)$ is not present (i.e., $h(n) = 0$ for all n), then the recurrence relation is said to be **homogeneous**; otherwise, it is said to be **nonhomogeneous**.

EXAMPLE 2.   The recurrence relation $a_n = 3a_{n-1} - 2a_{n-2}$ is a second-order homogeneous linear recurrence relation with constant coefficients (the coefficients are 3 and -2). The constant coefficient, linear homogeneous recurrence relation $a_n = 3a_{n-3} - 2a_{n-5}$ has order 5 (the coefficients are 0, 0, 3, 0, and -2). The recurrence relations $a_n = 3a_{n-1} - 2a_{n-2} - 4$ and $a_n = 3a_{n-1} - 2a_{n-2} + 7n^2$ are linear (i.e., linear expressions in the variables a_n, a_{n-1}, $\ldots$, a_{n-k}) and have constant coefficients, but they are nonhomogeneous. The recurrence relation $a_n = 3a_{n-1}^2 - 2a_{n-2}$ is not linear (because of the term a_{n-1}^2). The recurrence relation $a_n = 3a_{n-1} - 2n^3a_{n-2}$ is linear and homogeneous but has the nonconstant coefficient $-2n^3$.   ◆

In the latter part of this section we discuss a technique for finding the general solutions of linear homogeneous recurrence relations with constant coefficients, as well as a method that often works for nonhomogeneous problems as well.

ITERATIVE SOLUTIONS OF RECURRENCE RELATIONS

In Example 4 of Section 7.2 we showed that we could model the amount A_n in an individual retirement account at the end of n years with the recurrence relation $A_n = 2100 + 1.05A_{n-1}$, together with the initial condition $A_0 = 0$. We will explain the iterative approach to solving first-order recurrence relations by applying it to this problem. What we seek is an explicit formula for the amount of money in the account at the end of n years (A_n) as a function of n. For simplicity of notation let us replace the constant 2100 by d and the constant 1.05 by i. We start with the recurrence relation itself,

$$A_n = d + iA_{n-1}.$$

Now $A_{n-1} = d + iA_{n-2}$ (this is just a restatement of the recurrence relation with $n-1$ in place of n). If we substitute $d + iA_{n-2}$ for A_{n-1} in the original equation, we obtain

$$A_n = d + i(d + iA_{n-2})$$
$$= d(1 + i) + i^2 A_{n-2}.$$

Next we substitute $d + iA_{n-3}$ for A_{n-2}, this time using the fact that the recurrence relation must hold for $n - 2$. This gives us

$$A_n = d(1 + i) + i^2(d + iA_{n-3})$$
$$= d(1 + i + i^2) + i^3 A_{n-3}.$$

At this point we begin to see the pattern. Let us continue the calculation "all the way to the end."

$$A_n = d(1 + i + i^2) + i^3 A_{n-3}$$
$$= d(1 + i + i^2) + i^3(d + iA_{n-4}) = d(1 + i + i^2 + i^3) + i^4 A_{n-4}$$
$$= d(1 + i + i^2 + i^3) + i^4(d + iA_{n-5}) = d(1 + i + i^2 + i^3 + i^4) + i^5 A_{n-5}$$
$$\vdots$$
$$= d(1 + i + i^2 + \cdots + i^{n-1}) + i^n A_{n-n}$$
$$= d(1 + i + i^2 + \cdots + i^{n-1}) + i^n A_0.$$

Thus we have found the explicit solution we are seeking: A_n expressed as a function of n. Furthermore, the final expression in parentheses is just a geometric series, whose sum we know from Example 8 in Section 5.3. Therefore, as long as $i \neq 1$ our solution is

$$A_n = \frac{d(i^n - 1)}{i - 1} + i^n A_0.$$

If we want to know, for example, how much money will be in the account in 30 years, we do not need to compute A_1 through A_{30}, but rather just plug $n = 30$ into this solution. For the parameters of Example 4 in Section 7.2 ($d = 2100$, $i = 1.05$, and $A_0 = 0$), we have

$$A_n = \frac{2100(1.05^n - 1)}{1.05 - 1} + 1.05^n \cdot 0 = 42000(1.05^n - 1),$$

so $A_{30} = 42000(1.05^{30} - 1) \approx 139522$. Thus there will be about \$139,522 in the account at the end of 30 years.

In summary, this iterative, brute-force method consists of successively rewriting the recurrence relation $a_n = f(a_{n-1})$ by replacing a_j with an expression involving a_{j-1}, as j goes from $n - 1$ down to 1, at which point the initial condition for a_0 leaves us with an explicit formula for a_n. The formula may at first contain a sum or product, as it did in the example we just worked out. If the recurrence relation is fairly simple (as it was in this example), we can often write a closed-form expression for the sum or product.

In our next example we determine the efficiency of merge sort.

EXAMPLE 3. In Section 7.2 we showed that the number of comparisons used in merge sort satisfies the recurrence relation $T(n) = T(\lfloor n/2 \rfloor) + T(\lceil n/2 \rceil) + n$ for $n > 1$, with initial condition $T(1) = 0$. In order to apply the iterative method easily here, we restrict ourselves to the case in which n is a power of 2, say $n = 2^k$. Then our recurrence relation simplifies to $T(n) = 2T(n/2) + n$ or $T(2^k) = 2^k + 2T(2^{k-1})$. The iterative approach gives us the following calculation

$$T(n) = T(2^k) = 2^k + 2T(2^{k-1})$$
$$= 2^k + 2(2^{k-1} + 2T(2^{k-2})) = 2 \cdot 2^k + 2^2 T(2^{k-2})$$
$$= 2 \cdot 2^k + 2^2(2^{k-2} + 2T(2^{k-3})) = 3 \cdot 2^k + 2^3 T(2^{k-3})$$
$$= 3 \cdot 2^k + 2^3(2^{k-3} + 2T(2^{k-4})) = 4 \cdot 2^k + 2^4 T(2^{k-4})$$
$$\vdots$$
$$= k \cdot 2^k + 2^k T(2^{k-k}) = k \cdot 2^k + 2^k T(1) = 2^k \cdot k.$$

Now since $n = 2^k$, we can write $k = \log n$, so we have found the desired explicit formula for $T(n)$, namely, $T(n) = n \log n$. This calculation is valid as long as n is a power of 2.

If n is not a power of 2, the formula is still approximately correct, so in any case $T(n) \in O(n \log n)$. To see this formally, we can argue as follows. Given n, let m be the smallest power of 2 greater than or equal to n; then certainly $m < 2n$. We can merge sort a list of m numbers using at most $m \log m$ comparisons, which is less than $2n \log(2n)$, which is certainly in $O(n \log n)$. Now sorting the smaller list using merge sort can never take more comparisons than sorting the larger one, so the number of comparisons needed to sort the list with n elements is also in $O(n \log n)$.

The difference between the running time of an algorithm using $n \log n$ comparisons and one using around $n^2/2$ comparisons (which is what naive sorting algorithms, such as bubble sort, require) can be substantial in practice, as we saw at the end of Section 4.3. ◆

The iterative approach, together with some messy algebra, gives us the following general theorem for recurrence relations of the type we have been considering. We will not include the proof.

THEOREM 1. *Let f be an increasing function. Suppose that f satisfies the recurrence relation $f(n) = af(n/b) + cn^d$ (or at least the inequality $f(n) \leq af(n/b) + cn^d$), where a, c, and d are nonnegative real numbers and b is an integer greater than 1. Then $f(n) \in O(n^d)$ if $a < b^d$; $f(n) \in O(n^d \log n)$ if $a = b^d$; and $f(n) \in O(n^{\log_b a})$ if $a > b^d$.*

With merge sort we had $a = b = 2$, $c = d = 1$. Thus $a = b^d$, so $f(n) \in O(n^1 \log n)$. Let us see what this theorem says about fast matrix multiplication.

EXAMPLE 4. At the end of Section 7.2 we found that the recurrence relation for the number of arithmetic operations needed by Strassen's fast matrix multiplication algorithm is $T(n) = 7T(n/2) + 4.5n^2$. Theorem 1 applies with $a = 7$, $b = 2$, $c = 4.5$, and $d = 2$. Since $a > b^d$, the solution satisfies $T(n) \in O(n^{\log_2 7}) \approx O(n^{2.81})$. Since the usual matrix multiplication algorithm requires n^3 multiplications, this fast algorithm may be an improvement if n is large. ◆

The iterative approach we have discussed here is somewhat limited in application. It does not apply to most recurrence relations of order greater than 1; the reader is invited to try it in such cases and discover the mess that results. To solve higher-order recurrence relation problems, we turn to a more systematic algorithm.

LINEAR HOMOGENEOUS RECURRENCE RELATIONS WITH CONSTANT COEFFICIENTS

We will present an algorithm for finding the general solutions of linear homogeneous recurrence relations with constant coefficients. It always works (in principle) by transforming the recurrence relation problem into an algebraic problem of finding roots of a polynomial. (The reader who has learned how to solve linear homogeneous differential equations with constant coefficients will be at an advantage in reading this subsection, since the theory and methodology are essentially identical in the two cases. The reader with some background in linear algebra will find that the discussion has a familiar ring to it, since what we do in this algorithm is to find a basis for the vector space of solutions of the recurrence relation. We will not, however, assume familiarity with either differential equations or linear algebra in our development.)

The simplest case of a linear homogeneous recurrence relation with constant coefficients is the first-order relation $a_n = ca_{n-1}$. This problem can be solved by the iterative scheme of the preceding subsection: $a_n = ca_{n-1} = c^2 a_{n-2} = c^3 a_{n-3} = \cdots = c^n a_{n-n} = c^n a_0$. In other words, the explicit solution turns out to be an exponential function of n. Although we cannot apply the iterative method to recurrence relations of higher order, let us see if an exponential function might be the solution in those cases as well.

We begin by tackling the recurrence relation $a_n = 3a_{n-1} - 2a_{n-2}$ considered in Example 1. Suppose that we search for solutions of the form $a_n = r^n$, where r is some nonzero constant. (The constant solution $a_n = 0$ always satisfies a linear homogeneous recurrence relation, but it is not very interesting; it is certainly not the general solution.) There is, of course, no reason to expect ahead of time that an exponential function will actually be a solution, but motivated by the first-order case, let us look nonetheless. The worst that can happen is that this approach will lead nowhere, and if so, we can try something else.

If we plug $a_n = r^n$ into the given recurrence relation, we get $r^n = 3r^{n-1} - 2r^{n-2}$. Dividing both sides by r^{n-2} and transposing everything to the left, we obtain $r^2 - 3r + 2 = 0$. Thus $a_n = r^n$ is a solution of the given recurrence relation if and only if r satisfies the algebraic equation $r^2 - 3r + 2 = 0$. This equation is called the **characteristic equation** (and $r^2 - 3r + 2$ is called the **characteristic**

polynomial) of the recurrence relation. By elementary algebra (factoring or the quadratic formula), we see that there are two roots, $r = 1$ and $r = 2$. Thus we have found two solutions of our recurrence relation, $a_n = 1^n = 1$ and $a_n = 2^n$. The reader should check that these are, indeed, solutions of $a_n = 3a_{n-1} - 2a_{n-2}$.

In order to continue our discussion and find not just two but all solutions of the given recurrence relation, we need to see exactly why the linearity and homogeneity of a recurrence relation make it especially nice to work with.

THEOREM 2.   *Suppose that $a_n = g_1(n)$ and $a_n = g_2(n)$ are both solutions of the recurrence relation*

$$a_n = c_1 a_{n-1} + c_2 a_{n-2} + \cdots + c_k a_{n-k},$$

*where the c_i's are constants. Then any **linear combination** $a_n = A g_1(n) + B g_2(n)$ of these solutions is also a solution of the recurrence relation, where A and B are arbitrary constants.*

Proof.   We simply take the statements that g_1 and g_2 are solutions, namely that

$$g_1(n) = c_1 g_1(n - 1) + c_2 g_1(n - 2) + \cdots + c_k g_1(n - k)$$

and

$$g_2(n) = c_1 g_2(n - 1) + c_2 g_2(n - 2) + \cdots + c_k g_2(n - k),$$

multiply by A and B, respectively, and add, to get

$$Ag_1(n) + Bg_2(n) = c_1(Ag_1(n - 1) + Bg_2(n - 1)) + c_2(Ag_1(n - 2)$$
$$+ Bg_2(n - 2)) + \cdots + c_k(Ag_1(n - k) + Bg_2(n - k)),$$

which says that $Ag_1 + Bg_2$ is also a solution.   ∎

It should be clear that this result extends to a sum of any number of solutions.

Because of Theorem 2, then, we know that $a_n = A \cdot 1 + B \cdot 2^n = A + B \cdot 2^n$ is a solution of the recurrence relation $a_n = 3a_{n-1} - 2a_{n-2}$. Furthermore, it can be shown that this is *the general solution*: Every solution to the recurrence relation is $a_n = A + B \cdot 2^n$ for some A and B. In Example 1, for instance, we found the specific solution of this recurrence relation that satisfies the initial conditions $a_0 = 3$ and $a_1 = 7$. We summarize the method inherent in our discussion so far in the following theorem (we omit the proof).

THEOREM 3.   *Suppose that $\{a_n\}$ satisfies the second-order recurrence relation $a_n = c_1 a_{n-1} + c_2 a_{n-2}$ where c_1 and c_2 are constants, and $c_2 \neq 0$. Let r_1 and r_2 be the two roots of $r^2 - c_1 r - c_2 = 0$, and assume that $r_1 \neq r_2$. Then $a_n = Ar_1^n + Br_2^n$*

for some constants A and B. Conversely, $a_n = Ar_1^n + Br_2^n$ does satisfy the given recurrence relation.

Recall from algebra that the second-degree equation $r^2 - c_1 r - c_2 = 0$ always has two roots, r_1 and r_2, which may be rational numbers, irrational numbers, or even complex numbers. Theorem 3 takes care of the case in which the two roots are not equal. The recurrence relation for the Fibonacci sequence leads to such a case.

EXAMPLE 5. Find an explicit formula for the terms of the Fibonacci sequence $\{f_n\}$, defined by $f_0 = f_1 = 1$ and $f_n = f_{n-1} + f_{n-2}$ for all $n \geq 2$.

Solution. The characteristic equation for this recurrence relation is $r^2 - r - 1 = 0$. By the quadratic formula, its roots are $r_1 = (1 + \sqrt{5})/2$ and $r_2 = (1 - \sqrt{5})/2$. Therefore, the general solution of the recurrence relation is

$$f_n = A\left(\frac{1 + \sqrt{5}}{2}\right)^n + B\left(\frac{1 - \sqrt{5}}{2}\right)^n.$$

To determine the values of A and B, we plug in the initial conditions:

$$1 = f_0 = A\left(\frac{1 + \sqrt{5}}{2}\right)^0 + B\left(\frac{1 - \sqrt{5}}{2}\right)^0 = A + B$$

$$1 = f_1 = A\left(\frac{1 + \sqrt{5}}{2}\right)^1 + B\left(\frac{1 - \sqrt{5}}{2}\right)^1 = A\left(\frac{1 + \sqrt{5}}{2}\right) + B\left(\frac{1 - \sqrt{5}}{2}\right).$$

This is a system of two linear equations in two unknowns; it happens to have messy irrational coefficients, but one solves it by the usual methods of elementary algebra, to yield

$$A = \frac{1}{\sqrt{5}}\left(\frac{1 + \sqrt{5}}{2}\right), \qquad B = -\frac{1}{\sqrt{5}}\left(\frac{1 - \sqrt{5}}{2}\right).$$

Therefore, the specific solution—the algebraic formula for the terms of the Fibonacci sequence—is

$$f_n = \frac{1}{\sqrt{5}}\left(\frac{1 + \sqrt{5}}{2}\right)^{n+1} - \frac{1}{\sqrt{5}}\left(\frac{1 - \sqrt{5}}{2}\right)^{n+1}.$$

This formula may seem awkward, but it is essentially the simplest way to express these numbers explicitly. If we want to find f_8, for example, we plug

$n = 8$ into the last display and do the algebra. Alternatively, we can do the arithmetic on a calculator and round to the nearest natural number (if we want to avoid the messy algebra and are willing to believe that the calculator's round-off error will not get us in trouble—which it will not). We find that $f_8 = 34$. In fact, since the second term in our formula is always less than $\frac{1}{2}$ in absolute value, we can compute f_n simply as the first term, $((1 + \sqrt{5})/2)^{n+1}/\sqrt{5}$, rounded to the nearest integer. A calculator that displays 10 digits and keeps 13 digits of accuracy internally can correctly compute the sequence through $f_{48} = 7{,}778{,}742{,}049$ in this way.

Finally, note that the explicit formula gives us important qualitative information that the recurrence relation alone does not. We can tell that the Fibonacci sequence grows exponentially, and we even know the base of the exponentiation, namely $(1 + \sqrt{5})/2 \approx 1.618$ (since the second term very quickly goes to 0 as n increases). (The number $(1 + \sqrt{5})/2$ is called the **golden ratio**; it arises repeatedly in pure and applied mathematics.) Thus for instance, we can easily find the exact number of decimal digits in f_{5000} without calculating f_{5000} (which would not be a pleasant task) (see Exercise 32). ◆

As we remarked above, the roots of the characteristic equation might be complex numbers. We can work with them just as we worked with irrational numbers in Example 5: Numbers are numbers, after all. Exercise 20, for example, involves complex roots. Alternatively, one can use trigonometric functions to handle the case of complex roots, using the fact that $e^{a+bi} = e^a(\cos b + i \sin b)$; see Exercise 42.

To finish the case left undone by the algorithm implied by Theorem 3, we need to handle the case of a repeated root $r_1 = r_2$ of the characteristic polynomial. In such a case simply letting $a_n = Ar_1^n + Br_1^n$ will not do, since this is the same as $a_n = (A + B)r_1^n$, and there is really only one arbitrary constant. Since there are two initial conditions, we need two arbitrary constants. Instead, we need a distinctly different, independent, solution of the recurrence relation. It turns out that $a_n = nr_1^n$ is a solution in such a case. Again we omit the proof.

THEOREM 4. *Suppose that $\{a_n\}$ satisfies the second-order recurrence relation $a_n = c_1 a_{n-1} + c_2 a_{n-2}$ where c_1 and c_2 are constants, and $c_2 \neq 0$. Let r_1 and r_2 be the two roots of $r^2 - c_1 r - c_2 = 0$, and assume that $r_1 = r_2$. Then $a_n = Ar_1^n + Bnr_1^n$ for some constants A and B. Conversely, $a_n = Ar_1^n + Bnr_1^n$ does satisfy the given recurrence relation.*

EXAMPLE 6. Solve the recurrence relation and initial conditions

$$a_n = -6a_{n-1} - 9a_{n-2}$$

$$a_0 = 2, \qquad a_1 = 3.$$

Solution. The characteristic equation is $r^2 + 6r + 9 = 0$, which has the repeated root $r = -3$. Therefore, the general solution of the recurrence relation is, by Theorem 4,

$a_n = A \cdot (-3)^n + Bn \cdot (-3)^n$. To solve for A and B we plug in the initial conditions, yielding the equations $2 = A$ and $3 = -3A - 3B$, whence $A = 2$ and $B = -3$. Therefore, the unique solution of the recurrence relation with the given initial conditions is $a_n = 2 \cdot (-3)^n - 3n \cdot (-3)^n$.

As a check on our calculations, let us find a_4 both from our solution and from the recurrence relation. Our solution says that $a_4 = 2 \cdot (-3)^4 - 12 \cdot (-3)^4 = -810$. On the other hand, calculating from the recurrence relation, we find that $a_2 = -6a_1 - 9a_0 = -36$, then $a_3 = -6a_2 - 9a_1 = 189$, and finally $a_4 = -6a_3 - 9a_2 = -810$. ◆

Finally, we need to see how to extend our algorithm to linear homogeneous recurrence relations

$$a_n = c_1 a_{n-1} + c_2 a_{n-2} + \cdots + c_k a_{n-k}$$

with constant coefficients having order greater than 2. As before, we form the characteristic equation, which now looks like

$$r^k - c_1 r^{k-1} - c_2 r^{k-2} - \cdots - c_{k-1} r - c_k = 0.$$

By the fundamental theorem of algebra, this equation has exactly k solutions (real or complex numbers), counting multiplicities. Let $r_1, r_2, \ldots, r_t$ be the *distinct* roots of the characteristic polynomial, and let m_i be the multiplicity of the root r_i, for $i = 1, 2, \ldots, t$. (We say that the root r_i has **multiplicity** m_i in the polynomial $p(r)$ if $(r - r_i)^{m_i}$ is a factor of $p(r)$ but $(r - r_i)^{m_i+1}$ is not.) In particular, the sum of the m_i's is k. Then $a_n = r_i^n$, $a_n = n r_i^n$, $a_n = n^2 r_i^n$, $\ldots$, $a_n = n^{m_i-1} r_i^n$ are all solutions of the recurrence relation for each i from 1 to t, giving us a total of k solutions. (Note that for nonrepeated roots, $m_i = 1$, so this list contains only $a_n = r_i^n$.) By Theorem 2, the general solution is then a linear combination of these k solutions (i.e., the sum of arbitrary constants times each of them). Let us restate this as a theorem. Once again we omit the proof.

THEOREM 5. *Suppose that $\{a_n\}$ satisfies the recurrence relation*

$$a_n = c_1 a_{n-1} + c_2 a_{n-2} + \cdots + c_k a_{n-k},$$

where the c_i's are constants, and $c_k \neq 0$. Let $r_1, r_2, \ldots, r_t$ be the distinct roots of

$$r^k - c_1 r^{k-1} - c_2 r^{k-2} - \cdots - c_{k-1} r - c_k = 0,$$

and denote the multiplicity of r_i by m_i. Then

$$a_n = A_{1,1}r_1^n + A_{1,2}nr_1^n + A_{1,3}n^2 r_1^n + \cdots + A_{1,m_1}n^{m_1-1}r_1^n$$
$$+ A_{2,1}r_2^n + A_{2,2}nr_2^n + A_{2,3}n^2 r_2^n + \cdots + A_{2,m_2}n^{m_2-1}r_2^n$$
$$+ \cdots + A_{t,1}r_t^n + A_{t,2}nr_t^n + A_{t,3}n^2 r_t^n + \cdots + A_{t,m_t}n^{m_t-1}r_t^n,$$

where the $A_{i,j}$'s are some arbitrary constants. Conversely, the sequence defined by this formula does satisfy the recurrence relation.

The general solution of a kth-order linear homogeneous recurrence relation with constant coefficients, then, can be written down immediately once we find all the roots (with multiplicities) of the characteristic equation. In practice this may be difficult. (Indeed, there is a deep theorem of algebra that says there is in general no way to find, or even to write down names of, the exact roots of polynomials of degree greater than 4.) Such techniques from algebra as the rational root test and synthetic division are often helpful. Alternatively, if we find approximations for the roots (which is usually not hard with numerical techniques like Newton's method), we can get approximations for the explicit formulas we seek. Once we have the general solution, we can find the specific solution of a kth-order problem with initial conditions by plugging in the initial conditions to obtain a system of k linear equations in k unknowns and solving the system.

EXAMPLE 7. Find the solution of the recurrence relation $a_n = 2a_{n-1} + 3a_{n-2} - 8a_{n-3} + 4a_{n-4}$ that satisfies the initial conditions $a_0 = 3$, $a_1 = 0$, $a_2 = -8$, and $a_3 = -26$.

Solution. We begin by writing down the characteristic equation, namely $r^4 - 2r^3 - 3r^2 + 8r - 4 = 0$. By the rational root test, the only possible rational roots are ± 1, ± 2, and ± 4 (factors of the constant term), so we try plugging these values into the equation. We see that $r = 1$ is a solution; this means that $(r - 1)$ is a factor of the characteristic polynomial, and we divide (by long or synthetic division) to find the other factor: $r^4 - 2r^3 - 3r^2 + 8r - 4 = (r - 1)(r^3 - r^2 - 4r + 4)$. We see that $r = 1$ is also a root of this third degree factor, so we divide again and get $r^4 - 2r^3 - 3r^2 + 8r - 4 = (r - 1)(r - 1)(r^2 - 4)$. This factors on inspection, so our characteristic equation is finally $(r - 1)^2(r - 2)(r + 2) = 0$.

The characteristic equation has three distinct roots ($t = 3$ in the notation of Theorem 5): $r_1 = 1$, $r_2 = 2$, and $r_3 = -2$. The multiplicities are 2, 1, and 1, respectively (i.e., $m_1 = 2$, $m_2 = 1$, and $m_3 = 1$). Thus we can immediately write down the general solution of the recurrence relation:

$$a_n = A \cdot 1^n + Bn \cdot 1^n + C \cdot 2^n + D \cdot (-2)^n$$
$$= A + Bn + C \cdot 2^n + D \cdot (-2)^n.$$

Note that we have named the arbitrary constants with plain letters rather than subscripted ones for a simpler notation.

Finally, we are ready to find the values of the four constants, using the initial conditions. We plug $a_0 = 3$ into our solution and obtain

$$A + C + D = 3.$$

Similarly, upon plugging in the other three initial conditions, we get

$$A + B + 2C - 2D = 0$$

$$A + 2B + 4C + 4D = -8$$

$$A + 3B + 8C - 8D = -26.$$

Then we use whatever methods we like (elimination of variables, for example) to solve this system, and we find that the unique solution is $A = 8$, $B = 2$, $C = -5$, and $D = 0$. Therefore, the solution to this problem is $a_n = 8 + 2n - 5 \cdot 2^n$.

The author actually constructed this example by starting with the solution, so that the algebra would turn out not to be too messy. If one writes down a "random" recurrence relation with integer coefficients, the characteristic polynomial is likely to have messy irrational and complex roots. ◆

The exercises include problems involving some recurrence relations that are not linear, or do not have constant coefficients; they can be transformed into linear recurrence relations with constant coefficients.

NONHOMOGENEOUS RECURRENCE RELATIONS (*OPTIONAL*)

Many applications, such as some we saw in Section 7.2, result in nonhomogeneous recurrence relations. In Example 7 of Section 7.2, for instance, we modeled the problem of counting bit strings of length n containing two consecutive 0's with the recurrence relation $b_n = b_{n-1} + b_{n-2} + 2^{n-2}$. We now discuss a method that is often effective in solving such nonhomogeneous recurrence relations. We will still assume that our relations are linear with constant coefficients, although this assumption is not needed in Theorem 6.

Let

$$a_n = c_1 a_{n-1} + c_2 a_{n-2} + \cdots + c_k a_{n-k} + h(n)$$

be a given nonhomogeneous recurrence relation with constant coefficients. The **associated homogeneous recurrence relation** is the same relation with the $h(n)$ term omitted:

$$a_n = c_1 a_{n-1} + c_2 a_{n-2} + \cdots + c_k a_{n-k}.$$

It turns out that we can find the general solution of our given recurrence relation with the following two-step procedure. First, we find the general solution of the associated homogeneous recurrence relation. This we already know how to do from the preceding subsection. Next, we find one **particular solution** of the given nonhomogeneous recurrence relation, using a technique known as the "method of undetermined coefficients." Then we add these two functions to obtain the general solution of the nonhomogeneous recurrence relation. Let us see why this procedure is valid.

THEOREM 6.   *Suppose that $a_n = f(n)$ is the general solution of the linear homogeneous recurrence relation*

$$a_n = c_1 a_{n-1} + c_2 a_{n-2} + \cdots + c_k a_{n-k}.$$

Suppose that $a_n = g(n)$ is a particular solution of a nonhomogeneous recurrence relation with the same coefficients:

$$a_n = c_1 a_{n-1} + c_2 a_{n-2} + \cdots + c_k a_{n-k} + h(n).$$

Then the general solution of this nonhomogeneous recurrence relation is $a_n = f(n) + g(n)$.

Proof.   If we plug $a_n = f(n) + g(n)$ into the right-hand side of the nonhomogeneous recurrence relation, we get

$$c_1(f(n-1) + g(n-1)) + c_2(f(n-2) + g(n-2))$$
$$+ \cdots + c_k(f(n-k) + g(n-k)) + h(n)$$
$$= (c_1 f(n-1) + c_2 f(n-2) + \cdots + c_k f(n-k))$$
$$+ (c_1 g(n-1) + c_2 g(n-2) + \cdots + c_k g(n-k) + h(n))$$
$$= f(n) + g(n),$$

by our assumptions on f and g. Therefore $a_n = f(n) + g(n)$ is indeed a solution of the given nonhomogeneous recurrence relation. Clearly, it has the same k arbitrary constants that f had; we can find values for these k constants to satisfy any set of initial conditions (a justification of this last claim would get us too deeply into linear algebra). Therefore, $a_n = f(n) + g(n)$ is the *general* solution of the given recurrence relation. ∎

EXAMPLE 8.   Find the general solution of $a_n = 3a_{n-1} - 2a_{n-2} + 12 \cdot (-1)^n$, using the hint that $a_n = 2 \cdot (-1)^n$ is a particular solution. Then find the specific solution of this recurrence relation that also satisfies the initial conditions $a_0 = 8$ and $a_1 = 5$.

Solution. Let us verify that the hint is true. Plugging $a_n = 2 \cdot (-1)^n$ into the recurrence relation gives

$$2 \cdot (-1)^n = 3 \cdot 2 \cdot (-1)^{n-1} - 2 \cdot 2 \cdot (-1)^{n-2} + 12 \cdot (-1)^n,$$

which, after dividing both sides by $(-1)^{n-2}$ and simplifying, says that $2 = 2$, certainly a true statement. Thus we have a particular solution of the nonhomogeneous recurrence relation (we will see shortly how to get such a solution if no one is around to give us such a hint).

Next we look at the associated homogeneous recurrence relation, namely $a_n = 3a_{n-1} - 2a_{n-2}$. The general solution of this recurrence relation was obtained earlier in this section; it is $a_n = A + B \cdot 2^n$. Therefore, by Theorem 6, the general solution of the given nonhomogeneous recurrence relation is $a_n = A + B \cdot 2^n + 2 \cdot (-1)^n$.

To determine the specific solution that has $a_0 = 8$ and $a_1 = 5$, we perform the usual steps: Plug these conditions into our general solution and solve for A and B. Doing this yields $A = 5$ and $B = 1$ (details are left to the reader), so the specific solution is $a_n = 5 + 2^n + 2 \cdot (-1)^n$. ◆

We turn now to the method of undetermined coefficients for finding particular solutions of nonhomogeneous linear recurrence relations with constant coefficients. The method works well only if the form of the **nonhomogeneous term** $h(n)$ is such that we can predict what the form of the particular solution will be. The general heuristic is that the form of the solution will look like the form of $h(n)$. Let us be more specific.

First, suppose that $h(n)$ has the form $P(n)s^n$, where $P(n)$ is a polynomial (possibly of degree 0—i.e., a constant) and s is a nonzero constant (possibly equal to 1). Suppose also that s is *not* a root of the characteristic equation of the associated homogeneous recurrence relation. Then *there will be a particular solution of the nonhomogeneous recurrence relation of the form $Q(n)s^n$, where $Q(n)$ is a polynomial of degree less than or equal to the degree of $P(n)$* (let us call this degree d). A procedure for finding this particular solution follows from the preceding statement (which we will not prove). Write

$$a_n = (u_0 + u_1 n + \cdots + u_d n^d)s^n,$$

and plug this into the given nonhomogeneous recurrence relation. Divide both sides by s^{n-k}, leaving an equation involving polynomials in n. Since we want the two sides of this equation actually to be equal, set the corresponding coefficients of like powers of n equal, resulting in $d+1$ linear equations in the $d+1$ unknowns $u_0, u_1, \ldots, u_d$. Solve these equations to obtain the desired particular solution. An example will make the procedure clear.

EXAMPLE 9. Find a particular solution of the recurrence relation $a_n = 3a_{n-1} - 2a_{n-2} + 12n \cdot (-1)^n$.

Solution.   Since the nonhomogeneous term is $12n \cdot (-1)^n$, the procedure outlined above applies, with $P(n) = 12n$, $d = 1$, and $s = -1$. Note that s is not a root of the characteristic equation $r^2 - 3r + 2 = 0$ for the associated homogeneous recurrence relation. Thus we try a solution of the form

$$a_n = (u + vn) \cdot (-1)^n.$$

We use unsubscripted variable names for ease of notation. Note that even though the constant term was not present in the coefficient of $(-1)^n$ in the nonhomogeneous term of the given recurrence relation, we needed to allow for it in our particular solution; the coefficient of s^n in the nonhomogeneous term must be viewed as a *polynomial*.

Plugging our trial solution into both sides of the recurrence relation, we obtain

$(u + vn) \cdot (-1)^n$

$$= 3(u + v(n-1)) \cdot (-1)^{n-1} - 2(u + v(n-2)) \cdot (-1)^{n-2} + 12n \cdot (-1)^n.$$

Dividing both sides by $(-1)^{n-2}$ and simplifying, we get

$$u + vn = -3(u + v(n-1)) - 2(u + v(n-2)) + 12n$$

$$= -5u + 7v + (-5v + 12)n.$$

Thus for this solution to work, we need to equate the constant terms on both sides and the coefficients of n on both sides, giving us the simultaneous equations

$$u = -5u + 7v$$

$$v = -5v + 12.$$

The solution of these equations is easily obtained: $v = 2$ and $u = \frac{7}{3}$. Therefore, the particular solution we seek to the nonhomogeneous recurrence relation is $a_n = \left(\frac{7}{3} + 2n\right)(-1)^n$. The reader should substitute this solution into the recurrence relation to verify that it is correct.   ◆

We need to modify the procedure if s (in our current notation) happens to be a root of the associated homogeneous recurrence relation's characteristic equation. *For this reason, it is necessary to solve the associated homogeneous recurrence relation first.* Suppose that s is a root with multiplicity m. Then we need to look for a particular solution of the form $n^m Q(n) s^n$, rather than $Q(n) s^n$. If we fail to do so, the system of equations in the u_i's we get may have no solution.

Finally, if $h(n)$ is the sum of several functions (say, j of them), each of which is of the form $P(n)s^n$ (with different s's), then we treat them one at a time. In other words, we find j particular solutions of the j recurrence relations in each of which the nonhomogeneous term is just one of these j functions. Then we add these j particular solutions to obtain a particular solution of the given nonhomogeneous recurrence relation. This procedure works because of the linearity of the recurrence relation, for much the same reason that Theorem 6 holds; the reader should state and prove a formalization of this discussion (see Exercise 30).

We have discussed the method of undetermined coefficients only for nonhomogeneous terms of a particular form. Fortunately, that form is the one that arises most of the time. It includes polynomials, exponential functions, and products of the two, so it really covers quite a range. We end with a comprehensive example.

EXAMPLE 10. Find the solution of the recurrence relation with initial conditions

$$a_n = 4a_{n-1} - 3a_{n-2} + 2^n + n + 3$$

$$a_0 = 0, \qquad a_1 = 0.$$

Solution. First we look at the associated homogeneous recurrence relation, namely $a_n = 4a_{n-1} - 3a_{n-2}$. Its characteristic equation is $r^2 - 4r + 3 = 0$, which factors as $(r - 1)(r - 3) = 0$. Therefore, the general solution of the homogeneous recurrence relation is $a_n = A + B \cdot 3^n$.

Next we look at the nonhomogeneous term, $2^n + n + 3$. We break this up into two pieces. The first is 2^n, thought of as $1 \cdot 2^n$, where 1 is a polynomial of degree 0. The second piece is $n + 3$, thought of as $(n + 3) \cdot 1^n$, where $n + 3$ is a polynomial of degree 1. We find particular solutions for these two pieces separately.

Since 2 is not a root of the characteristic equation, we look for a solution of $a_n = 4a_{n-1} - 3a_{n-2} + 2^n$ of the form $a_n = u \cdot 2^n$. Plugging this proposed solution into both sides, we obtain

$$u \cdot 2^n = 4u \cdot 2^{n-1} - 3u \cdot 2^{n-2} + 2^n;$$

upon dividing by 2^{n-2} we have

$$4u = 8u - 3u + 4 = 5u + 4,$$

whence we find $u = -4$. Thus $a_n = -4 \cdot 2^n$ is a particular solution of $a_n = 4a_{n-1} - 3a_{n-2} + 2^n$.

In order to find a particular solution of $a_n = 4a_{n-1} - 3a_{n-2} + n + 3$, we first note that 1 is a root of multiplicity 1 of the characteristic polynomial. Therefore, rather than looking for a particular solution of the form $Q(n)$, where $Q(n)$ is a polynomial of degree 1, we must look for one of the form $nQ(n)$. Thus we set $a_n = un + vn^2$, and plug this into $a_n = 4a_{n-1} - 3a_{n-2} + n + 3$, obtaining

$$un + vn^2 = 4(u(n-1) + v(n-1)^2) - 3(u(n-2) + v(n-2)^2) + n + 3$$

$$= (4un - 4u + 4vn^2 - 8vn + 4v)$$

$$- (3un - 6u + 3vn^2 - 12vn + 12v) + n + 3$$

$$= (2u - 8v + 3) + (u + 4v + 1)n + vn^2.$$

Equating constant terms on both sides gives us $2u - 8v + 3 = 0$; equating coefficients of n gives us $u + 4v + 1 = u$; and equating coefficients of n^2 gives us the identity $v = v$. Solving the first two of these simultaneously, we obtain $v = -\frac{1}{4}$ and $u = -\frac{5}{2}$. Therefore, a particular solution of $a_n = 4a_{n-1} - 3a_{n-2} + n + 3$ is $a_n = -(5n/2) - (n^2/4)$.

Adding our two particular solutions and our general solution to the associated homogeneous recurrence relation, we obtain the general solution of the given nonhomogeneous recurrence relation:

$$a_n = A + B \cdot 3^n - 4 \cdot 2^n - \frac{5n}{2} - \frac{n^2}{4}.$$

Finally, plugging in the initial conditions $a_0 = a_1 = 0$, we obtain the equations $A + B - 4 = 0$ and $A + 3B - \frac{43}{4} = 0$, whence we find $A = \frac{5}{8}$ and $B = \frac{27}{8}$. Thus our final answer, the explicit formula for the terms of the sequence defined by the given recurrence relation and initial conditions, is

$$a_n = \frac{5}{8} + \frac{27}{8} \cdot 3^n - 4 \cdot 2^n - \frac{5n}{2} - \frac{n^2}{4}.$$

We can confirm our calculations (and there is surely ample opportunity for algebraic and arithmetic mistakes in a calculation such as this) by computing a_2 both from the formula and from the recurrence relation, obtaining the answer 9 in each case.   ◆

SUMMARY OF DEFINITIONS

arbitrary constant in a solution of a recurrence relation: a constant such that the solution is valid no matter what value the constant has (e.g., the constant A in the solution $a_n = A \cdot 2^n$ of the recurrence relation $a_n = 2a_{n-1}$).

associated homogeneous recurrence relation for the nonhomogeneous relation $a_n = c_1 a_{n-1} + c_2 a_{n-2} + \cdots + c_k a_{n-k} + h(n)$, with $h(n)$ not the zero function: the relation obtained by omitting $h(n)$ (e.g., $a_n = 3a_{n-1}$ is the homogeneous relation associated with $a_n = 3a_{n-1} + 2^n$).

characteristic equation of a linear homogeneous recurrence relation with constant coefficients: the equation obtained by setting the characteristic polynomial equal to 0 (e.g., $r^3 - r^2 + 5 = 0$ for the recurrence relation $a_n = a_{n-1} - 5a_{n-3}$).

characteristic polynomial of the kth-order linear homogeneous recurrence relation with constant coefficients $a_n = c_1 a_{n-1} + c_2 a_{n-2} + \cdots + c_k a_{n-k}$, $c_k \neq 0$: the polynomial $r^k - c_1 r^{k-1} - c_2 r^{k-2} - \cdots - c_{k-1} r - c_k$ (e.g., $r^3 - r^2 + 5$ for the recurrence relation $a_n = a_{n-1} - 5a_{n-3}$).

general solution of a kth-order recurrence relation: a solution of the recurrence relation that contains k arbitrary constants, such that proper choices of the constants will produce the specific solution of the recurrence relation with any given set of initial conditions $a_0 = b_0$, $a_1 = b_1$, ..., $a_{k-1} = b_{k-1}$, where the b_i's are constants (e.g., the solution $a_n = A \cdot 2^n$ of the recurrence relation $a_n = 2a_{n-1}$).

golden ratio: $(1 + \sqrt{5})/2 \approx 1.618$.

homogeneous linear kth-order recurrence relation: a linear recurrence relation $a_n = c_1 a_{n-1} + c_2 a_{n-2} + \cdots + c_k a_{n-k}$, where the c_i's are (perhaps constant) functions of n, $c_k \neq 0$ (e.g., $a_n = 3a_{n-2} - n^2 a_{n-3}$).

linear combination of the collection of functions f_1, f_2, ..., f_n: the sum of constants times the functions in the collection, namely $A_1 f_1 + A_2 f_2 + \cdots + A_n f_n$, where the A_i's are constants (e.g., $A \cdot 2^n + B \cdot (-2)^n + C$ is a linear combination of the functions 2^n, $(-2)^n$, and 1).

lincar kth-order recurrence relation: a recurrence relation in which the variables a_i appear only linearly, namely, a recurrence relation of the form $a_n = c_1 a_{n-1} + c_2 a_{n-2} + \cdots + c_k a_{n-k} + h(n)$, where the c_i's are (perhaps constant) functions of n, $c_k \neq 0$ (e.g., $a_n = 3a_{n-2} - n^2 a_{n-3} - n3^n$).

linear kth-order recurrence relation with constant coefficients: a linear recurrence relation $a_n = c_1 a_{n-1} + c_2 a_{n-2} + \cdots + c_k a_{n-k} + h(n)$, in which the c_i's are constants, $c_k \neq 0$ (e.g., $a_n = 3a_{n-2} - 4a_{n-3} - n3^n$).

multiplicity of the root r_i of the polynomial $p(r)$: the natural number m_i such that $(r - r_i)^{m_i}$ is a factor of $p(r)$ but $(r - r_i)^{m_i+1}$ is not (e.g., the root -3 has multiplicity 2 in the polynomial $p(x) = x(x - 1)^3 (x + 3)^2 (x - 3) = x^7 - 15x^5 + 8x^4 + 51x^3 - 72x^2 + 27x$).

nonhomogeneous: not homogeneous.

nonhomogeneous term in the nonhomogeneous linear recurrence relation $a_n = c_1 a_{n-1} + c_2 a_{n-2} + \cdots + c_k a_{n-k} + h(n)$: $h(n)$.

particular solution of a recurrence relation: a solution without arbitrary constants.

(explicit) solution of a recurrence relation for the sequence $\{a_n\}$: an explicit formula for a_n as a function of n for which the recurrence relation holds (e.g., $a_n = A \cdot 2^n$ is a solution of $a_n = 2a_{n-1}$).

specific solution of a kth-order recurrence relation and set of initial conditions $a_0 = b_0$, $a_1 = b_1$, ..., $a_{k-1} = b_{k-1}$, where the b_i's are constants: the solution of the recurrence relation that satisfies the initial conditions (e.g., the solution $a_n = 3 \cdot 2^n$ of the recurrence relation $a_n = 2a_{n-1}$ with initial condition $a_0 = 3$).

EXERCISES

1. By substituting the proposed solutions into the recurrence relation, determine whether the following functions are solutions of the recurrence relation

$$a_n = \frac{n-1}{n}a_{n-1} + \frac{2n-4}{n}a_{n-2}.$$

   (a) $a_n = 3$.
   (b) $a_n = 2^n$.
   (c) $a_n = 0$.
   (d) $a_n = 2^n/n$.
   (e) $a_n = 7 \cdot 2^n/n$.

2. Consider the recurrence relation $T(n) = 8T(n/2) + n^2$ with initial condition $T(1) = 1$, where n is restricted to being a power of 2. (This recurrence relation arose in analyzing matrix multiplication in Section 7.2.)
   (a) By plugging it into the recurrence relation and initial condition, show that $T(n) = 2n^3 - n^2$ is the solution.
   (b) Compute $T(16)$ both from the recurrence and from the explicit solution, and note that the same answer is obtained in each case.

3. Determine which of the following recurrence relations are homogeneous linear recurrence relations with constant coefficient. Determine the order of those that are, and explain why the others are not.
   (a) $a_n = a_{n-3}$.
   (b) $a_n = \sqrt{a_{n-1}}$.
   (c) $a_n = 5$.
   (d) $a_n = 3n^2 a_{n-1} - 3n^2 a_{n-2}$.

4. Use the iterative approach to solve the following recurrence relations and initial conditions.
   (a) $T(n) = 2T(n-1)$, $T(0) = 1$.
   (b) $T(n) = 2T(n-1)$, $T(0) = 0$.
   (c) $T(n) = 2T(n-1) + 1$, $T(0) = 0$.
   (d) $A_n = 3 + 2A_{n-1}$, $A_0 = 0$.

5. Use Theorem 1 to estimate $f(n)$ in big-oh terms if f satisfies the given recurrence relation. Assume that f satisfies the hypotheses of the theorem.
   (a) $f(n) = 3f(n/3) + 7n$.
   (b) $f(n) = 4f(n/2) + n^2$.
   (c) $f(n) = f(n/2) + n$.
   (d) $f(n) = 2f(n/2) + \sqrt{n}$.

6. Find the general solutions of the following recurrence relations.
   (a) $a_n = a_{n-1} + 6a_{n-2}$.
   (b) $a_n = -4a_{n-1} - 4a_{n-2}$.

(c) $a_n = 3a_{n-1} - 2a_{n-2}$.

(d) $a_n = 2a_{n-1} - a_{n-2}$.

7. Find the specific solutions of the recurrence relations given in Exercise 6, satisfying the initial conditions $a_0 = 2$ and $a_1 = -1$. [*Hint*: To check your work, compute a_2 (and maybe a_3 for good measure) both from the recurrence relation and from the formula you obtain to see that the two answers agree; if they do not, then at least one of them is wrong.]

8. By substituting the proposed solutions into the recurrence relation, determine whether the following functions are solutions of $a_n = -4a_{n-2}$. Recall that $i = \sqrt{-1}$.

(a) $a_n = i^n$.

(b) $a_n = (2i)^n$.

(c) $a_n = 3 \cdot (2i)^n + 4(-2i)^n$.

(d) $a_n = 2^n \sin(\pi n/2)$.

9. Consider the recurrence relation

$$a_n = \frac{(1-n)a_{n-1}}{a_{n-1} - n} \text{ for } n \geq 2.$$

(a)   Show that $a_n = 1/(1 + Cn)$ is a solution for all positive constants C.

(b)   Find the specific solution of the recurrence relation that satisfies the initial condition $a_1 = 3$.

(c)   Compute a_2, a_3, and a_4 directly from the recurrence relation and the initial condition $a_1 = 3$, and compare the values to those obtained using your answer to part (b).

(d)   Find a solution of the recurrence relation that is not in the form given in part (a). What initial condition does it correspond to? [*Hint*: Let $C \to \infty$.]

10. Use the iterative approach to solve the following recurrence relations and initial conditions.

(a) $a_n = -a_{n-1}$, $a_0 = 5$.

(b) $a_n = na_{n-1}$, $a_0 = 1$.

(c) $a_n = 2/a_{n-1}$, $a_0 = 1$.

(d) $a_n = n + a_{n-1}$, $a_0 = 1$.

11. Solve the recurrence relation and initial condition obtained in Exercise 1 of Section 7.2.

12. Solve the recurrence relation and initial condition obtained in Exercise 8 of Section 7.2.

13. In Exercise 9 of Section 7.2 we looked at what would happen if an investor in an IRA deposited money at the end of the year rather than at the beginning.

(a) Solve the recurrence relation and initial condition obtained in Exercise 9 of Section 7.2. (Recall that we solved the recurrence relation in the case of the investment at the beginning of the year in this section.)

(b) Compare the balance in the account under the two methods after 5 years, after 15 years, and after 30 years.

14. Find the indicated solutions of the following recurrence relations.
    (a) $a_n = 5a_{n-1} - 6a_{n-2} + 4^n$, general solution.
    (b) $a_n = 5a_{n-1} - 6a_{n-2} + 3^n$, general solution.
    (c) $a_n = 5a_{n-1} - 6a_{n-2} + 3^n + 4^n$, solution in which $a_0 = 7$ and $a_1 = 36$.
    (d) $a_n = 5a_{n-1} - 6a_{n-2} + n + 4^n$, solution in which $a_0 = 10$ and $a_1 = 35$.
    (e) $a_n = 5a_{n-1} - 6a_{n-2} + n^2 \cdot 4^n$, general solution.
    (f) $a_n = 5a_{n-1} - 6a_{n-2} + (n^2 + 1) \cdot 3^n$, general solution.

15. A certain fast algorithm for multiplying large integers (say each n digits long) works by reducing the problem to three problems of multiplying integers half as long, plus a number of operations on digits that is proportional to n. Apply Theorem 1 to determine the time complexity of this algorithm.

16. Estimate in big-oh terms the number of multiplications and divisions needed in Algorithm 7 of Section 5.2. [*Hint*: Use Theorem 1.]

17. The Lucas sequence is like the Fibonacci sequence, satisfying the same recurrence relation but with the initial conditions $L_1 = 1$ and $L_2 = 3$.
    (a) Find the first eight terms of the Lucas sequence.
    (b) Find an explicit formula for the Lucas sequence. [*Hint*: Use the work already done in the first part of Example 5; all that remains is plugging in the initial conditions.]
    (c) Use the formula obtained in part (b), and a calculator, to compute L_8 (noting that the numbers must be integers).
    (d) Compute L_3 exactly from the formula obtained in part (b); that is, use algebraic manipulation.

18. Consider the recurrence relation $a_n = 2a_{n-1} + a_{n-2} - 2a_{n-3}$.
    (a) Find the general solution.
    (b) Find the specific solution that satisfies the initial conditions $a_0 = a_1 = a_2 = 1$.
    (c) Find the specific solution that satisfies the initial conditions $a_0 = a_1 = a_2 = 0$.
    (d) Find the specific solution that satisfies the initial conditions $a_i = i$ for $0 \le i \le 2$.
    (e) Find the specific solution that satisfies the initial conditions $a_i = k$ for $0 \le i \le 2$, where k is a constant.

19. Consider the recurrence relation $a_n = a_{n-1} + 3a_{n-2}$.

(a)   Find the general solution.
(b)   Find the specific solution that satisfies the initial conditions $a_0 = 3$ and $a_1 = 1$.

20. Consider the recurrence relation $a_n = 2a_{n-1} - 10a_{n-2}$.
    (a)   Find the general solution.
    (b)   Find the specific solution that satisfies the initial conditions $a_0 = a_1 = 1$.
    (c)   Compute a_2 and a_3 directly from the recurrence relation, as well as from the formula in part (b), to verify that the formula is correct.

21. Find the general solution of the recurrence relation $a_n = 8a_{n-2} + 9a_{n-4}$, using the methods discussed in this section. Note that this is a fourth-order recurrence relation and that two of the roots of the characteristic polynomial are complex.

22. The recurrence relation in Exercise 21 can be viewed as a second-order recurrence relation on two intertwined sequences—the subsequence of terms with even indices and the subsequence of terms with odd indices. Find the general solution from this point of view, and reconcile your solution with the solution obtained in Exercise 21.

23. Consider the sequence defined by $a_n = 2a_{n-1} - 2a_{n-2} + 2a_{n-3} - a_{n-4}$, $a_0 = a_1 = a_2 = 1$, $a_3 = 2$.
    (a)   Write out the first 12 terms of the sequence.
    (b)   Find an explicit formula for the nth term of the sequence. [*Hint*: The solution to the characteristic equation involves complex numbers.]
    (c)   Verify that a_7 and a_{10} as given by your formula agree with the terms as computed in part (a).

24. Solve the following recurrence relations with initial conditions. [*Hint*: In both cases there is an alternative to using the method of this section (although you may want to use the method anyway): Compute some terms of the sequence, guess an answer, and then justify it.]
    (a) $a_n = a_{n-1} - a_{n-2} + a_{n-3}$, $a_i = i$ for $i = 1, 2, 3$.
    (b) $a_n = a_{n-1} + a_{n-2} - a_{n-3}$, $a_0 = a_1 = 1$, $a_2 = 2$.

25. Consider the sequence defined by $a_n = a_{n-1}^3 / a_{n-2}^2$, with $a_0 = 1$ and $a_1 = 2$.
    (a)   Write down the first five terms of the sequence.
    (b)   Find an explicit formula for the terms of the sequence. [*Hint*: Transform the problem into a linear one, using logarithms.]

26. Consider the sequence defined by $a_n = \left(\sqrt{a_{n-1}} + 2\sqrt{a_{n-2}} \right)^2$, with $a_0 = a_1 = 1$.
    (a)   Write down the first five terms of the sequence.

(b)   Find an explicit formula for the terms of the sequence. [*Hint*: Transform the problem into a linear one.]

27. Explain what the difficulty is in obtaining a nice formula for the number of bit strings of length n that do not contain three consecutive 0's (see Exercise 11 in Section 7.2).

28. Find the specific solution of the recurrence relation $a_n = 4a_{n-1} - 4a_{n-2} + n \cdot 2^n$, with initial conditions $a_0 = 1$ and $a_1 = 2$.

29. Find an explicit formula for the number of bit strings of length n that contain at least two consecutive 0's (see Example 7 in Section 7.2).

30. State and prove a generalization of Theorem 6 to handle recurrence relations of the form

$$a_n = c_1 a_{n-1} + c_2 a_{n-2} + \cdots + c_k a_{n-k} + h_1(n) + h_2(n) + \cdots + h_j(n).$$

(Omit the part of the proof that corresponds to the omitted part of the proof of Theorem 6.)

31. State precisely the special case of Theorem 5 in which the roots of the characteristic polynomial are all distinct.

32. Determine the exact number of decimal digits in f_{5000}.

33. The sequence 1, 5, 14, 30, 55, ..., is obtained by starting with 0 and adding successive squares ($0 + 1^2 = 1$, $1 + 2^2 = 5$, $5 + 3^2 = 14$, and so on).
    (a)   Find a formula for the nth term in this sequence, by writing down and solving an appropriate nonhomogeneous recurrence relation and initial condition.
    (b)   Compare your answer to the formula obtained in Example 2 in Section 5.3.

34. Let a sequence be defined in the following way. The first two terms are arbitrary; call them x and y. Thereafter, each term is the arithmetic average of the two preceding terms (their sum divided by 2).
    (a)   Write down a recurrence relation and initial conditions to model this problem.
    (b)   Find an explicit formula for the nth term in the sequence.
    (c)   What is the limit of this sequence (to what real number, if any, are the terms converging)?

35. According to the familiar Christmas carol, on the nth day of Christmas my true love gave to me $1 + 2 + \cdots + n$ presents. Suppose that there are d days of Christmas in all (traditionally, $d = 12$).
    (a)   Write down an explicit formula for the number of presents my true love gave to me on the nth day of Christmas.

(b)   Write down a recurrence relation and initial condition for P_n, the total number of presents I received through day n. Thus P_d is the total number of presents I received over the entire holiday.

(c)   Solve the recurrence relation and initial condition in part (b), obtaining a formula for P_d.

(d)   Plug $d = 12$ into the formula obtained in part (c) to find out how many presents I got in all.

36. Solve the recurrence relation

$$a_n = a_{n-1} + \frac{1}{n} - \frac{1}{n+1}$$

with initial condition $a_0 = 2$.

Challenging Exercises

37. Find a formula for the number of regions in the plane determined by n lines, no two of which are parallel and no three of which pass through a common point (see Exercise 36 in Section 7.2).

38. Find a formula for each of the following sums.

(a) $\sum_{i=1}^{n} i^4$.

(b) $\sum_{i=4}^{n} P(i, 4)$.

39. Repeat Exercise 34 with geometric mean replacing arithmetic average. (The geometric mean of x and y is defined to be $\sqrt{xy}$.)

40. Make an appropriate transformation to turn the recurrence relation in Exercise 1 into a linear recurrence relation. Solve the resulting recurrence relation and use the solution to find the general solution of the recurrence relation in Exercise 1. Assume that the recurrence relation holds for $n \geq 3$.

Exploratory Exercises

41. Compare and contrast the methods we have discussed for solving linear recurrence relations with constant coefficients and the methods for solving linear differential equations with constant coefficients.

42. Develop the theory for solving second-order linear homogeneous recurrence relations with constant coefficients in which the characteristic roots are complex numbers using trigonometric functions. Consult Liu [198] for assistance.

SECTION 7.4
THE INCLUSION–EXCLUSION PRINCIPLE

I wouldn't belong to any club
that would have me as a member.

—Groucho Marx

The addition principle (Theorem 2 in Section 6.1) told us how to find the cardinality of the union of two disjoint finite sets: The cardinality of the disjoint union is the sum of the cardinalities of the sets. Clearly, the same principle applies to more than two sets: If A_1, A_2, ..., A_n are pairwise disjoint sets, then $|A_1 \cup A_2 \cup \cdots \cup A_n| = |A_1| + |A_2| + \cdots + |A_n|$. What if the sets in the union are not necessarily pairwise disjoint? A corollary of the overcounting principle (Theorem 4 in Section 6.1) gave us a way to find the cardinality of the union in the case of two sets:

$$|A \cup B| = |A| + |B| - |A \cap B|.$$

The extension of this result to more than two sets is not so obvious. In this section we state and prove the corresponding counting principle for more than two sets, and then we apply it to some fairly interesting situations: counting the number of surjective functions from one set to another; determining the number of prime numbers in an interval of integers; and finding the probability that a random permutation leaves no object in its original position. All the sets in this section are assumed to be finite.

Let us recall the proof of Theorem 4 in Section 6.1. To count the number of elements in the union of sets A and B, we include each element of A and each element of B, for a total of $|A| + |B|$ elements. Unfortunately, in doing so we have counted each element in $A \cap B$ twice. Therefore, to get the right answer—the cardinality of $A \cup B$—we need to subtract the overcount, $|A \cap B|$. In other words, we need to exclude, once, the elements in $A \cap B$, since they have been included twice. (Note that we are not subtracting $|A \cap B|$ because we do not wish to count elements in the intersection: We are subtracting this amount because we do not want to count them more than once.) Thus $|A \cup B| = |A| + |B| - |A \cap B|$. To generalize this result to more than two sets, we will generalize this argument.

Let us begin with the case of three sets. How can we determine $|A \cup B \cup C|$, making no assumption about disjointness among these sets? The situation is pictured in the Venn diagram shown in Figure 7.4, where we have shaded the pairwise intersections using three different patterns.

Let us begin our count by including each element in each of the sets, obtaining $|A| + |B| + |C|$. To the extent that these sets overlap (all the shaded regions in Figure 7.4), we have overcounted. In particular, each element in $A \cap B$ that is not in C was included twice (once for being in A and once for being in B); each element

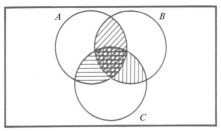

FIGURE 7.4   Intersection pattern for three arbitrary sets.

in $A \cap C$ that is not in B was included twice (once for being in A and once for being in C); each element in $B \cap C$ that is not in A was included twice (once for being in B and once for being in C); and each element in all three sets was included three times. Can we correct the overcount by subtracting $|A \cap B| + |A \cap C| + |B \cap C|$? Not quite; if we subtract that quantity, we will have excluded a bit too much. To see this, let us see what the calculation $|A| + |B| + |C| - (|A \cap B| + |A \cap C| + |B \cap C|)$ counts. Each element in exactly one of the three sets is counted correctly, for it is counted in exactly one of $|A|$, $|B|$, or $|C|$, and not at all in the cardinalities of the intersections. Furthermore, each element in exactly two of the sets A, B, and C will be counted correctly, since it will have been included twice (once for each of the sets it is in) and excluded once (for the intersection it is in). Unfortunately, the elements in all three sets (the most heavily shaded area in Figure 7.4) will have been included three times (once for each set) and excluded three times (once for each intersection), the net result being that they are not being counted at all. Hence to get the correct answer we need to include them once more, by adding $|A \cap B \cap C|$. Thus we have shown that

$$|A \cup B \cup C| = |A| + |B| + |C| - \big(|A \cap B| + |A \cap C| + |B \cap C|\big) + |A \cap B \cap C|.$$

Before stating and proving the general result for n sets, let us look at an application.

EXAMPLE 1.   Speaking before a crowd of pet owners, a mathematician asks some questions of the audience. "How many of you own a dog?" she asks, and 88 people raise their hands. Then she asks "How many of you own a cat?" and 99 people raise their hands. "And how many of you own a bird?" This time 33 people raise their hands. Realizing that she does not have enough information to figure out the exact pet ownership patterns of her audience, she inquires further and finds the following information: 15 people own both a dog and a cat; 11 own both a dog and a bird; seven own both a cat and a bird; and five own all three kinds of animal. Determine how many people in the audience own a dog, cat, or bird.

Solution.   If we let D, C, and B be the sets of people in the audience who own a dog, a cat, and a bird, respectively, then the given information tells us that $|D| = 88$, $|C| = 99$, $|B| = 33$, $|D \cap C| = 15$, $|D \cap B| = 11$, $|C \cap B| = 7$, and $|D \cap C \cap B| = 5$. We are asked to find $|D \cup C \cup B|$. By the formula we have just derived, the answer must be $88 + 99 + 33 - (15 + 11 + 7) + 5 = 192$.

The Venn diagram in Figure 7.5 shows the cardinalities of each of the sets in this problem. Each number in the diagram is the number of people in the region of the plane in which the number lies. For instance, there are 20 people who own a bird but no dog or cat ($|B - D - C| = 20$), whereas there are six people who own a dog and a bird but not a cat ($|(D \cap B) - C| = 6$).

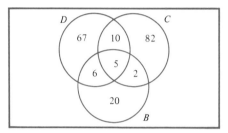

FIGURE 7.5   Cardinalities of various intersections of sets of pet owners.

This picture was constructed from the inside out, using the information given in the statement of the problem. First we were told that $|B \cap C \cap D| = 5$. Then since we were told that $|D \cap B| = 11$, we conclude that $|(D \cap B) - C| = 11 - 5 = 6$, and similarly for the regions with 10 and two people in them. Finally, since we were told that $|B| = 33$, we conclude that $|B - D - C| = 33 - (6 + 5 + 2) = 20$, and similarly for the regions with 82 and 67 people in them.

Note that the formula can also be used to find the cardinality of any one of the seven sets in the expression, given the cardinalities of the other six. For instance, if we had been told that 192 people in the audience owned a dog, cat, or bird, but had not been told how many owned all three, then we could have computed how many owned all three by subtracting $88 + 99 + 33 - (15 + 11 + 7)$ from 192.   ◆

The situation for four sets gets a little messier, but the idea is the same. To find the cardinality of $A \cup B \cup C \cup D$, we can add the cardinalities of the four sets; then subtract the cardinalities of the six intersections of pairs of these sets (there are six of them since $C(4, 2) = 6$) to account for the fact that the elements in more than one of the sets were included too often; then add back the cardinalities of the four intersections of three of these sets at a time to account for the fact that those elements in more than two of the sets were excluded too often by the previous step; and finally, subtract the cardinality of the intersection of all four sets, since the previous inclusion step went too far in reincluding the elements common to all four sets. Thus we obtain the formula

$|A \cup B \cup C \cup D|$

$$= |A| + |B| + |C| + |D|$$
$$- (|A \cap B| + |A \cap C| + |A \cap D| + |B \cap C| + |B \cap D| + |C \cap D|)$$
$$+ (|A \cap B \cap C| + |A \cap B \cap D| + |A \cap C \cap D| + |B \cap C \cap D|)$$
$$- |A \cap B \cap C \cap D|.$$

We do not claim that the argument given in the paragraph above was a proof of this formula. Instead, we will formulate the general principle as a theorem, and give a proof that works for any number of sets. For the general statement we need to name our sets with subscripts, and the sums in the formula need to be represented with summation notation. With these conventions, the preceding display becomes

$$|A_1 \cup A_2 \cup A_3 \cup A_4| = \sum_{1 \le i \le 4} |A_i| - \sum_{1 \le i < j \le 4} |A_i \cap A_j| + \sum_{1 \le i < j < k \le 4} |A_i \cap A_j \cap A_k|$$
$$- |A_1 \cap A_2 \cap A_3 \cap A_4|.$$

The general result then looks like this.

THEOREM 1. The Inclusion–Exclusion Principle. *If $A_1, A_2, \ldots, A_n$ are arbitrary finite sets, then*

$$|A_1 \cup A_2 \cup \cdots \cup A_n| = \sum_{1 \le i \le n} |A_i| - \sum_{1 \le i < j \le n} |A_i \cap A_j|$$
$$+ \sum_{1 \le i < j < k \le n} |A_i \cap A_j \cap A_k| - \cdots + (-1)^{n+1} |A_1 \cap A_2 \cap \cdots \cap A_n|.$$

Proof. It is enough to show that each element of the union gets counted exactly once by the expression on the right-hand side. Suppose that x is an element of $A_1 \cup A_2 \cup \cdots \cup A_n$, and suppose that x is in exactly t of the sets A_i. Then x will be in exactly $C(t, 2)$ sets $A_i \cap A_j$, with $1 \le i < j \le n$, in exactly $C(t, 3)$ of the sets $A_i \cap A_j \cap A_k$, with $1 \le i < j < k \le n$, and so on. Thus x will be counted (included) t times by the first summation on the right-hand side, counted (excluded) $C(t, 2)$ times by the second summation on the right-hand side, counted (included) $C(t, 3)$ times by the third summation on the right-hand side, and so on. Note that x will not be counted at all by the mth summation if $m > t$. Therefore, the total number of times that x will be counted is

(∗) $C(t, 1) - C(t, 2) + C(t, 3) - \cdots \pm C(t, t).$

Now Corollary 2 to the binomial theorem (Theorem 1 in Section 7.1) tells us that $-C(t, 0) + C(t, 1) - C(t, 2) + C(t, 3) - \cdots \pm C(t, t) = 0.$ Therefore, the sum (∗) must equal $C(t, 0) = 1.$ In other words, each element of the union was counted exactly one time. This completes the proof. ∎

EXAMPLE 2. The freshman class at a certain college consists of 240 students. Suppose that 130 members of the class are taking English, 100 members are taking history, 90 are taking calculus, and 50 are taking biology. Furthermore, suppose that 57 of them are taking both English and history, 15 are taking both English and calculus, 16 are taking both English and biology, 25 are taking both history and calculus, 19 are taking both history and biology, and 24 are taking both calculus and biology. Also, suppose that one person is taking English, history, and calculus; six are taking English, history, and biology; four are in English, calculus, and biology; and five are in history, calculus, and biology. Nobody is taking all four of these subjects. How many students are taking none of these four subjects?

Solution. If we let E, H, C, and B be the sets of students taking the subjects beginning with these letters, then the inclusion–exclusion principle tells us that

$$|E \cup H \cup C \cup B| = 130 + 100 + 90 + 50 - (57 + 15 + 16 + 25 + 19 + 24)$$
$$+ (1 + 6 + 4 + 5) - 0 = 230.$$

Since the class (the universal set for this problem) has 240 members in all, our answer is $|\overline{E \cup H \cup C \cup B}| = 240 - 230 = 10.$ ◆

In the remainder of this section, we turn to some interesting applications of the inclusion–exclusion principle.

COUNTING SURJECTIVE FUNCTIONS

If X is a set with k elements, and Y is a set with n elements, then there are n^k different functions from X to Y. This follows from the multiplication principle, because each such function f is determined by choosing, for each of the k elements of X, an element of Y to be its image under f; and for each element of X there are n such choices. Furthermore, if $k \leq n$, then there are

$$P(n, k) = n(n - 1)(n - 2) \cdots (n - k + 1)$$

one-to-one (injective) functions from X to Y. This again follows from the multiplication principle: There are n ways to choose the image of the first element of X for such a function, $n - 1$ ways to choose the image of the second element of

X after the image of the first element has been chosen, and so on, for the k elements of X. It is natural to ask the corresponding combinatorial question for onto (surjective) functions: *How many surjective functions are there from a k-set X onto an n-set Y?* The inclusion–exclusion principle will give us the answer. (For other approaches to counting surjective functions, see Exercise 32 in this section and Exercise 32 in Section 7.2.)

Before we state and prove the theorem that answers our question, let us look at an example in order to get a feeling for what is going on and how the inclusion–exclusion principle fits in.

EXAMPLE 3. Determine the number of surjective functions from $X = \{1, 2, 3, 4, 5, 6, 7\}$ to $Y = \{a, b, c, d\}$.

Solution. Certainly, there are some surjective functions, such as the function that sends 1, 2, and 3 to a, sends 4 to b, sends 6 to d, and sends 5 and 7 to c. And certainly some functions from X to Y are not surjective, such as the function that sends every element in X to c.

A function is surjective if it has each element of the codomain Y in its range. We can count the number of surjective functions by looking at the sets of functions that fail to have specific elements of Y in their ranges. In particular, let A be the set of functions from X to Y that do not have a in their range; let B be the set of functions from X to Y that do not have b in their range; and similarly for C and D. Then $A \cup B \cup C \cup D$ is the set of functions that fail to have at least one element of Y in their range; in other words, $|A \cup B \cup C \cup D|$ is the number of functions from X to Y that are *not* surjective. Since there are 4^7 functions from X to Y, our answer is $4^7 - |A \cup B \cup C \cup D|$. Thus we need to compute $|A \cup B \cup C \cup D|$. We will use the inclusion–exclusion principle to do so.

A function from X to Y that does not have a in its range is really the same thing as a function from X to $\{b, c, d\}$. Since there are 3^7 functions from the 7-set X to the 3-set $\{b, c, d\}$, we conclude that $|A| = 3^7$. By exactly the same reasoning, $|B| = |C| = |D| = 3^7$. These are the first four quantities we need to know in order to apply the inclusion–exclusion principle to computing $|A \cup B \cup C \cup D|$. Next we need to compute the cardinalities of the intersections of two of these sets, such as $A \cap B$. Now $A \cap B$ is the set of functions from X to Y that have neither a nor b in their range; thus $A \cap B$ is the same as the set of functions from the 7-set X to the 2-set $\{c, d\}$. Therefore, $|A \cap B| = 2^7$. In all there are $C(4, 2) = 6$ intersections of two of the sets among A, B, C, and D, and clearly the cardinality of each such intersection is this same number, 2^7.

Next we need to compute the cardinalities of the intersections of three of these sets, such as $A \cap B \cap C$. Since $A \cap B \cap C$ is the set of functions from X to Y that have neither a nor b nor c in their range, $A \cap B \cap C$ is the same as the set of functions from the 7-set X to the 1-set $\{d\}$. Therefore, $|A \cap B \cap C| = 1^7$. In all there are $C(4, 3) = 4$ intersections of three of the sets among A, B, C, and D, and clearly the cardinality of each such intersection is this same number, 1^7. Finally, there are no elements in $A \cap B \cap C \cap D$, since every function from X to Y must have at least one element in its range, as long as X is not empty.

If we combine all we have so far by the inclusion–exclusion principle, we obtain

$$|A \cup B \cup C \cup D| = 4 \cdot 3^7 - 6 \cdot 2^7 + 4 \cdot 1^7 - 0 = 8748 - 768 + 4 = 7984.$$

In other words, here are 7984 functions from X to Y that have one or more elements of Y missing from their range. Therefore there are $4^7 - 7984 = 16384 - 7984 = 8400$ functions that have no missing elements in their range; that is, there are 8400 surjective functions from $\{1, 2, 3, 4, 5, 6, 7\}$ to $\{a, b, c, d\}$.   ◆

Generalizing the argument given in this example, we obtain the following theorem.

THEOREM 2.   *If X is a set with k elements, and Y is a set with n elements, where $k \geq n \geq 1$, then there are*

$$n^k - C(n, 1)(n - 1)^k + C(n, 2)(n - 2)^k - \cdots + (-1)^{n-1}C(n, n - 1)1^k$$

different surjective functions from X onto Y.

Proof.   The proof follows the line of reasoning given in Example 3. For each i from 1 to n, let A_i be the set of functions from X to Y that do not contain the ith element of Y in their range. Now for each r from 1 to $n - 1$, there are $C(n, r)$ intersections involving exactly r of the sets A_i, and the cardinality of each of them is $(n - r)^k$, since a function from X to Y that leaves out (at least) r specified elements of Y from its range is the same thing as a function from X to the $(n - r)$-set consisting of Y with these r elements removed.

Therefore by the inclusion–exclusion principle, the cardinality of $\bigcup_{i=1}^{n} A_i$ is $C(n, 1)(n - 1)^k - C(n, 2)(n - 2)^k + \cdots + (-1)^n C(n, n - 1)1^k$; there are this many functions from X to Y that are not surjective. Since there are n^k functions in all from X to Y, the formula in the statement of the theorem follows.   ■

EXAMPLE 4.   Determine the number of functions from $\{a, b, c, d, e, f\}$ to $\{a, b, c, d, e\}$; how many of them are one-to-one and how many are onto?

Solution.   There are $5^6 = 15{,}625$ functions from this 6-set to this 5-set. None of them are one-to-one, since there are more elements in the domain than in the codomain. (This is an application of the pigeonhole principle (Section 6.4).) The number of onto functions among these 15,625 functions is given by Theorem 2; there are $15625 - C(5, 1)4^6 + C(5, 2)3^6 - C(5, 3)2^6 + C(5, 4)1^6 = 15625 - 5 \cdot 4096 + 10 \cdot 729 - 10 \cdot 64 + 5 \cdot 1 = 1800$ of them.   ◆

COUNTING PRIME NUMBERS

Since antiquity, mathematicians have been fascinated with the distribution of the prime numbers. The primes seem to occur rather randomly: Every now and then there will be a prime number among the composites. There seems to be a higher density of primes among small natural numbers than among larger ones (e.g., there are five primes between 100 and 120, whereas there is only one prime between 10,000 and 10,020). Euclid proved that there are an infinite number of primes (see Exercise 8a in Section 1.3). In the nineteenth century the great German mathematician C. Gauss conjectured (and later mathematicians proved) that the number of prime numbers less than N is approximately equal to $N/(\ln N)$, where $\ln N$ is the natural logarithm of N; in other words, about one number out of every $\ln N$ numbers from 1 to N is prime. This fact quantifies the statement that the density of the prime numbers decreases as the size of the numbers grows. In this subsection, using the inclusion–exclusion principle, we develop a method for determining the precise number of prime numbers less than N.

Suppose that we want to find all the prime numbers no larger than 50. We can do so with the following algorithm, known as the **sieve of Eratosthenes,** named after the ancient Greek mathematician who developed it. First we write down the numbers from 2 to 50. Then we cross out all the multiples of 2 except for 2 itself, namely 4, 6, 8, and so on. Next we cross out all the multiples of 3, except for 3: 6, 9, 12, and so on. Then we cross out the multiples of the next prime, 5, starting with $2 \cdot 5 = 10$. We do the same with the multiples of 7 starting with 14. At this point, all the composite numbers have been crossed out, since any number from 2 to 50 must have a prime factor less than or equal to $\lfloor\sqrt{50}\rfloor = 7$ (see Exercise 8 in Section 4.2). Thus all the numbers that have not been crossed out are prime (see Figure 7.6).

	2	3	~~4~~	5	~~6~~	7	~~8~~	~~9~~	~~10~~	11	~~12~~
13	~~14~~	~~15~~	~~16~~	17	~~18~~	19	~~20~~	~~21~~	~~22~~	23	~~24~~
~~25~~	~~26~~	~~27~~	~~28~~	29	~~30~~	31	~~32~~	~~33~~	~~34~~	~~35~~	~~36~~
37	~~38~~	~~39~~	~~40~~	41	~~42~~	43	~~44~~	~~45~~	~~46~~	47	~~48~~
~~49~~	~~50~~										

FIGURE 7.6 The Sieve of Eratosthenes: The prime numbers not exceeding 50 are not crossed out.

Let us use the inclusion–exclusion principle to see how many numbers have been crossed out. We need to observe first that if k is a positive integer, then there are $\lfloor 50/k \rfloor$ positive multiples of k no larger than 50, since every kth number is a multiple of k. Thus we crossed out $\lfloor 50/2 \rfloor - 1 = 25 - 1 = 24$ numbers as being multiples of 2 other than 2 itself; we crossed out $\lfloor 50/3 \rfloor - 1 = 16 - 1 = 15$

numbers as being multiples of 3 other than 3 itself; we crossed out $\lfloor 50/5 \rfloor - 1 = 10 - 1 = 9$ numbers as being multiples of 5 other than 5 itself; and we crossed out $\lfloor 50/7 \rfloor - 1 = 7 - 1 = 6$ numbers as being multiples of 7 other than 7 itself.

Now the four sets of numbers that we have just counted are not disjoint; for example, we crossed out the number 30 as a multiple of 2, again as a multiple of 3, and again as a multiple of 5. Thus to count the number of numbers we have crossed out, we need to apply the inclusion–exclusion principle. There are $\lfloor 50/(2 \cdot 3) \rfloor = 8$ multiples of both 2 and 3 (i.e., multiples of $2 \cdot 3 = 6$), there are $\lfloor 50/(2 \cdot 5) \rfloor = 5$ multiples of both 2 and 5 (i.e., multiples of $2 \cdot 5 = 10$); there are $\lfloor 50/(2 \cdot 7) \rfloor = 3$ multiples of both 2 and 7; there are $\lfloor 50/(3 \cdot 5) \rfloor = 3$ multiples of both 3 and 5; there are $\lfloor 50/(3 \cdot 7) \rfloor = 2$ multiples of both 3 and 7; and there is $\lfloor 50/(5 \cdot 7) \rfloor = 1$ multiple of both 5 and 7. We need to subtract all of these numbers to correct our count. Furthermore, there is one common multiple of 2, 3, and 5 (since $\lfloor 50/(2 \cdot 3 \cdot 5) \rfloor = 1$), and one common multiple of 2, 3, and 7, so we need to add back 2. (There are no multiples of $2 \cdot 5 \cdot 7$ or $3 \cdot 5 \cdot 7$ in this range; nor is there any number that is a multiple of all four of the primes 2, 3, 5, and 7.)

Putting this all together, we know that the number of numbers between 2 and 50 inclusive that are not prime is $24 + 15 + 9 + 6 - (8 + 5 + 3 + 3 + 2 + 1) + (1 + 1) = 34$. Since there are $50 - 1 = 49$ numbers in this range, we conclude that $49 - 34 = 15$ of them are prime. Indeed, Figure 7.6 shows this to be so: 15 numbers are not crossed out.

Obviously, there was nothing special about the number 50. Let us summarize this discussion as a theorem.

THEOREM 3.   *Let n be a positive integer, and let P be the set of prime numbers less than or equal to $\sqrt{n}$. The number of prime numbers less than or equal to n is given by*

$$(n - 1) - \sum_{i \in P} \left(\left\lfloor \frac{n}{i} \right\rfloor - 1 \right) + \sum_{\substack{i,j \in P \\ i < j}} \left\lfloor \frac{n}{i \cdot j} \right\rfloor - \sum_{\substack{i,j,k \in P \\ i < j < k}} \left\lfloor \frac{n}{i \cdot j \cdot k} \right\rfloor + \cdots.$$

DERANGEMENTS

We end this section by counting certain kinds of permutations. A permutation $x_1 x_2 \ldots x_n$ of the set $\{1, 2, \ldots, n\}$ is called a **derangement** if $x_i \neq i$ for each i from 1 to n. In other words, a derangement is a permutation in which no element occupies its original position. We denote the number of derangements for a given value of n by d_n. For example, the permutations 231 and 312 are the only derangements for $n = 3$; in neither of them is the number i in the ith position ($i = 1, 2, 3$). Each of the other permutations of $\{1, 2, 3\}$, however, leaves at least one element fixed in its original position: 132 leaves 1 fixed, 213 leaves 3 fixed, 321 leaves 2 fixed, and 123 leaves all three elements fixed. Thus there are two derangements for $n = 3$, so $d_3 = 2$.

A graphic illustration of the notion of a derangement is given by a tradition at the U.S. Naval Academy. At the end of the commencement ceremony, the graduating midshipmen throw their caps high into the air. If each midshipman retrieves a cap, we have a permutation of the caps. Clearly, there are $P(n, n) = n!$ such permutations if there are n members of the graduating class. Some of the midshipmen might end up with their own caps. If no midshipman ends up with his or her own cap, the resulting permutation is a derangement. If we assume that this operation results in a random permutation, we can ask for the probability that no midshipman retrieves his or her own cap, namely $d_n/n!$.

Let us use the inclusion–exclusion principle to find a formula for d_n. For each i from 1 to n, let A_i be the set of permutations $x_1 x_2 \ldots x_n$ of the set $\{1, 2, \ldots, n\}$ in which i is left fixed (i.e., for which $x_i = i$). Since a permutation is a derangement if and only if it leaves no element fixed, the set of permutations that are *not* derangements is $\bigcup_{i=1}^{n} A_i$. In order to apply the inclusion–exclusion principle, we need to calculate the cardinality of each intersection of some of the A_i's. Let us see what $|A_{j_1} \cap A_{j_2} \cap \cdots \cap A_{j_m}|$ is. A permutation is in $A_{j_1} \cap A_{j_2} \cap \cdots \cap A_{j_m}$ if it leaves all the numbers $j_1, j_2, \ldots, j_m$ in their original positions. Thus the permutation is free to permute the remaining $n - m$ elements, so there are $(n - m)!$ such permutations. Therefore, $|A_{j_1} \cap A_{j_2} \cap \cdots \cap A_{j_m}| = (n - m)!$. Note that this number is independent of the subscripts appearing in the intersection; it depends only on the fact that there are m of them. Next we determine how many such intersections there are for a fixed value of m. Since the intersection is determined by the choice of the m subscripts appearing in the expression, there are $C(n, m)$ such intersections.

Therefore, by the inclusion–exclusion principle the number of permutations in $A_1 \cup A_2 \cup \cdots \cup A_n$, that is, the number of permutations that are *not* derangements, is

$$C(n, 1)(n-1)! - C(n, 2)(n-2)! + C(n, 3)(n-3)! - \cdots + (-1)^{n-1} C(n, n)(n-n)!.$$

Recalling that $C(n, m) = n!/(m!(n - m)!)$, we can simplify this to

$$\frac{n!}{1!} - \frac{n!}{2!} + \frac{n!}{3!} - \cdots + (-1)^{n-1} \frac{n!}{n!}.$$

Finally, we obtain d_n by subtracting this last expression from $n!$, since there are $P(n, n) = n!$ permutations in all. Thus we have proved the following theorem.

THEOREM 4. *The number of derangements of n objects, $n \geq 1$, is given by*

$$d_n = n! \left(1 - \frac{1}{1!} + \frac{1}{2!} - \frac{1}{3!} + \cdots + (-1)^n \frac{1}{n!} \right).$$

For example, Theorem 4 tells us that

$$d_3 = 3! \left(1 - \frac{1}{1!} + \frac{1}{2!} - \frac{1}{3!}\right) = 6\left(1 - 1 + \frac{1}{2} - \frac{1}{6}\right) = 6 - 6 + 3 - 1 = 2,$$

in agreement with our explicit listing of the derangements of $\{1, 2, 3\}$ given above.

Recall from calculus that the Taylor series for the exponential function e^x (where $e \approx 2.71828$ is the base of the natural logarithm) is given by

$$e^x = 1 + \frac{x}{1!} + \frac{x^2}{2!} + \frac{x^3}{3!} + \cdots,$$

and that the convergence is quite rapid, especially for moderate values of x. In particular,

$$e^{-1} = 1 - \frac{1}{1!} + \frac{1}{2!} - \frac{1}{3!} + \cdots.$$

Thus our formula for d_n is approximately $n!e^{-1}$. In other words, if n is at all large, then $d_n/n! \approx e^{-1} \approx 0.368$. Even for n as small as 4 the estimate is quite good; we have $d_4 = 9/24 = 0.375$. Returning, then, to the midshipmen's caps, we see that the probability of no midshipman's retrieving his or her own cap is about 0.368, and the size of the graduating class has essentially no influence on the answer.

SUMMARY OF DEFINITIONS

derangement: a permutation of $\{1, 2, \ldots, n\}$ such that for each i from 1 to n, the number i is not in the ith position of the permutation; the number of derangements is denoted d_n (for example, 53421 is a derangement, but 51342 is not).

inclusion–exclusion principle: The cardinality of the union of a finite number of finite sets equals the sum of the cardinalities of the sets, minus the sum of the cardinalities of all pairwise intersections of the sets, plus the sum of the cardinalities of all three-way intersections of the sets, and so on.

sieve of Eratosthenes: an algorithm for finding the set of prime numbers up to a fixed bound.

EXERCISES

1. In a group of 60 college students, 37 like to play chess, 31 like to play bridge, and 19 like to play backgammon. Furthermore, 11 like to play chess and bridge, 16 like to play chess and backgammon, and five like to play bridge and backgammon. If all the students like at least one of these three games, how many like to play all three games?

2. Draw the Venn diagram showing the situation in Exercise 1, with the cardinalities of the various intersections indicated.

3. Three kinds of prizes were awarded to some of the 100 competitors in a race: blue ribbons for speed, red ribbons for endurance, and green ribbons for form. Suppose that 13 blue ribbons, 25 red ribbons, and 23 green ribbons were given out. Also suppose that the total number of people who won two ribbons (of different colors) was 17; but that no one won three ribbons. How many people won no ribbons?

4. Write out the inclusion–exclusion principle in full for five sets. In other words, express $|A \cup B \cup C \cup D \cup E|$ in terms of the cardinalities of these five sets and various intersections of them (there will be 31 terms).

5. Write out the expression in Theorem 2 using summation notation.

6. Consider the set $\mathcal{F}$ of functions from $\{1, 2, 3, 4\}$ to $\{1, 2, 3\}$.
   (a) Determine $|\mathcal{F}|$.
   (b) Write down a function in $\mathcal{F}$ that is not surjective.
   (c) Write down a function in $\mathcal{F}$ that is surjective.
   (d) Determine the number of elements of $\mathcal{F}$ that are surjective.

7. List all the prime numbers from 1 to 40 to determine how many there are. Then use Theorem 3 to calculate this number.

8. Write down all the derangements of $\{1, 2, 3, \ldots, n\}$ and verify that Theorem 4 gives the correct values for d_n, for the following values of n.
   (a) $n = 1$.
   (b) $n = 2$.
   (c) $n = 4$.

9. Explain what is wrong with the following story. Twenty-six foreign-language majors were comparing notes on what languages they had studied. It turned out that 12 had studied German, nine had studied French, and 12 had studied Russian. Also, one of them had studied both French and Russian, three of them had studied both German and Russian, and two of them had studied both French and German.

10. Draw a Venn diagram showing the sizes of the various sets of interest in Example 2, similar to Figure 7.4. (There will be 16 regions, including $\overline{E \cup H \cup C \cup B}$.)

11. Suppose that four sets intersect in the following ways: Each pair of them has 20 elements in common; each triple of them has three elements in common; and there is one element in all four of the sets. If each set has 250 elements in it, how many elements are there in at least one of the sets?

12. During a certain month, 15 of the cars checked at a state inspection facility had defective brakes, 20 had defective emissions control, and 25 had defective

lights. Furthermore, two of these cars had all three of these problems, and seven cars had exactly two of the three problems. How many cars had at least one defect?

13. Among a group of picture painters, 14 use oil, and of these 14, six use no other medium. Also, 17 use water color, and of these 17, seven use nothing else; and 18 use acrylic, including nine who use only acrylic. Furthermore, only one of these artists uses all three media.
    (a)   How many of the artists use exactly two of the three media in their work?
    (b)   How many of the artists use at least one of the three media?

14. Suppose that six employees are available to do 10 tasks. Each employee is capable of doing all of the tasks, and each task must have exactly one person assigned to do it. Recall in this exercise our convention that people are always to be considered distinct.
    (a)   Assuming that the tasks are to be considered distinct, how many ways are there for the employees to be assigned the tasks?
    (b)   Assuming that the tasks are to be considered distinct, how many ways are there for the employees to be assigned the tasks if each employee must receive at least one task?
    (c)   Assuming that the tasks are to be considered indistinguishable, how many ways are there for the employees to be assigned the tasks?
    (d)   Assuming that the tasks are to be considered indistinguishable, how many ways are there for the employees to be assigned the tasks if each employee must receive at least one task?

15. Repeat Exercise 14, assuming that there are 10 employees and six tasks.

16. Determine the number of prime numbers less than 100.

17. Determine the parity (oddness or evenness) of d_n as a function of n.

18. Compute d_5 and d_6 exactly, and compute $d_5/5!$ and $d_6/6!$ to five decimal places. Compare these latter values to $1/e$.

19. Show that there are no permutations of n objects that leave exactly $n - 1$ objects in their original positions.

20. Give a convincing argument as to the correct value of d_0.

21. Fifteen gentlemen check their hats before entering a formal dance. By the end of the dance, each has lost his hatcheck, and the hats are returned in a random order.
    (a)   How likely is it that Mr. Rosen will receive his own hat back?
    (b)   How likely is it that Mr. Rosen will receive his own hat back but nobody else will?

- (c) How likely is it that both Mr. Rosen and Mr. Roberts will receive their own hats back?
- (d) How likely is it that both Mr. Rosen and Mr. Roberts will receive their own hats back but nobody else will?
- (e) How likely is it that the five gentlemen who live in the Zeta Zeta Zeta fraternity will collectively (but not necessarily individually) receive the correct set of hats back?
- (f) How likely is it that each of the 15 gentlemen will receive his own hat back?
- (g) How likely is it that exactly one of the 15 gentlemen will receive his own hat back?
- (h) How likely is it that none of the 15 gentlemen will receive his own hat back?
- (i) How likely is it that at least one of the 15 gentlemen will receive his own hat back?

22. Spelling bees are to be held in a class of 10 students, on each of two consecutive days. In a spelling bee, the order in which the participants are arranged is important; for example, the first student in line must answer the first question, while the last student gets to relax for the first nine questions. Determine how many different orderings of the students for the two bees are possible under each of the following conditions.
    - (a) There are no restrictions.
    - (b) No student may be in the same position on both days.
    - (c) The students in the first half of the line the first day must be in the second half of the line the second day.

23. Consider permutations of the set $\{1, 2, 3, \ldots, 10\}$.
    - (a) How many such permutations leave all the odd numbers in their original positions?
    - (b) How many such permutations leave all the odd numbers, and only the odd numbers, in their original positions?
    - (c) How many such permutations leave exactly five numbers in their original positions?
    - (d) How many such permutations leave at least two numbers in their original positions?

24. Suppose that a cereal manufacturer decorates boxes of its cereal with pictures of sports stars. Assume that there are five different personalities featured, so that each personality appears randomly on one-fifth of the boxes. If you choose eight boxes at random, how likely is it that you will see all five stars? [*Hint*: Consider the function that sends a cereal box to the person who appears on it.]

25. Find the number of positive integers no greater than 300 that are either odd or perfect squares but not both.

26. Find the number of permutations of the 26 letters of the English alphabet (i.e., strings of length 26 using each letter exactly once) meeting the following conditions, where "contains" means "has as a substring."
    (a)   The permutation does not contain the string *math*.
    (b)   The permutation contains both of the strings *math* and *bio*.
    (c)   The permutation contains neither of the strings *math* or *bio*.
    (d)   The permutation contains both of the strings *bio* and *housing*.
    (e)   The permutation contains neither of the strings *bio* or *housing*.
    (f)   The permutation contains both of the strings *math* and *housing*.
    (g)   The permutation contains neither of the strings *math* or *housing*.
    (h)   The permutation contains none of the strings *math*, *bio*, or *housing*.

27. Determine how many bit strings of length 7 do not contain 00000 as a substring; use the inclusion–exclusion principle explicitly.

Challenging Exercises

28. State concisely (as a function of k and n) how many functions there are from a k-set to an n-set
    (a)   That are both one-to-one and onto.
    (b)   That are neither one-to-one nor onto.
    (c)   That are one-to-one but not onto.
    (d)   That are onto but not one-to-one.

29. Determine the number of numbers between 1 and 500 that are divisible by some perfect square greater than 1. (For example, we want to count 270, since 270 is divisible by the perfect square $3^2 = 9$; but we do not want to count 42, since $42 = 2 \cdot 3 \cdot 7$ has no perfect squares as factors.)

30. Find a recurrence relation and initial conditions for d_n, the number of derangements of the numbers from 1 to n. Your answer must be based on the definition of derangement, not on the formula obtained in Theorem 4. Then solve the recurrence relation to obtain a formula for d_n.

31. Consider the figure obtained by drawing a regular pentagon and its five diagonals. Determine the number of triangles (of all sizes and shapes) in this figure. [*Hint*: Formulate this problem in terms of the inclusion–exclusion principle.]

32. Let $S(k, n)$ be the number of partitions of a k-set into exactly n nonempty parts (see Exercise 20 in Section 7.2). Find a formula involving $S(k, n)$ for the number of surjective functions from a k-set to an n-set.

Exploratory Exercises

33. See how the number of primes from 1 to N compares to $N/\ln N$ for various values of N, such as 100, 200, 500, ... Then find out more about the prime number theorem by consulting Shapiro [271].

34. Investigate the relationship between the number of derangements of $\{1, 2, \ldots,$ $n\}$ and the number of permutations of this set that leave exactly one element in its original position. Consult Wilf [321] for more details.

35. Find out about **rook polynomials** by consulting Liu [198] and A. Tucker [306].

36. Find out about Carl Friedrich Gauss by consulting Bell [15].

SECTION 7.5
GENERATING FUNCTIONS

Mathematicians are like Frenchmen: whatever you say
to them they translate into their own language,
and forthwith it is something entirely different.

—**Goethe,** *Maxims and Reflexions*

In this section we introduce a powerful and valuable tool that enables us to encode a sequence of numbers as a function of one variable. This function is called a generating function for the sequence. By manipulating generating functions, according to familiar rules of algebra and calculus, we can gain information about the sequences they represent. We will see how to use generating functions to model the kinds of combinatorial problems that we encountered in Section 7.2. Finally, we will show how to solve recurrence relations using generating functions.

Suppose that $a_0, a_1, a_2, \ldots, a_k, \ldots$ is a sequence of numbers. (For our applications, these will usually be natural numbers, but in general they can be arbitrary real numbers or even complex numbers.) The **generating function** for the sequence $\{a_k\}$ is the function f defined by

$$f(x) = \sum_{k=0}^{\infty} a_k x^k = a_0 + a_1 x + a_2 x^2 + a_3 x^3 + \cdots + a_k x^k + \cdots.$$

The numbers in the given sequence form the coefficients of the powers of x in the power series that defines this function. We will think of x as a real variable. Recall from calculus that this infinite series does indeed define a function. The domain of f is the set of values of x for which the series converges, and the series is the Taylor series for f. The domain might be the set of all real numbers, as is the case for the exponential function

$$e^x = \sum_{k=0}^{\infty} \frac{x^k}{k!} = 1 + x + \frac{x^2}{2} + \frac{x^3}{6} + \frac{x^4}{24} + \cdots$$

(here $a_k = 1/k!$). The domain might be some finite interval centered at $x = 0$; for example, the constant sequence in which each $a_k = 1$ has as its generating function the rational function

$$\frac{1}{1-x} = \sum_{k=0}^{\infty} x^k = 1 + x + x^2 + x^3 + \cdots,$$

which converges if and only if $|x| < 1$, since the series is a geometric series (see Example 9 in Section 2.2). At worst, the domain might consist only of $\{0\}$, as will occur if the coefficients grow too fast; consider, for example, the function

$$f(x) = \sum_{k=0}^{\infty} k!\, x^k = 1 + x + 2x^2 + 6x^3 + 24x^4 + \cdots.$$

Most of the generating functions we will encounter turn out to be rational functions (the quotient of two polynomials), whose domains are always intervals of positive length centered at 0. It may come as somewhat of a surprise, but we will not concern ourselves at all with the intervals of convergence for these series. Instead, we will think of these as **formal** power series, in essence just a symbolic trick to gather all the coefficients $\{a_k\}$ together into one entity, which we can manipulate. In other words, x is just a formal variable, and we never will think of replacing x by numerical values.

EXAMPLE 1. We saw above that the generating function for the sequence $1, 1, 1, \ldots$ is the function $1/(1-x)$.

The generating function for the finite sequence $1, 3, 3, 1$, which we can think of as an infinite sequence by extending it with zeros (to $1, 3, 3, 1, 0, 0, 0, \ldots$), is by definition

$$f(x) = 1 + 3x + 3x^2 + 1 \cdot x^3 + 0 \cdot x^4 + 0 \cdot x^5 + \cdots$$
$$= 1 + 3x + 3x^2 + x^3 = (1 + x)^3.$$

The trigonometric function $\sin(x)$ has as its Taylor series

$$\sin(x) = x - \frac{x^3}{3!} + \frac{x^5}{5!} - \frac{x^7}{7!} + \cdots,$$

so $\sin(x)$ is the generating function for the sequence $0, 1, 0, -1/6, 0, 1/120, 0, -1/5040, \ldots$. ◆

Let us now study certain generating functions more systematically. Our goal here is to compile a catalog of functions that arise as we look at the applications in the remainder of this section.

We begin by looking at what the binomial theorem says, in the language of generating functions. Let n be a fixed natural number. Recall from Section 7.1 that

$$(a + b)^n = \sum_{k=0}^{n} C(n, k) a^{n-k} b^k.$$

If we let $a = 1$ and $b = x$, we obtain

$$(1 + x)^n = \sum_{k=0}^{n} C(n, k) x^k,$$

which tells us that *the generating function for the sequence of binomial coefficients $C(n, k)$, thought of as a sequence in which n is fixed and k is the index, is $f(x) = (1 + x)^n$.* Since the sequence is finite (or, equivalently, contains only zeros beyond the term $C(n, n)$), the generating function is just a polynomial: All the coefficients of x^k for $k > n$ are zero. We noted the example $(1 + x)^3 = 1 + 3x + 3x^2 + x^3$ above.

Next let us turn to the infinite geometric series, which we mentioned in Example 9 in Section 2.2. The simplest case we saw above, namely

$$\frac{1}{1 - x} = \sum_{k=0}^{\infty} x^k = 1 + x + x^2 + x^3 + \cdots.$$

If we substitute ax for x in this equation, where a is a constant, we obtain

$$\frac{1}{1 - ax} = \sum_{k=0}^{\infty} (ax)^k = \sum_{k=0}^{\infty} a^k x^k = 1 + ax + a^2 x^2 + a^3 x^3 + \cdots.$$

Thus we have proved the following fact.

THEOREM 1. *Let a be a fixed real number. Then the generating function for the sequence $1, a, a^2, a^3, \ldots$ is $1/(1 - ax)$.*

For instance, the generating function for the sequence $1, -3, 9, -27, \ldots$ is $1/(1 - (-3)x) = 1/(1 + 3x)$.

If we substitute x^r for x in the equation giving the power series for $1/(1 - x)$, where r is a positive integer, we obtain

$$\frac{1}{1 - x^r} = \sum_{k=0}^{\infty} (x^r)^k = \sum_{k=0}^{\infty} x^{rk} = 1 + x^r + x^{2r} + x^{3r} + \cdots.$$

Note that most of the coefficients here are 0; every rth coefficient is 1. Thus we obtain the following fact.

THEOREM 2.   *Let r be a fixed positive integer. Then the generating function for the sequence $1, 0, 0, \ldots, 0, 1, 0, 0, \ldots, 0, 1, 0, 0, \ldots, 0, \ldots$ (with $r - 1$ 0's between each pair of 1's) is $1/(1 - x^r)$.*

Operations from algebra and calculus can be applied to functions defined by power series: They can be added, multiplied by constants or powers of x, and differentiated or integrated. In each case the resulting power series is obtained by performing these operations term by term.

EXAMPLE 2.   Find the sequence whose generating function is $x/(1 - 2x^2)$.

Solution.   We have

$$\frac{x}{1 - 2x^2} = x \cdot \frac{1}{1 - 2x^2} = x \cdot (1 + 2x^2 + (2x^2)^2 + (2x^2)^3 + (2x^2)^4 + \cdots)$$

$$= x \cdot (1 + 2x^2 + 4x^4 + 8x^6 + 16x^8 + \cdots)$$

$$= x + 2x^3 + 4x^5 + 8x^7 + 16x^9 + \cdots$$

$$= \sum_{k=0}^{\infty} 2^k x^{2k+1}.$$

Therefore, the sequence is $0, 1, 0, 2, 0, 4, 0, 8, 0, 16, \ldots$.   ◆

EXAMPLE 3.   Find the generating function for the sequence $1, 1, 0, 1, 1, 0, 1, 1, 0, 1, 1, 0, \ldots$.

Solution.   The answer is, by definition, $f(x) = 1 + x + x^3 + x^4 + x^6 + x^7 + x^9 + x^{10} + \cdots$, but what we seek is a closed form for this function. Let us write f as the sum of two functions:

$$f(x) = (1 + x^3 + x^6 + x^9 + \cdots) + (x + x^4 + x^7 + x^{10} + \cdots).$$

From the discussion above we know nice forms for each of these: $1/(1 - x^3)$ and $x \cdot [1/(1 - x^3)]$, respectively. Therefore, we have $f(x) = (1 + x)/(1 - x^3)$.   ◆

EXAMPLE 4.   Find the generating function for the sequence $1, 2, 3, 4, \ldots$.

Solution.   If we differentiate both sides of the equation

$$\frac{1}{1 - x} = \sum_{k=0}^{\infty} x^k = 1 + x + x^2 + x^3 + \cdots,$$

we obtain

$$\frac{1}{(1-x)^2} = \sum_{k=1}^{\infty} kx^{k-1} = \sum_{k=0}^{\infty} (k+1)x^k = 1 + 2x + 3x^2 + 4x^3 + \cdots.$$

(The middle equality comes from a change of dummy variable, substituting $k+1$ for k.) Therefore, the generating function of $1, 2, 3, 4, \ldots$ is $1/(1-x)^2$.

If we multiply both sides of the last equation by x, we obtain

$$\frac{x}{(1-x)^2} = \sum_{k=1}^{\infty} kx^k,$$

so $x/(1-x)^2$ is the generating function for the sequence $0, 1, 2, 3, 4, \ldots$. ◆

Let us generalize the last example. If we differentiate the original left-hand side, $1/(1-x)$, r times, we obtain $r!/(1-x)^{r+1}$. If we differentiate the right-hand side, $\sum_{k=0}^{\infty} x^k$, r times, we obtain

$$\sum_{k=r}^{\infty} \frac{k!}{(k-r)!} x^{k-r} = \sum_{k=0}^{\infty} \frac{(k+r)!}{k!} x^k;$$

this equality follows from replacing the dummy index of summation k by $k+r$. If we equate these two derivatives, divide by $r!$, and recall that $C(a, b) = a!/(b!(a-b)!)$, we obtain

$$\frac{1}{(1-x)^{r+1}} = \sum_{k=0}^{\infty} C(k+r, r)x^k.$$

Thus we obtain the generating function for the binomial coefficients, this time thought of as a sequence in which the second argument is fixed, as summarized in the following theorem.

THEOREM 3. *Let r be a fixed natural number. Then the generating function for the sequence $\{C(k+r, r)\}$ is $1/(1-x)^{r+1}$.*

MODELING COMBINATORIAL PROBLEMS

Generating functions are a natural setting in which to model combinatorial problems. The addition principle and the multiplication principle (Section 6.1) correspond very closely to the addition and multiplication of the generating functions whose coefficients are the number of ways to perform tasks.

Let us focus on the following setting for our initial discussion. Suppose that we have a bag containing colored balls, and we wish to determine the number of ways to select various combinations of balls. To begin, suppose that the bag contains four identical red balls, five identical white balls, and two identical blue balls. For each natural number k, we wish to find the number of ways to select k balls, ignoring order; let us denote this quantity by a_k. For example, $a_2 = 6$, since our choice can be any of RR, RW, RB, WW, WB, or BB, using the obvious notation. We claim that the generating function for $\{a_k\}$ is

$$f(x) = (1 + x + x^2 + x^3 + x^4)(1 + x + x^2 + x^3 + x^4 + x^5)(1 + x + x^2).$$

In other words, we claim that the coefficient of x^k in $f(x)$, when it is multiplied out, is the number of ways to choose k balls from this bag. To explain why, let us see what the expression shown here for $f(x)$ represents. The first factor in $f(x)$ corresponds to choosing red balls. More precisely, $1 + x + x^2 + x^3 + x^4$ is the generating function for the number of ways to choose red balls only, in that the coefficient of x^i is the number of ways to choose i red balls: Since there are four red balls in the bag, this coefficient is 1 for $0 \le i \le 4$ and 0 for $i > 4$. Note that the statement that we could choose zero red balls *or* one red ball *or* two red balls *or* three red balls *or* four red balls was reflected in *adding* powers of x. Similarly, the second and third factors of $f(x)$ are the generating functions for the number of ways to choose white balls and blue balls, respectively. In each case, the exponent of x gives the number of balls chosen.

Now if we want to choose k balls in all, then we must select some number of red balls (say i_1 of them) *and* some number (i_2) of white balls *and* some number (i_3) of blue balls, so that $i_1 + i_2 + i_3 = k$. In other words, to solve our problem, we need to find the number of solutions to the equation $i_1 + i_2 + i_3 = k$, with the restrictions that $0 \le i_1 \le 4$, $0 \le i_2 \le 5$, and $0 \le i_3 \le 2$. Notice how this equation was translated into a generating function: Each factor in $f(x)$ corresponds to the possible choices for one of these integer variables. Therefore, the coefficient of x^k when $f(x)$ is multiplied out tells us how many solution there are to this equation, because multiplying powers of x corresponds to adding the exponents. We obtain a term x^k in the expanded expression whenever we can find a term x^{i_1} in the first factor *and* a term x^{i_2} in the second factor *and* a term x^{i_3} in the third factor such that $i_1 + i_2 + i_3 = k$.

In summary, the coefficient of x^k after $f(x)$ has been multiplied out is the number of ways to choose the k balls. When we carry out the algebra of expanding $f(x)$, as we can do either by hand or with a computer algebra package, we find that

$$f(x) = 1 + 3x + 6x^2 + 9x^3 + 12x^4 + 14x^5 + 14x^6 + 12x^7 + 9x^8 + 6x^9 + 3x^{10} + x^{11}.$$

Thus, for example, there are 14 ways to choose five balls and no ways to choose 17 balls. Note that once we wrote down the generating function, we transformed the problem from the realm of combinatorics to pure mechanical algebraic manipulation.

EXAMPLE 5. Find the generating function for the number of ways to choose k balls from a bag containing an unlimited number of identical red, white, and blue balls.

Solution. By the same reasoning as above, the generating function is

$$f(x) = (1 + x + x^2 + x^3 + \ldots)(1 + x + x^2 + x^3 + \ldots)(1 + x + x^2 + x^3 + \ldots)$$

$$= (1 + x + x^2 + x^3 + \ldots)^3$$

$$= \frac{1}{(1-x)^3}.$$

Knowing the generating function does not by itself solve the problem of computing the answer. To obtain the numerical answer for a given value of k, we have at least two options. On the one hand, once k is fixed, we may as well assume that the bag holds only k of each colored ball. Then the generating function becomes $f(x) = (1 + x + x^2 + x^3 + \ldots + x^k)^3$, and we can expand this polynomial and look at the coefficient of x^k. Alternatively, we can use Theorem 3, which tells us that the coefficient of x^k in $1/(1-x)^{r+1}$ is $C(k+r, r)$. Here $r = 2$, so we see that the answer is $C(k+3, 3)$.

This is the same answer we would obtain using the techniques of Section 6.3. With either approach we are counting the number of natural number solutions to the equation $i_1 + i_2 + i_3 = k$. In Section 6.3 we applied Theorem 3 of that section; here we translate the problem into a problem in generating functions and apply Theorem 3 of the present section. ◆

The generating function approach allows us to add various constraints to the problem, as the following example shows.

EXAMPLE 6. Find the generating function for the number of ways to choose k balls from a bag containing an unlimited number of identical red, white, and blue balls, if we must choose an even number of red balls, an odd number of white balls, and between three and five blue balls, inclusive. Then determine the number of ways to choose 50 balls under these conditions.

Solution. We are asked to find a generating function for counting the number of solutions to $i_1 + i_2 + i_3 = k$, where i_1 is an even natural number, i_2 is an odd natural number, and i_3 equals 3, 4, or 5. The generating function for choosing the red balls is $1 + x^2 + x^4 + \cdots$. Similarly, the generating function for choosing the white balls is $x + x^3 + x^5 + \cdots$. Finally, the generating function for choosing the blue balls is $x^3 + x^4 + x^5$. Thus the generating function for the number of ways to choose balls with these constraints is the product

$$f(x) = (1 + x^2 + x^4 + \cdots)(x + x^3 + x^5 + \cdots)(x^3 + x^4 + x^5).$$

We can write this in closed form by using Theorem 2:

$$f(x) = \frac{1}{1-x^2} \cdot \frac{x}{1-x^2} \cdot (x^3 + x^4 + x^5)$$

$$= \frac{x^4 + x^5 + x^6}{(1-x^2)^2}.$$

To find the number of ways to choose 50 balls under these conditions, we need to read off the coefficient of x^{50} in $(x^4+x^5+x^6)/(1-x^2)^2$. Thus we need to find the coefficients of x^{46}, x^{45}, and x^{44} in $1/(1-x^2)^2$ and add them together. By Example 4, $1/(1-x)^2$ is the generating function for the sequence $1, 2, 3, 4, \ldots$. It follows that $1/(1-x^2)^2$ is the generating function for the sequence $1, 0, 2, 0, 3, 0, 4, \ldots$. Thus we see that the coefficient of x^{46} is 24, the coefficient of x^{45} is 0, and the coefficient of x^{44} is 23. Our answer is therefore $24 + 0 + 23 = 47$.   ◆

EXAMPLE 7.   How many ways are there to make change for one dollar, using pennies, nickels, dimes, quarters, and half-dollar coins, if no more than 10 pennies may be used?

Solution.   We solve this problem by generalizing it a bit. Let a_k be the number of ways to obtain a total of k cents in coins, using no more than 10 pennies; for simplicity, we will assume that k is a multiple of 5, not exceeding 100 (since we need the answer only for $k = 100$). For instance, $a_{10} = 4$, since we can use a dime, two nickels, one nickel and five pennies, or 10 pennies to achieve a total of 10 cents. The problem asks for a_{100}.

The exponent on x represents the number of cents in the total. Half-dollars contribute 50 cents to the total, so the factor in our generating function that corresponds to choosing half-dollars will be $1+x^{50}+x^{100}$: We can choose no half-dollars and achieve 0 cents, one half-dollar and achieve 50 cents, or two half-dollars and achieve 100 cents. A similar analysis applies for quarters, dimes, and nickels, with generating functions $1 + x^{25} + x^{50} + x^{75} + x^{100}$, $1 + x^{10} + x^{20} + x^{30} + \cdots + x^{100}$, and $1 + x^5 + x^{10} + x^{15} + \cdots + x^{100}$, respectively. Finally, the generating function for choosing pennies under the given constraints and simplifying assumptions is $1 + x^5 + x^{10}$, once we observe that the change must necessarily contain either no pennies, five pennies, or 10 pennies.

Putting this all together, we see that the generating function for this problem is

$$f(x) = (1 + x^{50} + x^{100})(1 + x^{25} + x^{50} + x^{75} + x^{100})$$

$$\cdot (1 + x^{10} + x^{20} + x^{30} + \cdots + x^{100})$$

$$\cdot (1 + x^5 + x^{10} + x^{15} + \cdots + x^{100})$$

$$\cdot (1 + x^5 + x^{10}).$$

Note that by using the formula for the sum of a finite geometric series (Example 8 in Section 5.3), this can also be written more compactly as

$$f(x) = (1 + x^{50} + x^{100}) \cdot \frac{1 - x^{125}}{1 - x^{25}} \cdot \frac{1 - x^{110}}{1 - x^{10}} \cdot \frac{1 - x^{105}}{1 - x^{5}} \cdot (1 + x^{5} + x^{10}).$$

Laborious calculation shows that

$$f(x) = \sum_{k=0}^{\infty} a_k x^k = 1 + 2x^5 + 4x^{10} + \cdots + 105x^{100} + \cdots + x^{410},$$

so the answer to our problem is $a_{100} = 105$.

Again we could have phrased this problem in terms of counting solutions to an equation, before translating it into generating functions. The problem asks for the number of natural number solutions to $50h + 25q + 10d + 5n + p = 100$, subject to the constraint that $0 \le p \le 10$, where h, q, d, n, and p stand for the number of half-dollars, quarters, dimes, nickels, and pennies to be used, respectively. The fact that the sum is to be 100 also allowed us to impose the constraints $0 \le h \le 2$; $0 \le q \le 4$; $0 \le d \le 10$; $0 \le n \le 20$; and $p = 0$, 5, or 10. Each factor in the generating function reflected the contribution to the sum of one of the summands in this equation. ◆

Finally, let us use generating functions to model the problem of counting partitions of natural numbers. Recall from Section 6.3 that a partition of k is a decomposition of k into a sum of positive integers, with order ignored. Let a_k be the number of partitions of k. For example, $a_4 = 5$, since the partitions of 4 are 4, $3 + 1$, $2 + 2$, $2 + 1 + 1$, and $1 + 1 + 1 + 1$.

Our approach will have the same flavor as Example 7. To determine a partition of k we first need to decide how many of the summands will be 1's. Each of them will contribute 1 to the sum. The generating function for this much is thus $1 + x + x^2 + x^3 + \cdots$. Then we need to decide how many of the summands will be 2's. Since each of them will contribute 2 to the sum, the generating function for this much is $1 + x^2 + x^4 + x^6 + \cdots$. Similarly, the generating function for the selection of 3's in the sum is $1 + x^3 + x^6 + x^9 + \cdots$, and so on. In other words, letting i_s be the number of summands that equal s, we want to find the number of natural number solutions to the equation $i_1 + 2i_2 + 3i_3 + \cdots = k$.

Putting the pieces together by the multiplication principle, we see that the generating function for $\{a_k\}$ is the product

$$f(x) = (1 + x + x^2 + x^3 + \cdots)(1 + x^2 + x^4 + x^6 + \cdots)$$

$$\cdot (1 + x^3 + x^6 + x^9 + \cdots) \cdots$$

$$= \frac{1}{1-x} \cdot \frac{1}{1-x^2} \cdot \frac{1}{1-x^3} \cdots$$

$$= \prod_{i=1}^{\infty} \frac{1}{1-x^i}.$$

Even though this product is infinite, it is possible actually to compute the coefficients x^k by noting that in any decomposition of k there can be no more than k 1's, no more than $k/2$ 2's, no more than $k/3$ 3's, $\ldots$, and no summand greater than k. Thus to compute a_k we only need to expand a finite portion of this product. To compute a_4, for instance, we expand

$$(1 + x + x^2 + x^3 + x^4)(1 + x^2 + x^4)(1 + x^3)(1 + x^4)$$

and find the coefficient of x^4. The expansion is

$$1 + x + 2x^2 + 3x^3 + 5x^4 + 5x^5 + 6x^6 + 7x^7$$

$$+ 7x^8 + 6x^9 + 5x^{10} + 5x^{11} + 3x^{12} + 2x^{13} + x^{14} + x^{15},$$

so $a_4 = 5$.

The generating function approach lets us compute the number of partitions with various constraints as well—partitions into odd parts, partitions into parts of unequal size, and so on. The reader is asked to find some of these in Exercise 12.

Our final example shows how more complex combinatorial problems can be solved using generating functions. This problem could also be solved using the techniques of Section 6.3.

EXAMPLE 8.   A bag contains 50 red balls, 40 white balls, and 30 blue balls. How many ways are there to select 100 balls from the bag?

Solution.   Let a_k be the number of ways to select k balls from the bag. We need to find a_{100}; we will do so by using the generating function for $\{a_k\}$.

It should be clear that the generating function for this sequence is

$$(1 + x + x^2 + \cdots + x^{50})(1 + x + x^2 + \cdots + x^{40})(1 + x + x^2 + \cdots + x^{30}).$$

This can be rewritten (using the formula for the sum of a geometric series) as

$$\frac{1 - x^{51}}{1 - x} \cdot \frac{1 - x^{41}}{1 - x} \cdot \frac{1 - x^{31}}{1 - x},$$

which is the same as

$$\frac{1 - x^{31} - x^{41} - x^{51} + x^{72} + x^{82} + x^{92} - x^{123}}{(1 - x)^3}.$$

Now by Theorem 3, the denominator expands as $\sum_{k=0}^{\infty} C(k + 2, 2)x^k$. Thus the generating function can be written finally as

$$\sum_{k=0}^{\infty} (1 - x^{31} - x^{41} - x^{51} + x^{72} + x^{82} + x^{92} - x^{123})C(k + 2, 2)x^k.$$

We want to find the coefficient of x^{100}. Seven terms in the series as written contribute to this coefficient (namely, for $k = 100, 69, 59, 49, 28, 18,$ and 8):

$$a_{100} = C(102, 2) - C(71, 2) - C(61, 2) - C(51, 2) + C(30, 2) + C(20, 2) + C(10, 2).$$

Therefore, $a_{100} = 5151 - 2485 - 1830 - 1275 + 435 + 190 + 45 = 231$: There are 231 possible selections. ◆

SOLVING RECURRENCE RELATIONS

We discussed one method for solving recurrence relations in Section 7.3. Now we will see that generating functions can also be used to obtain explicit formulas for sequences satisfying recurrence relations.

We will explain the basic method by means of an example, namely one of the recurrence relations and initial conditions featured in Section 7.3:

$$a_k = 3a_{k-1} - 2a_{k-2}$$

$$a_0 = 3, \qquad a_1 = 7.$$

Let $f(x)$ be the generating function for the sequence $\{a_k\}$; in other words, let

$$f(x) = \sum_{k=0}^{\infty} a_k x^k.$$

Our goal is to use the recurrence relation and initial conditions to write down an equation that f must satisfy, then solve that equation to find f explicitly, and finally use the facts about specific generating functions that we compiled at the beginning of this section (Theorems 1, 2, and 3) to write down an explicit formula for a_k.

To apply generating functions to the recurrence relation $a_k = 3a_{k-1} - 2a_{k-2}$, we need to find the generating function whose kth coefficient is $3a_{k-1}$ and the generating function whose kth coefficient is $2a_{k-2}$. For the first we see that

$$3xf(x) = \sum_{k=0}^{\infty} 3xa_k x^k = \sum_{k=0}^{\infty} 3a_k x^{k+1} = \sum_{k=1}^{\infty} 3a_{k-1} x^k,$$

the last equality coming from a change of dummy variable (replacing k by $k-1$). Similarly,

$$2x^2 f(x) = \sum_{k=0}^{\infty} 2x^2 a_k x^k = \sum_{k=0}^{\infty} 2a_k x^{k+2} = \sum_{k=2}^{\infty} 2a_{k-2} x^k.$$

Now we have

$$f(x) - 3xf(x) + 2x^2 f(x)$$

$$= \sum_{k=0}^{\infty} a_k x^k - \sum_{k=1}^{\infty} 3a_{k-1} x^k + \sum_{k=2}^{\infty} 2a_{k-2} x^k$$

$$= a_0 + a_1 x + \sum_{k=2}^{\infty} a_k x^k - 3a_0 x - \sum_{k=2}^{\infty} 3a_{k-1} x^k + \sum_{k=2}^{\infty} 2a_{k-2} x^k$$

$$= a_0 + a_1 x - 3a_0 x + \sum_{k=2}^{\infty} (a_k - 3a_{k-1} + 2a_{k-2}) x^k$$

$$= a_0 + a_1 x - 3a_0 x.$$

The last equality is where we use the recurrence relation, namely $a_k - 3a_{k-1} + 2a_{k-2} = 0$. Factoring $f(x)$ out of the left-hand side, and using the initial conditions, we obtain

$$(1 - 3x + 2x^2)f(x) = 3 + 7x - 3 \cdot 3x = 3 - 2x,$$

or

$$f(x) = \frac{3 - 2x}{1 - 3x + 2x^2}.$$

Thus we have obtained the generating function for the solution $\{a_k\}$, so in principle we have obtained the solution.

But we can go even further. Let us expand the right-hand side of the preceding equation, using partial fractions (partial fraction decomposition is usually covered in a calculus course). We want to rewrite $(3 - 2x)/(1 - 3x + 2x^2)$, which factors as $(3 - 2x)/((1 - 2x)(1 - x))$, in the form $A/(1 - 2x) + B/(1 - x)$. To do so, we set

$$\frac{3 - 2x}{(1 - 2x)(1 - x)} = \frac{A}{1 - 2x} + \frac{B}{1 - x},$$

multiply through by the common denominator, and simplify, to obtain $3 - 2x = (A + B) + (-A - 2B)x$. Thus we must have the constant terms equal ($A + B = 3$), and the coefficients of x equal ($-A - 2B = -2$). Solving these equations simultaneously, we obtain $A = 4$ and $B = -1$. Thus $f(x) = 4/(1 - 2x) + (-1)/(1 - x)$.

Now we know from Theorem 1 that the generating function for $1/(1 - 2x)$ is $\sum_{k=0}^{\infty} 2^k x^k$, and the generating function for $1/(1 - x)$ is $\sum_{k=0}^{\infty} x^k$. Thus we have

$$f(x) = \frac{4}{1 - 2x} + \frac{-1}{1 - x}$$

$$= \sum_{k=0}^{\infty} 4 \cdot 2^k x^k + \sum_{k=0}^{\infty} -1 \cdot x^k = \sum_{k=0}^{\infty} (4 \cdot 2^k - 1)x^k.$$

Since $f(x) = \sum_{k=0}^{\infty} a_k x^k$, it follows that $a_k = 4 \cdot 2^k - 1$, and we have the desired explicit solution.

Let us apply the same technique to a nonhomogeneous problem from Section 7.3 (Example 10). The algebra becomes quite messy, but in principle the method is straightforward.

EXAMPLE 9. Solve the recurrence relation and initial conditions

$$a_k = 4a_{k-1} - 3a_{k-2} + 2^k + k + 3$$

$$a_0 = 0, \qquad a_1 = 0.$$

Solution. Let $f(x) = \sum_{k=0}^{\infty} a_k x^k$ be the generating function for the sequence $\{a_k\}$ that we wish to find. First we will use the recurrence relation and initial conditions to find a formula for $f(x)$. Then we will use partial fractions to write $f(x)$ in a form to which Theorems 1 and 3 apply. This will enable us to write down an explicit formula for a_k.

To convert the recurrence relation to an equation involving generating functions, we need the generating function for the sequence $\{2^k + k + 3\}$ which appears there. From Theorem 1 we know that $1/(1 - 2x) = \sum_{k=0}^{\infty} 2^k x^k$ and $3/(1-x) = \sum_{k=0}^{\infty} 3x^k$, and from Example 4 we know that $x/(1-x)^2 = \sum_{k=0}^{\infty} kx^k$. Thus $1/(1 - 2x) + x/(1 - x)^2 + 3/(1 - x)$ is the desired generating function.

Now proceeding as in the preceding problem, we see that

$$f(x) - 4xf(x) + 3x^2 f(x) = \sum_{k=0}^{\infty} a_k x^k - \sum_{k=1}^{\infty} 4a_{k-1}x^k + \sum_{k=2}^{\infty} 3a_{k-2}x^k$$

$$= a_0 + a_1 x - 4a_0 x + \sum_{k=2}^{\infty} (a_k - 4a_{k-1} + 3a_{k-2})x^k$$

$$= 0 + \sum_{k=2}^{\infty} (2^k + k + 3)x^k$$

(from the recurrence relation and initial conditions)

$$= \sum_{k=0}^{\infty} (2^k + k + 3)x^k - (2^0 + 0 + 3) - (2^1 + 1 + 3)x$$

$$= \frac{1}{1 - 2x} + \frac{x}{(1 - x)^2} + \frac{3}{1 - x} - 4 - 6x.$$

Factoring the left-hand side and dividing to isolate $f(x)$, we obtain

$$f(x) = \frac{1}{(1 - 2x)(1 - 3x)(1 - x)} + \frac{x}{(1 - 3x)(1 - x)^3}$$

$$+ \frac{3}{(1 - 3x)(1 - x)^2} + \frac{-4 - 6x}{(1 - 3x)(1 - x)}.$$

Now a tedious partial fraction exercise allows us to rewrite this as

$$f(x) = \frac{27/8}{1 - 3x} + \frac{-4}{1 - 2x} + \frac{23/8}{1 - x} + \frac{-7/4}{(1 - x)^2} + \frac{-1/2}{(1 - x)^3}.$$

Applying Theorems 1 and 3 yields

$$a_k = \frac{27}{8} \cdot 3^k - 4 \cdot 2^k + \frac{23}{8} - \frac{7}{4}(k + 1) - \frac{1}{2} \cdot \frac{(k + 2)(k + 1)}{2}$$

$$= \frac{27}{8} \cdot 3^k - 4 \cdot 2^k - \frac{1}{4}k^2 - \frac{5}{2}k + \frac{5}{8}.$$

Note that this is the same answer obtained using the methods explained in Section 7.3. ◆

SUMMARY OF DEFINITIONS

formal power series: an infinite series of powers of a variable, $\sum_{k=0}^{\infty} a_k x^k$, in which the variable is not thought of as representing a number; in particular, convergence is not an issue.

generating function for the sequence $a_0, a_1, a_2, \ldots, a_k, \ldots$: the function defined by $f(x) = \sum_{k=0}^{\infty} a_k x^k$.

EXERCISES

1. Find the sequence represented by each of the following generating functions.
   (a) $f(x) = (2 + 3x)^2$.

    (b) $f(x) = 1/(1 + x)$.
    (c) $f(x) = 3/(1 - 4x)$.
    (d) $f(x) = 3/(1 - x^4)$.
    (e) $f(x) = 2x/(1 + x^2)$.

2. Find the generating function (in closed form) for each of the following sequences (the obvious patterns are intended).
    (a) $2, 2, 2, \ldots$.
    (b) $1, 2, 4, 8, 16, \ldots$.
    (c) $1, -1, 1, -1, 1, -1, \ldots$.

3. Write out the first five terms of the sequences generated by the following generating functions.
    (a) $1/(1 - x)^2$.
    (b) $1/(1 - x)^3$.
    (c) $1/(1 - x)^4$.
    (d) $1/(1 - x)^5$.

4. Find the generating functions for the following combinatorial problems.
    (a)   The number of ways to choose k balls from a bag containing two red balls and five white balls.
    (b)   The number of ways to choose k balls from a bag containing two red balls and five white balls, if an odd number of white balls must be selected.
    (c)   The number of ways to choose k balls from a bag containing two balls of each of 10 different colors.
    (d)   The number of ways to choose k balls from a bag containing 10 balls of each of two different colors.

5. Use the answers to Exercise 4 (parts (a) and (b), multiplied out) to find the number of ways to choose four balls from a bag containing two red balls and five white balls
    (a)   With no restrictions.
    (b)   If an odd number of white balls must be selected.

6. Find the sequence represented by each of the following generating functions.
    (a) $f(x) = (2 + 3x)^4$.
    (b) $f(x) = (1 - x)/(1 + x)$.
    (c) $f(x) = (1 - x)/(1 - x^2)$.

7. Find the generating function (in closed form) for each of the following sequences (the obvious patterns are intended). [*Hint*: In many cases, it is best to write the sequence as a sum or difference of sequences.]
    (a) $2, -2, 2, -2, \ldots$.
    (b) $4, 8, 16, 32, 64, \ldots$.
    (c) $2, 4, 6, 8, \ldots$.
    (d) $0, 0, 0, 1, 1, 1, 1, 1, 1, 1, 1, 1, 1, 1, 1 \ldots$.

  (e) $0, 1, 0, -1, 0, 1, 0, -1, \ldots$.
  (f) $2, 3, 4, 5, 6, \ldots$.
  (g) $1, 3, 5, 7, 9, \ldots$.
  (h) $2, 1, 4, 3, 6, 5, 8, 7, 10, 9, \ldots$.

8. State and prove a theorem that combines the information in Theorems 1 and 3. In other words, what sequence is generated by $1/(1-ax)^{r+1}$, for a fixed natural number r and real number a?

9. Solve the problem posed in Example 3 by thinking of the given power series as $(1 + x + x^2 + x^3 + x^4 + \cdots) - (x^2 + x^5 + x^8 + x^{11} + \cdots)$. Show that the closed form you obtain actually represents the same function as the answer obtained in Example 3.

10. Find the generating function, written in closed form, for the problem of finding the number of ways to choose k balls from a bag containing an unlimited number of balls of each of 10 different colors
  (a) With no restrictions.
  (b) If an odd number of balls of each color must be selected.
  (c) If you may not choose exactly one ball of any color.
  (d) If you must choose the same number of balls of each color.

11. Consider the problem of finding the number of ways to choose among a collection of a lot of 3-cent stamps, 4-cent stamps, and 5-cent stamps to obtain k cents in postage.
  (a) Find the generating function that solves this problem. Write out explicitly the terms of this function up through x^{10}.
  (b) Determine how many ways there are to obtain 10 cents in postage from this collection.

12. Find the generating function for the problem of finding the number of partitions of the natural number k
  (a) With no restrictions.
  (b) Into odd parts (only odd numbers are allowed as summands).
  (c) Into parts of unequal size (at most one 1, at most one 2, at most one 3, etc.).
  (d) Into parts no larger than 4 (no parts of size $5, 6$, etc.).

13. Determine the coefficient of x^6 in each of the generating functions in Exercise 12.

14. Write down in closed form the generating function for the number of ways to hold k cents in present-day U.S. coins. (Coins come in denominations of 1, 5, 10, 25, 50, and 100 cents.)

15. Use generating functions to solve the following problems.
  (a) How many solutions are there to the equation $a + b + c + d = 30$, where the variables range over the natural numbers?

(b)   How many solutions are there to the equation $a+b+c+d = 30$, where the variables range over the natural numbers less than 10?

(c)   How many solutions are there to the equation $a+b+c+d = 30$, where the variables range over the positive integers?

(d)   How many solutions are there to the equation $a+b+c+d = 30$, where the variables range over the positive integers less than 10?

16. Use generating functions to solve the following problem: In how many ways can $2m+1$ balls be distributed into three distinct boxes, so that no box contains more than m balls?

17. Find the generating function for the number of ways for n distinct dice to show a total of k (each die can show a number from 1 to 6).

18. Use generating functions to solve the following recurrence relations with initial conditions.

    (a) $a_k = -a_{k-1}, \quad a_0 = 5.$

    (b) $a_k = a_{k-1} + 6a_{k-2}, \quad a_0 = 2, a_1 = -1.$

    (c) $a_k = -4a_{k-1} - 4a_{k-2}, \quad a_0 = 2, a_1 = -1.$

    (d) $a_k = 2a_{k-1} - a_{k-2}, \quad a_0 = 2, a_1 = -1.$

19. Use generating functions to solve the following recurrence relations with initial conditions.

    (a) $a_k = 5a_{k-1} - 6a_{k-2} + 4^k, \quad a_0 = 1, a_1 = 2.$

    (b) $a_k = a_{k-1} + a_{k-2}, \quad a_0 = 1, a_1 = 1$ (the Fibonacci sequence).

20. Use generating functions to solve the following recurrence relations with initial conditions.

    (a) $a_k = a_{k-1} + 4a_{k-2} - 4a_{k-3}, \quad a_0 = 2, a_1 = -1, a_2 = 0.$

    (b) $a_k = 3a_{k-1} - 3a_{k-2} + a_{k-3}, \quad a_0 = 0, a_1 = 1, a_2 = 2.$

    (c) $a_k = 2a_{k-1} + 3a_{k-2} - 8a_{k-3} + 4a_{k-4}, \quad a_0 = 3, a_1 = 0, a_2 = -8, \ a_3 = -26.$

21. Use generating functions to solve the following recurrence relations with initial conditions.

    (a) $a_k = -4a_{k-2}, \quad a_0 = 1, a_1 = 2.$

    (b) $a_k = -a_{k-4}, \quad a_0 = 1, a_1 = 2, a_2 = 3, a_3 = 1.$

Challenging Exercises

22. State and prove a theorem that combines the information in Theorems 1, 2 and 3. In other words, what sequence is generated by $1/(1 - ax^r)^s$, for fixed positive integers r and s?

23. Argue that the number of partitions of k into parts of unequal size is the same as the number of partitions of k into odd parts, by showing with formal algebraic manipulation that the generating functions for the two problems are equal. [*Hint*: $(1 + x^s)(1 - x^s) = 1 - x^{2s}$.]

24. Determine the number of strings of length n over the set $\{a, b, c, d\}$ that contain an odd number of a's (e.g., *caaba* is one such string of length 5). First write down a recurrence relation and initial conditions which the answer must satisfy, then use generating functions to solve the recurrence relation with initial conditions.

25. Let a_k be the number of ways to obtain a sum of k by rolling a die as often as desired. For example, $a_3 = 4$, since you can roll a 3, a 2 followed by a 1, a 1 followed by a 2, or three 1's. Show that the generating function for $\{a_k\}$ is $1/(1 - x - x^2 - x^3 - x^4 - x^5 - x^6)$. [*Hint*: You can roll the die no times, or one time, or two times, or three times, ...]

26. Show that if $f(x)$ is the generating function for the sequence $b_0, b_1, b_2, \ldots$, then $f(x)/(1 - x)$ is the generating function for the sequence $b_0, b_0 + b_1, b_0 + b_1 + b_2, \ldots$.

27. Use generating functions to find a formula for $1^2 + 2^2 + 3^2 + \cdots + k^2$. [*Hint*: Show that the generating function for the sequence whose kth term is $1^2 + 2^2 + 3^2 + \cdots + k^2$ is $(x + x^2)/(1 - x)^4$.]

28. Use generating functions to determine the number of ways to choose $3m$ balls from a total of $6m$ balls, a third of which are red, a third of which are white, and a third of which are blue.

29. How many ways are there to select 30 balls from a collection of sufficiently many red, white, and blue balls, under each of the following conditions? Use generating functions to solve these problems.
(a)   The number of red balls selected must be even.
(b)   The number of white balls selected must be odd.
(c)   The number of red balls selected must be even and the number of white balls selected must be odd.
(d)   The number of red balls selected must be even or the number of white balls selected must be odd.

Exploratory Exercises

30. Find out more about generating functions by consulting Liu [198], Pólya, Tarjan, and Woods [237], F. S. Roberts [255], A. Tucker [306], and Watkins [317].

31. Find out more about applying generating functions to partitions by consulting Alder [6,7].

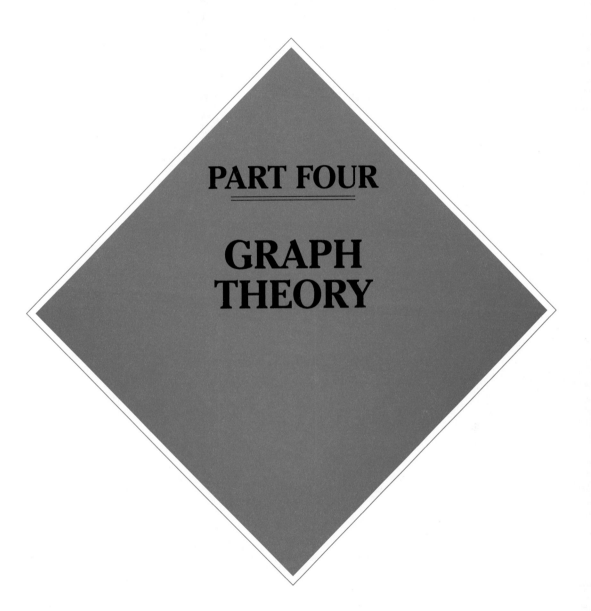

PART FOUR

GRAPH
THEORY

GRAPHS

Perhaps the most useful object in discrete mathematics, especially for computer science and other applications, is a structure called a graph. (Graphs in this sense have nothing to do with graphs of functions $y = f(x)$ in the xy-coordinate plane.) Graphs were first introduced in the eighteenth century, and active research in both pure and applied graph theory continues today. We will study several topics in graph theory, but we can only scratch the surface of this part of discrete mathematics.

In Section 8.1 we describe what a graph is and discuss several applications in which graph models are useful. (Applications continue to play a major role throughout this last part of the book.) We prove some interesting theorems about how one can "walk" through a graph, and look at some algorithmic questions about such walks, in Section 8.2. In Section 8.3 we deal with issues of how best to represent a graph and how to tell whether two graphs are really "different." Section 8.4 deals with drawing pictures of graphs in the plane. Finally, in Section 8.5 we discuss "coloring" a graph, from a theoretical, algorithmic, and applied point of view.

There are many textbooks and other good resources on graph theory that can serve as general references for this chapter and the next two. See Biggs, Lloyd, and Wilson [19] (which provides an interesting historical perspective and includes many of the original papers in the field), Bollobás [25], Bondy and Murty [27], Busacker and Saaty [32], Chachra, Ghare, and Moore [36], Chartrand and Lesniak [41], Christofides [44], Deo [67], Even [78,79], Gibbons [114], Gould [122], Harary [140], Ore

[226], Swamy and Thulasiraman [300], Trudeau [305] (a charming book), Wilson [326], and Wilson and Beineke [327]. In addition, many of the general references listed in Chapter 6 have extensive material on these topics.

SECTION 8.1
BASIC DEFINITIONS IN GRAPH THEORY

One picture is worth more than ten thousand words.

—Chinese proverb

We will begin by defining graphs abstractly, and then we will look at a few of the problems that graphs can model. We will be presenting an inordinate number of definitions for the reader to learn; indeed, some skeptics have called graph theory the "theory of definitions." This situation is unfortunate but unavoidable, since the very versatility of graph models requires that many new concepts and variations on these concepts be defined precisely. Even worse, the terminology in graph theory is not at all standard, so the reader may find some differences in definitions from author to author.

A **graph** $G = (V, E)$ consists of a finite nonempty set V, called the set of **vertices** (the singular is not "vertice" but **vertex**) and a set E of *unordered* pairs of *distinct* vertices, called the set of **edges**. If $e = \{u, v\}$ is an edge of G, we say that e **joins** u and v, that u and v are **endpoints** of e, that e is **incident** to u and to v, and that u and v are **adjacent** (to each other). We usually write the edge $\{u, v\}$ as uv or vu. A **picture** of the graph G is a drawing in the plane in which each vertex is represented by a distinct point, and each edge is represented by a curve or line segment joining the points representing its endpoints.

EXAMPLE 1. A picture of the graph with vertex set $\{a, b, c, n, w\}$ and edge set $\{an, nw, cn, aw\}$ is shown in Figure 8.1. In this graph, vertex a is adjacent to vertices n and w. Edge cn joins c and n, which are its endpoints. Vertex b is incident to no edges. ◆

Some authors call a graph, as we have defined it, a "simple undirected graph," to distinguish it from pseudographs and directed graphs, which we will define shortly. Synonyms for "vertex" include "node" and "point." Synonyms for "edge" include "arc" and "line."

Although we have defined a graph G as an ordered pair (V, E), this is just a formalism to enable us actually to say what a graph *is*, in a mathematically precise

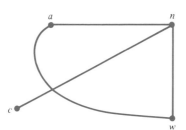

FIGURE 8.1   A graph with five vertices and four edges.

way. The reader should think of a graph, however, not with this formalism, but as a certain *structure*—either as the set of vertices, together with the additional information as to which vertices are adjacent, or as the set of vertices and the set of edges together with the additional information as to which edges are incident to which vertices. The picture enables us to see this structure clearly. (In sections to come we will sometimes impose even more structure on a graph—for example, designating one vertex as a "root" or assigning real numbers called "weights" to the edges. We will modify the picture accordingly to display the extra structure.)

In our definition of graph, edges have a symmetry about them: An edge is the *unordered* pair of its endpoints. We also need to define a structure in which edges are *directed* from one vertex to another.

A **directed graph** (or **digraph**) $G = (V, E)$ consists of a finite nonempty set V, called the set of **vertices**, and a set E of *ordered* pairs of (not necessarily distinct) vertices, called the set of (**directed**) **edges**. (Some authors always call directed edges "arcs" to distinguish them from the undirected edges found in undirected graphs.) If $e = (u, v)$ is a directed edge of G, we say that e **joins** u to v, that u and v are **endpoints** of e (more specifically that u is the **tail** of e and v is the **head** of e), that e is an **incoming** edge to v and e is an **outgoing** edge from u (and **incident** to both u and v), and that u and v are **adjacent** (to each other). We usually write the edge (u, v) as uv. A **picture** of the digraph G is a drawing in the plane in which each vertex is represented by a distinct point, and each edge is represented by a curve or line segment joining the points representing its endpoints, with an arrow on the curve or line segment pointing toward the head of the edge.

EXAMPLE 2.   A picture of the digraph with vertex set $\{a, b, c, d, e, f\}$ and edge set $\{aa, ab, cd, dc, ec\}$ is shown in Figure 8.2. In the digraph, edge ab joins vertex a to vertex b. The head of this edge is b and its tail is a. There is an edge from a to itself; such an edge is called a **loop**. There are a pair of **antiparallel** edges, cd and dc, joining the same endpoints but in the opposite order. Vertices e and c are adjacent, with ec as the only outgoing edge from e. Vertex a has two outgoing edges and one incoming edge. ◆

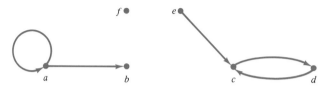

FIGURE 8.2 A digraph with six vertices and five edges.

Directed graphs should be familiar to the reader, since we used pictures of them to represent relations in Section 3.3. Indeed, we could rephrase the definition of digraph and say simply that *a digraph is a finite set together with a relation on that set.*

Graph theory is the study of directed and undirected graphs, both unadorned, as we have seen them so far, and also generalized or endowed with additional structure. There are some profound theorems that can be proved about graphs, and some useful applications of graphs to model phenomena in computer science, the sciences, engineering, the social sciences, and even the humanities.

GRAPH MODELS

The set of problems that can be modeled with a graph or a digraph (often with some additional structure) is virtually endless (see Exercise 43). The vertices can represent, for example, points in space or time, people, animals, species, tasks, states of a machine, atoms, words, or sports teams. Edges might represent roads, telephone lines, communications channels, cause, acquaintance, dependence, precedence, chemical bonds, encounters, flights, tasks, and so on.

EXAMPLE 3. The main roads of the interstate highway system can be modeled with a graph. The vertices of the graph represent major cities, and the edges represent portions of interstate highways between adjacent cities (i.e., with no other major cities on those stretches of highway). A small portion of this rather large graph is shown in Figure 8.3.

The vertices are cities, such as Detroit. There is an edge between Detroit and Toledo, for example, because I-75 runs between them. There is no edge between Flint and Toledo, because no interstate highway directly joins them.

We have here, as in most cases of modeling complex real-world situations, the problem of deciding exactly how to model facets of the situation that do not fit nicely into the intended pattern. For example, the intersection of I-80 and

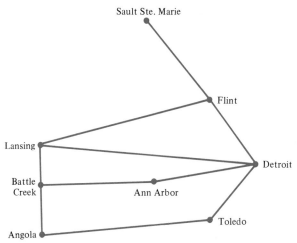

FIGURE 8.3 A portion of the graph model of the interstate highway system.

I-69 occurs in a rural area, not too far from Angola, Indiana. Since there is an interchange there, we probably want a vertex at that point in our model. Therefore, both of the words "major" and "between" in our description of what the vertices and edges should represent have to be compromised if we want a useful model. ◆

EXAMPLE 4.   Theorem 4 of Section 6.4 addressed a problem involving six people at a party. (It is not necessary to have read Section 6.4 to understand this example.) The theorem told us that in any group of six people at a party we would always be able to find three who were mutual acquaintances or three who were mutual strangers. A graph model is ideal here. Let each of the six people at the party be represented by a vertex, and let there be an edge between two distinct vertices if the two people are acquainted with each other. A possible graph that might result is shown in Figure 8.4.

Perhaps A is the host of the party; he knows everyone except B, who was brought along as the "significant other" of C. In the graph theory model, three mutual acquaintances are represented by three vertices, each adjacent to the other two (such as A, C, and D in Figure 8.4)—in short, a "triangle" in the graph. Three mutual strangers are represented by three vertices, none of which is adjacent to either of the others (such as B, D, and E). The graph model lets us "see" the situation in our minds much better than the verbal description. ◆

EXAMPLE 5.   The flow through a computer program can be modeled with a digraph, together with one piece of additional structure, namely an indication as to where the program starts. Vertices represent the executable statements in the program, and there is a directed edge from vertex s_1 to vertex s_2 if statement s_2 can be executed

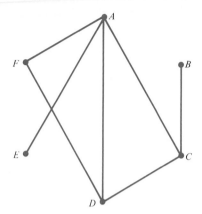

FIGURE 8.4 Graph model of acquaintances at a party.

immediately after statement s_1, as the program runs. One vertex is designated to represent the first executable statement in the program. For example, the following BASIC program can be modeled with the digraph shown in Figure 8.5, in which we have labeled the vertices with the line numbers of the statements they represent.

```
10   REMARK. THIS IS A SILLY PROGRAM TO COMPUTE 55.
20   LET N=1
30   LET X=0
40   IF N>10 THEN 80
50   LET X=X+N
60   LET N=N+1
70   GOTO 40
80   PRINT X
90   END
```

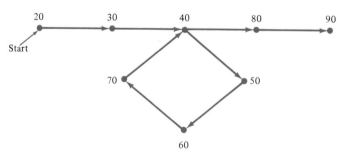

FIGURE 8.5 Graph model of the flow through a BASIC program.

Some programmers like to make such digraphs, which they call "flowcharts" (although the practice of documenting programs with flowcharts has largely been superseded by the widespread use of well-structured programming languages). Typically, the vertices are drawn as very large boxes, rather than as small points, and the program statements themselves are written inside the boxes. Additional information is sometimes indicated along the edges. See Figure 4.2 for another example of a flowchart. ◆

VARIATIONS OF GRAPHS

Sometimes the concepts of graph and digraph are not sufficiently rich to provide the model we need for a particular problem. For example, there may be six flights each day from Detroit to Washington. A digraph in which a directed edge from the vertex for Detroit to the vertex for Washington represents a flight might be an inadequate model; we might find that we want one directed edge for each flight. We need more complicated definitions to make precise the mathematical structures we want to use in cases like this. The concepts are really simpler than the definitions, as the reader will discover when looking at the explanations and examples that follow; nevertheless we need formal definitions in order to describe these concepts precisely.

A **pseudograph** $G = (V, E, i)$ consists of a finite nonempty set V of **vertices**, a finite set E of **edges** (disjoint from V) and a function $i : E \to \left\{ P \subseteq V \mid |P| = 1 \text{ or } 2 \right\}$, called the **incidence function**. If e is an edge, we call the elements of $i(e)$ the **endpoints** of e. If e has just one endpoint, e is called a **loop**. If e_1 and e_2 are two different edges that have the same endpoints (i.e., $i(e_1) = i(e_2)$), then we call e_1 and e_2 **parallel edges**. The words "joins," "adjacent," "incident," and "picture" are defined as they are for graphs. If a pseudograph has no loops, it is called a **multigraph**.

The incidence function makes explicit the relationship between edges and their endpoints. We did not need an incidence function when we defined "graph," since the edge itself, being a pair of vertices, indicated what its endpoints were. Here we cannot use a pair of vertices to represent an edge, because we want *different* edges to be allowed to have the *same* endpoints. Indeed, *we can (and will) consider a graph to be a special case of a pseudograph (or a multigraph) by taking the incidence function to be the obvious one:* $i(uv) = \{u, v\}$ *for each edge uv in the graph*. The only real difference, then, between graphs and these more general structures is that a graph contains neither loops nor parallel edges, a multigraph may contain parallel edges but not loops, and a pseudograph (as its derogatory name suggests) may contain either of these aberrations.

EXAMPLE 6.   Figure 8.6 shows a pseudograph with four vertices and nine edges. Edges e_1 and e_2 are parallel loops at vertex u. Edge e_3 is a single edge joining vertices u and v. Edges e_5, e_6, and e_7 are all parallel. ◆

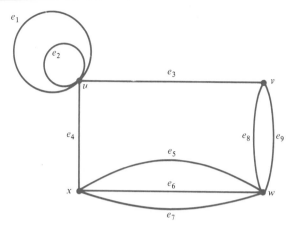

FIGURE 8.6 A pseudograph with four vertices and nine edges.

A **multidigraph** $G = (V, E, h, t)$ consists of a finite nonempty set V of **vertices**, a finite set E of (**directed**) **edges** (disjoint from V) and functions h and t from E to V, called the **head function** and **tail function**, respectively. If e is an edge, then $h(e)$ and $t(e)$ are the **endpoints** (the **head** and **tail**, respectively) of e. If e_1 and e_2 are two different edges that have the same head and the same tail (i.e., $h(e_1) = h(e_2)$ and $t(e_1) = t(e_2)$), then we call e_1 and e_2 **parallel edges**. The words "joins," "adjacent," "incident," "incoming," "outgoing," and "picture" are defined as they are for digraphs.

As in the undirected case, we can think of a digraph as being a multidigraph with no parallel edges, by defining $h(uv) = v$ and $t(uv) = u$ for each edge uv in the digraph. Two edges e_1 and e_2 such that $h(e_1) = t(e_2)$ and $h(e_2) = t(e_1)$ are antiparallel, but they are not parallel unless they are loops.

EXAMPLE 7. Over the course of a football season, the 28 teams in the National Football League each play 16 games, including two games against every other team in their own division. A multidigraph for representing the results of a season of play (assuming that there are no ties) can be formed by letting the 28 vertices represent the teams, and by putting a directed edge from u to v for each game in which the team represented by vertex u beat the team represented by vertex v. If, for example, the Detroit Lions beat the Chicago Bears twice during the season, then there will be parallel edges from the Lions' vertex to the Bears'. Since no team plays itself, there are no loops in this multidigraph. (In reality, ties usually do occur in at least one game during the season, so our model is not quite powerful enough. One possible enhancement to the model would be to allow both directed and undirected edges, and then represent tied games with undirected edges.)

An interesting problem, for which there is no clear-cut solution, is to decide on a ranking of the teams (a total order for the 28 vertices of our model), based on the season's results (as embodied in the multidigraph). ◆

MORE GRAPH THEORY VOCABULARY

Sometimes the number of edges incident to a vertex is important. In an acquaintance graph (see Example 4), the number of incident edges might be a reasonable measure of the popularity of the person represented by a vertex.

The **degree** of a vertex $v \in V$ in a graph $G = (V, E)$, denoted $d(v)$, is the number of edges incident to v, or, equivalently, the number of vertices adjacent to v. In a pseudograph, as well, the degree of a vertex v is the number of edges incident to v, *except that each loop at v is counted twice*. The **degree sequence** of a graph (or pseudograph) is a list, in nonincreasing order, of the degrees of all its vertices.

EXAMPLE 8. In the pseudograph shown in Figure 8.7, vertex a has degree 1, since it is incident to just one edge. Since vertex b is incident to five edges, $d(b) = 5$. Similarly, $d(f) = 0$, since there are no edges incident to f. Note that $d(g) = 3$ because the loop at g counts twice. The degree sequence of this pseudograph is $(5, 3, 3, 3, 3, 1, 0)$. ◆

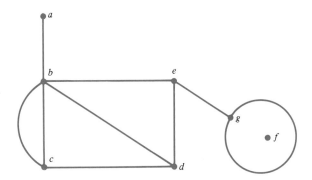

FIGURE 8.7 A pseudograph with degree sequence $(5, 3, 3, 3, 3, 1, 0)$.

So far in this section we have only stated definitions. Let us now prove something about these objects that we have defined and then use the result in an application. We want to see how the degrees of the vertices in a graph relate to the number of edges. Each edge has two endpoints, so each edge contributes 2 to the degrees of vertices. Therefore, if we add up all the degrees, we will have counted each edge twice, so the total will be $2|E|$. This proves our first theorem of graph theory.

THEOREM 1.   *Let G be a graph or a pseudograph with vertex set V and edge set E. Then*

$$\sum_{v \in V} d(v) = 2|E|.$$

COROLLARY.   *Let G be a graph or a pseudograph. Then G has an even number of vertices of odd degree.*

Proof.   By the theorem, the sum of *all* the degrees must be even. This sum is made up of the sum of the degrees of the vertices of even degree (which necessarily is even), and the sum of the degrees of the vertices of odd degree. Since the total is even, the latter sum must be even as well. Now the sum of an *odd* number of odd numbers is odd (see Exercise 35 in Section 5.3); therefore, there must be an *even* number of vertices of odd degree. ◼

EXAMPLE 9.   A new football league will have 14 teams, in two divisions of seven teams each. The league organizers would like to set up a schedule such that each team plays five of the other teams in its own division and three of the teams in the other division. Is such a schedule possible? If so, construct one.

Solution.   First let us model this problem with a graph containing 14 vertices, one for each team; an edge joins each pair of teams that play against each other. Since each team plays eight other teams, the degree of each vertex in our graph will be 8. Theorem 1 and its corollary do not prevent such a graph from existing, so at first glance there appears to be no reason that we should not be able to construct a schedule of the desired type.

Let us look at the problem more closely, though. We have not yet taken into account the division structure that is to be imposed on this league. Suppose that a schedule of the desired type exists. Then we can construct a graph model of one division: a graph with seven vertices, one for each team in that division, again with an edge joining each pair of teams in that division that are scheduled to play against each other. In this model each vertex has degree 5, since each team in the division plays exactly five of the other teams in the division. But this gives us an odd number of vertices (7) of odd degree (5), a contradiction to the corollary. Therefore, such a schedule is impossible. ◆

The number of incident edges is an important concept for digraphs, too. In this case we need to distinguish between incoming edges and outgoing edges.

Let $G = (V, E)$ be a digraph, or let $G = (V, E, h, t)$ be a multidigraph. The **in-degree** of a vertex $v \in V$, denoted $d^-(v)$, is the number of edges incoming to v; and the **out-degree** of v, denoted $d^+(v)$, is the number of edges outgoing from v.

EXAMPLE 10.   In the multidigraph shown in Figure 8.8, vertex A has in-degree 0 and out-degree 1. Similarly, $d^-(B) = d^+(B) = 3$, $d^-(C) = 2$, and $d^+(C) = 1$. ◆

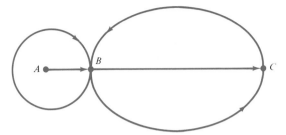

FIGURE 8.8   A multidigraph with in-degrees 0, 3, and 2, and out-degrees 1, 3, and 1.

Analogous to Theorem 2 is the following theorem about the in-degrees and out-degrees in a digraph; the proof is left to the reader (Exercise 12).

THEOREM 2.   *Let G be a digraph or a multidigraph with vertex set V and edge set E. Then*

$$\sum_{v \in V} d^-(v) = \sum_{v \in V} d^+(v) = |E|.$$

We next introduce a class of graphs that are useful in applications that involve objects of two different types.

A graph $G = (V, E)$ is **bipartite** if V can be partitioned into two subsets V_1 and V_2 (i.e., $V = V_1 \cup V_2$ and $V_1 \cap V_2 = \emptyset$), such that every edge $e \in E$ joins a vertex in V_1 and a vertex in V_2. Equivalently, we can say that in a bipartite graph with **parts** V_1 and V_2, no edge joins two vertices in V_1, and no edge joins two vertices in V_2.

It is worth pointing out that the definition does not say that a bipartite graph G must contain every possible edge joining vertices in different parts. Indeed, a bipartite graph need not contain any edges at all, and any graph with no edges is vacuously a bipartite graph. A generalization of the notion of bipartiteness is discussed in Exercise 16, as wel! as (at length) in Section 8.5. We illustrate the notion of a bipartite graph with an application to resource allocation.

EXAMPLE 11.   Four workers—A, B, C, and D—need to be assigned tasks from the set $\{I, J, K, L, M\}$, one task per worker, with no more than one worker per task. (One of the tasks will necessarily go begging.) Not all the workers are trained for all the tasks, however. Suppose that the bipartite graph in Figure 8.9 models the abilities of the workers with respect to the tasks.

The vertex set $V = \{A, B, C, D, I, J, K, L, M\}$ has been partitioned into $V = W \cup T$, where $W = \{A, B, C, D\}$ represents the set of workers and $T = \{I, J, K, L, M\}$ represents the set of tasks. The edge set

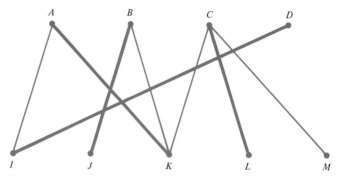

FIGURE 8.9 Bipartite graph showing which workers are capable of performing which tasks.

$$E = \{\, wt \mid \text{worker } w \in W \text{ can perform task } t \in T \,\},$$

contains only (but not all) edges joining vertices in W with vertices in T. As Figure 8.9 shows, for example, worker A can perform tasks I and K, but not tasks J, L, or M.

A typical problem here would be to find a set of assignments such that each worker is given a task. One such assignment is indicated by the heavy lines in Figure 8.9. In graph-theoretic terms we seek a subset of edges such that each vertex $w \in W$ is incident to exactly one edge in the subset, and each vertex $t \in T$ is incident to at most one edge in the subset. Whether such an assignment is possible, and how to find such an assignment efficiently, is one version of the "matching problem" (see Exercises 47–49). ◆

Next, we need to define the notion of one graph's appearing as part of another graph. We say that the graph $G = (V, E)$ is a **subgraph** of the graph $G' = (V', E')$ if $V \subseteq V'$ and $E \subseteq E'$. In this case we also call G' a **supergraph** of G. This definition implies in particular that whenever an edge of a graph is included in a subgraph, both of the endpoints of that edge are included as well.

Similar definitions apply to digraphs, pseudographs, and multigraphs, as well. Details are left to the reader (Exercise 32). We will use the terms "subgraph" and "supergraph" in all these situations.

EXAMPLE 12. The graph shown on the left in Figure 8.10 is called the **Petersen graph** (we will refer to it several times in this chapter). The graph shown on the right is a subgraph of the Petersen graph. It contains seven of the 10 vertices and six of the 15 edges of the Petersen graph. Note that the subgraph we have drawn does not contain, for example, edge ab: Not every edge in the supergraph that joins two vertices in the subgraph need be in the subgraph. Note also that in the picture of the subgraph, the positions of the vertices have been changed; the way a graph is drawn has nothing to do with the graph per se. ◆

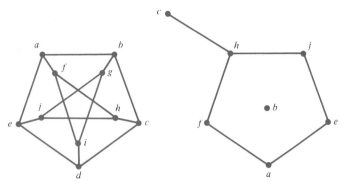

FIGURE 8.10   The Petersen graph (on the left) and one of its subgraphs.

Again let $G = (V, E)$ be a graph. Then the **complement of** G, denoted $\overline{G}$, is the graph $\overline{G} = (V, E')$ with the same vertex set as G whose edges are precisely the edges missing from G; that is, $uv \in E' \leftrightarrow uv \notin E$ for each pair of distinct vertices u and v in V.

EXAMPLE 13.   A graph and its complement are shown in Figure 8.11. Note that if a graph has n vertices and e edges, then its complement has n vertices and $[n(n-1)/2] - e$ edges, since there are $C(n, 2) = n(n-1)/2$ unordered pairs of distinct vertices in G.   ◆

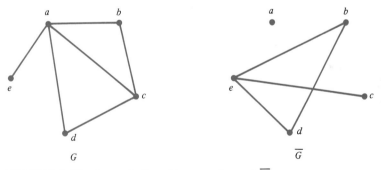

FIGURE 8.11   A graph G and its complement $\overline{G}$.

SOME SPECIAL GRAPHS

Some graphs will occur repeatedly in our discussions. We catalog them here, and give them names and notations, in order to have a store from which to draw.

The **complete graph** on n vertices, for $n \geq 1$, which we denote K_n, is a graph with n vertices and an edge joining every pair of distinct vertices. (Note that we have called this *the* complete graph on n vertices. In fact there are many different complete graphs on n vertices, one for each possible set of vertices, that is, one for each different set of cardinality n. The *structure* of all complete graphs on n vertices, however, is the same, and the structure is all we care about. Thus there is essentially only *one* K_n, and we are justified in our use of the word "the." We will return to this point, in depth, in Section 8.3.) The complete graph K_n has $C(n, 2) = n(n - 1)/2$ edges. Figure 8.12 is a picture of K_5.

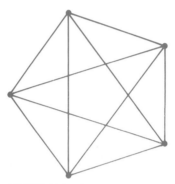

FIGURE 8.12 The complete graph on five vertices, K_5.

The *n*-cycle, for $n \geq 3$, denoted C_n, is the graph with n vertices, v_1, v_2, ..., v_n, and edge set $E = \{v_1 v_2, v_2 v_3, \ldots, v_{n-1} v_n, v_n v_1\}$. Note that C_n has n edges. Figure 8.13 is a picture of C_9.

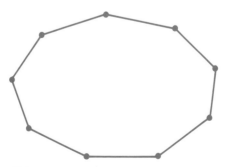

FIGURE 8.13 The 9-cycle.

The **complete bipartite graph** $K_{m,n}$, where m and n are positive integers, is the graph whose vertex set is the union $V = V_1 \cup V_2$ of disjoint sets of cardinalities m and n, respectively, and whose edge set is $\{\, uv \mid u \in V_1 \wedge v \in V_2 \,\}$. In other words, just as a complete graph is a graph that contains all possible edges, so a complete bipartite graph is a bipartite graph that contains all possible edges. Clearly, there are mn edges in $K_{m,n}$. Figure 8.14 is a picture of $K_{5,4}$.

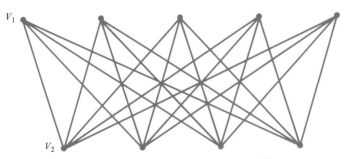

FIGURE 8.14   The complete bipartite graph $K_{5,4}$.

Finally, the **n-cube**, for $n \geq 0$, denoted Q_n, is the graph whose vertex set consists of all bit strings of length n, with an edge joining two vertices if and only if the bit strings disagree in exactly one bit. (For example, in Q_5 the vertex 00110 is adjacent precisely to 10110, 01110, 00010, 00100, and 00111.) Figure 8.15 contains pictures of Q_0, Q_1, Q_2, and Q_3. The name comes from the fact that this graph can be thought of as the vertices and edges of an n-dimensional "cube." The reader should label the vertices in these graphs to show that the pictures conform to the definitions (see Exercise 3). It is not hard to show (Exercise 18) that Q_n has 2^n vertices and $n2^{n-1}$ edges. Exercise 19 asks for a recursive definition of Q_n.

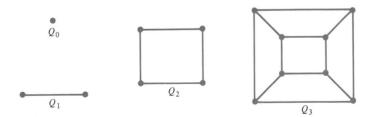

FIGURE 8.15   The n-cubes for $n = 0, 1, 2,$ and 3.

Several additional terms in graph theory and a few other special graphs are introduced in the exercises.

SUMMARY OF DEFINITIONS

adjacent: Two vertices are adjacent if they are joined by an edge.

antiparallel edges in a digraph or multidigraph: a pair of edges e_1 and e_2 such that the head of e_1 is the tail of e_2 and the head of e_2 is the tail of e_1.

bipartite graph G: a graph whose vertex set V can be partitioned into two subsets such that no edge of G joins vertices in the same subset, or a graph together with such a partition.

complement of the graph G: the graph, denoted $\overline{G}$, whose vertex set is the same as the vertex set of G, and whose edge set consists of all unordered pairs of vertices that are not edges in G.

complete bipartite graph on m and n vertices: the graph, denoted $K_{m,n}$, whose vertex set consists of two disjoint sets V_1 and V_2 whose cardinalities are m and n, respectively, and whose edge set consists of all unordered pairs $\{v_1, v_2\}$, where $v_1 \in V_1$ and $v_2 \in V_2$.

complete graph on n vertices: the graph, denoted K_n, whose vertex set has cardinality n and whose edge set consists of all unordered pairs of its vertices.

n-cube, where n is a natural number: the graph, denoted Q_n, whose vertex set consists of all bit strings of length n, and whose edge set consists of all unordered pairs of strings that differ in exactly one position.

n-cycle, where $n \geq 3$: the graph, denoted C_n, with n vertices, $v_1, v_2, \ldots, v_n$, and edge set $E = \{v_1v_2, v_2v_3, \ldots, v_{n-1}v_n, v_nv_1\}$.

degree of the vertex v in a graph or pseudograph: the number of edges incident to v, with each loop counted twice; denoted $d(v)$.

degree sequence of the graph or pseudograph G: a list, in nonincreasing order, of the degrees of the vertices of G.

digraph: a directed graph.

directed edge: an edge in a digraph or multidigraph.

directed graph: a pair $G = (V, E)$ consisting of a nonempty set V (the vertices) and a set E of ordered pairs of vertices (the edges).

edge: an element of the set E in a graph, digraph, pseudograph, or multidigraph, usually denoted uv, where u and v are its endpoints; in the case of a digraph or multidigraph (directed edge), the tail is listed before the head in the notation uv.

endpoints of the edge e: in a graph, the vertices comprising the unordered pair e; in a digraph, the vertices comprising the ordered pair e; in a pseudograph, the elements of $i(e)$; in a multidigraph, $h(e)$ and $t(e)$.

(undirected) graph: a pair $G = (V, E)$ consisting of a finite nonempty set V (the vertices) and a set E of unordered pairs of distinct vertices (the edges).

graph theory: the study of graphs and their variations.

head of edge uv: the vertex v.

head function in the multidigraph (V, E, h, t): the function h.

incidence function in the pseudograph (V, E, i): the function i.

incident: An edge is incident to its endpoints.

incoming edge to the vertex v: an edge whose head is v.

in-degree of the vertex v in a digraph or multidigraph: the number of edges incoming to v, denoted $d^-(v)$.

join: Edge uv joins vertex u to vertex v.

loop: an edge whose two endpoints are the same.

multidigraph: a 4-tuple $G = (V, E, h, t)$ consisting of a finite nonempty set V (the vertices), a finite set E disjoint from V (the edges), a function $h: E \to V$ (the head function), and a function $t: E \to V$ (the tail function).

multigraph: a pseudograph with no loops.

out-degree of the vertex v in a digraph or multidigraph: the number of edges outgoing from v, denoted $d^+(v)$.

outgoing edge from the vertex v: an edge whose tail is v.

parallel edges in a multidigraph: edges that have the same head and the same tail.

parallel edges in a pseudograph: edges that have the same endpoints.

part of the bipartite graph G (whose vertex set has been partitioned into two subsets such that no edge of G joins vertices in the same subset): one of these two subsets.

Petersen graph: the graph pictured in the left-hand side of Figure 8.10.

picture of the graph, digraph, pseudograph, or multidigraph G: a drawing in the plane, in which distinct vertices of G are represented by distinct points, and each edge uv of G is represented by a curve or line segment from u to v (with an arrow pointing toward the head of the edge in the case of digraphs and multidigraphs).

pseudograph: a triple $G = (V, E, i)$ consisting of a finite nonempty set V (the vertices), a finite set E disjoint from V (the edges), and a function $i : E \to \{ P \subseteq V \mid |P| = 1 \text{ or } 2 \}$ (the incidence function).

subgraph: The graph $G = (V, E)$ is a subgraph of the graph $G' = (V', E')$ if and only if $V \subseteq V'$ and $E \subseteq E'$.

supergraph: G is a supergraph of G' if and only if G' is a subgraph of G.

tail of edge uv: the vertex u.

tail function in the multidigraph (V, E, h, t): the function t.

vertex: an element of the set V in a graph, digraph, pseudograph, or multidigraph.

vertices: plural of "vertex."

EXERCISES

1. Draw a picture of each of the following graphs.
   (a) $V = \{p, q, r, s, t\}$, and each pair of distinct vertices is joined by an edge, except for $\{p, q\}$ and $\{r, s\}$.

(b) $V = \{0, 1, 2, 3, 4, 5, 6, 7, 8\}$ and $E = \{\, uv \mid u \neq v \wedge \exists w \in \mathbf{N} \colon u \cdot v = 3w \,\}$.

(c) $V = \{0, 1, 2, 3, 4, 5, 6, 7, 8\}$ and two distinct vertices are adjacent if and only if their sum is a multiple of 3.

(d)   The complete graph on four vertices.

(e) K_6.

(f) $K_{3,4}$.

2. Draw a picture of a multidigraph to model the following situation. Detroit, Washington, Boston, and Miami are served by Prime Airlines. Each day there are two flights from Detroit to each of the other cities, two flights from each of the other cities to Detroit, one flight from Miami to Washington, one from Washington to Boston, and one from Boston to Miami; and there is a sightseeing flight over the Everglades, leaving from and returning to Miami.

3. Verify that the pictures of Q_0, Q_1, Q_2, and Q_3 shown in Figure 8.15 are correct, by labeling their vertices with bit strings as described in the definition.

4. Determine the degrees of all the vertices in the following graphs and pseudographs.

(a)   Figure 8.1.

(b)   Figure 8.6.

(c) K_{22}.

(d) $K_{10,20}$.

5. Determine the in-degrees and out-degrees of all the vertices in the following digraphs and multidigraphs.

(a)   Figure 8.2.

(b)   The answer to Exercise 2.

(c)   The "less than" relation on the set $\{1, 2, 3, 4, 5, 6, 7\}$.

6. The complete bipartite graph $K_{1,n}$ is called an **n-pointed star**. Draw a picture of $K_{1,n}$, showing why this name was applied to this graph.

7. What is the relationship between a graph G and $\overline{\overline{G}}$ (the complement of the complement of G)?

8. Draw a picture of a pseudograph with five vertices and 12 edges, such that there are exactly three loops, exactly two of which are parallel, and such that four of the degrees of the vertices are 8, 7, 5, and 3.

9. Draw a picture of the bipartite graph G that models the "is an element of" relation on the set $\{1, 2, 3\}$ and its power set. Specifically, one part of G is $V_1 = \{1, 2, 3\}$, the other part of G is $V_2 = \mathcal{P}(\{1, 2, 3\})$, and there is an edge from $x \in V_1$ to $A \in V_2$ if and only if $x \in A$.

10. Draw a picture of a digraph to model the possible flows through the following bit of pseudocode. Include one vertex for each numbered statement.

```
(1)   N ← 1
(2)   while true do
        begin
(3)       N ← N + 1
(4)       X ← N
(5)       while X ≠ 1 do
            begin
(6)           if X is even
(7)             then X ← X/2
(8)             else X ← 3X + 1
(9)           if X = N
(10)            then print(N)
            end
        end
```

11. An **isolated vertex** in a graph or pseudograph is a vertex with degree 0.
    (a) Draw a picture of a graph with six vertices, one of which is isolated, and nine edges.
    (b) Prove or disprove that a graph with six vertices and 11 edges cannot have any isolated vertices.
    (c) Prove or disprove that a pseudograph with six vertices and 11 edges cannot have any isolated vertices.

12. Prove Theorem 2.

13. Indicate whether each of the following statements is true or false.
    (a) C_5 is a subgraph of the Petersen graph (see Figure 8.10).
    (b) C_4 is a subgraph of the Petersen graph.
    (c) K_5 is a subgraph of the Petersen graph.
    (d) K_3 is a subgraph of $K_{10,10}$.

14. Prove that if $(d_1, d_2, \ldots, d_n)$ is any nonempty sequence of n nonincreasing natural numbers whose sum is even, then this n-tuple is the degree sequence of some pseudograph with n vertices. [*Hint*: Use induction on n or on $\sum_{i=1}^{n} d_i$.]

15. For each of the following sequences, either draw a graph with this degree sequence or show that no such graph exists.
    (a) $(4, 4, 4, 3, 2, 1)$.
    (b) $(4, 4, 3, 3, 2, 1)$.
    (c) $(4, 4, 4, 4, 1, 1)$.
    (d) $(5, 4, 3, 2, 2)$.
    (e) $(1, 1, 1, 1, 1, 1, 1, 1)$.
    (f) $(2, 2, 2, 2, 2, 2, 2)$.
    (g) $(3, 3, 3, 3, 3, 3, 3, 3)$.

16. Shown here is a picture of the **complete tripartite graph** $K_{2,2,1}$ with parts of sizes 2, 2, and 1.

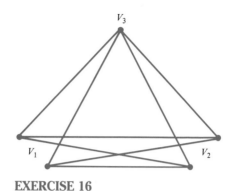

EXERCISE 16

Using this picture and the discussion of bipartite graphs as guides,
(a) Define what it should mean for a graph G to be tripartite.
(b) Define the complete tripartite graph $K_{l,m,n}$.
(c) Determine how many edges there are in $K_{l,m,n}$.

17. Show that in any graph with at least two vertices there are two vertices with the same degree. Is the same conclusion true for multigraphs?

18. Show that the n-cube Q_n has 2^n vertices and $n2^{n-1}$ edges.

19. Give a recursive definition of the n-cubes. [*Hint*: Show how to construct Q_{n+1} from Q_n.]

20. Draw a picture of the graph whose vertex set consists of the bit strings of length 5 that represent prime numbers in binary notation (00010, 00011, 00101, ..., 11111), such that two distinct vertices are adjacent if and only if the bit strings differ in at most one bit. (A graph like this, in which an edge joins two distinct vertices if and only if the vertices are not too different, is called a **similarity graph**.)

21. The **n-spoked wheel** is a graph consisting of an n-cycle with one additional vertex, which is joined to every vertex of the n-cycle.
(a) Draw a picture of the five-spoked wheel.
(b) How many edges does the n-spoked wheel have?
(c) What is the degree sequence of the n-spoked wheel?

22. An **intersection graph** is a graph, each vertex of which is a set; two distinct vertices are adjacent if and only if the sets are not disjoint.
(a) Draw a picture of the intersection graph whose vertices are all the two-element subsets of $\{1, 2, 3, 4\}$.
(b) Draw a picture of the complement of the intersection graph in which the vertices are all the two-element subsets of the set $\{1, 2, 3, 4, 5\}$. [*Hint*: Put appropriate labels on the vertices of the Petersen graph.]

23. Describe in English the following graphs.
   (a) $\overline{K}_n$
   (b) $\overline{K}_{m,n}$

24. What is the maximum possible number of edges in a bipartite graph with n vertices? Justify your answer.

25. Consider the following graph model of one aspect of the game of chess. The vertices in G are the 64 squares of the chessboard. Two distinct vertices (i.e., squares) are adjacent in G if and only if a queen placed on one of the squares attacks the other square (in the absence of intervening pieces). In other words, two distinct squares are adjacent in G if and only if they are in the same row, column, or diagonal of the chessboard. Determine the degree sequence of this graph and, thereby, the number of edges.

26. Repeat Exercise 25 for the knight.

27. Projects, consisting of several tasks, can be modeled by digraphs or multidigraphs in at least two ways. Consider the project of completing the fall term for Sophomore Sam. He has to perform the following tasks: (1) do the research for a term paper in history, (2) rent from a local store a word processor with which to write his history paper, (3) write a draft of the history paper, (4) edit the draft of the history paper and prepare the final copy, (5) return the word processor, (6) study for his history exam, (7) take his history exam, (8) do the last exercise set for discrete math, (9) study for his discrete math exam, (10) take his discrete math exam, (11) celebrate at the local pub after finishing all his work (papers and tests), and (12) fly home for the holidays.
   (a)   Draw a digraph model of this project by letting the vertices be the tasks, with an edge uv to represent the fact that task u must be finished before task v can be started. Omit edges whose existence is implied by the transitivity of the "must be finished before" relation.
   (b)   Draw a multidigraph model of this project by letting the vertices be certain points in time, with an edge for each task, where the tail of the edge represents the point in time at which the task may be started, and the head of the edge represents the point in time by which the task must be finished. You may need edges representing "dummy tasks" to make the logic come out right.

28. It is possible to define infinite graphs, simply by removing the word "finite" from the definition of graph that we have given. An infinite graph is called **locally finite** if the degree of each vertex is finite. Consider the infinite graph whose vertex set is the set of integers, with an edge between two distinct vertices u and v if and only if $|u - v| < 4$.
   (a)   Draw a picture of a portion of this graph.
   (b)   Determine whether this graph is locally finite.
   (c)   Determine the values of n for which K_n is a subgraph of this graph.

(d)   Determine the values of n for which C_n is a subgraph of this graph.

29. Repeat parts (b), (c), and (d) of Exercise 28 for the following infinite graphs.
    (a)   The vertices are all the points in the plane; two vertices are adjacent if and only if they are exactly 1 inch apart.
    (b)   The vertices are all the finite subsets of **N**; two distinct vertices are adjacent if and only if the subsets intersect.
    (c)   The graph consists of the union of disjoint copies of K_1, K_2, K_3, K_4, K_5, ...

30. A bipartite graph G can be used to model the relationship between students and courses at a certain university. Let the vertices in one part (call it S) represent the students enrolled this semester, and let the vertices in the other part (call it C) represent the courses being offered this semester. Vertex $s \in S$ is adjacent to $c \in C$ if s is taking c this semester. What is the real significance of each of the following graph theoretic aspects of G?
    (a)   The degree of vertex $s \in S$.
    (b)   The degree of vertex $c \in C$.
    (c)   The fact that two particular vertices $s_1 \in S$ and $s_2 \in S$ are both adjacent to the same vertex $c \in C$.
    (d)   The fact that no edge incident to $c_1 \in C$ has an endpoint in common with any edge incident to $c_2 \in C$.
    (e)   The maximum degree of the vertices in this graph (realistically).

31. The illustration shows the **double star** $S(5, 4)$.

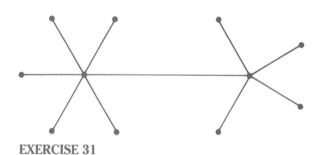

EXERCISE 31

    (a)   Write out a complete and precise definition of the double star $S(m, n)$, where m and n are natural numbers.
    (b)   Show that every double star is a bipartite graph.

32. Write out a careful definition of "subgraph" for each of the following categories.
    (a)   Digraphs.
    (b)   Pseudographs.

    (c)   Multidigraphs.

33. Show that the n-cube Q_n is bipartite.

34. The **union** of two graphs $G_1 = (V_1, E_1)$ and $G_2 = (V_2, E_2)$ is the graph, denoted $G_1 \cup G_2$, whose vertex set is $V_1 \cup V_2$ and whose edge set is $E_1 \cup E_2$.
    (a)   Draw a picture of the union of the 5-cycle whose vertices are (a, b, c, d, e) in that order and the 4-cycle whose vertices are (a, f, c, d) in that order.
    (b)   What can be said about $G \cup \overline{G}$ for any graph G?
    (c)   Prove or disprove that the union of two bipartite graphs is bipartite.

35. In each case, draw a picture to illustrate a graph with the given property, and write out the property symbolically, making sure to use quantifiers correctly.
    (a)   The graph $G = (V, E)$ has an isolated vertex (i.e., a vertex incident to no edges).
    (b)   The complete graph K_3 is a subgraph of $G = (V, E)$, but K_4 is not.
    (c)   The graph $G = (V, E)$ is bipartite with parts V_1 and V_2.
    (d)   The graph $G = (V, E)$ has a subgraph with the same vertex set as G in which every vertex has degree 1.

Challenging Exercises

36. A **tournament** is a digraph whose edge set consists of exactly one member of $\{uv, vu\}$ for each pair of distinct vertices u and v. (Thus a tournament can be used to model the results of a round-robin tournament in which every team plays every other team, with no ties possible; a directed edge uv indicates that u beat v.)
    (a)   Draw a picture of a tournament on five vertices with out-degrees 4, 2, 2, 1, and 1.
    (b)   A **king** in a tournament G is a vertex u such that for every other vertex v, either uv is an edge of G, or, for some vertex w, both uw and wv are edges of G. Show that every tournament has a king.
    (c)   A **triangular standoff** in a tournament G is a subgraph of G consisting of three vertices and three edges, each vertex having in-degree 1 in the subgraph. Suppose that G is a tournament in which for each vertex v we have $d^-(v) = d^+(v) = 6$. Determine how many triangular standoffs there are in G.

37. Characterize in a clear and satisfying way all graphs whose vertices all have degree 2.

38. Find a formula (it need not be in closed form) for the number of subgraphs that the complete graph K_n has. [*Hint*: K_4 has 112 subgraphs.]

39. Write a recursive algorithm to find, given a graph G, the largest r such that K_r is a subgraph of G.

40. Suppose that n children holding loaded water pistols are standing in an open field, no three of them in a line, such that all the distances between pairs of

them are distinct. At a given signal, each child shoots the child closest to him
or her with water.
(a)   Show that if n is even, then it is possible for every child to get wet.
(b)   Show that if n is odd, then at least one child must remain dry.

Exploratory Exercises

41. If G is a graph, then the **line graph** of G, denoted $L(G)$, is the graph whose
    vertex set is the set of *edges* of G, with two distinct vertices in $L(G)$ adjacent
    if and only if the edges of G they represent share a common endpoint.
    (a)   Draw pictures of the line graphs of several graphs.
    (b)   Find a formula for the number of edges in $L(G)$, in terms of some
          parameters of G.
    (c)   Find some other interesting facts about line graphs. (For example, is
          there any subgraph that they must or must not have?)

42. Investigate the problem of determining when a sequence of natural numbers
    is the degree sequence of a graph, and of constructing such a graph. Consult
    Wolfe [332] for a solution.

43. Find out more about mathematical models using graphs by consulting Char-
    trand [39], Gardner [107], Grecos [127], Johnsonbaugh and Murato [165], and
    F. S. Roberts [256,257].

44. Find out more about kings in tournaments (see Exercise 36b) by consulting
    Maurer [208].

45. Find out more about the Petersen graph by consulting Chartrand and Wilson
    [42].

46. A **regular graph** is a graph whose degree sequence is constant.
    (a)   Draw a picture of a regular graph other than K_{n+1} or $K_{n,n}$ in which
          the degree of each vertex is n, for $n = 1, 2, 3, 4, 5$.
    (b)   Consult Chartrand, Erdős, and Oellermann [40] to find out about graphs
          at the other extreme in terms of the degrees of the vertices.

47. A **matching** in a graph G is a subset of the edges of G such that every vertex
    of G is incident to exactly one edge in the subset.
    (a)   Draw pictures of two graphs with six vertices and eight edges, one that
          has a matching and one that does not.
    (b)   Show that when n is an even positive integer the edge set of K_n is the
          union of $n-1$ disjoint matchings, and explain what this result has to do
          with round-robin tournaments; or consult Freund [91] and Wallis [315].
    (c)   Delve into some of the references on graph theory to find out about
          Tutte's theorem, which gives necessary and sufficient conditions for a
          graph to have a matching.

48. See Exercise 47 for the definition of a matching. Let G be a bipartite graph with parts of equal cardinality. Delve into some of the references on graph theory to find necessary and sufficient conditions for G to have a matching.

49. Find out about applications of graph theory (in particular, matchings in bipartite graphs—see Exercise 48) to marriage and to college admissions by consulting Gale and Shapley [94].

50. Find out how graphs can be used to model algebraic structures by consulting Grossman and Magnus [128].

51. An **interval graph** is the intersection graph (see Exercise 22) of a set of intervals of real numbers.
   (a)   Show that $K_{1,3}$ is an interval graph.
   (b)   Show that C_4 cannot be an interval graph.
   (c)   Find out more about interval graphs by consulting Harary and Kabell [141].

52. Explore the following two-person game, in which the players play alternately. A graph is given initially. When it is his or her turn to move, a player chooses a vertex v of the current graph, erases every edge incident to v, and draws an edge from v to every vertex to which v was not adjacent in the current graph. The resulting graph then becomes the current graph. Each vertex can be chosen at most once in a game. The winner is the player who can create an isolated vertex in the graph (not necessarily the vertex he or she chose). Consult Ringeisen [250] for an analysis.

SECTION 8.2
TRAVELING THROUGH A GRAPH

"The road to the City of Emeralds is paved with yellow brick," said the Witch; "so you cannot miss it."

—L. Frank Baum, *The Wonderful Wizard of Oz*

On one level a graph or digraph can be viewed as a set of connections: An edge uv connects vertex u with vertex v. If another edge vw joins vertex v to vertex w, then vertex u is also, in a sense, connected to vertex w, via vertex v. This idea is particularly easy to appreciate if the vertices of a graph represent physical locations and the edges represent roads. In this section we make precise these ideas of "traveling" along the edges of a graph. As in Section 8.1, there are many new words to learn, and the terminology varies considerably from book to book. The terminology we use here is widely used among mathematicians engaged in research in graph theory.

Let $G = (V, E)$ be a graph (or, more generally, let $G = (V, E, i)$ be a pseudograph). Then a **walk** W of length $n \geq 0$ in G is a sequence

$$v_0, e_1, v_1, e_2, v_2, \ldots, v_{n-1}, e_n, v_n,$$

such that each $v_k \in V$, each $e_k \in E$, and for each k from 1 to n, edge e_k joins vertices v_{k-1} and v_k. The walk W is said to **join** v_0 and v_n, going from v_0 to v_n. If W satisfies certain other conditions, then W is given additional descriptive names:

- If the edges in W are all distinct, then W is called a **trail**.
- If the vertices in W are all distinct, then W is called a **path**.
- If $v_0 = v_n$ and $n \geq 1$, then W is said to be **closed**.
- If W is a closed trail in which $v_1, v_2, \ldots, v_n$ are all distinct, then W is called a **cycle**.

Since an edge in a graph is uniquely determined by its endpoints, a walk $v_0, e_1, v_1, e_2, v_2, \ldots, v_{n-1}, e_n, v_n$ *in a graph* can be denoted simply as $v_0, v_1, v_2, \ldots, v_n$.

EXAMPLE 1. Let G be the graph pictured in Figure 8.16.

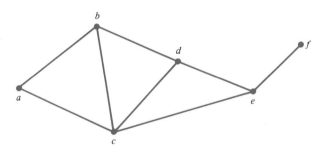

FIGURE 8.16 A graph G in which to walk.

(a) a, b, e, d is not a walk, since be is not an edge of G.

(b) b, d, e, d is a walk of length 3 from b to d. It is not a trail, since edge de occurs twice; and it is not closed since $b \neq d$.

(c) f, e, f is a closed walk of length 2, but it is not a trail.

(d) a, b, d, c is a path of length 3.

(e) a, b, c, e, d, c, a is a closed trail of length 6, but it is not a cycle (note that closed trails are allowed to "cross" themselves).

(f) b, c, d, b is a cycle of length 3. Note that viewed as a subgraph of G, this is just a 3-cycle, or **triangle**. ◆

The definitions of "walk," "trail," "path," "closed," and "cycle" can be extended to digraphs and multidigraphs, with travel permitted along a directed edge only from its tail to its head (in other words, in the direction indicated by the arrow in a picture of a digraph or multidigraph).

EXAMPLE 2.   Consider the multidigraph in Figure 8.17.

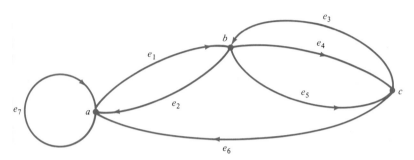

FIGURE 8.17   A multidigraph in which to walk.

(a) b, e_2, a, e_6, c is not a walk, since edge e_6 goes from c to a, not from a to c.

(b) $a, e_1, b, e_2, a, e_7, a$ is a closed trail of length 3 from a to a.

(c) $a, e_1, b, e_4, c, e_3, b, e_5, c$ is a trail of length 4, but it is not a path. Note that it would be ambiguous if we left out the names of the edges and wrote simply a, b, c, b, c; the latter notation does not specify either time which edge (e_4 or e_5) we are to use in traveling from b to c.

(d) a, e_7, a is a 1-cycle, and b, e_4, c, e_3, b is a 2-cycle.   ◆

Sometimes we consider a walk (especially when it is a cycle) to *be* its set of edges. For example, we say that C_n (the graph we called the n-cycle in Section 8.1) has a unique cycle (namely all of itself), even though it really has $2n$ distinct cycles, since a cycle in C_n can start at any of the n vertices and proceed in either of two directions.

CONNECTEDNESS

Now that we have defined the basic concepts of traveling along the edges of a graph, we can say what it means for a graph to be connected. In a communications network application, for example, it is important to know whether it is possible for each object in the network to communicate with all the others, possibly by going through some intermediaries; to discuss this problem, we need the notion of connectedness. We say that a graph (or pseudograph) G is **connected** if, whenever

u and v are vertices of G, there is a walk from u to v. A **component** of G is a *maximal* connected subgraph of G, that is, a connected subgraph of G that is not a subgraph of any other connected subgraph of G. For example, a communications network in which every object is ultimately able to communicate with every other object is connected; it has exactly one component.

EXAMPLE 3. The graph in Figure 8.18 is not connected. In fact, it has three components, one consisting of just the vertex h, another consisting of vertices f and g and the edge joining them, and the third consisting of the remaining five vertices and six edges. The subgraph consisting of the triangle (vertices a, b, and c and the three edges joining pairs of these vertices) is a connected subgraph, but it is not a component since it is not maximal: It is properly contained in the third component mentioned above. ◆

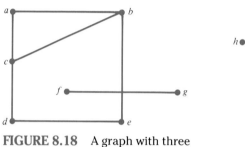

FIGURE 8.18　A graph with three components.

Intuitively, a graph is connected if, when you pick up a vertex v and leave the room, the whole graph comes with you; in any case, that portion of the graph that goes with you is the component containing v. Another way to look at the notions of connectedness and components is to observe that the relation R defined on the set of vertices of a graph G by

$$uRv \text{ if and only if there is a walk from } u \text{ to } v$$

is an equivalence relation (see Exercise 8). Each equivalence class of vertices (together with all the edges joining pairs of vertices in the class) is a component of G.

The reader should, at this point, be asking, "But how does one go about finding out whether a graph is connected? It seems rather cumbersome to go around looking for a walk between each pair of vertices." We will postpone answering this question until Sections 8.3 and 9.2. For now, we will look only at graphs that are small enough so that the "pick up a vertex and leave the room" heuristic enables one to determine connectedness.

The notion of connectedness becomes more complex for digraphs, since the relation R given above need no longer be an equivalence relation: It need not be symmetric. We say that a digraph (or multidigraph) G is **strongly connected** if, whenever u and v are vertices of G, there is a walk from u to v and a walk from v to u. A **strong component** of G is a maximal strongly connected subgraph of G.

EXAMPLE 4. The graph pictured in Figure 8.19 is not strongly connected, even though the multigraph one obtains by ignoring the directions on the edges is connected. This digraph has four strong components. Vertices a, b, c, and d (and the edges incident to pairs of these vertices) form one component, since it is possible to travel from each of these vertices to each of the other vertices in this component. Vertex e and vertex f are each in a component by themselves. Finally, vertices g, h, and i and the three edges gi, ih, and hg form a component. Note that some edges (ce, for example) are not in any strong component. ◆

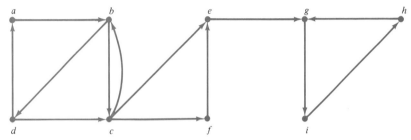

FIGURE 8.19 A digraph with four strong components.

Algorithms for determining the strong components of a digraph are rather complicated (see Exercise 34 in Section 9.2).

Let us prove a few theorems about walks, both to gain more experience in working with graph-theoretic concepts and to provide us with some lemmas for later use. These theorems hold for all varieties: graphs, pseudographs, digraphs, and multidigraphs.

THEOREM 1. *If G contains a walk from u to v, then G contains a path from u to v.*

Proof. Let $u = v_0, e_1, v_1, \ldots, e_n, v_n = v$ be a given walk from u to v. If all the vertices in this walk are distinct, then we are done. If not, we will show how to find a shorter walk from u to v. By repeating this shortening process as long as necessary (but only finitely many times, since the walk had only finite length to begin with), we obtain the desired path.

The shortening is easily accomplished: Let v_i and v_j, with $i < j$, be a pair of equal vertices in the walk. Then by excising the portion of the walk from v_i to v_j (excising v_i but not v_j), we obtain the walk

$$u = v_0, e_1, v_1, \ldots, v_{i-1}, e_i, v_j, e_{j+1}, v_{j+1}, \ldots, e_n, v_n = v.$$

Since at least one edge (e_j) has been deleted, this walk is shorter than the original. (The reader should draw a picture of what is going on and verify that this is indeed still a walk from u to v, even if $u = v$ or $i = 0$ or $j = n$.)

This has really been a proof by contradiction, in the spirit of a proof by mathematical induction. In particular we have shown that the *shortest* walk from u to v must necessarily be a path.   ■

THEOREM 2.   *If G contains a closed trail starting and ending at v, then G contains a cycle starting and ending at v.*

Proof.   The proof is similar to the proof of Theorem 1 and is left as Exercise 15 for the reader.   ■

EULER TOURS

If the streets of a town are modeled with a graph (streets are the edges and intersections are the vertices), an everyday concern for the town's safety officer can be formulated as a graph-theoretic problem. The safety officer would like to patrol every street, starting and ending her tour at the police station. To avoid traveling more than necessary, she wants to traverse each street only once. The problem is to determine if such a tour is possible and, if so, to find such a tour. We will show how to solve this problem in graph-theoretic terms. (See Exercise 35 for other important applications.)

We need another definition in order to deal with this problem. Let G be a connected graph (or pseudograph). An **Euler tour** in G is a closed trail that contains each edge exactly once.

EXAMPLE 5.   Consider the multigraph G pictured in Figure 8.20. There is an Euler tour in G as follows (we abuse the notation somewhat by listing only the vertices and omitting the edges—the ambiguity is minor, however, since we obviously mean to traverse one of the edges between a and g the first time and the other edge between a and g the second time):

$$a, h, c, b, a, g, f, e, d, c, f, j, h, i, j, d, i, c, g, a.$$

On the other hand, K_4 (the complete graph with four vertices) has no Euler tour, as the reader should verify by an ad hoc argument.   ◆

The problem of finding Euler tours is one of the oldest problems in graph theory. It turns out to be a very satisfying problem because there is an easily checked condition which is necessary and sufficient for the existence of an Euler

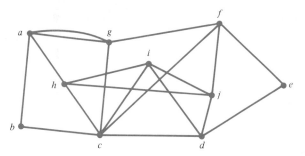

FIGURE 8.20   A multigraph with an Euler tour.

tour, and the proof of the sufficiency of the condition gives an algorithm for finding an Euler tour when one exists.

THEOREM 3.   *A connected graph (or pseudograph) G with at least one edge has an Euler tour if and only if the degree of every vertex in G is even.*

Proof.   Let us prove the necessity of this condition first. This half of the proof dates back to the great Swiss mathematician L. Euler in 1736 (see Exercise 20).

Suppose that we are given an Euler tour C in G, say starting and ending at a vertex u. Imagine a counter at each vertex, initially set to zero for each vertex. Let us travel along C, adding one to the counter at a vertex whenever we find an edge–vertex incidence in our travels. Thus we begin by adding one to the counter at u as we begin our tour. Thereafter for each vertex v in the tour (before we come to the final vertex in C), we add two to the counter at v, since we find one incidence as we come into v along some edge and another incidence as we leave v along a different edge. Finally, we add one more to the counter at u when we return to u for the completion of the tour. Thus, for every vertex other than u, we incremented the counter by 2 whenever we incremented it at all, so the counter must end up even; and the counter at u must also end up even, since it was incremented by 1 twice (once at the beginning of the tour and once at the end) and perhaps by 2 some number of times. Since the Euler tour C contained each edge exactly once, the counters clearly give the degrees of the vertices, and the "only if" part of the theorem is proved.

The "if" part is a little harder. [We will follow along an example, using the graph in Figure 8.20, as we go.] Let us try to construct an Euler tour C in G, given that G is connected, has at least one edge, and contains only vertices of even degree. We start at any vertex u and travel along the edges of the graph *until we cannot travel any further*, never using an edge we have already used. In doing so, we have obviously constructed a trail. Since G is connected and has at least one edge, this trail has length at least 1. Furthermore, since each vertex has even degree, we were never without a new edge to leave a vertex by, once we had come to that vertex, unless we had returned to our starting point u. Thus the trail we formed is actually a closed trail starting and ending at u; let us call that closed trail C_1. [In Figure 8.20 suppose that we start at vertex a and choose the following trail: a, h, c, b, a, g, a. At this point, we are stuck, because all the edges incident to a have been used. Thus C_1 is the closed trail a, h, c, b, a, g, a.]

If C_1 contains all the edges of G, we are done: C_1 is the desired Euler tour. If not, there must be vertices incident to edges not in C_1, and since G is connected, one of these vertices must also be incident to edges that are in C_1. Let v be such a vertex. We now repeat the process of constructing a trail, starting at v, using no edge more than once, and using no edge already in C_1. Again, since the degree of every vertex is even, we will not be forced to stop until we have returned to v, completing a closed trail C'. Now—and this is the key idea in the proof—we splice closed trail C' into closed trail C_1 to obtain a new closed trail C_2. Specifically, C_2 begins at u, proceeds along C_1 until it comes to vertex v (for the first time), then traverses C' completely, returning to v, and finally continues with C_1, returning to u. [In our example, suppose that we chose g as the starting point of C', and found the closed trail $g, f, e, d, c, f, j, d, i, c, g$. We splice this into C_1 to obtain C_2, namely

$$a, h, c, b, a, g, (f, e, d, c, f, j, d, i, c, g), a,$$

where we have indicated the insertion in parentheses.]

Again, if all the edges have been used, C_2 is the desired Euler tour; if not, we repeat the previous step, finding another closed trail, starting and ending at some vertex on C_2, to splice into C_2, thereby obtaining an even larger closed trail C_3. [In our example, we splice j, h, i, j into C_2 to obtain the Euler tour given in Example 5.] This step is repeated as often as necessary until all the edges have been included. This completes the constructive proof of the sufficiency part of the theorem. (Like the proof of Theorem 1, this proof is in some sense a proof by contradiction: If the *longest* closed trail in the graph does not use every edge (i.e., is not an Euler tour), we have derived a contradiction by finding a longer one.) ◼

It is easy enough to turn the proof into an algorithm for finding Euler tours, presented as Algorithm 1. To make it slightly easier to write down the closed trails we seek, we will assume that G is a graph, containing no loops or parallel edges. (It should be clear that this restriction is not really necessary, if we are willing to list our trails by including the names of the edges as well as the names of the vertices.) We assume that the vertices are labeled 1 through n; this makes it easy for us to state how the algorithm decides in a deterministic way which vertex to use next, among many possible choices, as it seeks to extend the closed trails it is "growing."

To analyze the efficiency of Algorithm 1 would require a detailed look at how some of the assignment statements are actually carried out. This, in turn, would

```
procedure euler(G : graph)
    {assume that G is connected, has at least one edge, and has vertices numbered
        1, 2, ..., n}
        H ← G {H will contain what remains of G}
        C ← 1 {initially, the tour C we are "growing" is trivial}
        while H contains edges do
            begin
                v ← lowest numbered vertex that appears in C and has edges in H
                        incident to it
                C' ← v {initially, the closed trail C' is the trivial walk with no edges}
                w ← v {w is the last vertex put into C'}
                while there are edges in H incident to w do
                    begin
                        x ← lowest numbered vertex adjacent to w in H
                        concatenate ", x" onto the end of C'
                        delete edge wx from H
                        w ← x {the new last vertex in C'}
                    end
                replace first occurrence of v in C with C' {splice C' into C}
            end
        return(C)
```

Algorithm 1.   Finding an Euler tour in a graph.

require a detailed examination of the data structures used to store the adjacency information for G. We will not go into these details here, but in fact the process can be carried out efficiently.

EXAMPLE 6.   Consider the children's game of dominoes (see Example 10 of Section 6.1 for a description of the set of dominoes). Is it possible to arrange all 28 dominoes end to end in a circle, so that each square of each domino is next to a square in the neighboring domino with the same number of dots on it?

Solution.   We use a pseudograph model in which the dominoes are represented by *edges*, not vertices. Let G be the pseudograph with seven vertices, labeled 0 through 6 (these stand for the possible numbers of dots on a square of a domino), with one edge joining each pair of vertices. Thus there are 28 edges (including the seven loops), one for each domino (the edge uv represents the domino with squares having u dots and v dots). A solution to our problem is precisely an Euler tour in this pseudograph; the edges in the tour, in order, determine the circular arrangement of the dominoes. Since each vertex in G has degree 8 (at each vertex there are edges to each of the other six vertices and a loop), Theorem 3 guarantees that such an Euler tour exists, and Algorithm 1 gives a method for finding one. The reader should obtain a set of dominoes (or make one out of paper) and find several actual solutions to this domino problem.   ◆

Variations on the theme of Euler tour, as well as some other interesting applications, are discussed in the exercises. These include Euler tours in directed graphs, Euler trails, and de Bruijn sequences.

HAMILTON CYCLES

Let us return to the public safety officer in our small town, who, in the last subsection, needed to patrol all the streets; she found that she needed an Euler tour of the town. Now suppose instead that she needs to visit each *intersection* once (perhaps to make sure that the traffic signals are functioning properly), rather than to traverse each *street* once. The questions are similar: Can she find a tour through the city, starting and ending at the police station, so that she visits each intersection exactly once? If so, how can she find such a tour? Again we will formulate and discuss these problems in graph theoretic terms. (For the same graph model arising in other applications, see Exercise 35.) Since this time we care only about the vertices of the graph, we may as well assume that there are no loops or parallel edges.

Let G, then, be a graph or digraph. A **Hamilton cycle** in G is a cycle that contains every vertex of G. In other words, a Hamilton cycle of G is a connected subgraph of G containing all the vertices of G in which every vertex has degree 2. Such cycles were investigated by the nineteenth-century British mathematician Sir William Rowan Hamilton.

EXAMPLE 7.   Consider the graph G pictured in Figure 8.21. There is a Hamilton cycle in G, shown with heavy lines. On the other hand, $K_{3,2}$ (the complete bipartite graph on parts of size 3 and 2) has no Hamilton cycle, as the reader should verify.   ◆

Although the problem of finding a Hamilton cycle looks similar to the problem of finding an Euler tour, it turns out to be much harder. No one knows a good (easily checked) necessary and sufficient condition for the existence of a Hamilton cycle in a graph or digraph, and, in contrast to the situation for Euler tours, there are no known efficient algorithms for finding them. (In particular, there is no known "good" algorithm (as defined in Section 4.4), that is, one whose worst-case running time is bounded by a polynomial function of the number of vertices in the graph. In fact, the question of whether a given graph with n vertices has a Hamilton cycle is an *NP*-complete problem (see Section 4.4). It is clearly in the class *NP*, since if a genie were to tell us which sequence of vertices to try, we could quickly prove that a graph with a Hamilton cycle indeed had one. On the other hand, it is not at all clear how we would go about finding the right sequence, short of trying all (or substantially all) of the $n!$ possibilities, and $n!$ is not polynomial in n.)

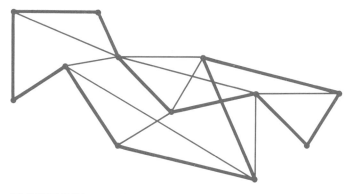

FIGURE 8.21   A graph with a Hamilton cycle.

We will look at two aspects of the Hamilton cycle problem: First we will show that a Hamilton cycle does exist in certain cases, and then we will show how it is sometimes possible to prove that a given graph does not have a Hamilton cycle.

THEOREM 4.   *Let G be a* **tournament** *on $n \geq 3$ vertices; that is, let G be a digraph without loops such that for every pair of distinct vertices u and v, exactly one of uv and vu is an edge of G. Further suppose that G is strongly connected. Then G has a Hamilton cycle.*

Proof.   Our proof is a proof by contradiction, really a type of proof by mathematical induction, with the induction being on the length of the longest cycle in G. First we claim that since G is strongly connected, G does contain some cycle. (The reader is asked to prove this assertion in Exercise 17.) Let C be such a cycle having the greatest possible length; denote the length by k. (Clearly $k \geq 3$, since G has no loops or antiparallel edges.) If $k = n$, we are done, for in this case C *is* the desired Hamilton cycle. If not, we will derive a contradiction by showing how to find a cycle of length greater than k. [This proof is fairly subtle, and you should draw appropriate pictures as you follow it along.]

Let C be $v_1, v_2, \ldots, v_k, v_1$, and let u be a vertex of G not in C. Look at the directions of the edges joining u and the v_i's (since G is a tournament, there *is* an edge, in one direction or the other, between u and each v_i). If these edges are not all directed from u or all directed to u, then there are vertices v_j and v_{j+1} in C (where we are considering v_{k+1} to be the same as v_1) such that $v_j u$ and $u v_{j+1}$ are edges of G. Then we can find a longer cycle: $v_1, v_2, \ldots, v_j, u, v_{j+1}, \ldots, v_k, v_1$. So assume that all the edges joining u and the v_i's are directed *from* u. (The case in which all these edges are directed *to* u is similar (see Exercise 29).) Since G is strongly connected, there is a walk from v_1 to u, and hence, by Theorem 1, a path from v_1 to u. This path leaves C for the last time at some vertex v_j and then continues $u_1, u_2, \ldots, u_l = u$, where the u_i's are distinct vertices not in C. But now we see a longer cycle in G, namely $v_1, v_2, \ldots, v_j, u_1, u_2, \ldots, u_l, v_{j+1}, \ldots, v_k, v_1$. Again we have derived a contradiction. Thus C must have been the desired Hamilton cycle. ■

Let us interpret Theorem 4 in terms of round-robin tournaments in sports. Suppose that each team in a league with n teams plays every other team exactly once. Construct a tournament in the graph-theoretic sense by representing each team by a vertex and representing the fact that team u beat team v by a directed edge from the vertex representing u to the vertex representing v. (We assume that there are no tied games.) To invoke Theorem 4, that this tournament is strongly connected, we have to verify that given any two teams u and v in the league, we can find a sequence of teams $u = v_1, v_2, \ldots, v_s = v$, such that v_i beat v_{i+1} for each i from 1 to $s - 1$. (For example, the Lions beat the Colts, who beat the Redskins, who beat the Bears.) The theorem tells us that if this condition is satisfied, then we can find a circular ordering of *all* the teams in the league, such that each team beat the next team in that ordering.

Note that Theorem 4 merely gives a sufficient condition for the existence of a Hamilton cycle in a digraph, not a necessary one. For example, the digraph consisting of vertices a, b, c, d, and e and edges ab, bc, cd, de, ea, dc and be is not a tournament yet still contains the Hamilton cycle a, b, c, d, e, a.

Finally, we turn to an example of a graph in which there is no Hamilton cycle. The techniques used in this example are typical of the kinds of ad hoc reasoning that can be helpful in proving the *nonexistence* of Hamilton cycles.

EXAMPLE 8. Consider the graph G pictured in Figure 8.22. We will argue (with a proof by contradiction) that G has no Hamilton cycle. Suppose, instead that G does have a Hamilton cycle C. Now vertex a must be in C; hence either edge xa or edge ya must be in C (perhaps both), since otherwise, the cycle would not be able to enter and leave a. If *both* edges xa and ya are in C, then edge ab is not, since in a Hamilton cycle, the degree of every vertex must be exactly 2. But that means that edges xb and yb must both be in C, in order to include vertex b in the cycle. This is clearly impossible, since the cycle cannot contain all four of these edges (xa, ya, xb, and yb) and still pick up the remaining four vertices of G.

Thus we have shown that exactly one of xa and ya is in C. By the symmetry of the graph, let us assume without loss of generality that xa is in C, but ya is not. Then edge ab is forced to be in C, since we need another edge incident in C to a. [The reader can darken the edges that have been forced into C and strike through those that have been forbidden, as the proof proceeds.] Of the remaining two edges at b, it would be impossible for xb to be in C, since that would form a cycle prematurely (including only vertices x, a, and b); hence yb must also be in C.

By a similar line of reasoning, we know that either edges xc, cd, and dy must be in C or edges xd, cd, and cy must be in C. In either case, the cycle has

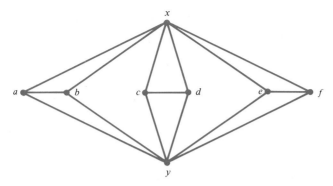

FIGURE 8.22   A graph G with no Hamilton cycle.

closed prematurely: vertices e and f have been left out. Hence G has no Hamilton cycle.   ◆

A traveling salesperson with many sites to visit on a sales trip needs to find Hamilton cycles in the appropriate graph model, to avoid needless backtracking. The salesperson also probably wants to minimize the actual distance traveled between sites (or cost incurred in such travel). We will return, in Section 10.1, to the problem of finding Hamilton cycles of minimum total cost in graphs in which the edges have costs associated with them.

SUMMARY OF DEFINITIONS

closed walk: a walk of length at least 1 that ends where it starts (i.e., in which the first vertex equals the last vertex).

component of the graph or pseudograph G: a maximal connected subgraph of G.

connected graph or pseudograph: a graph or pseudograph with the property that every two vertices are joined by a walk.

cycle: a closed trail in which all the vertices are distinct, except that the first vertex equals the last vertex.

Euler tour: a closed trail in a connected graph or pseudograph that contains each edge exactly once.

Hamilton cycle: a cycle in a graph or digraph that contains every vertex.

join: A walk $v_0, e_1, v_1, e_2, v_2, \ldots, v_{n-1}, e_n, v_n$ joins vertex v_0 to vertex v_n.

path: a walk in which all the vertices are distinct.

strong component of the digraph or multidigraph G: a maximal strongly connected subgraph of G.

strongly connected digraph or multidigraph: a digraph or multidigraph with the property that every two vertices are joined by walks in both directions.

tournament: a digraph G without loops such that for every pair of distinct vertices u and v in G, exactly one of uv and vu is an edge of G.

trail: a walk in which all the edges are distinct.

triangle: a 3-cycle.

walk of length $n \geq 0$ in the graph, digraph, pseudograph, or multidigraph G: a sequence $v_0, e_1, v_1, e_2, v_2, \ldots, v_{n-1}, e_n, v_n$ such that each v_i is a vertex of G, each e_i is an edge of G, and $e_i = v_{i-1}v_i$ for all $1 \leq i \leq n$; this walk goes from v_0 to v_n.

EXERCISES

1. In the graph shown, determine which of the terms "walk," "closed walk," "trail," "closed trail," "path," and "cycle" apply to the following sequences.

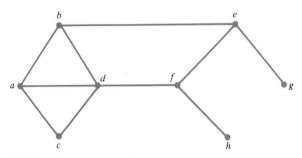

EXERCISE 1

(a) b, e, f, g.
(b) $a, b, e, f, d, a, c, d, b$.
(c) d, f, d.
(d) h.
(e) a, b, e, f, d, c, a.
(f) a, c, d, f, e, b, d, a.
(g) a, b, d, f, e, b, d, c.

2. In the digraph shown, give an example of each of the following.
(a) A closed walk that is not a trail.
(b) A path of length 3.
(c) A cycle of length 3.
(d) A closed trail that is not a cycle.
(e) A sequence that is not a walk but that would specify a walk in the graph obtained from this digraph by ignoring the directions on the edges.

3. Find two Euler tours in the graph depicted in Figure 8.20 that are substantially different from the one given in Example 5, that is, different by more than just

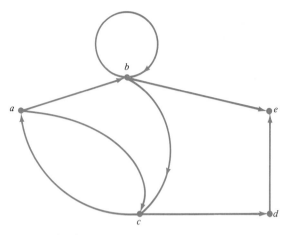

EXERCISE 2

a circular permutation (the "same" tour starting at a different point) and/or a reversal of direction.

4. Find two Hamilton cycles in the graph depicted in Figure 8.21 that are different (as sets of edges) from the one given in Example 7.

5. Give an example of each of the following.
    (a)   A connected graph with at least three edges that has a Hamilton cycle but does not have an Euler tour.
    (b)   A connected graph with at least three edges that has an Euler tour but does not have a Hamilton cycle.
    (c)   A connected graph with at least three edges that has both an Euler tour and a Hamilton cycle.
    (d)   A connected graph with at least three edges that has neither an Euler tour nor a Hamilton cycle.

6. Determine, with justification, whether each of the following statements is necessarily true. The answers may depend on the context (graph, digraph, multigraph, pseudograph, or multidigraph)—explain fully.
    (a)   Every path is a trail.
    (b)   Every cycle is a closed trail.
    (c)   Every cycle must have length at least 2.
    (d)   Every cycle must have length at least 3.

7. Suppose that G is a subgraph of H. Determine, with justification, whether each of the following statements is necessarily true.
    (a)   If G is connected, then H is connected.
    (b)   If H is connected, then G is connected.

(c)   The number of components of G is no greater than the number of components of H.

(d)   The number of components of G is no less than the number of components of H.

8. Let G be a graph. Prove that the relation R defined on the set of vertices of G by

$$uRv \text{ if and only if there is a walk from } u \text{ to } v$$

is an equivalence relation.

9. Suppose that we make a physical model of a graph using beads and string. We use one bead for each vertex, and we tie a piece of string from bead u to bead v for each edge uv in the graph. Does the heuristic for connectedness given in this section ("a graph is connected if and only if, when you pick up a vertex and leave the room, the whole graph comes with you") necessarily work for this model?

10. Show that if a graph G is not connected, then its complement $\overline{G}$ is connected.

11. What is the maximum number of edges a graph with n vertices can have and yet not be connected? Prove that your answer is correct.

12. Determine the number of components in the *complements* of the following graphs. (Your answers may depend on n.)
(a) K_n.
(b) C_n.
(c)   The n-spoked wheel (see Exercise 21 of Section 8.1).
(d) Q_n.

13. Identify the strong components of the following digraphs and multidigraphs depicted in Section 8.1.
(a)   Figure 8.2.
(b)   Figure 8.5.
(c)   Figure 8.8.

14. We will say that a walk W_1 **contains** walk W_2 if W_2 appears as a consecutive subsequence of W_1. Determine, with proof, whether each of the following statements is necessarily true.
(a)   In a graph, a closed trail contains a cycle.
(b)   In a graph, a closed walk contains a closed trail.
(c)   In a graph, a trail that is not a path contains a cycle.
(d)   In a digraph, a closed walk contains a cycle.

15. Prove Theorem 2.

16. Let u and v be vertices of a graph G. Prove that if there are two different trails from u to v in G, then G contains a cycle.

17. Show that a strongly connected digraph with at least two vertices contains a cycle.

18. Show that if a graph has exactly two vertices of odd degree (and an arbitrary number of vertices of even degree), then there is a path joining these two vertices.

19. Let G be a graph that has exactly three vertices of odd degree (and an arbitrary number of vertices of even degree). [This problem is not a misprint.]
    (a)   Show that G is connected.
    (b)   Show that G is not connected.

20. Euler first studied the concept of what is now called an Euler tour in the context of finding a walk through the city of Königsberg. The accompanying diagram shows that the city was situated on a river, and that the river branched, forming two islands. Bridges joining the islands and the banks of the river were as shown in the figure. The citizens of Königsberg were wont to take strolls, and they wished to arrange their walks in such a way that they crossed each of the seven bridges exactly once, before returning home.

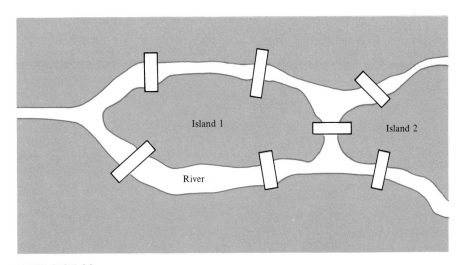

EXERCISE 20

    (a)   Model the problem of the citizens of Königsberg with a multigraph. Draw a picture of the multigraph, and state the problem in graph-theoretic terms. [*Hint*: The four land masses (two islands and two banks) can be the vertices, and the seven bridges can be the edges.]

    (b)   Show that there is no way for a citizen of Königsberg to take a walk with the desired properties.

21. For which values of n (or m and n) do the following graphs contain Euler tours? Justify your answers.
    (a) K_n.
    (b) $K_{m,n}$.
    (c) C_n.
    (d) Q_n.

22. An **Euler trail** in a connected graph (or pseudograph) G is a trail that contains every edge of G. Thus an Euler trail is the same as an Euler tour, except that it may end at a vertex different from where it began.
    (a) Determine, with proof, nice necessary and sufficient conditions for a connected pseudograph to contain an Euler trail. [*Hint*: Add an extra edge back to the start of the trail, if it ends at a vertex different from the one at which it began.]
    (b) Show that there is no Euler trail in the model of Königsberg's bridges (see Exercise 20).

23. Explain why a line figure in the plane can be drawn without lifting the pencil nor retracing any line if and only if the figure, looked at as a picture of a pseudograph, contains an Euler trail (see Exercise 22). Then, for each of the following figures, either find a drawing made without lifting the pencil or retracing any line, or prove that no such drawing is possible.

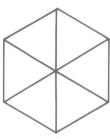

(a)   (b)   (c)

EXERCISE 23

24. Show that every connected graph with at least one edge contains a closed walk using each edge exactly twice.

25. Explain how in the proof of Theorem 3 we used the hypothesis that the graph had at least one edge. Is the conclusion not valid without this hypothesis?

26. Suppose that we are given a complete set of dominoes (see Example 6) on which the number of spots per square can range from 0 to n, rather than from

0 to 6, where n is a positive integer. For which values of n can we arrange all these dominoes in a circle as in Example 6?

27. Determine the set of values of m and n for which $K_{m,n}$ contains a Hamilton cycle.

28. For each of the graphs pictured here, either find a Hamilton cycle or give a proof that none exists. (Hamilton turned the problem of finding a Hamilton cycle in the graph in part (d) into a game called Around the World.)

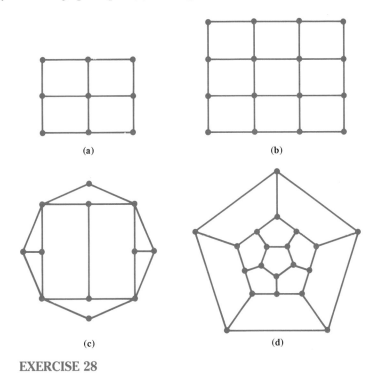

(a)

(b)

(c)

(d)

EXERCISE 28

29. Fill in the details of the proof of Theorem 4 for the case in which all the edges are directed *to* the vertex u.

Challenging Exercises

30. Determine the maximum number of edges that a graph with c components and n vertices can have, where $n \geq c \geq 1$.

31. Show that a graph is bipartite if and only if it contains no cycles of odd length.

32. Develop a definition of Euler tour for a multidigraph, and determine, with proof, nice necessary and sufficient conditions for a multidigraph to contain an Euler tour.

33. A **de Bruijn sequence** of order n is a circular arrangement of 0's and 1's such that each of the 2^n bit strings of length n appears exactly once as n consecutive digits in the arrangement. (The figure shows a de Bruijn sequence of order 3; reading consecutive three-bit strings, clockwise, starting from the top and proceeding clockwise, we find 000, 001, 011, 111, 110, 101, 010, and 100.)

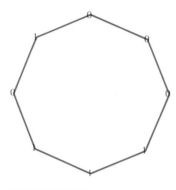

EXERCISE 33

(a) Find a de Bruijn sequence for $n = 2$.
(b) Find a de Bruijn sequence for $n = 4$.
(c) Give a general method for constructing de Bruijn sequences. [*Hint*: Construct a digraph whose vertices are all the bit strings of length $n-1$. Put two outgoing edges at each vertex, one labeled 0 and one labeled 1: An edge with label l goes from $a_1 a_2 \ldots a_{n-1}$ to $a_2 a_3 \ldots a_{n-1} l$. Find an Euler tour and look at the labels on the edges in the tour.]

34. Show that the n-cube Q_n contains a Hamilton cycle for all $n \geq 2$. Such a cycle is called a **Gray code**.

Exploratory Exercises

35. Explain how the notions of Euler tour or Hamilton cycle are relevant to the following applications. Be specific as to the models involved. Explicitly consider any additional assumptions you may need for an accurate model. Also, indicate how the problem changes if the assumptions are violated.
(a) The head of the food service for an airline company wants to try out the meals on every flight.
(b) The head of customer relations for an airline company wants to inspect the first-class passenger lounges at each of the airports the airline serves.
(c) A road crew needs to salt all the city's streets during a snowstorm.
(d) An automated drill press needs to be programmed to drill several screw holes in metal plates.

(e)   A postal carrier needs to deliver mail to all the houses in a neighborhood.

36. Find out about **Fleury's algorithm** for finding Euler tours by consulting Bondy and Murty [27].

37. Find out about a nonconstructive proof of the sufficiency half of Theorem 3, using induction on the number of vertices in the graph, by consulting Fowler [87].

38. Find out about Euler tours in graphs that can be drawn in the plane without edges crossing by consulting Kidwell and Richter [170].

39. Give reasonable definitions for the **radius** and **diameter** of a graph.
    (a)   What can be said about the relationship between radius and diameter in general?
    (b)   Prove that if the diameter of G is at least 3, then the diameter of $\overline{G}$ is at most 3.
    (c)   Consult Bloom, Kennedy, and Quintas [22] and Harary and Robinson [144] for further information.

40. Find out more about Hamilton cycles by consulting Nash-Williams [221] and Ore [228].

41. Find out more about de Bruijn sequences (see Exercise 33) by consulting Ralston [241].

42. Find out about Sir William Rowan Hamilton by consulting Bell [15] and Hankins [138].

SECTION 8.3
GRAPH REPRESENTATION
AND GRAPH ISOMORPHISM

A rose by any other name would smell as sweet.

—William Shakespeare, *Romeo and Juliet*

We have defined a graph as an abstract mathematical object—a pair of sets (vertices and edges), related in a certain way. Our first goal in this section is to explore ways in which we might represent graphs so that human beings and computers can work with them. This much is pretty straightforward, and we will not go into the subject very deeply, for it is really more a problem of computer science (data structures) than of mathematics. In the latter part of this section, we deal with the problem of determining whether two graphs have the same structure.

We have already indicated one method of representing a graph, namely as a picture in the plane, in which distinct points represent the vertices, and line segments or curves joining the points represent edges. We saw that with minor modifications, this pictorial method of representation applies to pseudographs, digraphs, and multidigraphs, as well. Pictures are very useful for human understanding of the structure of a small graph, especially if the drawing exploits symmetries in the graph. Pictorial representations have at least three disadvantages, however. First, in the current state of computer technology, it is not easy for a computer to work with a picture. Second, if the graph is at all large, the human mind and visual apparatus will boggle at a picture. Third, a picture does not easily lend itself to algorithmic calculations; imagine, for example, trying to count the number of 3-cycles in a fairly large graph drawn in the plane. We will not say any more about pictorial representations here, but we will return to the subject, with a crucial modification, in the next section.

ADJACENCY MATRICES

One of the most straightforward ways to represent a graph is with an adjacency matrix, which is just a systematic way to record which edges are present and which are absent. The following discussion will sound familiar, since we used matrices in essentially just this way to represent relations in Section 3.3.

Let $G = (V, E)$ be a graph, and fix an ordering $v_1, v_2, \ldots, v_n$ of its vertex set V. Then the **adjacency matrix** for G with respect to this ordering, denoted A_G (or simply A if the graph under discussion is clear), is the n by n matrix $A = (a_{ij})$ given by

$$a_{ij} = \begin{cases} 1 & \text{if } v_i v_j \in E \\ 0 & \text{if } v_i v_j \notin E. \end{cases}$$

In other words, $a_{ij} = 1$ if and only if v_i is adjacent to v_j. The matrix depends on the chosen ordering of the vertices, but we will usually think of the ordering as fixed and refer to "the" adjacency matrix of G.

EXAMPLE 1.   Consider the graph G shown in Figure 8.23, with the ordering on the vertices implied by alphabetical order (this is the ordering we usually use when constructing adjacency matrices).

Explicitly, $G = (V, E)$, where $V = \{a, b, c, d, e\}$ and $E = \{ab, ad, ae, bc, de\}$. The adjacency matrix for this graph is shown below. We will usually label the left and top margins of the matrix with the names of the vertices, as we have done here, in order to make it easier to read the information contained in the matrix.

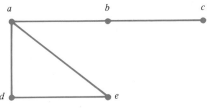

FIGURE 8.23   A graph G.

$$A = A_G = \begin{array}{c} \\ a \\ b \\ c \\ d \\ e \end{array} \begin{array}{ccccc} a & b & c & d & e \\ \left[\begin{array}{ccccc} 0 & 1 & 0 & 1 & 1 \\ 1 & 0 & 1 & 0 & 0 \\ 0 & 1 & 0 & 0 & 0 \\ 1 & 0 & 0 & 0 & 1 \\ 1 & 0 & 0 & 1 & 0 \end{array}\right] \end{array}.$$

For instance, the entry 1 in the fourth row, first column ($a_{41} = 1$) indicates that there is an edge from the fourth vertex (d) to the first vertex (a). On the other hand, the entry 0 in the second row, fifth column ($a_{25} = 0$) indicates that there is no edge from the second vertex (b) to the fifth (e). ◆

Note that since a graph has *undirected* edges, the adjacency matrix of a graph is symmetric about its main diagonal (from upper left to lower right); that is, $a_{ij} = a_{ji}$ for all i and j. Furthermore, since a graph has no loops, all the elements on the main diagonal, a_{ii}, are zero.

The matrix representation is easy to store in a computer and easy to work with in a high-level computer language that allows "random access" to elements of arrays (we can access the (i, j)th entry of a matrix simply by referring to it). If we need to tell whether two vertices are adjacent, we simply look at the appropriate element of the matrix. Also, if we want to consider all the vertices adjacent to a particular vertex v_i, we can simply look along the ith row of the adjacency matrix, and note which entries are 1's.

Furthermore, certain parameters and properties of the graph can be found easily from the matrix. The number of vertices in the graph is simply the number of rows (or, equivalently, the number of columns) of the matrix. The number of edges in a graph is equal to the number of 1's in the matrix divided by 2, since each edge $v_i v_j$ "appears" twice in the matrix, once as a 1 in the (i, j)th location and once as a 1 in the (j, i)th location. The row sums (or column sums) give the degrees of the vertices:

$$d(v_i) = \sum_{j=1}^{n} a_{ij} = \sum_{j=1}^{n} a_{ji}.$$

Adjacency matrices really come into their own when we want to count the number of walks in a graph, as the next theorem shows. Recall from Section 4.5 the definition of matrix multiplication: If A and B are n by n matrices, the product $A \times B$ is the n by n matrix (c_{ij}) whose entries are given by

$$c_{ij} = \sum_{k=1}^{n} a_{ik} b_{kj}.$$

Recall also that A^l denotes the product $A \times A \times \cdots \times A$, with l factors, where A^0 is the identity matrix I_n.

THEOREM 1. *Let G be a graph with n vertices, and let A be the adjacency matrix for G. Then for every integer $l \geq 0$, the (i, j)th entry of A^l is the number of walks of length l in G from v_i to v_j.*

Proof. We prove the theorem by induction on l. If $l = 0$, then $A^l = I_n$, and the statement reduces to the statements that the number of walks of length 0 in G from a vertex to itself is 1 (which is true), and the number of walks of length 0 in G from a vertex to a different vertex is 0 (which is also true). Similarly, if $l = 1$, then $A^l = A$, and the statement reduces to the true statements that the number of walks of length 1 in G from a vertex v_i to a vertex v_j is 1 if there is an edge from v_i to v_j and 0 if there is not. So we assume the inductive hypothesis that for each pair of vertices v_s and v_t in G, the (s, t)th entry of A^l gives the number of walks of length l in G from v_s to v_t, and try to prove that the (i, j)th entry of A^{l+1} gives the number of walks of length $l + 1$ in G from v_i to v_j.

Now a walk of length $l + 1$ from v_i to v_j consists of a walk of length l from v_i to *some* vertex v_k, followed by the edge $v_k v_j$ to reach v_j. By the addition principle, the number of such walks is the sum, over all possible choices of v_k, of the number of walks from v_i to v_j that have v_k as their next to last vertex. By the observation in the first sentence of this paragraph, the latter quantity is the product of the number of walks of length l from v_i to v_k and the number of edges (1 or 0) from v_k to v_j. Putting this all together and invoking the inductive hypothesis, we see that the number of walks of length $l + 1$ from v_i to v_j is $\sum_{k=1}^{n} b_{ik} a_{kj}$, where b_{ik} is the (i, k)th entry of A^l. But by the definition of matrix multiplication and the fact that $A^{l+1} = A^l \times A$, this sum is just the (i, j)th entry of A^{l+1}, and our theorem is proved. ∎

EXAMPLE 2. In Example 1 we displayed the adjacency matrix A for the graph shown in Figure 8.23. A straightforward calculation shows that the first few powers of this matrix are as follows.

$$A^2 = \begin{array}{c} \\ a \\ b \\ c \\ d \\ e \end{array} \begin{array}{ccccc} a & b & c & d & e \\ \left[\begin{array}{ccccc} 3 & 0 & 1 & 1 & 1 \\ 0 & 2 & 0 & 1 & 1 \\ 1 & 0 & 1 & 0 & 0 \\ 1 & 1 & 0 & 2 & 1 \\ 1 & 1 & 0 & 1 & 2 \end{array}\right] \end{array}, \qquad A^3 = \begin{array}{c} \\ a \\ b \\ c \\ d \\ e \end{array} \begin{array}{ccccc} a & b & c & d & e \\ \left[\begin{array}{ccccc} 2 & 4 & 0 & 4 & 4 \\ 4 & 0 & 2 & 1 & 1 \\ 0 & 2 & 0 & 1 & 1 \\ 4 & 1 & 1 & 2 & 3 \\ 4 & 1 & 1 & 3 & 2 \end{array}\right] \end{array},$$

$$A^4 = \begin{array}{c} \\ a \\ b \\ c \\ d \\ e \end{array} \begin{array}{ccccc} a & b & c & d & e \\ \left[\begin{array}{ccccc} 12 & 2 & 4 & 6 & 6 \\ 2 & 6 & 0 & 5 & 5 \\ 4 & 0 & 2 & 1 & 1 \\ 6 & 5 & 1 & 7 & 6 \\ 6 & 5 & 1 & 6 & 7 \end{array}\right] \end{array}.$$

Thus, for instance, since the $(2,4)$th entry of A^4 is 5, there are five walks of length 4 from the second vertex (b) to the fourth (d). Indeed, these walks are precisely b, c, b, a, d; b, a, b, a, d; b, a, d, e, d; b, a, d, a, d; and b, a, e, a, d. ◆

Let us carry the analysis one step further. Since a walk of length *at most l* is either a walk of length 0 or a walk of length 1 or a walk of length 2 or ... or a walk of length l, and since these possibilities are disjoint, the addition principle tells us that the number of walks of length at most l from vertex v_i to vertex v_j is the sum of the (i, j)th entries in A^0, A^1, A^2, ..., A^l, which by the definition of matrix addition (see Section 4.5) is the (i, j)th entry in the sum of these matrices. Thus we have proved the following theorem.

THEOREM 2. *Let G be a graph with n vertices, and let A be the adjacency matrix for G. Then for every integer $l \geq 0$ the (i, j)th entry of*

$$I_n + A + A^2 + \cdots + A^l$$

is the number of walks in G, of length less than or equal to l, from v_i to v_j.

Looking at Theorem 2 in the right light, we now have an algorithm for determining whether a graph is connected. Indeed, a graph G with n vertices is connected if and only if there is a walk between every pair of vertices in G. By Theorem 1 of Section 8.2, there is such a walk if and only if there is a path between every pair of vertices, and paths can have length at most $n - 1$ (since they do not use any vertex more than once). We have, therefore, the following corollary to Theorem 2.

COROLLARY.   *Let G be a graph with n vertices, and let A be the adjacency matrix for G. Then G is connected if and only if every entry of $I_n + A + A^2 + \cdots + A^{n-1}$ is nonzero.*

Let us analyze the computational complexity of the algorithm for determining connectedness implied by this corollary. One matrix multiplication can be performed in time proportional to n^3 (see Section 4.5), and here we need to perform at most $n - 2$ matrix multiplications at most n times (and do some adding). Thus this algorithm for determining whether G is connected has time complexity in $O(n^5)$. However, we can perform this matrix calculation along the lines of Horner's method for evaluating polynomials: Set $B \leftarrow I_n$ and then $n - 1$ times perform $B \leftarrow I_n + (B \times A)$; at the end, B is the desired sum. This reduces the time to $O(n^4)$. Although this algorithm may be feasible for fairly small graphs, it would hardly do for a graph with 1000 vertices (not to mention the fact that a million words of storage would be required to hold each matrix). We will see in Section 9.2 that there are much better algorithms for checking the connectedness of a graph.

EXAMPLE 3.   For the graph in Example 1, we find that already

$$I_5 + A + A^2 + A^3 = \begin{array}{c} \\ a \\ b \\ c \\ d \\ e \end{array} \begin{array}{ccccc} a & b & c & d & e \\ \begin{bmatrix} 6 & 5 & 1 & 6 & 6 \\ 5 & 3 & 3 & 2 & 2 \\ 1 & 3 & 2 & 1 & 1 \\ 6 & 2 & 1 & 5 & 5 \\ 6 & 2 & 1 & 5 & 5 \end{bmatrix} \end{array},$$

none of whose entries is 0, so certainly $I_5 + A + A^2 + A^3 + A^4$ is also totally nonzero. Hence the graph is connected.   ◆

Adjacency matrices for directed graphs can be defined in a completely analogous fashion. The following definition is really just a restatement of the definition of the matrix representing a relation, given in Section 3.3.

Let $G = (V, E)$ be a digraph, and fix an ordering $v_1, v_2, \ldots, v_n$ of its vertex set. Then the **adjacency matrix** for G with respect to this ordering, denoted A_G (or simply A if the digraph under discussion is clear), is the n by n matrix $A = (a_{ij})$ given by

$$a_{ij} = \begin{cases} 1 & \text{if } v_i v_j \in E \\ 0 & \text{if } v_i v_j \notin E. \end{cases}$$

EXAMPLE 4.   Consider the digraph G shown in Figure 8.24, with the ordering on the vertices implied by alphabetical order. Its adjacency matrix, with respect to the alphabetical order of the vertices, is

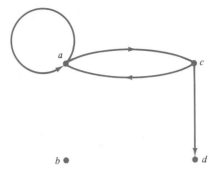

FIGURE 8.24   A digraph G.

$$A = A_G = \begin{array}{c}\ \\ a \\ b \\ c \\ d \end{array} \begin{array}{cccc} a & b & c & d \\ \left[\begin{array}{cccc} 1 & 0 & 1 & 0 \\ 0 & 0 & 0 & 0 \\ 1 & 0 & 0 & 1 \\ 0 & 0 & 0 & 0 \end{array}\right]\end{array}.$$

Note, for instance, that the second and fourth rows of the matrix are all 0's, since vertices b and d have no outgoing edges. There are four edges in this digraph, corresponding to the four 1's in the adjacency matrix. The out-degree of vertex c is 2 ($d^+(c) = 2$), since there are two 1's in the third row of A_G.   ◆

It is easy to see that Theorems 1 and 2 and the corollary (with "strongly connected" in place of "connected") hold for digraphs as well as for graphs.

EXAMPLE 5.   For the digraph in Figure 8.24, we can compute that

$$I_4 + A + A^2 + A^3 = \begin{array}{c}\ \\ a \\ b \\ c \\ d \end{array} \begin{array}{cccc} a & b & c & d \\ \left[\begin{array}{cccc} 7 & 0 & 4 & 2 \\ 0 & 1 & 0 & 0 \\ 4 & 0 & 3 & 2 \\ 0 & 0 & 0 & 1 \end{array}\right]\end{array}.$$

This tells us, for example, that there are four walks of length at most 3 from a to c, which are in fact a, c; a, a, c; a, a, a, c; and a, c, a, c. Since there are zero entries in this matrix, the digraph is not strongly connected. For example, since the $(1, 2)$th entry is zero, there is no walk from a to b.   ◆

It is also possible to define adjacency matrices for pseudographs and multidigraphs; we explore these ideas in Exercises 18 and 19.

ADJACENCY LISTS

There is a serious drawback to the use of adjacency matrices to represent graphs and digraphs: They may waste space, and the time required to look through that space, *if the graph or digraph does not have very many edges*. We call a graph or digraph with few edges **sparse**. (There is no precise definition of the word "sparse"; in general, if the number of edges is at most a reasonably small constant times the number of vertices, then we would call the graph or digraph sparse. For example, a graph with 1000 vertices and 5000 edges would be considered sparse; only a tiny fraction of the 499,500 possible edges are present in this graph.) The adjacency matrix for a sparse graph or digraph consists mainly of 0's, with only a relatively few 1's. Time and space can be saved in such cases, if we record only the information conveyed by the 1's. We record this information as a sequence of lists: For each vertex, we maintain a list of its neighbors.

Formally, let $G = (V, E)$ be a graph. The **neighborhood** of each vertex v is the set of **neighbors** of v, that is, those vertices adjacent to v. The **adjacency list** representation of G consists of the set of neighborhoods, indexed by the set of vertices.

We think of the adjacency list representation of G as a list of lists, rather than as a set of sets, because in practice the sets are ordered by some ordering of the vertices of the graph. In our examples, we use alphabetical order for vertices named by letters.

EXAMPLE 6.   For the graph shown in Figure 8.23, we can display the adjacency lists as follows:

a:	b, d, e
b:	a, c
c:	b
d:	a, e
e:	a, d

Thus, for instance, the neighborhood of a is $\{b, d, e\}$.   ◆

Just as with the adjacency matrices, the adjacency lists provide much easily accessible information about the graph. In many of the graph-theoretic algorithms in the remainder of this book, it will be helpful to imagine that the graphs are stored in terms of their adjacency lists. In particular, adjacency lists make it extremely easy to look through the neighbors of a vertex, a task we will be doing repeatedly in various graph algorithms.

A similar structure can be defined for digraphs. Depending on the application, we may want only to record, for each vertex v, the set of vertices u for which there is an edge *from v to u*; or we may want to maintain two adjacency lists for each vertex v, both the set of heads of edges outgoing from v, and the set of tails of edges incoming to v. In the case of a pseudograph or multidigraph, our adjacency lists could indicate the identities of the edges themselves. Again, we do not wish to go into too much detail about the variety and richness of the possible data structures that arise, since our aim is primarily mathematical.

EXAMPLE 7. For the digraph shown in Figure 8.24, we can display the outgoing adjacency lists as

$$
\begin{array}{ll}
a: & a, c \\
b: & - \\
c: & a, d \\
d: & -
\end{array}
$$

and the incoming adjacency lists as

$$
\begin{array}{ll}
a: & a, c \\
b: & - \\
c: & a \\
d: & c
\end{array}
$$

Together, these lists of lists allow easy access to whatever information about the digraph one may need. ◆

OTHER REPRESENTATIONS

There are other ways of recording the information contained in a graph (or a graph with additional structure). For example, the edges of a graph are often labeled with "weights"—numbers that might indicate, for example, the physical distance between two points, the cost of communicating between two locations, the capacity of a pipe between two pumping stations, or the duration of a task. In some applications, we might want to put the weights into an adjacency matrix, so that the (i, j)th entry of the matrix will be the weight of the edge between v_i and v_j (perhaps with a weight of 0 or ∞, depending on the application, if there is no edge). Alternatively, we could incorporate the weights into adjacency lists, so that each row becomes, rather than a list of vertices adjacent to v, a list of ordered pairs (u, w), where u is a vertex adjacent to v and w is the weight of edge vu.

EXAMPLE 8.   A system of roads is represented by the weighted graph shown in Figure 8.25. The weights indicate the distances between towns in miles. Using the alphabetical order of vertices (Eastville, Midtown, Southville, Suburb, Westville), we can construct the adjacency matrix for this weighted graph:

	Eastville	Midtown	Southville	Suburb	Westville
Eastville	0	50	60	∞	∞
Midtown	50	0	30	∞	30
Southville	60	30	0	36	50
Suburb	∞	∞	36	0	40
Westville	∞	30	50	40	0

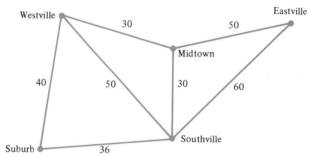

FIGURE 8.25   A road system modeled by a weighted graph G.

Note that we have put 0 as the weight from each vertex to itself; this makes sense if we are interested in travel along roads, since the distance from a town to itself is 0. Similarly, we have put ∞ in the other spaces where there are no edges (e.g., from Eastville to Suburb), since direct travel is impossible between those vertices. A typical application might be to find the shortest route from one town to another. This matrix contains all the information needed by any algorithm that will solve that problem. ◆

We will have much more to say about weighted graphs in Chapter 10.

In other applications it might be useful to maintain only a list of the edges, perhaps in order of increasing weight. Sometimes it might be useful to have a matrix that indicates the incidence relation between vertices and edges. If a graph G has vertices $v_1, v_2, \ldots, v_n$ and edges $e_1, e_2, \ldots, e_m$, then the **incidence matrix** of G is the n by m matrix $C = (c_{ij})$ such that

$$c_{ij} = \begin{cases} 1 & \text{if } v_i \text{ is incident to } e_j \\ 0 & \text{if } v_i \text{ is not incident to } e_j. \end{cases}$$

Often algorithms require that various labels or flags be assigned to vertices or edges; the labels can change as the algorithm proceeds. For example, we might need to keep track of whether or not we have visited a vertex as we walk through a graph. These labels or flags can be stored in arrays.

THE ISOMORPHISM PROBLEM

We turn now to the second subject of this section, one that is more mathematical in nature, and less an exercise in computer science. As we said in Section 8.1, a graph $G = (V, E)$ is a mathematical structure—a set V of vertices together with an adjacency relation on V. We want to consider only the *essence of the structure*, we do not care much about the actual set used for V. For example, the graphs $G = (\{a, b, c\}, \{ab, ac\})$ and $H = (\{x, y, z\}, \{xy, xz\})$ are not *equal*, since they have different vertex sets and different edge sets; but structurally, they are essentially "the same." Both G and H consist of three vertices, one of which is joined by an edge to each of the others, with no edge between the other two. In G we called the vertex of degree 2 a, while in H we called it x. By simply *renaming* the members of the vertex set of G, from a, b, and c, to x, y, and z, respectively, we can convert one of the graphs into the other. Our goal in the rest of this section is to make this notion of structural sameness precise and to consider the question of how to determine whether a pair of graphs are "the same" in this sense.

Let $G = (V, E)$ and $H = (W, F)$ be two graphs or digraphs. Then G and H are **isomorphic**, written $G \cong H$, if there is a bijective (one-to-one and onto) function

$$\varphi : V \to W$$

such that

$$\forall u \in V : \forall v \in V : (uv \in E \longleftrightarrow \varphi(u)\varphi(v) \in F).$$

The function φ is called an **isomorphism** between the two graphs or digraphs.

We can restate this definition in the case of graphs by saying that two graphs are isomorphic if there is a one-to-one correspondence between their vertex sets that *preserves adjacencies and nonadjacencies*. (In the case of digraphs, the directions of the edges must be preserved as well.) The isomorphism φ in the definition is precisely what does the "renaming" that we desired.

We will concentrate on graphs in our discussion, but most of what we say applies to digraphs as well, when suitably interpreted. We leave to the reader the task of defining isomorphism for pseudographs and multidigraphs (Exercise 38).

EXAMPLE 9. Consider the two graphs depicted in Figure 8.26. We will show that these two graphs are isomorphic. To do so, we need to exhibit a one-to-one correspondence

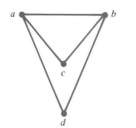 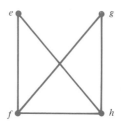

FIGURE 8.26 Two isomorphic graphs.

φ between $\{a, b, c, d\}$ (the vertex set of the graph on the left) and $\{e, f, g, h\}$ (the vertex set of the graph on the right) that induces a correspondence between the adjacency structures in the two graphs. There are several such φ's that will work (but not *all* one-to-one and onto functions from $\{a, b, c, d\}$ to $\{e, f, g, h\}$ will do); one is defined as follows:

$$\varphi(a) = f \qquad \varphi(b) = h \qquad \varphi(c) = e \qquad \varphi(d) = g.$$

Clearly, φ is one-to-one and onto. It remains to be checked that for each pair of vertices u and v in $\{a, b, c, d\}$, there is an edge between u and v if and only if there is an edge between $\varphi(u)$ and $\varphi(v)$. If $u = v$, there is no problem, since in neither graph are there any loops. In the graph on the left, there is an edge between *every* pair of distinct vertices, with the exception of cd. Thus for the graphs to be isomorphic, there must be an edge between every pair of distinct vertices in the graph on the right, except that there must be no edge between $\varphi(c)$ and $\varphi(d)$, that is, no edge between e and g. We see that the graph on the right has precisely the correct set of edges. Therefore, the graphs are isomorphic.

Another way to restate the definition is to say that two graphs are isomorphic if there exist orderings on their vertex sets such that their adjacency matrices are the same. The orderings induce the isomorphism φ. Thus in this example, if we order the vertices in the first graph a, b, c, d and the vertices in the second graph f, h, e, g, then both graphs have the adjacency matrix

$$\begin{bmatrix} 0 & 1 & 1 & 1 \\ 1 & 0 & 1 & 1 \\ 1 & 1 & 0 & 0 \\ 1 & 1 & 0 & 0 \end{bmatrix}.$$

It is easy to show that "is isomorphic to" is an equivalence relation on the class of all graphs. Indeed, any graph is isomorphic to itself via the identity function on its vertex set. If φ is an isomorphism from G to H, then φ^{-1} is an isomorphism from H to G. And if φ and ψ are isomorphisms from G_1 to G_2 and from G_2 to G_3, respectively, then $\psi \circ \varphi$ is an isomorphism from G_1 to G_3.

So far we have been talking in the positive vein. The last example indicated how one can show that two graphs *are* isomorphic; in fact, it is the only way to do so without using advanced mathematics beyond the scope of this book. On the other hand, how can one show that two graphs *are not* isomorphic? The definition of isomorphism begins with the existential quantifier: Two graphs are isomorphic if *there exists* an isomorphism φ between them. Recall from Section 1.2 how to refute an existentially quantified proposition, in this case to show that two graphs are not isomorphic: We must show that no possible choice of φ can preserve the adjacency–nonadjacency structure. To do this directly would be tedious for small graphs and quite infeasible for large ones, since there are $n!$ possible one-to-one and onto functions between two vertex sets of the same cardinality. We would have to show that each such function failed to preserve at least one adjacency or nonadjacency.

Fortunately, there is usually a better way to establish that two nonisomorphic graphs are not isomorphic—essentially a proof by contradiction. We find some graph-theoretic property that one of the graphs has and the other lacks. If the graphs were isomorphic, this could not happen; therefore, the graphs are not isomorphic. To make this precise, we begin with the definition of what kind of "graph theoretic property" we need to find. A (**graphical**) **invariant** is a parameter θ of a graph or an assertion P about a graph, such that if two graphs G and H are isomorphic, then $\theta(G) = \theta(H)$ or $P(G) \leftrightarrow P(H)$. In other words, an invariant is a parameter or property that has the same value or truth value on all isomorphic graphs.

For example, the number of edges in a graph is a graphical invariant. If two graphs are isomorphic, then the fact that the adjacency–nonadjacency structure must be preserved guarantees that there must be a one-to-one correspondence between the edges of the two graphs; therefore, any two isomorphic graphs must have the same number of edges. The property of being connected is a graphical invariant: If two graphs are isomorphic, then they are either both connected or both not connected. The reason in this case is that if two graphs are isomorphic via an isomorphism φ, then any walk from a vertex u to a vertex v in the first graph is sent by φ to a walk from $\varphi(u)$ to $\varphi(v)$ in the second, and any walk in the second is sent by φ^{-1} to a walk in the first. Indeed, it is hard to come up with a property of a graph that is *not* an invariant. Having *Fred* as an element of the vertex set is not an invariant. Neither is an assertion that depends on the way in which a graph is drawn; for example, whether edges cross is not a graphical invariant (see Figure 8.26).

Here is a (certainly not complete) list of useful invariants:

- The number of vertices.
- The number of edges.

- The number of components (or, more simply, whether the graph is connected).
- The existence of a 3-cycle as a subgraph (or, more generally, the existence of a subgraph with any particular graphical property).
- The length of the longest cycle (which is called the **circumference** of the graph) or the shortest cycle (which is called the **girth** of the graph).
- The degree sequence.
- Being bipartite.
- The existence of vertices of certain degrees adjacent to vertices of certain degrees.
- Any invariant of the complement.

Note that graphical invariants can be of two varieties—either numbers that can be calculated (the degree sequence, for example) or properties (containing a Hamilton cycle, for example).

The definition says that sharing graphical invariants is a *necessary* condition for isomorphism between graphs. For example, if a graph G contains a Hamilton cycle, then so does any other graph isomorphic to G. The sharing of a set of invariants is not *sufficient*, however. For example, it is possible for two nonisomorphic connected graphs each to have exactly eight vertices, 12 edges, degree sequence $(3, 3, 3, 3, 3, 3, 3, 3)$, girth 4, and circumference 8; see Figure 8.27 and Exercise 33.

 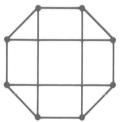

FIGURE 8.27   Two nonisomorphic graphs.

A good practical strategy for solving the problem of whether two given graphs are isomorphic is to look first for some graphical invariant that differs between the given graphs, particularly the degree sequence. (It makes sense to start with the easiest invariants.) If this fails to prove that the graphs are not isomorphic, then it may be wise to look for an isomorphism, using the information gained so far to narrow down the choices (e.g., any isomorphism must send a vertex of degree 4 to a vertex of degree 4). If this, too, fails, it may make sense to try again to prove that the graphs are not isomorphic, trying all the invariants listed above and using cleverness to find more. Often, more than one good invariant demonstrates that two given graphs are not isomorphic.

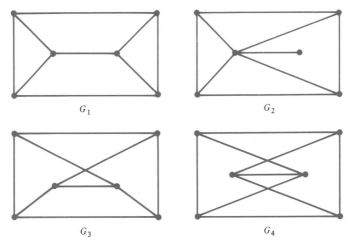

FIGURE 8.28 Some of these graphs are isomorphic.

EXAMPLE 10. Determine which pairs of graphs in Figure 8.28 are isomorphic.

Solution. We begin by looking for differences in graphical invariants. In this case, all four graphs have six vertices and nine edges, so these trivial tests fail to tell us anything. Let us look at the degree sequences. Graphs G_1, G_3, and G_4 are all **regular** of degree 3 (i.e., each vertex has degree 3); on the other hand, G_2 has degree sequence $(5, 3, 3, 3, 3, 1)$ rather than $(3, 3, 3, 3, 3, 3)$. Hence we can conclude that G_2 is not isomorphic to any of the other three.

At this point let us look for small cycles. Graph G_1 certainly contains some triangles (3-cycles)—two of them, in fact. So does G_4, although that might be harder to see, since the drawing does not display the triangles so clearly. On the other hand, we can convince ourselves after a few minutes that G_3 contains no triangles: The shortest cycle has length 4. (In fact, G_3 is bipartite, as can be readily demonstrated, so by Exercise 31 in Section 8.2, it has no cycles of odd length.) Since containing a triangle is a graphical invariant, we now know that G_3 is not isomorphic to either of the other graphs.

Finally, we need to compare G_1 and G_4. After several minutes, perhaps, of looking for more graphical invariants, we might switch tactics and try to find an isomorphism between them. Fortunately, the composer of this problem had some compassion for the solvers, because the drawing suggests an isomorphism. Suppose that we take the picture of G_4 and mentally "untangle" the middle two vertices by having them change places (without disturbing any of the adjacencies). With just a little mental effort (move the vertex on the right to the left, and move the vertex on the left to the right), we can see that the picture then becomes identical to the picture of G_1. Hence G_1 and G_4 are isomorphic; the isomorphism is given by the function that sends each point on the outer rectangle of G_1 to point on the outer rectangle of G_4 in the same relative position (upper left to upper left, etc.), the inside point on the left in G_1 to the inside point on the right in G_4, and the inside point on the right in G_1 to the inside point on the left in G_4. Under this function, all adjacencies and nonadjacencies are preserved. ◆

No one knows of a complete set of graphical invariants that are easy to compute, that is, one fixed set of easily computed invariants that will *always* distinguish between nonisomorphic graphs. If there were such a complete set of invariants for graphs, the isomorphism problem could be solved easily and algorithmically. In fact, there are no known simple, efficient algorithms for determining whether two graphs are isomorphic, although the search for one is currently an important area of research. (Exercise 32 asks for a proof that graph isomorphism is a problem in the class *NP* (see Section 4.4), but no one knows whether it is in *P*. Many people suspect that it is *not NP*-complete.)

We end this section by looking at the classification problem for various classes of graphs. Before stating the problem in its precise mathematical form, let us look at an important application to chemistry (see Exercise 42 for more details). We can view the structure of a molecule as a graph: The atoms are the vertices and the chemical bonds between atoms are the edges. (The actual model is usually more complicated than this, since, for instance, there are different kinds of chemical bonds, but we will keep our discussion simple.) It sometimes happens that two different molecules contain the same numbers of atoms of each element (and hence have the same chemical formula) but have different structures, and hence are compounds with different chemical and physical properties. The graph models for the two molecules are not isomorphic. Chemists call such pairs "structural isomers." For example, both butane and isobutane have four carbon atoms and 10 hydrogen atoms, and hence have the chemical formula C_4H_{10}; but the arrangement of the atoms is different: The graphs modeling butane and isobutane are not isomorphic. A chemist who wants to find all the different compounds with a given formula needs to find all the different—nonisomorphic—graphs that can model these compounds. In fact the nineteenth-century British mathematician A. Cayley predicted the existence of certain structural isomers from the graph theoretic models, before chemists discovered them in the laboratory. (Cayley looked at chemicals containing n carbon atoms and $2n + 2$ hydrogen atoms, called alkanes.)

The general mathematical problem is this:

> given a description of a collection of graphs, determine,
> up to isomorphism, all graphs that fit the description.

In other words, we want a complete catalog of all the graphs in the collection, except that we want each graph in the catalog to be really different from—not isomorphic to—each of the others. We sometimes express this by saying that we want to enumerate all the nonisomorphic graphs in the collection (or, more precisely, all the *equivalence classes* of graphs in the collection under the isomorphism relation, with each equivalence class represented by one graph, isomorphic to all the other graphs in that equivalence class). The subject of graphical enumeration concerns itself with listing and counting the number of nonisomorphic graphs of various types (see Exercise 44).

EXAMPLE 11.   Find all nonisomorphic connected graphs that have five vertices and six edges. How many are there?

Solution.   We need to proceed systematically if we want to be sure not to leave out any graphs, nor include two isomorphic graphs. One fairly effective way to organize our work in this case is by the graph's circumference: We concentrate on the longest cycle in such a graph. (Exercise 10 in Section 9.1 asks you to prove that a graph with at least as many edges as vertices must contain a cycle; for now it should be fairly clear after a little experimentation that a graph with five vertices and six edges must contain a cycle.)

First, suppose that the graph has a 5-cycle. This accounts for five of the edges, and there is only one more edge to add. No matter which edge is added, the result is always the same, up to isomorphism: a pentagon with a diagonal joining one of its vertices to one of the two vertices across from it. Thus there is only one nonisomorphic graph with five vertices and six edges that contains a 5-cycle; it is shown in Figure 8.29a.

Second, suppose that the graph contains no 5-cycle but does contain a 4-cycle, say a, b, c, d, a. The graph is connected and there is a fifth vertex e, so there must be an edge from e to one of the other four vertices. Without loss of generality, we may suppose that it is edge ae. Now we need to determine where the sixth edge can go. It cannot go from e to either b or d, since that would give us a 5-cycle, and we have already considered all such graphs with 5-cycles. It may go from e to c, however, and, since the resulting graph contains no 5-cycle, this graph is different from our earlier one (see Figure 8.29b). Alternatively, the sixth edge may join two vertices on the 4-cycle; it may be either ac (Figure 8.29c) or bd (Figure 8.29d). The two graphs we obtain are not isomorphic (the former has a vertex of degree 4 while the latter does not), and neither is isomorphic to our other graph without a 4-cycle, since the latter does not contain a vertex of degree 1, as these do. Thus we have four graphs in our catalog so far.

Finally, what if there is no 4-cycle or 5-cycle? With a little ad hoc reasoning, similar to what we have just done, the reader can show that there is exactly one more graph meeting the criteria: two triangles sharing one common vertex (Figure 8.29e). Thus the class in question contains exactly five graphs.   ◆

In general, graph classification problems are quite difficult, and there are usually no simple formulas for the number of nonisomorphic graphs with a given set of properties. Some of the exercises allow the reader to get his or her hands dirty in some fairly messy cases.

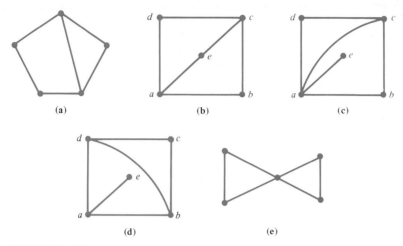

FIGURE 8.29 All nonisomorphic connected graphs with five vertices and six edges.

The notion of isomorphism is absolutely central to modern mathematics. We will encounter it again when we define isomorphism for various types of trees in Section 9.1. Much of the research done by pure mathematicians is, in a sense, trying to solve classification problems in various settings.

SUMMARY OF DEFINITIONS

adjacency list representation of the graph or digraph G: a list of the neighborhoods of each vertex of G.

adjacency matrix for the graph or digraph $G = (V, E)$ with respect to the ordering $v_1, v_2, \ldots, v_n$ of V: the matrix $A_G = (a_{ij})$ given by $a_{ij} = 1$ if $v_i v_j \in E$ and $a_{ij} = 0$ if $v_i v_j \notin E$.

circumference of the graph G: the length of the longest cycle in G (e.g., a graph on n vertices that has a Hamilton cycle has circumference n).

girth of the graph G: the length of the shortest cycle in G (e.g., if a bipartite graph has a cycle, then its girth is at least 4).

incidence matrix of the graph $G = (V, E)$, where $V = \{v_1, v_2, \ldots, v_n\}$ and $E = \{e_1, e_2, \ldots, e_m\}$: the n by m matrix $C = (c_{ij})$ whose entries are given by $c_{ij} = 1$ if v_i is incident to e_j and $c_{ij} = 0$ if v_i is not incident to e_j.

(graphical) invariant: a parameter of a graph G that has the same value for all graphs isomorphic to G, or a proposition about a graph G that has the same truth value for all graphs isomorphic to G.

isomorphic graphs or digraphs: G and H are isomorphic if and only if there is an isomorphism between them; written $G \cong H$.

isomorphism between the graphs or digraphs $G = (V, E)$ and $H = (W, F)$: a bijective function φ from V to W such that for all vertices u and v of G, $uv \in E$ if and only if $\varphi(u)\varphi(v) \in F$.

neighbor of the vertex v: a vertex adjacent to v.

neighborhood of the vertex v: the set of neighbors of v.

regular graph: A graph G is regular of degree k if and only if every vertex of G has degree k (e.g., the n-cycle is regular of degree 2 for all $n \geq 3$).

sparse graph or digraph: a graph or digraph having relatively few edges, compared to the number of edges it might have with the same set of vertices (e.g., a graph with 100 vertices and 150 edges—out of the 4950 possible edges).

EXERCISES

1. For the graph and digraph shown, find the adjacency matrix. Use the alphabetical order of the vertices.

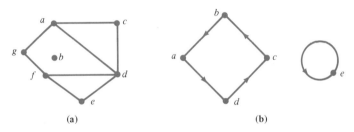

(a)   (b)

EXERCISE 1

2. Draw a picture of the graph whose adjacency matrix is

$$
\begin{array}{c c}
 & \begin{array}{ccccc} a & b & c & d & e \end{array} \\
\begin{array}{c} a \\ b \\ c \\ d \\ e \end{array} &
\left[\begin{array}{ccccc}
0 & 1 & 1 & 0 & 1 \\
1 & 0 & 0 & 1 & 0 \\
1 & 0 & 0 & 1 & 1 \\
0 & 1 & 1 & 0 & 1 \\
1 & 0 & 1 & 1 & 0
\end{array}\right].
\end{array}
$$

3. Draw a picture of the digraph whose adjacency matrix is

$$
\begin{array}{c c}
 & \begin{array}{c c c c c} a & b & c & d & e \end{array} \\
\begin{array}{c} a \\ b \\ c \\ d \\ e \end{array} &
\left[\begin{array}{c c c c c}
1 & 0 & 1 & 1 & 0 \\
1 & 1 & 0 & 1 & 0 \\
0 & 1 & 1 & 0 & 1 \\
1 & 0 & 1 & 0 & 1 \\
1 & 0 & 0 & 0 & 0
\end{array} \right].
\end{array}
$$

4. Determine the number of edges in each of the following, looking only at the adjacency matrix.
   (a) The graph whose adjacency matrix is given in Exercise 2.
   (b) The digraph whose adjacency matrix is given in Exercise 3.

5. Determine the following degrees of vertices, looking only at the adjacency matrix.
   (a) $d(c)$ for the graph whose adjacency matrix is given in Exercise 2.
   (b) $d^+(b)$ for the digraph whose adjacency matrix is given in Exercise 3.
   (c) $d^-(b)$ for the digraph whose adjacency matrix is given in Exercise 3.

6. Determine the number of walks from a to b in each of the following cases.
   (a) Walks of length 3 in the graph whose adjacency matrix is given in Exercise 2.
   (b) Walks of length at most 3 in the graph whose adjacency matrix is given in Exercise 2.
   (c) Walks of length 4 in the digraph whose adjacency matrix is given in Exercise 3.
   (d) Walks of length at most 4 in the digraph whose adjacency matrix is given in Exercise 3.

7. Write down the adjacency list representation of each of the following graphs.
   (a) The graph depicted in Exercise 1a.
   (b) The graph whose adjacency matrix is given in Exercise 2.

8. Write down the adjacency list representation of each of the following digraphs.
   (a) The digraph depicted in Exercise 1b.
   (b) The digraph whose adjacency matrix is given in Exercise 3.

9. Write down the adjacency matrix and draw a picture of the graph whose adjacency lists are

$$
\begin{array}{c l}
a: & c, d \\
b: & d \\
c: & a \\
d: & a, b
\end{array}
$$

10. Write down the adjacency matrix and draw a picture of the digraph whose incoming adjacency lists are

$$
\begin{array}{ll}
a: & a, b \\
b: & - \\
c: & b, c
\end{array}
$$

11. Construct an incidence matrix for the graph in Exercise 1a.

12. Refer to Figure 8.26.
   (a)   Find an isomorphism different from the one given in Example 9.
   (b)   Find a one-to-one correspondence between the vertex sets of these two graphs that is *not* an isomorphism.

13. Determine the number of trails from a to b (as opposed to arbitrary walks) in the situations given in Exercise 6.

14. Determine the number of paths from a to b (as opposed to arbitrary walks) in the situations given in Exercise 6.

15. For each of the following graphs, indicate what the adjacency matrix looks like (with appropriate use of 0's, 1's, and dots).
   (a) K_n.
   (b) C_n.
   (c) $K_{m,n}$.

16. Without doing any matrix multiplication, write down the matrix A^{423}, where A is the adjacency matrix of the digraph depicted in Exercise 1b.

17. What do the following quantities in the adjacency matrix for a digraph tell about the digraph?
   (a)   The sum of the entries in the ith row.
   (b)   The sum of the entries in the jth column.
   (c)   The sum of the entries on the main diagonal.

18. We define the adjacency matrix $A = (a_{ij})$ of a pseudograph by letting a_{ij} be the *number* of edges joining v_i and v_j.
   (a)   Draw a picture of the pseudograph whose adjacency matrix is

$$
\begin{bmatrix}
0 & 2 & 1 & 0 \\
2 & 2 & 0 & 1 \\
1 & 0 & 1 & 3 \\
0 & 1 & 3 & 0
\end{bmatrix}.
$$

   (b)   Is Theorem 1 still valid for pseudographs?

(c)   How can the degrees of the vertices be computed from the adjacency matrix of a pseudograph?

19. We define the adjacency matrix $A = (a_{ij})$ of a multidigraph by letting a_{ij} be the *number* of edges from v_i to v_j.
    (a)   Draw a picture of the multidigraph whose adjacency matrix is

$$\begin{bmatrix} 0 & 2 & 1 & 0 \\ 1 & 0 & 2 & 1 \\ 0 & 3 & 1 & 0 \\ 0 & 1 & 1 & 0 \end{bmatrix}.$$

    (b)   Is Theorem 1 still valid for multidigraphs?
    (c)   How can the in-degrees and out-degrees of the vertices be computed from the adjacency matrix of a multidigraph?

20. Explain how to determine the number of edges of a graph from the adjacency list representation of the graph.

21. Devise an efficient algorithm for constructing the incoming adjacency list representation of a digraph from its outgoing adjacency list representation. Write the algorithm in pseudocode.

22. The accompanying diagram shows traffic flow capacities on the major freeways of a fictitious metropolitan area, in hundreds of cars per minute. Write down a matrix for representing this weighted graph, using appropriate entries for the missing edges.

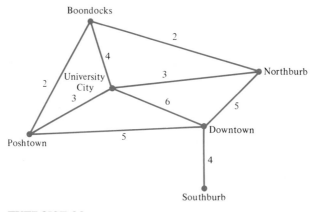

EXERCISE 22

23. Construct an edge list representation for the weighted graph in Exercise 22. Write the list in nonincreasing order of capacities.

24. Prove in detail that if G and H are isomorphic graphs (i.e., $G \cong H$), then G has a Hamilton cycle if and only if H has a Hamilton cycle.

25. Fill in the details of the proof, which was outlined in this section, that the relation "is isomorphic to" is an equivalence relation on graphs.

26. Determine whether the pairs of graphs or digraphs shown on p. 588 are isomorphic: Either label the vertices and exhibit an isomorphism, or find a graphical invariant on which the graphs or digraphs differ. (Some of these are easy; some are quite difficult.)

27. Find all of the following, up to isomorphism.
    (a) All graphs with four vertices.
    (b) All graphs with five vertices and four edges.
    (c) All graphs with six vertices and at most eight edges that contain a Hamilton cycle.
    (d) All digraphs with five vertices and two edges.
    (e) All regular graphs of degree 2 with 15 vertices.
    (f) All regular graphs of degree 3 with 15 vertices.

28. Prove that if $G \cong H$, then $\overline{G} \cong \overline{H}$.

29. A graph is called **self-complementary** if it is isomorphic to its complement.
    (a) Find a self-complementary graph with four vertices.
    (b) Find two nonisomorphic self-complementary graphs with five vertices.
    (c) Show that there is no self-complementary graph with three vertices.

30. Prove that Q_n is not isomorphic to any $K_{i,j}$ if $n \geq 3$.

31. Determine whether the digraphs that have the following adjacency matrices are isomorphic.

$$
A_1 = \begin{array}{c} \\ a \\ b \\ c \\ d \\ e \\ f \end{array}
\begin{array}{c} \begin{array}{cccccc} a & b & c & d & e & f \end{array} \\
\left[\begin{array}{cccccc}
1 & 0 & 0 & 1 & 0 & 1 \\
0 & 0 & 1 & 0 & 1 & 1 \\
1 & 0 & 1 & 1 & 0 & 1 \\
0 & 0 & 0 & 1 & 1 & 0 \\
0 & 0 & 0 & 0 & 0 & 1 \\
1 & 1 & 0 & 0 & 0 & 0
\end{array} \right] \end{array},
\qquad
A_2 = \begin{array}{c} \\ u \\ v \\ w \\ x \\ y \\ z \end{array}
\begin{array}{c} \begin{array}{cccccc} u & v & w & x & y & z \end{array} \\
\left[\begin{array}{cccccc}
1 & 0 & 0 & 1 & 0 & 0 \\
1 & 1 & 0 & 0 & 1 & 0 \\
0 & 0 & 0 & 1 & 1 & 1 \\
0 & 0 & 0 & 0 & 1 & 0 \\
0 & 1 & 1 & 0 & 0 & 0 \\
1 & 1 & 0 & 0 & 1 & 1
\end{array} \right] \end{array}.
$$

32. Explain in detail why the graph isomorphism problem is in the class NP.

33. Show that the two graphs in Figure 8.27 are not isomorphic.

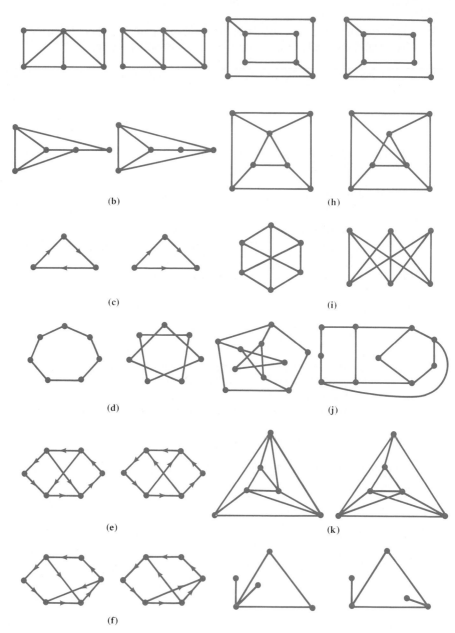

(b)

(h)

(c)

(i)

(d)

(j)

(e)

(k)

(f)

EXERCISE 26

Challenging Exercises

34. Find all of the following, up to isomorphism.
    (a)   All graphs with six vertices and six edges.
    (b)   All digraphs with four vertices and three edges, but no loops.

35. Show that there *is* a "universal" graphical invariant that will work for all pairs of graphs or digraphs (i.e., an invariant that will always distinguish between nonisomorphic graphs or digraphs). The catch is that this invariant is very hard to compute. [*Hint*: Define a distinguished adjacency matrix for a graph or digraph; that will be the invariant.]

36. Prove that for every positive integer n, there exists a self-complementary graph (see Exercise 29) with n vertices if and only if $n \equiv 0$ or $1 \pmod 4$. [*Hint*: For the necessity of this condition, look at the number of edges. For the sufficiency, proceed by induction, showing how to go from n to $n + 4$.]

37. Devise an algorithm for determining whether the graph G contains a triangle (i.e., whether G has K_3 as a subgraph). Assume that the graph is stored in terms of its adjacency lists, and have your algorithm run in at most $O(\max(ev, v^2))$ steps, where G has v vertices and e edges.

Exploratory Exercises

38. Define isomorphism for pseudographs and for multidigraphs. The answers are not unique.

39. Write in pseudocode an algorithm (as inefficient as need be) for determining whether two given graphs are isomorphic.

40. Implement on a computer the algorithm for determining connectedness discussed in this section.

41. Write computer programs to convert from matrix representation of a graph or digraph to adjacency list representation, and vice versa.

42. Find out more about chemical structure graphs by consulting Balaban [13], Balaban, Kennedy, and Quintas [14], and Hansen and Jurs [139].

43. Since the adjacency matrix of a graph is a matrix of real numbers, a person familiar with linear algebra can compute its eigenvalues. Find out what the eigenvalues reveal about the graph by consulting A. J. Hoffman [150].

44. Find out more about graphical enumeration by consulting Harary and Palmer [142] and Read [242].

45. Find out about **random graphs** by consulting Bollobás [26] and Palmer [231]. For example, suppose that we generate a graph with 100 vertices by flipping a coin for each of the 4950 pairs of distinct vertices in the graph and joining the pair by an edge if and only if the coin lands with heads up. Is it very

likely that the graph will have a Hamilton cycle, an Euler tour, or K_{10} as a subgraph? Is the graph likely to be connected? Is there likely to be a vertex of degree 7?

46. One of the hardest open problems in graph theory is the reconstruction problem for graphs. Suppose that someone has a graph G with n vertices. For each vertex v of G, he deletes vertex v and all incident edges, obtaining a graph G_v. Now he presents to you the n graphs G_v, with no labels on the vertices. You must reconstruct G. The twentieth-century mathematician Stanislaw Ulam has conjectured that the graph you obtain will be isomorphic to G.

   (a)   Suppose that the graphs given to you are two C_3's and two $K_{1,2}$'s. Find G.

   (b)   Explain how to determine the number of edges in G from the graphs given to you.

   (c)   Consult O'Neil [225] for more information about this problem.

   (d)   Consult Albers and Alexanderson [4] to find out about Stanislaw Ulam.

SECTION 8.4
PLANARITY OF GRAPHS

Whilst my physicians by their love are grown
Cosmographers, and I their map, who lie
Flat on this bed.

—John Donne

Many of our pictures of graphs contain edges that cross. The resulting clutter can make the graphs confusing and unpleasing to look at. Figure 8.30, for example, shows two pictures of K_4, the one on the left with a pair of edges crossing in the middle, the one on the right redrawn so as to remove the crossing.

In some applications it is important to have a physical model of a graph drawn in the plane without such crossings. For example, an electrical circuit etched onto a printed circuit board can be thought of as a graph: The electronic components are the vertices, and the connections between the electronic components are the edges. If two edges were to cross on the board, an unintended "short circuit" would result.

Sometimes crossings can be eliminated, as we saw in Figure 8.30, and, as we will see in this section, sometimes they cannot. We will study the properties of graphs drawn in the plane without edges crossing, and we will give a characterization of exactly which graphs can be so drawn. Essentially everything that we say in this section except Theorem 2 applies to pseudographs as well as to graphs. Also, since our concern in this chapter is geometric, there would be no point in worrying about directions on edges; hence we will not be referring to digraphs or multidigraphs here.

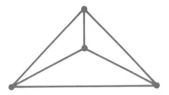

FIGURE 8.30   Two pictures of K_4.

Let $G = (V, E)$ be a graph (or pseudograph). A **planar embedding** of G is a picture of G in the plane such that the curves or line segments that represent edges intersect only at their endpoints. A graph that *has* a plane embedding is called a **planar** graph.

We will compromise the terminology somewhat and refer to "a graph embedded in the plane": The curves or line segments in the plane that represent edges of the graph are thought of *as* the edges of the graph, and the points in the plane that represent vertices of the graph are thought of *as* the vertices of the graph. Thus in a plane embedding, an edge cannot pass through itself, another edge, or a vertex, except at its endpoints.

We must be careful, throughout this section, to distinguish between graphs and their drawings in the plane. A planar graph can be drawn in the plane in many ways, and different embeddings may have different geometric properties. In Figure 8.31, for example, we see two embeddings of the same graph in the plane. In the embedding on the right, it is possible to draw a curve between the two vertices of degree two, without crossing any of the edges or touching the vertices; this is not possible for the embedding on the left.

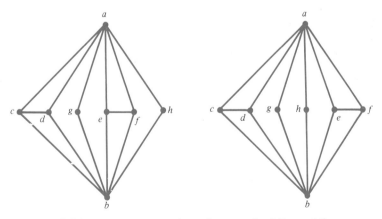

FIGURE 8.31   Two nonequivalent planar embeddings of the same graph.

We begin our study with some definitions and a major theorem about graphs embedded in the plane.

EULER'S FORMULA

When a graph is embedded in the plane, the portion of the plane *not* occupied by the points and lines (or curves) representing the graph consists of one or more connected **regions**. The best way to visualize this is to imagine drawing the graph on a sheet of paper and then cutting, with a razor blade, along all the edges. Each separate piece of paper that results from this operation is a region. Since every graph is finite, precisely one of the regions is **unbounded** (in a sense, "goes off to infinity").

We want to derive a relationship that exists between the *number* of regions formed by a graph embedded in the plane and some numerical invariants of the graph. Let $G = (V, E)$ be a graph (or pseudograph) embedded in the plane. Then we will let

$v = $ the number of vertices of G (i.e., $|V|$),
$e = $ the number of edges of G (i.e., $|E|$),
$c = $ the number of components of G (see Section 8.2), and
$r = $ the number of regions determined by the embedding of G.

It turns out that the number of regions does not depend on the way in which the graph is embedded in the plane. For example, both embeddings of the graph depicted in Figure 8.31 produced eight regions, including the unbounded region. We have $v = 8$, $e = 14$, $c = 1$, and $r = 8$ in both cases. The next theorem tells us how each of these four invariants can be determined from the other three.

THEOREM 1. Euler's Formula. *For every graph (or pseudograph) embedded in the plane,*

$$v - e + r = c + 1.$$

Proof. Our proof will proceed by induction on e, the number of edges of the graph. In the base case, $e = 0$. If there are no edges, then each vertex of the graph is in a component by itself. Therefore, $v = c$. Also, there is clearly only one region (intuitively, a piece of paper with v pinholes in it). Thus the equation $v - e + r = c + 1$ reads $v - 0 + 1 = v + 1$, which is true. We now assume the inductive hypothesis that the theorem is true for all graphs with e edges and try to verify it for a graph with $e + 1$ edges. (Everything we say in this proof applies to pseudographs as well as to graphs, but for simplicity, we will refer only to graphs.)

Any embedding of a graph with $e + 1$ edges can be obtained from an embedding of a graph with e edges: Simply draw the entire graph except for one edge, and then draw in the last edge. By our inductive hypothesis, we can assume that

$v - e + r = c + 1$ before the $(e + 1)$th edge is added. There are two cases to consider, depending on whether the new edge joins vertices in separate components or vertices in the same component.

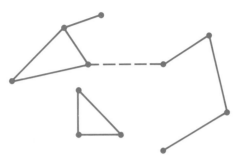

FIGURE 8.32   The $(e + 1)$th edge (dashed) joins vertices in separate components.

If the new edge joins vertices in separate components (see Figure 8.32), then the number of regions does not change: The points of the plane just to one side of the new edge are still in the same region as the points of the plane just to the other side. Thus r remains the same. On the other hand, the number of components (c) decreases by one, since two components coalesce into one. Also, of course, e increases by one when we add the new edge, but v remains unchanged. Therefore, in the equation $v - e + r = c + 1$, which was valid before the new edge was added (by the inductive hypothesis), both the left-hand side and the right-hand side decrease by one. Thus the formula still holds when the new edge is added, as desired.

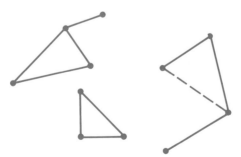

FIGURE 8.33   The $(e + 1)$th edge (dashed) joins vertices in the same component.

On the other hand, if the new edge joins vertices in the same component (see Figure 8.33), then a new region is formed: The old region into which the new edge was drawn is split into two. Thus both e and r increase by one; but v and c remain unchanged. Since $v - e + r = c + 1$ holds before the addition, and these changes leave both sides of the equation with no net change, the equation still holds. This completes the proof by induction. ■

EXAMPLE 1. Consider the pseudograph embedded in the plane as depicted in Figure 8.34.

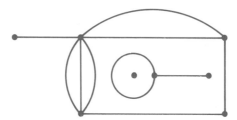

FIGURE 8.34 A pseudograph embedded in the plane with $v = 8$, $e = 10$, $r = 6$, and $c = 3$.

This pseudograph clearly has eight vertices and 10 edges. There are three components: the isolated vertex inside the loop, the piece that looks like a lollipop lying on its side, and the rest of the pseudograph. There are six regions: the inside of the lollipop (except for the isolated vertex), the inside of the rectangle (outside the lollipop), three regions formed between parallel edges, and the unbounded region. Euler's formula $v - e + r = c + 1$ becomes $8 - 10 + 6 = 3 + 1$ or $4 = 4$. ◆

In many cases our graphs are connected, that is, there is only one component. Setting $c = 1$ in Theorem 1, we obtain the following special case of Euler's formula as a corollary.

COROLLARY. *For every connected graph (or pseudograph) embedded in the plane,*

$$v - e + r = 2.$$

Euler's formula is a profound and powerful truth about the plane (see also Exercise 27). We now use the formula in a proof by contradiction to show that certain graphs *cannot* be embedded in the plane. To carry out this plan, we need two consequences of Euler's formula, which tell us that a planar graph cannot have too many edges.

THEOREM 2. *Let G be a planar graph (no loops or parallel edges are permitted) with at least two edges. Then*

$$e \leq 3v - 6.$$

Proof. Since G is a planar graph, we can consider a fixed embedding of G in the plane. We first derive an inequality relating the number of edges of G and the number of regions formed by this embedding. Each region can be considered as having a border made up of edges. For each region, we can count the number of edges forming its border, counting an edge twice if both of its "sides" border on the region. [Figure 8.35 shows a graph embedded in the plane with four regions. The triangular region is bordered by three edges. The large rectangular ring-shaped region has eight bordering edges—the four sides of the outer rectangle, plus the four sides of the inner rectangle. The small rectangular region has six bordering edges—the four sides of the rectangle plus the stray edge sticking into the rectangle counted twice (both of its sides abut the region). Finally, the unbounded region has 11 bordering edges, counting in this way.]

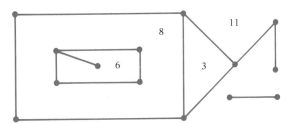

FIGURE 8.35 A graph embedded in the plane, showing the number of edges bordering each region.

Now since the graph has no loops or parallel edges, every region except the unbounded one is bounded by a polygon with at least three sides, and hence each such region has at least three bordering edges. The unbounded region has at least three bordering edges as well (here is where we need the fact that G has at least two edges). Thus if we add up the number of bordering edges for each region, we will obtain a number that is at least $3r$. On the other hand, since each edge contributes a border to exactly two regions (possibly the same region twice), this sum must equal $2e$. [For example, in Figure 8.35 the sum is $3 + 8 + 6 + 11 = 28$, and $e = 14$.] Therefore, we have $2e \geq 3r$ or $r \leq 2e/3$. Combining this with Euler's formula, which guarantees that $v - e + r = c + 1 \geq 2$ and solving for e, we obtain the advertised inequality. (The reader is asked to perform this routine algebra in Exercise 11.) ∎

Theorem 2 says that planar graphs are extremely sparse. A graph with n vertices could have as many as $n(n-1)/2 \in O(n^2)$ edges; a planar graph can have only $3n - 6 \in O(n)$ edges. Thus in some sense most graphs are not planar. Our next theorem, whose proof is similar to that of Theorem 2 and is left to the reader (Exercise 12), gives an even stronger inequality for planar graphs without triangles. In particular, it gives us a smaller bound for the number of edges in any planar bipartite graph. (Recall from Exercise 31 in Section 8.2 that a bipartite graph has no odd cycles, hence no triangles.)

THEOREM 3.   *Let G be a planar graph with at least two edges, in which no cycle has length less than 4. Then*

$$e \leq 2v - 4.$$

Note that each of these two theorems only gives a necessary condition for planarity, not a sufficient condition. It is definitely *not* a theorem that any graph that satisfies the inequalities has to be planar. An example is discussed in Exercise 13.

KURATOWSKI'S THEOREM

We observed that "most" graphs are nonplanar. Let us find some examples. The simplest graphs to study are the complete graphs and the complete bipartite graphs. We saw in Figure 8.30 that K_4 is planar (and hence so are K_1, K_2, and K_3). The complete graph on five vertices is not planar, however. Similarly, it is easy to see that the complete bipartite graphs $K_{1,n}$ and $K_{2,n}$ are planar for all n (see Exercise 10), but the complete bipartite graph with three vertices in each part is not planar. Let us prove the two nonplanarity claims we have made here.

THEOREM 4.   *The complete graph K_5 is not planar, and the complete bipartite graph $K_{3,3}$ is not planar.*

Proof. K_5 has five vertices and $C(5,2) = 10$ edges. If K_5 were planar, Theorem 2 would imply that $10 \leq 3 \cdot 5 - 6 = 9$, which is obviously false. Therefore, K_5 is not planar. Similarly, $K_{3,3}$ has six vertices and $3 \cdot 3 = 9$ edges. If $K_{3,3}$ were planar, Theorem 3 would imply that $9 \leq 2 \cdot 6 - 4 = 8$, again a contradiction. Therefore, $K_{3,3}$ is not planar.   ∎

We could have demonstrated that other graphs are nonplanar, but we will see that these two graphs occupy a special place among all nonplanar graphs. For one thing, they are minimally nonplanar: If any edge is removed from them, the result is planar. (The reader is asked in Exercise 2 to draw planar embeddings of K_5 with one edge removed, and of $K_{3,3}$ with one edge removed.) Even more, in a sense that we will make precise shortly, any nonplanar graph is nonplanar precisely because of K_5 or $K_{3,3}$.

Our goal is to characterize planar graphs (or, equivalently, to characterize nonplanar ones). Theorem 5 gives us a sufficient condition for a graph to be nonplanar, namely that the graph be K_5 or $K_{3,3}$. We would like a kind of converse to this theorem, a nice *necessary* condition for nonplanarity.

Certainly these are not the only nonplanar graphs, so *being* K_5 or $K_{3,3}$ is not a necessary condition for nonplanarity. Any graph that contained K_5 or $K_{3,3}$ as a subgraph would have to be nonplanar as well, since an embedding of a graph that contained K_5 or $K_{3,3}$ would provide an embedding of K_5 or $K_{3,3}$. So a first guess at a characterization of nonplanar graphs might be to conjecture that a graph is nonplanar if and only if it contains K_5 or $K_{3,3}$ as a subgraph. Unfortunately, this conjecture is not quite true; the graph shown in Figure 8.36 is also not planar.

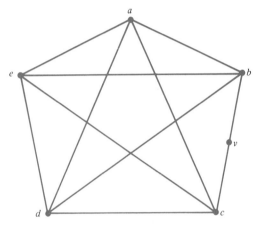

FIGURE 8.36   A subdivision of K_5.

The reason that this graph is not planar is *not* that it is drawn with edges crossing. The issue is whether it *can be* drawn so that edges do not cross. To see that this graph is not planar, we observe that *geometrically* this graph is just like K_5. The path b, v, c acts exactly as an edge bc would, insofar as the problem of drawing the graph in the plane is concerned.

On the other hand, it turns out that the *only* thing wrong with our conjecture is what is illustrated in Figure 8.36. Let us define precisely what that is. The graph $G' = (V', E')$ is said to be obtained from the graph $G = (V, E)$ by an **edge bisection** if $V' = V \cup \{x\}$, for some $x \notin V$, and $E' = (E - \{uv\}) \cup \{ux, xv\}$, for some edge $uv \in E$. A graph obtained from G by a sequence of zero or more edge bisections is called a **subdivision** of G.

Thus G' is obtained from G by placing a new vertex x in the middle of some edge uv, that is, replacing the edge uv by two edges, ux and xv. For example, the graph in Figure 8.36 is obtained from K_5 by bisecting the edge bc with the new vertex v; thus the graph in Figure 8.36 is a subdivision of K_5.

It should be clear that any graph that contains (as a subgraph) a subdivision of K_5 or $K_{3,3}$ cannot be planar, since geometrically a subdivided edge behaves just like an edge. What is not so clear—the proof is rather involved, and we omit it—is that the converse also holds. This fact was discovered and proved around 1930 by the Polish mathematician K. Kuratowski.

THEOREM 5.   *A graph is nonplanar if and only if it contains a subdivision of K_5 or $K_{3,3}$ as a subgraph.*

EXAMPLE 2.   The Petersen graph (see Example 12 in Section 8.1) is nonplanar. The Petersen graph looks something like K_5 in its usual drawing. Indeed, if there were a planar embedding of the Petersen graph, then we could shrink to a point each edge af, bg, ch, di, and ej (using the labels shown in the left-hand part of Figure 8.10) and obtain an embedding of K_5, a contradiction to Theorem 4. However, the Petersen graph does not contain K_5 or a subdivision of K_5. Instead, it contains a subdivision of $K_{3,3}$, as Figure 8.37 shows. We omit vertex x from the Petersen graph (and the three edges incident to x), leaving a subgraph with nine vertices and 12 edges. We claim that this subgraph is a subdivision of $K_{3,3}$. Indeed, the two parts are labeled $\{a, b, c\}$ and $\{1, 2, 3\}$ in Figure 8.37, and the subdivided edges are shown with heavy lines.   ◆

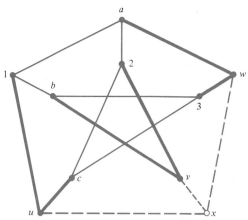

FIGURE 8.37   The Petersen graph contains a subdivision of $K_{3,3}$.

ALGORITHMIC CONSIDERATIONS

In practice, how can one actually determine whether a given graph is planar? If the graph is presented to us already drawn in the plane without edge crossings, then of course it is planar. If it is initially drawn with one or more crossings (or if a first attempt at drawing a picture of the graph from its adjacency matrix or adjacency list representation results in some edge crossings), then one can try mentally (or with paper and pencil, or, better yet if one is available, with an interactive graph theory computer graphics package) to move vertices and edges around until the crossings are eliminated. If this does not succeed, one can look for subgraphs that are subdivisions of K_5 or $K_{3,3}$; finding such a subgraph proves that the graph is not planar. This approach should be used in the exercises that ask you to determine whether a given graph is planar.

However, let us briefly consider the question from a more systematic point of view, less dependent on trial and error. Theorem 5 gives a necessary and sufficient condition for a graph to be planar: A graph G is planar if and only if G does not contain a subdivision of K_5 or $K_{3,3}$ as a subgraph. This suggests the following algorithm for checking planarity. For each subgraph H of G, successively replace all vertices of degree 2 by edges (in effect, "undo" any edge bisections), and then see whether the result is K_5 or $K_{3,3}$. Although the procedure is well defined, a little thought will convince the reader that it is not very efficient, since any graph has a very large number of subgraphs (at least $2^{|E|} - 1$). Perhaps a more serious flaw for practical applications is that this algorithm (or, indeed, the theorem itself) gives no indication of *how* to draw the graph in the plane without edge crossings, once it has been determined that it is planar.

Fortunately, there are better algorithms for determining planarity, and the algorithms are constructive in that they actually produce an embedding of a planar graph in the plane. What is more, the embeddings can be made to have the nice property that all the edges are represented by straight line segments. In fact, *there is an algorithm with time complexity in $O(n)$ for determining whether a graph G with n vertices is planar and, if it is, producing a straight-line planar embedding of G.* We cannot go into the details, but the reader is urged to look at Exercise 34.

SUMMARY OF DEFINITIONS

bisect edge e of graph or pseudograph G: to form a new graph or pseudograph G' which is identical to G except that edge $e = uv$ is replaced by one new vertex w and two new edges uw and wv; a picture for G' can be obtained from a picture for G simply by drawing a new point on edge e.

planar graph or pseudograph: a graph or pseudograph that has a planar embedding.

planar embedding of the graph or pseudograph G: a picture of G in the plane such that the line segments or curves that represent the edges of G intersect only at their endpoints.

region of a planar embedding of a graph or pseudograph: a maximal connected portion of the plane that remains when the curves, line segments, and points representing the graph are removed.

subdivision of the graph or pseudograph G: a graph or pseudograph obtained from G by a sequence of zero or more edge bisections.

unbounded region in the planar embedding of a graph or pseudograph: the one region that contains points arbitrarily far away from the graph or pseudograph.

EXERCISES

1. For each of the following pseudographs, determine the values of v, e, r, and c, and verify that Euler's formula holds.
    (a)   Figure 8.31, considered as one graph with two components (ignore the labels on the vertices).
    (b)   The pseudograph depicted in Figure 8.35.
    (c) and (d) See the accompanying figure.

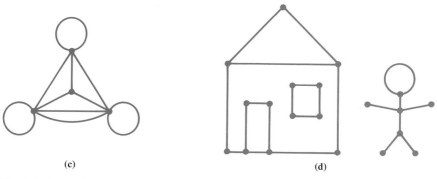

(c)   (d)

EXERCISE 1

2. Exhibit planar embeddings of the following graphs.
    (a) K_5 with one edge deleted.
    (b) $K_{3,3}$ with one edge deleted.

3. Determine the set of values of n for which K_n is planar.

4. Determine which of the graphs shown are subdivisions of K_4.

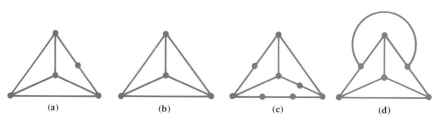

(a)   (b)   (c)   (d)

EXERCISE 4

5. An old brain-teaser asks whether it is possible to draw nonintersecting curves joining each of three houses to each of three utility companies (gas, water, and electricity), so that the curves do not cross. Explain what this brain-teaser has to do with this section, and solve the puzzle.

6. We showed in this section that not every graph can be embedded in the plane (i.e., two-dimensional space) without edges crossing. Can every graph be embedded in three-dimensional space without edges crossing?

7. Draw a planar embedding of the following graph such that all the edges are straight line segments.

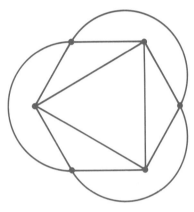

EXERCISE 7

8. Exhibit planar embeddings of the following graphs such that each edge is represented by a straight line segment.
   (a) K_5 with one edge deleted.
   (b) $K_{3,3}$ with one edge deleted.

9. Let G be a pseudograph, and let H be the graph obtained from G by deleting all loops and by replacing all sets of parallel edges by single edges. Show how a planar embedding of H can be used to construct a planar embedding of G.

10. Determine the set of values of the pair (m, n) for which $K_{m,n}$ is planar.

11. Verify the algebraic step in the proof of Theorem 2: $(2e \geq 3r \land v - e + r \geq 2) \rightarrow e \leq 3v - 6$.

12. Prove Theorem 3.

13. Consider the graph G, which consists of K_5 together with five new vertices and five new edges, one edge joining each of the new vertices to one of the

vertices of the K_5, with no two of the new vertices joined to the same vertex of K_5.
(a)   Draw a picture of G.
(b)   Determine v and e, the number of vertices and the number of edges of G, respectively.
(c)   Verify that $e \leq 3v - 6$ holds for G.
(d)   Show that G is not planar.
(e)   Explain why G is not a counterexample to Theorem 2.

14. Give an example of a connected graph (no loops or parallel edges allowed) with eight vertices and 18 edges that is
(a)   Planar.
(b)   Nonplanar.

15. Find two embeddings of a connected planar graph with five vertices that have substantially different geometric properties (as was the case with the embeddings shown in Figure 8.31); explain how they differ.

16. Show in each case that any graph satisfying the conditions cannot be planar.
(a)   A graph with 23 vertices and 65 edges.
(b)   A graph with 23 vertices, 62 edges, and two components.
(c)   A graph with 23 vertices, 40 edges, three components, and no triangles.

17. Determine whether the 4-cube Q_4 is planar.

18. Show that Theorem 2 is the best possible inequality obtainable if $v \geq 3$. In other words, show that for each $v \geq 3$ there exists a planar graph for which $e = 3v - 6$. (We express this fact by saying that the inequality given in Theorem 2 is **sharp** for all $v \geq 3$.)

19. Show that the inequality given in Theorem 3 is sharp (see Exercise 18) for all $v \geq 4$.

20. A **triangulation of the plane** is an embedding of a planar graph (no loops or parallel edges allowed) in such a way that every region (including the unbounded region) is a triangle (bounded by exactly three edges).
(a)   Give examples of triangulations of the plane with two regions, with four regions, and with six regions.
(b)   In any triangulation of the plane, any one of the following numbers determines the other two: v, e, and r. Find the six formulas implicit in this statement.
(c)   Show that there is no triangulation of the plane with an odd number of regions.
(d)   Show that for every even integer $r \geq 2$ there is a triangulation of the plane with r regions.

21. Give an example to show that the restriction in Theorems 2 and 3 that G have at least two edges is necessary, in the sense that there exist graphs with fewer than two edges for which the inequalities do not hold.

22. Suppose that a connected planar graph has degree sequence $(4, 4, 4, 3, 3, 3, 3, 3, 3, 3, 3, 3, 3, 2, 2, 2, 2, 2)$.
    (a)   Without drawing any pictures, calculate the number of regions that are formed when this graph is embedded in the plane.
    (b)   Draw a picture of such a graph embedded in the plane.

23. Suppose that a connected graph G in which all vertices have degree 4 is embedded in the plane so that it has 10% more regions than it has vertices.
    (a)   Without drawing any pictures, calculate the number of vertices in G.
    (b)   Draw a picture of such a graph embedded in the plane.

24. In each case, either produce a planar embedding of the graph whose adjacency matrix is shown, or show that the graph is not planar.

(a)

$$
\begin{array}{c}
 \\ a \\ b \\ c \\ d \\ e \\ f \\ g \\ h
\end{array}
\begin{array}{cccccccc}
a & b & c & d & e & f & g & h \\
\left[\begin{array}{cccccccc}
0 & 1 & 1 & 0 & 1 & 1 & 0 & 0 \\
1 & 0 & 1 & 0 & 1 & 1 & 0 & 1 \\
1 & 1 & 0 & 1 & 0 & 1 & 1 & 1 \\
0 & 0 & 1 & 0 & 0 & 1 & 0 & 1 \\
1 & 1 & 0 & 0 & 0 & 1 & 0 & 0 \\
1 & 1 & 1 & 1 & 1 & 0 & 0 & 0 \\
0 & 0 & 1 & 0 & 0 & 0 & 0 & 1 \\
0 & 1 & 1 & 1 & 0 & 0 & 1 & 0
\end{array}\right]
\end{array}.
$$

(b)

$$
\begin{array}{c}
 \\ a \\ b \\ c \\ d \\ e \\ f \\ g
\end{array}
\begin{array}{ccccccc}
a & b & c & d & e & f & g \\
\left[\begin{array}{ccccccc}
0 & 1 & 1 & 0 & 1 & 1 & 1 \\
1 & 0 & 0 & 1 & 1 & 1 & 0 \\
1 & 0 & 0 & 0 & 1 & 1 & 1 \\
0 & 1 & 0 & 0 & 0 & 1 & 1 \\
1 & 1 & 1 & 0 & 0 & 1 & 1 \\
1 & 1 & 1 & 1 & 1 & 0 & 1 \\
1 & 0 & 1 & 1 & 1 & 1 & 0
\end{array}\right]
\end{array}.
$$

25. Determine whether each of the given graphs (see p. 604) is planar.

Challenging Exercises

26. Determine whether each of the given graphs (see p. 604) is planar.

27. Show that Euler's formula is valid, as well, for the sphere. (The formula needs to be modified to be valid for other surfaces, such as the surface of a donut (see Exercise 37).)

28. State and prove an analogue of Theorem 3 for graphs with girth g (recall that the girth of a graph is the length of the shortest cycle in the graph).

29. If a graph is not planar, it cannot be drawn in the plane. Graph theorists have devised several graphical invariants to study the "degree of nonplanarity." The **thickness** of a graph $G = (V, E)$ is the minimum number of planes needed to embed G; more precisely, it is the minimum number k such that the edge set E can be partitioned as $E = E_1 \cup E_2 \cup \cdots \cup E_k$ so that each of the subgraphs $G_1 = (V, E_1), G_2 = (V, E_2), \ldots, G_k = (V, E_k)$ is planar. (The designer of

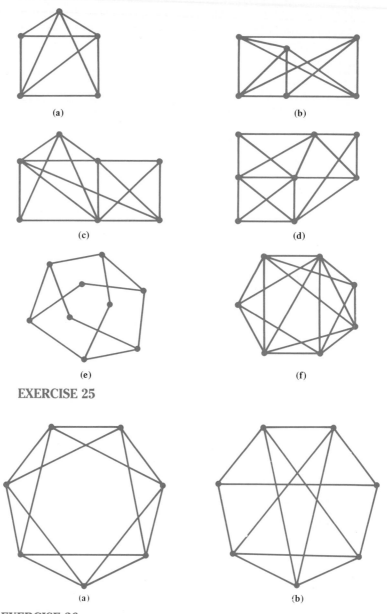

(a)

(b)

(c)

(d)

(e)

(f)

EXERCISE 25

(a)

(b)

EXERCISE 26

printed circuits might like to know the thickness of a graph being etched on the circuit boards, for this number would tell the number of boards needed for the circuit.)

(a)   Show that K_{11} has thickness 3. [*Hint*: There are two things to prove. First, you must find three planar graphs on an 11-element vertex set that together have an edge between every pair of distinct vertices. Second, you must show that it is impossible to do so with two planar graphs (try using Euler's formula in a proof by contradiction).]

(b)   Show that $K_{7,7}$ has thickness 3.

(c)   Show that $K_{7,7}$ with one edge deleted has thickness 2.

30.   A **Platonic** graph is a connected planar graph in which every vertex has the same degree and every region has the same number of bordering edges, each of these quantities being at least 3 (but not necessarily equal to each other). Find all Platonic graphs.

31.   A two-person game is played as follows. First $n \geq 3$ points are drawn in the plane, such that no three of them are collinear. The players play alternately, and a play consists of drawing a straight line segment between two of the points in such a way that it does not intersect any previously drawn line segments. The first player who cannot draw a line segment in this manner loses. Show that the outcome of this game depends only on the arrangement of points, not on the strategies adopted by the players, and determine the winner of the game (first player or second player) as a function of the arrangement.

32.   Solve Exercise 49 in Section 6.2 by applying Euler's formula. [*Hint*: Use Exercise 44 in Section 6.2.]

Exploratory Exercises

33.   In a two-person game called **sprouts** (invented by J. Conway) the players construct a graph embedded in the plane as follows. First n points are drawn in the plane. The players play alternately, and a play consists of drawing a new vertex in the plane (not on any previously drawn curve), and joining that vertex with a curve (an edge) to two of the vertices already present (possibly to the same vertex twice), in such a way that no edges ever intersect except at their endpoints, and the degree of each vertex never exceeds 3.

(a)   Play the game with a friend, starting with about 5 points.

(b)   Determine the winner of the game (first player or second player), assuming that the players adopt optimal strategies, for $n = 1$ and $n = 2$.

(c)   Determine a good bound on how many moves the game can last.

(d)   Find out more about sprouts (and a variant called Brussels sprouts) by consulting Gardner [96] and Prichett [240].

34.   Find out more about efficient algorithms for determining whether a graph is planar and finding embeddings of planar graphs (including straight-line embeddings) by consulting Bondy and Murty [27], Gibbons [114], Hopcroft and Tarjan [158], and Reingold, Nievergelt, and Deo [243].

35. In Exercise 29 we discussed a measure of nonplanarity of a graph, its thickness. Another good measure of the nonplanarity of a graph is its **crossing number**, defined to be the minimum number of places at which edges must cross in a picture of the graph drawn in the plane, if no edge may pass through a vertex except at its endpoint, and no point (other than a vertex) can have three or more edges pass through it.

(a) Find three graphs without vertices of degree 2 that have crossing number 1. [*Hint*: You need to show two things: that the graph is not planar and that it can be pictured with only one crossing.]

(b) Find out more about crossing numbers by consulting Chartrand and Lesniak [41] and Gardner [101].

36. Still another good measure of the nonplanarity (see Exercise 29) of a graph is its **book number**, defined to be the minimum number of "pages" needed in a book in order to draw the graph, if the vertices are placed along the "spine" of the book and the edges extend out into the pages.

(a) Find the book numbers of K_4 and K_5.

(b) Find out more about book numbers by consulting Chung, Leighton, and Rosenberg [46].

37. Nonplanar graphs can be embedded on surfaces other than the plane; see Exercise 27, for example. The complexity of the surface needed to embed a graph, called the **genus** of the surface, is another measure of the degree to which the graph is nonplanar (see Exercise 29). Find out about genus by consulting Chartrand and Lesniak [41].

38. The phenomenon exhibited in Kuratowski's theorem occurs in other areas of graph theory: a small collection of graphs being responsible for a graph's not having some property (K_5 and $K_{3,3}$ in the case of planarity). Find out more about these "obstructions" by consulting Wilf [322].

SECTION 8.5
COLORING OF GRAPHS

Four colors suffice!

**—University of Illinois postage
meter message in 1976**

In this section we consider the graph-theoretic model of a fairly common scheduling problem. The model will involve assigning "colors" to the vertices of a graph. We will see that this model also applies to coloring geographical maps.

Suppose that $n \geq 1$ committees have been appointed from a group of people. The committees need not be disjoint. Each committee needs to meet once a week.

How can a schedule be set up so that no two committees with a member in common meet at the same time?

One solution, obviously, is to have each committee meet at a different time. This requires n meeting times. Can the number of meeting times t be made smaller than n? If so, what is the minimum possible number $t_{\min}$ of meeting times? And how can a schedule with only $t_{\min}$ meeting times be produced? If all the committees are disjoint, then of course there is no problem: All the committees can meet simultaneously, so $t_{\min} = 1$. If the committees overlap in complicated ways, however, the problem is nontrivial.

A graph model works well for this problem. Let us represent the problem with a graph $G = (V, E)$. The vertices of G will be the committees. Thus $|V| = n$. Two vertices will be adjacent if the committees are *not* disjoint. Thus each edge indicates an obstruction to the scheduling process that must be overcome. Let us call the meeting times $1, 2, \ldots, t$. A scheduling then consists of a function from V to $\{1, 2, \ldots, t\}$ such that two adjacent vertices have different function values (two committees that are not disjoint have different meeting times). The first use of a model such as this was to the problem of coloring the countries on a geographical map so that adjacent countries had different colors (a problem we will study at the end of this section), so historically the codomain of this function was thought of as a set of colors, and the function was called a coloring. Thus our problem is, given a graph G, to find a coloring with the smallest possible number of colors.

CHROMATIC NUMBER

Let us be a bit more formal. Let $G = (V, E)$ be a graph. A *t*-**coloring** of G is a function $C : V \to \{1, 2, \ldots, t\}$ such that $C(u) \neq C(v)$ whenever $uv \in E$. A graph that has a t-coloring is said to be *t*-**colorable**. Clearly, any graph that is t-colorable is also t'-colorable for all $t' > t$. The minimum value of t for which there exists a t-coloring of G is called the **chromatic number** of G, and it is denoted $\chi(G)$. Thus G is t-colorable if and only if $\chi(G) \leq t$.

Informally, we will call the value $C(v)$ the **color** of vertex v under the coloring C, and we will often use "red," "blue," and so on, for elements of the codomain of C, rather than numbers from 1 to t. Note that we have referred to a graph here rather than a digraph or pseudograph. Clearly, directions on edges have no bearing on our problem. If a graph had a loop at a vertex v, then the graph would have no t-coloring for any t, since $C(v) = C(v)$. Parallel edges do not change adjacencies, so any problem about coloring a multigraph is equivalent to a problem about coloring the graph obtained by deleting extra parallel edges. Thus we have not lost any generality in working only with graphs.

EXAMPLE 1. Consider the graph G depicted in Figure 8.38 (the five-spoked wheel). We have indicated a 4-coloring of G in Figure 8.38, calling the colors red, blue, green, and yellow. Therefore the chromatic number of G is at most 4, that is, $\chi(G) \leq 4$. In fact $\chi(G) = 4$. To prove this, we must show that there is no coloring with three colors. We proceed by contradiction. Suppose that there were a 3-coloring of G, say with the colors red, blue, and green. Without loss of generality, let us assign the middle vertex m the color red. Since m is adjacent to all the other vertices, we will need to 2-color the vertices around the rim of the wheel, using the colors blue and green. Again without loss of generality, let us assume that a receives the color blue. This forces b to be green, which in turn forces c to be blue and therefore d to be green. But now vertex e is adjacent to a red vertex (m), a blue vertex (a), and a green vertex (d) and hence cannot receive any of these colors. This contradiction shows that G is not 3-colorable. ◆

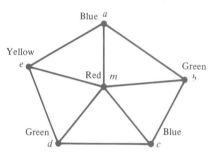

FIGURE 8.38   A 4-coloring of the five-spoked wheel.

Let us see how to apply the concept of graph coloring to a scheduling problem of the type we introduced above.

EXAMPLE 2.   Suppose that Adam, Beverly, Charles, Deborah, Evan, Felicity, and Greta are planning a ball. Adam, Charles, and Deborah constitute the publicity committee. Charles, Deborah, and Felicity are the refreshment committee. Felicity and Adam make up the facilities committee. Beverly, Charles, Deborah, and Evan form the decorations committee. Evan, Adam, and Greta are the music committee. Evan, Felicity, and Charles form the cleanup committee. How many meeting times are necessary in order for each committee to meet once?

Solution.   We draw a graph G with one vertex for each committee, as shown in Figure 8.39. Each vertex is labeled with the first letter of the name of the committee that it represents. Two vertices are adjacent whenever the committees have at least one member in common. For instance, P is adjacent to R, since Charles is on both the publicity committee and the refreshment committee.

To answer the question posed, we need to determine the chromatic number of G. Note that each of the four vertices P, D, M, and C is adjacent to each of the others. Thus in any coloring of G these four vertices must receive different colors, say 1, 2, 3, and 4, respectively, as shown in Figure 8.39. Now R is adjacent to all

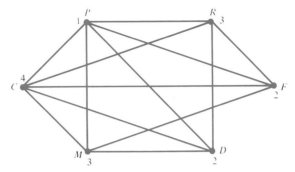

FIGURE 8.39   Graph model for the ball committee problem.

of these except M, so it cannot receive the colors 1, 2, or 4. Similarly, vertex F cannot receive colors 1, 3, or 4, nor can it receive the color that R receives. But we could color R with color 3 and F with color 2. This completes a 4-coloring of G. Since we also saw that *at least* four colors were required, we have found that $\chi(G) = 4$. Thus four meeting times are required, and we can read off a possible schedule from Figure 8.39: The publicity committee meets in the first time slot, the facilities committee and the decorations committee (which have no common members) meet in the second time slot, the refreshment and music committees meet in the third time slot, and the cleanup committee (appropriately enough) meets last. Our analysis showed that no fewer time slots will suffice for this problem.   ◆

The chromatic number of a graph is another graphical invariant. Let us state and prove a few easy facts about chromatic number.

THEOREM 1.   *A graph is bipartite if and only if its chromatic number is at most* 2.

Proof.   If the graph G is bipartite, with parts V_1 and V_2, then a coloring of G can be obtained by assigning red to each vertex in V_1 and blue to each vertex in V_2. Since there is no edge from any vertex in V_1 to any other vertex in V_1, and no edge from any vertex in V_2 to any other vertex in V_2, no two adjacent vertices can receive the same color. Conversely, if we have a 2-coloring C of a graph G, let V_1 be the set of vertices that receive color 1 (i.e., $C^{-1}(\{1\})$, in the notation of Section 3.2) and let V_2 be the set of vertices that receive color 2 (i.e., $C^{-1}(\{2\})$). Since adjacent vertices must receive different colors, there is no edge between any two vertices in the same part.   ■

Thus the concept of bipartiteness is really a special case of the concept of graph coloring, or the latter is a generalization of the former.

THEOREM 2.   *If K_n is a subgraph of a graph G, then $\chi(G) \geq n$.*

Proof. In any coloring, the vertices of the K_n contained in G must receive different colors, since they are all adjacent to each other. Thus at least n colors are required. ▪

The converse of Theorem 2 is not true, as Example 1 illustrated. There the chromatic number was 4, but, as is easily seen, K_4 is not a subgraph of the five-spoked wheel.

By an analysis similar to that carried out in Example 1, which we leave to the reader (Exercise 10), we have the following formula for the chromatic number of a cycle.

THEOREM 3. *The chromatic number of the n-cycle is given by*

$$\chi(C_n) = \begin{cases} 2 & \text{if } n \text{ is even} \\ 3 & \text{if } n \text{ is odd.} \end{cases}$$

Finally, we consider another application of graph coloring, this time to a problem faced by communications regulators.

EXAMPLE 3. Television channels are assigned to broadcasting stations by a governmental agency. Obviously, two stations in geographic proximity must get different channels, to avoid reception interference. Suppose that the rule has been adopted that stations within 140 miles of each other (as the crow flies) must have different channels. The grid in Figure 8.40 shows the locations of 15 hypothetical stations. Each square is 50 miles on a side. How many channels are required, and how can they be assigned to comply with the rule?

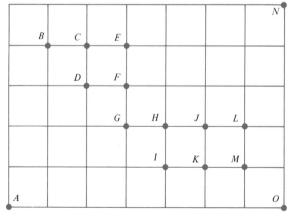

FIGURE 8.40 Locations of television stations A to O.

Solution. We will construct a graph model. The vertices of the graph are the stations; two vertices are adjacent in the graph if the stations are within 140 miles of each other. Using the Pythagorean theorem to compute distances, we obtain the graph G shown in Figure 8.41. For example, $d(HM) = \sqrt{50^2 + 100^2} \approx 112 \leq 140$, so HM is an edge of our graph, but $d(GL) = 150 > 140$, so no edge joins vertex G and vertex L.

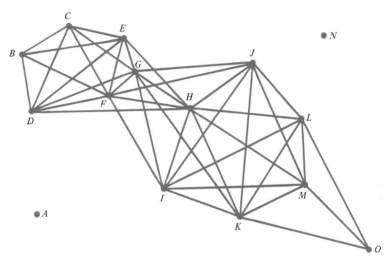

FIGURE 8.41 Graph model of the channel assignment problem.

Now an assignment of channels to stations is precisely a coloring of G: The channels are the colors. Thus to solve this problem, we have to find a coloring of G that uses as few colors as possible. Since $\{H, I, J, K, L, M\}$ form a complete subgraph of G, we know by Theorem 2 that $\chi(G) \geq 6$. On the other hand, it is not difficult to find a 6-coloring of G (as the reader is asked to do in Exercise 11). Thus six channels are both necessary and sufficient, and any 6-coloring of G determines an allowable assignment of channels to stations. ◆

THE FOUR-COLOR THEOREM

In the nineteenth century it was conjectured that the chromatic number of any *planar* graph is at most 4. (Clearly, four colors are required for some planar graphs, such as K_4.) This conjecture, although remarkably easy to state, is apparently very hard to prove. As we mentioned in Section 1.3, a false proof was accepted for several years in the latter part of the nineteenth century, but its error was eventually pointed out. Much of the motivation for graph theory research for the past hundred years has been the quest for a proof of this so-called four-color conjecture. Finally, in 1976, W. Haken and K. Appel from the University of Illinois produced a proof. Their proof was by contradiction. They were able to show that if the four-color conjecture were false, there would be a counterexample (a planar graph that is not 4-colorable) of one of about 1000 specific types. Then, with the aid of several hundred hours of computer calculations, they showed that there could be no counterexample of any of these types. Thus the conjecture became a theorem.

THEOREM 4. The Four-Color Theorem.   *If G is a planar graph, then $\chi(G) \leq 4$.*

Although the proof of the four-color theorem is beyond the scope of this book, we will prove that the chromatic number of any planar graph is at most 5. This proof provides a good example of the kind of reasoning that occurs in graph theory. We begin with a lemma, which says that any planar graph has to have a vertex with degree at most 5. This will allow us to give an inductive proof of the five-color theorem.

LEMMA.   *If G is a planar graph, then G contains a vertex whose degree is at most 5.*

Proof.   Suppose, to the contrary, that the degree of every vertex of G is at least 6. Then the sum of the degrees of the vertices would be at least $6v$, where v is the number of vertices in G. By Theorem 1 in Section 8.1, this sum is equal to $2e$, where e is the number of edges of G; thus we would have $2e \geq 6v$, or $e \geq 3v$. But this contradicts Theorem 2 in Section 8.4, which states that $e \leq 3v - 6$ for any planar graph with at least two edges. Thus G must contain a vertex of degree at most 5. ∎

THEOREM 5. The Five-Color Theorem.   *If G is a planar graph, then $\chi(G) \leq 5$.*

Proof.   We proceed by induction on the number of vertices of G. For the base case, if G has five or fewer vertices, we can assign each vertex a different color to obtain a 5-coloring of G. So assume that the theorem is true for graphs with n vertices, and let G be a graph with $n + 1$ vertices. Embed G in the plane without edges crossing. We must show that G has a 5-coloring.

By the lemma above, we can find a vertex u whose degree is at most 5. We delete u (and all adjacent edges) from G. The result is a planar graph G' with n vertices, so by the inductive hypothesis there is a 5-coloring of G'. We want to extend this coloring to a coloring of G by assigning a color to u in such a way that u is colored differently from all vertices adjacent to it. If the degree of u is strictly less than 5, then there is at least one color not assigned to a neighbor of u, and we obtain the desired coloring of G by assigning that color to u. Even if the degree of u is 5, it might happen that the neighbors of u do not use up all five colors, leaving a color free for u. The only difficulty arises if $d(u) = 5$ and the vertices adjacent to u all have different colors in the given coloring of G'. We assume without loss

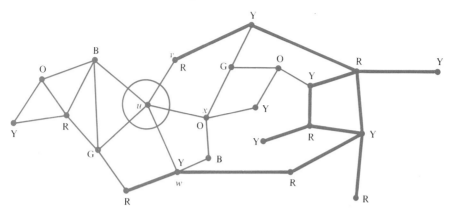

FIGURE 8.42   Vertex u is adjacent to vertices of five different colors.

of generality that the colors are blue, red, orange, yellow, and green, clockwise in that order around u. [The situation is illustrated in Figure 8.42.]

In this case we cannot extend the coloring of G' to a coloring of G, since there is no color available for u. Instead, we will have to modify the coloring of G' first. To this end, let v be the vertex adjacent to u that is colored red in the given coloring of G'. Look at the subgraph of G' consisting of all vertices colored either red or yellow and connected to v by a path using only red or yellow vertices, together with the edges incident to pairs of such vertices. [This subgraph is indicated by the heavy lines in Figure 8.42.] Now we can recolor every vertex in this subgraph, changing red to yellow and yellow to red, and obtain another coloring of G'. If vertex w, adjacent to u and originally colored yellow, is *not* in this subgraph, then after we have performed this recoloring, vertex u will no longer be adjacent to a red vertex, and we can color u red to obtain the desired coloring of G. However, it might happen that w *is* in this subgraph [as we indicate in Figure 8.42], in which case the color of w changes to red, and we are no better off than we were with the original coloring of G'.

Now comes the key part of the proof, exploiting the planarity of the graph G. If w is in the red-yellow subgraph emanating from v, then there is a path from v to w, using only vertices colored red or yellow. This path, followed by the path w, u, v back to v, separates the plane into two regions, one of which contains the vertex x, adjacent to u and colored orange [again, see in Figure 8.42]. Now recolor every green or orange vertex in the region containing x, switching green to orange and orange to green. Since there is a red–yellow "fence" surrounding this region, this recoloring produces another coloring of G' and does not change the colors of any of the other vertices adjacent to u. [In our picture, the region containing x was inside the fence; alternatively, it may be outside the fence, in which case the other two vertices adjacent to u would be inside, and the same conclusion would follow.] Thus the vertices adjacent to u are now all colored either red, green, yellow, or blue: We have freed up the color orange. Then we can color u orange. Having obtained a coloring of G, we have completed the inductive step of our proof. ■

In the remainder of this subsection we will see how to apply graph coloring to the coloring of geographical maps. We will see that the four-color theorem guarantees that any map can be colored with at most four colors.

Imagine a map of countries occupying a certain land area, such as Europe. We will suppose that each country is connected (all in one piece). Some countries will share a common border. An example of such a map is shown in Figure 8.43.

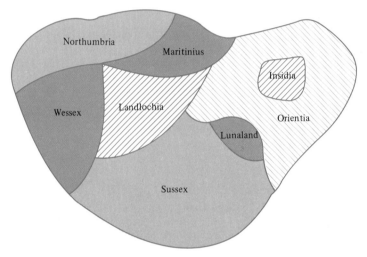

FIGURE 8.43   A geographical map.

Here "common border" means a stretch of some positive length, not just isolated points. In Figure 8.43, for example, we will not say that Wessex and Maritinius share a common border (just as Arizona and Colorado do not share a common border in the United States).

A mapmaker would like to color each country in the map so that the countries will stand out better to the human viewer. Countries with common borders should receive different colors. We have indicated a possible coloring of the map in Figure 8.43 by shading; we used four colors here. How many colors does the mapmaker need, and how can the mapmaker determine a coloring that uses this minimum number of colors? We will transform this map-coloring problem into a problem about coloring *graphs*.

A map, such as that shown in Figure 8.43, is a figure in the plane. We will show how to associate a multigraph embedded in the plane to any such map in such a way that the map-coloring problem is transformed into a coloring problem for this associated multigraph. Then, by ignoring the extra parallel edges, we reduce the problem to a coloring problem for a graph.

Let M be a map drawn in the plane. A multigraph G, embedded in the same plane, called the **multigraph associated with** M, is obtained as follows. A vertex is placed inside each country of M. For each connected stretch of common border between two countries, a curve is drawn from the vertex of the first country to the vertex of the second, passing through this stretch of common border.

EXAMPLE 4. The map drawn in Figure 8.43 is reproduced in Figure 8.44. A vertex has been placed inside each country, and edges have been drawn as explained in the definition. The result is the associated multigraph. ◆

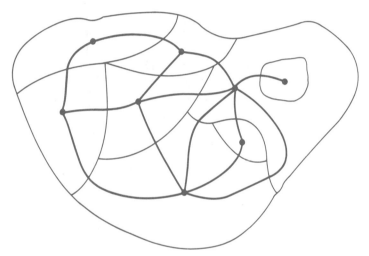

FIGURE 8.44 Associated multigraph of the geographical map in Figure 8.43.

By this definition, two countries share a common border in a map if and only if the vertices representing those countries are adjacent in the associated multigraph. Thus the coloring problem for maps reduces to the coloring problem for multigraphs, and hence to the coloring problem for graphs. Furthermore, it is clear from the construction that the graph we obtain is planar (indeed, we obtain it embedded in the plane). Therefore, we can apply the four-color theorem to obtain the following useful fact for mapmakers.

THEOREM 6. *Any map can be colored with at most four colors.*

COLORING ALGORITHMS

The algorithmically oriented reader (and the reader should, by this time, be at least somewhat algorithmically oriented) will ask the obvious question: *Given an arbitrary graph G, how can one determine its chromatic number $\chi(G)$ and find a $\chi(G)$-coloring of G?* Unfortunately, the answer is: *not very efficiently*. Even determining if an arbitrary graph is 3-colorable is an *NP*-complete problem (see Section 4.4), so it is unlikely that there is an efficient algorithm for determining chromatic number. If a graph is small, we can determine an optimum coloring (one using as few colors as possible) visually by trial and error and ad hoc reasoning, as we have done in Examples 1, 2, and 3. If a graph is large, as it would be, for example, in many real scheduling problems, it would be nice to have an algorithm available. We present here two algorithms for graph coloring. One is not very efficient in the worst case, and the other may not produce a coloring with as few colors as possible, but in practice these algorithms may be satisfactory.

Our first algorithm uses a recursive "backtracking" approach. Essentially, it tries all possible colorings, choosing the best. It is not efficient, requiring, in the worst case, time that is exponential in the number of vertices of the graph. Both the procedure and its recursive subprocedure return two things: a coloring of the graph received as input and the number of colors that that coloring uses. The subprocedure *extend_coloring*, moreover, takes three inputs in addition to G: a partial coloring of G (called C_in in the argument list), the number of colors that the partial coloring uses (called nc_in in the argument list), and the number of the first vertex that is uncolored in the partial coloring (v in the argument list).

The procedure *extend_coloring* takes the partial coloring of G given in C_in, which requires nc_in colors, and returns the best coloring that agrees with this partial coloring (best in the sense of requiring the fewest colors). Its method is to try every possible assignment of a new color to the first uncolored vertex and recursively call itself to determine the best extension of the incoming coloring with that additional assignment. It keeps track of the best among all such colorings (C_out) and returns it. The procedure *color* simply initializes all vertices to be uncolored and then calls *extend_coloring* on G with the empty coloring, which uses no colors, at vertex 1. Because of the exhaustive search that Algorithm 1 conducts, it is clear that it is correct, that is, that it produces a coloring of G using only $\chi(G)$ colors.

EXAMPLE 5.   Apply Algorithm 1 to the graph G (with $n = 6$ vertices) shown in Figure 8.45.

Solution.   We will call the colors blue, green, and red, in that order, to avoid confusion between color numbers and vertex numbers. Initially, the graph is uncolored, and *extend_coloring* is called, with $nc\_in = 0$ and $v = 1$. Vertex 1 is colored blue (it is not already adjacent to a blue vertex), and *extend_coloring* is called with $v = 2$. Since $nc\_in = 1$ for this call, the algorithm will try letting vertex 2 be first blue, and then green (because of the **for** loop). It colors vertex 2 blue and invokes *extend_coloring* again. After several more recursive calls, the algorithm discovers

procedure *color*(*G* : graph)
  {assume that *G* has vertices numbered 1, 2, ..., *n*}
    for *i* ← 1 **to** *n* **do**
      C_initial(*i*) ← 0 {initially all vertices are uncolored}
    return(*extend_coloring*(*G*, *C_initial*, 0, 1))
    {we obtain a minimum coloring by extending the empty coloring}
procedure *extend_coloring*(*G*, *C_in*, *nc_in*, *v*)
  {*C_in* is a partial coloring of a graph *G*, using *nc_in* colors; *v* is the vertex to
  be colored next; the procedure finds a coloring extending *C_in*, using as few
  colors as possible}
    if *v* = *n* + 1 **then** {base case: graph is already completely colored}
      return(*C_in*, *nc_in*)
    nc_out ← ∞
    for *j* ← 1 **to** *nc_in* + 1 **do** {try color *j* on vertex *v*}
      if *v* is not adjacent in *G* to a vertex *u* with *C_in*(*u*) = *j* **then**
        begin
          C_pass ← *C_in* {copy the given coloring}
          C_pass(*v*) ← *j* {assign color *j* to *v*}
          (*C_temp*, *nc_temp*) ← *extend_coloring*(*G*, *C_pass*,
                                    max(*nc_in*, *j*), *v* + 1)
           {recursive call to get complete coloring}
          if *nc_temp* < *nc_out* **then** {we found a new best coloring}
            begin
              nc_out ← *nc_temp*
              C_out ← *C_temp*
            end
        end
    return(*C_out*, *nc_out*)

Algorithm 1. Recursive backtracking procedure to color a graph.

that the best coloring with this partial coloring uses three colors: Vertices 3 and 5 are green, and vertices 4 and 6 are red. Thus *nc_out* is 3 at this point. Next it tries letting vertex 2 be green and invokes *extend_coloring* once more. Again after several additional recursive calls, the algorithm discovers that the best coloring with this partial coloring uses two colors: Vertices 3 and 6 are green, and vertices 4 and 5 are blue. Since this coloring is better than the one obtained with vertex 2 colored blue, the values of *nc_out* and *C_out* are updated, and this coloring is then returned to the first invocation of *extend_coloring*, and from there back to the main program. ◆

    Our second algorithm for coloring an arbitrary graph is not guaranteed to yield a coloring with the fewest colors, but it usually works reasonably well in practice.

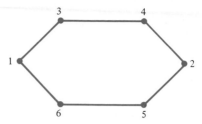

FIGURE 8.45 A 2-colorable graph.

procedure *approximately_color*(G : graph)
  {assume that G has vertices numbered 1, 2, ..., n, listed in order of nonincreasing degrees; $C(i)$ will be the color assigned to vertex i; nc will be the number of colors used}
  for $i \leftarrow 1$ **to** n **do**
    $C(i) \leftarrow 0$ {initially all vertices are uncolored}
  $nc \leftarrow 0$ {initially no colors are used}
  while uncolored vertices remain **do**
    begin
      $nc \leftarrow nc + 1$
      for $i \leftarrow 1$ **to** n **do**
        if $C(i) = 0$ and i is not adjacent to any j such that $C(j) = nc$
          then $C(i) \leftarrow nc$
    end
  return(C, nc) {a not necessarily optimum coloring}

Algorithm 2. An algorithm for coloring a graph, not necessarily with as few colors as possible.

The idea behind Algorithm 2 is quite simple. First color 1 is assigned to all vertices to which it can be assigned, then color 2 is assigned to all vertices to which it can be assigned, and so on. Vertices are considered in an order v_1, v_2, ..., v_n for which $d(v_1) \geq d(v_2) \geq \cdots \geq d(v_n)$; this often helps the algorithm to avoid running into unnecessary conflicts.

EXAMPLE 6. Let us apply Algorithm 2 to the graph G shown in Figure 8.46. The vertices are numbered 1 through 8, in nonincreasing order of degree. Let us call the colors blue, green, and red, in that order. The algorithm first assigns blue to vertex 1. It cannot assign blue to vertices 2 through 6, since they are all adjacent to vertex 1, but vertices 7 and 8 can (and do) receive the color blue. Next the color green ($nc = 2$ now) is assigned to the first uncolored vertex (2), and also to vertices 4 and 5, which are uncolored and not adjacent to vertices already colored green.

Finally, the third color (red) is assigned to vertices 3 and 6, and the coloring is complete.

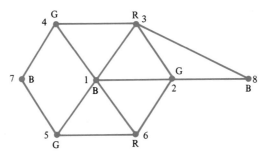

FIGURE 8.46 A 3-colorable graph.

Algorithm 2 found a 3-coloring of this graph, and since the graph contains a triangle, it is clear that it can do no better in this case. ◆

Algorithm 2 is our first example of a "greedy" algorithm. It does what seems best at each step (i.e., color another vertex with the smallest-numbered color available). Unfortunately, *there is no general theorem that says that optimizing at each step achieves an optimum overall result*. Indeed, the number of colors required by Algorithm 2 to color a graph G may be greater than $\chi(G)$; the reader is asked to find an example (Exercise 14). Another greedy coloring algorithm is discussed in Exercise 28; it too can fail to produce a $\chi(G)$-coloring. We will see other examples of greedy algorithms in subsequent chapters. Some of them work (i.e., find optimum solutions to the problems they are trying to solve), and some of them do not. It is usually nontrivial to demonstrate that a greedy algorithm produces the optimum solution if it actually does so.

The reader is asked in Exercise 25 to analyze the time complexity of Algorithm 2; it is quite efficient, especially if the data structures are such that the **for** loop can be implemented without having to search through the entire vertex set.

SUMMARY OF DEFINITIONS

associated multigraph G for the geographical map M in the plane: the multigraph whose vertices are the countries of M, with an edge joining two vertices for every stretch of common border between the two countries those vertices represent; the multigraph can be drawn in the plane by placing a vertex of G inside each country of M and drawing an edge between two vertices passing through each stretch of common border.

chromatic number of the graph G: the smallest t such that G is t-colorable; denoted $\chi(G)$.

color of the vertex v in a graph with the coloring C: $C(v)$—see t-coloring.

t-colorable graph: a graph that has a t-coloring (e.g., any bipartite graph is 2-colorable).

t-coloring of the graph $G = (V, E)$: a function C from V to $\{1, 2, \ldots, t\}$ that sends adjacent vertices to distinct colors; the elements of the codomain are called colors and are often referred to with color names instead of positive integers.

EXERCISES

1. An exam schedule needs to be set up for the following courses: calculus, data structures, discrete mathematics, European history, French, physics, psychology, and Shakespeare. The following pairs of courses (and only these) have students in common: calculus and French (i.e., there is at least one student who is taking both calculus and French), calculus and psychology, data structures and European history, discrete mathematics and French, discrete mathematics and physics, discrete mathematics and psychology, discrete mathematics and Shakespeare, European history and French, European history and Shakespeare, French and psychology, and physics and psychology. The exams must be scheduled in such a way that no student is required to take two exams on the same day. The problem is to determine the minimum number of examination days necessary, and to schedule the examinations.

   (a) Develop a graph model to solve this problem. Write down explicitly what the vertices and edges of the graph represent.

   (b) Solve the problem using the graph model.

2. Either show that the coloring shown in Figure 8.43 is optimum (i.e., that a coloring with fewer colors is not possible), or find a 3-coloring of this map.

3. Draw the map determined by the six New England states together with New York and draw the associated multigraph. Exhibit a coloring of the map using the minimum number of colors.

4. Repeat Exercise 3 for Missouri, Iowa, Illinois, Indiana, Ohio, Pennsylvania, West Virginia, Virginia, North Carolina, Kentucky, Tennessee, Arkansas, Oklahoma, Kansas, and Nebraska.

5. Find colorings of the given maps (p. 621), using as few colors as possible.

6. Draw the planar multigraphs associated with the maps given in Exercise 5.

7. Apply Algorithm 2 to the following graphs. Use the alphabetical order of the vertex names to break ties in the initial ordering of the vertices by degree.

   (a) The graph in Figure 8.39.

   (b) The graph in Figure 8.41.

8. Carry out the modeling exercise discussed in Example 3 (i.e., draw the graph model, determine its chromatic number, and interpret what this means in terms

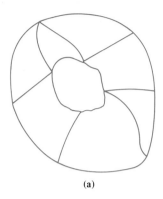

(a)

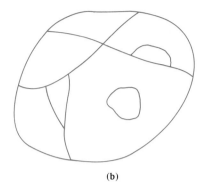

(b)

EXERCISE 5

of the original problem) if it is desired to assign channels to broadcasting stations whose coordinates in the xy-coordinate plane (in miles) are as follows: $(100, 50)$, $(100, 250)$, $(150, 50)$, $(150, 100)$, $(200, 150)$, $(200, 200)$, $(200, 250)$, $(250, 50)$, $(250, 250)$, $(300, 150)$, and $(300, 200)$. Assume that stations assigned the same channel cannot be closer than 140 miles.

9. For each of the following graphs, find colorings using the fewest possible colors.
   (a)   The Petersen graph (see Example 12 in Section 8.1).
   (b) and (c) See the accompanying figure.

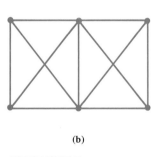

(b)

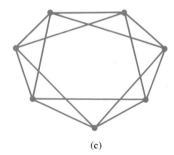

(c)

EXERCISE 9

10. Prove Theorem 3.

11. Find a 6-coloring of the graph in Figure 8.41.

12. Find an efficient algorithm for determining, given a graph G, whether $\chi(G) = 2$.

13. Apply Algorithm 1 to the accompanying graph, using the numbering of the vertices as shown. Organize your answer to convey the recursive nature of the execution of the algorithm.

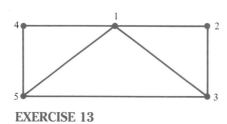

EXERCISE 13

14. Give an example of a graph for which Algorithm 2 does not produce a coloring with the fewest possible colors.

15. Characterize the class of all graphs G for which $\chi(G) = 1$.

16. Prove that if G is a subgraph of H, then $\chi(G) \leq \chi(H)$.

17. An **edge coloring** of a multigraph G is an assignment of "colors" to the edges of G so that edges that have a common endpoint receive different colors. The minimum number of colors required to edge-color G is called the **edge-chromatic number** of G, denoted $\chi'(G)$. For this problem, let d be the maximum degree of the vertices of G.
   (a)   Show that $\chi'(G) \geq d$.
   (b)   Compute the edge-chromatic number of K_4 to show that equality sometimes holds in part (a), that is, that it is possible to have $\chi'(G) = d$.
   (c)   Compute the edge-chromatic number of the graph shown here to prove that the inequality in part (a) may be strict, that is, that it is possible to have $\chi'(G) > d$.

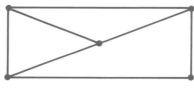

EXERCISE 17

18. Vizing's theorem states that if G is a graph, then $\chi'(G) \leq d+1$, where $\chi'(G)$ is the edge-chromatic number of G and d is the maximum degree of the vertices of G (see Exercise 17). Does the same inequality hold for multigraphs with parallel edges?

19. State and prove a useful proposition about the chromatic number of a graph with more than one component.

20. The graph depicted in Exercise 13 is called **uniquely 3-colorable**.
    (a) In what sense is the 3-coloring of that graph unique, and in what sense is it not unique?
    (b) Give a rigorous definition of what it should mean for a graph to be **uniquely k-colorable**.
    (c) Find a graph with chromatic number 3 that is not uniquely 3-colorable.

21. Determine whether each of the following statements is true or false.
    (a) Every graph can be colored with at most four colors.
    (b) If a graph has chromatic number 6, then it is not planar.
    (c) $\chi(K_5) > 4$.
    (d) $\chi(K_{3,3}) > 4$.
    (e) Every 4-colorable graph is planar.
    (f) Every 2-colorable graph is planar.
    (g) Every 1-colorable graph is planar.

22. The lemma in this section stated that a planar graph has to have a vertex of degree at most 5.
    (a) Prove that the number "5" cannot be reduced, by giving an example of a planar graph that has no vertices of degree less than 5.
    (b) Show that if a planar graph has no vertices of degree less than 5, then it must have at least 12 vertices of degree 5.
    (c) Discuss the situation for multigraphs that can be embedded in the plane. Does the lemma still hold?

23. Apply the ideas in the proof of the five-color theorem to use the given partial coloring of each of the accompanying graphs (p. 624) to construct a coloring of the entire graph (only the circled vertex, which is adjacent to vertices with all five colors, is not colored).

24. Find a formula for the chromatic number of the complement of the n-cycle, $\chi(\overline{C}_n)$.

25. Analyze the time complexity of Algorithm 2. State your answer in terms of the number of vertices, the number of edges, and the chromatic number of the graph.

26. Explain why determining whether $\chi(G) \leq 3$ is in the class NP.

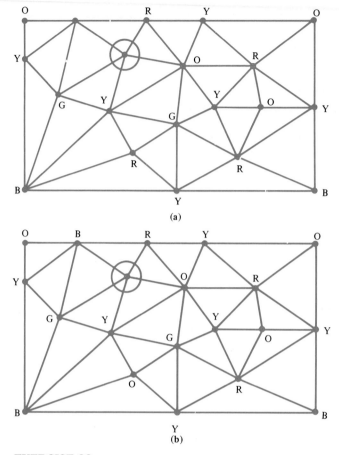

(a)

(b)

EXERCISE 23

Challenging Exercises

27. Give an example to show that in Algorithm 2 the initial ordering of the vertices can change the number of colors required to color the graph. (Both orderings must be nonincreasing with respect to degree.)

28. Another greedy algorithm for graph coloring proceeds as follows. Go through the vertices in arbitrary numerical order and assign to each vertex the smallest color number possible at the time the vertex is processed (i.e., the smallest number that is not already the color of an adjacent vertex).
    (a)   Write this algorithm in pseudocode.
    (b)   Apply the algorithm to the graphs in Figures 8.39 and 8.41, using the alphabetical order of the vertices.

(c)   Find a graph G with the fewest possible number of vertices for which the algorithm may fail to produce a $\chi(G)$-coloring.

29. A graph G is called **k-critical** if its chromatic number is k but the chromatic number of every proper subgraph is less than k. In other words, G requires k colors to color, but if even one edge, or an isolated vertex, is removed from G, then the resulting graph can be colored with fewer than k colors.
(a)   Show that K_n is n-critical.
(b)   Find a graph different from K_4 that is 4-critical.
(c)   Prove that every graph with chromatic number k contains a k-critical subgraph.
(d)   Show that every k-critical graph is connected.
(e)   Show that if G is k-critical, then the graph obtained by deleting exactly one edge from G has chromatic number $k - 1$.

Exploratory Exercises

30. Implement one or more of the coloring algorithms given in this section on a computer, and apply them to some fairly large graphs. Notice how the running times increase with graph size (number of vertices, number of edges) and how close the approximate algorithms come to finding an optimum coloring.

31. Find out more about edge colorings (see Exercises 17 and 18) by consulting Fiorini and Wilson [82].

32. Find out more about the history and the proof of the four-color theorem by consulting Appel and Haken [10,11], Gardner [106], and Saaty and Kainen [264].

33. Find out more about coloring algorithms by consulting Manvel [206].

34. Find out about coloring graphs embedded on other surfaces, such as the surface of a donut, by consulting Stahl [291].

35. Find out about how to count the number of colorings of a graph using t colors by looking up **chromatic polynomial** in Bondy and Murty [27].

36. A map drawn in the plane can be viewed as a multigraph—the borders between countries form the edges of the graph. The multigraph we associated with a map is usually called the **dual** of the map (viewed as a multigraph). Find out more about duals of graphs by consulting Bondy and Murty [27].

TREES

One particular kind of graph, called a tree, is very important in applications to computer science and other areas. It is also quite useful for studying graphs in general. In this chapter we prove some interesting facts about trees and explore some of their many applications.

Our general discussion of trees occupies Section 9.1. We encounter several different categories of trees, and we begin to look at some of the different settings in which tree models apply. In Section 9.2 we present efficient algorithms for finding trees as subgraphs of connected graphs. In many situations it is necessary to search through a tree systematically; we explore different ways of doing this in Section 9.3, with applications to evaluating algebraic and arithmetic expressions. Section 9.4 concerns applications of trees to searching (a topic we first encountered in Section 4.3) and data transmission. Finally in Section 9.5 we see how trees can be applied to finding optimal strategies for games.

The general graph theory references listed in Chapter 8 contain material on trees.

SECTION 9.1
BASIC DEFINITIONS FOR TREES

*How can it be that mathematics, being after all a
product of human thought independent of experience,
is so admirably adopted to the objects of reality?*

—Albert Einstein

Trees come in several varieties. A garden variety tree, so to speak, is just a connected graph with no closed trails; in particular, every tree is a graph. Often, though, it is useful to supply a tree with more structure, such as a distinguished vertex called the root, and a labeling or ordering among certain neighbors of each vertex. Whether two trees are to be considered the same (i.e., isomorphic) depends on how much structure we have endowed the trees with.

More generally, a **forest** is a graph that contains no closed trails. A **tree**, then, is a connected forest, that is, a connected graph with no closed trails. Just as we usually use the letter G to stand for a graph, we will usually use T to stand for a tree. A **rooted tree** is a tree with the additional structure of having one vertex designated as the **root**. (Formally, we could say that a rooted tree is a pair (T, r), where T is a tree and r is an element of the vertex set of T.) Isomorphisms in mathematics must preserve all relevant structure, so *two rooted trees are isomorphic if and only if there is a one-to-one correspondence between their vertex sets that preserves adjacencies, nonadjacencies, and the root.*

Our definition could just as well have said "no cycles" as "no closed trails," since by Theorem 2 in Section 8.2 a graph contains a closed trail if and only if it contains a cycle. We will usually find it easier to use this equivalent formulation. Trees cannot contain loops or parallel edges, not only because our definition specified "graph" and not "multigraph" or "pseudograph," but also because loops or parallel edges would automatically produce closed trails. Each component of a forest must clearly be a connected graph without cycles, so a forest is just the disjoint union of trees (appropriately, enough).

EXAMPLE 1. Determine the following sets.

    (a) All nonisomorphic trees with five vertices.
    (b) All nonisomorphic forests with five vertices.
    (c) All nonisomorphic rooted trees with four vertices.

Solution. For part (a), we can organize our search in terms of the length of the longest path in the tree. The tree could be nothing but a path of length 4 (Figure 9.1a); it could have longest path of length 3, in which case the other vertex must be adjacent to one of the inner vertices of that path (Figure 9.1b); or it could have longest path of

627

length 2, in which case it must look like Figure 9.1c (the four-pointed **star** $K_{1,4}$). Thus there are exactly three nonisomorphic trees with five vertices.

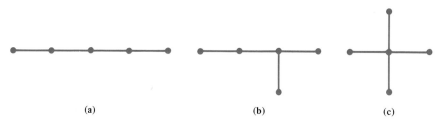

(a)   (b)   (c)

FIGURE 9.1   All nonisomorphic trees with five vertices.

For part (b) we have, first of all, the three graphs in Figure 9.1, since every tree is a forest. In addition, we have the seven additional forests with more than one component, shown in Figure 9.2. These are organized by number of components; a little thought should convince the reader that we have shown them all. Thus there are a total of 10 nonisomorphic forests with five vertices.

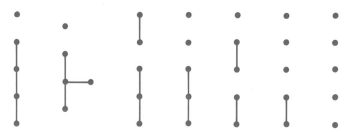

FIGURE 9.2   All nonisomorphic forests with five vertices and more than one component.

Finally, for part (c) we can organize our investigation in terms of the longest path with the root as one of its ends. For now, let us encircle the root in our pictures of rooted trees (we will have a better schematic shortly). Figure 9.3a shows the only rooted tree with four vertices with a path of length 3 coming from the root. If the longest path from the root has length 2, then the other vertex can be "attached" either to the root or to the vertex adjacent to the root (Figure 9.3b and c). Finally, every other vertex might be adjacent to the root (Figure 9.3d). Thus there are four nonisomorphic rooted trees with four vertices. Note that Figure 9.3a and b are isomorphic *as trees* but not *as rooted trees*. As a tree each is simply a path of length 3, but in one case the root has degree 1, while in the other case the root has degree 2. Similarly, Figure 9.3c and d are isomorphic as trees. This illustrates that the notion of "isomorphic" depends on the context (or, as mathematicians call it, the "category") in which we are working. ◆

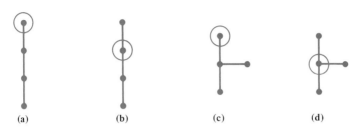

(a)   (b)   (c)   (d)

FIGURE 9.3   All nonisomorphic rooted trees with four vertices.

Any tree other than the trivial tree with only one vertex must contain vertices of degree 1. This fact is useful for giving inductive proofs of theorems about trees, because we are able to remove a vertex of degree 1 and produce a smaller tree. We give a constructive proof here; other proofs are suggested in Exercise 24.

LEMMA.   *A tree with more than one vertex contains at least two vertices of degree* 1.

Proof.   Let v be any vertex of a given tree T with more than one vertex. Then v must be incident to an edge (otherwise, T would not be connected). Construct a trail from v by starting with this edge and continuing as far as possible (recall that a trail may not use the same edge twice). Since T contains no cycles, this trail will never return to a vertex it has visited previously. Hence, because the graph is finite, this trail must eventually end at some vertex $u_1 \neq v$, where it is impossible to continue. The only reason it can be impossible to continue is that there is only one edge incident to u_1, namely the one by which the trail entered u_1. Therefore, u_1 has degree 1.

We still need to find another vertex of degree 1. If $d(v) = 1$, then v is the desired vertex. If not, then there is an edge incident to v other than the edge with which we began the trail above. Construct another trail from v starting with this edge, obtaining a vertex $u_2 \neq v$ of degree 1. Again since T contains no cycles, this trail cannot share any vertices with the first trail, other than v. Thus we know that $u_1 \neq u_2$, and our proof is complete.   ■

We next give some useful alternative characterizations of trees.

THEOREM 1. *Let G be a graph with n vertices. Then the following statements about G are equivalent.*

1. *G is a tree; that is, G is connected and has no closed trails.*
2. *G has exactly n − 1 edges and no closed trails.*
3. *G is connected and has exactly n − 1 edges.*
4. *G is connected, but if any edge is removed from G, then the resulting graph is not connected.*
5. *Given any two vertices u and v in G, there is a unique trail from u to v (and furthermore, this trail is a path).*
6. *G contains no closed trails, but if any edge in $\overline{G}$ is added to G, then the resulting graph contains a cycle.*

Proof. Theorems of this form are typical in mathematics. We defined a tree in a certain way, yet we could just as easily have given other, equivalent definitions; this theorem gives some of those equivalent statements. Note that the theorem is really saying 30 different things, namely that $i \to j$, for each different i and j from 1 to 6. For example, it is saying that there is a unique trail between any two vertices of a tree ($1 \to 5$), and it is saying that if a graph has this unique trail property, then it is a tree ($5 \to 1$). Fortunately, we do not have to prove 30 things in order to prove this theorem, since certain implications will follow logically from others. In fact, it suffices to prove six implications: $1 \to 2 \to 3 \to 4 \to 5 \to 6 \to 1$. The remaining 24 implications then follow by repeated applications of part (h) of Theorem 2 in Section 1.1. We will do four of these six proofs and leave the other two as Exercise 9.

$1 \to 2$. Suppose that G is a tree with n vertices; we want to show that G has exactly $n-1$ edges. We prove this by induction on n. If $n = 1$, then G must consist of just a single vertex and no edges, so G does have exactly $n - 1 = 1 - 1 = 0$ edges. Now let $n > 1$ and assume the inductive hypothesis that all trees with $n-1$ vertices have exactly $(n - 1) - 1 = n - 2$ edges. By the lemma, G contains a vertex x of degree 1. Remove x (and its one incident edge) from G to obtain a graph G' with $n-1$ vertices. We claim that G' is a tree. It certainly has no cycles, since any cycle in G' would have been a cycle in G. To see that G' is connected, let u and v be two vertices in G'. Then there is a walk between them in G, and therefore (by Theorem 1 of Section 8.2) a path from u to v in G. Obviously, this path cannot have used the deleted edge (since the path would have had nowhere to go after reaching x), so this path is a path in G' as well. Thus G' is a tree, so by the inductive hypothesis, G' has $n - 2$ edges. But G contains all the edges of G' plus the one edge we removed; hence G contains $(n - 2) + 1 = n - 1$ edges, as desired.

$2 \to 3$. This time we are given a graph G which has n vertices, $n - 1$ edges, and no closed trails; we must show that G is connected. By definition, G is a forest (i.e., the union of disjoint trees). Thus we know by the previous paragraph that G has $\sum(n_i - 1)$ edges, where the n_i's are the numbers of vertices in the components of G. If there are c components, then this sum equals $(\sum n_i) - c = n - c$. But we are assuming that there are $n - 1$ edges; hence $n - c = n - 1$, or $c = 1$. In other words, G is connected.

$3 \rightarrow 4$. Left to the reader.

$4 \rightarrow 5$. Suppose that G is a connected graph, the removal of any edge of which produces an unconnected graph. Let u and v be two vertices of G. We must show that there is exactly one trail from u to v. Since G is connected, there is a walk from u to v, and hence there is at least one trail (in fact, a path) from u to v by Theorem 1 of Section 8.2. Now suppose that there were two different trails from u to v. By Exercise 16 in Section 8.2, this implies that G contains a closed trail, say $v_0, v_1, \ldots, v_k, v_0$. If we delete edge $v_0 v_k$ of this closed trail, we do not disconnect G, since any walk that had used this edge can use $v_0, v_1, \ldots, v_k$ (in one direction or the other) in its place. This contradicts our hypothesis. Hence there is only one trail from u to v, and as we observed above, this trail is in fact a path.

$5 \rightarrow 6$. Assume that G has the unique trail property. First we must show that G has no closed trails. By Theorem 2 in Section 8.2, it suffices to show that G has no cycles. If G had a cycle, there would be two different paths joining every pair of vertices on the cycle—one obtained by going around part of the cycle forward, the other obtained by going around the rest of the cycle backward; this contradicts the hypothesis. Next we must show that if any edge uv is added to G, then the resulting graph does contain a cycle. This is clear: We follow the path from u to v in G (guaranteed by the hypothesis) and then travel along uv back to u.

$6 \rightarrow 1$. Left to the reader. ■

ROOTED TREES

Although plain (unrooted) trees have many applications (e.g., modeling the molecular structure of many families of chemicals), rooted trees are much more important for computer science and other applications. Once we endow a tree with a root, we can say a lot more about its structure.

By Theorem 1 there is a unique path between every pair of vertices in a tree. In particular, there is a unique path from the root of a rooted tree to every other vertex. We define the **level** of a vertex as the length of this path. Thus the root has level 0, the vertices adjacent to the root have level 1, the vertices adjacent to vertices at level 1 (other than the root) have level 2, and so on. In general, we can assign levels recursively by saying that the root is at level 0, and a vertex is at level l if it has not yet been assigned a level but is adjacent to a vertex at level $l - 1$.

We exhibit this level structure when we draw pictures of rooted trees. Henceforth we will draw rooted trees using the following convention. The root is placed at the top of the picture. Below the root, all in a horizontal row, we place the vertices at level 1. We draw the edges from the root to the vertices at level 1 as straight line segments. Below the vertices at level 1, we place the vertices at level 2, all in the same horizontal row, grouped under the vertices at level 1 to which they are adjacent (since paths are unique, each vertex at level 2 is adjacent to exactly one vertex at level 1). Again we use straight line segments for edges. We continue in this manner until we have drawn the entire tree.

Since the root is always at the top, we do not need to circle it in order to know that it is the root. See Figure 9.4 for a typical picture of a rooted tree. The root in this tree is vertex a.

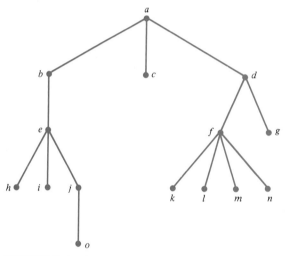

FIGURE 9.4 A rooted tree drawn in the conventional manner.

We define the **height** of a rooted tree to be the largest level number of a vertex occurring in the tree. For example, the tree in Figure 9.4 has height 4. Vertex o is at level 4, and no vertices are at a level greater than 4. If vertex o were removed, the resulting tree would have height 3.

We have defined trees as undirected graphs. We can think of a *rooted* tree, however, as a directed graph, by declaring that all edges are to be directed away from the root. Thus every edge will be directed from a vertex at level l to a vertex at level $l + 1$, for some natural number l. Under this convention, there is a unique directed path from the root to every other vertex, but there are not necessarily *directed* paths between every pair of vertices (e.g., between vertices e and k in

Figure 9.4). We will feel free to think of a rooted tree, then, as either a graph or a digraph, whichever is convenient in a given context.

A **subtree** of a tree T is by definition just a subgraph of T that is also a tree (which is equivalent to saying that it is a connected subgraph of T). More important, certain subtrees of *rooted* trees play a crucial role, as we will see. If T is a rooted tree, thought of as directed away from the root, and if v is a vertex of T, then the **subtree** of T **rooted at** v is the rooted tree containing v (as root) and all vertices that can be reached by a directed path from v, together with the edges of T joining pairs of these vertices. A subtree rooted at a vertex adjacent to the root is called an **immediate subtree** of T.

We can imagine obtaining the subtree rooted at v by snipping the picture just above v and taking the piece that contains v. For example, the subtree of the tree in Figure 9.4 rooted at e is the rooted tree shown in Figure 9.5. Clearly, a rooted tree with n vertices has exactly n subtrees of this type, one rooted at each vertex.

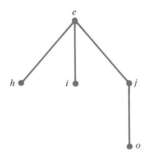

FIGURE 9.5 Subtree of the tree in Figure 9.4, rooted at e.

EXAMPLE 2. Rooted trees can be used to model organizations. Part of a typical university's academic personnel structure is given by the rooted tree shown in Figure 9.6.

Each person is represented by a vertex (drawn here as a box, to make room for the name). The president is the root; she is the chief academic officer of the university. Reporting directly to the president is the provost, so the provost is adjacent to the president, one level lower. The deans report to the provost, so they are shown at the next level, adjacent to the provost. Each dean oversees several department chairpersons, so there are edges from each dean to the department chairpersons he oversees. Our tree also shows some faculty members within departments and even a graduate student being supervised by a faculty member. The subtree rooted at a person is the part of the structure ultimately under the control of that person. ◆

EXAMPLE 3. We can use rooted trees to represent the structure of organized sets. In many computer operating systems, "files" (data or programs) are collected into

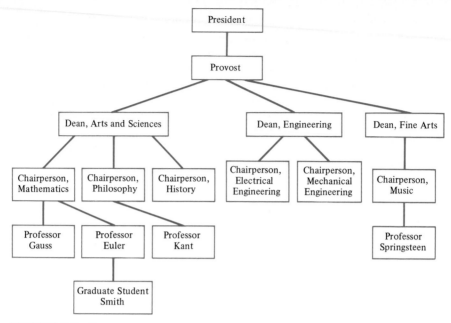

FIGURE 9.6 Rooted tree showing part of the organizational hierarchy at a typical university.

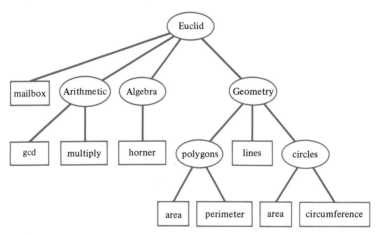

FIGURE 9.7 File and directory organization for a computer user.

"directories," directories collected into other directories, and so on. For example, in Figure 9.7 we show the subtree of the entire structure on a large computer system, the subtree controlled by one user, Euclid.

Directories are shown as ovals, files as rectangles. The user Euclid controls the directory named *Euclid*. At level 1 in this subtree is the data file *mailbox*, which consists of the messages that other users have sent to Euclid, and three directories in which Euclid has chosen to group his programs. He has two arithmetic programs, an algebra program, and five geometry programs; he has chosen to further organize the geometry programs by storing in separate directories those dealing with polygons and those dealing with circles. Note that it is possible to have programs with the same name in two different directories; they are presumably different programs. To specify a program, Euclid would give the path from the root of his subtree to the desired program, using whatever notation the system requires. For example, the program dealing with the area of a circle might be specified by *Geometry > circles > area*. ◆

Rooted trees can be looked at from a totally different, and very useful, point of view. We give a recursive definition of a rooted tree. This definition is the one that the reader should keep in mind as we discuss trees and their applications. In fact, one might say that a rooted tree is the canonical example of a recursive object.

Base case: A single vertex is a rooted tree, with this vertex as root.

Recursive part: If $k \geq 1$ and $T_1, T_2, \ldots, T_k$ are pairwise disjoint rooted trees with roots $v_1, v_2, \ldots, v_k$, then the following graph is also a rooted tree: a new vertex v, which is its root, together with $T_1, T_2, \ldots, T_k$, with v_i adjacent to v for each i from 1 to k (see the schematic in Figure 9.8).

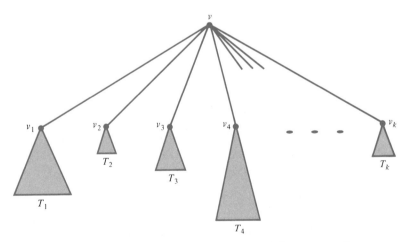

FIGURE 9.8 Rooted tree determined recursively by its immediate subtrees.

Let us use the "days of creation" metaphor (Section 5.1) to see a tree evolve. A rooted tree T with root v is constructed from its immediate subtrees. Each of these subtrees is born prior to the day on which T is born. For example, the tree in Figure 9.4 comes into existence as follows: Vertex o is a tree (born on day 0, by the base case), and so the subtree rooted at j is a tree (born on day 1) by the recursive step. Also, h and i by themselves are both trees, so the subtree rooted at e is a tree born on day 2. By similar reasoning, the subtree rooted at f exists by day 1, and the subtree rooted at d is born on day 2. The subtree rooted at b comes from the recursive step applied to the subtree rooted at e, so it comes into existence on day 3. The single vertex c was already around on day 0, so all three immediate subtrees are in existence by day 3. The entire tree comes into existence on day 4.

The taller a tree, the later its birthday. We use in the following proof the observation (whose proof is left to the reader in Exercise 13) that *the height of a tree is one more than the height of its tallest immediate subtree.*

THEOREM 2.   *A rooted tree has height h if and only if it is born on day h under the recursive definition of rooted tree.*

Proof.   Our proof is by induction on h. Clearly, the only trees born on day 0 are those consisting of a single vertex, and these are the only trees of height 0. Now fix $h > 0$ and assume that the theorem is true for values less than h.

For the "if" half of the theorem, let T be a tree born on day h. Then each of the immediate subtrees of T must have been born on a day prior to day h, and hence by the inductive hypothesis have height less than h. Since the height of T is exactly one more than the maximum of the heights of its immediate subtrees, the height of T is at most h. On the other hand, if the height of T were actually less than h, then by the inductive hypothesis, T would have been born before day h, a contradiction to our assumption. Hence the height of T is exactly h, as desired.

Conversely, suppose that T has height h. Then the tallest of its immediate subtrees must have height $h - 1$. Hence by the inductive hypothesis one of T's subtrees was not born until day $h - 1$, but all of its subtrees were born on or before day $h - 1$. It follows that T was born on day h.   ■

We have already introduced the botanical metaphor for the objects we are studying in this chapter, calling them trees and calling their distinguished vertices roots. We need to extend this terminology a little. For this purpose we will draw both on botany and on genealogy. The latter should make sense if you think of your own family tree or the family trees of European royalty.

Let T be a rooted tree with root r, viewed as a directed graph with edges directed away from the root. If uv is a directed edge of T, then we say that v is a **child** of u, and that u is the **parent** of v. Two different vertices with the same parent are said to be **siblings**. A child of a child of a vertex is that vertex's **grandchild**. The parent of the parent of a vertex is that vertex's **grandparent**.

More generally, all the vertices on the unique path from the root r to a vertex v, other than v itself, are called the **ancestors** of v. If u is an ancestor of v, then v is a **descendant** of u. A vertex with out-degree 0 is called a **leaf**; vertices that are not leaves are called **internal vertices**.

The terminology is quite natural and easy to get used to. Note that the descendants of u are all the vertices (other than u itself) in the subtree rooted at u. Vertices may have no children, one child, or more than one child. Vertices with no children are the leaves; internal vertices have children. The root is an internal vertex unless the tree contains only one vertex. Every vertex other than the root has exactly one parent. Every vertex at level 2 or greater has exactly one grandparent. If a tree has height h, all the vertices at level h are necessarily leaves, but not all leaves are necessarily at level h.

EXAMPLE 4.   Referring to Figure 9.4, we see that e is the parent of h, i, and j, and that these three vertices are e's children. Vertices h, i, and j are each other's siblings. The root a has three grandchildren: e, f, and g. The leaves are c, g, h, i, k, l, m, n, and o; the other vertices are internal vertices.   ◆

The distinction between leaves and internal vertices is often significant in applications. For example, in an organizational chart, the leaves correspond to people who supervise nobody. In a file and directory system on a computer (see Figure 9.7), internal vertices have to be directories, and the leaves are either files or empty directories (directories that have been set up but have nothing in them yet).

ORDERED TREES, BINARY TREES, AND m-ARY TREES

For many applications a rooted tree does not have enough structure. Often we need to impose an *order* on the children of each vertex. The genealogical metaphor helps here: A person's children have a natural ordering by age.

We define an **ordered** tree to be a rooted tree together with the additional structure of, for each internal vertex v, a linear ordering of the children of v. In the picture of an ordered tree, the children of each vertex are shown in order, from left to right. Two ordered trees are isomorphic if there is a one-to-one correspondence between their vertex sets that preserves adjacencies, nonadjacencies, the root, and order among children.

Note that all of our drawings of rooted trees could equally well have been drawings of ordered trees: The very act of drawing a rooted tree imparts a left-to-right order on the children of each vertex.

EXAMPLE 5.   Find all nonisomorphic ordered trees with four vertices.

Solution.   We already found all the rooted trees with four vertices in Example 1. There were four of them, as drawn (not according to our convention) in Figure 9.3. The trees in Figure 9.3a, c, and d have only one possible order, up to isomorphism, but the tree in Figure 9.3b has two, since either the first child or the second child of the root may be the vertex at level 1 with a child. The five nonisomorphic ordered trees are shown in Figure 9.9. ◆

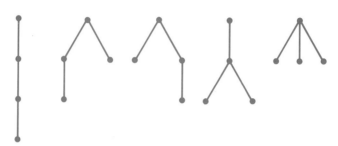

FIGURE 9.9   All nonisomorphic ordered trees with four vertices.

Next we define a binary tree, which has more structure yet. Binary trees are one of the most useful structures in computer science, not only in their own right, but also because they can be used in the representation of other data structures. Binary trees are ordered trees in which each vertex can have at most two children, together with one additional piece of structure.

To be precise, a **binary tree** is a rooted tree together with a designation, for each vertex other than the root, as to whether that vertex is the **left child** or the **right child** of its parent (it must be one or the other, but not both), such that each vertex has at most one left child and at most one right child. The **left subtree** of a binary tree is the subtree rooted at the left child of the root (if there is one), and the **right subtree** of a binary tree is the subtree rooted at the right child of the root (if there is one). A binary tree is called **full** if every internal vertex has both a left child and a right child. In drawing binary trees, we place a left child below and to the left of its parent and a right child below and to the right of its parent. Two binary trees are isomorphic if there is a one-to-one correspondence between vertices that preserves adjacencies, nonadjacencies, the root, left childhood, and right childhood.

Note that for a binary tree, even when a vertex has only one child, that child must be designated as being a left child or a right child. For every vertex v of a binary tree, the subtree rooted at v is again a binary tree. We can therefore speak of the **left subtree** or **right subtree** of any vertex v in a binary tree. If v has no child of the specified type, then we say that the corresponding subtree is **empty**.

EXAMPLE 6.   A typical binary tree is shown in Figure 9.10. The height of this tree is 3. The left subtree contains vertices b (its root), d, g, and h; the right subtree contains

c (its root), e, f, and i. The left subtree of e is empty. This is not a full binary tree, since, for instance, vertex b has a left child (d) but no right child. If we were to add two new leaves—a right child for b and a left child for e—then the resulting binary tree would be full. ◆

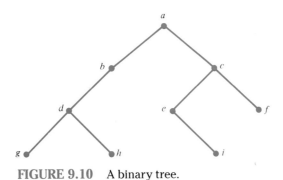

FIGURE 9.10 A binary tree.

EXAMPLE 7. Find all the nonisomorphic binary trees with three vertices.

Solution. There is only one unrooted tree with three vertices, and only two rooted trees (ordered or not), but as Figure 9.11 shows, there are five nonisomorphic binary trees with three vertices. The only one of these that is full is the one of height 1. ◆

FIGURE 9.11 All nonisomorphic binary trees with three vertices.

To emphasize the recursive nature of trees, let us give two recursive definitions involving binary trees.

> Base case: A single vertex is a full binary tree, with itself as root and no children.
>
> Recursive part: If B is a full binary tree with root r, and if v is a leaf of B (a vertex with no children), then the tree obtained by giving v a left child and a right child (new vertices, each with no children) is a full binary tree, still with r as its root.

This first definition lets us watch a full binary tree grow from the top, with each vertex that is destined to be an internal vertex eventually sprouting its two children.

Second, we define an **extended binary tree**, which is just like a binary tree except that the empty set is considered to be an extended binary tree.

> Base case: The empty set is an extended binary tree (it has no vertices).
>
> Recursive part: If B_1 and B_2 are disjoint extended binary trees, then an extended binary tree can be formed by taking a new vertex as root, making B_1 its left subtree and B_2 its right subtree.

Under this definition, left and right subtrees always exist, but they may be empty. Here we can watch a tree being built from the bottom up, similar to the way we viewed rooted trees recursively. A binary tree can be considered to be an extended binary tree simply by imagining the empty set as the left and right children of each leaf, and the empty set as the other child of each internal vertex with only one child. We will view binary trees as extended binary trees when it is convenient to do so.

EXAMPLE 8. The binary tree in Figure 9.10 can be thought of as an extended binary tree, built from the bottom up, as follows. Pairs of empty trees are attached to vertices f, g, h, and i, creating four trees with one vertex each. The trees rooted at g and h are made the left and right subtrees of a new vertex d. The tree thus formed and the empty tree are attached to vertex b to create the left subtree of Figure 9.10. A somewhat similar sequence gives the subtree rooted at c. Finally, the two extended binary trees rooted at b and c are attached to the new root a to form the entire tree. ◆

Interesting relationships exist among the height, the number of vertices, the number of leaves, and the number of internal vertices in a rooted tree. Let us start with a special case.

THEOREM 3. *Let B be a full binary tree with l leaves and i internal vertices. Then* $l = i + 1$.

Proof. We use induction on the recursive definition of full binary tree. The equality holds for the tree with one vertex ($i = 0$ and $l = 1$). Whenever the recursive step occurs, one leaf is replaced by two leaves and one internal vertex, for a net increase of one in both i and l; thus the equality remains true. ■

Any binary tree can be turned into a full binary tree by adding to each vertex with only one child a leaf as its other child. This increases the number of leaves but does not change the number of internal vertices. Therefore, we have the following corollary.

COROLLARY.   *Let B be a binary tree with l leaves and i internal vertices. Then $l \leq i+1$, and equality holds if and only if B is full.*

A rooted tree (perhaps with additional structure) is called an **m-ary tree**, $m \geq 0$, if every vertex has at most m children. (However, 0-ary and 1-ary trees are not very interesting.) A **full** m-ary tree is a rooted tree in which every internal vertex has exactly m children. In particular, a binary tree can be viewed as a 2-ary tree, although our definition of binary tree imposes the additional structure that every child has to be either *left* or *right*. We obtain the following analogue of our previous results for binary trees. The proof is left to the reader (Exercise 14).

THEOREM 4.   *Let T be an m-ary tree with l leaves and i internal vertices. Then $l \leq (m-1)i + 1$, and equality holds if and only if T is full.*

Next we relate the number of vertices in a tree to its height; our goal is to apply this information shortly to the analysis of algorithms.

THEOREM 5.   *Let T be an m-ary tree with n vertices, l of which are leaves. Suppose that the height of T is h. Then $l \leq m^h$. Also, as long as $m \geq 2$,*

$$n \leq \frac{m^{h+1} - 1}{m - 1}.$$

Equality holds in both cases if and only if T is full and all the leaves are at level h.

Proof.   We sketch the proof here and leave the details to the reader (Exercise 16c). Since the tree can fan out by no more than a factor of m at each level, there can be at most m^r vertices at level r. The maximum number of leaves would occur when all the leaves were at level h, whence $l \leq m^h$. The second inequality is obtained by finding the sum of the geometric series $\sum_{r=0}^{h} m^r$, which, as long as $m \geq 2$, is $(m^{h+1} - 1)/(m - 1)$ (see Example 9 in Section 2.2).   ∎

The upshot of this theorem is that a short tree cannot have too many leaves or vertices. Turning these results around algebraically (a task left to the reader in Exercise 15), we obtain the following corollary, which says that a tree with many leaves or vertices cannot be too short.

COROLLARY.   *A binary tree with l leaves has height at least $\lceil \log l \rceil$. A binary tree with n vertices has height at least $\lceil \log(n + 1) \rceil - 1$. For any fixed $m \geq 2$, an m-ary tree with l leaves [respectively, n vertices] has height at least $\lceil \log_m l \rceil$ [respectively, $\lceil \log_m((m - 1)n + 1) \rceil - 1$].*

We now apply this corollary to obtain an important lower bound on how fast a sorting algorithm can work. We saw in Section 4.3 that bubble sort takes time proportional to n^2 to sort n items. In Section 7.3 (Example 3) we saw that merge sort has running time in $O(n \log n)$, quite an improvement. We will now show that in some sense we cannot do any better than merge sort. The "information-theoretic" approach we will use here is often useful in obtaining lower bounds on the efficiency of algorithms.

Suppose that we have a list of n distinct numbers (or other objects that come with a natural order), and we wish to determine their relative order by comparing various pairs of numbers in the list. (Equivalently, suppose that we wish to sort the numbers into increasing order, using comparisons of numbers as our only tool.) Let us compute how many comparisons are required in the worst case.

We base our analysis on the notion of a decision tree. A **decision tree** is a binary tree, the internal vertices of which represent the yes/no questions that the algorithm asks as it executes. (We assume here that all questions are of this yes/no form. One can also study decision processes involving questions with more than two possible answers, and in that case the resulting tree will not be binary.) Execution starts at the root. If the answer to a question is "yes," the algorithm proceeds to the right subtree; if the answer is "no," the algorithm proceeds to the left subtree. Leaves represent terminations of the execution, presumably with the desired output. (We can imagine all the other computation that the algorithm does as also being included in the vertices, prior to the question, if any.) In Figure 9.12 we see a decision tree for an algorithm for determining the order of three distinct numbers a, b, and c.

Now let A be an algorithm that determines the order of (or sorts) a list of n numbers. Let T be the decision tree for A. (Thus Figure 9.12 is one possibility if

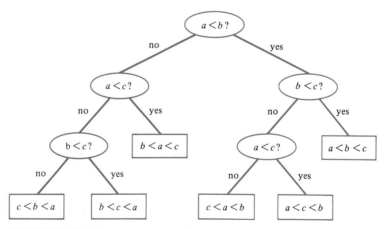

FIGURE 9.12 Decision tree for determining the order of three distinct numbers.

$n = 3$.) There are clearly $n!$ possible orders that the numbers might appear in, so if algorithm A is to work correctly, its decision tree will need to have at least $n!$ leaves. By the corollary above, the decision tree will therefore have to have height at least $\log(n!)$. This in turn means that in the worst case (that case in which the algorithm terminates at a leaf at the bottom level), at least $\log(n!)$ comparisons are required. So we conclude that *no* sorting algorithm whatsoever—past, present, or future—that uses comparisons of pairs of numbers as its method of operation can do better than $\log(n!)$ steps in the worst case. A consequence of the results of Exercise 26 in Section 4.3 is that $O(\log(n!)) = O(n \log n)$. Thus no such sorting algorithm is, asymptotically, any better than merge sort.

SUMMARY OF DEFINITIONS

ancestor of the vertex v in a rooted tree: a vertex other than v on the unique path from the root to v.

binary tree: a rooted tree together with a designation for each child as to whether it is the left child or the right child of its parent (it cannot be both), such that each vertex has at most one child of each type.

child of the vertex v in the rooted tree T: a vertex u such that vu is a directed edge in T; necessarily, u's level is one greater than v's level.

decision tree: a rooted tree (usually a binary tree) used to model an algorithm, in which each internal vertex represents a point in the algorithm at which a decision is made (with execution continuing in the appropriate subtree rooted at that vertex) and each leaf represents an outcome of the algorithm.

descendant of the vertex v in a rooted tree: a vertex that has v as an ancestor.

empty binary tree: the empty set; the (extended) binary tree having no vertices.

extended binary tree: a binary tree in which every vertex has both a left child and a right child, accomplished by considering a missing child as being the empty tree.

forest: a graph that has no closed trails.

full binary tree: a binary tree in which every internal vertex has both a left child and a right child.

full m-ary tree: an m-ary tree in which every internal vertex has exactly m children.

grandchild of the vertex v in a rooted tree: a child of a child of v.

grandparent of the vertex v in a rooted tree: the vertex that has v as its grandchild (will not exist if v is at level 0 or 1).

height of a rooted tree: the maximum level of all the vertices in the rooted tree.

immediate subtree of the rooted tree T: a subtree rooted at a child of the root of T.

internal vertex in a rooted tree: a vertex that has at least one child.

leaf in a rooted tree: a vertex that has no children.

left child of a vertex in a binary tree: a child that has been so designated, drawn on the left in the conventional drawing of the tree.

left subtree of the binary tree T: the subtree rooted at the left child of the root of T.

left subtree of the vertex v in a binary tree: the subtree rooted at the left child of v (will be empty if v has no left child).

level of the vertex v in a rooted tree: the length of the unique path from the root to v; thus the root is at level 0, the vertices adjacent to the root are at level 1, and so on.

m-ary tree: a rooted tree or ordered tree in which every vertex has at most m children.

ordered tree: a rooted tree in which the children of each vertex have been given an order; in the drawing, the children of each vertex are shown in order from left to right.

parent of the vertex v in a rooted tree: the vertex that has v as its child (will not exist if v is the root).

right child of a vertex in a binary tree: a child that has been so designated, drawn on the right in the conventional drawing of the tree.

right subtree of the binary tree T: the subtree rooted at the right child of the root of T.

right subtree of the vertex v in a binary tree: the subtree rooted at the right child of v (will be empty if v has no right child).

root of a tree: a designated vertex.

rooted tree: a tree with a root; sometimes a rooted tree is viewed as a digraph, in which the edges are directed away from the root (from level l to level $l + 1$).

siblings in a rooted tree: vertices with the same parent.

star: the graph $K_{1,n}$ for some natural number n.

subtree rooted at the vertex v in the rooted tree T: the rooted tree with v as root containing v and all its descendants in T.

subtree of the tree T: a subgraph of T that is a tree.

tree: a connected forest.

EXERCISES

1. Show that the average of all the degrees of the vertices in a tree is less than 2. [*Hint*: Use Theorem 1 in Section 8.1.]

2. Determine the following sets.
   (a) All nonisomorphic trees with three or fewer vertices.
   (b) All nonisomorphic nonconnected forests with three or fewer vertices.
   (c) All nonisomorphic rooted trees with three or fewer vertices.
   (d) All nonisomorphic binary trees with two or fewer vertices.

3. Compute the following quantities.
   (a) The number of edges in a forest with 72 vertices and seven components.

   (b)   The number of leaves in a full 3-ary tree with seven internal vertices.

   (c)   The number of internal vertices in a full 5-ary tree with 101 leaves.

   (d)   The maximum number of vertices possible in a binary tree of height 5.

4. Referring to Figure 9.4, draw each of the following.
   (a)   The immediate subtrees.
   (b)   The subtree rooted at h.
   (c)   The subtree rooted at f.

5. In Figure 9.4, identify each of the following.
   (a) d's children.
   (b) m's children.
   (c) b's grandchildren.
   (d) k's siblings.
   (e) d's descendants.
   (f) o's ancestors.

6. Draw a binary tree with vertices a, b, c, d, e, and f, such that b has no parent; a, d and f are the leaves; c is b's right child; d and e are c's children; and e and f are left children.

7. Determine the following sets.
   (a)   All nonisomorphic forests with four vertices.
   (b)   All nonisomorphic trees with six vertices.
   (c)   All nonisomorphic rooted trees with six vertices.
   (d)   All nonisomorphic ordered trees with five vertices.
   (e)   All nonisomorphic binary trees with four vertices.
   (f)   All nonisomorphic 2-ary trees with four vertices.
   (g)   All nonisomorphic ordered 2-ary trees with four vertices.

8. Give a recursive definition of a plain (unrooted) tree, capturing the idea that a tree can be built up by starting with a single vertex and repeatedly adding a new edge with one new vertex attached to it.

9. Prove the following parts of Theorem 1.
   (a) $3 \rightarrow 4$.
   (b) $6 \rightarrow 1$.

10. Let G be a graph that has at least as many edges as it has vertices. Prove that G contains a cycle.

11. Investigate the relationship between trees and bipartite graphs in the following ways.
   (a)   Show that every tree is a bipartite graph.
   (b)   Determine necessary and sufficient conditions on n and m so that the complete bipartite graph $K_{m,n}$ is a tree.

12. Show that every tree is planar. (There are many ways to do this; try to give more than one proof.)

13. Give a formal proof that the height of a rooted tree is one greater than the maximum of the heights of its immediate subtrees.

14. Prove Theorem 4.

15. Prove the corollary to Theorem 5.

16. An m-ary tree is called **complete** if it is full and all of its leaves are at the same level.
    (a)   Draw the complete 3-ary tree of height 2.
    (b)   Give a recursive definition of a complete m-ary tree.
    (c)   Give a rigorous inductive proof of Theorem 5.

17. Draw a full binary tree that has nine vertices, height 3, and one leaf at level 1. Explain how this tree "grows" by the recursive definition of full binary tree given in this section.

18. Draw a binary tree that has eight vertices, height 3, and one leaf at level 1. Explain how this tree can be viewed as an extended binary tree, and explain how it "grows" according to the bottom-up recursive definition of extended binary tree given in this section.

19. Determine the maximum possible number of descendants that a vertex at level l in a binary tree of height h can have.

20. Suppose that you tell five of your friends that you have been selected for the varsity basketball team. Suppose that each person who hears the news either tells five of his or her friends or tells nobody. Assume that no one hears the news more than once.
    (a)   Set up an appropriate graph-theoretic model.
    (b)   If there are 10 gossips (other than yourself), how many people know the news? [*Hint*: What does "gossip" mean in terms of the model?]
    (c)   If 75 people other than you know the news, how many gossips were there?
    (d)   If there are 85 people who heard the news but told nobody, how many people know the news?

21. Let T be a full m-ary tree, $m \geq 2$. Let n, i, and l be the number of vertices, internal vertices, and leaves of T, respectively. Obviously, $n = i + l$, and by Theorem 4, $l = (m - 1)i + 1$. This gives us two equations involving four variables, which means that any two of the four variables determine the other two. Determine the 12 equations implied by this last statement (simplified as much as possible).

22. Draw a decision tree for sorting (i.e., determining the order of) five distinct numbers a, b, c, d, and e, given that you already know that $a < b < c$ and $d < e$. Try to make it as small (in height) as possible.

23. Suppose that you are given a set of nine coins, eight of which are the same weight and one of which is a counterfeit that weighs less than the others. You are also given a balance scale, consisting of two pans capable of holding coins and an arm that indicates whether the contents of the two pans have the same weight, or, if not, which is heavier. Your task is to determine the counterfeit coin.
    (a) Explain how a 3-ary decision tree can be used to model any algorithm for solving this problem.
    (b) Compute the minimum height the decision tree can have, and interpret this in terms of the number of weighings that are needed to solve the problem.
    (c) Figure out a way to determine the counterfeit coin using as few weighings as possible.

Challenging Exercises

24. Prove the lemma proceding Theorem 1 in each of the following ways.
    (a) By induction on the number of edges in the tree.
    (b) By induction on the number of vertices in the tree.
    (c) By concentrating on a longest trail in the tree.

25. Give a rigorous, nonrecursive definition of what the following genealogical concepts should mean for vertices u and v in a tree (see also Exercise 28 in Section 5.1).
    (a) u is v's uncle.
    (b) u is v's first cousin.
    (c) u is v's nth cousin, where $n \geq 1$.
    (d) u is v's nth cousin, r times removed, where $n \geq 1$ and $r \geq 0$.

26. Let b_n be the number of nonisomorphic extended binary trees with n vertices, for each $n \geq 0$.
    (a) Write down a recurrence relation and initial condition that the sequence $\{b_n\}$ satisfies.
    (b) Compute b_n for all n less than 7.

27. Figure out a good method to determine, given an unrooted tree T, the length of the longest path in T.

28. An **AVL-tree** (also called a **height-balanced tree**) is a binary tree with the property that for each vertex v, the height of the left subtree at v and the height of the right subtree at v differ by at most 1. (Consider an "empty tree" to have height -1, so that, for example, there can be at most one vertex in

the left subtree of a vertex that has no right child.)
(a) Find all the AVL-trees of height less than 3.
(b) Draw two nonisomorphic AVL-trees of height 3 and one tree of height 3 that is not an AVL-tree.
(c) Let a_n be the number of nonisomorphic AVL-trees of height n. Find a recurrence relation with initial conditions for a_n.
(d) Use part (c) to determine the number of nonisomorphic AVL-trees of heights 3 and 4.

29. Define a set F of binary trees recursively as follows. The tree with one vertex is in F_0; and the two binary trees with two vertices are in F_1. For $n > 1$, if $T \in F_{n-1}$ and $S \in F_{n-2}$, then the binary tree that has S as its left subtree and T as its right subtree is in F_n; and the binary tree that has T as its left subtree and S as its right subtree is in F_n. The set F is the union of all the F_n's.
(a) Draw all the nonisomorphic trees in F_2 and F_3.
(b) Show that all the trees in F_n have height n.
(c) Let a_n be the number of nonisomorphic trees in F_n. Find a recurrence relation with initial conditions for a_n.
(d) Solve the recurrence relation from part (c).

30. Consider the following alternative definition of a rooted tree. A **rooted tree** T consists of a set V of vertices; a function l from V to the set of natural numbers ($l(v)$ is called the **level** of v in the tree); an element $r \in V$ (called the **root**) such that $l(r) = 0$; and a function p from $V - \{r\}$ to V such that $l(p(v)) = l(v) - 1$ for all $v \in V$ ($p(v)$ is called the **parent** of v). Say that two vertices in a rooted tree are **adjacent** if one is the parent of the other. Define a **path** to be a sequence of distinct vertices $\{v_n\}_{n=1}^m$ or $\{v_n\}_{n=1}^\infty$ (finite or infinite), such that v_n is adjacent to v_{n-1} for all $n \geq 2$. The advantage of this definition is that if we do not require V to be a finite set, then we can have infinite rooted trees.
(a) Use the definition to prove that the root is the only vertex with level 0.
(b) Give an example of a rooted tree with an infinite number of vertices that has an infinite path; give an example of a rooted tree with an infinite number of vertices that has paths of all finite lengths but no infinite path; give an example of a rooted tree with an infinite number of vertices that has no path of length greater than 3.
(c) Show that the lemma preceding Theorem 1 is not valid for infinite trees (i.e., infinite rooted trees with the root no longer distinguished).
(d) Define the set of children of a vertex v in a rooted tree, using the definitions given here.
(e) A rooted tree is called **locally finite** if every vertex has only finitely many children. Show that an infinite, locally finite rooted tree has an infinite path. (This result is known as König's lemma.) [*Hint*: An infinite version of the pigeonhole principle is called for.]

Exploratory Exercises

31. Explain why a family tree is not necessarily a tree in the graph-theoretic sense. Draw a (not too small) portion of your family tree that actually is a tree in the graph-theoretic sense.

32. Explain carefully in a paragraph, citing an example or two, how a binary tree B_T can be used to model an arbitrary ordered tree T on the same set of vertices. This model is useful for computer representation of trees. [*Hint*: Make the left child of a vertex in B_T its oldest child in T, and make the right child of a vertex its next younger sibling.]

33. The **center** of a tree is defined to be the set of vertices of minimum eccentricity, where the **eccentricity** of a vertex is the length of the longest path that starts at that vertex.
    (a)   Find the center of all the trees in Figure 9.1.
    (b)   After experimenting with several other trees, come up with a conjecture as to what the center of a tree can be. (How many vertices can be in the center, and what can be said about the vertices in the center?)
    (c)   Prove your conjecture from part (b).

34. A **graceful labeling** of a tree with n vertices is a bijective function L from the vertices of the tree to $\{1, 2, 3, \ldots, n\}$ (we call $L(v)$ the label on v) such that the set of $n - 1$ absolute values of the differences of the labels on the endpoints of the $n - 1$ edges are precisely $\{1, 2, \ldots, n - 1\}$. For example, we can gracefully label the tree shown in Figure 9.1b by assigning the numbers 2, 3, 1, and 4 to the vertices along the top, from left to right, and the number 5 to the vertex hanging down. It is conjectured that all trees have a graceful labeling, but no one has been able to prove this conjecture.
    (a)   Find a graceful labeling of the tree shown in Figure 9.4.
    (b)   Show that every tree that is just a path (e.g., the tree in Figure 9.1a) has a graceful labeling.
    (c)   Show that every caterpillar has a graceful labeling, where a **caterpillar** is a tree consisting of a path (the "body" of the caterpillar) together with any number of other vertices adjacent to vertices on this path (the "legs" of the caterpillar). For example, the tree shown in Figure 9.4 is a caterpillar, with body consisting of the path o, j, e, b, a, d, f, k.
    (d)   Find out more about graceful labeling by consulting Gardner [111] and Golomb [118].

35. Find out more about solving counterfeit coin problems (see Exercise 23) by consulting Manvel [205].

SECTION 9.2
SPANNING TREES

Under the spreading chestnut tree
The village smithy stands.

—Henry Wadsworth Longfellow

Trees and forests are important in their own right; in addition, finding trees and forests as subgraphs of other graphs will give us a way to get a better hold on the original graphs. We define a **spanning tree** of a connected graph G to be a subgraph of G that contains every vertex of G and is a tree. More generally, a **spanning forest** of a graph G is a subgraph of G consisting of a spanning tree for each component of G. In this section we prove that every connected graph has a spanning tree (and every graph has a spanning forest), and we give two efficient algorithms for finding spanning trees. The "search" techniques at the core of these algorithms turn out to be useful in other areas of graph theory and computer science, as well. (We discuss other algorithms for finding spanning trees—for weighted graphs—in Section 10.2.)

EXAMPLE 1.   Figure 9.13a shows a graph, with a spanning tree in heavy lines. Figure 9.13b shows an unconnected graph with a spanning forest, consisting of five spanning trees (two of which consist of just one vertex each).   ◆

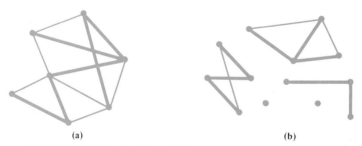

(a)   (b)

FIGURE 9.13   Graphs and their spanning trees or spanning forests.

A graph can certainly have more than one spanning tree, but every connected graph has at least one, as our first theorem shows.

THEOREM 1.   *Every connected graph (or pseudograph) contains a spanning tree.*

Proof. Let G be the given connected graph or pseudograph. If G has no cycles, then G is itself a tree, so we are done. If G does contain a cycle C, we delete one edge e in C. We claim that the resulting subgraph G' must still be connected. To see this, we must show that any two vertices in G' are joined by a walk. Since G is connected, they are joined by a walk in G. If that walk did not use e, then it is still a walk in G', so we are done. If e was included in the walk, then a new walk can be obtained by replacing e with the walk consisting of the rest of the cycle C, traversed backward (in other words, to avoid using e simply go around the cycle in the other direction).

Now if G' has no cycles, then it is the desired spanning tree of G. If it does have a cycle, then we repeat the process, deleting an edge in a cycle of G' to obtain a subgraph G''. By the same argument as before, G'' is connected. If it is cycle-free, then it is the desired spanning tree. If not, we repeat. Eventually we must obtain a connected subgraph that has no cycles (since there were only a finite number of edges to begin with, and we delete one edge at each stage). This subgraph is the desired spanning tree. (Our somewhat informal argument here can be made more rigorous by using mathematical induction rather than the notions of "repeat" and "eventually" (see Exercise 5).) ∎

COROLLARY. *Every graph (or pseudograph) contains a spanning forest.*

Proof. Left to the reader (Exercise 6). ∎

Unfortunately, the proof of Theorem 1 did not give us a useful and efficient way to *find* a spanning tree for a given connected graph. That proof required us to find cycles in graphs, and it is not at all clear how to go about doing so efficiently.

The thrust of the two efficient algorithms to be presented in this section actually is not to find a spanning tree in a graph. Rather, it is to explore the graph, systematically visiting all its vertices. The spanning trees—which are actually built as rooted trees—come, in a sense, as a by-product.

DEPTH-FIRST SEARCH

Depth-first search is a recursive algorithm for visiting all the vertices of a connected graph G. We mark each vertex of G as we visit it, so as not to visit a vertex more than once. Simultaneously, we construct a rooted spanning tree T of G (rooted at the vertex where the exploration begins) by including an edge uv in T whenever we move from a vertex u to a new vertex v. The key idea of this algorithm—what distinguishes it from the breadth-first search algorithm given later in this section—is that whenever we move on, say from a vertex u to a new vertex v, we finish exploring from v before we return to u to finish exploring from u. The recursive procedure *DFS* in Algorithm 1 does the exploring from the vertex given to it, and it returns the portion of the spanning tree constructed so far, along with the information as to which vertices have been visited. The main program (*depth_first_search*) initializes the arrays that keep track of which vertices have been visited and what has been added to the tree, and then calls *DFS* to start the search at the first vertex.

> **procedure** *depth_first_search*(*G* : connected graph)
> {assume that the vertices of *G* are numbered $1, 2, \ldots, n$; *visited*(*i*) is true if
> and only if vertex *i* has been visited; *T* is the spanning tree}
> *visited*(1) ← **true** {exploration starts at vertex 1}
> **for** *i* ← 2 **to** *n* **do**
> *visited*(*i*) ← **false** {no other vertices have been visited}
> *T* ← ({1}, ∅) {spanning tree initially contains vertex 1 and no edges}
> (*T*, *visited*) ← *DFS*(*G*, *T*, *visited*, 1)
> **return**(*T*)
>
> **procedure** *DFS*(*G* : graph, *T* : tree, *visited* : array, *i* : vertex of *G*)
> {*T* is a subgraph of *G*; *visited* indicates which vertices of *G* have been visited
> so far; this recursive procedure explores from vertex *i*}
> **for** *j* ← 1 **to** *n* **do**
> **if** (*j* is adjacent to *i*) ∧ not *visited*(*j*) **then**
> **begin** {*j* is a new, as yet unvisited, neighbor; continue exploring
> from *j*}
> *visited*(*j*) ← **true**
> add vertex *j* and edge *ij* to *T*
> (*T*, *visited*) ← *DFS*(*G*, *T*, *visited*, *j*)
> **end**
> **return**(*T*, *visited*)

Algorithm 1. Depth-first search of a graph.

This algorithm is mysterious enough that we should follow its execution on a small graph. In our example, we will suppress references to the arguments *G*, *visited*, and *T*, and just refer to calling *DFS*(*j*), where *j* is a vertex number. (In effect, we can just think of *visited* and *T* as global variables that are being updated continually as the program executes.)

EXAMPLE 2. Consider the graph *G* shown in Figure 9.14. The depth-first search proceeds as follows. [The reader should follow this discussion by circling the vertices of *G* as they are visited, and darkening edges as they are added to *T*.] We start at vertex 1, put it into the tree (it will be the root), and call *DFS*(1). Then we look at vertex 1's first neighbor, namely vertex 2. Since vertex 2 has not yet been visited, we visit it, put it and edge 12 into the tree, and call *DFS*(2), recursively. Now *i* = 2 and we look at the neighbors of vertex 2. (We will return to look at the other neighbors of vertex 1 later.) Of the neighbors of vertex 2, the first, vertex 1, has already been visited. Vertex 4 is the first *unvisited* neighbor of vertex 2, so we visit it, put it and edge 24 into T, and call $DFS(4)$.

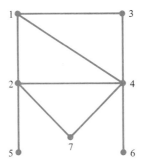

FIGURE 9.14 A graph on which to apply search algorithms for generating spanning trees.

Now in the subprocedure DFS with $i = 4$, we find the first unvisited neighbor to be vertex 3, so we visit vertex 3, add vertex 3 and edge 43 to T, and call $DFS(3)$. Since all the neighbors of vertex 3 have already been visited, $DFS(3)$ returns without taking any action (the condition on the **if** statement fails for each j). Thus we continue with $DFS(4)$. The next unvisited vertex adjacent to vertex 4 is vertex 6, so we visit vertex 6, put it and edge 46 into T, and call $DFS(6)$. Since vertex 6 has no unvisited neighbors, we return to DFS with $i = 4$. Finally, we reach $j = 7$, and we visit vertex 7, put vertex 7 and edge 47 into T, and call $DFS(7)$, during which again nothing new happens. That completes the execution of DFS with $i = 4$, so we return to DFS with $i = 2$ (the procedure that called $DFS(4)$).

At this point the next neighbor of vertex 2, namely vertex 5, is still unvisited, so we visit vertex 5, put it and edge 25 into T, and call $DFS(5)$. The latter procedure returns with no changes, and that completes $DFS(2)$. Finally, we return to the patiently waiting first invocation of DFS, when $i = 1$. Now vertices 3 and 4 have already been visited, so vertex 1 has no more unvisited neighbors, and our search is finished.

In Figure 9.15 we exhibit T in its conventional form with the root at the top. The edges of T are shown here with heavy lines; the edges *not* in T are also shown, with thin lines. Note that all the edges not in T join a vertex to an ancestor or descendant of that vertex. The reader is asked to show (Exercise 17) that this is always the case for depth-first search spanning trees. Such edges are called **back edges**. ◆

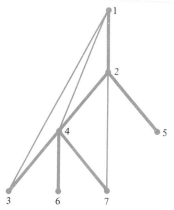

FIGURE 9.15 Depth-first search spanning tree of the graph in Figure 9.14, with back edges.

Let us see why Algorithm 1 works. At each stage of the algorithm, we have a subgraph T. Initially, T is a single point, and T changes by having a *new* vertex (thanks to the array *visited*) added to it, together with an edge from that new vertex to a vertex already in T (again thanks to *visited*). Thus T remains connected and without cycles—that is, is a tree—throughout. Furthermore, since G is connected, no vertex can escape inclusion in T. This follows by induction on the length of the shortest path from vertex 1 to the other vertices, since once a neighbor u of a vertex v is included and $DFS(u)$ called, it is only a matter of time before v gets included as well (certainly not later than the point at which the **return** from $DFS(u)$ occurs). We conclude, therefore that T is indeed a spanning tree of G.

Next let us see how efficient this algorithm is. Every time a vertex is first visited (and this occurs n times since the graph has n vertices), DFS is called. The amount of calculation that occurs just *in that invocation* of DFS is clearly in $O(n)$, because of the **for** loop. Thus the total time involved is in $O(n^2)$. By being a little more careful with the way we represent the graph and look at adjacencies, we can cut this down to $O(m)$, where m is the number of *edges* in G. (Specifically, we need to represent G by its adjacency lists (see Section 8.3), and replace the **for** loop by a loop through the *adjacent* vertices only. We could write this as "**for** each unvisited vertex j adjacent to i **do**.") The overhead involved in implementing the recursion is insignificant, since there are only $O(n)$ recursive calls. Thus we have the following theorem.

THEOREM 2. *The depth-first search algorithm finds a spanning tree of a connected graph with n vertices and m edges in $O(m)$ steps, which is at worst in $O(n^2)$.*

COROLLARY. *The depth-first search algorithm provides a way to determine whether a graph with n vertices and m edges is connected in $O(m)$ steps, which is at worst in $O(n^2)$.*

Proof. Given the not necessarily connected graph G, we begin exactly as in Algorithm 1. However, when the return from the call to *DFS* with vertex 1 occurs, we scan the array *visited*. If it contains only the value **true**, we conclude that G is connected (since all vertices have been included in the tree T). If it contains at least one value **false**, we conclude that G is not connected. This scanning takes only $O(n)$ steps, so the bounds given in Theorem 4 still hold. ■

This corollary gives a much more efficient method for determining connectedness than the matrix multiplication methods discussed in Section 8.3. The algorithm described in that section requires on the order of n^4 basic operations to determine whether a given graph with n vertices is connected, as opposed to the $O(n^2)$ operations here. In effect, almost all the calculations in the previous algorithm were wasted—providing no substantive new information. Thus the depth-first search method is greatly to be preferred.

BREADTH-FIRST SEARCH

Algorithm 1 is called *depth*-first search because it tries to delve as *deeply* into the graph as it can, following a path from the root (vertex 1) as far as it can go until forced to backtrack. The other approach to visiting the vertices of a graph systematically is to *fan out* from each vertex as *broadly* as possible. The resulting tree will likely be wider (have more children at each internal vertex) and not as deep (i.e., high) as the depth-first search spanning tree.

In **breadth-first search**, when we first encounter a vertex, we do not proceed to search further from that vertex immediately. Instead, we put that vertex into a queue to wait its turn for further processing. (A **queue** is a dynamic (changing) list in which items are removed in the same order as that in which they were added. The rule is "first in, first out" for this data structure.) When we do process a vertex v, we gather *all* of its unvisited neighbors at once and put them into the tree as children of v; the children then must wait their turn for further processing.

In Algorithm 2 we implement the queue by means of a list L, to which vertices are added at the end and removed for processing from the front. Initially, L contains only vertex 1 (to be the root of the breadth-first search spanning tree). When L becomes empty, all the vertices have been processed, and our spanning tree is complete.

EXAMPLE 3. We will again step through the algorithm for the graph in Figure 9.14. Initially, the tree T contains only vertex 1, and the list L contains only vertex 1. Since L is not empty, we remove the first vertex from L (leaving L temporarily

procedure *breadth_first_search*(G : connected graph)
{assume that the vertices of G are numbered $1, 2, \ldots, n$; *visited*(i) is true if and only if vertex i has been visited; T is the spanning tree; L is the list of visited but unprocessed vertices}
  visited(1) ← **true** {exploration starts at vertex 1}
  for i ← 2 **to** n **do**
    visited(i) ← **false** {no other vertices have been visited}
  T ← ({1}, ∅) {spanning tree initially contains vertex 1 and no edges}
  L ← (1) {vertex 1 awaits processing}
  while L is not empty **do**
    begin
      i ← first element of L
      L ← L with i removed
      for j ← 1 **to** n **do**
        if (j is adjacent to i) ∧ not *visited*(j) **then**
          begin {j is a new, as yet unvisited, neighbor}
            visited(j) ← **true**
            add vertex j and edge ij to T
            add vertex j to the end of L
          end
    end
  return(T)

Algorithm 2. Breadth-first search of a graph.

empty), setting i equal to 1. Then we look at vertex 1's neighbors. Vertices 2, 3, and 4 all qualify, so successively we put these vertices and the edges 12, 13, and 14 into T, and add these vertices (which have now been visited) to the end of L. At this point (the end of the first pass through the **while** loop), $L = (2, 3, 4)$.

Again we remove the first vertex in L, leaving L as $(3, 4)$, and setting i equal to 2. The neighbors of vertex 2 are vertices 1, 4, 5, and 7, but vertices 1 and 4 have already been visited. Hence we add to the tree only vertices 5 and 7, together with edges 25 and 27; and we add 5 and 7 to L, making $L = (3, 4, 5, 7)$. This completes the second pass through the **while** loop.

When vertex 3 is processed, no new vertices get added, since vertex 3 has no unvisited neighbors. Thus $L = (4, 5, 7)$ at the end of this third pass, and the tree contains vertices 1, 2, 3, 4, 5, and 7 at this point. Next vertex 4 is processed: Vertex 6 and edge 46 get added to the tree, and L is updated to be $(5, 7, 6)$. The final three passes generate no new additions to the tree. After the last pass, L is empty, and we are finished.

In Figure 9.16 we exhibit T in its conventional form. The edges of T are shown here with heavy lines; the edges *not* in T are also shown, with thin lines. Note that all the edges not in T join vertices at the same level or one level apart.

The reader is asked to show (Exercise 18) that this is always the case for breadth-first search spanning trees. Such edges are called **cross edges**. ◆

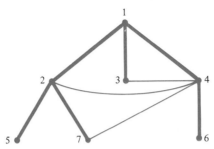

FIGURE 9.16   Breadth-first search spanning tree of the graph in Figure 9.14, with cross edges.

Algorithm 2 works for precisely the same reasons that Algorithm 1 works (see Exercise 21).

Next let us analyze the efficiency of breadth-first search. The **while** loop is executed as many times as there are vertices of G, since i assumes each vertex number exactly once. The amount of calculation that occurs in each pass is clearly in $O(n)$, because of the **for** loop. Thus the total time involved is in $O(n^2)$. Just as was the case with depth-first search, by looking only at the neighbors of each vertex, we can cut this time down to $O(m)$, where m is the number of edges in G. Thus we have the following analogue of Theorem 2 and its corollary.

THEOREM 3.   *The breadth-first search algorithm finds a spanning tree of a connected graph with n vertices and m edges in $O(m)$ steps, which is at worst in $O(n^2)$.*

COROLLARY.   *The breadth-first search algorithm provides a way to determine whether a graph with n vertices and m edges is connected in $O(m)$ steps, which is at worst in $O(n^2)$.*

SUMMARY OF DEFINITIONS

back edge: an edge in a graph that is not in its depth-first search spanning tree (from a given starting vertex); it will always join a vertex to an ancestor other than its parent.

breadth-first search: a "first in, first out" technique for visiting all the vertices of a connected graph and building a rooted spanning tree; whenever a new vertex is first encountered, that vertex is placed in a queue for later processing and the search continues from the previous vertex (see Algorithm 2).

cross edge: an edge in a graph that is not in its breadth-first search spanning tree (from a given starting vertex); it will always join two vertices that are at most one level apart in the tree and are not in the parent–child relation.

depth-first search: a recursive technique for visiting all the vertices of a connected graph and building a rooted spanning tree; whenever a new vertex is first encountered, the search continues from that vertex before being completed at the previous vertex (see Algorithm 1).

queue: a list in which elements are added at the tail and removed from the head, effecting a "first in, first out" order of processing; used in breadth-first search.

spanning forest of the graph G: a forest that consists of one spanning tree in each component of G.

spanning tree of the connected graph G: a tree that contains all the vertices of G.

EXERCISES

1. Find two nonisomorphic spanning trees for each of the following graphs.
   (a) Q_3.
   (b) $K_{2,5}$.
   (c) The six-spoked wheel (see Exercise 21 in Section 8.1).

2. Find a spanning forest for each of the following.
   (a) The pseudograph in Exercise 1d in Section 8.4.
   (b) The graph in Figure 9.10.

3. Find the depth-first search spanning tree of each of the following graphs, darkening the edges in the original picture as they are put into the tree. Then redraw the spanning tree in its conventional form, and show the back edges with dashed lines.
   (a) The graph in Figure 9.13a, assuming that the vertices are numbered from left to right.
   (b) The graph in Figure 9.13a, assuming that the vertices are numbered from right to left.

4. Find the breadth-first search spanning tree of each of the following graphs, darkening the edges in the original picture as they are put into the tree. Then redraw the spanning tree in its conventional form, and show the cross edges with dashed lines.
   (a) The graph in Figure 9.13a, assuming that the vertices are numbered from left to right.
   (b) The graph in Figure 9.13a, assuming that the vertices are numbered from right to left.

5. Prove Theorem 1 using induction on the number of edges in the graph.

6. Prove the corollary to Theorem 1.

7. Determine the number of different (not just nonisomorphic) spanning forests that the graph in Figure 9.13b has.

8. For the complete bipartite graph $K_{2,n}$, determine
   (a) The number of nonisomorphic spanning trees.
   (b) The number of different (not just nonisomorphic) spanning trees.

9. Find all nonisomorphic spanning trees of $K_{3,3}$.

10. Let G be a connected graph with n vertices and n edges.
    (a) Show that G contains exactly one cycle.
    (b) Determine the number of different (not just nonisomorphic) spanning trees that G has, in terms of the length of its cycle.

11. A **cut** of a connected graph is a minimal set of edges whose removal will disconnect the graph. Show that the intersection of a cut and the set of edges in any spanning tree is nonempty.

12. Prove that if a graph has 100 vertices and 98 edges, then it is not connected.

13. Find breadth-first search spanning trees of the following graphs. Draw them in the conventional manner.
    (a) K_5.
    (b) $K_{3,3}$.
    (c) Q_3, with the numbering implied by the vertex labels, viewed as numbers in base 2.

14. Repeat Exercise 13 for depth-first search.

15. Generalize from part (a) in Exercises 13 and 14 and state a theorem as to what the breadth-first search and depth-first search spanning trees for the complete graph look like.

16. Generalize from part (b) in Exercises 13 and 14 and state a theorem as to what the breadth-first search and depth-first search spanning trees for the complete bipartite graph look like.

17. Let T be a depth-first search spanning tree for a connected graph G. Show that every edge of G not in T joins a vertex in T to an ancestor or descendant.

18. Let T be a breadth-first search spanning tree for a connected graph G. Show that every edge of G not in T joins vertices in T that are either at the same level or differ in level by 1.

19. Find a graph G and an unrooted spanning tree T of G, such that no matter what vertex order is used, it is impossible to obtain T using either depth-first search or breadth-first search and ignoring the root.

20. The breadth-first search spanning tree in Figure 9.16 had cross edges joining either siblings or vertices in the uncle-nephew relation (see Exercise 25 in

Section 9.1). Give an example of a graph and a breadth-first search spanning tree in which there are cross edges not of these two types.

21. Show that the breadth-first search algorithm (Algorithm 2) necessarily produces a spanning tree. [*Hint*: Mimic the proof for Algorithm 1.]

22. Modify Algorithm 1 to produce a "depth-first search spanning forest" of an arbitrary (not necessarily connected) graph G.

23. Repeat Exercise 22 for breadth-first search (modify Algorithm 2).

24. We do not need a total ordering on the vertices of a connected graph in order to define depth-first search or breadth-first search unambiguously. All we need is a **local** ordering, in the following sense: We need to know where to start the exploration (the root of the resulting tree), and we need to know an order among the neighbors of each vertex. Given such a local ordering, depth-first search or breadth-first search produces an **induced** total ordering, given by the order in which the vertices are visited. Carry out the algorithm and find this induced ordering for each of the following. The graph is the one in Figure 9.14, and the root is vertex 1 in each case.
    (a) Depth-first search, where the neighbors are ordered clockwise as drawn, starting at noon (e.g., the neighbors of vertex 2 are, in order, 1, 4, 7, and 5; the neighbors of vertex 7 are, in order, 4 and 2).
    (b) Breadth-first search, where the neighbors are ordered clockwise as drawn, starting at noon.
    (c) Depth-first search, where the neighbors are ordered counterclockwise as drawn, starting at noon (e.g., the neighbors of vertex 2 are, in order, 1, 5, 7, and 4; the neighbors of vertex 7 are, in order, 2 and 4).
    (d) Breadth-first search, where the neighbors are ordered counterclockwise as drawn, starting at noon.

25. Explain why the level of a vertex v in the breadth-first spanning tree of a connected graph G is equal to the length of the shortest path in G from v to vertex 1 (the root of the tree).

Challenging Exercises

26. For the five-spoked wheel (see Exercise 21 in Section 8.1), determine the number of
    (a) Different (not just nonisomorphic) spanning trees.
    (b) Nonisomorphic spanning trees.

27. Let G be a connected graph.
    (a) Let T be a spanning tree of G. Let e be an edge in G but not in T. Show that there exists an edge f in T such that the graph obtained by adjoining e to T and deleting f is again a spanning tree of G.
    (b) Show that by using repeatedly the process described in part (a), we can transform any spanning tree T_0 of G into any other spanning tree T of

G; in other words, there is a sequence of spanning trees $T_0, T_1, T_2, \ldots,$ $T_k = T$ such that T_i and T_{i-1} differ by only one edge, for each i from 1 to k.

28. Prove or disprove each of the following.

(a) Unless G is a tree, the depth-first search spanning tree of G cannot be isomorphic as an unrooted tree to the breadth-first search spanning tree of G (assume that the ordering of the vertices is fixed).

(b) Unless G is a tree, a depth-first search spanning tree of G cannot be isomorphic as a rooted tree to any breadth-first search spanning tree of G (possibly based on a different ordering of the vertices).

29. Rewrite Algorithm 1 nonrecursively. [*Hint*: You will need to use a list called a **stack**, in which the most recently added element is removed and processed first ("last in, first out"). This data structure is in some sense the opposite of a queue, which was used in Algorithm 2.]

30. The graph shown in the figure consists of two paths of length $n - 1$ (called $u_1, u_2, \ldots, u_n$ and $v_1, v_2, \ldots, v_n$ in the picture), with each pair of corresponding vertices (u_i and v_i) joined by an edge.

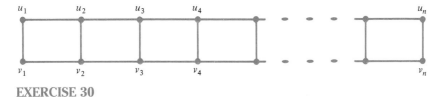

EXERCISE 30

(a) Let a_n be the number of different spanning trees that this graph has. Compute a_1, a_2, and a_3.

(b) Let b_n be the number of different spanning trees of this graph that include edge $u_1 v_1$. Compute b_1, b_2, and b_3.

(c) Write down a recurrence relation giving a_n in terms of a_{n-1} and b_n.

(d) Write down a recurrence relation giving b_n in terms of a_{n-1} and b_{n-1}.

(e) Use parts (c) and (d) to confirm your calculations in parts (a) and (b) and to calculate a_4, a_5, a_6, b_4, b_5, and b_6.

(f) Eliminate b from your system of two recurrence relations to obtain one recurrence relation for the sequence $\{a_n\}$. As a check, confirm your calculations in part (e).

(g) Solve the recurrence relation obtained in part (f) to give an explicit formula for a_n. [*Hint*: It involves irrational numbers.]

31. Implement on a computer the algorithms discussed in this section and apply them to some large graphs.

32. Modify Algorithm 2 so that the next vertex to be removed from the list and processed is not the vertex that has waited the longest in the list, but rather the vertex whose number is the smallest. Compare the resulting tree (it is still considered to be a breadth-first tree) with the tree obtained by Algorithm 2.

33. This exercise concerns the number of different (not just nonisomorphic) spanning trees of K_n.
    (a) Compute this quantity, for $1 \leq n \leq 5$ (the last of these requires a bit of elbow grease).
    (b) Make a conjecture as to a formula for the number of different spanning trees K_n has (the correct formula is fairly simple).
    (c) Find out more about this result, discovered by the British mathematician A. Cayley in 1889, by consulting Biggs, Lloyd, and Wilson [19], and Wilson [326].

34. A procedure similar to the searching algorithms discussed in this chapter can be used to find the strong components of digraphs (see Section 8.2). Consult Reingold, Nievergelt, and Deo [243] for details.

SECTION 9.3
TREE TRAVERSAL

Let not thy left hand know
what thy right hand doeth.

—Matthew 6:3

In this section we lay the groundwork for considering applications of trees, and we see how trees can be applied to arithmetic, algebraic, or other mathematical expressions. The latter material is important in the design of compilers for high-level computer languages. All of the trees in this section will be rooted, usually ordered, and sometimes binary.

There are several ways to list systematically the vertices of an ordered tree. The way that probably first occurs to most people is to list the root first, then list the vertices at level 1, in order from left to right as the tree is drawn, then the vertices at level 2 from left to right, and so on. The alphabetical order of the vertex labels in Figures 9.17 and 9.18 is this **level order**. Level order corresponds to breadth-first search of the tree, starting at the root. Surprisingly, this order is not very useful. Much more important are some orders related to depth-first search. Not surprisingly, these orders have recursive definitions.

PREORDER

If we apply depth-first search to an ordered tree, starting at the root and using the given (left to right) order of the children of each vertex to determine the order in which to visit them, then we get the preorder traversal of the tree. We will think of this process of systematically visiting all its vertices as a **traversal** of the tree. Let us make this idea more formal with a recursive definition, and then implement it with a recursive algorithm.

Let T be an ordered tree with root r. The **preorder** of the vertices of T is given by the following conditions:

If T contains just the one vertex r, then the preorder is r.

Otherwise, the preorder is r, followed by the preorder of the vertices in the immediate subtrees of T, in order.

EXAMPLE 1.   Let T be the ordered tree shown in Figure 9.17. The preorder of the vertices of T is obtained, according to the definition, by starting with the root a, following it with the vertices of the subtree rooted at b in preorder, followed by the vertices of the subtree rooted at c in preorder, followed by the vertices of the subtree rooted at d in preorder. The vertices of the subtree rooted at b, in preorder, are the root b, followed by the vertices of the subtree rooted at e, in preorder. The latter tree contains only one vertex, so the base case of the definition says that the preorder is just (e). Thus our preorder of the entire tree begins (a, b, e). Next since the subtree rooted at c has only one vertex, the preorder continues with c, giving (a, b, e, c) so far. Finally, we need to list the vertices in the subtree rooted at d, in preorder. This would be d, followed by f, then the preorder of the vertices rooted at g, then h. The preorder of the vertices of the subtree rooted at g is, again recursively from the definition, (g, i, k, j). Putting this all together, we have the preorder for the entire tree: $(a, b, e, c, d, f, g, i, k, j, h)$.

We have no trouble turning the definition into a recursive algorithm. We phrase this algorithm in terms of "processing" the vertices; what that means can vary with the context. For now let us suppose that process means **print**, in which case this algorithm writes out the names of the vertices, in preorder. Also, we will think of passing *the root* of a tree as the argument, rather than the entire tree. Since we also assume that each vertex comes with information as to who its children are, knowing the root means knowing—recursively—the whole tree. (This view corresponds to programming practice, since usually a tree is represented in a computer as a pointer to its root, and each vertex of the tree contains pointers to its children.)

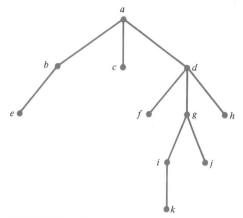

FIGURE 9.17 An ordered tree to be traversed.

procedure *preorder*(r : root of ordered tree)
    process r
    for each child v of r, in order, **do**
      call *preorder*(v)
    return

Algorithm 1. Preorder traversal of an ordered tree.

Conceptually, even if the tree is not ordered, merely rooted, preorder still makes sense as a partial order on the tree. Vertices still precede all of their children, but the order in which the subtrees are considered is not defined. In practice, usually a rooted tree receives a de facto order by the way it is represented, so we may as well assume that we are dealing with ordered trees.

In the special (and most important) case of a binary tree, this algorithm could be rewritten explicitly as follows.

procedure *binary_preorder*(r : root of binary tree)
    process r
    if r has a left child **then call** *binary_preorder*(*left_child*(r))
    if r has a right child **then call** *binary_preorder*(*right_child*(r))
    return

Algorithm 2. Preorder traversal of a binary tree.

With preorder, then, we process the rooted tree from the top down. This approach is useful whenever we need to deal with a vertex before we deal with its children.

EXAMPLE 2. Consider a book such as this one (just the text proper—ignore the preface, list of references, etc.). Its structure can be viewed as an ordered tree. The root is the title of the entire book. The children of the root are the names of the parts that the book is divided into, in order. In this book, we have four parts (*Foundations of Discrete Mathematics*, *Tools of Discrete Mathematics*, *Combinatorics*, and *Graph Theory*). The children of each part are the titles of the chapters comprising that part; in this book, the first part has three children, the second part has two, the third part has two, and the fourth part has three. One level farther down are the sections within a chapter, and below that are the subsections.

We can view this tree as existing in its own right, in the abstract, contemporaneously with the physical book itself. Now consider the detailed table of contents of the book. It is nothing but a preorder listing of this tree. Before section names are given, for example, we give the title of the chapter in which those sections are contained; and following each section name is the ordered list of its subsections. Nothing could be more natural than this preorder listing—indeed, it is hard to see how we would ever consider any other linear arrangement of this information. ◆

POSTORDER

In many applications of rooted trees, we need to process the tree from the bottom up. We do so with a postorder traversal of the vertices, processing the children of each vertex *before* we process the vertex itself. The following definition and Algorithms 3 and 4 are the obvious modifications of the corresponding items for preorder; all we have done is to move the processing of each vertex from *before* the processing of its children to *after* the processing of its children.

Let T be an ordered tree with root r. The **postorder** of the vertices of T is given by the following conditions:

If T contains just the one vertex r, then the postorder is r.

Otherwise, the postorder is the postorder of the vertices in the immediate subtrees of T, in order, followed by r.

EXAMPLE 3. Let T be the ordered tree shown in Figure 9.17. The postorder of the vertices of T is obtained, according to the definition, by ending with the root a, preceded by the vertices of the subtree rooted at b in postorder, the vertices of the subtree rooted at c in postorder, and the vertices of the subtree rooted at d in postorder. The vertices of the subtree rooted at b, in postorder, are the vertices of the subtree rooted at e, in postorder, followed by b. The latter tree contains only one vertex, so the base case of the definition says that the postorder is just (e). Thus our postorder of the entire tree begins (e, b) and ends with a. Next since the subtree rooted at c has only one vertex, c follows b in postorder. Finally, we need to list the vertices in the subtree rooted at d, in postorder. This would be f, then the postorder of the vertices rooted at g, then h, then d. The postorder of the vertices of the subtree rooted at g is, again recursively from the definition, (k, i, j, g). Putting this all together, we have the postorder for the entire tree: $(e, b, c, f, k, i, j, g, h, d, a)$. ◆

Again we have no trouble turning the definition into recursive algorithms.

procedure *postorder*(r : root of ordered tree)
    for each child v of r, in order, **do**
        call *postorder*(v)
    process r
    return

Algorithm 3. Postorder traversal of an ordered tree.

procedure *binary_postorder*(r : root of binary tree)
    if r has a left child **then call** *binary_postorder*(*left_child*(r))
    if r has a right child **then call** *binary_postorder*(*right_child*(r))
    process r
    return

Algorithm 4. Postorder traversal of a binary tree.

Section 9.5 contains an application of rooted trees to game theory; there we view the tree in its postorder arrangement. Let us look at a simpler application in which we need to process a tree from the bottom up.

EXAMPLE 4. Write a recursive algorithm to compute the height of a binary tree.

Solution. Let the desired function—and the procedure that computes it—be called *height*. More generally, *height*(v) will be the height of the subtree rooted at v. The base case occurs if v is a leaf; in that case *height*(v) = 0. Otherwise, we can find *height*(v) if we know the heights of the immediate subtrees of v, since clearly *height*(v) equals 1 plus the larger of those heights. Thus to find *height*(r), where r is the root of a binary tree T, we need to process T, in effect, in postorder.

We can make our algorithm even neater if we think of the input as the root of an *extended* binary tree, so that every nonempty vertex has both a left and a right child, even if one or both of these is empty. To be consistent with our recursive step, we need to set *height*($\emptyset$) equal to -1. The pseudocode is given as Algo-

procedure *height*(r : root of extended binary tree)
   if $r = \emptyset$ **then return**(-1)
     else return($1 + \max(height(left\_child(r)), height(right\_child(r)))$)

Algorithm 5.   Computing the height of an extended binary tree.

rithm 5. We leave to the reader the slightly longer version for ordinary binary trees (see Exercise 17). Similar algorithms can be used, for example, to count the number of leaves in a tree. To know how many leaves there are in a tree, we only need to know how many leaves there are in the immediate subtrees. Exercises such as this are left to the reader (see Exercises 15 and 16).   ◆

INORDER

Preorder for binary trees involves processing the root before processing the children; postorder involves processing the root after processing the children. The obvious third choice is to process the root *between* processing the children—in other words, to process the left child, then process the root, then process the right child. We make the obvious formal definition of inorder traversal of a binary tree, analogous to our previous ones.

Let T be a binary tree with root r. The **inorder** of the vertices of T is given by the following conditions:

> If T contains just the one vertex r, then the inorder is r.
>
> Otherwise, the inorder is the inorder of the vertices in the left subtree of T (if any), followed by r, followed by the inorder of the vertices in the right subtree of T (if any).

EXAMPLE 5.   Let T be the binary tree shown in Figure 9.18. Following the definition exactly as we did in Examples 1 and 3, we find that the inorder is $(d, b, g, e, k, m, h, a, c, l, i, f, j)$. After practicing on several exercises, the reader will become proficient at "reading" the vertices of a tree in any of the three orders that we have discussed.   ◆

The inorder traversal algorithm (see Algorithm 6) is exactly what we expect.

In Section 9.4 we will see how the inorder of binary trees allows us to use trees to represent ordered lists efficiently. We will not give any other applications here.

EXPRESSION TREES

We can tie all these traversal orders for trees together, and gain some insight into the structure of computations, with an application of trees to the evaluation of

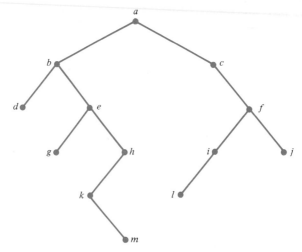

FIGURE 9.18   A binary tree to be traversed.

procedure *binary_inorder* (*r* : root of binary tree)
    if *r* has a left child **then call** *binary_inorder* (*left_child*(*r*))
    process *r*
    if *r* has a right child **then call** *binary_inorder* (*right_child*(*r*))
    return

Algorithm 6.   Inorder traversal of a binary tree.

mathematical expressions. Recall from Example 5 in Section 5.1 the definition of a fully parenthesized algebraic expression, or FPAE for short. It consists of strings of numerals, variable names, arithmetic **operator** symbols such as + (denoting addition) or ∗ (denoting multiplication), and parentheses, obeying certain rules of syntax. Each FPAE has a meaning: The arithmetic operations are applied in a particular order to certain numbers—the numbers whose numerals appear in the expression, the values of the variables whose names appear in the expression, and the values of subexpressions. We denote the value obtained when performing this calculation on a FPAE α by *EVAL*(α). This was the upshot of Example 6 in Section 5.1. For example, ((5 − 2) ∗ 7) "means" that 5 and 2 are to be subtracted (in that order), resulting in the number 3, and then 3 and 7 are to be multiplied, giving the value 21 for the entire expression. More formally, *EVAL*(((5 − 2) ∗ 7)) = 21.

Let us see how an ordered tree can be used to represent such a calculation. The leaves of the tree represent the numerals or variables, and the value of a leaf is the value of the corresponding number or the current value of the variable. Each internal vertex represents an arithmetic operator, and its children are the **operands** on which the operation is to be performed. In our example above, "−" would be a vertex with two children, 5 and 2, in order. The value of such an internal vertex is the answer to the calculation it expresses: the result of applying the operation to the operands. In particular, such a vertex can then be considered as an operand to a further operation, and our tree can grow to arbitrary heights. For example, the tree representing $((5 − 2) * 7)$ is shown in Figure 9.19.

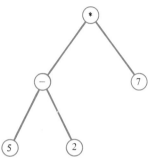

FIGURE 9.19 Expression tree for the calculation $((5 − 2) * 7)$.

We now give a formal recursive definition of **expression tree** and its **value**:

If x is a variable or a constant, then the ordered tree with just one vertex, x, is an expression tree, whose value is the value of x.

If o is an operator taking k operands, and if $T_1, T_2, \ldots, T_k$ are expression trees, then the ordered tree with root o and immediate subtrees $T_1, T_2, \ldots, T_k$ is an expression tree, whose value is the result of applying the operation specified by o to the values of these immediate subtrees, in order.

We sometimes speak of the "value" of a vertex; we just mean the value of the subtree rooted at that vertex. Note that we have not restricted the operators to being binary (although they usually are); to the extent that they are binary, we view expression trees as binary trees. Also, we have not restricted the context to arithmetic and algebra; expression trees can arise in Boolean (logical) calculations, set theoretic calculations, and other contexts.

Expression trees are really more fundamental than expressions: They capture the essence of a calculation, without regard to any particular linear arrangement of the symbols involved. We will see in this subsection that several different strings can represent the same calculation, even though there is essentially only one tree.

EXAMPLE 6. Find the expression tree for the following calculations:

(a) $(5 - 2) * 7 + (4 - 3)$.
(b) $(A \cup B) \cap (A \cup C)$.

Solution. Two things are worth noting before we begin. First, the expressions given here are not fully parenthesized, but they have an intended interpretation because of the usual conventions (in particular that multiplication has priority over addition). As fully parenthesized expressions, they would be $(((5 - 2) * 7) + (4 - 3))$ and $((A \cup B) \cap (A \cup C))$, respectively. Thus each represents a unique computation and hence we can write down their expression trees. Second, although the setting of the first is arithmetic, the setting of the second is set theory, and the same ideas apply.

(a) This expression is the sum of two quantities, namely $(5 - 2) * 7$ and $4 - 3$. Hence the expression tree has the binary operator "+" as its root, and its subtrees are the expression trees for these two summands. The first we already found, in Figure 9.19. For the second, the binary operation of subtraction is being applied to two operands: the number 4 and the number 3. Hence the expression tree for $4 - 3$ has "−" as its root, 4 as the left child, and 3 as the right child. Putting this all together, we obtain Figure 9.20 as the desired expression tree.

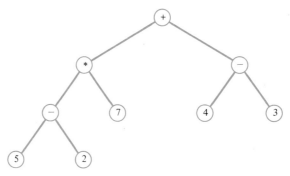

FIGURE 9.20 Expression tree for the arithmetic expression $(5 - 2) * 7 + (4 - 3)$.

To find the value of this expression tree, which is the value of the corresponding arithmetic expression, we need to work from the bottom up; in other words, we need to do a postorder traversal of the tree. (The use of the word "we" is perhaps ill-advised, since we will write down shortly a recursive algorithm for finding the value of an expression tree, and *we* always view recursive algorithms as top-down creatures. It would be more precise to say that *the computer* needs to work from the bottom up to evaluate an expression tree.) The value of the subtree rooted at the leftmost minus sign is $5 - 2 = 3$, that is, the value obtained when the operation "$-$" is applied to the values of the subtrees, 5 and 2. Next the vertex containing the multiplication sign has value 21, as we saw earlier. The value of the right subtree is the result of applying subtraction to 4 and 3, namely $4 - 3 = 1$. Finally, then, the value of the entire tree is the result of applying the operation at the root (addition) to the values 21 and 1, namely 22.

(b) We cannot find the value here, because we are not told what specific sets A, B, and C are. The expression tree is shown in Figure 9.21. ◆

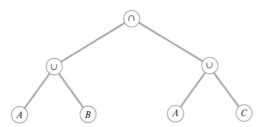

FIGURE 9.21 Expression tree for the set theoretic expression $(A \cup B) \cap (A \cup C)$.

Let us express in pseudocode the algorithm for evaluating an expression tree. We write it as a recursive subprocedure, *EVAL*, which returns the value; as usual, we pass the root of the tree as argument. For simplicity, we are thinking of each vertex both as a vertex in a tree (it is either a leaf or has children) *and* as the variable, constant, or operator that it represents. See Algorithm 7.

procedure *EVAL*(r : root of expression tree)
    if r is a constant **then return**(r)
        else if r is a variable **then return**(current value of r)
            {otherwise, r is an operator}
            else return(r applied to *EVAL*(c_1), *EVAL*(c_2), ..., *EVAL*(c_k))
                {where $c_1, c_2, \ldots, c_k$ are the children of r}

Algorithm 7. Evaluating an expression tree.

So far we have an algorithm for evaluating an expression tree. Next we want to consider explicitly the process of converting between expression trees and linear expressions (strings) that represent the same calculations. When a compiler for a high-level computer language such as Pascal encounters an algebraic, arithmetic, or logical expression in a program, it needs to translate that expression into a sequence of unary or binary computations. To do so, it will, either implicitly or explicitly, build the expression tree. We will see that each of the possible traversals of an expression tree corresponds to a way of writing the calculation as a string. Inorder traversal corresponds to the usual algebraic notation. We need to add parentheses to avoid ambiguity, and the resulting string is called the **fully parenthesized infix expression** for the calculation. In this form operators come between their operands. With preorder traversal, as we will see, no parentheses are necessary; the string that results from listing the vertices of an expression tree in preorder is called the **prefix** or **Polish expression**. Similarly, the string that results from listing the vertices in postorder is unambiguous without parentheses; it is called the **postfix** or **reverse Polish expression**. In a prefix expression the operators come before their operands, and in a postfix expression the operators come after their operands. (The name "Polish" comes from the fact that the twentieth-century Polish logician J. Łukasiewicz first used these alternative notations.)

We will restrict our discussion here to binary operators (forbidding something like the unary minus or the logical negation). Exercises 26–29 discuss how to handle unary operators, and Exercise 30 deals with operators that take more than two operands.

We are already familiar with infix expressions. We will define the infix expression formally with Algorithm 8: The result of applying this algorithm to an expression tree *is* the infix expression for that calculation. We assume that the **print** instruction in this algorithm (as well as the next two) does not issue a "carriage return"; thus the output will be a string all on one line. Also, we can assume that sufficient space is left around the printed variables and constants to enable someone looking at the output to tell where they begin and end. This is not crucial with infix expressions, since there will always be operators between them, but it is important for prefix and postfix expressions. (We need to distinguish between 42 3 and 4 23, for example.)

For example, we can think of Algorithm 8 executing on the expression tree given in Figure 9.20 in the following manner. At the highest level, the output will be the following string:

(subexpression from left subtree + subexpression from right subtree)

Recursively, the two subexpressions will be correctly produced at the appropriate time, resulting in the final answer:

$$(((5-2)*7)+(4-3))$$

Note that since each operator is printed *between* the printing of the left subtree and the right subtree, the infix expression prints all the symbols in the expression tree in their *inorder*. Indeed, the infix expression is nothing but an inorder listing of the contents of the expression tree, with pairs of parentheses added appropriately— a left parenthesis before each left subtree and a right parenthesis after each right subtree.

procedure *infix*(*r* : root of expression tree)
   {assume that all the operations are binary; the output of this procedure is
   a printed string, giving the fully parenthesized infix expression for the
   calculation of the tree}
     if *r* is a constant or variable **then print**(*r*)
       else {otherwise, *r* is an operator and has two children}
        begin
          print("(") {left parenthesis}
          call *infix*(*left_child*(*r*)) {which will print first operand}
          print(*r*) {the operator symbol}
          call *infix*(*right_child*(*r*)) {which will print second operand}
          print(")") {right parenthesis}
        end
    return

Algorithm 8.   Producing the infix expression from an expression tree.

What about the opposite problem, converting a FPAE into an expression tree? The key ingredient is to be able to recognize the outermost operator. The reader can show (Exercise 24) that the outermost operator is the unique operator that has exactly one more left parenthesis to its left than it has right parentheses to its left. For example, in $(((5 - 2) * 7) + (4 - 3))$, the plus sign is the outermost operator, because it is the only operator in the string that meets this condition (having 3 left parentheses and 2 right parentheses to its left). This operator becomes the root of the tree, the substring between the initial left parenthesis and this operator is translated (recursively) into the left subtree, and the substring between this operator and the final right parenthesis is translated (recursively) into the right subtree. The base case is a string without an operator or parentheses; it must be a constant or variable and is represented by a vertex with no children. (A somewhat finer analysis would enable us to write an algorithm for processing the infix expression totally from left to right, without the need for finding the outermost operator first and then backing up to process the first operand.)

Let us stand back for a minute and see what we have accomplished so far. Given a fully parenthesized arithmetic expression, we can algorithmically construct its expression tree by the method outlined in the preceding paragraph. Algorithm 7 enables us to find its value. Thus we have a two-stage process for mechanically evaluating such an expression. (A third step, or a modification of this process, would be needed if the original expression omitted some parentheses; we will not go into the details.) When we first studied algebra, we learned to evaluate an expression by essentially this process, because working "from the inside out" corresponds to evaluating the expression tree from the bottom up.

Next we turn to prefix notation, defined as the output of Algorithm 9. The algorithm is practically the same as Algorithm 8, except that there are no parentheses, and the symbols in the tree are printed in their preorder.

The prefix expression, then, is just the symbols in the expression tree, listed in preorder. (By "symbol" here we mean an entire variable name, an entire numeral for a number, or an operator. These are called "tokens" by computer scientists.) For the expression tree of Figure 9.20, the prefix expression is the string:

$$+ * - 5\,2\,7 - 4\,3$$

procedure *prefix*(r : root of expression tree)
   {assume that all the operations are binary; the output of this procedure is a
    printed string, giving the prefix (Polish) expression for the calculation of the
    tree}
    print(r) {whether r is a leaf or not}
    if r is not a leaf **then**
      begin
        call *prefix*(*left_child*(r)) {which will print first operand}
        call *prefix*(*right_child*(r)) {which will print second operand}
      end
    return

Algorithm 9.   Producing the prefix expression from an expression tree.

Conversely, given a prefix expression, we can construct its expression tree recursively as follows. If the first symbol is a constant or variable, then the expression tree consists of just that symbol as its root, with no other vertices (and the prefix expression necessarily has no other symbols). Otherwise, the first symbol is an operator symbol, which becomes the root of the expression tree. As much of the rest of the string is processed as is necessary to complete (recursively) a tree, which becomes the left subtree, and then the rest of the string is recursively translated into the right subtree. The reader can try this procedure on the string displayed above; the result should be Figure 9.20.

The fact that both the tree-to-prefix translation (Algorithm 9) and this prefix-to-tree translation are well defined (and inverses of each other) shows, as we claimed above, and in contrast to the situation with infix expressions, that no parentheses are needed to avoid ambiguity in priority of operations. For example, an infix expression such as $3 + 4 * 7$ is ambiguous without a convention on priorities. Written in prefix notation, it is either $* + 3\,4\,7$ or $+ 3 * 4\,7$, whichever is intended; we do not encounter the ambiguity of having one expression standing for more than one calculation. (The usual mathematical convention, incidentally, is not universal—many inexpensive electronic pocket calculators will give the answer 49 when the user enters $3 + 4 \times 7$, while more expensive ones will give the answer 31.)

As with infix expressions, then, we have a two-stage process for evaluating a prefix expression. There is also a simple one-pass process that humans can use.

We work from right to left. Whenever we encounter an operator symbol, we apply the operator to the two numbers currently to its immediate right (in order from left to right), and replace all three (the operator and the two numbers) with the answer. When we have finished we will be left with exactly one number, the final answer.

EXAMPLE 7. Evaluate the prefix expression $+ * - 5\,2\,7 - 4\,3$.

Solution. Working from the right, we encounter first the minus sign preceding the 4 and the 3. Carrying out this calculation, we obtain the answer 1, so our string becomes $+ * - 5\,2\,7\,1$. Next we find the minus sign preceding the 5 and the 2, so we replace $- 5\,2$ by the answer 3, obtaining $+ * 3\,7\,1$. We then multiply 3 by 7, obtaining the string $+ 21\,1$, and finally, add 21 and 1 to get the answer of 22. ◆

Finally, we turn the process for prefix expressions around to obtain analogous procedures for postfix expressions. The postfix expression for a given expression tree is by definition the output of Algorithm 10.

procedure *postfix* (r : root of expression tree)
    {assume that all the operations are binary; the output of this procedure is a
      printed string, giving the postfix (reverse Polish) expression for the calculation
      of the tree}
    if r is not a leaf **then**
      begin
        call *postfix* (*left_child* (r)) {which will print first operand}
        call *postfix* (*right_child* (r)) {which will print second operand}
      end
    print (r) {whether r is a leaf or not}
    return

Algorithm 10. Producing the postfix expression from an expression tree.

The postfix expression is just the symbols of the expression tree in postorder. For the expression tree of Figure 9.20, the postfix expression is the string:

$$5\,2 - 7 * 4\,3 - +$$

Given a postfix expression, we can construct its expression tree recursively as follows. If there is just one symbol, then the expression tree consists of just that symbol (which must be a constant or variable) as its root, with no other vertices. Otherwise, we work from right to left. The last symbol (necessarily an operator symbol) is the root. Then as much of the string immediately preceding the last symbol is processed as is necessary to complete (recursively) a tree, which becomes the right subtree. Final! he rest of the string becomes the left subtree. The reader can try this procedure on the string displayed above; again the result should be Figure 9.20.

Once again, no ambiguity results from the lack of parentheses. The procedure for evaluating postfix expressions by hand is similar to the procedure for evaluating prefix expressions by hand. We work from left to right. Whenever we encounter an operator symbol, we apply the operator to the two numbers currently to its immediate left (in order from left to right), and replace all three (the operator and the two numbers) with the answer. When we have finished we will be left with exactly one number, the final answer.

EXAMPLE 8. Evaluate the postfix expression $5\,2-7*4\,3-+$.

Solution. Working from the left, we encounter first the minus sign following the 5 and the 2. Carrying out this calculation, we obtain the answer 3, so our string becomes $3\,7*4\,3-+$. We then find the multiplication sign following the 7 and the 3, so we replace $3\,7*$ by the answer 21, obtaining $21\,4\,3-+$. Next we subtract 3 from 4, obtaining the string $21\,1+$, and add 21 and 1 to arrive at the final answer, 22. ◆

SUMMARY OF DEFINITIONS

expression tree: an ordered tree in which leaves represent variables or constants, and internal vertices represent operations being applied to the values of their children.

fully p enthesized infix expression for a calculation: an expression in which each operator is placed between its operands and each subexpression operand–operator–operand is enclosed in a pair of parentheses.

inorder of the vertices in a binary tree: the order consisting of the inorder of the vertices in the left subtree, followed by the root, followed by the inorder of the vertices in the right subtree.

level order of the vertices in an ordered tree: the order in which the root comes first, followed by all the vertices at level 1, from left to right, as the tree is drawn, followed by all the vertices at level 2 from left to right, and so on.

operand: a quantity on which an operator acts (e.g., 3 and $(x - y)$ are the operands of + in the infix expression $2 * (3 + (x - y))$ and in the equivalent prefix expression $*2 + 3 - x\,y$).

operator: a k-ary function from a set to itself (e.g., "$-$" is a binary $(k = 2)$ operator on the set of real numbers—it performs the operation of subtraction, taking the operands x and y to their difference $x - y$).

Polish expression: a prefix expression.

postfix expression for a calculation: an expression in which each operator is placed immediately after its operands; no parentheses are needed.

postorder of the vertices in an ordered tree: the order consisting of the postorders of the vertices in each immediate subtree, from left to right, followed by the root.

prefix expression for a calculation: an expression in which each operator is placed immediately before its operands; no parentheses are needed.

preorder of the vertices in an ordered tree: the order in which the root comes first, followed by the preorders of the vertices in each immediate subtree, from left to right.

reverse Polish expression: a postfix expression.

traversal of a graph: a systematic visiting of all the vertices of the graph.

value of an expression tree: The value of the trivial tree with just one vertex (the root) is the value of the constant or variable the root represents, and (recursively) the value of an expression tree with an operator at the root is the result of applying that operator to the values of the immediate subtrees.

EXERCISES

1. List the vertices of the tree in Figure 9.4 in the following orders.
   (a) Preorder.
   (b) Postorder.

2. List the vertices of the tree in Figure 9.10 in the following orders.
   (a) Preorder.
   (b) Postorder.
   (c) Inorder.

3. Draw the expression tree for each of the following expressions. Assume the usual priority conventions, in particular that multiplication has precedence over addition and subtraction, and that operations with the same priority are performed from left to right. Thus your tree will have only binary operators in it; $1 + 2 + 3$, for example, means $(1 + 2) + 3$.
   (a) $X * Y + (A - B)$.
   (b) $(X + Y) * Z + Q - (X - Y) * B$.
   (c) $(P \wedge (P \rightarrow Q)) \rightarrow R$.
   (d) $(A \cap (B \cup C)) - (A \cap B \cap C)$.

4. Write each of the calculations in Exercise 3 as a prefix expression.

5. Write each of the calculations in Exercise 3 as a postfix expression.

6. Write each of the calculations in Exercise 3 as a fully parenthesized infix expression.

7. Evaluate each of the following prefix expressions, without drawing the expression tree.
   (a) $* + 3 - 4\,2 + 6\,3$.

(b) $\cup \cap \{1,2,3,4\} \cup \{1\} \{3,5\} \{6,7\}$.

(c) $\vee F \wedge \wedge \vee F \wedge \rightarrow T F T F T$.

8. Draw the expression tree for each of the calculations in Exercise 7.

9. Write each of the calculations in Exercise 7 as a postfix expression.

10. Write each of the calculations in Exercise 7 as a fully parenthesized infix expression.

11. Evaluate each of the following postfix expressions, without drawing the expression tree.

(a) $3\,3*4\,4*+$.

(b) $\{1,2,3\} \{4,5\} \cap \{5,6\} \cup$.

(c) $F F F F \rightarrow \rightarrow \rightarrow$.

12. Draw the expression tree for each of the calculations in Exercise 11.

13. Write each of the calculations in Exercise 11 as a prefix expression.

14. Write each of the calculations in Exercise 11 as a fully parenthesized infix expression.

15. Write a recursive algorithm to print the leaves only of an ordered tree, in preorder.

16. Write recursive algorithms to determine the following parameters of binary trees. You will probably find it simpler to work with extended binary trees.

(a)   The number of vertices.

(b)   The number of leaves.

(c)   The number of internal vertices.

17. Rewrite Algorithm 5 so that it deals with binary trees, not with extended binary trees.

18. Show that the leaves have the same order, relative to each other, in all three listings of ordered trees (preorder, postorder, and, in the case of binary trees, inorder).

19. Find a binary tree whose preorder is given by $ABCDEFGHIJKLMNOP$, while its inorder is given by $BFEGDCAJIKHLONPM$.

20. Suppose that a binary tree has three vertices, whose preorder is ABC. Determine which of the six permutations of these three letters can possibly be the inorder listing of the vertices of the tree.

21. Show that knowing both the preorder and postorder of the vertices of a binary tree does not uniquely determine the tree.

22. Suppose that we were to define "Hebrew postorder" similarly to postorder, except that we process the immediate subtrees from right to left (before pro-

cessing the root), rather than from left to right. (The fanciful name comes from the fact that the Hebrew language is read from right to left.)
(a)   Find the Hebrew postorder for the tree in Figure 9.10.
(b)   State and prove a relationship between preorder and Hebrew postorder.

23. Given a binary tree T, let T^R be the tree which has the same vertices as T and in which each vertex has the same parent as it had in T, except that every left child is now a right child, and every right child is now a left child.
(a)   Give a recursive definition of T^R.
(b)   Write a recursive algorithm to construct T^R, given T.

24. Show that the outermost operator symbol in a fully parenthesized infix expression with at least one operator symbol is the unique operator symbol with the property that the number of left parentheses to its left is equal to one more than the number of right parentheses to its left.

25. Give a recursive definition of prefix and postfix expressions, without reference to an expression tree (i.e., mimicking the definition of FPAE in Example 5 of Section 5.1).

26. Our definition of an expression tree and its value did not require that the operators involved be binary. Many unary operators occur in mathematics. Draw expression trees for the following calculations.
(a) $((4 - (-3)) * 7)$.   (Make up a notation for unary minus.)
(b) $\sqrt{A * A + B * B}$.
(c) $\sqrt{A^2 + B^2}$.   (Make up a notation for squaring.)
(d) $(A \cap \overline{B}) \cup C$.   (Make up a notation for complementation.)
(e) $(P \vee (\neg P)) \rightarrow Q$.

27. Write each of the calculations in Exercise 26 as a prefix expression.

28. Write each of the calculations in Exercise 26 as a postfix expression.

29. Explain the difficulty with writing the calculations in Exercise 26 as fully parenthesized infix expressions.

30. Consider the set-theoretic expression $(A \cup B \cup C) \cap (A \cup D) \cap (B \cup (D \cap E \cap F \cap G))$. Because the union and intersection operations are associative (part (a) of Theorem 1 in Section 2.3), this expression is unambiguous. In fact, we can think of its outermost level, for example, as a ternary intersection: one operator ($\cap$) taking three arguments.
(a)   Draw the expression tree for this expression under this interpretation. Use $\cup_n$ and $\cap_n$, respectively, as the symbols for n-ary union and intersection.
(b)   Write the expression in prefix notation.
(c)   Write the expression in postfix notation.

31. A verbal statement of the Pythagorean theorem is that "in a right triangle, the square of the hypotenuse is equal to the sum of the squares of the other two sides." Is this calculation expressed in prefix, postfix, or infix?

32. Suppose that the tax code for a certain country states that "the tax shall be the larger of 20 percent of the taxpayer's earned income and 10 percent of the sum of his earned income and his unearned income."
    (a) Translate the tax into a prefix expression. Use L for the binary operator "larger of"; Y and N for the unary operators "20 percent of" and "10 percent of", respectively; and e and i for earned and unearned income, respectively.
    (b) Draw the expression tree for the expression obtained in part (a).

33. Write an algorithm for processing the vertices of an ordered tree in level order. [*Hint*: Look at Algorithm 2 in Section 9.2.]

34. An ordered tree can be written in **preorder list form**, defined as follows. The preorder list form of a tree with only one vertex r is just r. If a tree T has root r and immediate subtrees $T_1, T_2, \ldots, T_k$, with $k \geq 1$, and if L_i is the preorder list form for T_i for each i from 1 to k, then the preorder list form for T is $r(L_1, L_2, \ldots, L_k)$; note that the parentheses and commas are part of the notation.
    (a) Give the preorder list form of the tree in Figure 9.7.
    (b) Write a recursive algorithm to print the preorder list form of an ordered tree.
    (c) Modify the recursive step of the definition, so that the form for T is just $(L_1, L_2, \ldots, L_k)$, without the r. Thus only leaves appear (together with parentheses and commas to show the structure). Repeat part (a) under this modification.
    (d) Write a recursive algorithm to print the preorder list form of an ordered tree, as modified in part (c).

Challenging Exercises

35. Recall the definition of an AVL-tree from Exercise 28 in Section 9.1. Write a recursive algorithm for determining whether a given binary tree T is an AVL-tree. Assume that $height(v)$ has already been computed (and is available to your algorithm) for each vertex v of T.

36. Recall from Exercise 32 in Section 9.1 how an ordered tree T can be represented by a binary tree B_T. Find, with justification, whether preorder and/or postorder in T, are the same as any of preorder, postorder, and inorder in B_T.

37. Suppose that you are given two lists of a set of vertices. Show that there exists at most one binary tree whose preorder is given by the first list and whose inorder is given by the second list, and explain how to construct that tree if it exists. [*Hint*: Think recursively. See Exercise 19 for an example.]

38. Consider a string consisting only of A's and $+$'s, which are a variable and a binary operator, respectively. Give nonrecursive necessary and sufficient conditions for the string to be

(a) A valid prefix expression.

(b) A valid postfix expression.

(c) A valid algebraic expression (infix without parentheses).

Exploratory Exercises

39. Model the rules of succession for the British monarchy as a particular order of the vertices in an ordered family tree. Assume that the root of the tree is the reigning king or queen. Assume (contrary to past practice) that there is no intermarriage among the descendants of the root. Give a general explanation as to how to set up your model. Then draw the appropriate tree for the royal family headed by Queen Elizabeth II of England, and list her descendants in their order of succession (Prince Charles is first, his oldest son Prince William is second, and so on).

40. Write an algorithm in pseudocode (or, better yet, in a real high-level programming language) to simulate a calculator that uses the usual mathematical conventions on priority of operations. Assume the four arithmetic operations, digits, the decimal point, the equals sign, parentheses, and a "clear" key. The keys the user presses should be **read** by the algorithm, and what appears on the calculator after each key has been pressed should be **printed** by the algorithm.

SECTION 9.4
FURTHER APPLICATIONS OF BINARY TREES

LEXICOGRAPHER—A writer of dictionaries, a harmless drudge.

—**Samuel Johnson,** *Dictionary*

Binary trees occur throughout computer science. In this section we touch on two more applications of them. In the first we use a binary tree to store an ordered list (such as a list of words in a dictionary). The tree has the advantage of allowing for fast searches (comparable to the binary search discussed in Section 4.3), as well as allowing for rapidly adding to (or deleting from) the list. Our second application, Huffman codes, uses binary trees as a method of encoding information into and decoding it from bit strings. This is the form in which information can be stored on a magnetic disk or tape, or sent electronically from one location to another. The Huffman codes tend to be more efficient than other, more naive ways of doing the encoding and decoding, so their use can save time or space.

BINARY SEARCH TREES

The inorder traversal of a binary tree T imposes a linear order on the vertices of T. It is not hard to see that the following statement characterizes this order: *For every vertex v in T, every vertex in the left subtree of v is less than v, and every vertex in the right subtree of v is greater than v.* This will always be the order we are referring to in this subsection. A **binary search tree** is a binary tree, at each of whose vertices is stored an element of a linearly ordered set, such that the order on the set is identical to inorder in the tree. We denote the order by $\preceq$.

Typically, the set will be a set of words or names, ordered alphabetically (i.e., strings over some alphabet, ordered lexicographically), perhaps with additional information attached to each item as well (such as a definition of the word, the translation of a word into some other language, or the address of a person).

EXAMPLE 1. Consider the following set of ordered pairs of English and French words, ordered alphabetically by the English word in the pair: (*cat*, *le chat*), (*dog*, *le chien*), (*eat*, *manger*), (*girl*, *la fille*), (*house*, *la maison*), (*red*, *rouge*), (*tiger*, *le*

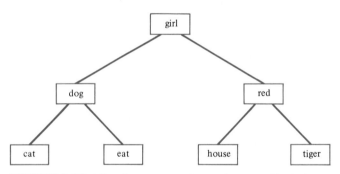

FIGURE 9.22   One binary search tree for some English words.

tigre). There are many binary search trees for this collection, two of which are shown in Figures 9.22 and 9.23. For simplicity, we show only the English word at each vertex.

In each case, the inorder of the vertices corresponds to alphabetical order of the English words. In other words, for each vertex v in the tree, all the words stored at vertices in v's left subtree precede (in alphabetical order) the word stored at v, and all the vertices in v's right subtree follow it.   ◆

We want to be able to find a word in a binary search tree. The recursive procedure given as Algorithm 1 for doing so should, by this point, seem completely natural.

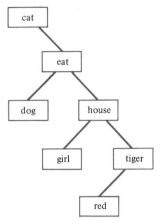

FIGURE 9.23 Another binary search tree for the same English words as in Figure 9.22.

procedure *search*(*r* : root of extended binary search tree, *x* : word)
  {*contents*(*v*) is the word stored at vertex *v*; the procedure will return the vertex
   whose *contents* is *x*, if there is one; otherwise, it will return 0}
  if *r* = ∅ **then return**(0)
    else if *x* = *contents*(*r*) **then return**(*r*)
      else if *x* ≺ *contents*(*r*) **then return**(*search*(*left_child*(*r*), *x*))
        else return(*search*(*right_child*(*r*), *x*))

Algorithm 1. Searching in an extended binary tree.

For simplicity, *we assume that our binary search trees are extended binary trees*, so that a missing child is thought of as the empty vertex.

EXAMPLE 2. Apply Algorithm 1 to search for "dog" and "lion" in the binary search tree shown in Figure 9.22.

Solution. To find "dog," we first compare "dog" to "girl." Since it is smaller (comes first alphabetically), we continue the search in the left subtree, whose root is "dog." Then when we compare "dog" to "dog" we find equality and hence return this vertex as the location of "dog" in the tree.

On the other hand, to attempt to find "lion," we first compare "lion" to "girl." Since it is larger (comes later alphabetically), we continue the search in the right subtree, whose root is "red." Then we compare "lion" to "red" and move to the left (since "lion" precedes "red" lexicographically), and then compare "lion" to "house" and move right. At this point the vertex we are looking at is empty, so we return 0 to indicate that the word we are seeking is not in the tree. ◆

Clearly, the algorithm works, since it directs its search to that subtree in which the sought-after item must be. Furthermore, the worst-case running time for this algorithm (counting, say, the number of comparisons of words it uses) is clearly in $O(h)$, where h is the height of the tree, since each recursive call to the algorithm sends us one level deeper into the tree, and the algorithm must terminate no later than when it reaches the bottom.

How efficient this is in terms of n, the number of vertices in the tree (which is the number of elements in the set—the natural parameter with which to measure efficiency here), depends on how "balanced" the tree is. If the tree is **complete** (i.e., full and having *all* of its leaves on level h), then the worst-case efficiency is in $O(\log n)$, since (by Theorem 5 in Section 9.1) a complete binary tree of height h contains $n = 2^{h+1} - 1$ vertices. If n is not 1 less than a power of 2, we cannot put all the leaves at level h, but we can put all the leaves either at level h or at level $h - 1$, in which case h is still approximately $\log n$.

On the other hand, if the tree is badly lopsided (e.g., if no vertex has a left child, so that $h = n - 1$), then searching in a binary search tree could be little better than linear search, with efficiency in $O(n)$. Since we are free to determine which tree to use, we can arrange to make it almost complete and thereby achieve the desired $O(\log n)$ efficiency. Note that this is the same as we achieved with the binary search discussed in Section 4.3, and the two algorithms are essentially the same in this case. In summary, we have the following theorem.

THEOREM 1. *Searching in a binary tree with n vertices has a worst-case running time in $O(h)$, where h is the height of the tree. By making the tree as balanced as possible (with all the leaves at level h or level h − 1), we can achieve a worst-case running time in $O(\log n)$.*

EXAMPLE 3. The maximum number of comparisons needed to search for a word in the binary search tree in Figure 9.22 is 3. We need three comparisons to find a word at the bottom level, such as "house," or to discover that a word such as "butterfly" is not in the tree. On the other hand, five comparisons are needed to find "red" or to fail to find "sugar" in the more unbalanced binary search tree for the same set shown in Figure 9.23. ◆

Actually, the problem is more complex than we are yet admitting to. Typically, we will be searching for some words much more often than others. If we are translating children's stories from English to French, we might expect to encounter the words "girl" or "to" fairly frequently, but rarely the words "balance" or "leaf." Thus not only should we try to make our binary search tree fairly balanced, so that the searches in the worst cases do not take too many steps, but we should also try to arrange that *frequently occurring words are near the top* (at low-numbered levels), so that the repeated searches for them will take less time. The ability to make such adjustments is one way in which the binary search tree is more useful and versatile than a simple list.

EXAMPLE 4.   Suppose that the relative frequency with which the words in our seven-word dictionary of Example 1 need to be looked up is given in the following table. (We are assuming that there will be no searches for words not in the table.) Compare the binary search trees shown in Figures 9.22 and 9.23 as to their average-case efficiency.

cat	0.35
eat	0.30
dog	0.10
house	0.10
girl	0.05
red	0.05
tiger	0.05

Solution.   Under these hypotheses, we can find an average-case efficiency by taking the weighted sum of the search times for the seven words in each tree, weighted by their relative frequencies.

In the tree in Figure 9.22, it takes one comparison to locate "girl," and that happens 5% of the time. It takes two comparisons to locate "dog" or "red"; these two words occur a total of 15% of the time. Similarly, the other four words, occurring the remaining 80% of the time, require three comparisons. Therefore the average number of comparisons will be $1 \cdot 0.05 + 2 \cdot 0.15 + 3 \cdot 0.80 = 2.75$. In other words, over the long run, after searching many times, we will have used an average of about 2.75 comparisons per search.

Similarly, we find that the average number of comparisons for the tree in Figure 9.23 is $1 \cdot 0.35 + 2 \cdot 0.30 + 3 \cdot 0.20 + 4 \cdot 0.10 + 5 \cdot 0.05 = 2.20$. Given the choice, we would probably prefer to use the tree in Figure 9.23, since over the long run we are saving about one half of a comparison per search.

If some other distribution of relative frequencies were given, then perhaps the first tree would be preferable. Also, if we were concerned with the *maximum* search time, rather than the average search time (perhaps for a "real-time" application), the first tree might be more desirable.   ◆

There are algorithms (see Exercise 31) for determining an optimum binary search tree for a given set of relative frequencies—one that minimizes the expected search time. The reader is asked to try (in Exercise 26) to find an optimum binary search tree for the frequencies given in Example 4.

Before leaving binary search trees, we consider another important feature that makes them preferable to linear lists in many circumstances. Suppose that we have a "dynamic" list, which, in addition to being searched, is periodically augmented (or perhaps reduced) through the addition of new items (or the deletion of old ones). We need to insert each new item into its correct place. With a linear list, this might involve moving practically every item in the list to make room for the addition. With a binary tree, insertion is as easy as searching.

To insert an element x into a binary search tree, we simply *search* for x in the tree. When we do not find it, we know that the last place we looked was an empty child of some vertex. By the way the search algorithm works, that is precisely where x should be placed. Hence we create a new child to replace the empty child and put x there; we assume that children of newly created vertices are automatically empty.

Algorithm 2 implements this strategy. It assumes that the word to be inserted is not already in the tree. It is easy to modify the algorithm to handle the case in which this word is already in the tree (see Exercise 13).

procedure *insert*(r : root of extended binary search tree, x : word not in the tree)
  {*contents*(v) is the word stored at vertex v; the procedure will create a new
  leaf for x, in the appropriate place; the algorithm assumes that the tree is not
  empty}
    if $x \prec contents(r)$ **then** {move into left subtree}
        if *left_child*(r) = $\emptyset$ **then**
            begin {create a new vertex for x}
                create a new vertex v, with empty children
                $contents(v) \leftarrow x$
                $left\_child(r) \leftarrow v$
            end
          else call *insert*(*left_child*(r), x)
      else {move into right subtree}
        if *right_child*(r) = $\emptyset$ **then**
            begin {create a new vertex for x}
                create a new vertex v, with empty children
                $contents(v) \leftarrow x$
                $right\_child(r) \leftarrow v$
            end
          else call *insert*(*right_child*(r), x)
    return

Algorithm 2.   Inserting a new item into a binary tree.

EXAMPLE 5.   Suppose that we want to create a binary search tree by successively inserting, in the given order, the words in the sentence "Little Miss Muffet sat on a tuffet." We are assuming that the words are to be ordered in the tree alphabetically. We begin by making a tree with just one vertex, whose contents is the word "little." Next we insert the word "miss." Since "miss" follows "little" alphabetically, "miss" is inserted as a new right child of the root. Next we insert "muffet" as the right child of "miss" (the algorithm taking us to the right from the root to "miss" and then to the right from "miss" to an empty vertex, where we add "muffet"). Similarly, "sat" gets inserted as the right child of "muffet." Next we insert "on"; the algorithm takes us to the right three times, but when we compare "on" to "sat," we move to the empty left child. Therefore, "on" becomes the left child of "sat." Next, "a" becomes the left child of the root, and finally, "tuffet" is inserted as the right child of "sat." The final tree appears in Figure 9.24.   ◆

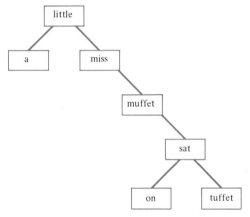

FIGURE 9.24   Binary search tree constructed from a nursery rhyme.

Deletions are not quite this simple, and we will not go into them here. You might try to find a deletion algorithm (Exercise 24). You might also consider the issue of how to keep a binary search tree balanced as insertions and deletions are made. The tree in Figure 9.24 has height 4; it is quite lopsided. Since there are only seven vertices in the tree, we could rearrange them into a tree of height just 2 (Exercise 6). Fortunately, there are algorithms for inserting into and deleting from binary search trees that maintain some kind of balance. Again we will not go into any details, other than to mention that certain data structures, such as AVL-trees and B-trees, provide ways to do this (see Exercises 29 and 30).

HUFFMAN CODES

When information is stored in a computer, on a disk, or on a tape, or when information is sent over communications lines, it is usually encoded as bit strings.

Frequently, this information is text—a string of characters from some alphabet, such as the alphabet of uppercase English letters, a period, a comma, and a space. (Numerical data can be viewed as text; the alphabet contains digits, a decimal point, etc.) Recall from Section 2.2 that V^* denotes the set of strings (words) over the alphabet V. An encoding method, then, is just a function from V^* to $\{0, 1\}^*$. One straightforward function simply encodes each character as a bit string of a fixed length and then concatenates the encodings of the characters in the text.

For example, to encode the alphabet mentioned above (which consists of 29 symbols), we could use a bit string of length 5 for each symbol (that gives us $2^5 = 32$ possible codes). We might let A be encoded by 00000, B by 00001, C by 00010, ..., Z by 11001, period by 11010, comma by 11011, and space by 11100. (The remaining three bit strings would not encode anything.) The translation of the word "cat," then, would be 000100000010011. Decoding a bit string is just as easy: We simply read five bits at a time and consult our translation scheme to find the character each chunk of five bits represents.

Standard codes have been developed for such fixed-length encoding. Most familiar is the ASCII code, which uses seven bits to encode upper- and lowercase letters, digits, punctuation, and some special symbols—a total of 128 symbols.

We will now see how using an encoding scheme based on a binary tree can sometimes be more efficient than fixed-length encoding, in the sense that on the average fewer bits will be needed to encode messages in this tree-based method than in the fixed-length method. We will translate each symbol of text into a bit string, as before, but the length of the bit strings will vary. If we can arrange for frequently occurring letters to have short codes, then we will be able to reduce the total number of bits needed for a typical message. The problem that we need to overcome, however, is that with variable-length codes, decoding becomes more complicated, since it is no longer clear when the bit string for one symbol ends and the bit string for the next begins.

We might draw an analogy with Morse code, which is also a variable-length code. Short codes are used for letters that occur frequently (such as *dot* for the letter E) and longer codes are used for letters that occur less frequently (such as *dot-dash-dash-dash* for the letter J). In Morse code, breaks between letters are indicated by short pauses (in effect, a third encoding character). We want to avoid having to introduce the analogue of pauses, since this is hard to do with only two symbols (0 and 1).

Instead, we will arrange for our code to have the following property: *No code for a letter is an initial substring of the code for another letter*. A code with this property is called a **prefix code**. For instance, we will not allow the relationship that exists between the codes for E and J in Morse code, since the code for E is the same as the beginning of the code for J. To ensure that this condition is met, and to provide ourselves with a practical means of encoding and decoding, we will create a binary tree, the leaves of which are the symbols to be encoded.

A **prefix code tree** for an alphabet V is a binary tree whose leaves are in one-to-one correspondence with the elements of V. If v is a symbol in V, then the encoding of v is the bit string that represents the path from the root of the tree to the leaf corresponding to v, where a 0 represents descent to a left child and a 1 represents descent to a right child. (Some authors use the opposite convention and draw their trees with the left-leading edges labeled 1 and the right-leading edges labeled 0.)

EXAMPLE 6. Consider the prefix code tree shown in Figure 9.25. The alphabet here is $\{B, D, E, F, I, O, P, S, T, W, Y\}$, and we have labeled the leaves with the elements of this set. Note that we have also labeled each edge of the tree either 0 or 1: Edges leading to left children are labeled 0 and edges leading to right children are labeled 1. Let us see how a message might be encoded.

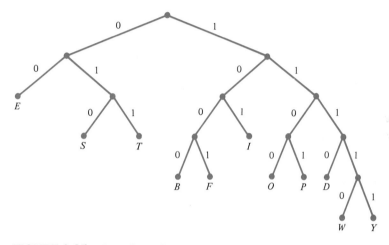

FIGURE 9.25   A prefix code tree.

Consider the message *STOP*. The symbol S is reached from the root by descending left, then right, then left; hence its code is 010. Similarly, the code for T is 011. The next symbol, O, is reached by the path right–right–left–left, so its code is 1100; and the code for P is 1101. Thus the encoding of the word *STOP* is 01001111001101. To implement this translation from letters to codes, we could store all the pairs consisting of a letter and its code in a list or binary search tree; thus this list would contain pairs such as $(S, 010)$ and $(P, 1101)$. Each encoding would require that a pair be looked up in this list or search tree.

For decoding, we need to use the prefix code tree. Suppose that we receive the message 100000001001. To decode it we start at the beginning and follow the indicated path from the root of the tree until we reach a leaf. Moving right (because of the first 1), we are not yet at a leaf; moving left (because of the second symbol, 0), we are still not at a leaf; moving left again (because of the third symbol, another 0), we are still not at a leaf; moving left again (because of the fourth symbol, 0), we arrive at the leaf labeled B. Hence the first symbol of the message is B. We have consumed the first four symbols in the bit string. Now we continue with the fifth symbol, starting again from the root of the tree. Two 0's take us to the leaf labeled E, so E is the second symbol of the message. Again we start from the root, working now on the seventh bit. Two 0's take us again to the leaf labeled E, so E is also the third symbol of the message. Returning to the root of the tree, and starting with the ninth bit, we find 1001, which leads us to F and exhausts the bit string. Thus the final symbol of the message is F, and the entire message is *BEEF*.   ◆

In analogy with binary search trees, we want to structure the prefix code tree in such a way that the *weighted average* length of the code for a letter is as small as possible, taking into account the frequencies with which the letters occur. Suppose in Example 6 that we knew, from past experience with the kinds of messages that we needed to encode, that the relative frequencies of the letters in this alphabet were given by Table 9.1. Then we can compute the weighted average number of bits per letter as follows. The letter E occurs 30% of the time and has a code of length 2. The letters S, T, and I occur 35% of the time and have codes of length 3. All the remaining letters except W and Y occur 30% of the time and have codes of length 4; and these two letters, occurring 5% of the time, have code length 5. Thus the expected code length is $2 \cdot 0.30 + 3 \cdot 0.35 + 4 \cdot 0.30 + 5 \cdot 0.05 = 3.10$. If we had configured the tree differently, or if the relative frequencies had been different, we might have obtained a different answer. On the other hand, a fixed-length code would need four bits for every symbol. This saves space for W and Y, but wastes much more space for the frequently occurring E, S, T, and I. The net gain for the prefix code is $4.00 - 3.10 = 0.90$ bit per character on the average.

Table 9.1.   Relative frequencies for the alphabet $V = \{B, D, E, F, I, O, P, S, T, W, Y\}$.

B	0.05	D	0.05	E	0.30
F	0.05	I	0.08	O	0.10
P	0.05	S	0.10	T	0.17
W	0.02	Y	0.03		

In the remainder of this section we look at a recursive algorithm for finding a prefix code tree for a given alphabet V that achieves the minimum weighted average length code per symbol (weighted by a given relative frequency for V). The code constructed by this algorithm is called a **Huffman code**.

To set notation, let $V = \{s_1, s_2, \ldots, s_n\}$, where $n \geq 2$, and suppose that the relative frequency of the symbol s_i is f_i. Thus $\forall i: 0 \leq f_i \leq 1$ and $\sum_{i=1}^{n} f_i = 1$.

Without loss of generality, we will assume that the symbols are ordered so that $f_1 \leq f_2 \leq \cdots \leq f_n$. It will be easiest to describe the algorithm in prose.

In the base case ($n = 2$), the "coding" is really nothing other than a change of name. We let the code for s_1 be 0 and the code for s_2 be 1. Trivially, the weighted average length of code words is 1. Thus in this case, we obtain the Huffman code tree shown in Figure 9.26. We have placed each symbol name in a circle representing the corresponding vertex.

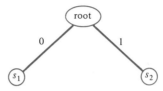

FIGURE 9.26   Huffman
code tree for an alphabet
with two symbols.

For the general (recursive) case, first we find the two symbols that occur least frequently, which by the notation agreement above are s_1 and s_2. We build the tree shown in Figure 9.27, and call the root of that tree t_1, where t_1 represents an artificial "symbol." Now we recursively apply the algorithm to the following problem: The alphabet is $V' = V - \{s_1, s_2\} \cup \{t_1\}$, and the relative frequencies are the same as they were, except that the relative frequency of t_1 is $f_1 + f_2$. In other words, we replace the two symbols s_1 and s_2 by the artificial symbol t_1, and we weight t_1 with the sum of the weights (relative frequencies) of the two symbols it is replacing. This problem has an alphabet with only $n - 1$ symbols. We take the solution to the problem for V' (some prefix code tree) and replace the occurrence of the leaf t_1 in that tree by the tree shown in Figure 9.27. That is the Huffman code tree for V.

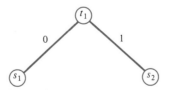

FIGURE 9.27   Combining
the two least frequently
occurring symbols into a
subtree.

For example, suppose that the prefix code tree returned by the recursive call to the algorithm is the one shown in Figure 9.28a. Then the Huffman code tree produced by the algorithm is the tree shown in Figure 9.28b.

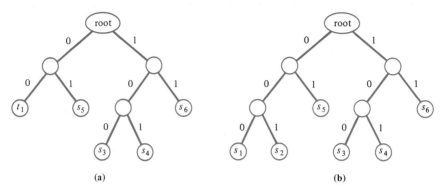

(a)   (b)

FIGURE 9.28   Huffman code tree on six symbols derived from one on five symbols.

It can be proved by induction that this algorithm produces a prefix code with smallest possible average code length (see Exercise 33).

EXAMPLE 7.   Implement the Huffman code tree algorithm on the alphabet and relative frequencies given in Table 9.1.

Solution.   The algorithm is stated recursively, but we will implement it iteratively. First we combine W and Y, the two symbols with the smallest frequencies, into the tree shown in Figure 9.29.

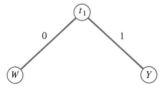

FIGURE 9.29   Combining W and Y into the artificial symbol t_1.

The frequency of t_1 is $0.02 + 0.03 = 0.05$. At this point our alphabet is (in order of nondecreasing frequency) $\{B, D, F, P, t_1, I, O, S, T, E\}$. We combine B and

D into a tree rooted at t_2, with frequency $0.05 + 0.05 = 0.10$. The third step is, similarly, to combine F and P into a tree rooted at t_3, with frequency 0.10. (These choices were not unique, since there were five symbols with frequency 0.05; we arbitrarily picked two of them each time.) At this point our alphabet is $\{t_1, I, O, S, t_2, t_3, T, E\}$. Now we combine t_1 and I into the tree shown in Figure 9.30 with root t_4, having frequency $0.05 + 0.08 = 0.13$.

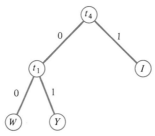

FIGURE 9.30   Combining t_1 and I into the artificial symbol t_4.

Now t_1 and I are replaced by t_4, so the alphabet is $\{O, S, t_2, t_3, t_4, T, E\}$; note that t_4 is listed in the correct order by relative frequency. The next step is to combine O and S into an artificial symbol t_5 with frequency $0.10 + 0.10 = 0.20$, reducing the alphabet to $\{t_2, t_3, t_4, T, t_5, E\}$. The artificial symbol t_6 is created out of t_2 and t_3; the tree rooted at t_6 is shown in Figure 9.31. Its frequency is $0.10 + 0.10 = 0.20$. The alphabet has become $\{t_4, T, t_5, t_6, E\}$. After t_4 and T are combined into t_7 with frequency $0.13 + 0.17 = 0.30$, the alphabet is $\{t_5, t_6, E, t_7\}$. Continuing, we let t_8 have children t_5 and t_6 (giving it a frequency of $0.20 + 0.20 = 0.40$) and let t_9 have children E and t_7 (giving it a frequency of $0.30 + 0.30 = 0.60$). Finally, we combine t_8 and t_9 into the final Huffman code tree. Recalling what subtrees of our artificial symbols stood for, we obtain the solution shown in Figure 9.32.

Note that this tree is a little better than the prefix code tree shown in Figure 9.25. The only change in code length is that O now has a code of length 3 rather than 4, and I now has a code of length 4 rather than 3. Since O occurred more frequently than I (0.10 compared to 0.08), the weighted average code length must have decreased by $0.10 - 0.08 = 0.02$, to 3.08. If we accept the claim that the algorithm really does produce an optimum code, then no prefix code can have a weighted average length of less than 3.08.   ◆

There are two other aspects to encoding and decoding, which we will not consider here. First, it is important to be able to protect against errors in the

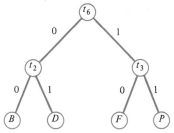

FIGURE 9.31 Combining t_2 and t_3 into the artificial symbol t_6.

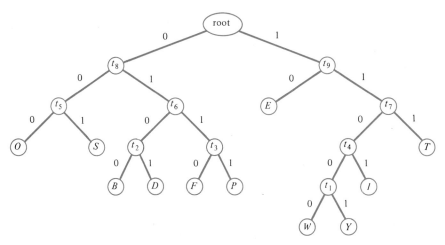

FIGURE 9.32 Huffman code tree for the alphabet in Table 9.1.

transmission of data. For example, if a compact disk, containing the code necessary for a compact disk player to reproduce Beethoven's Fifth Symphony, is scratched, we would like the player to be able to ignore the scratch and reproduce the music as originally recorded. If we build enough redundancy into our code, this will be possible. The theory of such **error-correcting codes** is an active area of mathematical research today, with applications not only to compact disks, but also to telecommunications and other fields (see Exercise 32). Second, we have already alluded in Section 4.4 to the problem of encoding messages so that an unintended receiver will not be able to figure out their meanings. The study of such secret codes, **cryptography**, is another important area of mathematical research, interesting in its own right and useful in political and financial applications (see Exercise 16 in Section 4.4).

SUMMARY OF DEFINITIONS

binary search tree: a binary tree with an element of an ordered set at each vertex, such that for each vertex v in the tree, all the elements at vertices in the left subtree of v precede (in the given order) the element at v, and all the elements at vertices in the right subtree of v succeed the element at v.

complete binary tree: a binary tree in which every internal vertex has both a left child and a right child, and in which all the leaves are at the same level; if this tree has height h, then it contains 2^h leaves and $2^{h+1} - 1$ vertices in all.

cryptography: the study of making and breaking codes for the transmission of messages that someone other than the intended receiver will not be able to decode.

error-correcting code: a code with enough built-in redundancy so that even if a coded message contains a small number of errors, the receiver will still be able to decode the message correctly.

Huffman code for the alphabet V with given relative frequencies: a prefix code for V in which the weighted average length of encoded symbols is a minimum.

prefix code for the alphabet V: a method of encoding the symbols in V as bit strings, such that for every two symbols x and y, the encoding of x is not an initial substring of the encoding of y.

prefix code tree for a prefix code for the alphabet V: a binary tree in which edges to each right child are labeled 1 and edges to each left child are labeled 0, and in which the leaves are labeled with the elements of V; the code for the symbol $x \in V$ is the bit string formed by the unique path from the root to the leaf labeled x.

EXERCISES

1. Draw two binary search trees for the set of words stored in the tree in Figure 9.22, different from the trees in Figures 9.22 and 9.23.

2. Explain (by indicating the comparisons made) how Algorithm 1 finds the word "girl" in the binary search tree in Figure 9.23.

3. Explain (by indicating the comparisons made) how Algorithm 1 finds that the word "boy" is not in the binary search tree
   (a)   In Figure 9.22.
   (b)   In Figure 9.23.

4. Insert, successively, the words "Eating her curds and whey" into the binary search tree given in Figure 9.24.

5. Build a binary search tree containing the same words as the tree in Figure 9.24, obtained in the same way as Figure 9.24 was obtained, except starting at the

end of the sentence (with "tuffet" as the root of the tree), and working toward the front.

6. Write down the complete binary tree for the words stored in the tree in Figure 9.24.

7. Using the prefix code tree in Figure 9.25, encode the following words as bit strings.
    (a)   *FIDO.*
    (b)   *SIT.*
    (c)   *BOWWOW.*

8. Using the prefix code tree in Figure 9.25, decode the following bit strings.
    (a)   1110110011010011111.
    (b)   110100010011.
    (c)   011101110101011111.

9. Find the average number of comparisons used for searching the binary search trees that you drew in Exercise 1, assuming the relative frequencies given in Example 4.

10. Find a more efficient binary search tree than the one in Figure 9.23, assuming the relative frequencies given in Example 4.

11. Assume that an element of a linearly ordered set is stored at each vertex of a binary tree T. Prove or disprove the following proposition: T is a binary search tree if and only if at each vertex v, if v has a left child, then *contents(left_child(v))* $\prec$ *contents(v)*, and if v has a right child, then *contents(v)* $\prec$ *contents(right_child(v))*.

12. If a dictionary is stored in a binary search tree, how can we obtain a printout of the dictionary in alphabetical order?

13. Combine Algorithms 1 and 2 into one algorithm which, given the root of a binary search tree T and an item x, inserts a vertex containing x into T if it is not already there, and in any case returns the vertex at which x is stored.

14. Write down explicit formulas (in terms of n) for the following.
    (a)   The minimum, over all possible binary search trees T with n vertices, of the maximum, over all words w in T, of the number of comparisons needed by Algorithm 1 to find w in T.
    (b)   The maximum, over all possible binary search trees T with n vertices, of the minimum, over all words w in T, of the number of comparisons needed by Algorithm 1 to find w in T.
    (c)   The minimum, over all possible binary search trees T with n vertices, of the minimum, over all words w in T, of the number of comparisons needed by Algorithm 1 to find w in T.

(d)   The maximum, over all possible binary search trees T with n vertices, of the maximum, over all words w in T, of the number of comparisons needed by Algorithm 1 to find w in T.

15. If all the items in a binary search tree are to be searched for with equal frequency, what will the optimum binary search tree—the tree that minimizes the average number of comparisons—look like?

16. Find a set of relative frequencies such that the binary search tree given in Figure 9.24 is exactly as efficient as the complete binary search tree containing the same seven words (see Exercise 6).

17. Determine whether the following statement is necessarily true: If 11111 is a substring of the bit string encoding a message using the prefix code tree in Figure 9.25, then the letter Y is in the message.

18. Explain why prefix code trees should be full binary trees.

19. How many prefix code trees are possible for an alphabet of four symbols, assuming that only full binary trees are allowed?

20. Suppose that letters occur with the following frequencies in messages to be encoded: A, 30%; B, 15%; C, 15%; D, 15%; E, 15%; F, 10%.
    (a)   Construct two substantially different prefix code trees for this alphabet.
    (b)   Compute the average code length for the two trees constructed in part (a).

21. Use the algorithm given in this section to build the Huffman code tree for the following alphabets with relative frequencies.
    (a)   The one given in Exercise 20.
    (b)   The alphabet in which one symbol occurs 40% of the time and five other symbols occur 12% of the time each.
    (c)   The alphabet in which one symbol occurs 60% of the time and five other symbols occur 8% of the time each.

22. Show that if a symbol s of an alphabet has relative frequency at least 50%, then in the Huffman code for that alphabet, the symbol s will have a code of length 1.

23. Show that if T is any optimum prefix code tree, and if u and v are symbols in the alphabet such that u occurs more frequently than v, then the level at which u occurs in the tree must be less than or equal to the level at which v occurs in the tree.

Challenging Exercises

24. Describe an algorithm for deleting an element from a binary search tree.

25. Show that the number of nonisomorphic binary trees with n vertices is less than or equal to $n!$, and is less than $n!$ if $n \geq 3$. [*Hint*: Consider the act of building a binary search tree.]

26. Find the most efficient binary search tree for the words stored in the tree in Figure 9.23, assuming the relative frequencies given in Example 4.

Exploratory Exercises

27. Implement Algorithms 1 and 2 and use them to construct some large binary search trees.

28. Implement the Huffman code tree algorithm and use it to construct some Huffman codes.

29. Find out about inserting and deleting from an AVL-tree (see Exercise 28 in Section 9.1) by consulting Reingold, Nievergelt, and Deo [243] and Standish [292].

30. Find out about **B-trees**—another form of fairly balanced trees—by consulting Reingold, Nievergelt, and Deo [243] and Standish [292].

31. Find out about constructing optimum binary search trees by consulting Reingold, Nievergelt, and Deo [243] and Standish [292].

32. Find out about error-correcting codes by consulting Cartwright [34], MacWilliams and Sloane [203], Pless [235], and Sloane [275].

33. Find out more about Huffman codes by consulting Meyer [212] and Standish [292].

SECTION 9.5
GAME TREES

The players all played at once, without waiting for turns,
quarrelling all the while, and fighting for the hedgehogs;
and in a very short time the Queen was in a furious passion,
and went stamping about and shouting, "Off with his head!" or
"Off with her head!" about once in a minute.

—Lewis Carroll, *Alice's Adventures in Wonderland*

Our final application of trees is to the study of games. In the games we consider, there are two players, whom we will call Blue (referred to in the masculine) and Red (referred to in the feminine). The players play alternately, according to certain rules as to what constitutes a legal move in a given situation. The game continues until a player whose turn it is to move has no legal moves. The rules for the two players need not be the same. Usually, Blue goes first, although since we will be analyzing games recursively, we will also need to consider the subgame beginning in the middle of another game, when it is Red's turn to move.

Each player has complete information about what moves the other player has made, and there are no random elements (such as a roll of dice) in these games. Games such as tic-tac-toe, chess, checkers, dots and boxes, nim (see Exercise 1), and sprouts (see Exercise 33 in Section 8.4) fall into this category. Most card games (in which the players conceal their holdings from each other), many board games (that depend on chance devices), and games such as scissors–rock–paper (in which the players play simultaneously, not alternately) do not meet our definition of a game.

We will model a game with a rooted tree, called the **game tree**. The vertices of the tree represent the positions of the "game board"—the state of the game—at each point in time during the game. The edges represent the moves. The root of the game tree is the game board at the outset of the game. The children of the root are the possible positions of the game board after Blue has made his first move. The children of one of these vertices, say the one representing the position that arose when Blue made move m, are the possible positions of the game board after Red has responded to m, and so on. Thus all possible positions that the game board might reach are included in the tree. The vertices at even-numbered levels represent positions of the game board when it is Blue's turn to move, and vertices at odd-numbered levels represent positions when it is Red's turn to move. In drawing game trees, we will follow the convention that *vertices at even-numbered levels are represented by rectangles or squares, and vertices at odd-numbered levels are represented by ovals or circles*.

Usually a game tree will be finite, since games that can potentially go on arbitrarily long tend not to be too practical (see Exercises 23 and 29, however). Even chess has special rules that can prevent arbitrarily long games. Note that if a position can arise through two different sequences of moves, then there will be two vertices representing that position, located in different parts of the game tree.

The leaves of a game tree represent final positions—positions at which the game is over. Since leaves have no children, no further moves are allowed. Each leaf has attached to it a value, which we can think of as a payoff to Blue and, simultaneously, a cost to Red, when the game ends in that position. (In what follows, we will be assigning values to *all* vertices, not just to the leaves.)

In the simplest games, one player wins and the other loses, so a winning final position for Blue could be represented by a 1, while a winning position for Red could be represented by a -1. In other cases ties are possible, representable as a payoff of 0. In still other games, a numerical value is associated with each final position. If this value is positive, it can be thought of as the number of dollars that Red must pay to Blue at the end of the game. If this value is a negative number $-v$, Blue pays Red v dollars.

We assume that Blue wants to maximize his winnings (or minimize his losses), so *Blue would like the game to end at a leaf with as large a value as possible*. Similarly, Red wants to minimize Blue's winnings (since they represent her losses) or maximize his losses (i.e., her winnings), so *Red would like the game to end at a leaf with as small a value as possible*, preferably one with a large negative value.

These observations will enable us to find optimal strategies for the two players and assign values to all vertices in the tree.

A complete **play** of the game consists of a path from the root to a leaf. Blue and Red determine, alternately, where the path goes. A **strategy** for Blue is a set of rules by which he will choose his moves. In terms of the game tree, a strategy is simply a function b from the set of internal vertices at even-numbered levels to the set of vertices at odd-numbered levels such that $b(v)$ is a child of v for each v at an even-numbered level. Simply put, $b(v)$ is the position that Blue moves to when confronted with position v. Similarly, a strategy for Red is a function r from the set of internal vertices at odd-numbered levels to the set of vertices at even-numbered levels such that $r(v)$ is a child of v for each v at an odd-numbered level. Suppose that Blue is using strategy b and Red is using strategy r. If we denote the root of the game tree by s, then a play of the game will consist of the path $s, b(s), r(b(s)), b(r(b(s))), r(b(r(b(s)))), \dots$, terminating at some leaf.

In this section we will see how this general framework applies to some specific games, and how the players can determine optimal strategies.

SOME SIMPLE GAMES

Let us begin by discussing the rules, and drawing all or part of the game tree, for three interesting games.

EXAMPLE 1.   A game that we will call pebbles is played with piles of stones. A position consists of a finite number (possibly 0) of stones, arranged in zero or more piles of one or more stones each. When it is a player's turn to move, he or she may either pick up and remove any *positive* number of stones from any *one* pile, or pick up no stones but merge any *two* piles into one. (Unlike some of the other games we will see in this section, this game is symmetric for the two players—the definition of a legal move for Blue is exactly the same as the definition of a legal move for Red.) We can represent positions by sequences of zero or more positive integers representing the number of stones in the piles, and since order does not matter, we will list the integers in nonincreasing order. For example, if the current position is $(7, 3, 3, 1)$, so that there are four piles, with 7, 3, 3, and 1 stones in them, respectively, then a player could move to the positions $(5, 3, 3, 1)$ or $(7, 6, 1)$, or $(7, 3, 3)$, among many others; but it would not be legal to move to positions $(7, 3, 3, 1)$ or $(11, 3)$ or $(4, 3, 3, 3, 1)$. If a player is faced with no stones (the empty sequence, which we will represent by "0"), he or she loses. (People playing the game might rephrase this by saying that the player who takes the last stone wins.)

For any initial configuration of stones (or, equivalently, for any nonincreasing sequence of zero or more positive integers), we can draw the game tree. The initial configuration is the root. Below that are all the configurations that Blue can move to on his first move. Under each vertex at level 1 are the configurations that Red can move to on her first move in response to Blue's first move. Since the game ends only when all the stones are gone, each leaf represents the empty sequence.

In Figure 9.33 we have drawn the game tree for the initial position $(1, 1, 1)$ (we have omitted the parentheses and commas to save space). The position is written inside the square or circle at each vertex. Above each vertex we have indicated the value of that vertex. Ignore for now the values at the internal vertices (we have not defined them yet). The values at the leaves are all either -1 or 1, since the game must end in either a win or a loss for Blue. The square leaves—at even-numbered levels, when it is Blue's turn to move—have the value -1, since our rules said that a player unable to move loses. Similarly, circular leaves—at odd-numbered levels, when it is Red's turn to move—have the value 1, since Red loses when faced with no stones (and therefore Blue wins).

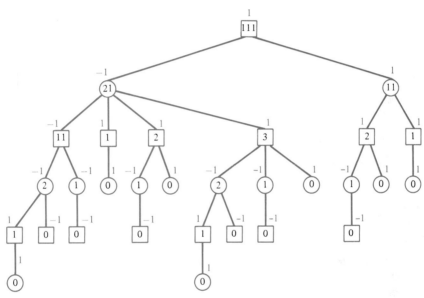

FIGURE 9.33 Game tree for the game of pebbles, with values at each vertex.

Suppose that Blue decides to adopt the strategy of always combining the two largest piles into one, if there are two or more piles, and taking all of the pile when only one pile remains. Suppose that Red decides to adopt the strategy of always taking one stone from the largest pile, if there are two or more piles, and taking all of the pile when only one pile remains. Then the game will begin with Blue moving from the initial position $(1, 1, 1)$ to the position $(2, 1)$, followed by Red moving to the position $(1, 1)$, followed by Blue moving to the position (2), followed by Red winning by moving to the position 0. ◆

EXAMPLE 2.   The familiar game of tic-tac-toe is played on an initially empty 3 by 3 array of squares. Blue draws an X when it is his turn to move, and Red draws an O when it is her turn to move, in any empty square, except that if there are three X's or three O's in a row (horizontally, vertically, or diagonally), then there are no legal moves—the game has just ended in a win for the other player.

Tic-tac-toe has too large a game tree for us to draw here (but not too large for a large computer to manage). We have drawn part of it. Because of the symmetry of the board, it really suffices to consider just the three initial moves for Blue shown in Figure 9.34 (corner, side, and center). Exploiting the symmetry of games is a trick that can often cut significantly the amount of work required in their analyses.

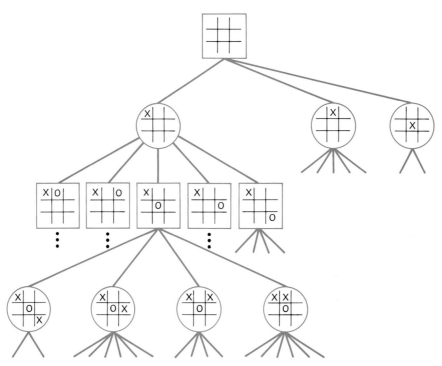

FIGURE 9.34   Beginning of the game tree for tic-tac-toe.

Red has responses to all of these; we have shown only the responses to Blue's corner move. Again we have exploited symmetry to cut down on the size of the tree. At level 3, we have shown all the positions that can face Red after Blue has replied to Red's response in the center to his initial corner move. The tree will have height 9, since the game can last for nine moves. There will be many leaves at level 9, and there will also be leaves at levels 5, 6, 7, and 8. Each leaf will be assigned the value 1, -1, or 0, depending on whether the position at that leaf has three X's in a row, three O's in a row, or neither, respectively. ◆

EXAMPLE 3. Another well-known childhood game is dots and boxes. The initial position consists of an I by J rectangular grid of dots, where I and J are fixed integers greater than 1. (Mathematically, we could say that it consists of all points (i, j) in the plane such that i and j are integers, $1 \le i \le I$, and $1 \le j \le J$.) A player moves by drawing a line segment between two adjacent dots (dots 1 unit apart), with the proviso that a line segment cannot be drawn between dots that have already been joined to each other. Furthermore, if this segment closes off a box (a 1 by 1 square), then the player who drew the line writes his or her initial in the box or boxes so formed and draws another line, continuing to draw lines as long as boxes are formed. (For this example, we will not insist that a player complete a box if able to do so, although that rule is usually also stipulated.) The game ends when all $I(J - 1) + J(I - 1)$ lines have been drawn. The payoff to Blue is the number of boxes marked B minus the number of boxes marked R.

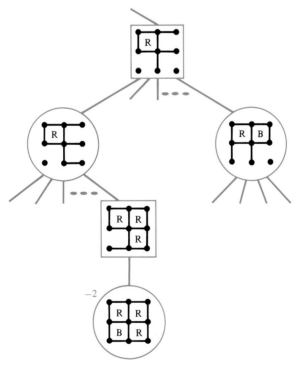

FIGURE 9.35 A small portion of the game tree for dots and boxes.

In Figure 9.35 we have drawn a very small portion of the game tree for dots and boxes with $I = J = 3$. The circle vertex on the left has the square vertex shown as its child, for instance, because Red can complete the box in the upper right-hand corner, then complete the box in the lower right-hand corner, and then draw the lower horizontal line on the left. She puts her R into the two boxes she completed, and her turn ends. This circle vertex has other children as well, corresponding to other moves by Red.

There is one leaf shown in Figure 9.35 and it has value -2, since Red has 2 more boxes than Blue has in the final position. ◆

Several other games are defined in the exercises.

VERTEX EVALUATION AND OPTIMAL STRATEGIES

In order to determine who will win a game (or by how much) if both players play optimally, we need only extend our definition of the value of a vertex to the internal vertices in a game tree. We will do so in a bottom up, recursive fashion. Thus a postorder traversal of the game tree can be used to find the value of each vertex.

The **value** of a vertex in a game tree is defined recursively as follows.

> Base case: The value of a leaf is the payoff to Blue if the game ends in the position represented by the leaf.
>
> Recursive part: If v is an internal vertex at an even-numbered level, then the value of v is the maximum of the values of the children of v; and if v is an internal vertex at an odd-numbered level, then the value of v is the minimum of the values of the children of v.

The **value** of a game is the value of the root of the game tree for that game. This definition is illustrated in Figure 9.36 and Figure 9.37.

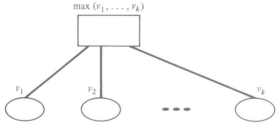

FIGURE 9.36   Recursive assignment of values to vertices at even-numbered levels.

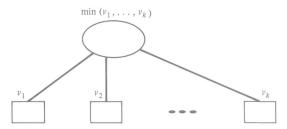

FIGURE 9.37 Recursive assignment of values to vertices at odd-numbered levels.

EXAMPLE 4. The values of all the vertices in the game tree for pebbles with starting position $(1, 1, 1)$ are shown in Figure 9.33. They were obtained by working from the bottom up, after all the leaves were assigned values. For example, the value of the square vertex representing position (1) at the far left on level 4 is 1 because $\max(1) = 1$. On the other hand, the value of the circle vertex representing position (2) at the far left on level 3 is -1 because $\min(1, -1) = -1$. Continuing in this manner we finally assign the value 1 to the root. Thus the value of this game is 1. A computer could, in a similar way, find the values of all the vertices in the game trees for tic-tac-toe or dots and boxes. ◆

The definition translates directly into an algorithm for finding the value of any game, as long as we can generate the game tree. Our algorithm needs two arguments—the position of the game being evaluated, which is, in effect, the root of the game tree for the game that starts at this position (or its opposite in which Red plays first), and the player whose turn it is to move. A call to *game_value*(*initial_position*, Blue) will find the value of the game. Note that we do not actually phrase the algorithm in terms of a tree, with vertices and edges. Instead, we deal directly with positions and moves. Nevertheless, what is really going on is that the computer executing this algorithm is traversing the game tree and assigning values to its vertices in postorder.

It remains to prove that the value of the game is precisely what we want it to be, namely the amount that Blue will win if both players play optimally. The following theorem practically proves itself by induction.

THEOREM 1. *The value of a game is the amount that Blue will win if both players play optimally—Blue playing to maximize the payoff to himself and Red playing to minimize the payoff to Blue.*

Proof. We prove a stronger proposition, namely that the value of *every* vertex in the game tree is the amount that Blue will win if both players play optimally and play starts from the position (and player to move) of that vertex. Theorem 1 is this proposition for the root of the game tree. Induction is on the recursive definition of value of a

procedure *game_value*(P : position of game, *player* : Red or Blue)
    {returns the value of the position to Blue}
        if P is a leaf **then return**(payoff of P to Blue)
            else if *player* = Blue **then**
                begin {compute maximum of values of children}
                    $v \leftarrow -\infty$
                    for each legal move m for Blue **do**
                        begin {compute value of game at resulting position}
                            $Q \leftarrow (P$ followed by move $m)$
                            $v' \leftarrow game\_value(Q, \text{Red})$
                            if $v' > v$ **then** $v \leftarrow v'$
                        end
                    return(v)
                end
            else {*player* = Red}
                begin {compute minimum of values of children}
                    $v \leftarrow \infty$
                    for each legal move m for Red **do**
                        begin {compute value of game at resulting position}
                            $Q \leftarrow (P$ followed by move $m)$
                            $v' \leftarrow game\_value(Q, \text{Blue})$
                            if $v' < v$ **then** $v \leftarrow v'$
                        end
                    return(v)
                end

Algorithm 1.   Evaluating a game.

vertex (or, equivalently, the distance from a vertex to a leaf). To avoid awkward wording, we will not keep repeating the phrase "if both players play optimally"; it is to be assumed throughout the rest of the proof.

If the vertex is a leaf, there is nothing to prove: The value of a leaf is the payoff to Blue by definition, and no further plays are made. The base case disposed of, we turn to the inductive step. Let P be an arbitrary internal vertex of the game tree (P standing for "position"), and assume the inductive hypothesis—that for each child Q of P, the value of Q is the payoff to Blue starting from position Q (appropriate player to move).

There are two cases to consider, depending on whose turn it is to move. First suppose that it is Blue's turn to move at P. Blue wants to maximize his payoff, so he will choose that position Q (a child of P) that will result in the largest payoff. By the inductive hypothesis, the values of all the children of P are precisely those payoffs, so Blue will choose a child Q whose value is the maximum among the values of the children of P. But this is, by definition, the value of P. Hence the claim holds for P, as desired. Similarly, if it is Red's turn to move at P, then she will choose that position Q (a child of P) that will result in the smallest payoff. By the inductive hypothesis, the values of all the children of P are precisely those payoffs, so Red will choose a child Q whose value is the minimum among the values of the children of P. But this is again, by definition, the value of P. Hence once more the claim holds for P, as desired, and our proof is complete. ∎

Theorem 1 and its proof tell us what the optimal strategies are. Blue's strategy, b, is the function that sends each square vertex v to the circle vertex with greatest value among the children of v. (If there is more than one child with the greatest value, b can arbitrarily choose one of them.) Similarly, Red's strategy, r, is given by $r(v) =$ (the square vertex with least value among the children of v), for each circle vertex v. Armed with a computer to evaluate the game tree, the players do not need to think at all.

EXAMPLE 5. Blue wins the game of pebbles starting with position $(1, 1, 1)$, since the value of the root of the game tree in Figure 9.33 is 1. To win, Blue must remove one stone, leaving the position $(1, 1)$, which has the value 1 according to Figure 9.33. (If he mistakenly combined two piles, to reach position $(2, 1)$, then Red could win by optimal play, since the value of that vertex is -1.) Now Red, facing position $(1, 1)$, is caught between a rock and a hard place, since both of the moves she might make lead to positions with value 1. Whichever move she makes (combining the stones into one pile or removing one stone), Blue then removes the rest of the stones and wins the game.

The reader is asked to show in Exercise 7 that Blue wins the game of pebbles starting with any odd number of piles of one stone each, whereas Red wins the game if it starts with any even number of piles with one stone each.

A computer analysis of the tic-tac-toe game tree will result in a value of 0 for the root, since, as every child who plays the game soon learns, optimal play results in a draw. Each player's optimal play, at each turn, will be to a vertex with value 0. ◆

PRUNING AND OTHER IMPROVEMENTS

Since game trees can grow quite large, blind application of Algorithm 1 to a game may result in a longer computation than can feasibly or economically be done. We will mention four techniques that can sometimes be used to shorten the computation.

First, we have already noted that the same position may occur many times in the game tree. In Figure 9.33, for example, the position (2) occurs four times. Algorithm 1 finds the value of each vertex "from scratch" each time. If we could somehow store the information that (2) has value 1 when it is Blue's turn to move and the value -1 when it is Red's turn after we find out these facts for the first time, we could cut down on some unnecessary computation. The difficulty, though, is that finding an efficient way to store and sift through the positions already evaluated may be extremely complex. We will not pursue this idea further here, other than to use it once in our next example.

Second, if the game tree is very deep, it may be infeasible to implement Algorithm 1 under any circumstances. Chess, for example, may last for several hundred moves, and there are scores of possible moves from most positions. A complete analysis of the game tree—although possible in theory—would be out of the question. Instead, various heuristic evaluation functions can be used to give approximations to the values of internal vertices in the game tree. In chess, for instance, we might look at the pieces remaining on the board, immediate potential captures, control of the center, and pawn formation. A heuristic evaluation function would assign a value close to 1 ("win") for positions that were very strong for Blue (if Blue were ahead by a rook, for example, and not in immediate danger) and a value close to -1 for positions that were very weak for Blue. Values between these extremes would be used for less clear-cut positions. Algorithm 1 would be modified to carry out the recursion only a small number of levels down the tree from the current position (say, six or so, the number perhaps depending on the position) and to use the heuristic evaluation function to find approximations to the values of any internal vertices encountered at this maximum depth. In fact, a version of this technique is what many chess-playing computer programs use. Furthermore, it is what human players do in most games, more or less, consciously or unconsciously. Again, we do not have time to pursue these ideas here.

Third, we can modify Algorithm 1 to take advantage of a technique called **pruning**. Suppose that we are trying to find the value of a square vertex v. To do so, we need to find the values of its children and take the largest. Suppose that we have already found the values of some of the children, and the largest currently known value is m. Then we know for certain that the value of v is at least m. Now suppose that we are evaluating another of v's children, say u. See Figure 9.38 for a diagram of the situation at hand.

To find the value of u, we need to look at the values of u's children. Suppose that as we do so, we find a value m' that is less than or equal to m. Now the value of u is no bigger than m', since the value of u is the minimum of the values of u's children. Since $m' \leq m$, we can already conclude that the value of u is not going to affect the value of v. Hence we can dispense with evaluating u. In some cases, this can mean cutting out quite a bit of work because we do not need to generate large portions of the game tree. Similarly, in finding the value of a circle vertex v, we can cut short the evaluation of a child of v whenever we come to a grandchild of v whose value is greater than or equal to the currently known minimum for v.

Finally, as a special case of pruning, we note that the evaluation process can be simplified if the outcome of a game is always win (1) or lose (-1). In this case, a square vertex will have value 1 if and only if at least one of its children does; hence we can stop evaluating its children if we find a child with value 1. This certainly makes sense: Blue does not care about other possible moves once he has

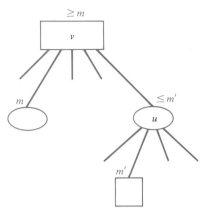

FIGURE 9.38 If $m' \leq m$, then the value of u need not be computed.

found one winning move from a given position. Dually, we can stop evaluating the children of a circle vertex when we find a child with value -1.

Our final example illustrates how pruning can cut down the amount of work in evaluating game trees. The amount of work saved will depend to a large extent on the luck of the order in which we consider the children of each vertex.

EXAMPLE 6. Consider the game that we will call arithmetic, played as follows. A position consists of a finite sequence of one or more positive real numbers. A move consists of replacing any one pair of adjacent numbers in the sequence by their sum or product. The game ends when the sequence consists of just one number, and that is the payoff to Blue. For instance, it is legal to move from the position $(2, 3, 4, 5)$ to positions $(2, 12, 5)$ or $(2, 3, 9)$, among others.

Since Blue wants to maximize the final value and Red wants to minimize it, we may as well assume that Blue will combine a pair of numbers using the operation (addition or multiplication) that results in the larger value (it can never hurt to do so), and that Red will use the operation that results in the smaller value. For instance, from $(2, 3, 4, 5)$, Blue would prefer $(2, 3, 20)$ to $(2, 3, 9)$ (and never choose the latter), whereas Red would never choose the former.

Let us determine the payoff if the game starts with the sequence $(1, 2, 3, 4, 5)$, assuming that both players play optimally. When the sequence contains n terms, there are effectively $n - 1$ moves available, since in light of the comments above, a player would contemplate only one operation on each pair of adjacent numbers in the list. Thus the game tree branches to four children of the root, $4 \cdot 3 = 12$ grandchildren, $4 \cdot 3 \cdot 2 = 24$ great-grandchildren, and 24 great-great-grandchildren (the leaves), a total of 65 vertices. We will see that we need to look at only 27 of them in order to prove that the game has value 46, if we use pruning and keep track of previously evaluated positions.

Figure 9.39 shows as much of the game tree as is required to find the value of the root. The values of the vertices other than the leaves are indicated; the values of the leaves are, by the rules of the game, the numbers in the boxes.

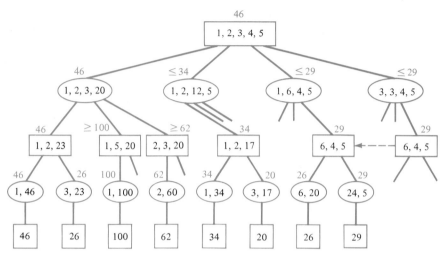

FIGURE 9.39 Game tree for a game of arithmetic.

The first pruning came when we were evaluating the children of the leftmost child of the root, $(1, 2, 3, 20)$. Once we found that its leftmost child, $(1, 2, 23)$, had value 46, we knew that the value of $(1, 2, 3, 20)$ was at most 46. Then we turned to the evaluation of its second child, $(1, 5, 20)$. When we found that *its* child $(1, 100)$ had value 100, we knew that the value of $(1, 5, 20)$ was at least 100. Therefore, the value of $(1, 5, 20)$ would have no influence on the value of its parent $(1, 2, 3, 20)$, whose value was already bounded above by 46. We show the fact that we did not need to look at the other child of $(1, 5, 20)$ by an edge leading nowhere in Figure 9.39. Similarly, we were able to dispense with the second child of $(1, 2, 3, 20)$'s third child, and concluded that the value of $(1, 2, 3, 20)$ was 46.

At this point, then, we knew that the value of the game (i.e., the value of the root) was at least 46. Thus we were able to dispense with evaluating each further child of the root once we found one of *its* children with value less than or equal to 46. In particular, when we found that the value of the root's fourth grandchild, $(1, 2, 17)$, was 34, we knew that the value of the root's second child, $(1, 2, 12, 5)$, would have no effect on the value of the root, so we were able to move on to the third child, $(1, 6, 4, 5)$. There, if we chose to look at the grandchild $(6, 4, 5)$ first (and found its value to be 29), we could dispense with further evaluation of the root's third child.

Finally, when we considered the root's last child, $(3, 3, 4, 5)$, we found that it had a child we had just evaluated, namely $(6, 4, 5)$; in other words, these two first cousins were identical. Therefore, we did not need to reproduce the rest of the subtree rooted there but rather immediately knew its value to be 29. Thus the value of $(3, 3, 4, 5)$ was known to be no bigger than 29 and hence not relevant to the value of the root, which was at least 46.

In summary, this game has value 46, and the path from the root to the leaf with payoff 46 shown in Figure 9.39 is the way the game will go if both players play optimally. If Blue does not make the choices given by this path, he risks winning less than 46; if Red does not make the choices given by this path, she risks losing more than 46. ◆

SUMMARY OF DEFINITIONS

game tree: a tree in which the vertices represent positions of the game, with the initial position at the root; the children of a vertex are the positions that can arise when one move is made from the position at that vertex; Blue moves from positions at even-numbered levels (where the vertices are denoted by squares), Red moves from positions at odd-numbered levels (where the vertices are denoted by circles).

play of a game: a path from the root of the game tree to a leaf; it represents a sequence of moves made by the players, playing alternately with Blue going first, starting at the initial position of the game, and ending when the game is over.

pruning a game tree: when evaluating a square vertex v, terminating the evaluation of a child u of v as soon as a child of u is found with value less than or equal to the value of some previously evaluated child of v (because the value of u is no longer relevant in determining the value of v); and dually when evaluating a circle vertex.

strategy for the player C in a game: a function f from the set of possible positions of the game that can arise for C to the set of possible positions of the game that can arise for C's opponent, such that for each position P that can arise for C, it is legal for C to move from position P to position $f(P)$; in terms of the game tree, a strategy for Blue (respectively, Red) is a function f on the set of vertices at even-numbered levels (respectively, odd-numbered levels) such that $f(P)$ is a child of P for all positions P in its domain.

value of a game: the value of the root of the game tree.

value of a vertex in a game tree (recursively defined): The value of a leaf is the payoff to Blue if the game ends in that position; the value of a vertex at an even-numbered level is the maximum of the values of its children, and the value of a vertex at an odd-numbered level is the minimum of the values of its children.

EXERCISES

In these exercises, exploiting symmetry where appropriate can reduce the sizes of the game trees. Using pruning when you can will cut down the amount of work involved. Keeping track of previously analyzed positions will further speed your solutions.

1. The game of **nim** is played with piles of stones. A position consists of a finite number (possibly 0) of stones, arranged in zero or more piles of one or more stones each. When it is a player's turn to move, he or she may pick up and remove any positive number of stones from any one pile. If a player is faced with no stones, he or she loses. (Our game pebbles was just a variation of nim.) Consider the game of nim with initial position consisting of two piles, one with three stones and one with two stones.
   (a) Draw the game tree. You do not need to repeat subtrees that occur elsewhere in the tree.
   (b) Find the value of the game by evaluating the vertices from the bottom up.
   (c) Determine who wins the game.

2. A game is described in Exercise 39 of Section 5.3. Repeat Exercise 1 for this game, starting with a pile of six stones.

3. A game is described in Exercise 40 of Section 5.3. Repeat Exercise 1 for this game, starting with a pile of six stones.

4. Repeat Exercise 1 for the game of pebbles, with starting position $(2, 1, 1)$. You may use the results obtained in Example 1.

5. Repeat Exercise 1 for the game of pebbles, with starting position $(2, 2, 1)$. You may use the results obtained in Exercise 4 and Example 1.

6. Consider the game of nim (Exercise 1) when there are two piles, say with m and n stones, respectively.
   (a) Show that Red (the second player) wins if $m = n$. [*Hint*: Use induction.]
   (b) Show that Blue (the first player) wins if $m \neq n$. [*Hint*: Use part (a).]

7. Show that in the game of pebbles, starting with n piles of one stone each, Blue wins if n is odd and Red wins if n is even. [*Hint*: Use induction.]

8. Draw the entire subtree (taking advantage of symmetry) of the game tree for tic-tac-toe rooted at the position in which there are X's in two diagonally opposite corners and an O in the center. Omit "obviously dumb moves" (failure to win immediately if possible, otherwise failure to block an opponent's two-in-a-row if such a threat exists). Include the values of all vertices in your tree.

9. Draw the entire subtree (taking advantage of symmetry) of the game tree for dots and boxes rooted at the top position shown in Figure 9.35, with the rule that a player *must* complete a box whenever able to do so. Include the values of all vertices in your tree.

10. Obtain an upper bound on the number of vertices in the game tree for tic-tac-toe.

11. In the complete game trees for the following games (with no simplifications made for symmetry), determine how many grandchildren the root has.
    (a)   Chess.
    (b)   Checkers.
    (c)   Dots and boxes (as a function of I and J).

12. Repeat parts (a) and (b) of Exercise 1 for the game of arithmetic, with starting position $(2, 3, 4, 5)$.

13. Repeat parts (a) and (b) of Exercise 1 for the game of arithmetic, with starting position $(1, 1, 1, 1, 1, 1, 1)$.

14. Find a path from the root to a leaf in the game tree for tic-tac-toe, in which every vertex has value 0. [*Hint*: Play tic-tac-toe with an intelligent friend and record your moves.]

15. An opposite version of nim is identical to nim, except that the player who takes the last stone loses, rather than wins (as is the case for nim). Repeat Exercise 1 for this game, starting with two piles, containing three stones and two stones, respectively.

16. The game of **hex** is played on an n by n board of hexagons, illustrated here for $n = 4$. Initially, the board is empty. At each turn a player places a marker of his or her own color in any unoccupied hexagon. If Blue creates a blue path from the lower left edge of the board to the upper right edge, he wins; if Red creates a red path from the lower right edge of the board to the upper left edge, she wins. (In the picture, Blue has won.) Analyze the game of hex for the following values of n.

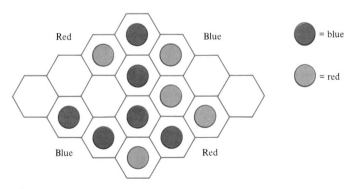

EXERCISE 16

(a) $n = 2$.
(b) $n = 3$.
(c) $n = 4$.

17. A game we will call give-away pebbles is played with a finite number of stones. The initial position consists of zero or more piles of stones, each pile containing one or more stones. A move consists either of removing one entire pile *and giving it to the other player to keep* or of splitting one pile into two piles (each containing at least one stone). The game ends when all the stones have been removed. The winner is the player with the greater number stones (the game is a draw if the two players have the same number of stones). Repeat Exercise 1 for the initial position consisting of one pile of five stones.

18. One version of the game of **hackenbush** is played as follows. An upper half-plane is given, bounded by a horizontal line, called the ground. A planar pseudograph is embedded (with no crossings) in the half-plane, each connected component of which contains at least one vertex on the ground; no edges may touch the ground. See the accompanying typical hackenbush playing board. A move consists of choosing an edge remaining in the picture and removing (erasing) that edge together with all vertices and edges that thereby become disconnected from the ground. For example, in the picture above, if Blue chooses to attack the left stick figure's neck, he erases that edge, the loop (the figure's head) and the vertex at the bottom of the loop. If a player is unable to move—because the figure has been completely erased, except for the vertices on the ground—he or she loses.

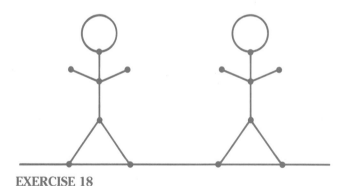

EXERCISE 18

(a)   Analyze the hackenbush position shown here. [*Hint*: The right observation makes the analysis trivial.]
(b)   Analyze the hackenbush position consisting of just one stick figure of the type shown. Show that the winning strategy is unique in this case.

Challenging Exercises

19. Analyze the game of pebbles when there are initially two piles. Who wins (with optimal play) under what conditions on the sizes of the piles?

20. Analyze the opposite version of nim (see Exercise 15), starting with two piles, say with m and n stones, respectively.

21. Discuss in detail how Algorithm 1 would need to be modified to incorporate each of the following improvements.
    (a)   Avoiding analyzing a position more than once.
    (b)   Heuristic evaluation functions.
    (c)   Pruning.
    (d)   Simplified analysis for win/lose games.

22. A **solitaire game** is just like the games we discuss in this section, except that there is only one player, Blue, who tries to have the game end at a leaf with maximum value. A general technique called **backtracking** is essentially equivalent to analyzing the resulting game tree. Many combinatorial problems can be phrased in these terms.
    (a)   Modify Algorithm 1 so that it finds the value of a solitaire game.
    (b)   Use this idea to solve the following problem for $n = 5$. The numbers from 1 to n are arranged around the circumference of a circle. The n sums of three consecutive numbers at a time are computed, and m is the smallest of these sums. Find the largest value that m can have. (For example, if the arrangement is $3-1-4-2-5$, then the five sums are 8, 7, 11, 10, and 9. Therefore, m is 7 for this arrangement.)

23. Find a simple (appealing and interesting, if possible) game (as defined in this section), that satisfies each of the following conditions.
    (a)   The game can go on forever, and it will if both players play optimally.
    (b)   The game can go on forever, but it will not if both players play optimally.
    (c)   The game can go on arbitrarily long, but not forever. In other words, the game tree is infinite (see Exercise 30 in Section 9.1), has a path from the root of length at least l for each positive integer l, but has no infinite paths.

Exploratory Exercises

24. Make up, play, and analyze some interesting games of your own that fall into the category of games we are considering.

25. Implement Algorithm 1 on some interesting games and determine who will win if both players play optimally.

26. Program a computer to play some games with a human opponent, playing optimally by using Algorithm 1.

27. The notion of game as we have defined it in this section is consistent with Conway's definition of game given in Example 2 of Section 5.1.
    (a) Explain this statement.
    (b) Find out more about game theory in Conway's sense by consulting the references mentioned in Exercise 40 in Section 5.1.

28. See Exercise 16 for the rules of hex.
    (a) Argue that it is impossible for there to be both a winning blue path and a winning red path on the board at the same time. Thus it does not really matter whether the game stops as soon as a path is formed or whether it continues for n^2 moves until all the hexagons are filled.
    (b) Argue that it is impossible to fill the entire hex board with red and blue markers without creating either a winning blue path or a winning red path on the board. Thus the game of hex cannot end in a draw.
    (c) Show that for all n, Blue has a winning strategy. [*Hint*: Do not attempt to find the strategy or use induction. Instead argue by contradiction, using the result of part (b).]
    (d) Find out more about hex by consulting Berman [18] and Gardner [109].

29. It is possible to define games that last forever, with the winner determined at the conclusion of play.
    (a) Consider the following game. Blue and Red alternately write down decimal digits of a real number between 0 and 1. For example, Blue might select 3 for the first digit, then Red might play 9 for the second digit, then Blue might play 9 for the third digit, and so on forever in whatever manner they choose. If this game is played forever, the players have constructed a unique real number (which begins 0.399 in the example). Blue wins the game if this number is a rational number; otherwise, Red wins. Find a winning strategy for Red.
    (b) Make up other interesting rules for determining the winner in the number-construction game in part (a), and analyze the corresponding game (as to who has a winning strategy).
    (c) It turns out that using the axiom of choice (Exercise 37 in Section 2.1) one can prove that there are infinite games of the type described here in which neither player has a winning strategy (quite different from the situation for finite games). Some set theorists reject the axiom of choice in favor of the axiom of determinacy—that every such game has a winning strategy for one player or the other. Consult Moschovakis [218] to get a feeling for the impact of these ideas on research in modern set theory.

30. Three-dimensional tic-tac-toe is fairly interesting if played on a 4 by 4 by 4 "board." The winner is the first player to get four-in-a-row.
    (a) Play the game with some friends.

    (b)   Argue that the value of the game must be either 0 or 1—that it cannot be -1.

    (c)   Consult Patashnik [234] to learn whether the first player has a winning strategy.

31. Find out about the theory of games in which the players play simultaneously, by consulting M. D. Davis [60], Packel [230], and J. D. Williams [325].

32. Find out more about nim (Exercise 1), hackenbush (see Exercise 18) and other games by consulting Gardner [96,100,109,111].

GRAPHS AND DIGRAPHS WITH ADDITIONAL STRUCTURE

The applications of graph models are almost endless. Many times, however, we need more than just a graph or digraph to represent all the information present in a problem. The extra information can usually be represented by one or more numerical values associated with each vertex and/or one or more numerical values associated with each edge. In this chapter we consider a few of these situations.

We look first at connected graphs or digraphs in which the edges have nonnegative real numbers attached to them, informally thought of as their lengths. Given such a structure, at least two questions naturally arise: How can we find shortest or longest paths in the graph meeting certain conditions (Section 10.1), and how can we find spanning trees of the graph with minimum total length (Section 10.2)? We present algorithms for these problems. Next (Section 10.3) we again assign nonnegative real numbers to the edges of a connected digraph, but this time the numbers represent the capacities of the edges to carry some commodity, such as oil or automobile traffic. A natural question now is how to find the maximum amount of the commodity that can flow from one point in the graph to another. Again, we present an algorithm for answering this question.

In addition to the general references listed at the beginning of Chapter 8, see Papadimitriou and Steiglitz [232] and Tarjan [302].

SECTION 10.1
SHORTEST PATHS AND LONGEST PATHS

A salesman is got to dream, boy.
It comes with the territory.

—Arthur Miller, *Death of a Salesman*

In this section we focus our attention on walks (usually paths or cycles) in graphs or digraphs. If a graph is connected (or strongly connected in the case of a directed graph), we know that for every pair of distinct vertices u and v there is a path between u and v. Breadth-first search from u will in fact find such a path for us, and the path it finds will be a shortest one, in the sense that it will contain the fewest possible number of edges. We want to generalize the notion of length to go beyond the mere counting of edges. For example, an edge in a digraph representing a flight from New York to Los Angeles should perhaps be considered "longer" in some applications than an edge representing a flight between Detroit and Chicago.

We begin with some definitions. In this section no generality would be gained by allowing loops or parallel edges, so let $G = (V, E)$ be either a graph or a digraph without loops. Suppose that G comes with the additional structure of a function w from the set of edges of G to the set of nonnegative real numbers. Such a function is called a **weight function**, and a graph or digraph together with such a function is called a **weighted** graph or digraph. For each $e \in E$ the value $w(e)$ is called the **weight** of e. We define the **weight** of a walk in a weighted graph or digraph to be the sum of the weights of the edges comprising the walk (as many times as they appear); in other words, the weight of the walk $v_0, e_1, v_1, e_2, v_2, \ldots, v_{l-1}, e_l, v_l$ is $w(e_1) + w(e_2) + \cdots + w(e_l)$. The weight of an edge is also sometimes called its **cost**, the function w is called the **cost function**, and the weight of a walk is called its **cost**. The two terms are used interchangeably. We can think of the weight of a walk as its (weighted) length. For weighted graphs or digraphs in which the weight of every edge is 1, the weighted length of a walk is just its length as we defined it in Section 8.2.

The reader should have no trouble thinking of applications of weighted graphs and digraphs. A graph representing a system of highways can be weighted by the lengths of the stretches of highway making up the edges of the graph. A graph representing a proposed communications network could be weighted by the cost of installing each communications line in the network; or the complete graph on the set of stations in the network could be weighted by the anticipated volume of calls between each pair of stations. A digraph representing an airline schedule could be weighted by the distance, the time, or the fare of each flight. Note that we are attaching weights to the edges, not to the vertices.

We can indicate the weights of the edges in a picture of a weighted graph or digraph simply by writing the weights next to the edges. Representing a weighted graph or digraph with adjacency lists also poses no problem, since for each vertex v we can store not only the list of vertices adjacent to v, but also the weights of the edges providing the adjacency. Furthermore, we can represent weighted graphs or digraphs by generalizing the adjacency matrix representation of the unweighted versions. The **cost matrix** of a weighted graph or digraph is the matrix $C = (c_{ij})$ in which $c_{ij} = w(i, j)$ is the weight of the edge from vertex i to vertex j, if there is such an edge; is 0 if $i = j$; and is ∞ if $i \neq j$ and there is no edge from i to j.

EXAMPLE 1. The cost matrix A of the weighted graph shown in Figure 10.1 is given below.

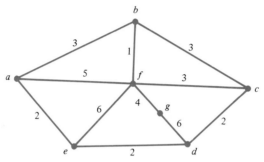

FIGURE 10.1 A weighted graph.

$$A = d \begin{array}{c@{\quad}ccccccc} & a & b & c & d & e & f & g \\ a & 0 & 3 & \infty & \infty & 2 & 5 & \infty \\ b & 3 & 0 & 3 & \infty & \infty & 1 & \infty \\ c & \infty & 3 & 0 & 2 & \infty & 3 & \infty \\ d & \infty & \infty & 2 & 0 & 2 & \infty & 6 \\ e & 2 & \infty & \infty & 2 & 0 & 6 & \infty \\ f & 5 & 1 & 3 & \infty & 6 & 0 & 4 \\ g & \infty & \infty & \infty & 6 & \infty & 4 & 0 \end{array}.$$

Since this is an undirected graph, the cost matrix is symmetric. ◆

We will consider two algorithms for finding the **(weighted) distance** between given vertices u and v in a weighted graph or digraph. This quantity is defined to be the weight of a walk of minimum weight from u to v if there is a walk from u to v, and ∞ otherwise. Since all weights are nonnegative, it does us no good, in looking for a minimum weight walk, to visit a vertex twice, so the minimum weight walk will always be a path.

The first algorithm uses a matrix operation related to matrix multiplication. Since this method turns out not to be the best one, there is no standard notation for the operation. Nevertheless, it is instructive to look at the method. Define the operation $\diamond$ as follows. Given the m by n matrix $A = (a_{ik})$ and the n by p matrix $B = (b_{kj})$, the m by p matrix $A \diamond B$ is the matrix whose (i, j)th entry is

$$\min_{1 \leq k \leq n} (a_{ik} + b_{kj}).$$

In other words, to determine the (i, j)th entry of $A \diamond B$, we look at the ith row of A and the jth column of B, and take the smallest sum of corresponding entries we can find. (Addition involving ∞ is defined by declaring $\infty + \infty = \infty$ and $\infty + a = \infty$ for every finite number a.) If A is a square matrix (in our case, the cost matrix of a graph or digraph), then recursively we let $A^{\diamond 1} = A$ and $A^{\diamond(l+1)} = A^{\diamond l} \diamond A$ for all $l \geq 1$.

EXAMPLE 2.   For the cost matrix A given in Example 1 we have

$$A^{\diamond 2} = A \diamond A = \begin{array}{c} \\ a \\ b \\ c \\ d \\ e \\ f \\ g \end{array} \begin{array}{c} \begin{array}{ccccccc} a & b & c & d & e & f & g \end{array} \\ \left[\begin{array}{ccccccc} 0 & 3 & 6 & 4 & 2 & 4 & 9 \\ 3 & 0 & 3 & 5 & 5 & 1 & 5 \\ 6 & 3 & 0 & 2 & 4 & 3 & 7 \\ 4 & 5 & 2 & 0 & 2 & 5 & 6 \\ 2 & 5 & 4 & 2 & 0 & 6 & 8 \\ 4 & 1 & 3 & 5 & 6 & 0 & 4 \\ 9 & 5 & 7 & 6 & 8 & 4 & 0 \end{array} \right] \end{array}.$$

Note that every entry in $A^{\diamond 2}$ is no bigger than the corresponding entry in A, since one of the sums in the expression $\min_{1 \leq k \leq n} (a_{ik} + b_{kj})$ is $a_{ij} + b_{jj} = a_{ij} + 0 = a_{ij}$.   ◆

It should be fairly clear (and can be proved rigorously by induction—see Exercise 7) that if A is a cost matrix, then the (i, j)th entry of $A^{\diamond l}$ gives the weight of a walk of length at most l (i.e., having at most l edges) between vertex i and vertex j having minimum weight. Since it suffices to consider paths when looking for the weighted distance between two vertices, and since a path can have length at most $n - 1$, where n is the number of vertices in the graph or digraph, *the matrix $A^{\diamond(n-1)}$ gives the weighted distances between all pairs of vertices in the weighted graph or digraph whose cost matrix is A.*

Let us analyze the efficiency of the algorithm implied by this discussion. The obvious algorithm for performing the operation $\diamond$, just like matrix multiplication, takes at least n^3 steps (for each of the n^2 entries of the answer, we have to look at n sums). The algorithm calls for $n - 2$ applications of this operation. Therefore, $O(n^3(n - 2)) = O(n^4)$ steps are required in all.

There are two drawbacks to this algorithm, aside from the fact that it appears not to be very efficient. First, all we obtain are the *weights* of the minimum weight paths, not the paths themselves. Second, we may be obtaining information that is not needed. We may only want to obtain the distance between two specific vertices; it seems a shame to compute the n^2 distances between *all* pairs of vertices in order to get this one distance.

DIJKSTRA'S ALGORITHM

We present now an algorithm for finding minimum weight paths that is more efficient than the algorithm just presented and overcomes the two objections just raised. It is known as Dijkstra's algorithm (named after the computer scientist who discovered it).

We display the procedure in pseudocode as Algorithm 1. Let us explain the notation and the ideas involved. A weighted graph or digraph G whose vertices are labeled $1, 2, \ldots, n$ is given, together with a vertex s called the **source** and a vertex t called the **sink**. We assume that G actually has a path from the source to the sink (see also Exercise 19). The cost matrix is part of the input to the algorithm; we denote the weight (cost) of the edge from i to j by $w(i, j)$. Recall that if there is no edge from i to j, then $w(i, j)$ is 0 (if $i = j$) or ∞ (if $i \neq j$). The algorithm finds a minimum weight path from the source s to the sink t, and the weight of this path.

The procedure needs to keep track of three things for each vertex. First, *final*(i) will be a Boolean variable; we call a vertex i "final" if *final*$(i) = $ **true**, and we call it "nonfinal" if *final*$(i) = $ **false**. Initially, all the vertices are nonfinal; one more vertex is marked final during each pass through the main loop of the algorithm (the source—vertex s—is always the first vertex to be marked final); and the algorithm stops once the sink—vertex t—has been marked final. Intuitively, at each stage, the algorithm knows all there is to know about the final vertices.

Second, *distance*(i) will be a numerical variable that tells us the weight of the minimum weight path from the source to vertex i that contains only final vertices (except possibly for i itself). Initially, since all vertices are originally nonfinal, we have *distance*$(i) = \infty$ for all $i \neq s$, and *distance*$(s) = 0$ since the path of length 0 from s to itself has weight 0. At the point at which i becomes final, *distance*(i) will cease to change and remains in fact the weighted distance from the source to i.

Finally, *previous*(i) for all $i \neq s$ will be a variable whose value is a vertex number (an integer from 1 to n); it tells us the next to the last vertex in a path from s to i that has weight *distance*(i). We do not define *previous*(s). Like *distance*(i), *previous*(i) will cease to change once vertex i becomes final. Therefore, when the algorithm terminates, the minimum weight path from s to t (which is what we are trying to find) can be found, in reverse order, by starting at t and following the *previous* pointers back to s; in other words, the path is $s = previous(previous(\cdots previous(t) \cdots)), \ldots, previous(previous(t)), previous(t),\ t$.

The key idea behind the algorithm is as follows. During each pass through the main loop, we find the nonfinal vertex v whose *distance* is smallest. We make this vertex final. Then, for each nonfinal vertex k, we see whether the path we just found from s to v, followed by the edge from v to k, has smaller weight than the minimum weight path we already knew about from s to k. If so, that is, if $distance(v) + w(v, k) < distance(k)$, then we update $distance(k)$ and $previous(k)$ accordingly.

procedure $dijkstra(G :$ weighted graph or digraph, $s, t :$ vertices of G)
   {this procedure finds the minimum weight path from s to t; it is assumed
     that there is a path from s to t; the vertices of G are labeled $1, 2, \ldots, n$;
     $C = (w(i, j))$ is the cost matrix of G}
   for $i \leftarrow 1$ **to** n **do** {initialize variables}
     begin
       $final(i) \leftarrow$ **false**
       $distance(i) \leftarrow \infty$
     end
   $distance(s) \leftarrow 0$
   while not $final(t)$ **do**
     begin
       $min \leftarrow \infty$ {find nonfinal v whose $distance$ is minimum
                       among all nonfinal vertices}
       for $j \leftarrow 1$ **to** n **do**
         if (not $final(j)$) and ($distance(j) < min$) **then**
           begin
             $v \leftarrow j$
             $min \leftarrow distance(j)$
           end
       $final(v) \leftarrow$ **true**
       for $k \leftarrow 1$ **to** n **do** {update $distance$ and $previous$}
         if not $final(k)$ **then**
           if $distance(v) + w(v, k) < distance(k)$ **then**
             begin
               $distance(k) \leftarrow distance(v) + w(v, k)$
               $previous(k) \leftarrow v$
             end
   end
   return$(distance, previous)$
   {$distance(t)$ is the weight of a minimum weight path from s to t; the path can
     be found, in reverse, by following $previous$ starting at t}

Algorithm 1. Dijkstra's algorithm for finding a minimum weight path.

EXAMPLE 3. Let us apply Dijkstra's algorithm to find the minimum weight path from a to g in the graph shown in Figure 10.1. We will display the values of the variables at each stage in the algorithm by listing each vertex v, followed by an asterisk if it is final, and the current value of $distance(v)$, with $previous(v)$ (if it has been set) as a superscript. Thus after initialization we have the following situation:

$$a : 0 \quad b : \infty \quad c : \infty \quad d : \infty \quad e : \infty \quad f : \infty \quad g : \infty.$$

The nonfinal vertex with minimum $distance$ at this point is (as is always the case) the source, namely vertex a. Thus we make a final and then update the $distance$'s from a to its neighbors, giving us

$$a^* : 0 \quad b : 3^a \quad c : \infty \quad d : \infty \quad e : 2^a \quad f : 5^a \quad g : \infty.$$

This time the nonfinal vertex with minimum $distance$ is e, so it becomes final. The $distance$ from a to d is now updated to 4 (since $distance(e) + w(e, d) = 2 + 2 < \infty = distance(d)$), and $previous(d) \leftarrow e$, since the path that achieves this smaller distance passes through e just before it arrives at d. On the other hand, $distance(f)$ and $previous(f)$ are not changed, since the path from a to f via e has greater weight than the path we already knew about, directly from a to f. Thus after two passes we have this situation:

$$a^* : 0 \quad b : 3^a \quad c : \infty \quad d : 4^e \quad e^* : 2^a \quad f : 5^a \quad g : \infty.$$

The process continues. Vertex b is the next vertex to be made final. Note that this time the value of $distance(f)$ decreases, since the path through b has less weight than the path directly from a to f. We now have

$$a^* : 0 \quad b^* : 3^a \quad c : 6^b \quad d : 4^e \quad e^* : 2^a \quad f : 4^b \quad g : \infty.$$

At this point there are two vertices tied for being "closest" to a (i.e., having smallest $distance$) among the unmarked vertices, namely d and f. We can choose either, although the algorithm actually picks the one with smaller vertex number. So let us finalize d next. The updated situation is as follows (note that $previous(c)$ has not changed, since the path from a to c through d has the same weight as the path from a to c through b that we already knew about):

$$a^* : 0 \quad b^* : 3^a \quad c : 6^b \quad d^* : 4^e \quad e^* : 2^a \quad f : 4^b \quad g : 10^d.$$

The next snapshot is

$$a^* : 0 \quad b^* : 3^a \quad c : 6^b \quad d^* : 4^e \quad e^* : 2^a \quad f^* : 4^b \quad g : 8^f,$$

and then

$$a^* : 0 \quad b^* : 3^a \quad c^* : 6^b \quad d^* : 4^e \quad e^* : 2^a \quad f^* : 4^b \quad g : 8^f,$$

and finally,

$$a^* : 0 \qquad b^* : 3^a \qquad c^* : 6^b \qquad d^* : 4^e \qquad e^* : 2^a \qquad f^* : 4^b \qquad g^* : 8^f.$$

Note that the sink g was the only nonfinal vertex before the last step, so in fact we have found not only the weighted distance from a to g, but also the weighted distance from a to *every* vertex in the graph. It would be simple enough to modify the algorithm to give all these distances, dispensing with a sink altogether (see Exercise 12).

We can recover the path that the algorithm found by looking at the *previous* values. Since $previous(g) = f$, we know that the path ended by going from f to g. Since $previous(f) = b$, we know that the path reached f from b. Finally, since $previous(b) = a$, we know that the path reached b from a. Thus the path is a, b, f, g, and indeed its weight is $3 + 1 + 4 = 8 = distance(g)$. The fact that Dijkstra's algorithm always finds minimum weight paths assures us that there is no path of weight less than 8 from a to g. ◆

It is not hard to verify that Dijkstra's algorithm always finds a minimum weight path from the source to the sink. (Recall that this does need to be verified: The mere *existence* of a terminating algorithm does not in the least guarantee that it provides the correct answer to the problem we are trying to solve.) The key observation is simply that when a vertex v is made final, the path from the source s to v known to the algorithm—namely, the path of minimum weight passing through only final vertices—is in fact the minimum weight path from s to v in the graph or digraph. Indeed, since the values of $distance(j)$ for nonfinal j are no less than the value of $distance(v)$ at this point, a path from s to v containing such a j could not have less weight than the path already known from s to v.

Finally, we consider the efficiency of Dijkstra's algorithm applied to a graph (or digraph) with n vertices. The initialization phase of the procedure takes $O(n)$ steps, which pales in comparison to the rest of the algorithm. There may be as many as n passes through the main **while** loop, since the sink may be the last vertex to become a final vertex. Within the **while** loop there are two **for** loops, one after the other. Each requires $O(n)$ steps. Therefore, each pass through the **while** loop requires $O(n)$ steps, so the entire algorithm requires $O(n^2)$ steps.

Thus for a given source s, we can find minimum weight paths from s to all other vertices in the graph in time $O(n^2)$. If we want to find minimum weight paths from all vertices, then we could apply the algorithm n times, letting the source successively be each vertex in the graph. This means that we can find least expensive paths between all pairs of vertices in time $O(n^3)$, better than the $O(n^4)$ steps required by the matrix algorithm discussed earlier (see Exercise 27, however).

PROJECT SCHEDULING (*OPTIONAL*)

Our next application involves the finding of maximum weight paths, rather than minimum weight paths. Suppose that a project (such as the American project to send human beings to the moon during the decade of the 1960s) consists of many subprojects (such as building and testing the required powerful rockets, perfecting docking maneuvers, and determining that humans can survive long periods of weightlessness). The entire project can be represented by a digraph: Each vertex represents a subproject, and there is a directed edge from vertex a to vertex b if and only if subproject a must be completed before subproject b can begin. For example, rockets must be built before they can be tested. In general, each vertex can have many successors (there may be many subprojects that cannot begin until a given subproject is finished) and many predecessors (there may be many subprojects that must be completed before a given project can begin). One property that such a project digraph must have, however, is that it is **acyclic**: There can be no cycles $v_1, v_2, \ldots, v_r, v_1$, since if there were, no subproject in the cycle could ever be started. In the terminology of Section 3.4, the set of subprojects forms a poset in which "precedes" means "must be completed prior to."

Now each subproject will require a certain amount of time. Let us assume that these durations or "completion times" are known. (In practice, completion times cannot usually be known exactly, but they can at least be estimated. Also, probabilistic techniques can be applied if we want to think of completion times as random variables.) The problem we wish to consider is this: *Given the digraph representing a project, together with completion times for each subproject, determine the duration (completion time) for the entire project.* Obviously, the answer is not simply the sum of the completion times for the subprojects, since many subprojects can be going on simultaneously (e.g., building the new rockets can take place while the tests of human responses to prolonged weightlessness are being carried out using older, less powerful rockets). We want an algorithm for finding the total completion time. As a by-product, we would like to identify the sequences of subprojects that cause the completion time to be as large as it is (these are called **critical paths**), and we would like to be able to determine when each subproject must be started in order to complete the entire project within the required time.

Project scheduling problems such as this are common in operations research and industrial management. Techniques for solving them are often known by such acronyms as CPM ("critical path method") and PERT ("program evaluation and review technique"). Let us see how completion times and critical paths can be found.

As we have formulated the problem, the vertices, not the edges, of the digraph model of a project have the weights, the weight of a vertex being the completion time of the subproject represented by that vertex. At the risk of complicating the model, and making it somewhat redundant, let us change the model slightly so that we can continue to work with digraphs in which the edges have weights. First, we need to adjoin one new vertex, called the **sink**, which represents the completion of the project. Any subproject that is not prerequisite for another subproject will be considered to be prerequisite for the sink. There are no edges leading from the sink. We define the weight $w(i, j)$ of the directed edge from vertex i to vertex j to be the time required for the completion of the subproject represented by vertex i *if the* *completion of subproject i is required for the beginning of subproject j*; otherwise, there will be no edge from i to j. Thus the weight of every edge leaving a given vertex will be the same, and this value is the time required to complete the subproject represented by that vertex. Furthermore, it is not necessary to have an edge from i to j if there is a vertex k with edges ik and kj. In such a case, the completion of the subproject at i is certainly necessary before the subproject at j can begin, but this information is already implied by the facts that j cannot begin before k is finished, and k cannot begin before i is finished. This simplification makes the digraph easier to work with. In the terminology of Section 3.4, it is enough to work with the Hasse diagram of our poset.

To make our model even cleaner, we will also assume that there is a unique vertex, called the **source**, which has no predecessors; this vertex represents the beginning of the entire project, and we may as well assume that it takes no time. If there is not already a source present, we can adjoin one, putting in edges from the source to each vertex in the original digraph having no predecessors; these new edges will have weight 0.

EXAMPLE 4. The acyclic digraph shown in Figure 10.2 represents a project having 10 subprojects, together with a source s and a sink t. The subproject represented by vertex d, for example, takes 2 units of time for completion, and subprojects e and g cannot be started until d is finished. (Neither can, for instance, subproject h, but there is no edge from d to h since the edges de and eh already contain this information.) Subproject d itself cannot start until subprojects a and b are finished, because of the presence of edges ad and bd. ◆

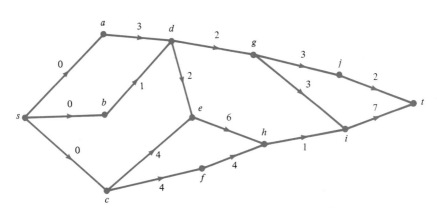

FIGURE 10.2 Digraph representing a project. Vertex s is the source.

We now describe an algorithm for finding the completion time of the entire project and a critical path, given a weighted acyclic digraph with a unique source s and a unique sink t. We want to find a path of maximum weight from s to t.

For a change we will sketch the algorithm in prose form, rather than presenting it in detail in pseudocode (see Exercise 13).

We will compute the completion time $T(v)$ of each vertex v in the given digraph, that is, the cost of the most expensive path from s to v. We do so in the following way. First, the completion time $T(s)$ of the source is 0. Then as long as there are vertices whose completion times have not been determined, we find one such vertex v such that the completion times of all of v's predecessors have already been determined. That there must be such a vertex follows from the fact that the given digraph is acyclic. (To find such a vertex, we can start with any vertex whose completion times have not yet been determined and, as long as the vertex we are currently considering still has predecessors whose completion times have not yet been determined, we consider one such predecessor; this process must terminate since there are only a finite number of vertices.) Then we set

$$T(v) = \max_{i} \left(T(i) + w(i, v) \right),$$

where the maximum is taken over all vertices i for which there is an edge iv in the digraph. The last vertex to have its completion time determined is the sink t. The number $T(t)$ is the completion time of the entire project—the weight of the most expensive path from s to t. Just as in Dijkstra's algorithm, we can keep track of the path itself, as well as its cost, by letting $previous(v)$ be a vertex i for which the maximum is achieved in the definition of $T(v)$.

EXAMPLE 5. This algorithm, applied to the digraph in Example 4, operates as follows. First we set $T(s) = 0$. Next we need to find a vertex all of whose predecessors have been processed; vertices a, b, and c all fall into this category, since their only predecessor is s. The edge from s to a has weight 0, so $T(a) = 0 + 0 = 0$. Similarly, $T(b) = 0$, and $T(c) = 0$. At this point we could process vertex d or vertex f. Suppose that we choose to process vertex d. We need to determine the larger of $0+3$ (considering the edge ad) and $0+1$ (considering the edge bd). Since the former value is larger, we set $T(d) = 0 + 3 = 3$, and $previous(d) = a$. If we process vertex e next, we find that its completion time is the larger of $3+2$ and $0+4$, namely $T(e) = 3+2 = 5$, and thus $previous(e) = d$. Continuing in this manner, we soon find that $T(f) = 4$, $previous(f) = c$, $T(g) = 5$, $previous(g) = d$, $T(h) = 11$, $previous(h) = e$, $T(i) = 12$, $previous(i) = h$, $T(j) = 8$, $previous(j) = g$, and $T(t) = 19$, $previous(t) = i$. Therefore, the project requires 19 units of time to complete, and a critical path (in fact, the only one in this example) is $s, a, d, e, h,$ i, t. We also find out, for example, that subproject j cannot be started until 8 units of time have elapsed.

Refinements of the algorithm would allow us to determine not only the earliest time at which a subproject *may* be started, but also the latest time at which a subproject *must* be started in order to finish the entire project by time $T(t)$ (see Exercise 23). For example, subproject j must be started by time $19 - 2 = 17$ in order that it be finished at time 19. ◆

THE TRAVELING SALESPERSON PROBLEM (*OPTIONAL*)

We end this section by considering a third problem of finding optimal paths in weighted graphs or digraphs. Let G be a complete graph on n vertices in which each edge has a nonnegative weight (possibly ∞). We seek a Hamilton cycle—a cycle containing every vertex (see Section 8.2)—with least weight. This problem is usually called the **traveling salesperson problem**, because it models the problem a salesperson faces in deciding the order in which to visit each of $n-1$ different locations and return home. Here the weight of edge uv is the cost (in dollars, time, mileage, or whatever) in traveling from vertex u to vertex v.

We can also use this model for the problem of programming an automatic drilling machine to drill several holes in sheets of metal, one sheet after another. If we want to minimize the amount of time required for drilling a sequence of sheets, we need to minimize the time required in having the sheets moved around as the successive holes are drilled. (In this case we may want to find a minimum weight Hamilton *path*—a path containing every vertex—rather than a minimum weight Hamilton cycle, if it is not necessary to return each sheet to its starting position before it is removed from the drilling machine.)

In contrast to the nice algorithms we had for finding optimal paths earlier in this section, there is no good solution to the traveling salesperson problem currently known. The problem (when suitably phrased as a yes/no question) is *NP*-complete (see Section 4.4). Indeed, a special case of the traveling salesperson problem is the problem of determining whether a graph H has a Hamilton cycle. (To see this, let G be the complete graph on the same vertex set as H, with the edge uv in G having weight 0 if the edge uv is in H, and weight 1 if uv is not in H. Then H has a Hamilton cycle if and only if G has a minimum weight Hamilton cycle of weight 0.)

There are some algorithms using backtracking (a systematic method of considering all possible potential solutions to a problem and eliminating those that cannot be optimal) that, when run on fast computers, are reasonably effective in finding minimum weight Hamilton cycles in graphs of up to about 50 vertices. There are also some algorithms that find Hamilton cycles that are not necessarily optimal but are guaranteed to come close to the optimal, say within a factor of 2, in certain cases. We will not go into the details (see Exercise 28). Instead, we content ourselves with looking at one small example and finding the minimum weight Hamilton cycle by ad hoc reasoning.

EXAMPLE 6.   Consider the weighted K_5 shown in Figure 10.3. A Hamilton cycle must contain two edges incident to each vertex. Let us get a lower bound on the weight of such a cycle by finding the sum of the weights of the two edges of least weight at every vertex, and then dividing by 2 (we need to divide by 2 because each edge is in effect counted twice in such an estimate). The minimum weight of the Hamilton cycle as it passes through vertex a is $2+3=5$; the minimum weight through b is $1+3=4$; the minimum weight through c is $2+3=5$; the minimum weight through d is $1 + 2 = 3$, and the minimum weight through e is $3 + 4 = 7$. Therefore, a lower bound (not necessarily achievable) on the weight of a Hamilton cycle is $(5 + 4 + 5 + 3 + 7)/2 = 12$.

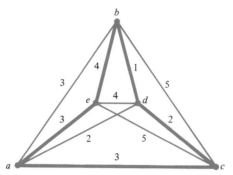

FIGURE 10.3 A weighted complete graph, with minimum weight Hamilton cycle darkened.

Next let us try to find a low-weight Hamilton cycle by inspection. The cycle a, e, b, d, c, a seems like a good candidate; its weight is $3 + 4 + 1 + 2 + 3 = 13$. Now we must argue that in fact this cycle is best possible—that there is no cycle of weight 12. First, let us consider what happens at vertex c. Our cycle uses the edges of weight 2 and 3, and unless both of these edges are used, the cost of passing through vertex d will be at least $2 + 5 = 7$, rather than 5. Therefore, the minimum weight Hamilton cycle that fails to use both of the edges cd and ca will be at least $(7-5)/2 = 1$ greater than the lower bound of 12 we found above. Since $12 + 1 = 13$, no such cycle can be better than the cycle we have already found. Thus we may assume that an optimal Hamilton cycle uses edges cd and ca.

At this point we can assume that edge da is not used, since its addition prematurely completes the cycle. Now at vertex d, if we used edge de rather than db, then we would have no hope of obtaining a Hamilton cycle of weight less than $12 + (3/2) = 13.5$, since we would be increasing the cost of passing through vertex d by $4 - 1 = 3$. Therefore, we can assume that edge db is in the optimal cycle as well. At this point, no more choices are possible: There is only one Hamilton cycle that uses edges cd, ca, and db, namely the one we chose. Thus we conclude that the Hamilton cycle of length 13 that we have found is a minimum weight Hamilton cycle. ◆

SUMMARY OF DEFINITIONS

acyclic graph or digraph: a graph or digraph without cycles.
cost of an edge or walk: the weight of the edge or walk.

cost function: weight function.

cost matrix of the weighted graph or loopless digraph G: the matrix $C = (c_{ij})$ in which c_{ij} is the weight of edge ij if edge ij is present, is 0 if $i = j$, and is ∞ if $i \neq j$ and there is no edge from i to j.

critical path in a weighted acyclic digraph: a path of maximum weight from source to sink.

Dijkstra's algorithm: a vertex labeling process for finding minimum weight paths in graphs or digraphs (Algorithm 1).

(weighted) distance from vertex u to vertex v in a weighted graph or digraph: the weight of a walk with smallest weight from u to v if such a walk exists, ∞ otherwise.

sink in a path problem: the vertex at which the path is to end.

source in a path problem: the vertex from which the path is to start.

traveling salesperson problem: the problem of finding a minimum weight Hamilton cycle in a weighted complete graph.

weight of the edge e of weighted graph or digraph with weight function w: the value of $w(e)$.

weight of a walk in a weighted graph or digraph with weight function w: the sum of the weights of the edges in the walk, with each edge counted as many times as it appears in the walk.

weight function w for the graph or loopless digraph G: a function from the set of edges of G to the nonnegative real numbers.

weighted graph or digraph: a graph or digraph G together with a weight function on G.

EXERCISES

1. Consider the weighted graph in Figure 10.1.
   (a)   List all paths of length at most 3 (i.e., having three or fewer edges) from a to f.
   (b)   Compute the weight of each of the paths found in part (a).

2. Draw a picture of a weighted graph to model the one-way air fares of an airline that flies between Detroit and Chicago for \$19, between Chicago and St. Louis for \$19, between Detroit and St. Louis for \$34, between St. Louis and Tulsa for \$49, and between Tulsa and Dallas for \$59. (Even though these are one-way fares, an undirected graph can be used because the fare is the same in each direction.)

3. Draw a picture of a weighted digraph to model the volume of mail among the different categories *consumers*, *retailers*, and *banks* in a community if each month there are 4000 pieces of mail from banks to consumers, 20,000 pieces from retailers to consumers, 2000 pieces from consumers to retailers,

300 pieces from consumers to banks, 500 pieces from banks to retailers, and 1000 pieces from retailers to banks.

4. Write down the cost matrices of the following weighted graph and digraph.
   (a)   The graph in Exercise 2.
   (b)   The digraph in Exercise 3.

5. Apply Dijkstra's algorithm to find a minimum weight path from b to e in the weighted graph in Figure 10.1.

6. Apply the algorithm for finding a critical path and completion time to the project consisting of the following subprojects and dependencies. Subprojects A through K take 1, 2, 4, 3, 4, 1, 7, 2, 3, 3, and 1 weeks, respectively. Subproject K cannot be started until H and I are finished; J cannot be started until F and G are finished; I cannot be started until F is finished; H cannot be started until E is finished; G cannot be started until both C and D are finished; F cannot be started until B is finished; and E cannot be started until A is finished.

7. Explain why the (i, j)th entry of $A^{\diamond l}$ gives the weight of a walk of length at most l between vertex i and vertex j having minimum weight. [*Hint*: Use mathematical induction.]

8. Let A be the matrix given in Example 1.
   (a)   Compute $A^{\diamond 3}$.
   (b)   Explain all the differences between $A^{\diamond 3}$ and $A^{\diamond 2}$ (computed in Example 2), in terms of paths in the graph.

9. Suppose that G is a graph or digraph whose edges are weighted by not necessarily positive numbers. Explain the difficulty in finding (or even defining) a minimum weight walk between two vertices in G.

10. Consider the weighted digraph on the set $\{1, 2, 3, 4, 5\}$ such that there is an edge of weight $i + (i - j)^2$ from i to j for all $i \neq j$.
    (a)   Write down the cost matrix for this digraph.
    (b)   Apply Dijkstra's algorithm to find the distances from vertex 5 to every other vertex.
    (c)   Use the result of part (b) to find a minimum weight path from vertex 5 to vertex 1.

11. Apply Dijkstra's algorithm to find a minimum weight path from a to z in the given graph.

12. Explicitly modify Algorithm 1 so that it computes distances from s to all other vertices.

13. Write down in pseudocode the algorithm for finding the completion time of an acyclic weighted digraph representing a project. Assume that the source and sink are given as part of the input. Be explicit as to how the procedure selects

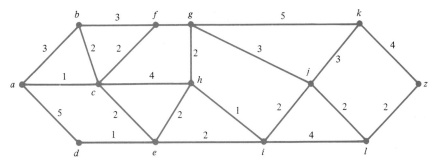

EXERCISE 11

the next vertex v whose completion time is to be computed. Your algorithm should also produce the information needed to find a critical path.

14. Explain how we can determine *all* the critical paths in an acyclic weighted digraph representing a project.

15. Find a maximum weight Hamilton cycle in the weighted complete graph in Figure 10.3.

16. How should the concept of isomorphism be defined for weighted graphs? Write down a precise definition.

17. A minimum weight Hamilton cycle in a complete weighted graph could be determined by generating all the Hamilton cycles and computing the weight of each. Assume that it takes 10^{-5} second to generate and compute the weight of one cycle.
    (a) How long would it take to find a minimum weight Hamilton cycle in a weighted K_{20} by this method?
    (b) What is the largest value of n for which a minimum weight Hamilton cycle in a weighted K_n could be found by this method in less than 1 hour?

18. Find a minimum weight Hamilton cycle for the weighted graph on p. 734.

19. In Dijkstra's algorithm we assumed that there was a path from the source to the sink.
    (a) What will happen if Algorithm 1 is applied to a graph in which there is no path from s to t?
    (b) Modify Algorithm 1 so that it returns the correct answer (∞) if no such path exists.

20. A weighted graph is said to satisfy the **triangle inequality** if for every three distinct vertices i, j, and k, we have $w(ij) + w(jk) \geq w(ik)$.
    (a) Determine whether the weighted graph in Figure 10.1 satisfies the triangle inequality.

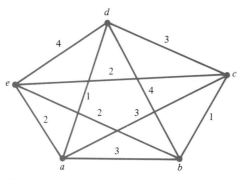

EXERCISE 18

(b)   Prove or disprove: If a weighted graph satisfies the triangle inequality, then a minimum weight path between vertices u and v is also a minimum length path between u and v (i.e., has the fewest possible number of edges in it).

(c)   Explain why weighted graphs that represent physical distances between locations must satisfy the triangle inequality.

21.   The **diameter** of a connected graph is the largest distance between two vertices in the graph, where the distance between vertices u and v is defined to be the length of the shortest path from u to v. (Here we are referring to unweighted graphs, but an analogous definition could be made for weighted graphs as well.) Explain how Dijkstra's algorithm could be used to find the diameter of a connected graph, and analyze the time complexity of finding the diameter by this method.

22.   Explain how to use the breadth-first search algorithm of Section 9.2 to find a minimum weight path between given vertices in a weighted graph in which all the weights are positive integers. [*Hint*: Construct a nonweighted graph to model the weighted one.]

Challenging Exercises

23.   Explain how to determine, given an acyclic weighted digraph representing a project, the latest time at which a subproject may be started in order to finish the entire project by time $T(t)$.

24.   Recall the monotone subsequence problem from Section 6.4. Model this problem as a pair of longest path problems and explain how to solve it.

Exploratory Exercises

25.   Model a project scheduling problem by letting edges, rather than vertices, represent the subprojects (see also Exercise 27 in Section 8.1). Develop an algorithm for finding the completion time and critical paths with this model.

26. Explain how to model a project scheduling problem by putting the weights on the vertices, rather than on the edges. Modify our algorithm for finding the completion time and critical paths with this model.

27. We saw that the matrix $A^{\circ(n-1)}$ gives the distances between all pairs of vertices in the weighted graph or digraph whose cost matrix is A, but that computing it naively as we did seemed not too efficient. Consult Lawler [193] to find better ways to solve the problem of finding distances between all pairs of vertices.

28. Find out more about the traveling salesperson problem by consulting Lawler, Lenstra, Kan, and Shmoys [194].

29. Find out more about project scheduling and related topics by consulting Graham [123,124].

SECTION 10.2
MINIMUM SPANNING TREES

A fool is he who is greedy when others possess.

—The Teaching of Merikane, 2100 B.C.

In Section 9.2 we saw how to find spanning trees in connected graphs, using depth-first or breadth-first search. Usually, a connected graph will have many different spanning trees, but every spanning tree in a graph with n vertices has $n-1$ edges. If we are dealing with a connected *weighted* graph, we can look at the **weight** or **cost** of a spanning tree, that is, the sum of the weights of its edges. The weights of different spanning trees may be quite different. Sometimes it is desirable to find a spanning tree with minimum weight; such a tree is called a **minimum spanning tree**. In this section we look at some efficient algorithms for finding minimum spanning trees.

EXAMPLE 1.   A bank has a number of automatic teller machines (ATMs) at various locations (branch offices, shopping malls, etc.) throughout a city. Each ATM, which allows customers to make deposits and withdrawals, must be able to communicate with the bank's main computer, located at its central office. Communication is accomplished through dedicated telephone lines, leased from the local phone company. The monthly charge for a line between point P and point Q is proportional to the actual straight-line distance between P and Q (say, 2 dollars per mile). It is not necessary to lease a line from every ATM to the central office, however. As long as there is some path of leased lines from each ATM to the central office, perhaps passing through several other ATM locations, communication is possible. The communications officer of the bank needs to decide which lines to lease to minimize the telephone bill.

We can model this problem with a complete weighted graph. The vertices are the locations of the ATMs and the central office. The weight of edge PQ is the monthly cost of leasing a dedicated phone line between points P and Q (or, equivalently, the straight-line distance between P and Q). A solution to the problem is a minimum spanning tree.

Suppose that Figure 10.4, drawn to scale, represents the ATM network for the bank. The central office is labeled a, and the ATMs are labeled b, c, d, e, f, g, and h. The tree shown appears to be the minimum spanning tree; it seems that no other tree on this set of eight vertices has smaller total length. For instance, if we deleted edge gh and replaced it by edge ah, the total length would certainly increase. ◆

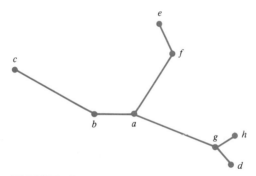

FIGURE 10.4   A network of automatic teller machines.

In this example we needed to find a minimum spanning tree for the complete graph. The complete graph on n vertices has many different spanning trees: It already has $(n-1)!$ of them consisting just of paths (see also Exercise 33 in Section 9.2). Thus even if we had an algorithm to generate all the spanning trees and compute their weights, it would be infeasible to find a minimum one by exhaustive search if n is at all large.

Both of the algorithms for finding minimum spanning trees that we will discuss in this section—and they are quite similar—are **greedy** algorithms. A greedy algorithm attempts to find an optimum structure (in this case a minimum spanning tree) by building it up piece by piece (in this case edge by edge), choosing at each stage in the construction the optimum piece (in this case the edge of minimum weight). In other words, a greedy algorithm tries to get at the global optimum through locally optimal steps. Before we look at these algorithms, let us see that this greedy approach does not always lead to an optimum structure. Recall that we already saw one example of a greedy algorithm that did not always work—a graph coloring algorithm in Section 8.5.

EXAMPLE 2.   Consider the following greedy algorithm for finding a minimum weight path from vertex s to vertex t in a connected weighted graph. Start at vertex s. At each step, move along the edge of minimum weight to a vertex not yet visited. Stop when you reach vertex t. Applied to the graph in Figure 10.5, for example, the algorithm will first choose edge su, since its weight (1) is less than the weight of the other edge incident to s. Next the algorithm will choose edge uv, of weight 3. Finally, it will choose edge vt. Thus it finds the path s, u, v, t of weight $1+3+2 = 6$ from s to t, when the minimum path is s, v, t, having weight $2+2 = 4$. Obviously, the algorithm does not work. It may seem silly to have expected it to, but on a naive level, it seems reasonable: If we want to find the path of minimum weight, why not use the edges of minimum weight? Unfortunately, this greedy approach of choosing the optimum piece at each stage is shortsighted in this problem; the optimum pieces (edges) did not lead to the optimum total structure (path).

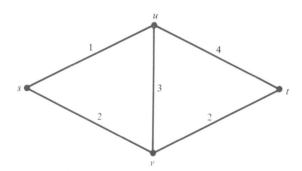

FIGURE 10.5   The greedy path from s to t is not the minimum path.

The point of this example is that a greedy algorithm does not need to work just because it is greedy. If a greedy algorithm works (as do the greedy algorithms we will discuss for finding minimum spanning trees), it is not merely because it chooses an optimum piece at each stage; a proof that this strategy leads to an optimum final structure in the problem at hand is necessary.   ◆

We will construct a minimum spanning tree for a connected weighted graph with n vertices one edge at a time. In other words, at each stage in the algorithm, we will add one more edge to what we already have. After $n-1$ steps we will have a spanning tree, and it will be one that has minimum weight.

At each stage we will add an edge of minimum weight, consistent with our intention to end up with a tree. Since a tree is connected and has no cycles, we cannot simply add *any* edge; some edges may make it impossible to end up with a tree. There are two ways in which we can determine the edge to add next. One is to make sure that we have a tree at each stage. Thus we can add an edge of minimum weight among all edges whose addition still gives us a tree, that is, leaves the graph connected and does not create any cycles. The algorithm that follows this strategy is called **Prim's algorithm**. The second approach is not to worry about connectedness as we proceed, simply adding an edge of minimum weight among all those edges whose addition does not create any cycles. At intermediate stages of the algorithm we may have a forest, rather than a tree, but after $n - 1$ edges have been added, we will have a tree. The algorithm that follows this strategy is called **Kruskal's algorithm**.

PRIM'S ALGORITHM

The input to Prim's algorithm is a connected weighted graph G with n vertices, assumed to be labeled $1, 2, \ldots, n$. We denote the weight of edge ij by $w(i, j)$. The algorithm constructs a tree T in stages. Initially, T consists of the vertex 1 and no edges. At the end, T is a minimum spanning tree. At each stage the algorithm finds an edge e with one endpoint in T and one endpoint not in T. If such an edge is added to T, the result will have no cycles (since it was not the case that both ends of e were already in T) and will be connected (since one end of e was already in T); thus it will again be a tree. Furthermore, since G is connected, there will always be at least one such edge. Among all such edges, the algorithm chooses one of minimum weight. Algorithm 1 gives Prim's algorithm in pseudocode.

EXAMPLE 3.   Use Prim's algorithm to find a minimum spanning tree of the weighted graph shown in Figure 10.6.

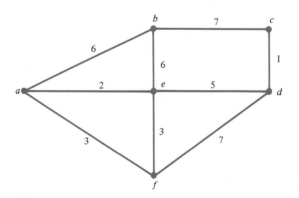

FIGURE 10.6   A connected weighted graph.

procedure *prim*(G : connected weighted graph)
{assume that the vertices of G are labeled $1, 2, \ldots, n$; $w(i, j)$ denotes the
weight of edge ij; a tree will be built one edge at a time; it is denoted T and
is thought of as a set of edges and vertices}
    $T \leftarrow \{1\}$ {initially T has no edges and one vertex}
    for $k \leftarrow 1$ **to** $n - 1$ **do**
      begin
        $min \leftarrow \infty$
        for each edge ij of G **do** {find edge to add to T}
          if exactly one of i and j is in T **then**
            if $w(i, j) < min$ **then**
              begin
                $\{i\_best, j\_best\} \leftarrow \{i, j\}$
                $min \leftarrow w(i, j)$
              end
        $T \leftarrow T \cup \{\{i\_best, j\_best\}, i\_best, j\_best\}$
      end
    return(T) {a minimum spanning tree}

Algorithm 1.   Prim's algorithm for finding a minimum spanning tree.

Solution.   We assume that the vertices are ordered alphabetically. Thus the tree starts as the single vertex a. There are three edges that meet the condition of the first **if** statement, with one endpoint in T and one endpoint not in T, namely ab, ae, and af. The one with smallest weight is ae, so that edge and its other endpoint, e, are added to T. Next we look among the edges that now have one endpoint in T, namely ab, af, be, de, and ef. Two of these—af and ef—are tied for having the smallest weight (3), so we choose one of them, say af, to add to the tree (if the edges are scanned in alphabetical order, this is the edge that the algorithm will choose). At this point, then, T consists of edges ae and af, together with their endpoints. The next edge to be added is de, since it has the smallest weight of all those edges with exactly one endpoint in T. Continuing in this manner, we then add edges cd and ab, obtaining the spanning tree shown with heavy lines in Figure 10.7.

The weight of this spanning tree is $2 + 3 + 5 + 1 + 6 = 17$. Since Prim's algorithm works (as we will show momentarily), we know that no other spanning tree has smaller weight. There are three other spanning trees with the same weight, which we could have obtained by making different choices when there were ties among the edges. One such tree contains the edges ae, ef, de, cd, and be.   ◆

Let us analyze the complexity of Prim's algorithm. In doing so, we need to be somewhat careful about how our data are stored. The procedure consists

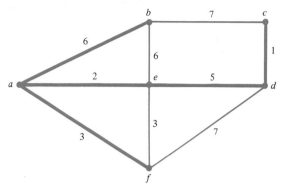

FIGURE 10.7 A minimum spanning tree of the weighted graph in Figure 10.6.

of an outer loop, which is executed $n - 1$ times. Within the outer loop is an inner loop; it is executed once for each edge of G. The statement within that loop thus is executed a total of $(n - 1)e$ times, where e is the number of edges of G. We claim that it takes $O(1)$ time to execute that statement. Certainly, the comparison $w(i, j) < min$ and the two assignment statements can be done in $O(1)$ steps. Ostensibly the determination as to whether "exactly one of i and j is in T" requires a search through T. In fact, it does not, since we can have a Boolean variable $T(v)$ for each vertex v in the graph; initially, each $T(v)$ is **false**, and we set $T(v)$ to be **true** when v is put into T. Thus checking whether i and j are in T amounts to nothing more than checking the values of $T(i)$ and $T(j)$. Since the statements within the inner loop take $O(1)$ steps, and since they are executed $(n - 1)e$ times, the algorithm has execution time in $O(ne)$. Since the number of edges in a graph is at most $n(n - 1)/2$ (this maximum occurs when the graph is complete), this estimate can also be given as $O(n^3)$.

Actually, it is possible to improve the algorithm to make it run faster. For instance, we could sort the edges by weight before we start (we can do this in time $O(e \log e)$, using an efficient sorting algorithm such as the merge sort of Section 5.2), and run through the list each time from least costly to most costly edge. The first acceptable edge (one with exactly one of its endpoints in T) is the one to be added to T. Edges that are added to T, and edges that are found to have both endpoints in T can be removed from the list. The complexity analysis becomes fairly complicated after implementing various such modifications to improve the efficiency; there are in fact implementations of Prim's algorithm with time complexity in $O(e \log n)$.

Finally, we come to the crucial question: How can we be sure that the algorithm works? The answer is the proof of the following theorem.

THEOREM 1. *Prim's algorithm produces a spanning tree of G having minimum weight.*

Proof. It is clear from our previous discussion that the algorithm produces a spanning tree; we need to show that the tree T it produces has minimum weight among all spanning trees of G. The proof is subtle.

Let $e_1, e_2, \ldots, e_{n-1}$ be the edges of T in the order they were added to T, and for each i from 0 to $n-1$, let T_i be the tree containing just the edges $\{e_1, e_2, \ldots, e_i\}$ (in other words, T_i is the tree formed after the ith stage of the algorithm). Note that T_0 is the tree containing just the starting vertex and no edges, and $T_{n-1} = T$. Now G certainly *has* at least one minimum spanning tree, since it has only a finite number of spanning trees, and one or more of them have the smallest weight. For each minimum spanning tree of G there is a largest natural number k (possibly 0) such that the tree contains all the edges $e_1, e_2, \ldots, e_k$ (i.e., contains T_k as a subgraph). (If $k = 0$, the spanning tree contains no edges of T.) Let T' be a minimum spanning tree of G that boasts the largest such k. If $k = n - 1$, we are finished, because then $T' = T$, so T is indeed a minimum spanning tree. Therefore, we assume that $k < n - 1$. Now we will show how to find another minimum spanning tree T'', which contains $e_1, e_2, \ldots, e_{k+1}$, contradicting the choice of T'. Then, in fact, k must equal $n - 1$ and we will be done.

Thus we assume that T' contains all of the edges $e_1, e_2, \ldots, e_k$, but not e_{k+1}. Consider the graph consisting of T' together with edge e_{k+1}. Since this graph has one too many edges to be a tree, it must contain a cycle (see Theorem 1 in Section 9.1), and since the cycle was not present in T', the cycle must contain e_{k+1}. Now not every edge in the cycle can belong to T_{k+1}, since T_{k+1} was also a tree, so there is an edge in the cycle that is in T' but not in T_{k+1}. Furthermore, we can choose such an edge so that one of its endpoints is in T_k. (To find such an edge, start at the endpoint of e_{k+1} that is in T_k, and follow the cycle in the direction away from e_{k+1} until it reaches an edge not in T_k.) Call this edge e'. See Figure 10.8, in which the edges in the set $\{e_1, e_2, \ldots, e_k\}$ are shown in heavy lines.

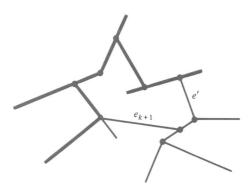

FIGURE 10.8 Edge e_{k+1} creates a cycle in T'.

Let T'' be the tree formed from T' by adjoining e_{k+1} and deleting e'. (That T'' is a tree follows from Theorem 1 in Section 9.1, since it is connected and has the right number of edges.) Now we come to the key point. In the execution of Prim's algorithm, when e_{k+1} was included in T, the edge e' was also a candidate for being included in T. Since e_{k+1} was included rather than e', it must be the case that $w(e_{k+1}) \leq w(e')$. Therefore, the weight of T'' can be no greater than the weight of T', since the weight of T'' equals the weight of T' plus $w(e_{k+1})$ minus $w(e')$. But T' was a minimum spanning tree, so its weight was the smallest possible. Thus the weight of T'' equals the weight of T', and T'' is another minimum spanning tree of G. Clearly, T'' contains all of the edges e_1, e_2, ..., e_{k+1}, as desired, and our proof is complete. ∎

KRUSKAL'S ALGORITHM

Our other approach to finding a minimum spanning tree is simpler to explain but harder to implement. We build a sequence of forests, adding one edge at a time, starting with the "empty forest" (containing no vertices or edges) and ending with a spanning forest (which, since G is connected, must be a spanning tree). At each stage we add an edge of minimum weight that still gives us a forest, that is, an edge of minimum weight that does not produce a cycle when added to the forest we already have.

The pseudocode looks very much like that for Prim's algorithm.

procedure *kruskal*(G : connected weighted graph)
  {assume that the vertices of G are labeled $1, 2, \ldots, n$; $w(i, j)$ denotes the weight of edge ij; a forest will be built one edge at a time; it is denoted F and is thought of as a set of edges}
    $F \leftarrow$ the empty tree
    for $k \leftarrow 1$ **to** $n - 1$ **do**
      begin
        $min \leftarrow \infty$
        for each edge ij of G **do** {find edge to add to T}
          if (ij is not in F) and ($F \cup \{ij\}$ contains no cycle) **then**
            if $w(i, j) < min$ **then**
              begin
                $\{i\_best, j\_best\} \leftarrow \{i, j\}$
                $min \leftarrow w(i, j)$
              end
        $F \leftarrow F \cup \{\{i\_best, j\_best\}\}$
      end
    return(F) {a minimum spanning tree}

Algorithm 2. Kruskal's algorithm for finding a minimum spanning tree.

EXAMPLE 4. Use Kruskal's algorithm to find a minimum spanning tree of the weighted graph shown in Figure 10.6.

Solution. The minimum weight edge is cd, so the first forest consists of just this edge. The next smallest weight edge is ae, so the forest after two passes through the outer loop consists of the two edges cd and ae. Note that F is not a tree at this point. Now there are two edges that have weight 3, and they are the edges of minimum weight that are not currently in F: af and ef. Neither will create a cycle when added to F. We choose af to include, so the forest now consists of cd, ae, and af. We cannot choose edge ef at the next step, since adding it to the current F would create a cycle. Therefore, we add the next least costly edge, de. Note that at this point F has become a tree again. Finally, we add edge ab (or be, since they have the same weight), obtaining the same minimum spanning tree as we found in Example 3. ◆

The complexity of Kruskal's algorithm is quite hard to analyze. First, note that we can speed up the process quite a bit by presorting the edges by weight. We only need to go through this sorted list of edges once. When we come to an edge, we either throw it into the forest F (because it is the minimum weight edge that does not form a cycle with what we already have) or else we can forget about it entirely (because it forms a cycle with what is already in F and hence will never be used in our spanning tree). Therefore in effect the statements within the inner loop of Algorithm 2 are executed only e times, where e is the number of edges of G. The problem with implementing Kruskal's algorithm is that it is hard to determine whether "$F \cup \{ij\}$ contains no cycle" when we are considering adding edge ij to the forest. It turns out that efficient algorithms for doing so are complicated and take slightly more on the average than the $O(1)$ time required to perform the same test in Prim's algorithm. Even so, as with Prim's algorithm, the best implementation currently known of Kruskal's algorithm uses $O(e \log n)$ steps.

Finally the proof that Kruskal's algorithm works is similar to the proof for Prim's algorithm. Thus the proof of the following theorem is left to the reader (Exercise 16).

THEOREM 2. *Kruskal's algorithm produces a spanning tree of G having minimum weight.*

SUMMARY OF DEFINITIONS

cost of a spanning tree: the weight of the spanning tree.

greedy algorithm: an algorithm that tries to achieve a global optimum by performing a sequence of steps, each of which is locally optimum (e.g., Kruskal's and Prim's algorithms).

Kruskal's algorithm: the algorithm for finding a minimum spanning tree in a connected weighted graph that builds up the tree by repeatedly selecting the minimum cost edge whose addition does not create a cycle.

minimum spanning tree in the connected weighted graph G: a spanning tree of G with smallest weight.

Prim's algorithm: the algorithm for finding a minimum spanning tree in a connected weighted graph that builds up the tree by repeatedly selecting the minimum cost edge incident to a vertex already in the tree whose addition does not create a cycle.

weight of a spanning tree: the sum of the weights of the edges in the tree.

EXERCISES

1. Suppose that a bank has its central computer at the origin of a hypothetical Cartesian coordinate system, with ATMs at the following locations: $(11, 11)$, $(11, -11)$, $(21, 20)$, $(28, 33)$, $(28, -22)$, $(30, -13)$, $(36, -27)$, $(41, -7)$, $(42, 32)$, and $(49, 0)$.
   (a) Plot the central computer and the ATMs accurately on graph paper.
   (b) Find a minimum spanning tree visually. [*Hint*: Keep Prim's or Kruskal's algorithm in mind while doing this.]

2. Apply Prim's algorithm to find a minimum spanning tree in the following weighted graphs.
   (a) Figure 10.1.
   (b) Figure 10.3.
   (c) The graph in Exercise 11 in Section 10.1.
   (d) The graph in Exercise 18 in Section 10.1.

3. Apply Kruskal's algorithm to find a minimum spanning tree in the weighted graphs listed in Exercise 2.

4. Let e be an edge of a connected weighted graph G whose weight is less than or equal to the weight of every other edge of G.
   (a) Show that G contains a minimum spanning tree containing e.
   (b) Is it necessarily the case that every minimum spanning tree of G contains e?

5. Let v be a vertex in a connected weighted graph G, and let e be an edge incident to v whose weight is less than or equal to the weight of every other edge incident to v.
   (a) Show that G contains a minimum spanning tree containing e.
   (b) Is it necessarily the case that every minimum spanning tree of G contains e?

6. Consider the following special case of the **knapsack problem**: Given a finite sequence $s_1, s_2, \ldots, s_n$ of positive integers, together with a positive integer g,

find the subsequence whose sum is the closest to g without exceeding g. (Intuitively think of g as the capacity of a knapsack and $s_1, s_2, \ldots, s_n$ as the sizes of objects that you want to put into the knapsack. The problem is to put as much as possible into the knapsack.) Consider the following greedy algorithm for solving this problem. Order the numbers in the sequence, so that without loss of generality we can assume that $s_1 \geq s_2 \geq \cdots \geq s_n$. Initialize $c \leftarrow g$ and $i \leftarrow 1$. Then carry out the following loop.

> **while** $i \leq n \wedge c > 0$ **do**
>   **begin**
>   **if** $s_i \leq c$ **then**
>   **begin**
>   put s_i into the subsequence
> $c \leftarrow c - s_i$
>   **end**
> $i \leftarrow i + 1$
>   **end**

   (a)   Apply this greedy algorithm to the sequence $10, 7, 7, 4, 3, 2$ with $g = 20$, and determine whether it gives the correct answer.

   (b)   Apply this greedy algorithm to the sequence $10, 7, 7, 4, 3, 2$ with $g = 15$, and determine whether it gives the correct answer.

7. This exercise deals with another example of a greedy algorithm that fails to find the optimum solution.

   (a)   Describe a greedy algorithm for finding a minimum weight Hamilton cycle in a complete weighted graph. (Your algorithm must produce a Hamilton cycle.)

   (b)   Apply your algorithm from part (a) to the graph shown in Figure 10.3, starting at vertex a and resolving all ties in favor of the vertex occurring first alphabetically. Does your algorithm produce a minimum weight cycle?

8. Explain how an algorithm for finding a minimum spanning tree in a connected weighted graph can be extended to an algorithm for finding a minimum spanning forest (defined in the obvious way) in an arbitrary weighted graph.

9. Rewrite Kruskal's algorithm, incorporating the idea of presorting the edges and going through the list of edges only once.

10. Consider the following algorithm for finding a minimum spanning tree. Initialize T to be the given connected weighted graph. If T is a tree, then we are done. If not, delete the edge of greatest weight that is in a cycle (equivalently, the edge of greatest weight whose deletion will not disconnect the graph), and call the result T. Repeat this process until T is a tree. Apply this algorithm to the weighted graph shown in Figure 10.6.

11. Sometimes an application might call for a maximum weight spanning tree.
    (a) Explain how to use the ideas of this section to find a maximum weight spanning tree of a given connected weighted graph.
    (b) Find a maximum weight spanning tree for the weighted graph shown in Figure 10.6.

12. Let G be a connected weighted graph, and let E_0 be a set of edges of G that do not form any cycles.
    (a) Explain how to find a spanning tree of G that has minimum weight among all those spanning trees that contain all the edges in E_0.
    (b) Find a spanning tree of the weighted graph in Figure 10.6 of minimum weight subject to the condition that the tree contains edges bc and be.

Challenging Exercises

13. Prove that the algorithm given in Exercise 10 is correct.

14. Suppose that a connected weighted graph G has an edge e whose weight is strictly less than the weight of every other edge of G.
    (a) Show that every minimum spanning tree of G must contain e.
    (b) Change the hypothesis so that the weight of e is just strictly less than the weight of every other edge incident to either endpoint of e. Either prove that the same conclusion follows, or else provide a counterexample.

15. Show that if the weights of the edges in a connected weighted graph G are all distinct, then G has a unique minimum spanning tree.

16. Prove Theorem 2. [*Hint*: The proof is very similar to the proof of Theorem 1, perhaps a little easier.]

17. Give a proof of Theorem 1 along the following lines. Prove by induction on k that the tree constructed after the kth pass through the outer loop of Prim's algorithm is contained in some minimum spanning tree of G.

Exploratory Exercises

18. Implement Prim's algorithm, making it as efficient as you can. Apply it to some large connected weighted graphs.

19. Find a good way to keep track of the components in the forest F that is constructed in Kruskal's algorithm to enable the algorithm to determine quickly whether adding edge ij would create a cycle.

20. Implement Kruskal's algorithm, making it as efficient as you can. You will need to make use of the ideas in Exercise 19. Apply it to some large connected weighted graphs.

21. Consult Graham and Hell [125] for a history of the development of minimum spanning tree algorithms. (Prim and Kruskal did not originate the ideas in the algorithms that bear their names.)

22. Consult Even [78] to find out about another algorithm for finding minimum spanning trees known as Sollin's algorithm.

23. In practice, the spanning trees that form the communications network for a bank's ATMs must satisfy certain constraints. For example, each "branch" of the tree emanating from the root (the central office) cannot contain too many vertices. Consult Elias and Ferguson [76] for approaches to dealing with this harder problem.

24. A generalization of the greedy approach to finding optima is based on the annealing process in chemistry. Consult Kirkpatrick, Gelett, and Vecchi [172] for information on this approach.

SECTION 10.3
FLOWS

Less is more.

—Ludwig Mies van der Rohe

Much of the activity of the business world involves the flow of things from one place to another, whether it is oil through a pipeline, automobiles along freeways, information via satellite, or work through an office. When designing or modifying the network along which these flows take place, we want to be able to determine the behavior of the flow under various constraints that are imposed on the network. In this section we model situations like these with a weighted directed graph, and we develop an algorithm for finding the maximum flow that can occur in such a network.

Let G be a directed graph without loops that has a nonnegative real number, called the **capacity**, assigned to each edge. The capacity of the edge ij will be denoted C_{ij}. As the name suggests, we will think of the capacity of an edge as representing the amount of a commodity that can flow through that edge per unit of time (thousands of cars per hour, cubic meters of water per minute, barrels of oil per day, etc.). Furthermore, we can assume that C_{ij} is defined for all pairs of vertices i and j by setting $C_{ij} = 0$ if there is no edge from i to j. (Conversely, we can assume that any edge with capacity 0 is not there, since none of the commodity can flow through it.) We assume that G has a distinguished vertex s whose in-degree is 0; this vertex is called the **source**. We assume also that G has a distinguished vertex t (different from s) whose out-degree is 0; this vertex is called the **sink**. The vertices other than the source and sink are called **interior vertices**. This entire structure—digraph with capacities, source, and sink—is called a **capacitated network**.

We are interested in the amount of the commodity that can "flow" from the source to the sink in a capacitated network. More precisely, a **flow** through G is a function F from the set of edges of G to the set of nonnegative real numbers, satisfying certain conditions. The value of F at edge ij is denoted F_{ij} and called the **flow** through ij. We think of F as being measured in the same units as capacities (such as cubic meters per minute). The conditions on F are that

$$0 \le F_{ij} \le C_{ij} \qquad \text{for each edge } ij$$

and

$$\sum_i F_{ik} = \sum_j F_{kj} \qquad \text{for all } k \ne s \text{ or } t.$$

Intuitively, the first condition simply says that the flow through an edge cannot exceed the capacity of that edge. Indeed, if $F_{ij} = C_{ij}$, we say that edge ij is **saturated**. The second condition says that the total flow into each interior vertex must equal the total flow out of that vertex: The **net flow** into (or out of) each interior vertex is 0. In other words, the flow goes *from* the source *to* the sink; it cannot pile up at an interior vertex.

As you might expect (and are asked to show in Exercise 4), the total flow out of the source, $\sum_j F_{sj}$, must equal the total flow into the sink $\sum_i F_{it}$, since there is nowhere else for it to go. This common value is called the **value** of the flow F. We will draw a picture of a capacitated network by writing the capacity next to each edge. Flows are shown by writing the flow through each edge immediately following the capacity and separated from it by a comma: C_{ij}, F_{ij}.

EXAMPLE 1. Figure 10.9 represents a flow through a capacitated network. It might represent water flowing through pipes (the edges are the pipes, the vertices are connections between pipes or pumping stations to help the flow along), or it might represent traffic flowing along one-way streets (the edges are the streets, the vertices are the intersections). The numbers are assumed to be in whatever units are appropriate for the application.

To check that this is a flow, we need to check that $0 \le F_{ij} \le C_{ij}$ for each edge, and that the net flow at each interior vertex equals 0. The former is easy to do by inspection: On each of the five edges, the first number, which represents the capacity, is greater than or equal to the second, which represents the flow. At the interior vertex u there is a flow of 4 in and a flow of $2 + 2 = 4$ out, so the net flow there is 0. Similarly, at vertex v there is a flow of $0 + 2 = 2$ in and 2 out, again a net flow of 0. The value of this flow is 4, since there are 4 units of flow out of s and also $2 + 2 = 4$ units of flow into t. Intuitively, we might imagine, say, 4 thousand gallons of water per minute going through a wide pipe su; at vertex u half the water continues through pipe ut and half continues through pipe uv. The latter then goes on through pipe vt to reach the sink.

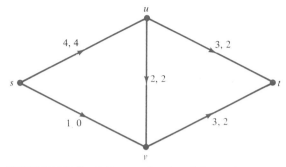

FIGURE 10.9 A flow with value 4 through a capacitated network.

Note that this flow is not the largest possible flow for this network. If we increased the flow in edge sv from 0 to 1, and increased the flow in edge vt from 2 to 3, we would have a flow with value 5. ◆

In general, a capacitated network will have many flows. It will always have the **zero flow**, in which each $F_{ij} = 0$. The problem we wish to solve is the following: *Given a capacitated network G, find a flow that has the greatest possible value*; such a flow, or its value, is called a **maximum flow** for the network. To solve this problem means to find an algorithm that produces the desired flow as output, given the network as input.

THE FORD–FULKERSON ALGORITHM

The elegant but fairly involved algorithm we will discuss was discovered in the 1950s by L. Ford and D. Fulkerson. Before going into the details of the **Ford–Fulkerson algorithm**, we outline its basic idea. The algorithm is a greedy one, but it works only little by little. We start with a flow (the zero flow, for instance), and then if possible **augment** it to a flow with a larger value. We repeat this augmentation process as long as we can; when no further augmentation is possible, we have a maximum flow.

Each augmentation operation consists of finding a certain path from the source to the sink in the underlying *undirected* graph (some of the edges in this path may be traversed backwards). Finding this path, called an **augmentation path**, is the hard part of the algorithm. The edges along the augmentation path that are directed in the direction of the path will have excess capacity not being used by the current flow, so we can increase the flow through these edges. Edges along the augmentation path that are directed in the direction opposite to the path will have some positive flow, so we can decrease the flow through these edges, in effect diverting some of the flow through these edges to other edges. More specifically, we find an amount $\Delta > 0$, and we increase the flow by Δ in each edge of the augmentation path that is directed in the direction of the path, and we decrease the flow by Δ in each edge of the augmentation path that is directed opposite to the direction of the path. The effect of these changes is to leave the net flow unchanged at each interior vertex, and to increase the flow out of the source (and into the sink) by Δ. If we are unable to construct an augmentation path, then we know (because of Theorem 2 below) that the current flow is maximum.

The augmentation path is constructed through a process of labeling the vertices of G, somewhat like the labeling process used in Dijkstra's algorithm (Section 10.1). The label at a vertex v is a pair (Δ, u), where Δ, which is always strictly positive, tells us how much more of the commodity could reach v, and u tells us which vertex it would come from. We start by labeling the source $(\infty, -)$: Any amount of the commodity would be more than willing to leave the source if it can find a place to go, and there is no vertex that it "comes from." The labeling process continues in a prescribed manner until we have labeled the sink, if that is possible. The augmentation path is then the reverse of the path obtained by following the second coordinates of the labels from the sink back to the source. The amount Δ by which we can augment the flow along this path is the first coordinate of the label on the sink. We modify the flow along this path and start over, trying to find another augmentation path. We repeat this process as long as possible. When the labeling process terminates without reaching the sink, we have a maximum flow.

In Example 1 above, the path s, v, t would be the augmentation path found by the algorithm when presented with the flow shown in Figure 10.9. The flow can be augmented by 1 unit along this path, yielding a larger flow.

It is time for the details of the algorithm. We present it in pseudocode, with many of the steps given in English. The procedure keeps track of a set L of the vertices labeled but not yet fully processed.

procedure *max_flow* (G : capacitated network)
  {assume that the vertices of G are $1, 2, \ldots, n$, two of which are the source s
  and the sink t; C_{ij} is the capacity of edge ij; and F_{ij}, which may change
  during each pass through the main loop, is the flow through edge ij}
  for $i \leftarrow 1$ **to** n **do**
    for $j \leftarrow 1$ **to** n **do**
      $F_{ij} \leftarrow 0$ {initial flow is zero}
  finished $\leftarrow$ **false**
  while not *finished* **do**
    begin
      label $s : (\infty, -)$
      $L \leftarrow \{s\}$
      while $L \neq \emptyset$ and $t \notin L$ **do** {find augmentation path}
        begin {labeling phase}
          $v \leftarrow$ a vertex in L (say the first one numerically)

{suppose that the label of v is (Δ, u); thus Δ is
the amount of extra flow that can reach v}
remove v from L
for each unlabeled vertex w in G **do**
    if vw is an edge of G with $F_{vw} < C_{vw}$ **then**
        begin
          $\Delta' \leftarrow \min(\Delta, C_{vw} - F_{vw})$
          {Δ' is necessarily greater than 0}
          label $w : (\Delta', v)$
          put w into L
        end
    else if wv is an edge of G with $F_{wv} > 0$ **then**
        begin
          $\Delta' \leftarrow \min(\Delta, F_{wv})$
          {Δ' is necessarily greater than 0}
          label $w : (\Delta', v)$
          put w into L
        end
end {of labeleing phase}
if $L = \emptyset$ **then** *finished* $\leftarrow$ **true**
else
    begin {augmenting phase}
      $\Delta \leftarrow$ first coordinate of label of t
      by following second coordinates of labels, starting
        at t, determine a path (in the underlying
        undirected graph) back to s;
      for each edge ij in this path **do**
        if ij is oriented in the "s to t" direction **then**
          $F_{ij} \leftarrow F_{ij} + \Delta$
        else
          $F_{ij} \leftarrow F_{ij} - \Delta$
    end {of augmenting phase}
end {of outer **while** loop}
return(F) {a maximum flow in G}

Algorithm 1. Ford–Fulkerson algorithm for finding a maximum flow in a capacitated network.

Notice the overall structure of the procedure. There are two phases to the algorithm (after some initialization), and the two phases are repeated until a maximum flow is obtained (the **while** not *finished* loop). The first phase (the inner **while** loop) is the labeling phase, which enables us to find an augmentation path. The second (augmenting) phase actually finds the path and augments the flow along it.

Let us see why the label (Δ', v) that we put on vertex w during the labeling phase tells us, as we claimed above, that the flow into w can be increased by an amount Δ', coming from v. As we process vertex v, the vertices adjacent to v may be labeled in the inside **for** loop of the labeling phase of this procedure. We know at that point that the flow into v can be increased by an amount Δ. If there is an edge from v to w, and if that edge is not saturated (has some unused capacity, i.e., if $C_{vw} - F_{vw}$ is a strictly positive number), then at least some of the extra flow that reached v can continue on to w. The amount that can reach w is the smaller of the amount that could reach v (namely, Δ) and the extra space available in edge vw (namely, $C_{vw} - F_{vw}$). Therefore, in this case the label assigned by the algorithm to w is correct. On the other hand, we can also get more flow into w if there is an edge from w to v that currently has a nonzero flow in it, since we can decrease the flow through wv. One way to see this is to think of the flow through wv as being a "negative flow" from v to w, which can be offset by an actual (positive) flow from v to w. Some of the extra flow Δ available at v can thus be added to this "negative flow"; in effect, there is "excess capacity" in the "negative edge" vw, since the "negative flow" can be increased from $-F_{wv}$ to 0. Therefore, in this case as well, the label assigned to w is correct.

Let us look at two examples. In the first we will go through just one pass of the algorithm, starting from a given flow, labeling the vertices, finding an augmentation path, and augmenting the flow accordingly. In the second example we will follow the algorithm through as many passes as it needs to find the maximum flow.

EXAMPLE 2. Apply the labeling procedure to find an augmentation path for the flow shown in Figure 10.10, and find the new, larger flow. (For convenience we use letters rather than numbers to name the vertices; we use alphabetical order to determine which vertex to process next if there is a choice.)

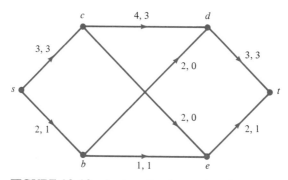

FIGURE 10.10 A given flow in a network.

Solution. [The reader should redraw the picture and write the labels next to the vertices as we find them.] We first label the source $s : (\infty, -)$, and put s into L. Since s is the only vertex in L, we remove it from L and look at the neighbors of s. There are two, both the head of an edge from s. Edge sc is saturated—has no unused capacity—so we do nothing with vertex c yet. There is $2 - 1 = 1$ unit of excess capacity in edge sb, however, so we label vertex $b : (1, s)$ and put b into L. The 1 in this label is the minimum of ∞ (the extra flow that could reach s) and 1 (the extra flow that could go through edge sb). This completes the processing of s.

At this point, L contains only the vertex b. We remove it from L and process it. There are three edges involving b, but edge sb is not looked at, since s is already labeled. Edge be has no excess capacity, so we do nothing with vertex e yet. Edge bd has an excess capacity of $2 - 0 = 2$ units, so all of the extra 1 unit of flow that could reach b can flow on to d through this edge if it wants to. Therefore, we label $d : (1, b)$, and put d into L. Again L has only one vertex in it, namely d, so we remove d from L and process d. Edge bd comes from an already labeled vertex, and edge dt is already saturated. There is edge cd, however, from c to d, that currently has 3 units of flow. We think of this as -3 units of flow from d to c. This flow can be increased by up to 3 units, from -3 to 0, by decreasing the flow from c to d (it cannot be increased beyond that since the flow from c to d cannot be negative). Now only 1 unit of flow can reach d (since the label of d is $(1, b)$), so, in fact, the amount of extra flow that can reach c is $\min(1, 3) = 1$. Therefore, we label $c : (1, d)$, and add c to L.

We next remove c from L and process c. It should be easy to see by now what this means: e gets the label $(1, c)$, since there is extra capacity in ce which can carry the extra flow that reached c, and e is put into L. Finally, we process e. Since there is enough extra capacity in et, we label $t : (1, e)$. The sink has now been labeled, so we have completed the labeling phase of this pass of the algorithm.

Next we need to find the augmentation path. We follow the second coordinates of the labels, starting with the label of t, and we are carried first to e, then to c, then to d, then to b, and finally to s. (It is just an accident that the augmentation path used all the vertices in the digraph.) Therefore, the augmentation path is s, b, d, c, e, t. We show it, with the flows currently in each of those edges, in Figure 10.11.

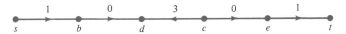

FIGURE 10.11 Augmentation path with the old flows.

The edges sb, bd, ce, and et are oriented in the "s to t" direction, so we *increase* the flow in each of these edges by $\Delta = 1$. The edge cd is oriented oppositely, so we *decrease* the flow in this edge by $\Delta = 1$ (or, equivalently, we increase the flow in the negative edge dc by 1, from -3 to -2). The new flows are shown in Figure 10.12.

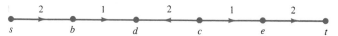

FIGURE 10.12   Augmentation path with the new flows.

Now the crucial point is that after these changes have been made, *we still have a flow*. This is because the net flow (the flow in minus the flow out) at each interior vertex has not changed: The flow in increased by Δ at each interior vertex, and the flow out also increased by Δ (provided that we think of the flow along the edge cd as a negative flow along the negative edge dc). The flow out of the source (and the flow into the sink) also increased by Δ, and that was our goal. Thus the value of the flow has increased from 4 to 5; see Figure 10.13 for the new flow.

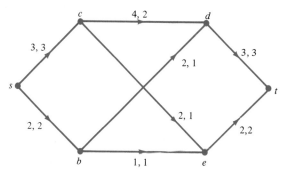

FIGURE 10.13   Increased flow in the network shown in Figure 10.10.

It is clear that the flow in Figure 10.13 is maximum, since all the edges leading out of the source are filled to capacity (as are all the edges leading into the sink). Indeed, if we were to try to apply another pass of the algorithm, no vertex other than s would be labeled, so the algorithm would terminate. As we will soon see, it is not always the case that the edges out of s or into t need to be saturated for a flow to be maximum.   ◆

EXAMPLE 3.   Use Algorithm 1 to find the maximum flow through the capacitated network shown in Figure 10.14. We show the zero flow; ignore temporarily the vertex labels (and the fact that some of the lines are heavier than others).

Solution.   We start with the zero flow. The reader should verify that one pass through the algorithm gives us the augmentation path s, b, d, t, which can be augmented by 2 units of flow (remember to use alphabetical order to decide which vertex to process next if L has more than one vertex in it). Figure 10.14 shows the labels and the

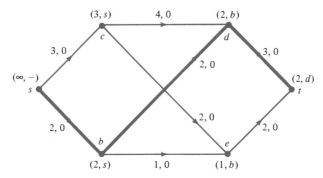

FIGURE 10.14   Algorithm 1 applied to the zero flow.

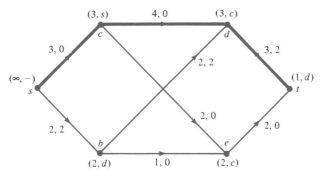

FIGURE 10.15   Algorithm 1 applied to the flow obtained after one pass.

path. Therefore, after one pass the flow is as shown in Figure 10.15 (again ignore the vertex labels for a moment).

Next the procedure starts over on this flow, finding the augmentation path s, c, d, t with 1 unit of excess capacity (see Figure 10.15). After making the augmentation, we have the flow shown in Figure 10.16. One more pass produces the flow shown in Figure 10.17. A final pass through the procedure results only in vertex s being labeled. Since t was not labeled, we are finished, and Figure 10.17 shows a maximum flow. Note that this is not the same flow as we obtained working with the same capacitated network in Example 2 (Figure 10.13), although both these maximum flows necessarily have the same value, 5. In other words, the flow that achieves the maximum value is not always unique.   ◆

The Ford–Fulkerson algorithm is too complicated for us to analyze its efficiency here. Indeed, it is not even immediately clear why the algorithm ever terminates; it might seem that we could go on increasing the flow forever. It turns out that, with suitable technical assumptions and modifications (e.g., that the ca-

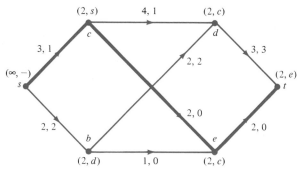

FIGURE 10.16 Algorithm 1 applied to the flow obtained after two passes.

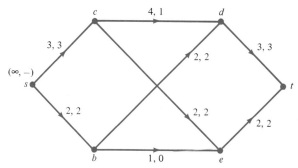

FIGURE 10.17 Algorithm 1 applied to the flow obtained after three passes.

pacities not be irrational numbers, or that the set L be processed as a queue—first in, first out), the Ford–Fulkerson algorithm will terminate in $O(n^5)$ steps. Let us give a proof that the algorithm terminates if the capacities are integers.

THEOREM 1. *Let G be a capacitated network in which the capacities are natural numbers. Then Algorithm 1 terminates, and the value of the flow it produces is a natural number.*

Proof. The algorithm begins with the flow whose value is 0, and each pass through the outermost **while** loop (the one controlled by *finished*) produces a strictly larger flow than the previous flow. Let us prove by induction that the flow $\{F_{ij}\}$ after each pass through this loop has only integer values. This statement is certainly true for the base case (no passes), since initially each $F_{ij} = 0$. We assume that the flow has an integer value (the inductive hypothesis) and must show that the flow obtained after one more iteration of this loop again has an integer value.

The amount by which the flow is augmented, Δ, is obtained by applying the min function to numbers of the form $C_{vw} - F_{vw}$ and F_{wv}. Since all of these numbers are integers (by the inductive hypothesis and the assumption that the capacities are natural numbers), Δ is an integer. Therefore, the value of the new flow is the sum of two integers, which is an integer. This completes the proof that the algorithm produces only flows whose values are natural numbers.

To show that the algorithm terminates, we first note that each pass through the outer **while** loop must terminate, because after at most n passes through the inner **while** loop (the one controlled by the contents of the set L), each vertex that was put into L will have been removed from L (and processed), so L will be empty. The value of the flow found by the algorithm can never exceed the sum of the capacities of all the edges leading out of the source, so there is a finite integer B, which bounds the value of the flow. Now on each pass through the outer **while** loop, the flow increases by an amount $\Delta > 0$; since Δ is an integer, the flow increases by at least 1. Therefore, the algorithm must terminate after at most B passes through this loop. ∎

THE MAXFLOW–MINCUT THEOREM

To give an indication of why the Ford–Fulkerson algorithm necessarily finds a maximum flow, we need to consider one more concept. Let G be a capacitated network with source s and sink t. A **cut** in G is a set C of edges whose removal will leave the digraph with no directed path from s to t. The **value** of a cut C is the sum of the capacities of the edges in the cut. A **minimum cut** is a cut whose value is the least among all cuts of G, or it is the value of such a cut.

EXAMPLE 4.   Let G be the capacitated network shown in Figure 10.14. The set $\{sc, sb\}$ is a cut, since if these edges are removed, there is no path from s to t (indeed, there is no path from s to any other vertex). In particular, there can be no flow from s to t if these edges are removed. The value of this cut is $3 + 2 = 5$. Similarly, the set of edges incident to t forms a cut, $\{dt, et\}$, and it also happens to have value 5. The cut $\{sc, bd, be\}$ has value $3 + 2 + 1 = 6$.   ◆

Note that G must have at least one cut, since if we remove *all* the edges, then certainly there will be no path from s to t. Furthermore, there are only a finite number of cuts (since there are only 2^e subsets of edges, where e is the number of edges), and every finite set of numbers has a minimum, so a minimum cut exists.

It is not at all obvious how to go about finding efficiently a minimum cut among the 2^e different subsets of the edge set of G. Remarkably, the Ford–Fulkerson algorithm for finding maximum flows also finds minimum cuts. First let us observe that *the value of any flow can be no greater than the value of any cut*, since by reducing the flow in each edge of a cut to 0 we totally stop the flow. It follows that *the value of a maximum flow can be no greater than the value of a minimum cut*. What is less clear is that there is a flow F and a cut C such that the value of F equals the value of C. Granting for a moment the truth of this statement (which we are about to prove), it follows that *the value of the maximum flow is equal to the value of the minimum cut*.

Let us see why the Ford–Fulkerson algorithm finds a minimum cut and a maximum flow. Suppose that we apply the algorithm to G. Let V denote the set of vertices of G. When the algorithm terminates (with F as the final, nonaugmentable flow), the set $S \subseteq V$ of vertices labeled during the final pass will be nonempty, since $s \in S$. Furthermore, if we denote by T the set $V - S$ of unlabeled vertices, then $T \neq \emptyset$, since $t \in T$ (if $t \in S$, this would not have been the final pass). The set of edges $C = \{\, uv \mid u \in S \wedge v \in T \,\}$ is clearly a cut for G (any path from s to t must at some point cross the boundary from S, where s lives, into T, where t lives). See the schematic diagram in Figure 10.18.

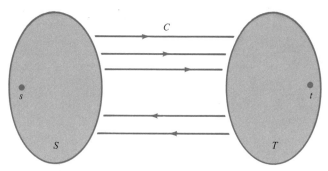

FIGURE 10.18   Schematic diagram of the cut determined by Algorithm 1.

Now the value of the cut C is the sum of the capacities of the edges of C. Since none of the vertices in T is labeled, it must be the case that the flow F fills each edge in the cut to capacity. Furthermore, each edge from T to S must have a flow of 0 in it, again because if any such edge had a positive flow, then the endpoint of that edge in T would have been labeled. Therefore, the value of the flow F must be the sum of the capacities of the edges in C, because that much flow leaves S for T and none comes back from T to S. Thus we have shown that the value of the cut C equals the value of the flow F. By the argument above, we conclude that F is a maximum flow and C is a minimum cut.

Combining this argument with the result of Theorem 1, we see that the Ford–Fulkerson algorithm works, and that it finds not only a maximum flow but also a minimum cut. We summarize the result we have just proved. (The technical assumption that the capacities are integers is not really a confining one; see Exercise 20.)

THEOREM 2.   *In a capacitated network with capacities that are natural numbers, the maximum value of a flow is equal to the minimum value of a cut. Algorithm 1* *produces a maximum flow, and the set of edges from labeled vertices to unlabeled vertices when the algorithm terminates is a minimum cut.*

EXAMPLE 5. In Example 3, when the algorithm terminated, only vertex s was labeled. Therefore, the minimum cut determined by the algorithm was $\{sc, sb\}$—the set of edges from labeled vertices (s) to unlabeled vertices (all the others). Its value, 5, is the same as the value of the maximum flow. ◆

The exercises include some useful variations of capacitated networks in which we can also find maximum flows.

SUMMARY OF DEFINITIONS

augment a flow through a capacitated network: to increase by a fixed positive amount the flow through each forward-oriented edge of an augmentation path from the source to the sink, and to decrease by the same amount the flow through each backward-oriented edge of this path.

augmentation path for a flow through a capacitated network: a path from the source to the sink in the underlying undirected graph in which no edge oriented in the direction of the path is saturated and no edge oriented in the opposite direction has zero flow; the amount of the augmentation is the smallest of the excess capacities ($C_{ij} - F_{ij}$ for the forward-oriented edges and F_{ij} for the backward-oriented edges).

capacitated network: a directed graph with nonnegative weights (capacities) on the edges, together with a unique source and a unique sink.

capacity of an edge: a nonnegative real number representing the amount of a commodity that can flow through the edge.

cut in a capacitated network: a set of edges whose removal will leave the digraph with no directed path from the source to the sink.

flow through an edge in a capacitated network: a nonnegative real number that does not exceed the capacity of the edge.

flow through a capacitated network: a flow through each edge of the network, such that the total flow into each interior vertex equals the total flow out of that vertex.

Ford–Fulkerson algorithm: a procedure for finding a maximum flow through a capacitated network by successively finding augmentation paths, using a labeling process (Algorithm 1).

interior vertex in a capacitated network: a vertex other than the source and the sink.

maximum flow through a capacitated network: a flow with the largest possible value (or the value of this flow).

minimum cut in a capacitated network: a cut with the smallest possible value (or the value of this cut).

net flow into an edge: the flow into the edge minus the flow out of the edge.

saturated edge in a capacitated network with a flow: an edge for which the flow equals the capacity.

sink in a capacitated network: the vertex with out-degree 0, which absorbs the flow.

source in a capacitated network: the vertex with in-degree 0, from which the flow emanates.

value of a cut: the sum of the capacities of the edges in the cut.

value of a flow: the flow out of the source (which equals the flow into the sink).

zero flow: the flow through a capacitated network in which the flow through each edge is 0.

EXERCISES

1. Find five different flows through the capacitated network shown in Figure 10.9, other than the flow shown there. Use only integers for the numbers F_{ij} in your flows.

2. Consider the capacitated network shown in Figure 10.9.
   (a) Draw the zero flow in this network.
   (b) Draw the flow obtained from the zero flow after augmenting it by 1 along the path s, u, v, t.

3. Find the flow obtained from the flow shown in Figure 10.9 after augmenting it by 1 along the path s, v, u, t.

4. Give an algebraic proof that for any flow in a capacitated network with source s and sink t, $\sum_j F_{sj} = \sum_i F_{it}$. [*Hint:* Start with the fact that $\sum_k \sum_j F_{kj} = \sum_k \sum_i F_{ik}$.]

5. Give an example of a maximum flow in a network such that not all of the edges from the source are saturated and not all of the edges to the sink are saturated.

6. Try to find as large a flow as possible by trial and error in the given network.

7. Try to find as large a flow as possible by trial and error in the given network.

8. Apply the Ford–Fulkerson algorithm to the flow you obtained in Exercise 6, to see if it is a maximum flow. If it is not, carry out the algorithm as many passes as necessary to obtain a maximum flow.

9. Apply the Ford–Fulkerson algorithm to the flow you obtained in Exercise 7, to see if it is a maximum flow. If it is not, carry out the algorithm as many passes as necessary to obtain a maximum flow.

10. Apply the Ford–Fulkerson algorithm to the network in Exercise 6, starting from the zero flow. Observe whether it is the same maximum flow obtained in Exercise 8.

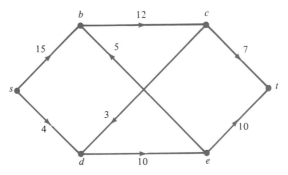

EXERCISE 6

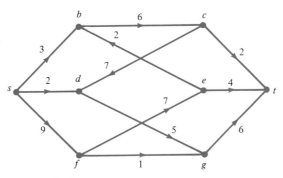

EXERCISE 7

11. Apply the Ford–Fulkerson algorithm to the network in Exercise 7, starting from the zero flow. Observe whether it is the same maximum flow obtained in Exercise 9.

12. What minimum cut corresponds to the maximum flow found in each of the following problems?
    (a) Exercise 8.
    (b) Exercise 9.
    (c) Exercise 10.
    (d) Exercise 11.

13. Make the weighted graph shown in Figure 10.1 into a digraph by directing each edge from its endpoint that comes first alphabetically to its endpoint that comes last alphabetically. Find a maximum flow and a minimum cut in the resulting network (the weights are the capacities; the source is a and the sink is g).

14. Apply the Ford–Fulkerson algorithm to the network consisting of the edges sb, sd, bc, be, dc, ct, and et, with a capacity of 1 on each edge.

15. Suppose that there are several sources in a capacitated network, rather than just one. Show how to transform the network into a network with just one source so that the maximum flow in the transformed network can be used to determine the maximum flow in the original network. [*Hint*: Add a "supersource."]

16. Repeat Exercise 15, allowing for both multiple sources and multiple sinks.

17. Find the maximum flow from the set of four vertices on the far left (a, b, c, d) to the set of three vertices on the far right (j, k, l) in the digraph shown (see Exercise 16). Assume that all edges are directed from left to right.

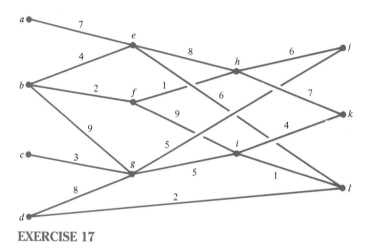

EXERCISE 17

18. Suppose that there are capacities on the vertices in a capacitated network as well as capacities on the edges (this might be the case if, for example, the vertices represent pumping stations). Show how to transform a network of this type into a network of the type we have considered in this section so that the maximum flow in the transformed network can be used to determine the maximum flow in the original network. [*Hint*: Expand each vertex into an edge.]

19. Find the maximum flow in the capacitated network shown in Exercise 7, under the assumption that the capacity of vertex d is 3, and the capacity of vertex e is 5 (see Exercise 18).

20. Show that Theorems 1 and 2 are valid if the capacities are allowed to be nonnegative rational numbers, rather than being restricted to being natural numbers.

Challenging Exercises

21. Suppose that some of the edges in a network are undirected—the commodity can flow in either direction along the edge. Show how to incorporate this possibility into the algorithm for finding maximum flows.

22. Apply the algorithm developed in Exercise 21 to find a maximum flow from a to g in the undirected graph shown in Figure 10.1 (the weights are the capacities).

23. Find a maximum flow from a to z in the undirected graph (see Exercise 21) shown in Exercise 11 in Section 10.1 (the weights are the capacities); be sure to justify that you have found a maximum flow. [*Hint*: Look for a minimum cut first.]

24. View the situation described in Exercise 22 of Section 8.3 as a problem of flow to downtown. Note that this network is undirected (see Exercise 21 of the current section). Find a flow with value 20 (measured in hundreds of cars per minute), if each vertex other than Downtown supplies 400 cars per minute.

Exploratory Exercises

25. Find out more about flows in networks by consulting Chachra, Ghare, and Moore [36], Ford and Fulkerson [86], Lawler [193], Lovász and Plummer [200], and Yao [333].

26. Flow problems in networks are related to reliability problems—how safe is the network from the breakdown of certain edges? Find out about these issues by consulting Frank and Frisch [88].

27. Find out about flows throughout an entire network (rather than with a fixed source and sink) by consulting Gomory and Hu [119].

28. In Example 11 in Section 8.1 we formulated the problem of matching jobs to workers who could perform them in terms of finding a certain set of edges in a bipartite graph.
    (a) Show how this "matching problem" can be solved by reformulating it as a network flow problem.
    (b) Apply your method to solve several large matching problems.
    (c) Find out more about this approach by consulting Lovász and Plummer [200].

REFERENCES

1. Acu, D., "Some Algorithms for the Sums of Integer Powers," *Mathematics Magazine*, **61** (1988), 189–191.

2. Aho, A. V., J. E. Hopcroft, and J. D. Ullman, *Data Structures and Algorithms*. Reading, Mass.: Addison-Wesley (1983).

3. Aho, A. V., J. E. Hopcroft, and J. D. Ullman, *The Design and Analysis of Computer Algorithms*. Reading, Mass.: Addison-Wesley (1974).

4. Albers, D. J., and G. L. Alexanderson, eds., *Mathematical People: Profiles and Interviews*. Basle, Switzerland: Birkhäuser (1985).

5. Albers, D. J., and L. A. Steen, "A Conversation with Don Knuth," *The Two-Year College Mathematics Journal*, **13** (1982), 2–18 and 128–141.

6. Alder, H. L., "Partition Identities—from Euler to the Present," *American Mathematical Monthly*, **76** (1969), 733–746.

7. Alder, H. L., "The Use of Generating Functions to Discover and Prove Partition Identities," *The Two-Year College Mathematics Journal*, **10** (1979), 318–329.

8. Andrews, G. E., *The Theory of Partitions*. Reading, Mass.: Addison-Wesley (1976).

9. Anton, H., *Elementary Linear Algebra*, 4th ed. New York: Wiley (1984).

10. Appel, K., and W. Haken, "The Four Color Problem," in *Mathematics Today: Twelve Informal Essays*, L. A. Steen, ed. New York: Springer-Verlag (1978), pp. 153–180 [reprinted in D. M. Campbell and J. C. Higgins, eds., *Mathematics: People, Problems, Results*, Vol. 2. Belmont, Calif.: Wadsworth (1984), pp. 154–173].

11. Appel, K., and W. Haken, "The Solution of the Four-Color-Map Problem," *Scientific American*, **237**(4) (Oct. 1977), 108–121.

12. Baase, S., *Computer Algorithms: Introduction to Design and Analysis*, 2nd ed. Reading, Mass.: Addison-Wesley (1988).

13. Balaban, A. T., ed., *Chemical Applications of Graph Theory*. New York: Academic Press (1976).

14. Balaban, A. T., J. W. Kennedy, and L. V. Quintas, "The Number of Alkanes Having *n* Carbons and a Longest Chain of Length *d*," *Journal of Chemical Education*, **65** (1988), 304–313.

15. Bell, E. T., *Men of Mathematics*. New York: Simon & Schuster (1937).

16. Benacerraf, P., and H. Putnam, eds., *Philosophy of Mathematics: Selected Readings*, 2nd ed. Cambridge: Cambridge University Press (1983).

17. Berlekamp, E. R., J. H. Conway, and R. K. Guy, *Winning Ways*. New York: Academic Press (1982).

18. Berman, D., "Hex Must Have a Winner: An Inductive Proof," *Mathematics Magazine*, **49** (1976), 85–86.

19. Biggs, N. L., E. K. Lloyd, and R. J. Wilson, *Graph Theory, 1736–1936*. Oxford: Clarendon Press (1976).

20. Bittinger, M. L., *Logic, Proof, and Sets*, 2nd ed. Reading, Mass.: Addison-Wesley (1982).

21. Blair, W. D., C. B. Lacampagne, and J. L. Selfridge, "Factoring Large Numbers on a Pocket Calculator," *American Mathematical Monthly*, **93** (1986), 802–808.

22. Bloom, G. S., J. W. Kennedy, and L. V. Quintas, "A Characterization of Graphs of Diameter Two," *American Mathematical Monthly*, **95** (1988), 37–38.

23. Bogart, K. P., *Introductory Combinatorics*. London: Pitman (1983).

24. Bogart, K. P., and P. G. Doyle, "Non-sexist Solution to the Ménage Problem," *American Mathematical Monthly*, **93** (1986), 514–518.

25. Bollobás, B., *Graph Theory: An Introductory Course*. New York: Springer-Verlag (1979).

26. Bollobás, B., *Random Graphs*. Orlando, Fla.: Academic Press (1985).

27. Bondy, J. A., and U. S. R. Murty, *Graph Theory with Applications*. New York: American Elsevier (1976).

28. Borwein, J. M., and P. B. Borwein, "Ramanujan and Pi," *Scientific American*, **258**(2) (Feb. 1988), 112–117.

29. Bose, R. C., and B. Manvel, *Introduction to Combinatorial Theory*. New York: Wiley (1984).

30. Brassard, G., and P. Bratley, *Algorithms: Theory and Practice*. Englewood Cliffs, N.J.: Prentice-Hall (1988).

31. Brualdi, R. A., *Introductory Combinatorics*. New York: North-Holland (1977).

32. Busacker, R. G., and T. L. Saaty, *Finite Graphs and Networks: An Introduction with Applications*. New York: McGraw-Hill (1965).

33. Campbell, D. M., "The Computation of Catalan Numbers," *Mathematics Magazine*, **57** (1984), 195–208.

34. Cartwright, M., "Crackly Phones and the Schoolgirl Problem," *New Scientist* (July 3, 1986), 36–40.

35. Castellanos, D., "The Ubiquitous Pi," *Mathematics Magazine*, **61** (1988), 67–98 and 148–163.

36. Chachra, V., P. M. Ghare, and J. M. Moore, *Applications of Graph Theory Algorithms*. New York: Elsevier North-Holland (1979).

37. Chaitin, G., "Randomness and Mathematical Proof," *Scientific American*, **232**(5) (May 1975), 47–52.

38. Charlesworth, A., "Infinite Loops in Computer Programs," *Mathematics Magazine*, **52** (1979), 284–291 [reprinted in D. M. Campbell and J. C. Higgins, eds., *Mathematics: People, Problems, Results*, Vol. 3. Belmont, Calif.: Wadsworth (1984), pp. 15–23].

39. Chartrand, G., *Introductory Graph Theory*. New York: Dover (1985) [reprint of *Graphs as Mathematical Models*. Boston: Prindle, Weber & Schmidt (1977)].

40. Chartrand, G., P. Erdős, and O. R. Oellermann, "How to Define an Irregular Graph," *College Mathematics Journal*, **19** (1988), 36–42.

41. Chartrand, G., and L. Lesniak, *Graphs and Digraphs*, 2nd ed. Belmont, Calif.: Wadsworth (1986).

42. Chartrand, G., and R. J. Wilson, "The Petersen Graph," in *Graphs and Applications*, F. Harary and J. S. Maybee, eds. New York: Wiley (1985), pp. 69–100.

43. Choe, B. R., "An Elementary Proof of $\sum_{n=1}^{\infty} 1/n^2 = \pi^2/6$," *American Mathematical Monthly*, **94** (1987), 662–663.

44. Christofides, N., *Graph Theory: An Algorithmic Approach*. New York: Academic Press (1975).

45. Chu, I. P., and R. Johnsonbaugh, "Tiling Deficient Boards with Trominoes," *Mathematics Magazine*, **59** (1986), 34–40.

46. Chung, F. R. K., F. T. Leighton, and A. L. Rosenberg, "Embedding Graphs in Books: A Layout Problem with Applications to VLSI Design," *SIAM Journal of Algebraic and Discrete Methods*, **8** (1987), 33–58.

47. Cohen, D. I. A., *Introduction to Computer Theory*. New York: Wiley (1986).

48. Cohen, P. J., and R. Hersh, "Non-Cantorian Set Theory," *Scientific American*, **217**(6) (Dec. 1967), 104–116.

49. Conway, J. H., "All Games Bright and Beautiful," *American Mathematical Monthly*, **84** (1977), 417–434.

50. Conway, J. H., "A Gamut of Game Theories," *Mathematics Magazine*, **51** (1978), 5–12.

51. Conway, J. H., *On Numbers and Games*. New York: Academic Press (1976).

52. Cormier, R. J., and R. B. Eggleton, "Counting by Correspondence," *Mathematics Magazine*, **49** (1976), 181–186.

53. Cull, P., and E. F. Ecklund, Jr., "Towers of Hanoi and Analysis of Algorithms," *American Mathematical Monthly*, **92** (1985), 407–420.

54. Dacić, R. M., "Mersenne Numbers and Binomial Coefficients," *Mathematics Magazine*, **54** (1981), 32.

55. Date, C. J., *An Introduction to Database Systems*, 4th ed., Vol. 1. Orlando, Fla.: Addison-Wesley (1986).

56. Dauben, J. W., "Georg Cantor and the Origins of Transfinite Set Theory," *Scientific American*, **248**(6) (June 1983), 122–131.

57. Dauben, J. W., *Georg Cantor—His Mathematics and Philosophy of the Infinite*. Cambridge, Mass.: Harvard University Press (1979).

58. Davis, M. "What Is a Computation?" in *Mathematics Today: Twelve Informal Essays*, L. A. Steen, ed. New York: Springer-Verlag (1978), pp. 241–267.

59. Davis, M., and R. Hersh, "Hilbert's 10th Problem," *Scientific American*, **229**(5) (Nov. 1973), 84–91 [reprinted in D. M. Campbell and J. C. Higgins, eds., *Mathematics: People, Problems, Results*, Vol. 2. Belmont, Calif.: Wadsworth (1984), pp. 136–148].

60. Davis, M. D., *Game Theory: A Nontechnical Introduction*. New York: Basic Books (1983).

61. Davis, P. J., "What Do I Know? A Study of Mathematical Self-Awareness," *College Mathematics Journal*, **16** (1985), 22–41.

62. Dawson, J. W., Jr., "Kurt Gödel in Sharper Focus," *The Mathematical Intelligencer*, **6**(4) (1984), 9–17.

63. DeLong, H., "Unsolved Problems in Arithmetic," *Scientific American*, **224**(3) (Mar. 1971), 50–60.

64. DeMillo, R. A., R. J. Lipton, and A. J. Perlis, "Social Processes and Proofs of Theorems and Programs," *The Mathematical Intelligencer*, **3** (1980), 31–40.

65. Deneen, L. L., "Secret Encryption with Public Keys," *UMAP Journal*, **8** (1987), 9–29.

66. Denning, D. E. R., *Cryptography and Data Security*. Reading, Mass.: Addison-Wesley (1982).

67. Deo, N., *Graph Theory with Applications to Engineering and Computer Science*. Englewood Cliffs, N.J.: Prentice-Hall (1974).

68. Dewdney, A. K., "Computer Recreations," *Scientific American*, **251**(5) (Nov. 1984), 19–28.

69. Dijkstra, E. W., "Programming as a Discipline of Mathematical Nature," *American Mathematical Monthly*, **81** (1974), 608–612.

70. Dixon, J. D., "Factorization and Primality Tests," *American Mathematical Monthly*, **91** (1984), 333–353.

71. Dornhoff, L. L., and F. E. Hohn, *Applied Modern Algebra*. New York: Macmillan (1978).

72. Dudley, U., "Formulas for Primes," *Mathematics Magazine*, **56** (1983), 17–22.

73. Duncan, D. R., and B. H. Litwiller, "Checkerboards and Sugar Cubes: Geometric Counting Patterns," *The Two-Year College Mathematics Journal*, **4**(2) (Spring 1973), 41–47.

74. Edwards, H. M., "Fermat's Last Theorem," *Scientific American*, **239**(4) (Oct. 1978), 104–122.

75. Eggleton, R. B., and R. K. Guy, "Catalan Strikes Again! How Likely Is a Function to Be Convex?" *Mathematics Magazine*, **61** (1988), 211–219.

76. Elias, D., and M. J. Ferguson, "Topological Design of Multipoint Teleprocessing Networks," *IEEE Transactions on Communications*, **22** (1974), 1753–1762.

77. Enderton, H. B., *A Mathematical Introduction to Logic*. New York: Academic Press (1972).

78. Even, S., *Algorithmic Combinatorics*. New York: Macmillan (1973).

79. Even, S., *Graph Algorithms*. Rockville, Md.: Computer Science Press (1979).

80. Feldman, J. A., "Programming Languages," *Scientific American*, **241**(6) (Dec. 1979), 94–117.

81. Fike, C., *Computer Evaluation of Mathematical Functions*. Englewood Cliffs, N.J.: Prentice-Hall (1968).

82. Fiorini, S., and R. J. Wilson, *Edge Colorings of Graphs*. London: Pitman (1977).

83. Fisher, J. C., E. L. Koh, and B. Grünbaum, "Diagrams Venn and How," *Mathematics Magazine*, **61** (1988), 36–40.

84. Flanders, H., "A Democratic Proof of a Combinatorial Identity," *Mathematics Magazine*, **44** (1971), 11.

85. Fletcher, P., and C. W. Patty, *Foundations of Higher Mathematics*. Boston: PWS-Kent (1988).

86. Ford, L. R., and D. R. Fulkerson, *Flows in Networks*. Princeton, N.J.: Princeton University Press (1962).

87. Fowler, P. A., "The Königsberg Bridges—250 Years Later," *American Mathematical Monthly*, **95** (1988), 42–43.

88. Frank, H., and I. T. Frisch, "Network Analysis," *Scientific American*, **223**(1) (July 1970), 94–103.

89. Freeman, J. W., "The Number of Regions Determined by a Convex Polygon," *Mathematics Magazine*, **49** (1976), 23–25.

90. Frei, G., "Leonhard Euler's Convenient Numbers," *The Mathematical Intelligencer*, **7**(3) (1985), 55–58.

91. Freund, J. E., "Round Robin Mathematics," *American Mathematical Monthly*, **63** (1956), 112–114.

92. Friedman, A. D., and P. R. Menon, *Theory and Design of Switching Circuits*. Rockville, Md.: Computer Science Press (1975).

93. Galda, K., "An Informal History of Formal Proofs: From Vigor to Rigor?" *The Two-Year College Mathematics Journal*, **12** (1981), 126–139.

94. Gale, D., and L. S. Shapley, "College Admissions and the Stability of Marriage," *American Mathematical Monthly*, **69** (1962), 9–14.

95. Gardner, M., *Knotted Doughnuts and Other Mathematical Entertainments*. New York: Freeman (1986).

96. Gardner, M., *Mathematical Carnival*. New York: Knopf (1975).

97. Gardner, M., *Mathematical Circus*. New York: Knopf (1979).

98. Gardner, M., "Mathematical Games," *Scientific American*, **210**(3) (Mar. 1964), 120–128.

99. Gardner, M., "Mathematical Games," *Scientific American*, **224**(3) (Mar. 1971), 106–109.

100. Gardner, M., "Mathematical Games," *Scientific American*, **226**(1) (Jan. 1972), 104–107.

101. Gardner, M., "Mathematical Games," *Scientific American*, **228**(6) (June 1973), 106–110.

102. Gardner, M., "Mathematical Games," *Scientific American*, **234**(6) (June 1976), 120–125.

103. Gardner, M., "Mathematical Games," *Scientific American*, **237**(2) (Aug. 1977), 120–124.

104. Gardner, M., "Mathematical Games," *Scientific American*, **237**(5) (Nov. 1977), 18–28.

105. Gardner, M., "Mathematical Games," *Scientific American*, **238**(5) (May 1978), 24–30.

106. Gardner, M., "Mathematical Games," *Scientific American*, **242**(2) (Feb. 1980), 14–22.

107. Gardner, M., "Mathematical Games," *Scientific American*, **242**(3) (Mar. 1980), 24–38.

108. Gardner, M., "Mathematical Games," *Scientific American*, **242**(5) (May 1980), 16–28.

109. Gardner, M., *The Scientific American Book of Mathematical Puzzles & Diversions*. New York: Simon & Schuster (1959).

110. Gardner, M., *The 2nd Scientific American Book of Mathematical Puzzles & Diversions*. New York: Simon & Schuster (1961).

111. Gardner, M., *Wheels, Life, and Other Mathematical Amusements*. New York: Freeman (1983).

112. Garey, M. R., and M. S. Johnson, *Computers and Intractability: A Guide to the Theory of NP-Completeness*. New York: Freeman (1979).

113. Genesereth, M. R., and N. J. Nilsson, *Logical Foundations of Artificial Intelligence*. Los Altos, Calif.: Kaufman (1987).

114. Gibbons, A., *Algorithmic Graph Theory*. Cambridge: Cambridge University Press (1985).

115. Gilbert, W. J., "Negative Based Number Systems," *Mathematics Magazine*, **52** (1979), 240–244.

116. Gödel, K., "What Is Cantor's Continuum Problem?" *American Mathematical Monthly*, **54** (1947), 515–525 [also reprinted (revised and expanded) in P. Benacerraf and H. Putnam, eds., *Philosophy of Mathematics: Selected Readings*, 2nd ed. Cambridge: Cambridge University Press (1983), pp. 470–485].

117. Golomb, S. W., "The Fifteen Billiard Balls—A Case Study in Combinatorial Problem Solving," *Mathematics Magazine*, **58** (1985), 156–159.

118. Golomb, S. W., "How to Number a Graph," in *Graph Theory and Computing*, R. C. Read, ed. New York: Academic Press (1972), pp. 23–37.

119. Gomory, R. E., and T. C. Hu, "Multi-terminal Flows in a Network," in *Studies in Graph Theory*, Part 1, D. R. Fulkerson, ed. Washington, D.C.: Mathematical Association of America (1975), pp. 172–199.

120. Goodman, A. W., "On Sets of Acquaintances and Strangers at Any Party," *American Mathematical Monthly*, **66** (1959), 778–783.

121. Goodman, N. D., "Mathematics as an Objective Science," *American Mathematical Monthly*, **86** (1979), 540–551.

122. Gould, R., *Graph Theory*. London: Benjamin/Cummings (1988).

123. Graham, R. L., "The Combinatorial Mathematics of Scheduling," *Scientific American*, **238**(3) (Mar. 1978), 124–132.

124. Graham, R. L., "Combinatorial Scheduling Theory," in *Mathematics Today: Twelve Informal Essays*, L. A. Steen, ed. New York: Springer-Verlag (1978), pp. 183–211.

125. Graham, R. L., and P. Hell, "On the History of the Minimum Spanning Tree Problem," *Annals of the History of Computing*, **7** (1985), 43–57.

126. Graham, R. L., B. L. Rothschild, and J. H. Spencer, *Ramsey Theory*. New York: Wiley (1980).

127. Grecos, A. P., "A Diagrammatic Solution to 'Instant Insanity' Problem," *Mathematics Magazine*, **44** (1971), 119–124.

128. Grossman, I., and W. Magnus, *Groups and Their Graphs*. Washington, D.C.: Mathematical Association of America (1965).

129. Grossman, J. W., and R. S. Zeitman, "An Inherently Iterative Computation of Ackermann's Function," *Theoretical Computer Science*, **57** (1988), 327–330.

130. Grünbaum, B., "The Construction of Venn Diagrams," *College Mathematics Journal*, **15** (1984), 238–247.

131. Grünbaum, B., and G. C. Shephard, *Tilings and Patterns*. New York: Freeman (1986).

132. Guy, R. K., "Conway's Prime Producing Machine," *Mathematics Magazine*, **56** (1983), 26–33.

133. Guy, R. K., "John Horton Conway: Mathematical Magus," *The Two-Year College Mathematics Journal*, **13** (1982), 290–299.

134. Guy, R. K., "John Isbell's Game of Beanstalk and John Conway's Game of Beans-Don't-Talk," *Mathematics Magazine*, **59** (1986), 259–269.

135. Guy, R. K., "The Strong Law of Small Numbers," *American Mathematical Monthly*, **95** (1988), 697–712.

136. Hall, M., *Combinatorial Theory*, 2nd ed. New York: Wiley (1986).

137. Halmos, P. R., *Naive Set Theory*. Princeton, N.J.: Van Nostrand (1960).

138. Hankins, T. L., *Sir William Rowan Hamilton*. Baltimore: Johns Hopkins University Press (1980).

139. Hansen, P. J., and P. C. Jurs, "Chemical Applications of Graph Theory," *Journal of Chemical Education*, **65** (1988), 661–664.

140. Harary, F., *Graph Theory*. Reading, Mass.: Addison-Wesley (1969).

141. Harary, F., and J. A. Kabell, "An Intuitive Approach to Interval Numbers of Graphs," *Mathematics Magazine*, **53** (1980), 39–44.

142. Harary, F., and E. M. Palmer, *Graphical Enumeration*. New York: Academic Press (1973).

143. Harary, F., and K. Plochinski, "On Degree Achievement and Avoidance Games for Graphs," *Mathematics Magazine*, **60** (1987), 316–321.

144. Harary, F., and R. W. Robinson, "The Diameter of a Graph and Its Complement," *American Mathematical Monthly*, **92** (1985), 211–212.

145. Hardy, G. H., and E. M. Wright, *An Introduction to Number Theory*, 5th ed. Oxford: Oxford University Press (1979).

146. Hellman, M. E., "The Mathematics of Public-Key Cryptography," *Scientific American*, **241**(2) (Aug. 1979), 146–157.

147. Henkin, L., "Are Logic and Mathematics Identical?" in *Mathematics: People, Problems, Results*, D. M. Campbell and J. C. Higgins, eds., Vol. 2. Belmont, Calif.: Wadsworth (1984), pp. 223–232.

148. Hilton, P., and J. Pedersen, "Looking into Pascal's Triangle: Combinatorics, Arithmetic, and Geometry," *Mathematics Magazine*, **60** (1987), 305–316.

149. Hodges, A., *Alan Turing: The Enigma*. New York: Simon & Schuster (1983).

150. Hoffman, A. J., "Eigenvalues of Graphs," in *Studies in Graph Theory*, Part 2, D. R. Fulkerson, ed. Washington, D.C.: Mathematical Association of America (1975), pp. 225–245.

151. Hoffman, P., "The Man Who Loves Only Numbers," *The Atlantic Monthly*, **260**(5) (Nov. 1987), 60–74.

152. Hofstadter, D. R., "Analogies and Metaphors to Explain Gödel's Theorem," *The Two-Year College Mathematics Journal*, **13** (1982), 98–114 [reprinted in D. M. Campbell and J. C. Higgins, eds., *Mathematics: People, Problems, Results*, Vol. 2. Belmont, Calif.: Wadsworth (1984), pp. 262–275].

153. Hofstadter, D. R., *Gödel, Escher, Bach: An Eternal Golden Braid*. New York: Basic Books (1979).

154. Hofstadter, D. R., *Metamagical Themas: Questing for the Essence of Mind and Pattern*. New York: Basic Books (1985).

155. Honsberger, R., *Mathematical Gems from Elementary Combinatorics, Number Theory, and Geometry*. Washington, D.C.: Mathematical Association of America (1973).

156. Honsberger, R., *Mathematical Gems III*. Washington, D.C.: Mathematical Association of America (1985).

157. Hopcroft, J. E., "Turing Machines," *Scientific American*, **250**(5) (May 1984), 86–98.

158. Hopcroft, J. E., and R. E. Tarjan, "Efficient Planarity Testing," *Journal of the ACM*, **21** (1974), 549–568.

159. Hopcroft, J. E., and J. D. Ullman, *Introduction to Automata Theory, Languages, and Computation*. Reading, Mass.: Addison-Wesley (1979).

160. Hu, T. C., *Combinatorial Algorithms*. Reading, Mass.: Addison-Wesley (1982).

161. Hwang, K., *Computer Arithmetic: Principles, Architecture and Design*. New York: Wiley (1979).

162. Jean, R. V., "The Fibonacci Sequence," *UMAP Journal*, **5** (1984), 23–47.

163. Jeffrey, R., *Formal Logic: Its Scope and Limits*. New York: McGraw-Hill (1967).

164. Jewett, R. I., and K. A. Ross, "Random Walks on **Z**," *College Mathematics Journal*, **19** (1988), 330–342.

165. Johnsonbaugh, R., and T. Murato, "Petri Nets and Marked Graphs—Mathematical Models of Concurrent Computation," *American Mathematical Monthly*, **89** (1982), 552–566.

166. Just, E., and N. Schaumberger, "A Curious Property of the Integer 38," *Mathematics Magazine*, **46** (1973), 221.

167. Kamke, E., *Theory of Sets*, trans. by F. Bagemihl. New York: Dover (1950).

168. Kelly, C., "An Algorithm for Sums of Integer Powers," *Mathematics Magazine*, **57** (1984), 296–297.

169. Kenner, H., "The Several Sorts of Sorts," *Discover*, **7**(5) (May 1986), 78–83.

170. Kidwell, M. E., and R. B. Richter, "Trees and Euler Tours in a Planar Graph and Its Relatives," *American Mathematical Monthly*, **94** (1987), 618–630.

171. Kirch, A. M., *Elementary Number Theory: A Computer Approach*. New York: Intext (1974).

172. Kirkpatrick, S., C. D. Gelett, Jr., and M. P. Vecchi, "Optimization by Simulated Annealing," *Science*, **220** (1983), 671–680.

173. Kneebone, G. J., *Mathematical Logic and the Foundations of Mathematics*. New York: Van Nostrand (1963).

174. Knuth, D. E., "Algorithms," *Scientific American*, **236**(4) (Apr. 1977), 63–80.

175. Knuth, D. E., *The Art of Computer Programming*, Vol. 1, *Fundamental Algorithms*, 2nd ed. Reading, Mass.: Addison-Wesley (1973).

176. Knuth, D. E., *The Art of Computer Programming*, Vol. 2, *Seminumerical Algorithms*, 2nd ed. Reading, Mass.: Addison-Wesley (1981).

177. Knuth, D. E., *The Art of Computer Programming*, Vol. 3, *Sorting and Searching*. Reading, Mass.: Addison-Wesley (1973).

178. Knuth, D. E., "Computer Science and Its Relation to Mathematics," *American Mathematical Monthly*, **81** (1974), 323–343.

179. Knuth, D. E., "Computer Science and Mathematics," *American Scientist*, **61** (1973), 707–713 [reprinted in D. M. Campbell and J. C. Higgins, eds., *Mathematics: People, Problems, Results*, Vol. 3. Belmont, Calif.: Wadsworth (1984), pp. 37–47].

180. Knuth, D. E., "Mathematics and Computer Science: Coping with Finiteness," *Science*, **194** (1976), 1235–1242 [reprinted in D. M. Campbell and J. C. Higgins, eds., *Mathematics: People, Problems, Results*, Vol. 2. Belmont, Calif.: Wadsworth (1984), pp. 209–222].

181. Knuth, D. E., "Supernatural Numbers," in *The Mathematical Gardner*, D. A. Klarner, ed. Boston: Prindle, Weber & Schmidt (1981), pp. 310–325.

182. Knuth, D. E., *Surreal Numbers*. Reading, Mass.: Addison-Wesley (1974).

183. Kohavi, Z., *Switching and Finite Automata Theory*, 2nd ed. New York: McGraw-Hill (1978).

184. Kolata, G., "Does Gödel's Theorem Matter to Mathematics?" *Science*, **218** (Nov. 19, 1982), 779–780.

185. Kolata, G. B., "A Profile of Ronald L. Graham," *The Two-Year College Mathematics Journal*, **12** (1981), 290–301.

186. Kolman, B., *Elementary Linear Algebra*, 4th ed. New York: Macmillan (1986).

187. Konečný, V., "A Recursive Formula for the Number of Partitions of an Integer N into m Unequal Parts," *Mathematics Magazine*, **45** (1972), 91–94.

188. Kreisel, G., "Observations on Popular Discussions of Foundations," in *Axiomatic Set Theory*. Washington, D.C.: American Mathematical Society (1971), pp. 189–198.

189. Kronsjö, L., *Algorithms: Their Complexity and Efficiency*. New York: Wiley (1979).

190. Kropa, J. C., "Calculator Algorithms," *Mathematics Magazine*, **51** (1978), 106–108.

191. Lagarias, J., "The $3x + 1$ Problem and Its Generalizations," *American Mathematical Monthly*, **92** (1985), 3–23.

192. Lakatos, I., "Proofs and Refutations," in *Mathematics: People, Problems, Results*, D. M. Campbell and J. C. Higgins, eds., Vol. 2. Belmont, Calif.: Wadsworth (1984), pp. 194–208.

193. Lawler, E. L., *Combinatorial Optimization: Networks and Matroids*. Orlando, Fla.: Holt, Rinehart and Winston (1976).

194. Lawler, E. L., J. K. Lenstra, A. H. G. Rinnooy Kan, and D. B. Shmoys, eds., *The Traveling Salesman Problem: A Guided Tour of Combinatorial Optimization*. New York: Wiley (1985).

195. Lehman, H., *Introduction to the Philosophy of Mathematics*. Oxford: Basil Blackwell (1979).

196. Lewis, H. R., and C. H. Papadimitriou, "The Efficiency of Algorithms," *Scientific American*, **238**(1) (Jan. 1978), 96–109.

197. Lewis, H. R., and C. H. Papadimitriou, *Elements of the Theory of Computation*. Englewood Cliffs, N.J.: Prentice-Hall (1981).

198. Liu, C. L., *Introduction to Combinatorial Mathematics*. New York: McGraw-Hill (1968).

199. Lovász, L., *Combinatorial Problems and Exercises*. New York: North-Holland (1979).

200. Lovász, L., and M. D. Plummer, *Matching Theory*. New York: North-Holland (1986).

201. Luciano, D., and G. Prichett, "Cryptology: From Caesar Ciphers to Public-Key Cryptosystems," *College Mathematics Journal*, **18** (1987), 2–17.

202. Lunnon, F., "Counting Polynominoes," in *Computers in Number Theory*, A. O. L. Atkins and B. J. Birch, eds. Orlando, Fla.: Academic Press (1971), pp. 347–372.

203. MacWilliams, F. J., and N. J. A. Sloane, *The Theory of Error-Correcting Codes*. New York: North-Holland (1978).

204. Malpas, J., *Prolog: A Relational Language and Its Applications*. Englewood Cliffs, N.J.: Prentice-Hall (1987).

205. Manvel, B., "Counterfeit Coin Problems," *Mathematics Magazine*, **50** (1977), 90–92.

206. Manvel, B., "Extremely Greedy Coloring Algorithms," in *Graphs and Applications*, F. Harary and J. S. Maybee, eds. New York: Wiley (1985), pp. 257–270.

207. Maor, E., *To Infinity and Beyond: A Cultural History of the Infinite*. Basle, Switzerland: Birkhäuser (1987).

208. Maurer, S. B., "The King Chicken Theorems," *Mathematics Magazine*, **53** (1980), 67–80.

209. Maurer, S. B., et al., "FORUM: The Algorithmic Way of Life Is Best," *College Mathematics Journal*, **16** (1985), 2–21.

210. Mendelson, E., *Introduction to Mathematical Logic*, 2nd ed. New York: Van Nostrand (1979).

211. Meyer, C. H., and S. M. Matyas, *Cryptography: A New Dimension in Computer Data Security*. New York: Wiley (1982).

212. Meyer, W., "Huffman Codes and Data Compression," *UMAP Journal*, **5** (1984), 278–296.

213. Monk, J. D., *Introduction to Set Theory*. New York: McGraw-Hill (1969).

214. Monk, J. D., *Mathematical Logic*. New York: Springer-Verlag (1976).

215. Monk, J. D., "On the Foundations of Set Theory," *American Mathematical Monthly*, **77** (1970), 703–711.

216. Montague, H. F., and M. D. Montgomery, "How Mathematicians Develop a Branch of Pure Mathematics," in *Mathematics: People, Problems, Results*, D. M. Campbell and J. C. Higgins, eds., Vol. 1. Belmont, Calif.: Wadsworth (1984), pp. 279–288.

217. Morash, R. P., *Bridge to Abstract Mathematics: Mathematical Proof and Structures*. New York: Random House (1987).

218. Moschovakis, Y. N., *Descriptive Set Theory*. New York: North-Holland (1980).

219. Nagel, E., and J. R. Newman, *Gödel's Proof*. New York: New York University Press (1958).

220. Naps, T. L., and B. Singh, *Introduction to Data Structures with Pascal*. St. Paul, Minn.: West (1986).

221. Nash-Williams, C. St. J. A., "Hamiltonian Circuits," in *Studies in Graph Theory*, Part 2, D. R. Fulkerson, ed. Washington, D.C.: Mathematical Association of America (1975), pp. 301–360.

222. Nijenhuis, A., and H. S. Wilf, *Combinatorial Algorithms for Computers and Calculators*, 2nd ed. New York: Academic Press (1978).

223. Niven, I., *Mathematics of Choice or How to Count Without Counting*. New York: Random House (1965).

224. Norwood, F. H., "Long Proofs," *American Mathematical Monthly*, **89** (1982), 110–112.

225. O'Neil, P. V., "Ulam's Conjecture and Graph Reconstructions," *American Mathematical Monthly*, **77** (1970), 35–43.

226. Ore, O., *Graphs and Their Uses*. Washington, D.C.: Mathematical Association of America (1963).

227. Ore, O., *Invitation to Number Theory*. Washington, D.C.: Mathematical Association of America (1967).

228. Ore, O., "Note on Hamilton Circuits," *American Mathematical Monthly*, **67** (1960), 55.

229. Oullette, H., and G. Bennett, "Stirling's Triangle of the First Kind—Absolute Value Style," *The Two-Year College Mathematics Journal*, **8** (1977), 195–202.

230. Packel, E., *The Mathematics of Games and Gambling*. Washington, D.C.: Mathematical Association of America (1981).

231. Palmer, E. M., *Graphical Evolution: An Introduction to the Theory of Random Graphs*. New York: Wiley (1985).

232. Papadimitriou, C. H., and K. Steiglitz, *Combinatorial Optimization: Algorithms and Complexity*. Englewood Cliffs, N.J.: Prentice-Hall (1982).

233. Park, S. K., and K. W. Miller, "Random Number Generators: Good Ones Are Hard to Find," *Communications of the ACM*, **31** (1988), 1192–1201.

234. Patashnik, O., "Qubic: $4 \times 4 \times 4$ Tic-Tac-Toe," *Mathematics Magazine*, **53** (1980), 202–216.

235. Pless, V., *Introduction to the Theory of Error-Correcting Codes*. New York: Wiley (1982).

236. Pólya, G., *How to Solve It*. New York: Doubleday (1957).

237. Pólya, G., R. E. Tarjan, and D. R. Woods, *Notes on Introductory Combinatorics*. Basle, Switzerland: Birkhäuser (1983).

238. Pomerance, C., "Recent Developments in Primality Testing," *The Mathematical Intelligencer*, **3** (1981), 97–105.

239. Pomerance, C., "The Search for Prime Numbers," *Scientific American*, **247**(6) (Dec. 1982), 136–147.

240. Prichett, G. D., "The Game of Sprouts," *The Two-Year College Mathematics Journal*, **7**(4) (Dec. 1976), 21–25.

241. Ralston, A., "De Bruijn Sequences—A Model Example of the Intersection of Discrete Mathematics and Computer Science," *Mathematics Magazine*, **55** (1982), 131–143.

242. Read, R., "The Graph Theorists Who Count—and What They Count," in *The Mathematical Gardner*, D. A. Klarner, ed. Boston: Prindle, Weber & Schmidt (1981), pp. 326–345.

243. Reingold, E. M., J. Nievergelt, and N. Deo, *Combinatorial Algorithms: Theory and Practice*. Englewood Cliffs, N.J.: Prentice-Hall (1977).

244. Renz, P., "Mathematical Proof: What It Is and What It Ought to Be," *The Two-Year College Mathematics Journal*, **12** (1981), 83–103.

245. Ribenboim, P., "The Early History of Fermat's Last Theorem," in *Mathematics: People, Problems, Results*, D. M. Campbell and J. C. Higgins, eds., Vol. 2. Belmont, Calif.: Wadsworth (1984), pp. 74–82.

246. Richards, I., "Impossibility," *Mathematics Magazine*, **48** (1975), 249–262 [reprinted in D. M. Campbell and J. C. Higgins, eds., *Mathematics: People, Problems, Results*, Vol. 2. Belmont, Calif.: Wadsworth (1984), pp. 251–261].

247. Richards, I., "Number Theory," in *Mathematics Today: Twelve Informal Essays*, L. A. Steen, ed. New York: Springer-Verlag (1978), pp. 37–64.

248. Riesel, H., "Modern Factorization Methods," *BIT*, **25** (1985), 205–222.

249. Riesel, H., *Prime Numbers and Computer Methods for Factorization*. Basle, Switzerland: Birkhäuser (1985).

250. Ringeisen, R. D., "Isolation, a Game on a Graph," *Mathematics Magazine*, **47** (1974), 132–138.

251. Riordan, J., *An Introduction to Combinatorial Analysis*. New York: Wiley (1958).

252. Riordan, J., *Combinatorial Identities*. New York: Wiley (1968).

253. Roberts, E. S., *Thinking Recursively*. New York: Wiley (1986).

254. Roberts, F. S., "Applications of Ramsey Theory," *Discrete Applied Mathematics*, **9** (1984), 251–261.

255. Roberts, F. S., *Applied Combinatorics*. Englewood Cliffs, N.J.: Prentice-Hall (1984).

256. Roberts, F. S., *Discrete Mathematical Models, with Applications to Social, Biological, and Environmental Problems*. Englewood Cliffs, N.J.: Prentice-Hall (1976).

257. Roberts, F. S., *Graph Theory and Its Applications to Problems of Society*. Philadelphia: Society for Industrial and Applied Mathematics (1978).

258. Rogers, J. B., *A Prolog Primer*. Reading, Mass.: Addison-Wesley (1986).

259. Rosen, K. H., *Elementary Number Theory and Its Applications*, 2nd ed. Reading, Mass.: Addison-Wesley (1988).

260. Rucker, R., *Infinity and the Mind: The Science and Philosophy of the Infinite*. Basle, Switzerland: Birkhäuser (1982).

261. Ruelle, D., "Strange Attractors," *The Mathematical Intelligencer*, **3** (1980), 126–137.

262. Russell, B., *Introduction to Mathematical Philosophy*. London: Allen & Unwin (1948).

263. Ryser, H., *Combinatorial Mathematics*. Washington, D.C.: Mathematical Association of America (1963).

264. Saaty, T. L., and P. C. Kainen, *The Four-Color Problem: Assaults and Conquest*. New York: Dover (1986).

265. Schensted, C., "Longest Increasing and Decreasing Subsequences," *Canadian Journal of Mathematics*, **13** (1961), 179–191.

266. Schroeder, M. R., "Number Theory and the Real World," *The Mathematical Intelligencer*, **7**(4) (1985), 18–26.

267. Schroeder, M. R., *Number Theory in Science and Communication*. New York: Springer-Verlag (1984).

268. Sedgewick, R., *Algorithms*, 2nd ed. Reading, Mass.: Addison-Wesley (1988).

269. Shader, L. E., "Another Strategy for SIM," *Mathematics Magazine*, **51** (1978), 60–63.

270. Shamir, A., R. L. Rivest, and L. M. Adleman, "Mental Poker," in *The Mathematical Gardner*, D. A. Klarner, ed. Boston: Prindle, Weber & Schmidt (1981), pp. 37–43.

271. Shapiro, H. N., *Introduction to the Theory of Numbers*. New York: Wiley (1983).

272. Simmons, G. J., "Cryptology: The Mathematics of Secure Communication," *The Mathematical Intelligencer*, **1** (1979), 233–246.

273. Singmaster, D., "Covering Deleted Chessboards with Dominoes," *Mathematics Magazine*, **48** (1975), 59–66.

274. Sloan, M. E., *Computer Hardware and Organization: An Introduction*. Chicago: Science Research Associates (1976).

275. Sloane, N. J. A., "Error-Correcting Codes and Cryptography," in *The Mathematical Gardner*, D. A. Klarner, ed. Boston: Prindle, Weber & Schmidt (1981), pp. 346–382.

276. Slowinski, D., "Searching for the 27th Mersenne Prime," *Journal of Recreational Mathematics*, **11** (1979), 258–261.

277. Small, C., "Waring's Problem," *Mathematics Magazine*, **50** (Jan. 1977), 12–16.

278. Smith, H. F., *Data Structures: Form and Function*. San Diego, Calif.: Harcourt Brace Jovanovich (1987).

279. Smith, K. J., "Pascal's Triangle," *The Two-Year College Mathematics Journal*, **4**(1) (Winter 1973), 1–13.

280. Smoryński, C., "Some Rapidly Growing Functions," *The Mathematical Intelligencer*, **2** (1980), 149–154.

281. Smullyan, R., *Forever Undecided*. New York: Knopf (1987).

282. Smullyan, R., *The Lady or the Tiger?* New York: Knopf (1982).

283. Smullyan, R., *What Is the Name of This Book?* Englewood Cliffs, N.J.: Prentice-Hall (1978).

284. Snapper, E., "What Is Mathematics?" *American Mathematical Monthly*, **86** (1978), 551–557.

285. Snapper, E., "The Three Crises in Mathematics: Logicism, Intuitionism, and Formalism," *Mathematics Magazine*, **2** (1979), 207–216 [reprinted in D. M. Campbell and J. C. Higgins, eds., *Mathematics: People, Problems, Results*, Vol. 2. Belmont, Calif.: Wadsworth (1984), pp. 183–193].

286. Solow, D., *How to Read and Do Proofs: An Introduction to the Mathematical Thought Process*. New York: Wiley (1982).

287. Spencer, D. D., *Computers in Number Theory*. Rockville, Md.: Computer Science Press (1982).

288. Spencer, J., "Large Numbers and Unprovable Theorems," *American Mathematical Monthly*, **90** (1983), 669–675.

289. Spencer, J., "Short Theorems with Long Proofs," *American Mathematical Monthly*, **90** (1983), 365–366.

290. Spitznagel, E. L., Jr., "Properties of a Game Based on Euclid's Algorithm," *Mathematics Magazine*, **46** (1973), 87–92.

291. Stahl, S., "The Other Map Coloring Theorem," *Mathematics Magazine*, **58** (1985), 131–145.

292. Standish, T. A., *Data Structure Techniques*. Reading, Mass.: Addison-Wesley (1980).

293. Stein, R. G., "A Combinatorial Proof that $\sum k^3 = \left(\sum k\right)^2$," *Mathematics Magazine*, **44** (1971), 161–162.

294. Stockmeyer, L. J., and A. K. Chandra, "Intrinsically Difficult Problems," *Scientific American*, **240**(5) (May 1979), 140–159.

295. Stubbs, D. F., and N. W. Webre, *Data Structures with Abstract Data Types and Pascal*, 2nd ed. Pacific Grove, Calif.: Brooks/Cole (1989).

296. Suppes, P., *Axiomatic Set Theory*. New York: Van Nostrand (1960).

297. Suppes, P., *Introduction to Logic*. New York: Van Nostrand (1957).

298. Sved, M., "Counting and Recounting," *The Mathematical Intelligencer*, **5**(4) (1983), 21–26.

299. Sved, M., "Counting and Recounting: The Aftermath," *The Mathematical Intelligencer*, **6**(4) (1984), 44–45.

300. Swamy, M. N. S., and K. Thulasiraman, *Graphs, Networks, and Algorithms*. New York: Wiley (1988).

301. Swart, E. R., "The Philosophical Implications of the Four-Color Problem," *American Mathematical Monthly*, **87** (1980), 697–707.

302. Tarjan, R. E., *Data Structures and Network Algorithms*. Philadelphia: Society for Industrial and Applied Mathematics (1983).

303. Tesler, L. G., "Programming Languages," *Scientific American*, **251**(3) (Sept. 1984), 70–78.

304. Tomescu, I., *Problems in Combinatorics and Graph Theory*, trans. by R. A. Melter. New York: Wiley (1985).

305. Trudeau, R. J., *Dots and Lines*. Kent, Ohio: Kent State University Press (1976).

306. Tucker, A., *Applied Combinatorics*, 2nd ed. New York: Wiley (1985).

307. Tucker, A. B., Jr., *Programming Languages*, 2nd ed. New York: McGraw-Hill (1986).

308. Tymoczko, T., "Computers, Proofs and Mathematicians: A Philosophical Investigation of the Four-Color Proof," *Mathematics Magazine*, **53** (1980), 131–138.

309. Tymoczko, T., "The Four-Color Problem and Its Philosophical Significance," *Journal of Philosophy*, **76** (1979), 57–83.

310. Usiskin, Z., "Perfect Square Patterns in the Pascal Triangle," *Mathematics Magazine*, **46** (1973), 203–208.

311. Wagon, S., "The Collatz Problem," *The Mathematical Intelligencer*, **7**(1) (1985), 72–76.

312. Wagon, S., "Perfect Numbers," *The Mathematical Intelligencer*, **7**(2) (1985), 66–68.

313. Wagon, S., "Primality Testing," *The Mathematical Intelligencer*, **8**(3) (1986), 58–61.

314. Wall, D. D., "Fibonacci Series Modulo m," *American Mathematical Monthly*, **67** (1960), 525–532.

315. Wallis, W. D., "One-Factorization of Graphs: Tournament Applications," *College Mathematics Journal*, **18** (1987), 116–123.

316. Wang, H., *Reflections on Kurt Gödel*. Cambridge, Mass.: MIT Press (1987).

317. Watkins, W., "Generating Functions," *College Mathematics Journal*, **18** (1987), 195–211.

318. Wetzel, J. E., "On the Division of the Plane by Lines," *American Mathematical Monthly*, **85** (1978), 647–656.

319. Wilder, R. L., *The Foundations of Mathematics*, 2nd ed. New York: Wiley (1965).

320. Wilf, H. S., *Algorithms and Complexity*. Englewood Cliffs, N.J.: Prentice-Hall (1986).

321. Wilf, H. S., "A Bijection in the Theory of Derangements," *Mathematics Magazine*, **57** (1984), 37–40.

322. Wilf, H. S., "Finite Lists of Obstructions," *American Mathematical Monthly*, **94** (1987), 267–271.

323. Wilf, H. S., "The 'Why-Don't-You-Just...?' Barrier in Discrete Algorithms," *American Mathematical Monthly*, **86** (1979), 30–36.

324. Williams, H. C., "Factoring on a Computer," *The Mathematical Intelligencer*, **6**(3) (1984), 29–36.

325. Williams, J. D., *The Compleat Strategyst*. New York: McGraw-Hill (1966).

326. Wilson, R. J., *Introduction to Graph Theory*, 2nd ed. Orlando, Fla.: Academic Press (1979).

327. Wilson, R. J., and L. W. Beineke, eds., *Applications of Graph Theory*. Orlando, Fla.: Academic Press (1979).

328. Winston, P. H., and B. Horn, *Lisp*. Reading, Mass.: Addison-Wesley (1981).

329. Wirth, N., *Algorithms and Data Structures*. Englewood Cliffs, N.J.: Prentice-Hall (1986).

330. Wirth, N., *Algorithms + Data Structures = Programs*. Englewood Cliffs, N.J.: Prentice-Hall (1976).

331. Wirth, N., "Data Structures and Algorithms," *Scientific American*, **251**(3) (Sept. 1984), 60–69.

332. Wolfe, D., "Making Connections: A Graphical Construction," *Mathematics Magazine*, **54** (1981), 250–255.

333. Yao, F., "Maximum Flows in Networks," in *The Mathematics of Networks*, Proceedings of Symposia in Applied Mathematics, **26**, S. A. Burr, ed. Providence, R.I.: American Mathematical Society (1982), pp. 31–43.

334. Yates, S., "The Mystique of Repunits," *Mathematics Magazine*, **51** (1978), 22–28.

335. Zippin, L., *Uses of Infinity*. Washington, D.C.: Mathematical Association of America (1962).

ANSWERS TO ODD-NUMBERED EXERCISES

CHAPTER 1: LOGIC

SECTION 1.1 PROPOSITIONS

1. (c), (h), and (j) are not propositions.

3. (b) F (f) T (i) T (k) T

5. (a) Paris is not the capital of Spain.
   (b) $4 < 7$ or 13 is prime.
   (c) If $4 < 7$, then 13 is prime and Paris is the capital of Spain.
   (d) $4 \not< 7$ or 13 is not prime.
   (e) It is not the case that both $4 < 7$ and 13 is prime.
   (f) Either $4 < 7$ implies that 13 is prime, or 13 is prime implies that Paris is the capital of Spain.
   (g) 13 is prime and Paris is the capital of Spain.
   (h) $4 \not< 7$
   (i) Either 13 is prime or Paris is not the capital of Spain.

7. In each case the final column of T's shows that the proposition is a tautology.

(a)

P	Q	$P \to Q$	$P \wedge (P \to Q)$	$(P \wedge (P \to Q)) \to Q$
T	T	T	T	T
T	F	F	F	T
F	T	T	F	T
F	F	T	F	T

(b)

P	Q	$P \to Q$	$\overline{P}$	$Q \vee \overline{P}$	$(P \to Q) \Longleftrightarrow (Q \vee \overline{P})$
T	T	T	F	T	T
T	F	F	F	F	T
F	T	T	T	T	T
F	F	T	T	T	T

(c)

P	Q	$\overline{P}$	$P \wedge \overline{P}$	$(P \wedge \overline{P}) \to Q$
T	T	F	F	T
T	F	F	F	T
F	T	T	F	T
F	F	T	F	T

9. (a) T if $x = 2$; F if $x = 3$
   (d) T if $p = 17$; F if $p = 18$
   (e) F if $x = 2$ and $y = 1$; never T
   (g) F if He $=$ Ronald Reagan; never T
   (l) T if $N = 11$; F if $N = 10$

11. (a) T   (b) F   (c) F   (d) F   (e) T
    (f) T   (g) T   (h) T   (i) T

13. If P is false, then $x \le 2$, which implies that $x < 7$; this makes Q true. Therefore, it is impossible to have both P and Q false. The other six possibilities are all actually possible. If $x = 5$ or $x = 6$, then P and Q are true, and R is true in the first case, false in the second; if $x = 11$ or $x = 12$, then P is true and Q is false; and if $x = 1$ or $x = 2$, then P is false and Q is true.

15. (b) and (c) are tautologies; (e) is a contradiction.

17. If Q and R are F; then $Q \wedge (P \vee \overline{R})$ is F, but $(Q \wedge P) \vee \overline{R}$ is T.

19. In the following truth table the second column has a T whenever the last column has a T. The proposition says: "If Q is true when P holds, and if Q is true when P does not hold, then Q is true."

P	Q	$P \to Q$	$\overline{P}$	$\overline{P} \to Q$	$(P \to Q) \wedge (\overline{P} \to Q)$
T	T	T	F	T	T
T	F	F	F	T	F
F	T	T	T	T	T
F	F	T	T	F	F

21. tautology

23. (a) $S \to (L \wedge P)$; $(\overline{L} \wedge \overline{P}) \to \overline{S}$

(b) The following truth table shows that in every case in which $S \to (L \wedge P)$ is true, $(\overline{L} \wedge \overline{P}) \to \overline{S}$ is also true; in the third line, however, the latter is true but the former is false.

L	P	S	$L \wedge P$	$S \to (L \wedge P)$	$\overline{L} \wedge \overline{P}$	$(\overline{L} \wedge \overline{P}) \to \overline{S}$
T	T	T	T	T	F	T
T	T	F	T	T	F	T
T	F	T	F	F	F	T
T	F	F	F	T	F	T
F	T	T	F	F	F	T
F	T	F	F	T	F	T
F	F	T	F	F	T	F
F	F	F	F	T	T	T

(c) She should have said "If you're not less than 6 years old or your parent isn't present, then you may not swim in the pool." This is $(\overline{L} \vee \overline{P}) \to \overline{S}$. By the contrapositive law, this is equivalent to $S \to \overline{\overline{L} \vee \overline{P}}$, which is in turn equivalent to $S \to (L \wedge P)$ by DeMorgan's law and the double negative law.

25. (a) always true

(b) For $x \cdot y$ to be positive, it is sufficient that x and y both be positive.

(c) For $x \cdot y$ to be positive, it is necessary that x and y both be positive. For x and y both to be positive, it is sufficient that $x \cdot y$ be positive.

(d) not always true

27. (a) Neither 42548 nor $4+2+5+4+8 = 23$ is divisible by 3. Both 121551 and $1 + 2 + 1 + 5 + 5 + 1 = 15$ are divisible by 3.

(b) A necessary and sufficient condition for a natural number to be divisible by 3 is that the sum of its digits be divisible by 3.

29. (a) The following table shows that Q is true whenever $P \wedge (P \to Q)$ is true, namely in line 1.

P	Q	$P \to Q$	$P \wedge (P \to Q)$
T	T	T	T
T	F	F	F
F	T	T	F
F	F	T	F

(b) The following table shows that $P \vee Q$ is true whenever P is true, namely in lines 1 and 2.

P	Q	$P \vee Q$
T	T	T
T	F	T
F	T	T
F	F	F

(c) The following table shows that P is true whenever $P \wedge Q$ is true, namely in line 1.

P	Q	$P \wedge Q$
T	T	T
T	F	F
F	T	F
F	F	F

(d) The following table shows that P is true whenever $\overline{P} \to P$ is true, namely in line 1.

P	$\overline{P}$	$\overline{P} \to P$
T	F	T
F	T	F

(e) The following table shows that P is true whenever $\overline{P} \to F$ is true, namely in line 1.

P	$\overline{P}$	$\overline{P} \to F$
T	F	T
F	T	F

(f) Since F is never true, it holds vacuously that P is true whenever F is true.

(g) Since T is always true, it holds trivially that T is true whenever P is true.

(h)   The following table shows that $P \to R$ is true whenever $(P \to Q) \land$ $(Q \to R)$ is true, namely in lines 1, 5, 7, and 8.

P	Q	R	$P \to Q$	$Q \to R$	$(P \to Q) \land (Q \to R)$	$P \to R$
T	T	T	T	T	T	T
T	T	F	T	F	F	F
T	F	T	F	T	F	T
T	F	F	F	T	F	F
F	T	T	T	T	T	T
F	T	F	T	F	F	T
F	F	T	T	T	T	T
F	F	F	T	T	T	T

31. (d) and (e) are logical equivalences.

33. (a) $(P \land \overline{Q}) \lor (\overline{P} \land Q)$ (b) $\overline{P} \land \overline{Q}$ (c) $\overline{P} \lor \overline{Q}$

35. $P \Longleftrightarrow Q$ means that every assignment of truth values to the propositional variables in P and Q results in the same truth value for P and for Q. By the definition of $\leftrightarrow$, this occurs if and only if $P \leftrightarrow Q$ is always true, that is, $P \leftrightarrow Q$ is a tautology.

37. (a) $(P \land Q) \lor R;\ \overline{P \lor Q} \land (T \land Q)$
   (b) If we make these replacements twice, then each of $\land$, $\lor$, T, and F returns to what it was originally.
   (c) We see that $(P \land Q) \lor (P \land \overline{Q}) \Longleftrightarrow P$ by comparing columns 1 and 5 in the following truth table.

P	Q	$P \land Q$	$P \land \overline{Q}$	$(P \land Q) \lor (P \land \overline{Q})$
T	T	T	F	T
T	F	F	T	T
F	T	F	F	F
F	F	F	F	F

We see that the dual propositions are logically equivalent

$$(P \lor Q) \land (P \lor \overline{Q}) \Longleftrightarrow P$$

by comparing columns 1 and 5 in the following truth table.

P	Q	$P \vee Q$	$P \vee \overline{Q}$	$(P \vee Q) \wedge (P \vee \overline{Q})$
T	T	T	T	T
T	F	T	T	T
F	T	T	F	F
F	F	F	T	F

39. (a) $\overline{P} \Longleftrightarrow P \uparrow P$ (b) $P \vee Q \Longleftrightarrow (P \uparrow P) \uparrow (Q \uparrow Q)$

(c) $P \wedge Q \Longleftrightarrow (P \uparrow Q) \uparrow (P \uparrow Q)$

41. (a)   See discussion of disjunctive normal form in Section 1.4.

(b)   Any use of $\vee$ in a proposition involving $\vee$, $\wedge$, and $^{-}$ can be replaced by substituting $\overline{\overline{P} \wedge \overline{Q}}$ whenever we see an expression of the form $P \vee Q$. Repeated applications of this substitution, from the inside of the expression outward, eliminates all uses of $\vee$. Thus $\{\wedge, ^{-}\}$ is complete. The dual reasoning applies to eliminating $\wedge$, so $\{\vee, ^{-}\}$ is also complete.

(c)   We know that $\{\vee, ^{-}\}$ is complete by part (b). From Exercise 39 we can replace each occurrence of $\overline{P}$ by $P \uparrow P$, one at a time, working from the inside out. This results in a proposition using only $\vee$ and $\uparrow$. Then by replacing each occurrence of $P \vee Q$ (one at a time, from the inside out) with $(P \uparrow P) \uparrow (Q \uparrow Q)$, we obtain a logically equivalent expression involving only $\uparrow$. A dual argument applies to $\downarrow$.

(d) $[(P \uparrow P) \uparrow (P \uparrow P)] \uparrow (Q \uparrow Q)$

(e) $[(P \downarrow P) \downarrow Q] \downarrow [(P \downarrow P) \downarrow Q]$

SECTION 1.2   LOGICAL QUANTIFIERS

1. (a) $\forall x \colon (x^2 > 4 \ \leftrightarrow \ (x > 2 \vee x < -2))$

(b) $\exists x \colon x = x^2$

(c) $\forall x > 1 \colon \exists y \colon (y > x \ \wedge \ y < 2x)$

(d) $\exists x \colon \forall y \colon y^2 > x$

3. (a)   There exists an x such that for every y, $x + y = y$.

(b)   For every x there exists a y such that $x + y = y$.

(c)   For every x and for every y, $x + y = y$.

(d)   There exist x and y such that $x + y = y$.

(e)   There exists an x such that for every y, $x - y = y$.

(f)   For every x there exists a y such that $x - y = y$.

5. (a)   There exist a and b, both greater than 1, whose product is x (i.e., x is composite).

(b)   For every x there exists a y such that $x < y^2$.

(c)   For every x there exists a y such that either $x = 3y$ or $x = 3y + 1$ or $x = 3y + 2$.

(d)   For every x, if $x < 2$, then $x^2 < 4$.
(e)   For every x, if $x^2 < 4$, then $x < 2$.
(f)   There exists an x such that $x < 5$ implies $x < 3$.
(g)   There exists an x such that $x^2 - 2x - 120 = 0$.
(h)   For every x, $x^2 > a$.

7. (a) $\exists x\colon \exists y\colon (x \neq y \,\wedge\, M(\text{Diana}, x) \,\wedge\, M(\text{Diana}, y) \,\wedge\, F(\text{Charles}, x) \,\wedge$
$F(\text{Charles}, y))$
(b) $\exists! x\colon (M(\text{Suzanne}, x) \wedge F(\text{Jerry}, x))$ (assuming that we are interested in their joint offspring)
(c) $\exists x\colon (M(x, \text{Pam}) \wedge F(\text{Sam}, x))$
(d) $\exists x\colon \exists y\colon \exists z\colon (M(x, y) \wedge M(x, z) \wedge F(y, \text{Pam}) \wedge F(z, \text{Conrad}))$
(e) $\exists x\colon \forall y\colon (\overline{M(x, y) \wedge F(x, y)})$

9. (a) $\exists n\colon 100 = 5 \cdot n$
(b) $\neg \exists n\colon 1000 = 8 \cdot n$
(c) $\forall x\colon ((\exists y\colon x = 6 \cdot y) \leftrightarrow \exists y\colon x = 2 \cdot y)$
(d) $\forall x\colon ((\exists y\colon x = 3 \cdot y) \rightarrow \neg \exists y\colon x = 2 \cdot y)$
(e) $\exists x\colon \forall y\colon \overline{x = y \cdot y}$

11. (a)   T   (b)   F   (c)   F   (d)   F   (e)   T

13. (a)   neither   (b)   T   (c)   T   (d)   F   (e)   T
(f)   T   (g)   T   (h)   neither

15. (a)   100 is not a multiple of 5.
(b)   1000 is a multiple of 8.
(c)   Some multiple of 6 is not a multiple of 2, or some multiple of 2 is not a multiple of 6.
(d)   Some multiple of 3 is even.
(e)   Every number has a square root.

17. (a)   Some perfect square is not less than 500.
(b)   Every perfect square is not less than 500.
(c)   Some perfect square is less than 500.
(d)   Every perfect square is less than 500.
(e)   Some perfect square is less than 500.
(f)   Every perfect square is less than 500.

19. (a)   F   (b)   F   (c)   T   (d)   F   (e)   T
(f)   T   (g)   T   (h)   F   (i)   F

21. (a) $7921 > 1 \wedge \neg \exists x\colon \exists y\colon (7921 = x \cdot y \,\wedge\, x > 1 \,\wedge\, y > 1)$; false
(b) $\forall x\colon [(\exists y\colon x = y^2) \rightarrow (x \leq 1 \,\vee\, \exists a\colon \exists b\colon (x = a \cdot b \,\wedge\, a > 1 \,\wedge\, b > 1))]$; true

23. (a) $P(1, 1) \wedge P(1, 2) \wedge P(2, 1) \wedge P(2, 2)$
(b) $P(1, 1) \vee P(1, 2) \vee P(2, 1) \vee P(2, 2)$

(c) $(P(1, 1) \lor P(1, 2)) \land (P(2, 1) \lor P(2, 2))$

(d) $(P(1, 1) \land P(1, 2)) \lor (P(2, 1) \land P(2, 2))$

25. $\forall x \colon \exists y \colon \exists t \colon L(x, y, t)$

27.   (a) $x \neq y \ \land \ \exists m \colon \exists f \colon (C(x, m, f) \land C(y, m, f))$

(b) $\exists p \colon \exists q \colon \exists r \colon (C(y, p, q) \land C(q, r, x))$

(c) $\forall p \colon \forall q \colon \forall r \colon \forall s \colon \neg [C(p, q, r) \ \land \ (C(q, x, s) \ \lor \ C(q, s, x) \ \lor \ C(r, x, s) \ \lor \ C(r, s, x))]$

(d) $\exists m \colon \exists f \colon \exists a \colon \exists b \colon [(C(y, m, f) \ \land \ C(x, a, b) \ \land \ C(m, a, b) \ \land \ m \neq x) \ \lor \ (C(y, m, f) \land C(x, a, b) \land C(f, a, b) \land f \neq x)]$

29.   (a)   equivalent   (b)   not equivalent   (c)   not equivalent

(d)   equivalent   (e)   not equivalent   (f)   not equivalent

31.   (a) $\forall n{>}2 \colon \forall x \colon \forall y \colon \forall z \colon (x^n + y^n \neq z^n)$

(b)   positive integers x, y, z, and n, with $n > 2$, such that $x^n + y^n = z^n$

33.   The quantified proposition has no free variables, so it cannot mean that $x = 0$ (a statement about x).

35. $(\exists x \colon P(x)) \land \forall x \colon \forall y \colon [(P(x) \land P(y)) \to x = y]$

37.   false

39.   (a) $\forall f \colon \forall a \colon (f \text{ is continuous at } a \longleftrightarrow \forall \epsilon > 0 \colon \exists \delta > 0 \colon \forall x \colon (|x - a| < \delta \to |f(x) - f(a)| < \epsilon))$

(b) $\forall x \colon \forall p \colon (x \text{ is a quadratic residue modulo } p \longleftrightarrow \exists y \colon \exists m \colon x - y^2 = pm)$

(c) $\forall f \colon \forall a \colon \forall b > a \colon [(f \text{ is continuous on } [a, b] \ \land \ f \text{ is differentiable on } (a, b) \ \land \ f(a) = 0 \ \land \ f(b) = 0) \to \exists c \colon (a < c < b \land f'(c) = 0)]$

(d) $\forall p \colon \forall q \colon (p \neq q \to \exists! l \colon (p \text{ is on } l \ \land \ q \text{ is on } l))$

SECTION 1.3   PROOFS

1.   Let $2n$ be the given even number. Then $(2n)^2 = 4n^2 = 2(2n^2)$. Since this is 2 times some number, it is even.

3.   (a)   Theorem. The product of two odd numbers is odd. *Proof*: Let $2n + 1$ and $2m + 1$ be the numbers. Their product is $(2n + 1)(2m + 1) = 4nm + 2n + 2m + 1 = 2(2nm + n + m) + 1$, which is odd by definition.

(b)   Theorem. The product of two even numbers is even. *Proof*: Let $2n$ and $2m$ be the numbers. Their product is $(2n)(2m) = 2(2nm)$, which is even by definition.

(c)   Theorem. The product of an even number and an odd number is even. *Proof*: Let $2n$ and $2m+1$ be the numbers. Their product is $(2n)(2m + 1) = 2(2nm + n)$, which is even by definition.

5.   (a)   Let $x = 6n$ be the given number. Then $x = 3(2n)$, so x is a multiple of 3.

   (b)   This is false; for example, 9 is a multiple of 3 but not a multiple of 6.

   (c)   We give an indirect proof, by proving the equivalent proposition, "If x is a multiple of 6, then x is a multiple of 2." Let $x = 6n$ be the given number. Then $x = 2(3n)$, so x is a multiple of 2.

7. Let x_1, x_2, ..., x_n be the real numbers, and let A be their average. This means that $A = (x_1 + x_2 + \cdots + x_n)/n$. If the statement is not true, then $x_i > A$ for all i. If we add the inequalities $x_i > A$ for $i = 1, 2, \ldots, n$, we obtain $x_1 + x_2 + \cdots + x_n > nA$, or $(x_1 + x_2 + \cdots + x_n)/n > A$, a contradiction to the definition of A above. Thus the statement is true.

9. (a)   This "proof" starts by assuming what we are trying to prove. The proposition is false; $n = 4$ provides a counterexample.

   (b)   One case has been omitted: n might be an even number not divisible by 4, such as 6. The proposition is false ($6^2 - 1$ is not a multiple of 4).

   (c)   This "proof" has things backward. Looking at $p = 17$ would only tell us something about the twinliness of 76 if 17 were a factor of 76, but $17 + 2$ were not. The proposition is true: 19 is a prime factor of 76, but $19 + 2$ is not, so 76 is not twinly.

   (d)   The "proof" erroneously assumes that there is some b such that $P(a, b)$ holds. The proposition is false. Let $P(x, y)$ be the proposition $x \neq x \wedge y \neq y$, which is always false. Then the two axioms are both vacuously true, but $\forall a\colon P(a, a)$ is false.

11. (a)   By the hypothesis we have $a = sb$ and $b = tc$ for some integers s and t. Thus $a = s(tc) = (st)c$, which shows that a is a multiple of c.

   (b)   Since $6083824773 = 13 \cdot 467986521$, the statement is true.

   (c)   We can write $n^2 + n$ as $n(n + 1)$. Now if n is even, then by Exercise 3, $n(n + 1)$ is even. Otherwise n is odd, so $n + 1$ is even, and again by Exercise 3, $n(n + 1)$ is even.

13. (a)   Let $2n + 1$ and $2m + 1$ be the given odd numbers. Then the difference of their squares is $(2n+1)^2 - (2m+1)^2 = 4n^2 + 4n + 1 - (4m^2 + 4m + 1) = 4(n^2 + n - m^2 - m)$, a multiple of 4.

   (b)   Let $2n + 1$ and $2m + 1$ be the given odd numbers. Then the sum of their squares is $(2n + 1)^2 + (2m + 1)^2 = 4n^2 + 4n + 1 + (4m^2 + 4m + 1) = 4(n^2 + n + m^2 + m) + 2$. Clearly, the remainder when dividing this by 4 is 2; therefore, it is not a multiple of 4.

15. Let $x < y$ be the two distinct real numbers. Let $z = (x + y)/2$. We claim that $x < z < y$. (For the first inequality, add x to both sides of $x < y$ and divide by 2; for the second, add y to both sides of $x < y$ and divide by 2.) Thus z is the desired number strictly between x and y.

17. Suppose that $5\sqrt{2}$ were rational. Since $\frac{1}{5}$ is also rational, the product $\left(\frac{1}{5}\right) \cdot (5\sqrt{2}) = \sqrt{2}$ must be rational, contradicting Theorem 5.

19. Lemma: If n^2 is a multiple of 3, then n is a multiple of 3. (Proof similar to proof of Theorem 4, since numbers that are not multiples of 3 are of the form $3k+1$ or $3k+2$.) Now suppose that $\sqrt{3} = a/b$ in lowest terms. Then $a^2 = 3b^2$. Therefore, a^2 is a multiple of 3, so a is as well, say $a = 3n$. Then we have $9n^2 = 3b^2$, or $3n^2 = b^2$. This implies that b^2, and hence b, is a multiple of 3. This has now contradicted our assumption that a/b was in lowest terms.

21. There must exist irrational numbers r and s such that r^s is rational. Look at $\sqrt{2}^{\sqrt{2}}$. If this is rational, we are finished—take $r = s = \sqrt{2}$. Otherwise, let $r = \sqrt{2}^{\sqrt{2}}$ and $s = \sqrt{2}$. Then $r^s = \left(\sqrt{2}^{\sqrt{2}}\right)^{\sqrt{2}} = \sqrt{2}^{\sqrt{2}\cdot\sqrt{2}} = 2$ is rational.

23. Let x and y be positive real numbers. Then $(\sqrt{x} - \sqrt{y})^2 \geq 0$, with equality if and only if $x = y$. But this is equivalent to $\sqrt{x}^2 - 2\sqrt{x}\sqrt{y} + \sqrt{y}^2 \geq 0$, that is, $x + y \geq 2\sqrt{x}\sqrt{y}$, or, as desired, $\sqrt{xy} \leq (x+y)/2$.

25. Among five consecutive positive integers, exactly one will be divisible by 5. Hence the product N is a multiple of 5. Similarly, the product is a multiple of 3, since at least one of any three consecutive integers must be divisible by 3. Furthermore, among the five consecutive numbers there are at least two consecutive even ones, say $2a$ and $2a + 2$. Their product is $2a(2a + 2) = 4a(a + 1)$. By Exercise 11c, $a(a + 1)$ is even, so $4a(a + 1)$ is divisible by $4 \cdot 2 = 8$. Thus the prime factorization of N contains at least 2^3. Putting this all together, we see that N is divisible by $2^3 \cdot 3 \cdot 5 = 120$.

SECTION 1.4   BOOLEAN FUNCTIONS

1. (a)   1   (b)   0

3. The following table shows that equality holds in all four cases: The first and fourth columns are identical.

x	y	$x + y$	$x(x+y)$
0	0	0	0
0	1	1	0
1	0	1	1
1	1	1	1

5. $x\overline{y} + \overline{x}y$

7. (a) $xyz + xy\overline{z} + x\overline{y}z + \overline{x}yz$
   (b) $xy + xz + yz$

9. (a) $x \cdot \overline{xy} = x \cdot (\overline{x} + \overline{y}) = x \cdot \overline{x} + x \cdot \overline{y} = 0 + x \cdot \overline{y} = x\overline{y}$
   (b) $x + \overline{x + y} = x + (\overline{x}\,\overline{y}) = (x + \overline{x}) \cdot (x + \overline{y}) = 1 \cdot (x + \overline{y}) = x + \overline{y}$

(c) $x + xy = x \cdot 1 + xy = x(1 + y) = x(y + 1) = x \cdot 1 = x$

(d) $x(x + y) = (x + 0)(x + y) = x + 0 \cdot y = x + y \cdot 0 = x + 0 = x$

11. (a)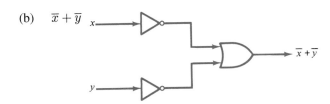

(b) $\overline{x} + \overline{y}$

(c) $\overline{x}\,y + x\,\overline{y} + \overline{x}\,\overline{y}$

13. (a) x (b)   1   (c) $y\,\overline{z} + \overline{y}\,z$ (d) $x(\overline{yz})$ or $x(\overline{y} + \overline{z})$

15. (a) $x\,y\,z + x\,y\,\overline{z} + x\,\overline{y}\,z + x\,\overline{y}\,\overline{z} + \overline{x}\,y\,z + \overline{x}\,y\,\overline{z} + \overline{x}\,\overline{y}\,z$

(b) $\overline{x}\,\overline{y}\,\overline{z}$ (c) $\overline{x}\,y\,z + \overline{x}\,y\,\overline{z}$

17.

$$x_1 x_2 + \overline{x}_1 x_2 + \overline{x}_1 \overline{x}_2$$

19. (a)   If $x = 1$ and $y = 0$, then $\overline{xy} = 1$ but $\overline{x}\,\overline{y} = 0$.

(b)   If $x = 1$ and $y = 0$, then $\overline{x + y} = 0$ but $\overline{x} + \overline{y} = 1$.

21. (a)   For each case in which $f(a_1, a_2, \ldots, a_n) = 0$, write down a sum $y_1 + y_2 + \cdots + y_n$ that will be 0 only in that case, by letting $y_i = x_i$ if $a_i = 0$, and $y_i = \overline{x}_i$ if $a_i = 1$. Then take the product of all such sums. This expression will have the value 0 if and only if $f(x_1, x_2, \ldots, x_n) = 0$.

(b) $(x + y + z)(x + y + \overline{z})(x + \overline{y} + z)(\overline{x} + y + z)$

(c)   Form the disjunction normal form of the complement of the desired function (i.e., write down all the minterms that are missing from the given disjunctive normal form expression), complement it, and then apply DeMorgan's law in two stages (and the double complement law if necessary), pushing the complementation operation inside.

CHAPTER 2:   SETS

SECTION 2.1   BASIC DEFINITIONS IN SET THEORY

1. {Ronald Reagan, Jimmy Carter, $\{25, 4, 48\}$, $\{\{4\}, \emptyset, \{1, 2\}, \mathbf{N}, \mathbf{Z}\}\}$

3. (a) no   (b) yes   (c) yes
    (d) yes   (e) no   (f) no

5. (a) $\{-99, -98, \ldots, -1, 0, 1, 2, \ldots, 98, 99\}$
    (b) $\{1, 2, \ldots, 67, 68, 70, 71, 72, \ldots, 100\}$
    (c) $\{0, 1, 4, 9, 16, 25, \ldots\}$

7. {red, blue}, {red, green}, {red, yellow}, {blue, green}, {blue, yellow}, {green, yellow}

9. (a) F   (b) F   (c) T

11. (a) $\exists x \colon (x \in A \land x \notin B)$
    (b) $\emptyset \not\subseteq A \iff \exists x \colon (x \in \emptyset \land x \notin A)$ This is clearly false, since there is no x such that $x \in \emptyset$.

13. (a) $\{\, x \mid x \in \mathbf{N} \land 0 < x < 100 \land \neg\exists m \in \mathbf{N} \colon x = 10m \,\}$
    (b) $\{\, 1/x^3 \mid x \in \mathbf{N} \land x > 0 \,\}$
    (c) $\{\, n(n+1) \mid n \in \mathbf{N} \land 1 \le n \le 99 \,\}$
    (d) $\{\, x \mid \exists m \in \mathbf{Z} \colon x = 3m \,\}$
    (e) $\{\, 6n + 1 \mid n \in \mathbf{Z} \,\}$

15. {Henry VIII, $\{2, 15\}$, $\{25, 4\}$, $\{11, 17\}\}$

17. Let x be an arbitrary element of A. Since $A \subseteq B$, it follows that $x \in B$. Then since $B \subseteq C$, we conclude that $x \in C$, as desired.

19. (a) $\{\emptyset\}$ (b) $\{1\}$ (c) $A = \{1\}$ and $B = \{1, \{1\}\}$

21. Let $C \in \mathcal{P}(A)$. Then $C \subseteq A$, and it follows directly from Exercise 17 that $C \subseteq B$, that is, $C \in \mathcal{P}(B)$.

23. 65536

25. By the observation made before Theorem 1, we know that $A = \emptyset \leftrightarrow (A \subseteq \emptyset \land \emptyset \subseteq A)$. By Theorem 1, we know that $\emptyset \subseteq A$ is always true. Hence $A = \emptyset \leftrightarrow A \subseteq \emptyset$.

27. (a) Let $n \in \mathbf{N}$ correspond to $2n + 2$ in the set of even positive integers.
    (b) This set is finite, hence countable.
    (c) We can list this set as $\{0, \frac{1}{2}, -\frac{1}{2}, 1, -1, 1\frac{1}{2}, -1\frac{1}{2}, 2, -2, 2\frac{1}{2}, -2\frac{1}{2}, \ldots\}$.

29. (a) $\forall A \colon \forall B \colon \exists C \colon \forall x \colon (x \in C \leftrightarrow (x = A \lor x = B))$

    (b) $\exists S \colon \forall x \colon \overline{\overline{x \in S}}$

    (c) $\forall A \colon \exists P \colon \forall C \colon (C \in P \leftrightarrow \forall x \colon (x \in C \to x \in A))$

31. The set of all rational numbers can be listed: $0, \frac{1}{1}, -\frac{1}{1}, \frac{1}{2}, -\frac{1}{2}, \frac{2}{1}, -\frac{2}{1}, \frac{1}{3}, -\frac{1}{3},$
$\frac{3}{1}, -\frac{3}{1}, \frac{1}{4}, -\frac{1}{4}, \frac{2}{3}, -\frac{2}{3}, \ldots$.

33. (a) Express all such reals as infinite decimals, possibly ending with a string of all 0's, but not ending with a string of all 9's.

    (b) Assume that the set of real numbers between 0 and 1 is countable, and can therefore be labeled as r_1, r_2, r_3, ... Let d_{ij} be the jth digit in the decimal for r_i.

    (c) Define a number $r = 0.d_1 d_2 d_3 \ldots$ by setting $d_i = 4$ if $d_{ii} \neq 4$ and setting $d_i = 5$ if $d_{ii} = 4$. Then $d_i \neq d_{ii}$ for all i, so that r is not in the list. On the other hand, r does not end in a string of 9's, so r must be in the list. This contradiction shows that our assumption that the set of real numbers between 0 and 1 is countable was wrong.

    (d) The set of real numbers between 0 and 1 is a subset of the set of all real numbers. If the latter were countable, the former would also have to be countable, and we just showed that this is not the case.

35. (a) $\omega + 2 = \{0, 1, 2, \ldots, \omega, \omega + 1\}$; $\omega + 3 = \{0, 1, 2, \ldots, \omega, \omega + 1, \omega + 2\}$

    (b) $\omega \cdot 2 = \{0, 1, 2, \ldots, \omega, \omega + 1, \omega + 2, \ldots\}$

    (c) $\omega^2 = \{0, 1, 2, \ldots, \omega, \omega + 1, \omega + 2, \ldots, \omega \cdot 2, \omega \cdot 2 + 1, \omega \cdot 2 + 2, \ldots, \omega \cdot 3, \omega \cdot 3 + 1, \omega \cdot 3 + 2, \ldots, \ldots\}$; $\omega^2 + 1 = \{0, 1, 2, \ldots, \omega, \omega + 1, \omega + 2, \ldots, \omega \cdot 2, \omega \cdot 2 + 1, \omega \cdot 2 + 2, \ldots, \omega \cdot 3, \omega \cdot 3 + 1, \omega \cdot 3 + 2, \ldots, \ldots, \omega^2\}$

SECTION 2.2 SETS WITH STRUCTURE

1. (a) $\{(1, 1), (1, 3), (2, 1), (2, 3)\}$

    (b) $\{(a, a, a), (a, a, b), (a, b, a), (a, b, b), (b, a, a), (b, a, b), (b, b, a), (b, b, b)\}$

    (c) $\{(1, 5), (2, 5), (3, 5), (4, 5)\}$

    (d) $\{(1, 4, 5), (2, 4, 5), (3, 4, 5)\}$

    (e) $\emptyset$

3. *uwvuuvu*

5. initial substrings: λ, *r*, *ro*, *rol*, *roll*; other substrings: *o*, *ol*, *oll*, *l*, *ll*

7. (a) $\begin{bmatrix} 2 & 3 & 4 & 5 \\ 3 & 4 & 5 & 6 \\ 4 & 5 & 6 & 7 \end{bmatrix}$ (b) $\begin{bmatrix} 1 & 1 & 1 & 1 & 1 \\ 0 & 1 & 0 & 1 & 0 \\ 0 & 0 & 1 & 0 & 0 \\ 0 & 0 & 0 & 1 & 0 \\ 0 & 0 & 0 & 0 & 1 \end{bmatrix}$

(c)
$$\begin{bmatrix} p & pa & pap & papa \\ \lambda & a & ap & apa \\ \lambda & \lambda & p & pa \\ \lambda & \lambda & \lambda & a \end{bmatrix}$$

9. (a) 38 (b) 278 (c) 700 (d) 208 (e) 12

11. Let $(a, c) \in A \times C$. Then $a \in A$ and $c \in C$. Since $A \subseteq B$, and $C \subseteq D$, we know that $a \in B$ and $c \in D$. Hence $(a, c) \in B \times D$, as desired.

13. If $A = B$, then we have the identity $A \times A = A \times A$. Conversely, suppose that $A \ne B$. Then there is an element in one of the sets but not in the other; without loss of generality, say an element a that is in A but not in B. Since $B \ne \emptyset$, we can also find an element $b \in B$. Then $(a, b) \in A \times B$, but $(a, b) \notin B \times A$ since $a \notin B$. Thus $A \times B \ne B \times A$.

15. 00001111, 00010111, 00011011, 00011101, 00100111, 00101011, 00101101, 00110011, 00110101, 01000111, 01001011, 01001101, 01010011, 01010101

17. $\{(0, 0, 3), (0, 3, 0), (3, 0, 0), (0, 1, 2), (0, 2, 1), (1, 0, 2), (1, 2, 0), (2, 0, 1), (2, 1, 0), (1, 1, 1)\}$

19. (a)
$$\begin{bmatrix} 1 & 1 & 1 & 6 \\ 1 & -1 & 0 & 5 \\ 1 & -1 & -3 & 4 \end{bmatrix}$$
(b)
$$\begin{bmatrix} 0 & 3 & 2 & 0 \\ 0 & 0 & 0 & 7 \\ 0 & 3 & 0 & 0 \\ 1 & 0 & 4 & 0 \end{bmatrix}$$

(c)
$$\begin{bmatrix} 0 & 1 & 1 & \sqrt{8} \\ 1 & 0 & \sqrt{2} & \sqrt{5} \\ 1 & \sqrt{2} & 0 & \sqrt{5} \\ \sqrt{8} & \sqrt{5} & \sqrt{5} & 0 \end{bmatrix}$$

21. (a) $\{\{x \mid a < x < b\} \mid a < b\}$, indexed by pairs of real numbers (a, b) with $a < b$
    (b) $\{$ last name of $x \mid x$ is a person in the United States $\}$
    (c) $\{\{mn \mid m \in \mathbf{N}\} \mid n \in \mathbf{N} \wedge n \ne 0\}$, indexed by the positive integers

23. Let A_n be the set of prime divisors of the natural number n. Then $\{A_n \mid n = 5, 10, 20\}$ is different from $\{A_n \mid n = 5, 10\}$ as indexed sets; but as sets, both are simply $\{\{5\}, \{2, 5\}\}$.

25. (a) 11 (b) 36

27. (a) $\displaystyle \sigma(n) = \sum_{\substack{d \mid n \\ 1 \le d < n}} d$

    (b) $\displaystyle \mathrm{trace}(A) = \sum_{i=1}^{n} a_{ii}$, where $A = (a_{ij})$ is an n by n matrix

29. (a) 5040 (b) 0 (c) 108 (d) 171

31. (a) yes (b) no (c) no

33. A palindrome is any string of the form $u_1u_2\ldots u_{n-1}u_nu_{n-1}\ldots u_2u_1$, where $n \geq 1$ and each $u_i \in U$, or of the form $u_1u_2\ldots u_{n-1}u_nu_nu_{n-1}\ldots u_2u_1$, where $n \geq 0$ and each $u_i \in U$.

SECTION 2.3   OPERATIONS ON SETS

1. (a) $\emptyset$ (b) $\{3,9\}$ (c) U (d) $\{2,3,4,6,8,9,10\}$
(e) $\{1,2,4,5,7,8,10\}$ (f) $\{1,5,7\}$ (g) $\{2,3,4,8,9,10\}$

3. (a) $M \cap C$ (b) $\overline{M} \cap C$ (c) $\overline{M \cap \overline{C}}$
(d) $A \cap \overline{C} \cap M$ (e) $\overline{A} \cap \overline{M} \cap C$

5. If $x \in A \cap B$, then $x \in A$ and $x \in B$. In particular, $x \in A$, so $x \in A \vee x \in B$. Thus $x \in A \cup B$.

7. (a) $\{1\}$ and $\{2,3\}$ (b) $\{0,4\}$ and $\{0,5,6,7,\ldots\}$
(c) $\{2,3\}$ and $\{3,4,5,\ldots\}$

9. if and only if the sets are disjoint

11. $\overline{\bigcup_{i \in I} A_i} = \left\{ x \mid x \notin \bigcup_{i \in I} A_i \right\} = \left\{ x \mid \neg\left(x \in \bigcup_{i \in I} A_i\right) \right\} = \left\{ x \mid \neg(\exists i \in I : x \in A_i) \right\} = \left\{ x \mid \forall i \in I : x \notin A_i \right\} = \left\{ x \mid \forall i \in I : x \in \overline{A}_i \right\} = \bigcap_{i \in I} \overline{A}_i$

13. $\{\{0,1\},\{2,3\},\{4,5\},\ldots\}$

15. $A \oplus B = \left\{ x \mid (x \in A \wedge x \notin B) \vee (x \in B \wedge x \notin A) \right\} = \left\{ x \mid (x \in A \vee x \in B) \wedge (x \in A \vee x \notin A) \wedge (x \notin B \vee x \in B) \wedge (x \notin B \vee x \notin A) \right\} = \left\{ x \mid (x \in A \vee x \in B) \wedge T \wedge T \wedge (\overline{x \in B \vee x \in A}) \right\} = \left\{ x \mid (x \in A \vee x \in B) \wedge \overline{x \in B \wedge x \in A} \right\} = \left\{ x \mid x \in A \cup B \wedge x \notin A \cap B \right\} = (A \cup B) - (A \cap B)$

17. (a) The double-hatched region is $A \cap (B \oplus C)$.

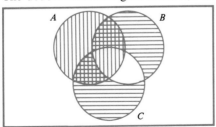

(b)   The entire shaded region is $(A - B) \cup (C - A)$.

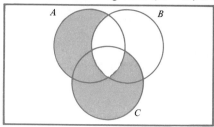

(c)   The entire shaded region is $A \oplus (B \oplus C)$.

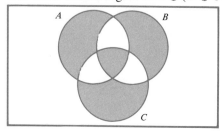

19. (a) $(A \cap \overline{B \cup C}) \cup (B \cap C \cap \overline{A})$

(b) $(A \cap B) \cup (A \cap C) \cup (B \cap C)$

(c) $(A \cup B \cup C) - (A \cap B) - (A \cap C) - (B \cap C)$

(d) $(B - A) \cup (A \cap C \cap \overline{B})$

21. (a)   Suppose that $x \in A \cup B$. Then $x \in A$ or $x \in B$. In either case, $x \in B \cup A$. The converse is proved in a similar fashion.

(b)   Suppose that $x \in A \cap (B \cup C)$. Then $x \in A$ and $x \in B \cup C$. The latter condition means that $x \in B$ or $x \in C$. If $x \in B$, then $x \in A \cap B$; and if $x \in C$, then $x \in A \cap C$. In either case $x \in (A \cap B) \cup (A \cap C)$. Conversely, suppose that $x \in (A \cap B) \cup (A \cap C)$. First, assume that $x \in A \cap B$. Then $x \in A$ and $x \in B$. Hence $x \in B \cup C$ as well, so $x \in A \cap (B \cup C)$. The other possibility is that $x \in A \cap C$. Then $x \in A$ and $x \in C$. Hence $x \in B \cup C$ as well, so again $x \in A \cap (B \cup C)$.

(c)   Suppose that $x \in A \cap U$. Then in particular $x \in A$. Conversely, if $x \in A$, then $x \in A \cap U$, since by convention x is always an element of U.

(d)   Suppose that $x \in A \cup A$. Then $x \in A$ or $x \in A$ (i.e., $x \in A$). Conversely, if $x \in A$, then $x \in A \cup A$.

(e)   Suppose that $x \in A \cap \overline{A}$. Then $x \in A$ and $x \in \overline{A}$ (i.e., $x \in A$ and $x \notin A$). This is impossible, so no such elements x exist. Therefore, $A \cap \overline{A} = \emptyset$.

(f)   Suppose that $x \in \overline{\overline{A}}$. Then it is not the case that $x \in \overline{A}$ (i.e., it is not the case that $x \notin A$). Therefore, $x \in A$. Conversely, if $x \in A$, then $x \notin \overline{A}$, whence $x \in \overline{\overline{A}}$.

(g)   Suppose that $x \in \overline{A \cup B}$. Then $x \notin A \cup B$. This means that it is not the case that $x \in A$ or $x \in B$, which implies that x is in neither A nor B. Hence $x \in \overline{A}$ and $x \in \overline{B}$, so $x \in \overline{A} \cap \overline{B}$. Conversely, if $x \in \overline{A} \cap \overline{B}$, then $x \notin A$ and $x \notin B$, so it is not the case that $x \in A \cup B$. Thus $x \in \overline{A \cup B}$.

23.   (a)   no   (b)   yes   (c)   no   (d)   yes
       (e)   yes   (f)   yes

25.   These operations correspond in the sense that $x \in A - B$ if and only if it is not the case that $x \in A$ implies $x \in B$.

27.   (a)   the $\subseteq$ relation
       (b)   Analogous to $P \Longrightarrow P \vee Q$ is $A \subseteq A \cup B$. Analogous to $P \wedge Q \Longrightarrow P$ is $A \cap B \subseteq A$. Both are clear.

29.   (a)   T   (b)   F   (c)   T   (d)   T

31.   (a)   **Q**   (b)   **R**   (c)   **R**   (d) $\emptyset$ (e) $\emptyset$ (f)   **R**

33.   One example is the collection of all sets $A_p = \{ pq \mid q \text{ is prime} \}$, where p is a prime number.

35.   (a)   If A is finite, then $\overline{A}$ has a finite complement (namely, A), and if $\overline{A}$ is finite, then A has a finite complement.
       (b)   Suppose that A and B are both in C. If either is finite, then $A \cap B$ is finite, hence in C. Otherwise, both $\overline{A}$ and $\overline{B}$ are finite, so $\overline{A} \cup \overline{B}$ is finite. But $\overline{A} \cup \overline{B} = \overline{A \cap B}$, so $A \cap B$ has a finite complement and hence is an element of C.
       (c)   Since $A \cup B = \overline{\overline{A} \cap \overline{B}}$, the fact that C is closed under complementation and intersection implies that C is closed under union.
       (d)   similar to part (c), since $A - B = A \cap \overline{B}$

CHAPTER 3: FUNCTIONS AND RELATIONS

SECTION 3.1   FUNCTIONS

1.   (a)   no   (b)   no   (c)   yes   (d)   no
      (e)   no   (f)   no

3.   (a) $\{(1,2),(2,3),(3,4),(4,5)\}$ (b) $\{(1,8),(2,8),(3,8),(4,8)\}$
      (c) $\{(1,2),(2,2),(3,4),(4,4)\}$

5.   (a)   2.32193   (b)   3.32193   (c)   166.09640

7. (a)   4   (b)   2   (c)   3   (d)   0   (e)   1

9.

+	0	1	2	3	4
0	0	1	2	3	4
1	1	2	3	4	0
2	2	3	4	0	1
3	3	4	0	1	2
4	4	0		2	3

×	0	1	2	3	4
0	0	0	0	0	0
1	0	1	2	3	4
2	0	2	4	1	3
3	0	3	1	4	2
4	0	4	3	2	1

11. (a)   Let q be Prince William of Great Britain.   (b)   It is always 1.
(c)   no   (d)   yes

13. $\forall a \in A: \exists! b \in B: (a, b) \in f$

15. $\{((\{1, 2, 3\}, 1), (\{1, 2\}, 1), (\{1, 3\}, 1), (\{2, 3\}, 2), (\{1\}, 1), (\{2\}, 2), (\{3\}, 3)\}$

17. $D: \mathbf{R} \times (\mathbf{R} - \{0\}) \to \mathbf{R}$ given by $D(x, y) = x/y$ is a function of two variables.

19. $\odot = \{((a, b), c) \mid (c = a \wedge a \geq b) \vee (c = b \wedge a \leq b)\}$

21. (a)   no   (b)   yes   (c)   yes   (d)   yes
(e)   yes   (f)   no

23. $\lfloor \log_{10} n \rfloor + 1$

25. (a) $1 + y$ (b) $2y$ (c) $y/2$ (d) x
(e) x^2 (f)   2

27. (a)   Write the real number x as $\lfloor x \rfloor + \epsilon$, with $0 \leq \epsilon < 1$. Then $0 \leq x - \lfloor x \rfloor < 1$; the desired inequalities follow algebraically.
(b)   Write $x = \lceil x \rceil - \epsilon$, with $0 \leq \epsilon < 1$. Then $0 \leq \lceil x \rceil - x < 1$; the desired inequalities follow algebraically.

29. $\lfloor x + 0.5 \rfloor$

31. If $x \equiv 0$ or 3   (mod 6), then $x = 6k$ or $x = 6k + 3$; in either case, x is divisible by 3, and therefore not prime. If $x \equiv 2$ or 4   (mod 6), then x is divisible by 2, hence not prime. Thus every prime number must be congruent to either 1 or 5, modulo 6. Numbers of the former form can be written as $6k + 1$, and numbers of the latter form can be written as $6k - 1$.

33. False; let $m = 5$, $a = b = 2$, $c = 1$, $d = 6$.

35. 2 has no multiplicative inverse modulo 6, since $2k$ is always congruent to 0, 2, or 4   (mod 6), never 1.

37. (a)   yes   (b)   no   (c)   yes

39. $\{(S, n) \mid S \in \mathcal{P}(\mathbf{N}) - \{\emptyset\} \wedge n \in S \wedge \forall m \in S: m \geq n\}$

SECTION 3.2   FUNCTIONS IN THE ABSTRACT

1. (a)

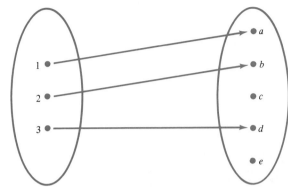

(b)

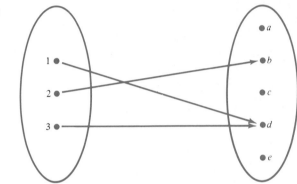

3. (a) $f(x) = x$ (b) $f(x) = x + 1$ (c) $f(x) = \lfloor x/2 \rfloor$
   (d) $f(x) = 3$

5. Let $f(0) = 1$, $f(1) = 0$, and $f(x) = x$ for all $x \geq 2$.

7. (a)   bijective   (b)   neither   (c)   bijective
   (d)   injective   (e)   surjective

9. If A is a finite set and $f\colon A \to A$, then f is injective if and only if f is surjective.

11. (a) $x = f(f(y)) \lor x = f(m(y))$
    (b) $y = m(f(x)) \lor y = m(m(x)) \lor y = f(f(x)) \lor y = f(m(x))$
    (c) $x \neq m(y) \land x \neq f(y) \land \exists u\colon (u = m(x) \land (u = m(f(y)) \lor u = m(m(y))))$
    (d) $\{ y \mid x \neq m(y) \land x \neq f(y) \land \exists u\colon (u = m(x) \land (u = m(f(y)) \lor u = m(m(y)))) \}$

13. (a) $(f + g)(x) = 3$ for all x
    (b) $(f \cdot g)(x) = 2$ for all x
    (c) $(f - g)(x) = -1$ if x is odd, 1 if x is even

(d) $(f \circ g)(x) = 2$ if x is odd, 1 if x is even

(e) $(f \circ f)(x) = 1$ if x is odd, 2 if x is even

(f) $(g \circ f)(x) = 2$ if x is odd, 1 if x is even

(g) $(g \circ g)(x) = 1$ if x is odd, 2 if x is even

15. (a) $\forall y \colon (y \in C \leftrightarrow \exists x \in A \colon f(x) = y)$

   (b) $\forall y \in B \colon \exists x_1 \in A \colon \exists x_2 \in A \colon (x_1 \neq x_2 \wedge f(x_1) = y \wedge f(x_2) = y \wedge \forall x \in A \colon [f(x) = y \rightarrow (x = x_1 \vee x = x_2)])$

17. (a) Suppose that $f(x_1) = f(x_2)$. Then $g(f(x_1)) = g(f(x_2))$. Since $g \circ f$ is one-to-one, this means that $x_1 = x_2$. Therefore, f is one-to-one.

   (b) False; let $A = \{a\}$, $B = \{b, d\}$, and $C = \{c\}$.

19. Let $f \colon A \rightarrow \mathbf{R}$ be the given function. If some horizontal line $y = b$ intersects the graph of f in two distinct points (x_1, b) and (x_2, b), then f is not injective, since $f(x_1) = f(x_2)$. Conversely, if f is not injective, then there exist distinct real numbers x_1 and x_2 such that $f(x_1) = f(x_2)$. If b is their common value, then the line $y = b$ intersects the graph in more than one point.

21. Take the graph of f and project it onto the vertical axis by moving each point (x, b) on the graph horizontally over to the point $(0, b)$. The resulting subset of the vertical axis represents the range.

23. (a) $f + g = \{ (x, u + v) \mid (x, u) \in f \wedge (x, v) \in g \}$

   (b) $i_A = \{ (a, a) \mid a \in A \}$

25. $f \circ i_A(a) = f(i_A(a)) = f(a)$ for all $a \in A$; $i_B \circ f(a) = i_B(f(a)) = f(a)$ for all $a \in A$

27. if and only if the domain is the empty set

29. (a) $f(\emptyset) = \{ f(x) \mid x \in \emptyset \} = \{ f(x) \mid \mathbf{F} \} = \emptyset$

   (b) $f^{-1}(\emptyset) = \{ x \in A \mid f(x) \in \emptyset \} = \{ x \in A \mid \mathbf{F} \} = \emptyset$

   (c) $f^{-1}(B) = \{ x \in A \mid f(x) \in B \} = A$ since $f(x) \in B$ for all $x \in A$

   (d) $f(A) = \{ f(x) \mid x \in A \}$, which is the range of f

31. (a) all multiples of 7   (b) $\{ 7k + 1 \mid k \in \mathbf{Z} \}$

   (c) same as (b)   (d) $\{ x \in \mathbf{Z} \mid x \bmod 7 = 0 \text{ or } 1 \}$

   (e) $\{0, 1, 2, 3, 4, 5, 6\}$ (f) $\{1\}$ (g) $\{0, 1\}$

33. $R = \{a\}$; $S = \{b, c\}$

35. Let $B = \{1\}$, $A = \{a, b\}$, $g(1) = a$, and $f(a) = f(b) = 1$.

37. 14

39. (a) yes   (b) yes   (c) no

41. Given a rational number a/b in lowest terms, with a an integer and b a positive integer, let $f(a/b) = 2^a 3^b$ if $a \geq 0$, and let $f(a/b) = 5^{-a} 3^b$ if $a < 0$.

SECTION 3.3 RELATIONS

1. (a) $\{(1,1),(1,2),(2,2)\}$
   (b) $\{(1,1),(1,2),(2,2)\}$
   (c) $\{(1,1),(2,1),(3,1),(2,2)\}$
   (d) $\{(1,1),(2,2)\}$
   (e) $\emptyset$
   (f) $\{(1,1),(1,2),(2,1),(2,2),(3,1),(3,2)\}$

3. (a) $(10,10),(10,11),(4,25) \in R; (0,10),(5,5) \notin R$
   (b) $(3,0),(3,1),(3,2) \in R; (2,2),(2,3) \notin R$
   (c) no elements in R; $(2,1),(0,0) \notin R.$

5. (a)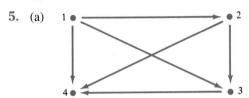

   (b)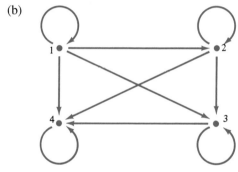

   (c)

7. (a) C is not a function.
   (b) because no one is his or her own child
   (c) "is a grandchild of"

9. (a) $(0,0), (0,5), (1,2), (1,7), (2,4), (3,1), (3,6), (4,3), (5,0), (5,5), (6,2),$
   $(6,7), (7,4)$

(b)

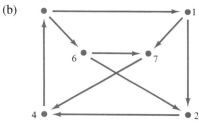
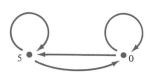

(c)

$$
\begin{array}{c c}
 & \begin{array}{cccccccc} 0 & 1 & 2 & 3 & 4 & 5 & 6 & 7 \end{array} \\
\begin{array}{c} 0 \\ 1 \\ 2 \\ 3 \\ 4 \\ 5 \\ 6 \\ 7 \end{array} &
\left[\begin{array}{cccccccc}
1 & 0 & 0 & 0 & 0 & 1 & 0 & 0 \\
0 & 0 & 1 & 0 & 0 & 0 & 0 & 1 \\
0 & 0 & 0 & 0 & 1 & 0 & 0 & 0 \\
0 & 1 & 0 & 0 & 0 & 0 & 1 & 0 \\
0 & 0 & 0 & 1 & 0 & 0 & 0 & 0 \\
1 & 0 & 0 & 0 & 0 & 1 & 0 & 0 \\
0 & 0 & 1 & 0 & 0 & 0 & 0 & 1 \\
0 & 0 & 0 & 0 & 1 & 0 & 0 & 0
\end{array} \right]
\end{array}
$$

11. (a) $\{(a, a), (a, b), (b, a), (b, c), (c, c)\}$

(b) $\{(1, 2), (1, 3), (1, 4), (3, 1), (3, 2), (3, 3), (3, 4), (4, 2), (4, 4)\}$

(c) $\{(1, b), (2, a), (2, b), (2, c), (3, c), (4, a), (4, b)\}$

(d) $\{(a, 1), (c, 1), (a, 3), (b, 3), (c, 4)\}$

(e) $\{(1, 3), (2, 1), (2, 3), (2, 4), (3, 1), (3, 3), (4, 1), (4, 3), (4, 4)\}$

(f) $\{(a, a), (a, b), (b, a), (c, b), (c, c)\}$

(g) $\{(a, a), (a, b), (b, a), (c, b), (c, c)\}$

13. (a) $R \circ \Delta_A = \{ (a, b) \mid \exists a' \in A : (a, a') \in \Delta_A \land (a', b) \in R \}$. For each $a \in A$ there is a unique $a' \in A$ such that $(a, a') \in \Delta_A$, namely $a' = a$. Therefore, $(a, b) \in R \circ \Delta_A$ if and only if $(a, b) \in R$, so $R \circ \Delta_A = R$.

(b) $\Delta_B \circ R = \{ (a, b) \mid \exists b' \in B : (a, b') \in R \land (b', b) \in \Delta_B \}$. For each $b \in B$ there is a unique $b' \in B$ such that $(b', b) \in \Delta_B$, namely $b' = b$. Therefore, $(a, b) \in \Delta_B \circ R$ if and only if $(a, b) \in R$, so $\Delta_B \circ R = R$.

(c) False; let $A = B = \{1, 2\}$, $R = A \times A$.

(d) False; let $A = \{1\}$, $B = \{2, 3\}$, $C = \{4\}$, $R = \{(1, 2)\}$, $S = \{(2, 4)\}$.

(e) Suppose that $(a, c) \in S \circ (R \cup R')$. Then $(a, b) \in R \cup R'$ and $(b, c) \in S$ for some $b \in B$. If $(a, b) \in R$, then $(a, c) \in S \circ R$, and if $(a, b) \in R'$, then $(a, c) \in S \circ R'$; in either case $(a, c) \in (S \circ R) \cup (S \circ R')$. Conversely, suppose that $(a, c) \in (S \circ R) \cup (S \circ R')$. Without loss of generality, suppose that $(a, c) \in S \circ R$. Then $(a, b) \in R$ and $(b, c) \in S$ for some $b \in B$. Thus $\exists b \in B : ((a, b) \in R \cup R' \land (b, c) \in S)$, so $(a, c) \in S \circ (R \cup R')$.

(f) False; let $A = \{1\}$, $B = \{2, 3\}$, $C = \{4\}$, $R = \{(1, 2)\}$, $R' = \{(1, 3)\}$, $S = \{(2, 4), (3, 4)\}$.

15. a subset of A

17. (a) Let a be any element of A. Then $(a, a) \in R$, so $(a, a) \notin \overline{R}$. Therefore, $\overline{R}$ is not reflexive.

(b) Suppose that $(x, y) \in \overline{R}$. Then $(x, y) \notin R$. If (y, x) were in R, then (x, y) would be in R as well; therefore, $(y, x) \notin R$. In other words, $(y, x) \in \overline{R}$. Thus $\overline{R}$ is symmetric.

(c) False; let $A = \{x, y\}$, $R = \emptyset$.

(d) False; let $A = \{x\}$, $R = \emptyset$.

19. Suppose that $\{R_i\}_{i \in I}$ is the given collection of relations on A, and let $R = \bigcap_{i \in I} R_i$ be their intersection.

(a) Let $a \in A$. Then since each R_i is reflexive, $(a, a) \in R_i$ for each i. Therefore, $(a, a) \in R$, so R is reflexive.

(b) Let $(x, y) \in R$, so that $(x, y) \in R_i$ for each i. Then since each R_i is symmetric, $(y, x) \in R_i$ for each i. Therefore, $(y, x) \in R$, so R is symmetric.

21. (a) symmetric
(b) antisymmetric, transitive
(c) antisymmetric, transitive
(d) reflexive, transitive
(e) none
(f) none
(g) symmetric
(h) antisymmetric, transitive
(i) reflexive, antisymmetric, transitive

23.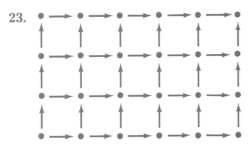

25. (a) Suppose that $(x, y) \in R \wedge (y, z) \in R$. By symmetry we know that $(y, x) \in R$ and hence by antisymmetry, $x = y$. Hence $(y, z) = (x, z)$, so $(x, z) \in R$.

(f) any subset of Δ_A

27. (a) Two authors a and b are related if there is a sequence of authors $a = a_0$, $a_1, a_2, \ldots, a_n = b$ such that a_{i-1} and a_i have written a joint paper, for $i = 1, 2, \ldots, n$.

(b) the relation that always holds

(c) "is an ancestor of"

(d) the relation that always holds, that assuming that every two people have a common ancestor (Adam and Eve?)

29. (a) all 1's along the main diagonal

(b) all 0's along the main diagonal

(c) The matrix is symmetric about the main diagonal.

(d) There are never two 1's symmetrically located around the main diagonal.

31. (a) Assume that R is symmetric. If $(x, y) \in R$, then $(y, x) \in R$, so $(x, y) \in R^{-1}$. Thus $R \subseteq R^{-1}$. Similarly, if $(x, y) \in R^{-1}$, then $(y, x) \in R$, which implies by symmetry that $(x, y) \in R$. Thus $R^{-1} \subseteq R$. Therefore, $R = R^{-1}$. Conversely, assume that $R = R^{-1}$. If $(x, y) \in R$, then $(x, y) \in R^{-1}$, so $(y, x) \in R$. In other words, R is symmetric.

(b) Assume that R is transitive. If $(x, y) \in R^2$, then there exists a $z \in A$ such that $(x, z) \in R$ and $(z, y) \in R$. Thus by transitivity, $(x, y) \in R$. This shows that $R^2 \subseteq R$. Conversely, assume that $R^2 \subseteq R$. Let $(x, y) \in R$ and $(y, z) \in R$. Then by definition, $(x, z) \in R^2$, so $(x, z) \in R$. This shows that R is transitive.

(c) R is reflexive $\leftrightarrow \forall a \in A: (a, a) \in R \leftrightarrow \Delta_A \subseteq R$

(d) Suppose that R is antisymmetric, and let $(x, y) \in R \cap R^{-1}$. Then $(x, y) \in R \wedge (x, y) \in R^{-1}$. The latter condition means that $(y, x) \in R$, so by antisymmetry, $x = y$. This shows that $R \cap R^{-1} \subseteq \Delta_A$. Conversely, suppose that $R \cap R^{-1} \subseteq \Delta_A$. If both $(x, y) \in R$ and $(y, x) \in R$, then $(x, y) \in R \cap R^{-1}$, so $(x, y) \in \Delta_A$. This means that $x = y$. Therefore, R is antisymmetric.

SECTION 3.4 ORDER RELATIONS AND EQUIVALENCE RELATIONS

1. (a) $\{(a, a), (a, b), (b, b), (c, c), (d, c), (d, d), (e, b), (e, c), (e, d), (e, e), (e, h),$ $(f, a), (f, b), (f, c), (f, d), (f, e), (f, f), (f, g), (f, h), (g, a), (g, b), (g, g),$ $(h, b), (h, c), (h, h)\}$

(b)

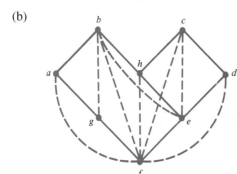

3. $(a, a), (a, b), (a, c), (b, a), (b, b), (b, c), (c, a), (c, b), (c, c)$

5. (a)

(b) It is reflexive, symmetric, and transitive.

(c) $\{\{1, 5\}, \{2, 6\}, \{3\}, \{4\}\}$

(d)

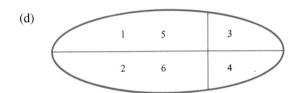

(e) $[1] = \{1, 5\} = [5]; [2] = \{2, 6\} = [6]; [3] = \{3\}; [4] = \{4\}$

7. (a) $\{(1, 1), (1, 2), (1, 3), (2, 1), (2, 2), (2, 3), (3, 1), (3, 2), (3, 3), (4, 4), (4, 5),$
$(5, 4), (5, 5), (6, 6)\}$

(b)

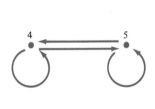 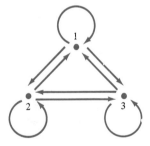 

9. (a) m is the minimum element of $(A, \preceq)$ if and only if $\forall x \in A : m \preceq x$

(b) m is a minimal element of $(A, \preceq)$ if and only if $\forall x \in A : (x \preceq m \rightarrow x = m)$

11. The empty relation on $\{1\}$ is antisymmetric and transitive, but not reflexive. The relation that always holds on $\{1, 2\}$ is reflexive and transitive, but not antisymmetric. The relation $\{(1, 1), (1, 2), (2, 2), (2, 3), (3, 3)\}$ on $\{1, 2, 3\}$ is reflexive and antisymmetric, but not transitive.

13.

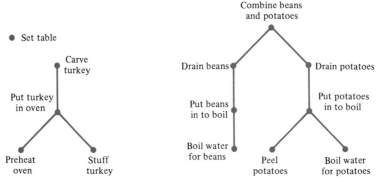

15. (a) $y = px$ for some prime p
    (b) the numbers of distinct prime divisors of y
    (c) an infinite number

17. (a) The proof of Theorem 1 in Section 3.3 showed that the intersection of transitive relations is transitive. Exercises 20d and e in Section 3.3 showed the same for reflexivity and antisymmetry.
    (b) False; the union of $\leq$ and $\geq$ is not antisymmetric.
    (c) False; let $R = \emptyset$.

19. (a) $\emptyset$ is minimal and minimum; **N** is maximal and maximum.
    (b) $(1, 1)$ is minimal and minimum; $(3, 3)$ is maximal and maximum.
    (c) a and b are minimal; no minimum; e is maximal and maximum.
    (d) a, b, c, h, and i are minimal; f, g, h and j are maximal; no minimum or maximum.
    (e) f is minimal and minimum; b and c are maximal; no maximum.
    (f) 1 is minimum, maximum, minimal and maximal.
    (g) 1 and 2 minimal and maximal; no minimum or maximum.

21. Replace "dominate" by "are dominated by," $u \preceq v$ by $v \preceq u$, "maximal" by "minimal," and so on in the rigorous proof of Theorem 2.

23. $\emptyset$ on $\{1\}$ is symmetric and transitive, but not reflexive; $\{(1, 1), (1, 2), (2, 2)\}$ on $\{1, 2\}$ is reflexive and transitive, but not symmetric; $\{(1, 1), (1, 2), (2, 1), (2, 2), (2, 3), (3, 2), (3, 3)\}$ on $\{1, 2, 3\}$ is reflexive and symmetric, but not transitive.

25. (b) and (f) are not equivalence relations.

27. the equivalence classes of R

29. Since $x - x = 0 = 0 \cdot m$ is a multiple of m, we know that the relation is reflexive. If $x - y = km$, then $y - x = (-k)m$; this tells us that the relation is symmetric. If $x - y = km$ and $y - z = lm$, then $x - z = (x - y) + (y - z) = km + lm = (k + l)m$; this tells us that the relation is transitive.

31. (a)   1   (b)   1   (c)   2

33. (a)   15   (b)   15

35. (a)   follows from the proof of Theorem 1 in Section 3.3
 (b)   False; let $R_1 = \{(1,1), (1,2), (2,1), (2,2), (3,3)\}$, $R_2 = \{(1,1), (2,2), (2,3), (3,2), (3,3)\}$.
 (c)   Reflexivity is clear; the inverse of any symmetric relation is itself; if $(x,y) \in R^{-1} \wedge (y,z) \in R^{-1}$, then $(y,x) \in R \wedge (z,y) \in R$, whence $(z,x) \in R$, and so $(x,z) \in R^{-1}$.
 (d)   False; the complement is never reflexive.

37. (a) $\{\{1,3,5,7,\ldots\},\ \{2,6,10,14,\ldots\},\ \{4,12,20,28,\ldots\},\ \{8,24,40,56,\ldots\},\ldots\}$
 (b) $\{\{1,2,4,8,16,\ldots\},\ \{3,6,12,24,48,\ldots\},\ \{5,10,20,40,80,\ldots\},\ \{7,14,28,56,112,\ldots\},\ldots\}$

39. Form its reflexive, symmetric, transitive closure (see Section 3.3).

41. (a)   Since the identity function on A is bijective, we have ARA. If ARB, then there is a bijective function $f\colon A \to B$. Then $f^{-1}\colon B \to A$ is a bijective function as well, so BRA. If ARB and BRC, then there are bijective functions $f\colon A \to B$ and $g\colon B \to C$. Then $g \circ f$ is a bijective function from A to C (Theorem 2 in Section 3.2), so ARC.
 (b)   For $i = 0,1,2,\ldots$, let C_i be the set of all finite subsets of $\mathbf{N}$ with cardinality i; and let C_∞ be the set of all infinite subsets of $\mathbf{N}$. The equivalence classes are all the sets C_i for $i = 0,1,2,\ldots,\infty$.

43. (a) $\leq$ on $\mathbf{R}$
 (b)   the relation R on $\mathbf{R} \times \mathbf{R}$ given by $(a,b)R(c,d) \longleftrightarrow (a \leq c \wedge b \leq d)$
 (c) $\leq$ on $\mathbf{Z}$
 (d)   Let $a \prec b$. Then there is an element x_1 such that $a \prec x_1 \prec b$. Using the definition of denseness again, we obtain an element x_2 such that $x_1 \prec x_2 \prec b$. Continuing in this way, we obtain an infinite sequence of distinct elements $x_1 \prec x_2 \prec x_3 \prec \cdots$.

45. (a) $\pi_1 = \{\{1\}, \{2,3\}, \{4,5\}\}$ and $\pi_2 = \{\{1,2,3\}, \{4,5\}\}$
 (b) $\pi_1 = \{\{1,2,3\}, \{4,5\}\}$ and $\pi_2 = \{\{1,2\}, \{3,4\}, \{5\}\}$
 (c)   For any partition, $\pi \preceq \pi$, since we can take $Y = X$ in the definition. Thus the refinement relation is reflexive. Next suppose that $\pi_1 \preceq \pi_2$ and $\pi_2 \preceq \pi_1$. Let X be an arbitrary element of π_1. Then $X \subseteq Y$ for some $Y \in \pi_2$. Furthermore, $Y \subseteq X'$ for some $X' \in \pi_1$. Hence $X \subseteq X'$. Since the elements of π_1 are nonempty and disjoint, this can only happen if $X = X'$. Thus we have $X \subseteq Y \subseteq X$, whence $X = Y$. This shows that every element of π_1 is also in π_2. By similar reasoning every element of π_2 is in π_1, so $\pi_1 = \pi_2$. Thus we have established that $\preceq$ is antisymmetric. Finally, suppose that $\pi_1 \preceq \pi_2$ and $\pi_2 \preceq \pi_3$. We

must show that $\pi_1 \preceq \pi_3$. Let $X \in \pi_1$. Then there is a $Y \in \pi_2$ such that $X \subseteq Y$, and hence also a $Z \in \pi_3$ such that $Y \subseteq Z$. But then $X \subseteq Z$, as desired.

(d) $\{\{1, 2, 3, 4, 5\}\}$

(e) $\{\{1\}, \{2\}, \{3\}, \{4\}, \{5\}\}$

CHAPTER 4: ALGORITHMS

SECTION 4.1 THE IDEA OF AN ALGORITHM

1. Suppose that a, b, and c are the three numbers. Compare a and b. If $a < b$, then compare a and c. If $c < a$, then give a as output; otherwise $(c > a)$, compare b and c, and give b as output if $b < c$ and c as output if $c < b$. On the other hand, if $b < a$, then compare b and c. If $c < b$, then give b as output; otherwise $(c > b)$, compare a and c, and give a as output if $a < c$ and c as output if $c < a$.

3. Set $i = 1$. As long as $i^2 \leq n$, continue to add 1 to i. Stop as soon as $i^2 > n$, giving $(i - 1)^2$ as output.

5.

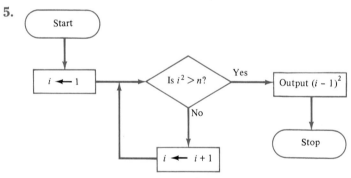

7. (a) 50   (b) 1   (c) 101

9. $d = \gcd(x, y) \longleftrightarrow [d \mid x \wedge d \mid y \wedge \forall a \colon ((a \mid x \wedge a \mid y) \rightarrow a \leq d)]$

11. Let the prime factorizations of x and y be $x = p_1^{x_1} p_2^{x_2} \cdots p_k^{x_k}$ and $y = p_1^{y_1} p_2^{y_2} \cdots p_k^{y_k}$, respectively, where we included all the prime factors of both x and y, so that some of the x_i's or y_i's might be 0. Then $\gcd(x, y) = p_1^{g_1} p_2^{g_2} \cdots p_k^{g_k}$, where $g_i = \min(x_i, y_i)$. Now suppose that $d \mid x \wedge d \mid y$. Then $d = p_1^{d_1} p_2^{d_2} \cdots p_k^{d_k}$, where each $d_i \leq x_i$ and $d_i \leq y_i$. Therefore, $\forall i \colon d_i \leq g_i$, so $d \mid \gcd(x, y)$.

13. (a) $m = \operatorname{lcm}(x, y) \longleftrightarrow [x \mid m \wedge y \mid m \wedge \forall z \colon ((x \mid z \wedge y \mid z) \rightarrow m \leq z)]$

(b) Let the prime factorizations of x and y be $x = p_1^{x_1} p_2^{x_2} \cdots p_k^{x_k}$ and $y = p_1^{y_1} p_2^{y_2} \cdots p_k^{y_k}$, respectively, where we included all the prime factors of both x and y, so that some of the x_i's or y_i's might be 0. Then $\text{lcm}(x, y) = p_1^{l_1} p_2^{l_2} \cdots p_k^{l_k}$, where $l_i = \max(x_i, y_i)$.

(c) If x and y are positive integers, and m is any positive common multiple of x and y, then $\text{lcm}(x, y) \mid m$. *Proof*: Write the prime factorizations of x, y, and their least common multiple as in part (b). Now suppose that $x \mid m$ and $y \mid m$. Then $m = p_1^{m_1} p_2^{m_2} \cdots p_k^{m_k}$, where each $m_i \geq x_i$ and $m_i \geq y_i$. Therefore, $\forall i \colon m_i \geq l_i$, so $\text{lcm}(x, y) \mid m$.

(d) Using the notation above and from Exercise 11, we can write $x = p_1^{x_1} p_2^{x_2} \cdots p_k^{x_k}$, $y = p_1^{y_1} p_2^{y_2} \cdots p_k^{y_k}$, $\text{lcm}(x, y) = p_1^{l_1} p_2^{l_2} \cdots p_k^{l_k}$, and $\gcd(x, y) = p_1^{g_1} p_2^{g_2} \cdots p_k^{g_k}$, where $l_i = \max(x_i, y_i)$ and $g_i = \min(x_i, y_i)$. Then the prime p_i appears to the power $x_i + y_i$ in $x \cdot y$, and to the power $l_i + g_i$ in $\text{lcm}(x, y) \cdot \gcd(x, y)$. Since for any two numbers u and v it is always true that $\max(u, v) + \min(u, v) = u + v$, these two exponents agree. The desired conclusion follows.

(e) Compute $\gcd(x, y)$ by the Euclidean algorithm. Then divide xy by $\gcd(x, y)$.

(f) 11163157

15. Since $\gcd(x, y) \mid x$ and $\gcd(x, y) \mid y$, this is just a special case of the transitivity of the "divides" relation (see Exercise 14 in Section 3.4).

17. Note that the remainder when $x^2 - 1$ is divided by x is $x - 1$, since $x^2 - 1 = x(x-1) + (x-1)$. Thus by Theorem 2 we have $\gcd(x^2 - 1, x) = \gcd(x, x-1) = \gcd(x - 1, 1) = \gcd(1, 0) = 1$.

19. (a) $34019/48848$ (b) $39/56$ (c) $238135/341936$

21. 1. Set i equal to 0.
    2. Set j equal to i.
    3. If $i^2 + 3j^2 > N$, then replace i by $i + 1$, return to step 2, and continue from there. Otherwise, continue with step 4.
    4. Set k equal to j.
    5. If $i^2 + j^2 + 2k^2 > N$, then replace j by $j + 1$, return to step 3, and continue from there. Otherwise, continue with step 6.
    6. Set l equal to k.
    7. If $i^2 + j^2 + k^2 + l^2 = N$, then output (i, j, k, l) and stop. If $i^2 + j^2 + k^2 + l^2 > N$, then replace k by $k + 1$, return to step 5, and continue from there. Otherwise (i.e., when $i^2 + j^2 + k^2 + l^2 < N$), replace l by $l + 1$ and repeat this step.

23. Set *largest* equal to s_1, and set *second* equal to $-\infty$. Now for each i from 2 to n, repeat the following steps: compare s_i to *second*, and if $s_i > second$,

then replace *second* by s_i, compare the new value of *second* to *largest*, and if *second* > *largest*, interchange *second* and *largest*. At the end, output *second*.

25. 1. Set i equal to 1.
2. If $i > N$, then output 0 and stop. Otherwise, continue with step 3.
3. If $w = w_i$, then output i and stop. Otherwise continue with step 4.
4. Replace i by $i + 1$, return to step 2, and continue from there.

27. In each case, let $A = \{a_1, a_2, \ldots, a_n\}$ and $B = \{b_1, b_2, \ldots, b_m\}$ be the input sets, with $a_1 < a_2 < \cdots < a_n$ and $b_1 < b_2 < \cdots < b_m$. Set a_{n+1} and b_{m+1} equal to ∞. The output is the set $C = \{c_1, c_2, \ldots, c_k\}$.

(a) 1. Set i equal to 1; set j equal to 1; set k equal to 0.
2. If $i = n + 1$ and $j = m + 1$, then output $\{c_1, c_2, \ldots, c_k\}$ and stop. Otherwise, continue with step 3.
3. If $a_i = b_j$, then continue with step 4. If $a_i < b_j$, then replace i by $i + 1$; otherwise (when $a_i > b_j$) replace j by $j + 1$. In either of these cases, return to step 2 and continue from there.
4. Replace k by $k + 1$; set c_k equal to a_i; replace i by $i + 1$; replace j by $j + 1$; and return to step 2 and continue from there.

(b) 1. Set i equal to 1; set j equal to 1; set k equal to 0.
2. If $i = n + 1$ and $j = m + 1$, then output $\{c_1, c_2, \ldots, c_k\}$ and stop. Otherwise, replace k by $k + 1$ and continue with step 3.
3. If $a_i = b_j$, then continue with step 4. If $a_i < b_j$, then skip to step 5 and continue from there. If $a_i > b_j$, then skip to step 6 and continue from there.
4. Set c_k equal to a_i; replace i by $i + 1$; replace j by $j + 1$; and return to step 2 and continue from there.
5. Set c_k equal to a_i; replace i by $i + 1$; and return to step 2 and continue from there.
6. Set c_k equal to b_j; replace j by $j + 1$; and return to step 2 and continue from there.

(c) 1. Set i equal to 1; set j equal to 1; set k equal to 0.
2. If $i = n + 1$, then output $\{c_1, c_2, \ldots, c_k\}$ and stop. Otherwise, continue with step 3.
3. If $a_i = b_j$, then replace i by $i + 1$, replace j by $j + 1$, return to step 2 and continue from there. Otherwise, if $a_i > b_j$, then replace j by $j + 1$ and repeat this step from its beginning. Otherwise (i.e., when $a_i < b_j$) continue with step 4.
4. Replace k by $k + 1$; set c_k equal to a_i; replace i by $i + 1$; and return to step 2 and continue from there.

29. Repeatedly bend down a corner of the rectangle, as shown in the following diagram, and discard square $AB'CB$ by cutting along AB. When only a square remains, you have found the common measure.

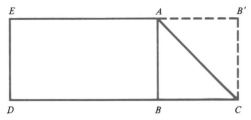

31. 1.   Set l equal to 1, and set u equal to 2.
2.   If $u - l < 10^{-n}$, then output l and stop. Otherwise, continue with step 3.
3.   Set x equal to $(u + l)/2$. If $x^2 < 2$, then replace l by x. Otherwise, replace u by x. In either case, return to step 2 and continue from there.

33. We need to check for reflexivity and transitivity in both cases, symmetry in part (a), and antisymmetry in part (b). Form the matrix representing R, that is, the n by n matrix $M_R = (m_{ij})$ in which $m_{ij} = 1$ if iRj and $m_{ij} = 0$ if not. To check for reflexivity, see if $m_{ii} = 1$ for all i. To check for transitivity, compute $M_{R \circ R}$, defined in Section 3.3 to be the Boolean matrix product of M with itself. The relation is transitive if and only if $M_{R \circ R} = M$ (given that R is reflexive). To check for symmetry, see that each below-the-diagonal entry has the same value as the above-the-diagonal entry symmetrically located across the main diagonal. To check for antisymmetry, see that for each below-the-diagonal entry that has the value 1, the above-the-diagonal entry symmetrically located across the main diagonal has the value 0.

SECTION 4.2   PSEUDOCODE
DESCRIPTION OF ALGORITHMS

1.   **procedure** *small_sum*(A: list of integers)
$\{A = (a_1, a_2, \ldots, a_n);$ *sum* is the sum of the elements of A less than 100$\}$
sum $\leftarrow 0$
for $i \leftarrow 1$ **to** n **do**
if $a_i < 100$ **then** *sum* $\leftarrow$ *sum* $+ a_i$
return(*sum*)

3. (a)   **procedure** *middle*(a, b, c : distinct natural numbers)
$\{$finds the middle value among a, b, and $c\}$
if $a < b$ **then**
if $c < a$ **then return**(a)
else if $b < c$ **then return**(b)
else return(c)
else if $c < b$ **then return**(b)
else if $a < c$ **then return**(a)
else return(c)

(b) **procedure** *add_fractions*(a, b, c, d : integers)
      {assumes that $b > 0$ and $d > 0$; returns the numerator and
      denominator of $(a/b) + (c/d)$}
        return$(ad + bc, bd)$

(c) **procedure** *large_square*(n : positive integer)
      {computes largest perfect square $\leq n$}
        $i \leftarrow 1$
        while $i^2 \leq n$ **do**
          $i \leftarrow i + 1$
        return$((i - 1)^2)$

(d) **procedure** *big_word*($w_1, w_2, \ldots, w_n$: strings)
      {finds a longest word in the list, and its length}
        maxword $\leftarrow w_1$
        maxlength $\leftarrow$ length(w_1)
        for $i \leftarrow 2$ **to** n **do**
          if length(w_i) > *maxlength* **then**
            begin
              maxword $\leftarrow w_i$
              maxlength $\leftarrow$ length(w_i)
            end
        return(*maxlength*, *maxword*)

5. (a) $i \leftarrow m$
        while $i \leq n$ **do**
          begin
            statement
            $i \leftarrow i + 1$
          end

(b) $i \leftarrow n$
        while $i \geq 1$ **do**
          begin
            statement
            $i \leftarrow i - 1$
          end

(c) $i \leftarrow m$
        while $i \leq n$ **do**
          begin
            statement
            $i \leftarrow i + k$
          end

7. Suppose that d is a common divisor of x, y, and z. Then d is a common divisor of x and y, so by Theorem 1 in Section 4.1, d is a common divisor of

gcd(x, y) and z. Conversely, if d is a divisor of gcd(x, y), then by Exercise 15 in Section 4.1, d divides both x and y. Hence d is a common divisor of x, y, and z. Thus we have shown that the sets of divisors about which the two sides of the equation speak are identical, and so their greatest elements are equal.

9. **procedure** *naive_gcd*(x, y : positive integers)
   {computes gcd(x, y) by brute force}
       $d \leftarrow \min(x, y)$
       while $\neg(d \mid x \wedge d \mid y)$ **do**
         $d \leftarrow d - 1$
       return(d)

11. **procedure** *small_prime*(N : integer > 1)
    {finds smallest prime factor of N}
        $d \leftarrow 2$
        while $d \nmid N$ **do**
          $d \leftarrow d + 1$
        return(d)

13. **procedure** *factorization*(N : integer > 1)
    {the output will be a set of pairs, $L = \{(p_1, e_1), (p_2, e_2), \ldots, (p_k, e_k)\}$, where $N = p_1^{e_1} p_2^{e_2} \cdots p_k^{e_k}$ is the prime factorization}
        $k \leftarrow 0$
        $p \leftarrow 2$
        while $N > 1$ **do**
          begin
            $e \leftarrow 0$
            while $p \mid N$ **do**
              begin
                $e \leftarrow e + 1$
                $N \leftarrow N/p$
              end
            if $e > 0$ **then**
              begin
                $k \leftarrow k + 1$
                $p_k \leftarrow p;\ e_k \leftarrow e$
              end
            $p \leftarrow p + 1$
          end
        return(L)

15. **procedure** *mode*(s : string of decimal digits)
    {finds the digit that occurs most often in s; assume that answer is unique; uses procedure *tabulate* from this section}
        $count \leftarrow tabulate(s)$

$maxdigit \leftarrow 0$
for $i \leftarrow 1$ **to** 9 **do**
    if $count(i) > count(maxdigit)$ **then** $maxdigit \leftarrow i$
return($maxdigit$)

17. **procedure** $order\_4\,(a, b, c, d$: distinct numbers)
    {returns the same numbers, put into increasing order}
        if $a > b$ **then** interchange a and b
        if $b > c$ **then** interchange b and c
        if $c > d$ **then** interchange c and d
        if $a > b$ **then** interchange a and b
        if $b > c$ **then** interchange b and c
        if $a > b$ **then** interchange a and b
        return(a, b, c, d)

19. **procedure** $exercise\_19\,(A$: list of nonzero real numbers)
    {assume that $A = (a_1, a_2, \ldots, a_n)$ has at least one positive and one
    negative number}
    $pos\_count \leftarrow 0$; $neg\_count \leftarrow 0$
    $pos\_sum \leftarrow 0$; $neg\_sum \leftarrow 0$
    for $i \leftarrow 1$ **to** n **do**
      if $a_i > 0$ **then**
          begin
             $pos\_count \leftarrow pos\_count + 1$
             $pos\_sum \leftarrow pos\_sum + a_i$
          end
        else
          begin
             $neg\_count \leftarrow neg\_count + 1$
             $neg\_sum \leftarrow neg\_sum - a_i$
          end
    return($(pos\_sum/pos\_count) - (neg\_sum/neg\_count)$)

21. **procedure** $circulant(a_1, a_2, \ldots, a_n)$
    {constructs circulant matrix ($circ$) whose first row is $(a_1, a_2, \ldots, a_n)$}
    for $i \leftarrow 1$ **to** n **do**
      for $j \leftarrow 1$ **to** n **do**
        begin
          $k \leftarrow i - 1 + j$
          if $k > n$ **then** $k \leftarrow k - n$
          $circ(i, j) \leftarrow a_k$
        end
    return($circ$)

23. **procedure** *phi*(*n* : positive integer)
    {computes the Euler phi function}
        count ← 0
        for *i* ← 1 **to** *n* **do**
          if *euclid*(*i*, *n*) = 1 **then** *count* ← *count* + 1
        return(*count*)

25. **procedure** *modes*(*s* : string of decimal digits)
    {finds the digits that occur most often in *s*; answer is given as a set *L* of
      digits; uses procedure *tabulate* from this section}
        count ← *tabulate*(*s*)
        maxdigit ← 0
        L ← {0}
        for *i* ← 1 **to** 9 **do**
          if *count*(*i*) = *count*(*maxdigit*) **then**
            L ← *L* ∪ {*maxdigit*}
          else if *count*(*i*) > *count*(*maxdigit*) **then**
            begin
              maxdigit ← *i*
              L ← {*i*}
            end
        return(*L*)

27. **procedure** *substring*(*s*, *t* : strings)
    {*s* = $s_1 s_2 \ldots s_n$ and *t* = $t_1 t_2 \ldots t_m$; determines whether *s* is a substring
      of *t*}
        for *j* ← 0 **to** *m* − *n* **do**
        begin
          matched ← **true**
          for *i* ← 1 **to** *n* **do**
            if $s_i \neq t_{j+i}$ **then** *matched* ← **false**
          if *matched* **then return**(**true**)
        end
        return(**false**)

29. **procedure** *length_of_repeat*(*n* : positive integer)
    {finds the length of the repeat in the decimal expansion of 1/*n*; the array
      R holds the successive remainders in the long division process}
        R(1) ← 1 {the numerator is 1}
        l ← 1
        while true do
          begin
            l ← *l* + 1
            R(*l*) ← 10 · *R*(*l* − 1) mod *n*
            if *R*(*l*) = 0 **then return**(0)

> **for** $i \leftarrow 1$ **to** $l - 1$ **do**
> > **if** $R(i) = R(l)$ **then return**$(l - i)$
>
> **end**

SECTION 4.3   EFFICIENCY OF ALGORITHMS

1. (a) $i \leftarrow 1$; *eating* $\neq$ *a*; $i \leftarrow 2$; *eating* $\neq$ *curds*; $i \leftarrow 3$; *eating* = *eating*; return 3.

    (b) $i \leftarrow 1$; *more* $\neq$ *a*; $i \leftarrow 2$; *more* $\neq$ *curds*; $i \leftarrow 3$; *more* $\neq$ *eating*; $i \leftarrow 4$; *more* $\neq$ *her*; $i \leftarrow 5$; *more* $\neq$ *little*; $i \leftarrow 6$; *more* $\neq$ *miss*; $i \leftarrow 7$; *more* $\neq$ *muffet*; $i \leftarrow 8$; *more* $\neq$ *on*; $i \leftarrow 9$; *more* $\neq$ *sat*; $i \leftarrow 10$; *more* $\neq$ *tuffet*; loop finished, return 0.

    (c)   *low* $\leftarrow 1$, *high* $\leftarrow 10$, *middle* $\leftarrow 5$; *eating* $\preceq$ *little*, so *high* $\leftarrow 5$, *middle* $\leftarrow 3$; *eating* $\preceq$ *eating*, so *high* $\leftarrow 3$, *middle* $\leftarrow 2$; *eating* $\npreceq$ *curds*, so *low* $\leftarrow 3$, loop finished; *eating* = *eating*, so return 3.

    (d)   *low* $\leftarrow 1$, *high* $\leftarrow 10$, *middle* $\leftarrow 5$; *more* $\npreceq$ *little*, so *low* $\leftarrow 6$, *middle* $\leftarrow 8$; *more* $\preceq$ *on*, so *high* $\leftarrow 8$, *middle* $\leftarrow 7$; *more* $\preceq$ *muffet*, so *high* $\leftarrow 7$, *middle* $\leftarrow 6$; *more* $\npreceq$ *miss*, so *low* $\leftarrow 7$, loop finished; *more* $\neq$ *muffet*, so return 0.

3. (a) $O(n^3)$ (b) $O(n^2)$ (c) $O(\sqrt{n})$ (d) $O(2^n)$

5. (a) $O(n)$ (b) $O(\log n)$ (c) $O(1)$

7. (a) $(high - low)/2$ or $((high - low)/2) - 1$ if $high - low$ is even; $(high - low - 1)/2$ if $high - low$ is odd

    (b)   If $(high - low)$ is odd, then $(high - low) \geq 1$, so that the new value of $(high - low)$ is still greater than or equal to 0 by the formula from part (a). If $(high - low)$ is even and not yet 0, then $(high - low) \geq 2$, so again the formula from part (a) guarantees that $(high - low)$ remains nonnegative. Thus $(high - low)$ never becomes less than 0 and so must equal 0 when the loop terminates.

9. $\log_a n = (\log_b n)/(\log_b a)$, and $1/(\log_b a)$ is a positive constant.

11. (a)   Let $C_1 = m^m/m!$, where $m = \lceil k \rceil$. Then for $n > m$ we have $k^n = k^{n-m} \cdot k^m \leq m^{n-m} \cdot m^m \leq n(n-1)\cdots(m+1) \cdot m^m = n(n-1)\cdots(m+1) \cdot C_1 \cdot m! = C_1 n!$.

    (b) $n! = n(n-1)\cdots 2 \cdot 1 \leq n \cdot n \cdots n \cdot n = n^n$

    (c)   Let $m = \lceil k \rceil$. Let $C_1 = m^m/m!$. Then if $n > m$,

    $$mC_1(n+1)! = mC_1(n+1)n(n-1)\cdots(m+1)m!$$
    $$= m(n+1)n(n-1)\cdots(m+1)m^m$$
    $$\geq m(n+1)m \cdot m \cdots m \cdot m^m$$
    $$= (n+1)m^{n+1} \geq (n+1)k^{n+1}.$$

Thus

$$(n+1)! \geq (n+1)\frac{1}{mC_1}k^{n+1}.$$

In particular,

$$\frac{(n+1)!}{k^{n+1}} \geq (n+1)\frac{1}{mC_1} \to \infty$$

as $n \to \infty$, so $n! \notin O(k^n)$.

(d) $n^n = n \cdot n \cdots n \geq n(n-1) \cdots 2 \cdot n \geq n! \cdot n$. Thus $n^n/n! \geq n \to \infty$ as $n \to \infty$. Therefore $n^n \notin O(n!)$.

13. Let C_1, C_2, k_1, and k_2 be such that $f_1(n) \leq C_1 g(n)$ for all $n \geq k_1$, and $f_2(n) \leq C_2 g(n)$ for all $n \geq k_2$. Let $C = C_1 + C_2$ and let $k = \max(k_1, k_2)$. Then for all $n \geq k$, we have $f_1(n) + f_2(n) \leq C_1 g(n) + C_2 g(n) = Cg(n)$.

15. (a) This follows from the fact that under the hypothesis, $f(n) + g(n) \leq g(n) + g(n) = 2g(n)$ for all large n.

(b) This follows from the definition and the fact that $mf(n) \leq mf(n)$ for all n.

17. (a) $O(n^2)$ (b) $O(n^2)$ (c) $O(n \log n)$ (d) $O(n)$

19. (a) We show the list at the end of each interchange. $(2, 4, 5, 8, 6, 10, 12)$, $(2, 4, 5, 8, 6, 10, 12)$, $(2, 4, 5, 6, 8, 10, 12)$, $(2, 4, 5, 6, 8, 10, 12)$, $(2, 4, 5, 6, 8, 10, 12)$, $(2, 4, 5, 6, 8, 10, 12)$

(b) After the outer loop is executed for a given value of *end*, a_{end} is the *end*th largest element in the list, in its correct place. Thus a_n, a_{n-1}, ..., a_2 are all correctly placed, and hence so is a_1.

(c) $O(n^2)$

21. **procedure** *speedy*(A, x)
    {searches A for x, as in *linear_search*}
        $a_0 \leftarrow x$ {the process will find x here if not before}
        $i \leftarrow n$
        while true do
            begin
                if $x = a_i$ **then return**(i)
                $i \leftarrow i - 1$
            end

23. Find an increasing, infinite sequence n_1, n_2, ..., such that $g(n_i)/f(n_i) \to \infty$ as $i \to \infty$. The function h defined by $h(n) = g(n)$ if $n = n_i$ for some even i, $h(n) = f(n)$ otherwise, is strictly between f and g.

25. (a) By L'Hôpital's rule,

$$\lim_{n \to \infty} \frac{n^a}{\log n} = \lim_{n \to \infty} \frac{an^{a-1}}{(1/n)\log e} = \lim_{n \to \infty} \left(\frac{a}{\log e}\right) n^a = \infty.$$

(b)　$n \approx 2^{996}$

27.　$f(n) = 2^{\sqrt{n}}$

SECTION 4.4　INTRACTABLE AND UNSOLVABLE PROBLEMS

1.　(a)　This is a nontautology; $\overline{T} \vee (F \wedge T)$ is F.

(b)　The truth table to show that this is a tautology will have eight rows.

3.　In each case we let n be the number of digits in N.

(a)　Given k, we can easily verify that $N = 2k$ by a multiplication of k by 2. Since k has about n digits, this only takes $O(n)$ steps (using the usual grade-school multiplication algorithm).

(b)　same as part (a), replacing 2 by 3

(c)　Given k, we can easily verify that $N = k^2$ by a multiplication of k by k. Since k has about $n/2$ digits, this only takes $O(n^2)$ steps (using the usual grade-school multiplication algorithm).

(d)　Given i, j, k, and l, we can easily verify that $N = i^2 + j^2 + k^2 + l^2$ in $O(n^2)$ steps, since each of the four numbers has no more than n digits.

5.　$193707721 \times 761838257287 = 147573952589676412927 = 2^{67} - 1$

7.　(a)　Given a hint as to which sets form the pairwise disjoint collection, we form their pairwise intersections (there are only $k(k-1)/2$ of them) to verify this.

(b)　Given a hint as to the partition, we sum each half to verify that the sums are equal.

(c)　Given such a C as a hint, which we can assume has no more than $2S$ elements, we can take each pair in S and verify the condition by searching C.

9.　Generate each subset of S (there are $2^{|S|}$ such subsets), and check whether its sum is g. Since this takes at least $2^{|S|}$ steps, it is a bad algorithm.

11.　Sort $S = \{s_1, s_2, \ldots, s_n\}$ so that $s_1 < s_2 < \cdots < s_n$. Then apply the following steps.

```
for i ← n down to 1 do
    if s_i < x then
        begin
            x ← x - s_i
            put s_i into the subset
        end
```

At this point $x = 0$ if and only if the original x was a sum of some subset of S. This works because $s_i > \sum_{j=1}^{i-1} s_j$, so that if $s_i < x$ when we are looking at s_i, then we know that we need to include s_i in the subset. Clearly, this algorithm has $O(n)$ complexity.

13. Construct the following algorithm for the one-variable case. The input is the formula for f, where $f(x) = 0$ is the given equation. Set n equal to 0. Repeatedly calculate $f(n)$ and $f(-n)$ and then add 1 to n. If you ever find that $f(n) = 0$ or $f(-n) = 0$, then halt. Now we encode this algorithm and its input (f) and give it to a purported algorithm for the halting problem. It answers yes if and only if the equation $f(x) = 0$ has an integral solution. Thus the solution to the halting problem gives a solution to the polynomial problem. In the multivariable case of looking for an integral solution to $f(x_1, x_2, \ldots, x_k) = 0$, the algorithm checks, during the nth pass through the loop, all k-tuples in which the sum of the absolute values of the coordinates is n, again halting if and when it finds a solution.

SECTION 4.5 ALGORITHMS FOR ARITHMETIC AND ALGEBRA

1. (a) $carry = 0$; $c_0 = 5 + 9 + 0 = 14$, $c_0 = 4$, $carry = 1$; $c_1 = 7 + 5 + 1 = 13$, $c_1 = 3$, $carry = 1$; $c_2 = 3 + 2 + 1 = 6$, $c_2 = 6$, $carry = 0$; $c_3 = 0$; answer is 0634.
   (b) $carry = 0$; $c_0 = 9 + 9 + 0 = 18$, $c_0 = 8$, $carry = 1$; $c_1 = 9 + 9 + 1 = 19$, $c_1 = 9$, $carry = 1$; $c_2 = 9 + 0 + 1 = 10$, $c_2 = 0$, $carry = 1$; $c_3 = 1$; answer is 1098.

3. (a) 88   (b) 24   (c) 96

5. (a) For the sum, $c_0 = -2 + 5 = 3$, $c_1 = 4 + 1 = 5$, $c_2 = 3 + 0 = 3$, so the sum is $3 + 5x + 3x^2$. For the product, $c_0 = 0 + (-2) \cdot 5 = -10$, $c_1 = 0 + (-2) \cdot 1 = -2$, $c_1 = -2 + 4 \cdot 5 = 18$, $c_2 = 0 + 4 \cdot 1 = 4$, $c_2 = 4 + 3 \cdot 5 = 19$, $c_3 = 0 + 3 \cdot 1 = 3$, so the product is $-10 + 18x + 19x^2 + 3x^3$.
   (b) For the sum, $c_0 = 0 + 0 = 0$, $c_1 = 0 + 0 = 0$, $c_2 = 3 + (-3) = 0$, so the sum is $0 + 0x + 0x^2$, i.e., the zero polynomial. For the product, $c_0 = 0 + 0 \cdot 0 = 0$, $c_1 = 0 + 0 \cdot 0 = 0$, $c_2 = 0 + 0 \cdot (-3) = 0$, $c_1 = 0 + 0 \cdot 0 = 0$, $c_2 = 0 + 0 \cdot 0 = 0$, $c_3 = 0 + 0 \cdot (-3) = 0$, $c_2 = 0 + 3 \cdot 0 = 0$, $c_3 = 0 + 3 \cdot 0 = 0$, $c_4 = 0 + 3 \cdot (-3) = -9$, so the product is $0 + 0x + 0x^2 + 0x^3 - 9x^4 = -9x^4$.

7. $A + B = \begin{bmatrix} 8 & 0 & -6 & 9 \\ 4 & 0 & 0 & 10 \end{bmatrix}$

9. (a) 16   (b) 109   (c) 2   (d) 1296

11. **procedure** $to\_base\_ten(\beta$: integer $\geq 2,\ d_{n-1}d_{n-2}\ldots d_1d_0$: string of digits)
    $\{$computes the value of the base β numeral $d_{n-1}d_{n-2}\ldots d_1d_0\}$
    $x \leftarrow d_{n-1}$
    for $i \leftarrow n-2$ **down to** 0 **do**
    $\quad x \leftarrow \beta \cdot x + d_i$
    return(x)

13. **procedure** $difference(x, y$: natural numbers)
    $\{x$ and y are represented by the decimal numerals $a_{n-1}\ldots a_0$ and
    $b_{m-1}\ldots b_0$, respectively; the answer $z = x - y$ will appear as the decimal
    numeral $c_{k-1}\ldots c_0$, where $k = \max(n, m)$; we assume that a_i is defined
    to be 0 for $i \geq n$, that b_i is defined to be 0 for $i \geq m$, and that $x \geq y\}$
    $k \leftarrow \max(n, m)$
    $borrow \leftarrow 0$
    for $i \leftarrow 0$ **to** $k - 1$ **do**
    $\quad$**begin**
    $\qquad c_i \leftarrow a_i - b_i - borrow$
    $\qquad$**if** $c_i < 0$ **then**
    $\qquad\quad$**begin**
    $\qquad\qquad c_i \leftarrow c_i + 10$
    $\qquad\qquad borrow \leftarrow 1$
    $\qquad\quad$**end**
    $\qquad$**else** $borrow \leftarrow 0$
    $\quad$**end**
    return(z)

    The time complexity of the algorithm is $O(k) = O(n + m)$.

15. $2^{340} = \left(2^{10}\right)^{34} = 1024^{34} \equiv 1^{34} = 1 \pmod{341}$

17. **procedure** $random\_integer(N$: positive integer)
    $\{$computes a random integer between 1 and N, inclusive; assumes that
    $rand$ returns a random real number between 0 and 1 each time it is
    called$\}$
    $x \leftarrow rand$
    return$(1 + \lfloor x \cdot N \rfloor)$

19. (a) **procedure** $matrix\_add(A, B)$
    $\{A = (a_{ij})$ and $B = (b_{ij})$ are m by n matrices; the sum is the m by
    n matrix $C = (c_{ij})\}$
    for $i \leftarrow 1$ **to** m **do**
    $\quad$**for** $j \leftarrow 1$ **to** n **do**
    $\qquad c_{ij} \leftarrow a_{ij} + b_{ij}$
    return(C)

(b) **procedure** *matrix_multiply*(A, B)
  $\{A = (a_{ik})$ is an m by n matrix and $B = (b_{kj})$ is an n by p matrix;
  the product is the m by p matrix $C = (c_{ij})\}$
  for $i \leftarrow 1$ **to** m **do**
    for $j \leftarrow 1$ **to** p **do**
      begin
        $c_{ij} \leftarrow 0$
        for $k \leftarrow 1$ **to** n **do**
          $c_{ij} \leftarrow c_{ij} + a_{ik}b_{kj}$
      end
  return(C)

21. the matrix all of whose entries are 0

23. Since $2^{24} = 16777216 \equiv 16 \not\equiv 1 \pmod{25}$, we know that 25 is not prime.

25. $s^{-1} + q^{-1} < p^{-1} + r^{-1}$

27. (a) **procedure** *slow*$(a_1, a_2, \ldots, a_n :$ sequence of real numbers)
  $\{$naively finds the sum of all products $a_i a_j$ for $i < j\}$
    $sum \leftarrow 0$
    for $i \leftarrow 1$ **to** $n - 1$ **do**
      for $j \leftarrow i + 1$ **to** n **do**
        $sum \leftarrow sum + a_i a_j$
    return(sum)

  The complexity of this algorithm is proportional to n^2.
  (b) Computing $((a_1 + a_2 + \cdots + a_n)^2 - a_1^2 - a_2^2 - \cdots - a_n^2)/2$ takes only $O(n)$ steps.

CHAPTER 5:   INDUCTION AND RECURSION

SECTION 5.1   RECURSIVE DEFINITIONS

1. 144

3. 105

5. (a) $\begin{bmatrix} -2 & -3 \\ 9 & 1 \end{bmatrix}$ (b) $\begin{bmatrix} 1 & 0 \\ 0 & 1 \end{bmatrix}$

  (c) $\begin{bmatrix} -11 & -4 \\ 12 & -7 \end{bmatrix}$ (d) $\begin{bmatrix} 1 & -1 \\ 3 & 2 \end{bmatrix}$

7. (a) yes (b) no (c) no (d) no

(e) yes (f) no

9. day 0: 1, s, 3; day 1: (-1), $(-s)$, $(3+1)$; day 2: $((-1)-s)$, $((-s)*(3+1))$; day 3: $(((-1)-s)+((-s)*(3+1)))$

11. (a) 1, 2, 2, 3, 3, 3, 4, 4, 4, 4, 5, 5, 5, 5, 5, 6
    (b) $n - P(n-1)$ might not be a positive integer less than n.
    (c) one 1, two 2's, three 3's, four 4's, etc.

13. (a) $x^0 = 1$, $x^{n+1} = x \cdot x^n$
    (b) Let $I(x, n) = x^{x^{\cdot^{\cdot^{x}}}}$, with n x's in the exponent ($n + 1$ x's in all). $I(x, 0) = x$; $I(x, n+1) = x^{I(x,n)}$.

15. (a) 9 (b) 9

17. (a) 5
    (b) $A(n, 0)$ is defined directly. For $i > 0$, $A(n, i)$ is defined in terms of values of $A(n', i')$, where $i' < i \ \lor \ (i' = i \land n' < n)$.

19. Variable names, 0, and 1 are Boolean expressions. If α and β are Boolean expressions, then so are (α), $\overline{\alpha}$, $\alpha\beta$, $\alpha \cdot \beta$, and $\alpha + \beta$.

21. (a) 1, 0.5, 0.6666667, 0.6, 0.625, 0.6153846, 0.6190476, 0.6176471, 0.6181818, 0.6179775, 0.6180556, 0.6180258, 0.6180371, 0.6180328, 0.6180344, 0.6180338
    (b) The sequence converges to $(-1 + \sqrt{5})/2$.

23. (a) There is clearly only one 4-nomino with four squares in a row. If a 4-nomino has three squares in a row, then the other square can only be attached in two locations (up to rotation and flipping), and the result of each of these possibilities is shown. Otherwise, the 4-nomino must start with three squares in an L shape, and there are only two different places to add the fourth square without creating four in a row; both of these pictures are shown.
    (b) There are twelve 5-nominoes, as shown here.

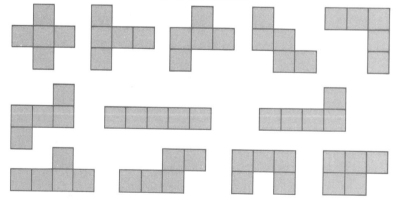

25.

(a)

New Lists	New Insides
Day 0 ( )	5
Day 1 (5)	( ); 5, 5
Day 2 (( )); (5, 5)	(5); ( ), ( ); ( ), 5, 5; 5, 5, ( );
	5, 5, 5, 5; ( ), 5; 5, ( ); 5, 5, 5

(b)

New Lists	New Insides
Day 0 ( )	3; 4; 7; 2
Day 1 (4)	( ); 7, 2
Day 2 (7, 2)	( ), ( ); (4)
Day 3 (( ), ( ))	(7, 2); 3, (4)
Day 4	(( ), ( )); 3, (4), (7, 2)
Day 5	3, (4), (7, 2), (( ), ( ))
Day 6 (3, (4), (7, 2), (( ), ( )))	

27. (a) day 0: 5, 7; day 1: 4, 6, 10, 12, 14; day 2: 3, 8, 9, 11, 13, 15, 16, 17, 18, 19, 20, 21, 22, 24, 26, 28

(b) Since $x \in E \to 2x \in E$, and since $5 \in E$, E is unbounded. Since $x \in E \to x - 1 \in E$, E has no gaps below any number. Therefore, $E = \mathbf{Z}$.

29. If d is a digit, then $d.$ and $.d$ are unsigned decimals. If d is a digit and α is an unsigned decimal, then $d\alpha$ and αd are unsigned decimals. If α is an unsigned decimal, then $+\alpha$ and $-\alpha$ are signed decimals.

31. (a) C_0, C_1, C_2, and C_3 are shown from top to bottom.

(b) The total length of the removed segments is $(\frac{1}{3}) \sum_{k=0}^{\infty} (\frac{2}{3})^k = 1$.

(c) The Cantor set contains the numbers $\frac{1}{3}, \frac{1}{9}, \frac{1}{27}, \ldots$, since these are never removed.

(d) Use base 3 notation. The removed segments are those values whose "decimals" (base 3) must contain a 1. Thus all those base 3 "decimals" containing only 0's and 2's remain. There are just as many of these as there are base 2 "decimals" with 0's and 1's, which represent all the numbers in the interval $[0, 1]$.

33. (a) A game is winning for Righty if and only if there exists a game in R that is either winning for Righty or winning for the second player and every game in L is either winning for Righty or winning for the first player. A game is winning for the first player if and only if there exists a game in L that is either winning for Lefty or winning for the second player and there exists a game in R that is either winning for Righty or winning for the second player. A game is winning for the second player if and only if every game in L is either winning for Righty or winning for the first player and every game in R is either winning for Lefty or winning for the first player.

(b) Second player wins 0; first player wins $*$, $(\{*, 1\}, \{-1\})$; Lefty wins 1, $(\{0\}, \{1\})$, $(\{1\}, \emptyset)$; Righty wins -1.

SECTION 5.2 RECURSIVE ALGORITHMS

1. Move disk 1 from peg A to peg B. Move disk 2 from peg A to peg C. Move disk 1 from peg B to peg C. Move disk 3 from peg A to peg B. Move disk 1 from peg C to peg A. Move disk 2 from peg C to peg B. Move disk 1 from peg A to peg B.

3. **procedure** *reverse*(s : string)
   {assume that $s = s_1 s_2 \ldots s_n$, with $n \geq 0$}
     if $n \leq 1$ **then return**(s)
     else return($s_n reverse(s_1 s_2 \ldots s_{n-1})$)

5. Output is $(1, 3, 5, 6, 6, 9, 9, 20, 24, 31, 49)$.

7. **procedure** *simpler_hanoi*(X, Y, Z : peg names, n : natural number)
   {assume that the peg names are A, B, and C in some order}
     if $n > 0$ **then**
       begin
         call *simpler_hanoi*($X, Z, Y, n - 1$)
         print("Move disk " n " from peg " X " to peg " Y ".")
         call *simpler_hanoi*($Z, Y, X, n - 1$)
       end
     return

9. (a) **procedure** *largest*(L : list of numbers)
       {assume that $L = (l_1, l_2, \ldots, l_n)$, with $n \geq 1$}
       if $n = 1$ **then return**(l_1)
       else
         begin
           $a \leftarrow largest((l_1, l_2, \ldots, l_{n-1}))$
           if $l_n > a$ **then return**(l_n)
           else return(a)
         end

  (b) **procedure** *largest_index*(L : list of numbers)
       {assume that $L = (l_1, l_2, \ldots, l_n)$, with $n \geq 1$}

if $n = 1$ **then return**(1)
  else
    begin
      $i \leftarrow largest\_index((l_1, l_2, \ldots, l_{n-1}))$
      if $l_n > l_i$ **then return**(n)
        else return(i)
    end

11. (a)   **procedure** *fib_3* (n : positive integer)
      if $n \leq 3$ **then return**(1)
        else return(*fib_3* $(n - 1) +$ *fib_3* $(n - 2) +$ *fib_3* $(n - 3)$)

    (b)   **procedure** *fib_3_iterative*(n : positive integer)
      $x \leftarrow 1;\ y \leftarrow 1;\ z \leftarrow 1$
      for $i \leftarrow 4$ **to** n **do**
        begin
          next $\leftarrow x + y + z$
          $x \leftarrow y;\ y \leftarrow z;\ z \leftarrow$ *next*
        end
      return(z)

    (c) $O(n)$ steps

13. **procedure** *valid*(s : string)
    {assume that $s = s_1 s_2 \ldots s_n$, with $n \geq 1$}
      if $n = 1$ and s_i is a letter **then return**(**true**)
        else if s_n is a letter or digit **then return**(*valid*($s_1 s_2 \ldots s_{n-1}$))
        else return(**false**)

15. **procedure** $F(n$: natural number)
    if $n = 0$ **then return**(1)
      else return($n - M(F(n - 1))$)

    procedure $M(n$: natural number)
    if $n = 0$ **then return**(0)
      else return($n - F(M(n - 1))$)

17. **procedure** *linear*(A, x)
    {assume setting of Algorithm 1 in Section 4.3}
      if $n = 0$ **then return**(0)
        else if $x = a_n$ **then return**(n)
        else return(*linear*$((a_1, a_2, \ldots, a_{n-1}), x)$)

19. $O(n + m)$

21. (a)   **procedure** *element_of* $(x$: number, B : list of numbers)
      {assume that $B = (b_1, b_2, \ldots, b_n)$, with $n \geq 0$}

if $n = 0$ **then return(false)**
    else if $x = b_n$ **then return(true)**
        else return($element\_of(x, (b_1, b_2, \ldots, b_{n-1}))$)

(b)   **procedure** $subset\_of(A, B : \text{lists of numbers})$
        {assume that $A = (a_1, a_2, \ldots, a_m)$, with $m \geq 0$}
        if $m = 0$ **then return(true)**
            else if $\neg element\_of(a_m, B)$ **then return(false)**
                else return($subset\_of((a_1, a_2, \ldots, a_{m-1}), B)$)

(c)   **procedure** $equals(A, B : \text{lists of numbers})$
        return($subset\_of(A, B) \land subset\_of(B, A)$)

(d)   **procedure** $adjoin(x : \text{number}, \ B : \text{list of numbers})$
        {assume that $B = (b_1, b_2, \ldots, b_n)$, with $n \geq 0$}
        return($(b_1, b_2, \ldots, b_n, x)$)

(e)   **procedure** $reduce(A : \text{list of numbers})$
        {assume that $A = (a_1, a_2, \ldots, a_n)$, with $n \geq 0$}
        if $n = 0$ **then return**(A)
            else if $element\_of(a_n, (a_1, a_2, \ldots, a_{n-1}))$
                then return($reduce((a_1, a_2, \ldots, a_{n-1}))$)
                else return($adjoin(a_n, reduce((a_1, a_2, \ldots, a_{n-1}))))$

23. **procedure** $subsets(n : \text{positive integer})$
      call $rec\_subsets(n, n, \lambda)$
      return

    procedure $rec\_subsets(n : \text{positive integer}, \ i : \text{natural number}, \ s : \text{string})$
        if $i = 0$ **then print**(s)
          else
            begin
                call $rec\_subsets(n, i - 1, s)$
                $s \leftarrow$ the number i followed by a blank, followed by s
                call $rec\_subsets(n, i - 1, s)$
            end
        return

25. **procedure** $EVAL(L : \text{list of items})$
      {assume that $L = (l_1, l_2, \ldots, l_n)$, with $n \geq 1$, represents a valid FPAE;
      each l_i is either an unsigned integer, a variable, an operator symbol, or a
      left or right parenthesis}
        if $n = 1$ **then return**($EVAL\_atom(l_1)$)
          else
            begin {need to find middle operator symbol}
                $count \leftarrow 1$
                for $i \leftarrow 2$ **to** $n - 1$ **do**
                    if $count = 1$ and l_i is an operator symbol θ **then**

$$\textbf{return}(EVAL((l_2, \ldots, l_{i-1})) \, \theta \, EVAL((l_{i+1}, \ldots, l_{n-1})))$$
$$\textbf{else if } l_i = \text{“(” } \textbf{then } count \leftarrow count + 1$$
$$\textbf{else if } l_i = \text{“)” } \textbf{then } count \leftarrow count - 1$$
end

27. (a) It computes an approximation for the so-called arithmetic-geometric mean of 1 and 10, returning approximately 4.250407.

(b) **procedure** $gauss\_recursive(x, y, \epsilon : \text{positive real numbers with } x \leq y)$
if $y - x \leq \epsilon$ **then return**(x)
else return$(gauss(\sqrt{xy}, (x + y)/2, \epsilon))$

SECTION 5.3 PROOF BY MATHEMATICAL INDUCTION

1.

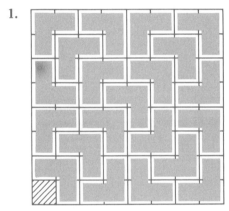

3. Base case: $1 \cdot 2 = 1 \cdot 2 \cdot 3/3$. Assume inductive hypothesis: $1 \cdot 2 + 2 \cdot 3 + \cdots + (k-1)k = [(k-1)k(k+1)]/3$. Then $1 \cdot 2 + 2 \cdot 3 + \cdots + (k-1)k + k(k+1) = [(k-1)k(k+1)]/3 + k(k+1) = [((k+1)-1)(k+1)((k+1)+1)]/3$.

5. $[S(1) \ \wedge \ \forall k \geq 1 : (S(k) \rightarrow S(k+1))] \rightarrow \forall n : S(n)$

7. Base case: $\frac{1}{2} \geq \frac{1}{2}$. Assume inductive hypothesis:

$$\frac{1}{2} \cdot \frac{3}{4} \cdot \frac{5}{6} \cdots \frac{2k-1}{2k} \geq \frac{1}{2k}.$$

Then

$$\frac{1}{2} \cdot \frac{3}{4} \cdot \frac{5}{6} \cdots \frac{2k-1}{2k} \cdot \frac{2k+1}{2k+2} \geq \frac{1}{2k} \cdot \frac{2k+1}{2k+2} = \frac{2k+1}{2k} \cdot \frac{1}{2k+2}$$

$$\geq \frac{1}{2k+2} = \frac{1}{2(k+1)}.$$

9. Base case: $1 \cdot 1! = 2! - 1$. Assuming inductive hypothesis,

$$\sum_{j=1}^{k+1} j \cdot j! = \sum_{j=1}^{k} j \cdot j! + (k+1)(k+1)!$$

$$= (k+1)! - 1 + (k+1)(k+1)! = (k+2)! - 1.$$

11. For $n = 0$ or 1, Algorithm 3 uses no additions, and $f(n) - 1 = 1 - 1 = 0$. Assume inductive hypothesis, that for all $i < k$, Algorithm 3 requires $f(i) - 1$ addition operations to compute $f(i)$. Then to compute $f(k)$, the algorithm recursively computes $f(k-1)$ and $f(k-2)$ and adds the answers. This requires $[f(k-1) - 1] + [f(k-2) - 1] + 1 = f(k-1) + f(k-2) - 1 = f(k) - 1$ additions.

13. $50000/100001$

15. $n = 0$, $n = 1$, $n \geq 5$

17. (a) Base case is false.
    (b) Inductive step is not proved for *all* $k \geq 1$.
    (c) $a_2 \notin B$ if $k = 2$.
    (d) Inductive hypothesis cannot be invoked for $i = k - 2$ unless $k \geq 2$, which requires checking two base cases.
    (e) Base case is false.

19. Suppose that $S(n)$ is a proposition with a free variable n that ranges over the positive integers. Then if $S(1)$ is true, and if for every $k > 1$ the implication $S(k-1) \rightarrow S(k)$ is true, then $S(n)$ is true for all n.

21. Suppose that $S(n)$ is a proposition with a free variable n that ranges over the natural numbers. Then if $S(0)$, $S(1)$, and $S(2)$ are true, and if for every $k > 2$ the implication $(\forall i < k\colon S(i)) \rightarrow S(k)$ is true, then $S(n)$ is true for all n.

23. By Example 6, $f_{1000} > 1.5^{1000}$. Thus $\log_{10} f_{1000} > 1000 \log_{10} 1.5 \approx 176.09$. Therefore, $f_{1000} > 10^{176}$, so f_{1000} has more than 176 decimal digits.

25. To prove: For all $n \geq 0$, if x is a FPAE born on day n, then $LP(x) = RP(x)$. If $n = 0$, then x must be an unsigned integer or variable name, so $LP(x) = 0 = RP(x)$. Let $k > 0$, and assume inductive hypothesis, that for all $i < k$, every FPAE y born on day i satisfies $LP(y) = RP(y)$. Let x be a FPAE born on day k. Then $x = (\alpha + \beta)$, $x = (\alpha - \beta)$, $x = (\alpha * \beta)$, $x = (\alpha/\beta)$, or $x = (-\alpha)$ for some α and β born prior to day k. The cases are similar, so we will assume that $x = (\alpha/\beta)$. By inductive hypothesis, $LP(\alpha) = RP(\alpha)$, and $LP(\beta) = RP(\beta)$. But then $LP(x) = 1 + LP(\alpha) + LP(\beta) = 1 + RP(\alpha) + RP(\beta) = RP(x)$.

27. If $\forall n: S(n)$ is not true, then there must be one or more counterexamples. Let n_0 be the smallest counterexample. By base case, $n_0 > 1$. By the choice of n_0, $S(n_0 - 1)$ is true, so by the inductive step $S(n_0)$ holds, a contradiction.

29. (a) $\quad n + 2$ (b) $\quad 2n + 3$ (c) $\quad 2^{n+3} - 3$

31. f_n is even $\leftrightarrow n \equiv 2 \pmod 3$

33. (a) $\quad f_{n+2}^2 - f_{n+1}^2 = (f_{n+2} - f_{n+1})(f_{n+2} + f_{n+1}) = f_n f_{n+3}$

    (b) To prove: $f_{n+1}^2 - f_n f_{n+2} = (-1)^{n+1}$ for all n. Base case: $f_1^2 - f_0 f_2 = 1 - 2 = (-1)^1$. Assume inductive hypothesis: $f_{k+1}^2 - f_k f_{k+2} = (-1)^{k+1}$. Then $f_{k+2}^2 - f_{k+1} f_{k+3} = f_{k+2}^2 - f_{k+1}(f_{k+2} + f_{k+1}) = f_{k+2}^2 - f_{k+1} f_{k+2} - f_{k+1}^2 = f_{k+2}^2 - f_{k+1} f_{k+2} - (f_k f_{k+2} + (-1)^{k+1}) = f_{k+2}^2 - f_{k+2}(f_{k+1} + f_k) - (-1)^{k+1} = f_{k+2}^2 - f_{k+2} f_{k+2} - (-1)^{k+1} = -(-1)^{k+1} = (-1)^{k+2}$.

35. (a) Base case: Empty sum is 0 by definition. Assume inductive hypothesis, that sum of k even numbers is even. Then $(a_1 + a_2 + \cdots + a_k) + a_{k+1}$ is the sum of two even numbers, which is even.

    (b) Base case: Empty sum is 0 by definition. Assume inductive hypothesis, that sum of $2k$ odd numbers is even. Then $(a_1 + a_2 + \cdots + a_{2k}) + a_{2k+1} + a_{2k+2}$ is the sum of an even number and two odd numbers, which is the sum of an even number and an even number, which is even.

    (c) Base case: Sum of one odd number is itself, which is odd. Assume inductive hypothesis, that sum of $2k + 1$ odd numbers is odd. Then $(a_1 + a_2 + \cdots + a_{2k+1}) + a_{2k+2} + a_{2k+3}$ is the sum of an odd number and two odd numbers, which is the sum of an odd number and an even number, which is odd.

37. (a) Base case: $3 \mid 6$. Assume inductive hypothesis, that $3 \mid k^3 - 4k + 6$. Then $(k + 1)^3 - 4(k + 1) + 6 = k^3 + 3k^2 + 3k + 1 - 4k - 4 + 6 = (k^3 - 4k + 6) + 3(k^2 + k - 1)$, the sum of two multiples of 3.

    (b) $n^3 - 4n + 6 = (n - 1)n(n + 1) - 3(n - 2)$. First term is divisible by 3 since one of the three consecutive numbers $n - 1$, n, and $n + 1$ is necessarily a multiple of 3; and second term is clearly divisible by 3.

39. (a) Prove by induction on m that if game table consists of n stones and $n - m$ piles, then exactly m plays remain. Base case ($m = 0$): If there are $n - 0 = n$ piles, then clearly each pile contains one stone, so no plays remain. Assume the inductive hypothesis, and suppose that there are $n - (m + 1) = n - m - 1$ piles. Whatever player P now does, the number of piles must increase by 1, leaving $n - m$ piles. By inductive hypothesis, m plays remain after that. Therefore, $m + 1$ plays remained when there were $n - (m + 1)$ piles left. Let $m = n - 1$ to obtain desired conclusion.

(b) This model gives the same game. Since there are $n - 1$ spaces between adjacent stones, the game lasts exactly $n - 1$ moves.

(c) If n is odd, then second player wins; if n is even, then first player wins.

41. Base case $(v = \lambda)$: $(uv)^R = (u\lambda)^R = u^R = \lambda u^R = \lambda^R u^R$. Assume inductive hypothesis, that the statement is true for second strings of length k, and let $v = \alpha x$, where α is a string of length k and x is a symbol. Then $(uv)^R = (u\alpha x)^R = ((u\alpha)x)^R = x(u\alpha)^R = x(\alpha^R u^R) = (x(\alpha^R))u^R = (\alpha x)^R u^R = v^R u^R$.

43. Let $S(n)$ be the statement that any solution to the problem of transferring n disks from one given peg to another (different) given peg requires at least $2^n - 1$ moves. Base case is clear, since if $n = 1$, then $2^1 - 1 = 1$ move is required. Assume inductive hypothesis. In order to solve problem with $k + 1$ disks, we must at some point move disk $k + 1$. This can happen only when the k other disks have been moved off of the peg on which this largest disk sits, onto one of the other pegs. By inductive hypothesis, this requires at least $2^k - 1$ moves. After disk $k + 1$ has been transferred to the desired peg for the last time, these other k disks must make their way back onto the peg on which disk $k + 1$ now sits, and again by the inductive hypothesis, this requires at least $2^k - 1$ moves. Thus the total number of moves required is at least $(2^k - 1) + 1 + (2^k - 1) = 2^{k+1} - 1$.

CHAPTER 6: ELEMENTARY COUNTING TECHNIQUES

SECTION 6.1 FUNDAMENTAL PRINCIPLES OF COUNTING

1. (a) 6   (b) $\{(a, a), (a, b), (a, c), (b, a), (b, b), (b, c)\}$

3. (a) 32   (b) 26

5. If $A_1, A_2, \ldots, A_n$ are finite sets, with $n \geq 1$, then $|A_1 \times A_2 \times \cdots \times A_n| = |A_1| \cdot |A_2| \cdots |A_n|$.

7. (a) $\{(a, a), (b, b), (c, c), (d, d)\}$, $\{(a, a), (b, a), (c, a), (d, a)\}$, $\{(a, a), (b, c), (c, c), (d, a)\}$

   (b) 625   (c) n^k

9. (a) 16384   (b) 4096   (c) 14197   (d) 9094

   (e) 2916   (f) 384   (g) 729   (h) 144

11. (a) $\aleph_0$ (b) 2^{32} (c) 2^{32}

13. 204

15. (a) 3^n (b) $[n(n-1)/2] \cdot 2^{n-2}$ (c) $3^n - n \cdot 2^{n-1} - 2^n$
(d) $[n(n-1)/2] \cdot (2^{n-2} - 2)$

17. (a) 32768 (b) 7776 (c) 24992 (d) 9031
(e) 25838 (f) 6930 (g) 7680 (h) 12005

19. (a) 52488 (b) 37512 (c) 13440 (d) 59049

21. (a) 3628800 (b) 17280 (c) 604800 (d) 0

23. (a) 2 (b) $2^n - 2^{n-2}$ (c) n
(d) $n(n-1)/2$ (e) 2^{n-1}

25. (a) $2^{|A|}$ (b) $2^{|B|+|C|}$ (c) $2^{|B|} \cdot 2^{|C|}$ (d) $2^{|A|} - |A| - 1$

27. 700

29. 382

31. 512

33. (a) 2 (b) 9 (c) 25 (d) 150

35. $29 \cdot 27 \cdot 25 \cdots 3 \cdot 1$

SECTION 6.2 PERMUTATIONS AND COMBINATIONS

1. (a) 210 (b) 7 (c) 5040 (d) 5040 (e) 1814400

3. (a) 20
(b) 13, 15, 17, 19, 31, 35, 37, 39, 51, 53, 57, 59, 71, 73, 75, 79, 91, 93, 95, 97
(c) 10
(d) 135, 137, 139, 157, 159, 179, 357, 359, 379, 579

5. (a) 45 (b) 210 (c) 1 (d) 10

7. 15

9. 282

11. 9310

13. (a) 211915132 (b) 17952000 (c) 24540 (d) 14677520

15. 1088430/10477677064400

17. (a) 1732104 (b) 369600 (c) 221760 (d) 4758039

19. 1

21. 1001

23. 39916800

25. (a) 239500800 (b) 119750400

27. $C(2n, n)$

29. $C(n, k + l)$

31. 2^{n^2}

33. 26333

35. (a) 635013559600 (b) 11404407300 (c) 136852887600
(d) 0 (e) 347373600; about once in 1828 hands

37. 119680/3838380

39. (a) $C(n - k, k)$ (b) $P(n - k, k)$ (c) $P(n - k, k) \cdot 2^k$

41. (a) 62403588 (b) 317506779800 (c) about 2.6×10^{11}

43. (a) $C(n, k)$ (b) 2^n (c) 1 if $k = 0$; $n - k + 1$ if $k > 0$
(d) $(n^2 + n + 2)/2$

SECTION 6.3 COMBINATORIAL PROBLEMS INVOLVING REPETITIONS

1. (a) 20160 (b) 19958400 (c) 6

3. (a) 11440 (b) 10000000

5. (a) 1296 (b) 126

7. (a) 27 (b) 10 (c) 4200 (d) 1
(e) 220 (f) 14

9. (a) 10 (b) 0 (c) 19324305 (d) 0
(e) $\displaystyle\sum_{r=2}^{(n-1)/2} \frac{n!}{r!(r - 2)!((n + 5)/2 - r)!((n - 1)/2 - r)!}$ if n is odd;
0 if n is even

11. (a) 12341 (b) 123410 (c) 6545 (d) 12341

13. 480

15. 1181952200

17. n^k

19. (a) $C(163, 24)$ (b) \$15,400,000 (c) \$14,000,000
(d) \$1,100,000

21. (a) $\lceil k/2 \rceil$ (b) $2^{k-1} - 1$

23. 2215220/3838380

25. (a)   1024   (b)   670442572800   (c)   234662231

27. (a)   336   (b)   231   (c)   666

SECTION 6.4   THE PIGEONHOLE PRINCIPLE

1. (a)   11   (b)   3

3. If every raise were at least $1200, then the average could not be its actual $10000/9 \approx \$1111.11$.

5. In Theorem 6, take $p = k$, $q = l$, $r = 2$, S equal to the set of people at the party, and T equal to the "is acquainted with" relation.

7. The number of subsets of S is 210. The largest possible sum is 206. Since the sums are all natural numbers, by the pigeonhole principle at least two of the subsets have the same sum.

9. Apply the pigeonhole principle to the partition $\{\{1,2\}, \{3,4\}, \{5,6\}, \ldots, \{2n-1, 2n\}\}$.

11. Divide square into four smaller squares as shown.

By the pigeonhole principle, two of the five points are in the same small square, which has diameter $\sqrt{2}$.

13. If each hole contained at least m pigeons, then there would be at least mn pigeons altogether, contradicting the hypothesis.

15. Let S be a sequence of length n. Let $m = \lceil \sqrt{n} \rceil - 1$. Then $m < \sqrt{n}$, so $m^2 < n$, that is, $m^2 + 1 \le n$. Look at the subsequence of S consisting of the first $m^2 + 1$ terms. By Theorem 3, this subsequence (and hence S, as well) has a monotone subsequence of length $m + 1$. But $m + 1 = \lceil \sqrt{n} \rceil \ge \sqrt{n}$, so there is a monotone subsequence of length at least $\sqrt{n}$.

17. We can show that $R(4) > 9$ with the people A, B, C, P, Q, R, X, Y, and Z such that each of A, B, and C knows each other, each of P, Q, and R knows each other, and each of X, Y, and Z knows each other.

19. $\forall r \ge 2 : \forall p \ge r : \forall q \ge r : \exists N : \forall S : \forall T : [(|S| = N \ \wedge \ T$ is a symmetric r-ary relation on $S) \rightarrow ([\exists P : (P \subseteq S \ \wedge \ |P| = p \ \wedge \ T$ holds on $P)] \ \vee \ [\exists Q : (Q \subseteq S \ \wedge \ |Q| = q \ \wedge \ \overline{T}$ holds on $Q)])]$, where the quantifiers on p, q, r, and N

range over natural numbers, and the quantifiers on S, P, and Q range over sets.

21. By the pigeonhole principle, two of the 128 numbers 7, 77, 777, ..., 77...7 must be congruent modulo 127. Hence their difference, which is of the form 77...700...0, is a multiple of 127.

23. (a) If there were 27 or fewer training sessions, then some robot would have received three or fewer sessions. Thus there would be at least seven workers who could not use this robot. If these seven workers need a robot at the same time, only six robots would be available to serve them.

   (b) Robot 1 is trained for workers 1, 2, 3, and 4; robot 2 is trained for workers 2, 3, 4, and 5; robot 3 is trained for workers 3, 4, 5, and 6; ...; robot 7 is trained for workers 7, 8, 9, and 10.

   (c) Assume that we have n workers and k robots. We need $k(n - k + 1)$ training sessions (otherwise some robot will serve at most $n - k$ masters, and the other k workers will have only $k - 1$ robots available to them). Furthermore, $k(n-k+1)$ sessions are sufficient, for we can train robot i for workers i, $i + 1$, ..., $i + n - k$.

25. 5 if $n = 3$; 3 if $n \geq 4$

27. **procedure** $long\_increase(x_1, x_2, \ldots, x_n : \text{distinct numbers})$
   $\{l_i$ will be the length of the longest increasing subsequence starting at x_i;
   w_i will be the index of the next term in such a subsequence$\}$
   $l_n \leftarrow 1$
   $w_n \leftarrow 0$
   for $i \leftarrow n - 1$ **down to** 1 **do**
     begin
       $l_i \leftarrow 1$
       $w_i \leftarrow 0$
       for $j \leftarrow i + 1$ **to** n **do**
         if $x_i < x_j \wedge l_i < l_j + 1$ **then**
           begin
             $l_i \leftarrow l_j + 1$
             $w_i \leftarrow j$
           end
     end
   $max \leftarrow -\infty$
   for $i \leftarrow 1$ **to** n **do**
     if $l_i > max$ **then**
       begin
         $max \leftarrow l_i$
         $start \leftarrow i$
       end
   return$(max, start, (w_1, w_2, \ldots, w_n))$

29. By the generalized pigeonhole principle, person A must have either 10 acquaintances or 10 nonacquaintances at the party. By symmetry we can without loss of generality assume that he has 10 acquaintances. By Exercise 28, among these 10 we can find either three mutual acquaintances or four mutual strangers. In the latter case we are done. In the former case, these three together with A form the desired set of four mutual acquaintances.

31. Person A must have either six nonacquaintances, six friends, or six enemies among the other party-goers (otherwise there would be only $5 + 5 + 5 = 15$ other people). Without loss of generality, assume that there are six nonacquaintances. If any two of these are not acquainted, then we have our three mutual strangers. Otherwise, we apply Theorem 4 to these six people (using the categories of friend and enemy) to get three mutual friends or three mutual enemies.

33. Among the $n!$ ways to pair red and blue points, the one having minimum total length of the line segments used has no intersections. (If there were an intersection between R_1B_1 and R_2B_2, then we could reduce the total length if we paired R_1 with B_2 and R_2 with B_1.)

CHAPTER 7: ADDITIONAL TOPICS IN COMBINATORICS

SECTION 7.1 COMBINATORIAL IDENTITIES

1. (a) $x^6 + 6x^5y + 15x^4y^2 + 20x^3y^3 + 15x^2y^4 + 6xy^5 + y^6$
   (b) $a^4 - 12a^3 + 54a^2 - 108a + 81$
   (c) $32x^5 + 240x^4y + 720x^3y^2 + 1080x^2y^3 + 810xy^4 + 243y^5$
   (d) $x^8 + 8x^6 + 28x^4 + 56x^2 + 70 + 56x^{-2} + 28x^{-4} + 8x^{-6} + x^{-8}$

3. Row 12: $1, 12, 66, 220, 495, 792, 924, 792, 495, 220, 66, 12, 1$; row 13: $1, 13, 78, 286, 715, 1287, 1716, 1716, 1287, 715, 286, 78, 13, 1$; row 14: $1, 14, 91, 364, 1001, 2002, 3003, 3432, 3003, 2002, 1001, 364, 91, 14, 1$. The sums of the rows are $2^{12} = 4096$, $2^{13} = 8192$, and $2^{14} = 16384$; the alternating sums are 0.

5. Each side equals 330.

7. 60

9. (a) 3^n (b) $(-1)^n$

11. A triangular arrangement with 1 dot in the top row, 2 dots in the second row, ..., $n - 1$ dots in the last row, has a total of $1 + 2 + \cdots + (n - 1) = (n - 1)n/2 = C(n, 2)$ dots in all.

13. **(a)** $C(2n, 2) = 2C(n, 2) + n^2$

(b)   Both sides simplify to $2n^2 - n$.

15. **(a)**   Both expressions equal 1260.

(b)   Both sides simplify to $n!/[(k-1)!(n-k)!]$.

(c)   If $n = 1$, then $k = 1$ and we have $1 \cdot C(1, 1) = 1 = 1 \cdot C(0, 0)$. Assume that the identity is true for $n - 1$; we must show it for n. If $k = n$, then both sides equal n. Otherwise, $k \le n - 1$, and

$$
\begin{aligned}
kC(n, k) &= k(C(n-1, k) + C(n-1, k-1)) \\
&= kC(n-1, k) + (k-1)C(n-1, k-1) + C(n-1, k-1) \\
&= (n-1)C(n-2, k-1) + (n-1)C(n-2, k-2) \\
&\quad + C(n-1, k-1) \\
&= (n-1)[C(n-2, k-1) + C(n-2, k-2)] + C(n-1, k-1) \\
&= (n-1)C(n-1, k-1) + C(n-1, k-1) \\
&= nC(n-1, k-1).
\end{aligned}
$$

(d)   To choose from a set of n people a committee of k people including a chairperson, we can either choose the committee members (in one of $C(n, k)$ ways) and then choose a chairperson from among them (in one of $C(k, 1) = k$ ways), or else we can choose the chairperson first (in one of $C(n, 1) = n$ ways), and then choose the other $k - 1$ members of the committee from among the other $n - 1$ people (in one of $C(n - 1, k - 1)$ ways).

17. We want to show that $\sum_{k=0}^{n} C(n, k)2^k = 3^n$. Let us count the number of ways to paint the elements of an n-set red, white, or blue. We can first decide how many of the elements are to be red or white: Let k be this number, so $0 \le k \le n$. There are $C(n, k)$ ways to select a subset of k elements to be red/white, and then there are 2^k ways to decide on the colors for the elements in our chosen set. Thus the left-hand side counts the desired quantity. Clearly the right-hand side does as well.

19. The right-hand side is the number of ways to choose n people from $2n$ people, who happen to consist of n men and n women. The left-hand side is the number of ways to choose from this collection k men and $n - k$ women [namely, $C(n, k) \cdot C(n, n - k) = C(n, k)^2$], summed over all possible values of k.

21. Both sides equal $(2m + 2n)!/[(m!)^2(n!)^2]$.

23. The number of permutations of k copies of each of n distinct things is

$$\frac{(kn)!}{k!k!\cdots k!} = \frac{(kn)!}{(k!)^n}.$$

Hence $(k!)^n$ must divide $(kn)!$.

25. (a) Choose a captain from among the n boys. Then choose the remaining $n - 1$ debaters from among the $2n - 1$ remaining people.

(b) The number of ways to choose the team with i boys and $n - i$ girls is $C(n, i) \cdot C(n, n - i) = C(n, i)^2$. The number of ways to choose the captain is i.

27. Consider the top row to be row 1. Let $B(m, k)$ be the weight borne by the kth person in row m. Then

$$B(1, 1) = 0$$

$$B(m, 1) = \frac{1}{2}(w + B(m - 1, 1))$$

$$B(m, m) = \frac{1}{2}(w + B(m - 1, m - 1))$$

$$B(m, k) = w + \frac{1}{2}(B(m - 1, k) + B(m - 1, k - 1)) \quad \text{for } 2 \leq k \leq m - 1.$$

In this triangle, we let $w = 1$ for simplicity.

```
                              0
                       0.5          0.5
                 0.75        1.5          0.75
           0.875      2.125       2.125    0.875
       0.9375    2.5        3.125        2.5      0.9375
   0.96875   2.71875    3.8125      3.8125    2.71875   0.96875
0.984375  2.84375   4.265625   4.8125   4.265625   2.84375  0.984375
```

29. $\sum_{i=0}^{\lfloor n/2 \rfloor} C(n - i, i) = f_n$

SECTION 7.2 MODELING COMBINATORIAL PROBLEMS WITH RECURRENCE RELATIONS

1. (a) $A_{n+1} = 1.05A_n$, with $A_0 = 1200$ (b) $1458.61

3. 203

5. $P_3(8) = 10$: $3 + 3 + 2$, $3 + 3 + 1 + 1$, $3 + 2 + 2 + 1$, $3 + 2 + 1 + 1 + 1$, $3 + 1 + 1 + 1 + 1 + 1$, $2 + 2 + 2 + 2$, $2 + 2 + 2 + 1 + 1$, $2 + 2 + 1 + 1 + 1 + 1$, $2 + 1 + 1 + 1 + 1 + 1 + 1$, and $1 + 1 + 1 + 1 + 1 + 1 + 1 + 1$

7. (a)   24   (b)   29   (c)   88   (d)   286

9. $A_n = 2000 + 1.05 A_{n-1}$, with $A_0 = 0$

11. $a_n = a_{n-1} + a_{n-2} + a_{n-3}$, with $a_0 = 1$, $a_1 = 2$, and $a_2 = 4$

13. $a_n = 2a_{n-1} + 2a_{n-2}$, with $a_0 = 1$ and $a_1 = 3$

15. (a) $a_n = a_{n-1} + a_{n-2}$, with $a_0 = a_1 = 1$ (b)   233

17. (a) $q(n) = \sum_{k=0}^{M} C(n-1, k) q(n - k - 1)$, where M is the smaller of $n - 1$ and 3, with $q(0) = 1$ (b)   196

19. We know that $p(n) = \sum_{k=0}^{n-1} C(n-1, k) p(n - k - 1)$. Replace the dummy variable k by $n - k - 1$.

21. $L(n, k) = L(n-1, k) + L(n-1, k-1)$ as long as $1 \le k \le n - 1$; $L(n, 0) = L(n, n) = 1$

23. (a) $a_n = a_{n-1} + a_{n-2}$, with $a_1 = 1$ and $a_2 = 2$ (b)   8

25. (a) $T(a, b) = T(a-1, b) + T(a, b-1)$, with $T(a, 0) = T(0, b) = 1$
    (b)   To reach (a, b), we need a sequence of a moves to the right and b moves up, that is, a string of length $a + b$ consisting of a R's and b U's. There are $C(a + b, a)$ such strings. Thus $T(a, b) = C(a + b, a)$.

27. (a) $T(4) = 156$; $T(8) = 1380$ (b) $n = 128$

29. (a) $T(n) = 1 + T(\lceil n/2 \rceil)$, with $T(1) = 1$ (b)   6

31. (a)   If the bit string of length $n \ge 3$ starts with a 0, then in order not to contain 101 it must continue as a string of length $n - 1$ with no 101 as a substring; and there are a_{n-1} such strings. Otherwise the string starts with k 1's, for some $k \ge 1$, followed by 00, unless it ends before the 00 appears. The remaining $n - k - 2$ bits must form a bit string without the substring 101. For each k from 1 to $n - 2$, then, there are a_{n-k-2} such strings. Finally, the string can be $11 \ldots 110$ or $11 \ldots 111$. Thus $a_n = a_{n-1} + a_{n-3} + a_{n-4} + a_{n-5} + \cdots + a_1 + a_0 + 2$. The initial conditions are $a_0 = 1$, $a_1 = 2$, and $a_2 = 4$.
    (b)   351
    (c) $a_n = 2a_{n-1} - a_{n-2} + a_{n-3}$, with $a_0 = 1$, $a_1 = 2$, and $a_2 = 4$

33. (a) Let x_0 be a fixed element of the k-set A. There are n ways to choose the image y_0 of x_0 under a surjective function from A to the n-set B. In order for this function to be surjective, we need to have either that it is still surjective when restricted to $A - \{x_0\}$ (and there are $f(k - 1, n)$ ways this can happen), or that it is surjective to $B - \{y_0\}$ when so restricted, and there are $f(k - 1, n - 1)$ ways this can happen.

(b) $f(k, 1) = 1$ and $f(k, k) = n!$ (c)   1560

35. (a) *RRRUUU, RRURUU, RRUURU, RURRUU, RURURU*

(b) $T'(a, b) = T'(a - 1, b) + T'(a, b - 1)$, with $T'(a, b) = 0$ if $b > a$, and $T'(a, 0) = 1$ (c)   42

37. (a) There are eight partitions of 7 into at most three parts $(7, 6 + 1, 5 + 2, 5 + 1 + 1, 4 + 3, 4 + 2 + 1, 3 + 3 + 1, 3 + 2 + 2)$ and eight partitions of 7 into parts no larger than 3 $(3 + 3 + 1, 3 + 2 + 2, 3 + 2 + 1 + 1, 3 + 1 + 1 + 1 + 1, 2 + 2 + 2 + 1, 2 + 2 + 1 + 1 + 1, 2 + 1 + 1 + 1 + 1 + 1, 1 + 1 + 1 + 1 + 1 + 1 + 1)$.

(b) Flipping the Ferrers diagram representing a partition of k into parts no bigger than m (i.e., k dots in at most m columns) around the line through the upper left corner having slope -1 produces a Ferrers diagram representing a partition of k into at most m parts (i.e., k dots in at most m rows), and vice versa.

SECTION 7.3   SOLVING RECURRENCE RELATIONS

1. (a) no   (b) no   (c) yes   (d) yes   (e) yes

3. (a) yes, order 3   (b) not linear   (c) not homogeneous
(d) coefficients not constant

5. (a) $O(n \log n)$ (b) $O(n^2 \log n)$ (c) $O(n)$ (d) $O(n)$

7. (a) $a_n = (7/5) \cdot (-2)^n + (3/5) \cdot 3^n$
(b) $a_n = 2(-2)^n - (3/2)n \cdot (-2)^n$
(c) $a_n = 5 - 3 \cdot 2^n$
(d) $a_n = 2 - 3n$

9. (a) $$\frac{(1 - n)a_{n-1}}{a_{n-1} - n} = \frac{(1 - n)/(1 + C(n - 1))}{1/(1 + C(n - 1)) - n} = \frac{1 - n}{1 - n(1 + C(n - 1))}$$

$$= \frac{1 - n}{1 - n + Cn(1 - n)} = \frac{1}{1 + Cn} = a_n$$

(b) $a_n = 3/(3 - 2n)$
(c) $a_2 = -3, a_3 = -1, a_4 = -\frac{3}{5}$
(d) $a_n = 0$

11. $A_n = 1200 \cdot (1.05)^n$

13. (a) $A_n = 40000 \cdot (1.05^n - 1)$
    (b)   Depositing the money at the beginning of the year results in a 5% greater balance at the end of every year.

15. $O(n^{\log 3})$

17. (a)   1, 3, 4, 7, 11, 18, 29, 47
    (b) $L_n = \left(\dfrac{1 + \sqrt{5}}{2}\right)^n + \left(\dfrac{1 - \sqrt{5}}{2}\right)^n$ (c)   47   (d)   4

19. (a) $a_n = A \cdot \left(\dfrac{1 + \sqrt{13}}{2}\right)^n + B \cdot \left(\dfrac{1 - \sqrt{13}}{2}\right)^n$

    (b) $a_n = \left(\dfrac{39 - \sqrt{13}}{26}\right) \cdot \left(\dfrac{1 + \sqrt{13}}{2}\right)^n + \left(\dfrac{39 + \sqrt{13}}{26}\right) \cdot \left(\dfrac{1 - \sqrt{13}}{2}\right)^n$

21. $a_n = A \cdot 3^n + B \cdot (-3)^n + C \cdot i^n + D \cdot (-i)^n$

23. (a)   1, 1, 1, 2, 3, 3, 3, 4, 5, 5, 5, 6
    (b) $a_n = \frac{1}{2} + \frac{1}{2}n + \frac{1}{4}i^n + \frac{1}{4}(-i)^n$ (c) $a_7 = 4$; $a_{10} = 5$

25. (a)   1, 2, 8, 128, 32768   (b) $a_n = 2^{2^n - 1}$

27. Characteristic equation has no rational roots.

29. $2^n - \dfrac{1}{\sqrt{5}}\left(\dfrac{1 + \sqrt{5}}{2}\right)^{n+2} + \dfrac{1}{\sqrt{5}}\left(\dfrac{1 - \sqrt{5}}{2}\right)^{n+2}$

31. Suppose that $\{a_n\}$ satisfies the recurrence relation $a_n = c_1 a_{n-1} + \cdots + c_k a_{n-k}$, with $c_k \neq 0$. If the characteristic equation has k distinct roots, $r_1, r_2, \ldots, r_k$, then $a_n = A_1 r_1^n + \cdots + A_k r_k^n$, where the A_i's are arbitrary constants. Conversely, $a_n = A_1 r_1^n + \cdots + A_k r_k^n$ satisfies the recurrence relation.

33. (a) $a_n = n/6 + n^2/2 + n^3/3$ (b)   algebraically equivalent

35. (a) $n(n + 1)/2$ (b) $P_n = P_{n-1} + n(n + 1)/2$, $P_0 = 0$
    (c) $P_n = n/3 + n^2/2 + n^3/6$ (d)   364

37. $(n^2 + n + 2)/2$

39. (a) $a_n = \sqrt{a_{n-1} a_{n-2}}$, with $a_0 = x$ and $a_1 = y$
    (b) $a_n = x^{1/3} y^{2/3} \left(x^{2/3} y^{-2/3}\right)^{(-1/2)^n}$
    (c) $x^{1/3} y^{2/3}$

SECTION 7.4 THE INCLUSION–EXCLUSION PRINCIPLE

1. 5

3. 56

5. $\sum_{i=0}^{n-1} C(n,i)(-1)^i (n-i)^k$

7. 12

9. $|G \cup F \cup R| = |G| + |F| + |R| - |F \cap R| - |G \cap R| - |F \cap G| + |F \cap G \cap R| = 27 + |F \cap G \cap R|$ contradicts the fact that only 26 people were present.

11. 891

13. (a) 12   (b) 35

15. (a) 1000000   (b) 0   (c) 5005   (d) 0

17. odd if n is even; even if n is odd

19. If $n - 1$ objects are in their original positions, then there is no place for the other object to go except its original position.

21. (a) $1/15$ (b) $d_{14}/15!$ (c) $1/210$ (d) $d_{13}/15!$
    (e) $1/3003$ (f) $1/15!$ (g) $d_{14}/14!$ (h) $d_{15}/15!$
    (i) $1 - (d_{15}/15!)$

23. (a) 120   (b) 44   (c) 11088   (d) 958879

25. 149

27. 120

29. 194

31. 35

SECTION 7.5 GENERATING FUNCTIONS

1. (a) $4, 12, 9, 0, 0, 0, \ldots$ (b) $a_k = (-1)^k$ (c) $a_k = 3 \cdot 4^k$
   (d) $3, 0, 0, 0, 3, 0, 0, 0, 3, 0, 0, 0, \ldots$ (e) $0, 2, 0, -2, 0, 2, 0, -2, \ldots$

3. (a) 1, 2, 3, 4, 5   (b) 1, 3, 6, 10, 15   (c) 1, 4, 10, 20, 35
   (d) 1, 5, 15, 35, 70

5. (a) 3   (b) 1

7. (a) $f(x) = 2/(1+x)$
   (b) $f(x) = 4/(1-2x)$
   (c) $f(x) = 2/(1-x)^2$

(d) $f(x) = x^3/(1-x)$
(e) $f(x) = x/(1+x^2)$
(f) $f(x) = (2-x)/(1-x)^2$
(g) $f(x) = (1+x)/(1-x)^2$
(h) $f(x) = (x^2+x+2)/(1-x^2)^2$

9. $1/(1-x) - x^2(1/(1-x^3)) = (1+x)/(1-x^3)$

11. (a) $f(x) = \dfrac{1}{1-x^3} \cdot \dfrac{1}{1-x^4} \cdot \dfrac{1}{1-x^5}$

$$= 1 + x^3 + x^4 + x^5 + x^6 + x^7 + 2x^8 + 2x^9 + 2x^{10} + \cdots$$

(b)   2

13. (a)   11   (b)   4   (c)   4   (d)   9

15. (a)   5456   (b)   84   (c)   3654   (d)   84

17. $f(x) = (x + x^2 + x^3 + x^4 + x^5 + x^6)^n$

19. (a) $a_k = 9 \cdot 2^k - 16 \cdot 3^k + 8 \cdot 4^k$
    (b) $a_k = -(1/\sqrt{5})((1-\sqrt{5})/2)^{k+1} + (1/\sqrt{5})((1+\sqrt{5})/2)^{k+1}$

21. (a) $a_k = (-1)^{\lfloor k/2 \rfloor} \cdot 2^k$
    (b) $a_k = (-1)^{k/4}$ if $k \equiv 0 \pmod 4$; $a_k = 2 \cdot (-1)^{(k-1)/4}$ if $k \equiv 1$ (mod 4); $a_k = 3 \cdot (-1)^{(k-2)/4}$ if $k \equiv 2 \pmod 4$; $a_k = (-1)^{(k-3)/4}$ if $k \equiv 3 \pmod 4$

23. To show that

$$(1+x)(1+x^2)(1+x^3)\cdots = \frac{1}{1-x} \cdot \frac{1}{1-x^3} \cdot \frac{1}{1-x^5} \cdots$$

rewrite the right-hand side as

$$\frac{1}{1-x} \cdot \frac{(1-x)(1+x)}{1-x^2} \cdot \frac{1}{1-x^3} \cdot \frac{(1-x^2)(1+x^2)}{1-x^4} \cdot \frac{1}{1-x^5} \cdot \frac{(1-x^3)(1+x^3)}{1-x^6} \cdots$$

and then cancel common factors to obtain the left-hand side.

25. The number of ways to get a sum of k rolling a die t times is the coefficient of x^k in $(x + x^2 + x^3 + x^4 + x^5 + x^6)^t$. Thus for the generating function we need to sum over all t from 0 to ∞, which yields the given formula.

27. $C(k+2, 3) + C(k+1, 3)$

29. (a)   256   (b)   240   (c)   120   (d)   376

CHAPTER 8:   GRAPHS

SECTION 8.1   BASIC
DEFINITIONS IN GRAPH THEORY

1.   (a)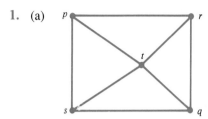

(b)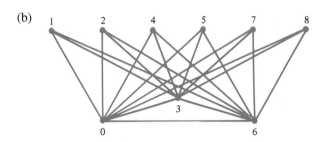

(c)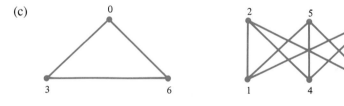

(d)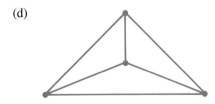

(e)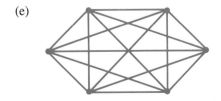

(f)

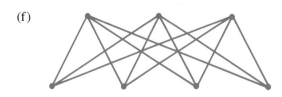

3.

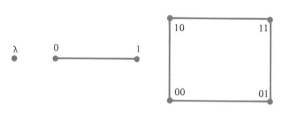

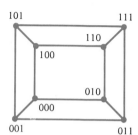

5. (a) $d^+(a) = 2, d^+(b) = 0, d^+(c) = 1, d^+(d) = 1, d^+(e) = 1, d^+(f) = 0,$
$d^-(a) = 1, d^-(b) = 1, d^-(c) = 2, d^-(d) = 1, d^-(e) = 0, d^-(f) = 0$

(b) $d^+(Detroit) = 6, d^+(Boston) = 3, d^+(Washington) = 3, d^+(Miami) = 4,$
$d^-(Detroit) = 6, d^-(Boston) = 3, d^-(Washington) = 3, d^-(Miami) = 4$

(c) $d^+(1) = 6, d^+(2) = 5, d^+(3) = 4, d^+(4) = 3, d^+(5) = 2, d^+(6) = 1,$
$d^+(7) = 0, d^-(1) = 0, d^-(2) = 1, d^-(3) = 2, d^-(4) = 3, d^-(5) = 4,$
$d^-(6) = 5, d^-(7) = 6$

7. $\overline{\overline{G}} = G$

9.

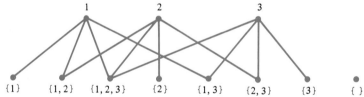

11. (a)

(b) If one vertex is isolated, there can be at most $C(5, 2) = 10$ edges. Thus if there are 11 edges, no vertex can be isolated.

(c) False; take a multigraph with six vertices, one pair of which has 11 parallel edges between them.

13. (a) T (b) F (c) F (d) F

15. (a)

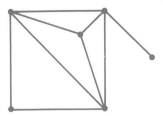

(b)   impossible
(c)   impossible
(d)   impossible
(e)

(f)

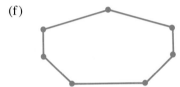

(g)

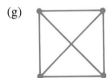

17. The possible degrees in a graph with n vertices are the n values $0, 1, 2, \ldots,$ $n - 1$. The only way to avoid having two vertices with the same degree is for the degree sequence to be precisely $(n - 1, n - 2, \ldots, 2, 1, 0)$. But this is impossible, since if one vertex has degree $n - 1$, then there can be no vertex with degree 0. The statement need not hold for multigraphs.

19. Q_0 is a single vertex. Q_{n+1} is obtained from two disjoint copies of Q_n, say with corresponding vertices $v_1, v_2, \ldots, v_{2^n}$ and $v'_1, v'_2, \ldots, v'_{2^n}$, together with edges $v_i v'_i$ for each i.

21. (a)

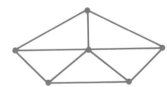

(b) $2n$ (c) $(n, 3, 3, \ldots, 3)$, with n 3's

23. (a)   the graph with n vertices and no edges
(b)   the disjoint union of K_m and K_n

25. 728 edges, with degree sequence $(4 \cdot 27, 12 \cdot 25, 20 \cdot 23, 28 \cdot 21)$, where $k \cdot d$ denotes k copies of the number d

27. (a)

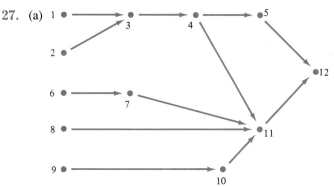

(b)

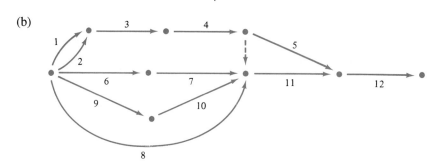

29. (a) Not locally finite; K_n is subgraph if and only if $n \leq 3$; C_n is subgraph for all $n \geq 3$.

(b) Not locally finite; K_n is subgraph for all n; C_n is subgraph for all $n \geq 3$.

(c) Locally finite; K_n is subgraph for all n; C_n is subgraph for all $n \geq 3$.

31. (a) The double star is a graph whose vertex set is $V = \{a, b, a_1, a_2, \ldots, a_m, b_1, b_2, \ldots, b_n\}$ with edges ab, aa_i for all i, and bb_j for all j.

(b) Let $V_1 = \{a, b_1, b_2, \ldots, b_n\}$ and $V_2 = \{b, a_1, a_2, \ldots, a_m\}$.

33. The parts are the set of all bit strings of length n with an odd number of 1's and the set of all bit strings of length n with an even number of 1's.

35. (a) $\exists v \in V : \forall e \in E : v \notin e$

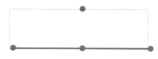

(b) $(\exists v_1 \in V : \exists v_2 \in V : \exists v_3 \in V : (\{v_1, v_2\} \in E \wedge \{v_1, v_3\} \in E \wedge \{v_2, v_3\} \in E)) \wedge \neg(\exists v_1 \in V : \exists v_2 \in V : \exists v_3 \in V : \exists v_4 \in V : (\{v_1, v_2\} \in E \wedge$

$\{v_1, v_3\} \in E \land \{v_1, v_4\} \in E \land \{v_2, v_3\} \in E \land \{v_2, v_4\} \in E \land \{v_3, v_4\}$
$\in E))$

(c) $\forall e \in E : \exists v_1 \in V_1 : \exists v_2 \in V_2 : e = \{v_1, v_2\}$

(d) $\exists E' \subseteq E : \forall v \in V : \exists! e \in E' : v \in e$

37. a disjoint union of cycles

39. **procedure** *maxclique*$(G$: graph with vertices $v_1, v_2, \ldots, v_n)$
    if $n = 1$ **then return**(1)
    $H_1 \leftarrow G$ with v_n deleted
    $H_2 \leftarrow G$ with v_n and all vertices not adjacent to v_n deleted
        {in each case, edges are kept if their endpoints remain}
    return$(\max(maxclique(H_1), 1 + maxclique(H_2)))$

SECTION 8.2   TRAVELING THROUGH A GRAPH

1. (a)   none   (b)   walk, trail   (c)   walk, closed walk
    (d)   walk, trail, path   (e)   walk, closed walk, trail, closed trail, cycle
    (f)   walk, closed walk, trail, closed trail   (g)   walk

3. $a, h, i, j, d, i, c, h, j, f, e, d, c, f, g, a, g, c, b, a$ and $a, h, j, i, h, c, i, d,$
$j, f, e, d, c, f, g, a, g, c, b, a$

5. (a) 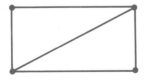   (b)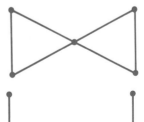

(c)    (d)

7. None are necessarily true.

9. no

11. $(n-1)(n-2)/2$

13. (a) $a; b; c, d; e; f$
    (b) 20; 30; 40, 50, 60, 70; 80; 90
    (c) $a; b, c$

15. Let $v = v_0, e_1, v_1, \ldots, e_n, v_n = v$ be a given closed trail starting and ending at v. If all the vertices $v_1, v_2, \ldots, v_n$ are distinct, then we are done. If not, let v_i and v_j, with $0 < i < j$ be a pair of equal vertices in the trail. Excise the portion of the trail from v_i to v_j (deleting v_i but not v_j) to obtain a shorter trail. Repeat until no equal vertices remain.

17. Let u and v be distinct vertices in the digraph. By definition there are walks from u to v and from v to u. Concatenating these gives a closed walk $u = u_0, u_1, \ldots, u_n = u$ from u to itself. Let j be the smallest natural number such that $u_j = u_i$ for some $i < j$; clearly $j \leq n$, and the i for which $u_i = u_j$ is uniquely determined. Then $u_i, u_{i+1}, \ldots, u_j$ is the desired cycle.

19. By the corollary to Theorem 1 in Section 8.1, the hypothesis is false. Hence any conclusion follows.

21. (a) all odd $n \geq 3$
    (b) all pairs for which m and n are both even positive integers
    (c) all $n \geq 3$
    (d) all even $n \geq 2$

23. Such a tracing is exactly an Euler trail. The graph in part (b) has no tracing.
    (a)    (c)

25. If there are no edges, we cannot begin to form the first closed trail C_1.

27. $m = n \geq 2$

29. Since G is strongly connected, there is a walk from u to v_1, and hence, by Theorem 1, a path from u to v_1. This path enters C for the first time at some vertex v_j, having begun $u = u_1, u_2, \ldots, u_l, v_j$, where the u_i's are distinct vertices not in C. But now we see a longer cycle in G, namely

$v_1, v_2, \ldots, v_{j-1}, u_1, u_2, \ldots, u_l, v_j, v_{j+1}, \ldots, v_k, v_1$. This contradiction shows that C must have been a Hamilton cycle.

31. If G is bipartite, then every cycle must have even length, since it constantly jumps from one part to the other and eventually ends up where it began. We prove the converse by induction. Suppose that G has no cycles of odd length. If G has one or two vertices, then clearly G is bipartite. Assume the inductive hypothesis, and let G be a graph with $n + 1$ vertices. Let v be a vertex of G, and let $v_1, v_2, \ldots, v_r$ be the vertices adjacent to v. Let G' be the graph obtained by removing v from G. By the inductive hypothesis, G' is bipartite. Choose the parts V_1 and V_2 so that for as large a k as possible, $v_1, v_2, \ldots, v_k$ are in V_1. If $k < r$, there must be a path from v_{k+1}, which is in V_2, to some $v_i \in V_1$, necessarily of odd length (otherwise, we could put v_{k+1} into V_1 by switching parts for all vertices in the same component as v_{k+1}). But then the cycle consisting of this path, followed by the edge to v, followed by the edge to v_{k+1}, is an odd cycle, contradicting the hypothesis. Therefore, $k = r$, and we can put v into V_2 to obtain a bipartition of G.

33. (a) 0011 (b) 0000111100101101
    (c) Construct a graph with 2^{n-1} vertices and 2^n edges as suggested in the hint. By Exercise 32, there is an Euler tour in this digraph, since each vertex has in-degree and out-degree 2. The labels on this tour are the desired de Bruijn sequence.

SECTION 8.3 GRAPH REPRESENTATION AND GRAPH ISOMORPHISM

1. (a)

	a	b	c	d	e	f	g
a	0	0	1	1	0	0	1
b	0	0	0	0	0	0	0
c	1	0	0	1	0	0	0
d	1	0	1	0	1	1	0
e	0	0	0	1	0	1	0
f	0	0	0	1	1	0	1
g	1	0	0	0	0	1	0

(b)

	a	b	c	d	e
a	0	0	0	1	0
b	1	0	0	0	0
c	0	1	0	0	0
d	0	0	1	0	0
e	0	0	0	0	1

3.

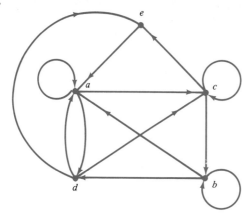

5. (a) 3 (b) 3 (c) 2

7.

(a)
a:	c, d, g
b:	
c:	a, d
d:	a, c, e, f
e:	d, f
f:	d, e, g
g:	a, f

(b)
a:	b, c, e
b:	a, d
c:	a, d, e
d:	b, c, e
e:	a, c, d

9.

$$\begin{bmatrix} 0 & 0 & 1 & 1 \\ 0 & 0 & 0 & 1 \\ 1 & 0 & 0 & 0 \\ 1 & 1 & 0 & 0 \end{bmatrix}$$

11.

$$\begin{array}{c} a \\ b \\ c \\ d \\ e \\ f \\ g \end{array} \begin{bmatrix} 1 & 1 & 1 & 0 & 0 & 0 & 0 & 0 \\ 0 & 0 & 0 & 0 & 0 & 0 & 0 & 0 \\ 1 & 0 & 0 & 1 & 0 & 0 & 0 & 0 \\ 0 & 1 & 0 & 1 & 1 & 1 & 0 & 0 \\ 0 & 0 & 0 & 0 & 1 & 0 & 1 & 0 \\ 0 & 0 & 0 & 0 & 0 & 1 & 1 & 1 \\ 0 & 0 & 1 & 0 & 0 & 0 & 0 & 1 \end{bmatrix}$$

13. (a) 2 (b) 3 (c) 7 (d) 12

15. (a)
$$\begin{bmatrix} 0 & 1 & 1 & 1 & \cdots & 1 & 1 \\ 1 & 0 & 1 & 1 & \cdots & 1 & 1 \\ 1 & 1 & 0 & 1 & \cdots & 1 & 1 \\ 1 & 1 & 1 & 0 & \cdots & 1 & 1 \\ \vdots & \vdots & \vdots & \vdots & \ddots & \vdots & \vdots \\ 1 & 1 & 1 & 1 & \cdots & 0 & 1 \\ 1 & 1 & 1 & 1 & \cdots & 1 & 0 \end{bmatrix}$$

(b)
$$\begin{bmatrix} 0 & 1 & 0 & 0 & \cdots & 0 & 1 \\ 1 & 0 & 1 & 0 & \cdots & 0 & 0 \\ 0 & 1 & 0 & 1 & \cdots & 0 & 0 \\ 0 & 0 & 1 & 0 & \cdots & 0 & 0 \\ \vdots & \vdots & \vdots & \vdots & \ddots & \vdots & \vdots \\ 0 & 0 & 0 & 0 & \cdots & 0 & 1 \\ 1 & 0 & 0 & 0 & \cdots & 1 & 0 \end{bmatrix}$$

(c)
$$\begin{bmatrix} 0 & 0 & \cdots & 0 & 1 & 1 & \cdots & 1 \\ \vdots & \vdots & \ddots & \vdots & \vdots & \vdots & \ddots & \vdots \\ 0 & 0 & \cdots & 0 & 1 & 1 & \cdots & 1 \\ 1 & 1 & \cdots & 1 & 0 & 0 & \cdots & 0 \\ \vdots & \vdots & \ddots & \vdots & \vdots & \vdots & \ddots & \vdots \\ 1 & 1 & \cdots & 1 & 0 & 0 & \cdots & 0 \end{bmatrix}$$

17. (a) the out-degree of v_i (b) the in-degree of v_j
    (c) the number of loops in the digraph

19. (a)

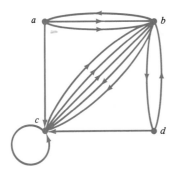

(b) yes   (c) $d^+(v_i) = \sum_{j=1}^{n} a_{ij}$; $d^-(v_j) = \sum_{i=1}^{n} a_{ij}$

21. **procedure** *in_list*(L : outgoing adjacency lists)
       {assume that vertices are labeled $v_1, v_2, \ldots, v_n$; L and L' are indexed by
          $\{v_1, v_2, \ldots, v_n\}$}

```
initialize L' to have all its entries empty
for i ← 1 to n do
    for each entry v_j in L(v_i) do
        adjoin v_i to L'(v_j)
    return(L') {incoming adjacency lists}
```

23. $(D,U,6)$, $(D,N,5)$, $(D,P,5)$, $(B,U,4)$, $(D,S,4)$, $(N,U,3)$, $(P,U,3)$, $(B,N,2)$, $(B,P,2)$

25. The identity function from V to itself is tautologically an isomorphism from $G = (V,E)$ to itself, so the relation is reflexive. Suppose that $\varphi: V \to V'$ is an isomorphism from $G = (V,E)$ to $G' = (V',E')$. Thus $\forall u \in V: \forall v \in V: (uv \in E \leftrightarrow \varphi(u)\varphi(v) \in E')$. Now since φ is bijective it has an inverse φ^{-1}, and the condition is equivalent to $\forall u' \in V': \forall v' \in V': (\varphi^{-1}(u')\varphi^{-1}(v') \in E \leftrightarrow u'v' \in E')$. Thus φ^{-1} is an isomorphism from G' to G. This shows that the relation is symmetric. Finally, to show transitivity, let $\varphi: V_1 \to V_2$ and $\psi: V_2 \to V_3$ be isomorphisms from $G_1 = (V_1,E_1)$ to $G_2 = (V_2,E_2)$, and from $G_2 = (V_2,E_2)$ to $G_3 = (V_3,E_3)$, respectively. Thus $\forall u \in V_1: \forall v \in V_1: (uv \in E_1 \leftrightarrow \varphi(u)\varphi(v) \in E_2)$ and $\forall u' \in V_2: \forall v' \in V_2: (u'v' \in E_2 \leftrightarrow \psi(u')\psi(v') \in E_3)$. This implies that $\forall u \in V_1: \forall v \in V_1: (uv \in E_1 \leftrightarrow \psi(\varphi(u))\psi(\varphi(v)) \in E_3)$, so $\psi \circ \varphi: V_1 \to V_3$ is an isomorphism from G_1 to G_3.

27. (a)

(b)

(c)

(d)

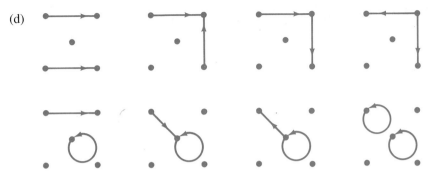

(e) C_{15}, $C_{12} \cup C_3$, $C_{11} \cup C_4$, $C_{10} \cup C_5$, $C_9 \cup C_6$, $C_8 \cup C_7$, $C_9 \cup C_3 \cup C_3$,
$C_8 \cup C_4 \cup C_3$, $C_7 \cup C_5 \cup C_3$, $C_7 \cup C_4 \cup C_4$, $C_6 \cup C_6 \cup C_3$, $C_6 \cup C_5 \cup C_4$,
$C_5 \cup C_5 \cup C_5$, $C_6 \cup C_3 \cup C_3 \cup C_3$, $C_5 \cup C_4 \cup C_3 \cup C_3$, $C_4 \cup C_4 \cup C_4 \cup C_3$,
$C_3 \cup C_3 \cup C_3 \cup C_3 \cup C_3$

(f) none

29. (a)   (b)

(c) There is no way for G and $\overline{G}$ to each have half of the edges, since K_3 has three edges.

31. They are isomorphic.

33. The graph on the right is bipartite, but the graph on the left is not.

35. Given G, let $M(G)$ be the n by n 0–1 matrix which can be the adjacency matrix for G and which comes first in lexicographic order among all such matrices, where we consider the n^2 entries of the matrix ordered row by row, from left to right in each row.

37. For each edge uu' of G (obtained from the adjacency lists), and for each vertex w, determine whether both u and u' are adjacent to w. It takes $O(\max(e, v))$ steps to peruse the adjacency lists for all the edges, and $O(v)$ steps to check out all possible w for each edge.

SECTION 8.4   PLANARITY OF GRAPHS

1. (a) $16 - 28 + 15 = 2 + 1$ (b) $14 - 14 + 4 = 3 + 1$
   (c) $4 - 10 + 8 = 1 + 1$ (d) $20 - 22 + 6 = 3 + 1$

3. all $n \le 4$

5. The construction cannot be done because $K_{3,3}$ is not planar.

7.

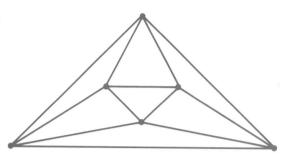

9. In the planar embedding of H, slightly fatten each edge and replace it by the required number of parallel edges. Put any required loops in empty space near the vertex.

11. $2e \geq 3r \implies r \leq 2e/3$; $v - e + r \geq 2 \implies 2 - v + e \leq r$. Therefore, $2 - v + e \leq 2e/3$, which is equivalent to $3v - 6 \geq e$.

13. (a)

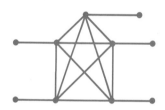

(b) $v = 10$, $e = 15$ (c) $15 \leq 24$ (d)   It contains K_5.
(e)   Theorem 2 said that if G is planar, then $e \leq 3v - 6$, not the converse.

15. The following two embeddings of the same graph differ in that in one case the two vertices of degree 1 are in the same region, whereas in the other they are not.

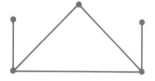 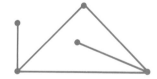

17. Not planar; it contains a subdivision of $K_{3,3}$.

19. For $v = 4$, the graph C_4 embedded in the plane satisfies $v = 2e - 4$. Note that all regions are four-sided. Given a planar embedding of a graph for which the equality holds, we can form a planar embedding of a graph with one more vertex for which the equality holds by putting a new vertex u inside a region $vwxy$ and adding edges uv and ux. Having added one more vertex

and two more edges, we have not disturbed the equality; and the resulting planar embedding still has only four-sided regions.

21. K_2 satisfies neither inequality.

23. (a) 20   (b)

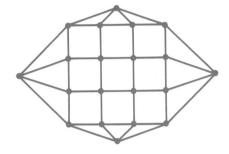

25. (a), (b), (c), and (d) are planar.

27. Suppose that a graph is embedded on the sphere. Pick a point in the interior of some region, puncture the sphere there, and stretch it open to form a plane. The region that was punctured becomes the unbounded region of the plane. All the other regions, as well as vertices, edges, and components, remain unchanged. Thus Euler's formula for the sphere follows from Euler's formula for the plane.

29. (a) Each plane can contain at most $3 \cdot 11 - 6 = 27$ edges. Thus two planes can contain at most 54 edges, one short of the 55 needed for K_{11}. Therefore, the thickness is greater than 2. The following figure shows that it is 3.

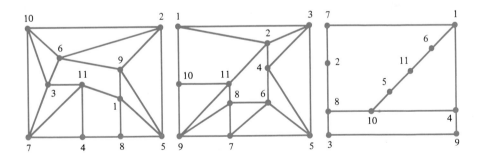

(b) Each plane can contain at most $2 \cdot 14 - 4 = 24$ edges. Thus two planes can contain at most 48 edges, one short of the 49 needed for $K_{7,7}$. Therefore, the thickness is greater than 2. The picture for part (c) shows that it is 3 (put the missing edge $a1$ in a third plane).

(c)

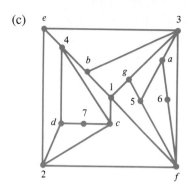 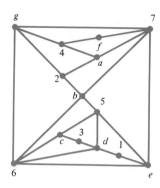

31. Suppose that the arrangement consists of d points forming a convex polygon, with $n - d$ points in its interior, where $3 \leq d \leq n$. When play is finished, the picture must consist of this d-gon with its interior completely triangulated. (Since every polygon of more than three sides contains at least one diagonal in its interior, play cannot stop until only triangles are left inside the d-gon.) Thus the number of moves is fixed in advance and does not depend on the players' strategies. The first player wins if $n - d$ is even; the second player wins if $n - d$ is odd.

SECTION 8.5   COLORING OF GRAPHS

1. (a)   Vertices represent courses. Two vertices are adjacent if the courses they represent have a student in common. We want a minimum coloring. The graph is shown below.

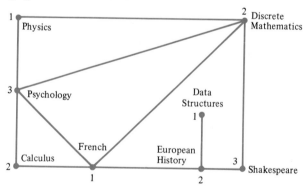

(b)   One minimum coloring gives rise to the scheduling of physics, data structures, and French exams on day 1; calculus, discrete math, and European history exams on day 2; and psychology and Shakespeare exams on day 3.

3. The chromatic number is 3.

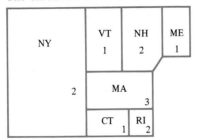 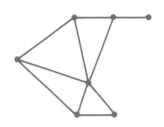

5. (a) 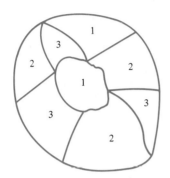 (b)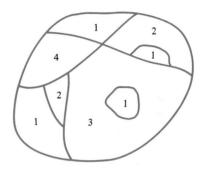

7. (a) color 1: C; color 2: P; color 3: D, F; color 4: M, R
(b) color 1: H, C, O, A, N; color 2: F, K; color 3: G, L, B; color 4: I, D; color 5: J, E; color 6: M

9. (a)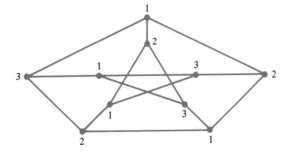

(b)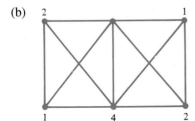

(c)

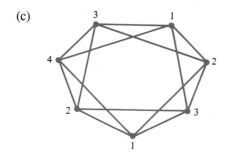

11. solution found in Exercise 7b

13. Assume that the colors are red, blue, green, and yellow.
  Color 1 red.
    Color 2 blue.
      Color 3 green.
        Color 4 blue.
        Color 5 yellow (4-coloring found).
      Color 4 green.
        Color 5 blue (3-coloring found).
      Color 4 yellow.
        Color 5 blue.
 done

  The algorithm colors 1 red, 2 blue, 3 green, 4 green, and 5 blue.

15. graphs with no edges

17. (a) There are d edges with a common endpoint, so at least d colors are required.
 (b) $\chi'(K_4) = 3$
 (c) The edge-chromatic number of the graph is 4.

19. The chromatic number of a graph with more than one component is the largest of the chromatic numbers of its components.

21. (a) F (b) T (c) T (d) F (e) F
  (f) F (g) T

23. (a)

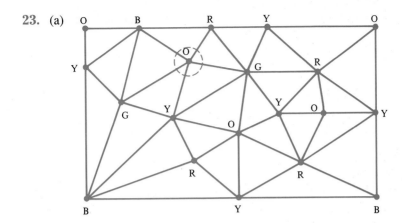

(b)

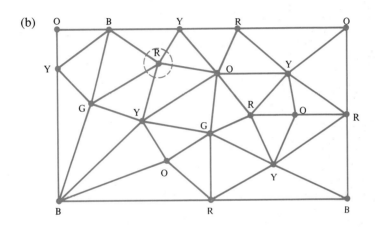

25. $O(\chi(G) \cdot (v + e))$

27. The algorithm will produce a 2-coloring of the graph with the numbering shown on the left, but not with the numbering shown on the right.

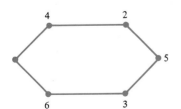
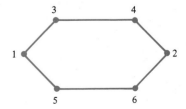

29. (a)   If we remove edge ab from K_n, we can $(n-1)$-color the resulting graph by using the same color for a and b.

(b)   the five-spoked wheel

(c)   Let $\chi(G) = k$. If G is k-critical, we are done. If not, G has a proper subgraph G_1 such that $\chi(G_1) = k$. If G_1 is k-critical, we are done. If not, we repeat this process, obtaining a proper subgraph G_2 with $\chi(G_2) = k$. We continue as long as possible. Since G has only finitely many vertices and edges, the process must eventually halt, giving us a k-critical graph.

(d)   By Exercise 19, the chromatic number of a graph equals the chromatic number of one of its components. Thus some proper subgraph of a nonconnected graph has the same chromatic number as the graph, so no nonconnected graph can be k-critical.

(e)   The graph G' obtained by deleting one edge uv from G has chromatic number less than k. If its chromatic number were less than $k-1$, then a $(k-2)$-coloring of G', together with color $k-1$ on vertex v, is a $(k-1)$-coloring of G, a contradiction.

CHAPTER 9:   TREES

SECTION 9.1   BASIC DEFINITIONS FOR TREES

1. The sum of the degrees must be twice the number of edges, namely $2(n-1)$. Thus the average is $2(n-1)/n < 2$.

3. (a)   65   (b)   15   (c)   25   (d)   63

5. (a) f, g (b)   none   (c) h, i, j (d) l, m, n
   (e) f, g, k, l, m, n (f) a, b, e, j

7. (a)

(b)

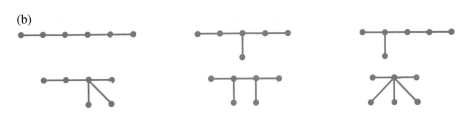

(c)

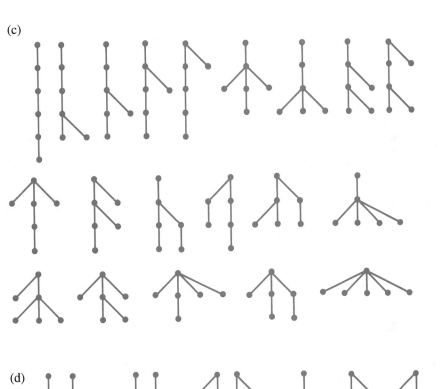

(d)

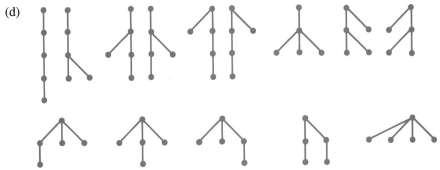

(e)

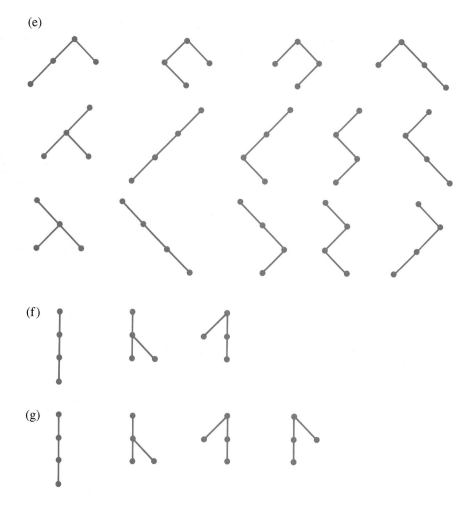

(f)

(g)

9. (a) Let G be a connected graph with n vertices and $n-1$ edges. Suppose (for an indirect proof) that G is still connected when some edge is removed. If this smaller graph has no closed trails, it is by definition a tree. If it has a closed trail, we can delete any edge in this closed trail without disconnecting the graph, giving us a smaller connected graph. We continue in this manner, deleting edges in closed trails, until we obtain a tree. But by $1 \to 2$, this tree must have $n-1$ edges, a contradiction to the fact that we deleted an edge at the first step.

   (b) Let G be a graph with no closed trails such that if any edge not in G is added to G, then the resulting graph contains a cycle. Suppose (for an indirect proof) that G is not a tree; thus G is not connected. Let u and v be in different components of G. Let G' be G with edge uv

added. By the hypothesis, G' contains a cycle C, necessarily using edge uv. But then C without uv must be a path connecting u and v in G, contradicting the choice of u and v. Therefore, G is connected.

11. (a) Draw the tree as a rooted tree. One part consists of all vertices at even-numbered levels; the other part consists of all vertices at odd-numbered levels.

(b) $m = 1$ or $n = 1$

13. The height of a tree is the maximum length of a path from the root to a leaf. Such a path must consist of an edge from the root to a child v_i of the root, followed by a path from this child to a leaf in its subtree T_i. Thus the longest path has length one greater than the maximum of the heights of its immediate subtrees T_i.

15. By Theorem 5 (with $m = 2$), $l \leq 2^h$. Thus $\log l \leq h$, so $h \geq \lceil \log l \rceil$ (since h is an integer). Also, by Theorem 5, $n \leq 2^{h+1} - 1$, so $h+1 \geq \lceil \log(n+1) \rceil$, or $h \geq \lceil \log(n+1) \rceil - 1$. Essentially the same inequalities hold using logarithms base m.

17. In the order the vertices are numbered, each vertex that will end up as an internal vertex sprouts its two children.

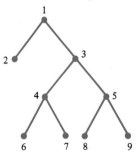

19. $2^{h-l+1} - 2$

21. $m = [(l-1)/i] + 1,\ n = i + l;$
$l = (m-1)i + 1,\ n = mi + 1;$
$l = n - i,\ m = (n-1)/i;$
$i = (l-1)/(m-1),\ n = [(l-1)/(m-1)] + l;$
$i = n - l,\ m = (n-1)/(n-l);$
$i = (n-1)/m,\ l = n - [(n-1)/m]$

23. (a) Any weighing can be modeled as an internal vertex of the tree, with three children. Leaves are "solutions"—decisions as to which coin is under weight. It is a 3-ary tree.
    (b) height 2, two weighings
    (c) Perform the weighings indicated in the following tree. The leaves list the counterfeit coin found.

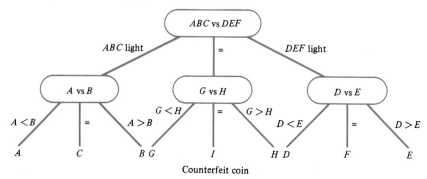

Counterfeit coin

25. (a) There exists a vertex w which is v's parent and u's sibling.
    (b) $u \neq v$, u and v are not siblings, and there exists a vertex w which is u's and v's grandparent.
    (c) There exist vertices u_i and v_i for $i = 1, 2, \ldots, n$ and vertex w such that both $w, u_1, u_2, \ldots, u_n, u$ and $w, v_1, v_2, \ldots, v_n, v$ are directed paths in the tree and $u_1 \neq v_1$.
    (d) Either there exist vertices u_i for $i = 1, 2, \ldots, n$, vertices v_i for $i = 1, 2, \ldots, n + r$, and vertex w such that both $w, u_1, u_2, \ldots, u_n, u$ and $w, v_1, v_2, \ldots, v_{n+r}, v$ are directed paths in the tree and $u_1 \neq v_1$, or else there exist vertices u_i for $i = 1, 2, \ldots, n + r$, vertices v_i for $i = 1, 2, \ldots, n$, and vertex w such that both $w, u_1, u_2, \ldots, u_{n+r}, u$ and $w, v_1, v_2, \ldots, v_n, v$ are directed paths in the tree and $u_1 \neq v_1$.

27. Make T a rooted tree from any vertex. Let v be a vertex at maximum level in this rooted tree. The height of the tree when rooted at v is the length of the longest path in the tree.

29. (a) F_2:

F_3:

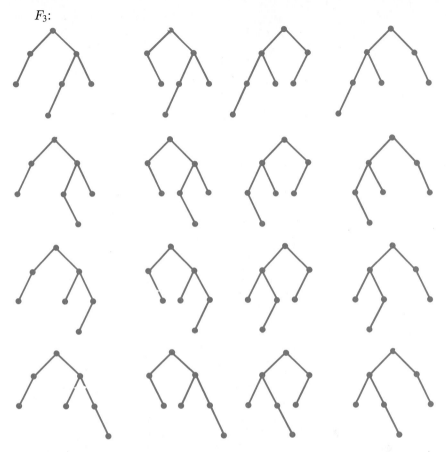

(b) Clearly true for F_0 and F_1. Assume inductive hypothesis. If $T \in F_n$ is formed with $T_1 \in F_{n-1}$ and $T_2 \in F_{n-2}$ as its immediate subtrees, then $\text{height}(T) = 1 + \max(\text{height}(T_1), \text{height}(T_2)) = n$.

(c) $a_n = 2a_{n-1}a_{n-2}$, $a_0 = 1$, $a_1 = 2$

(d) $a_n = 2^{f_{n+1}-1}$, where $\{f_n\}_{n=0}^{\infty}$ is the Fibonacci sequence.

SECTION 9.2   SPANNING TREES

1. (a)

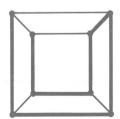 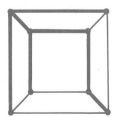

(b)

(c)

3. (a)

(b)

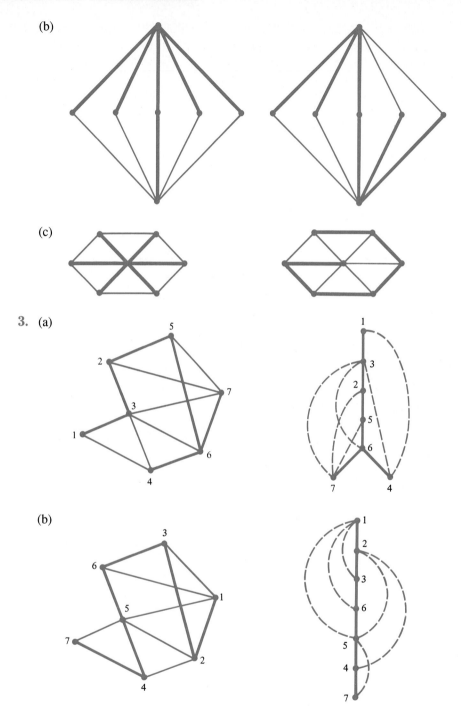

5. If G has no edges, then G must be K_1, which is itself a tree. Assume that the statement is true for graphs with k edges. Let G be a graph with $k + 1$ edges. If G is a tree, we are done. Otherwise, G must have a cycle. Remove an edge from the cycle. The resulting graph is still connected, but it has k edges. By the inductive hypothesis, it has a spanning tree, which perforce is also a spanning tree for G.

7. 32

9.

11. Suppose that C is a cut and T is a spanning tree. If the edges of C are removed, then the resulting graph is disconnected, hence cannot contain T. Therefore, at least one edge of T was in C.

13. (a)    (b)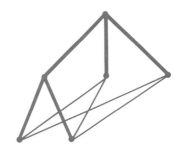

(c)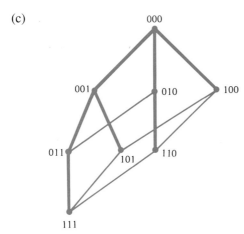

15. The breadth-first search spanning tree for K_n is $K_{1,n-1}$; the depth-first search spanning tree for K_n is a path.

17. Suppose that an edge uv of G is not used in the depth-first search spanning tree. Suppose without loss of generality that vertex u is encountered first during the search process. Since edge uv is not used, it must be the case that the search proceeded with other neighbors of u, whose numbers were less than v, and while searching from one of them, say v', encountered v. Thus v is in the subtree rooted at v', hence a descendant of v'. But v' is a descendant of u, so v is a descendant of u.

19. the tree on five vertices in K_5 that is not a path or $K_{1,4}$

21. Since no edge is ever put into T that leads to a previously visited vertex, T can contain no cycles. Every vertex v will eventually be visited, however (and an edge to it included in the tree): We see this by induction on the length of the shortest path from vertex 1 to vertex v.

23. **procedure** *breadth_first_search*(G : graph)
    {assume setting of Algorithm 2}
        $F \leftarrow \emptyset$
        for $i \leftarrow 1$ **to** n **do**
            visited(i) $\leftarrow$ **false**
        next $\leftarrow 1$
        while *next* > 0 **do**
            begin
                visited(*next*) $\leftarrow$ **true**
                $T \leftarrow (\{next\}, \emptyset)$
                $L \leftarrow (next)$
                while L is not empty **do**
                    begin
                        $i \leftarrow$ first element of L
                        $L \leftarrow L$ with i removed
                        for $j \leftarrow 1$ **to** n **do**
                            if (j is adjacent to i) $\wedge$ not *visited*(j) **then**
                                begin
                                    visited(j) $\leftarrow$ **true**
                                    add vertex j and edge ij to T
                                    add vertex j to the end of L
                                end
                    end
                $F \leftarrow F \cup T$
                next $\leftarrow 0$
                for $i \leftarrow n$ **down to** 1 **do**
                    if not *visited*(i) **then** *next* $\leftarrow i$
            end
        return(F)

25. Vertices at distance 1 from vertex 1 are at level 1, since they get visited from vertex 1; vertices at distance 2 from vertex 1 are adjacent to vertices at distance 1 from vertex 1, so they must appear at level 2 as children of vertices at level 1, and so on.

27. (a) Add e to T. The result has a cycle. Let f be any edge of the cycle other than e. Then $T \cup \{e\} - \{f\}$ is again a tree, since it is connected and has the right number of edges.

(b) We need only show how, given a spanning tree T and another spanning tree $T' \neq T$, we can find a spanning tree T'' which is the same as T except that one edge of $T - T'$ has been replaced by an edge $e \in T' - T$. Apply part (a), being careful to pick the edge f to be deleted from T not to lie in T' (this is possible, because not every edge in the cycle can lie in T').

29. **procedure** *iterative_DFS* (G : connected graph)
$\quad$ *visited*(1) $\leftarrow$ **true**
$\quad$ **for** $i \leftarrow 2$ **to** n **do**
$\quad\quad$ *visited*(i) $\leftarrow$ **false**
$\quad$ $T \leftarrow (\{1\}, \emptyset)$
$\quad$ $L \leftarrow (1)$
$\quad$ **while** $L \neq \emptyset$ **do**
$\quad\quad$ **begin**
$\quad\quad\quad$ $i \leftarrow$ first element of L
$\quad\quad\quad$ $j \leftarrow 1$
$\quad\quad\quad$ *found* $\leftarrow$ **false**
$\quad\quad\quad$ **while** $j \leq n \wedge$ not *found* **do**
$\quad\quad\quad\quad$ **begin**
$\quad\quad\quad\quad\quad$ **if** j is adjacent to $i \wedge$ not *visited*(j) **then**
$\quad\quad\quad\quad\quad\quad$ **begin**
$\quad\quad\quad\quad\quad\quad\quad$ *visited*(j) $\leftarrow$ **true**
$\quad\quad\quad\quad\quad\quad\quad$ *found* $\leftarrow$ **true**
$\quad\quad\quad\quad\quad\quad\quad$ add j to the front of L
$\quad\quad\quad\quad\quad\quad\quad$ add j and ij to T
$\quad\quad\quad\quad\quad\quad$ **end**
$\quad\quad\quad\quad\quad$ $j \leftarrow j + 1$
$\quad\quad\quad\quad$ **end**
$\quad\quad\quad$ **if** not *found* **then** remove first element from L {namely, i}
$\quad\quad$ **end**
$\quad$ **return**(T)

SECTION 9.3   TREE TRAVERSAL

1. (a) $a, b, e, h, i, j, o, c, d, f, k, l, m, n, g$
   (b) $h, i, o, j, e, b, c, k, l, m, n, f, g, d, a$

3. (a)

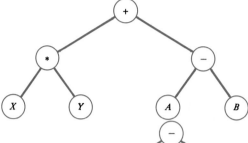

(b)

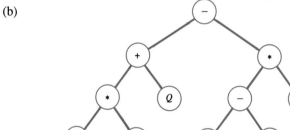

(c)

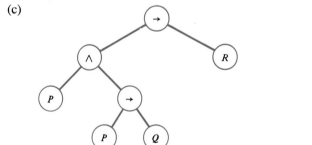

(d)

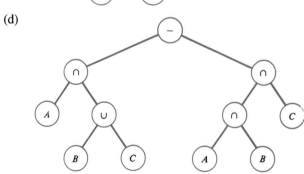

5. (a) $XY*AB-+$ (b) $XY+Z*Q+XY-B*-$
   (c) $PPQ\rightarrow\wedge R\rightarrow$
   (d) $ABC\cup\cap AB\cap C\cap-$

7. (a) 45   (b) $\{1,3,6,7\}$ (c) F

9. (a) $342-+63+*$ (b) $\{1,2,3,4\}\{1\}\{3,5\}\cup\cap\{6,7\}\cup$
   (c) $FFTF\rightarrow T\wedge\vee F\wedge T\wedge\vee$

11. (a) 25   (b) $\{5,6\}$ (c) T

13. (a) $+*33*44$ (b) $\cup\cap\{1,2,3\}\{4,5\}\{5,6\}$
    (c) $\rightarrow F\rightarrow F\rightarrow FF$

15. **procedure** *leaves*(*r* : root of ordered tree)
        if *r* has no children **then print**(*r*)
            else for each child *v* of *r*, in order, **do**
                call *leaves*(*v*)
        return

17. **procedure** *height*(*r* : root of binary tree)
        if *r* has a left child **then** *left_height* $\leftarrow$ *height*(*left_child*(*r*))
            else *left_height* $\leftarrow -1$
        if *r* has a right child **then** *right_height* $\leftarrow$ *height*(*right_child*(*r*))
            else *right_height* $\leftarrow -1$
        return(1 + max(*left_height*, *right_height*))

19.

21. The tree with root *A* and left child *B* has the same preorder and postorder as the tree with root *A* and right child *B*.

23. (a) If *T* is empty, then T^R is empty. If *T* has root *r*, then T^R has root *r*, with left subtree T_1^R, where T_1 is the right subtree of *T* (if any), and right subtree T_2^R, where T_2 is the left subtree of *T* (if any).

    (b) **procedure** *reverse*(*r* : root of extended binary tree)
        if $r = \emptyset$ **then return**($\emptyset$)

> **else**
> > **begin**
> > > create a new vertex r_0
> > > make *reverse*(*right_child*(r)) the left child of r_0
> > > make *reverse*(*left_child*(r)) the right child of r_0
> > > **return**(r_0)
> > **end**

25. A variable name or a constant is a prefix expression. If α_1 and α_2 are prefix expressions and θ is an operator, then $\theta\,\alpha_1\,\alpha_2$ is a prefix expression. A variable name or a constant is a postfix expression. If α_1 and α_2 are postfix expressions and θ is an operator, then $\alpha_1\,\alpha_2\,\theta$ is a postfix expression.

27. (a) $* - 4 \ominus 3\,7$ (where $\ominus$ is unary minus)
  (b) $\sqrt{\ } + * A\,A * B\,B$
  (c) $\sqrt{\ } + \Uparrow A \Uparrow B$ (where $\Uparrow$ is squaring)
  (d) $\cup \cap A \ominus B\,C$ (where $\ominus$ is complementation)
  (e) $\rightarrow \vee P \neg P\,Q$

29. There is no place "between" the operands to put the operator.

31. prefix

33. **procedure** *level_process*(r : root of ordered tree)
  {L is a list of vertices waiting to be processed}
  > $L \leftarrow (r)$
  > **while** L is not empty **do**
  > > **begin**
  > > > $v \leftarrow$ first element of L
  > > > remove v from L
  > > > process v
  > > > **for** each child c of v from left to right **do**
  > > > > add c to the end of L
  > > **end**
  > **return**

35. **procedure** *AVL*(r : root of binary tree)
  > **if** r has no children **then return**(**true**)
  > **else if** r has no left child **then**
  > > **return**(truth value of *height*(*right_child*(v) $= 0$))
  > **else if** r has no right child **then**
  > > **return**(truth value of *height*(*left_child*(v) $= 0$))
  > **else return**(truth value of
  > > $|height(left\_child(v)) - height(right\_child(v))| \le 1$
  > > > $\wedge\ AVL(left\_child(v)) \wedge AVL(right\_child(v)))$

37. Let x be the first vertex in the first list: It needs to be the root of the tree.

Find x in the second list, and let L_1 and L_2 be the lists of vertices that precede and follow x in the second list, respectively. A necessary condition that the tree exists is that the first list consist of the vertices in L_1, in some order, followed by the vertices in L_2, in some order. If the condition is met, then the left and right subtrees of x are determined recursively by L_1 and L_2 and their preorder arrangements given in the first list.

SECTION 9.4   FURTHER APPLICATIONS OF BINARY TREES

1.

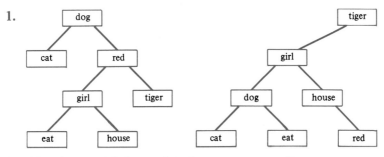

3. (a) $boy \prec girl$; $boy \prec dog$; $boy \prec cat$; return 0.
   (b) $boy \prec cat$; return 0.

5.

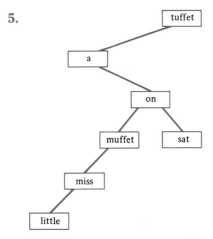

7. (a)   100110111101100   (b)   010101011
   (c)   10001100111101111 0110011110

9. 2.8 for the first tree, 3.55 for the second

11. False; put *hello* at the root, *goodbye* as the left child of the root, and *mayday* as the right child of *goodbye*.

13. **procedure** *find_or_insert*(*r* : root of nonempty binary search tree, *x* : word)
    if *x* = *contents*(*r*) **then return**(*r*)
      else if *x* ≺ *contents*(*r*) **then**
        if *left_child*(*r*) = ∅ **then**
          begin
            create a new vertex *v*
            contents(*v*) ← *x*
            left_child(*r*) ← *v*
            return(*v*)
          end
        else return(*find_or_insert*(*left_child*(*r*), *x*))
      else {*x* ⊀ *contents*(*r*) in this case}
        if *right_child*(*r*) = ∅ **then**
          begin
            create a new vertex *v*
            contents(*v*) ← *x*
            right_child(*r*) ← *v*
            return(*v*)
          end
        else return(*find_or_insert*(*right_child*(*r*), *x*))

15. as balanced as possible, with all leaves at level h or at level $h - 1$

17. False; the message could begin TW.

19. five, but only two essentially different ones

21. (a)

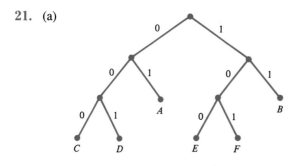

(b)

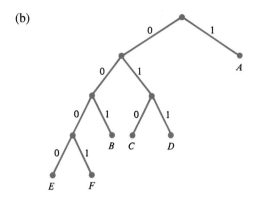

(c)

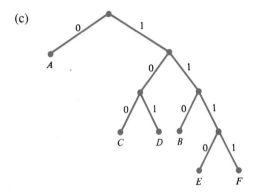

23. If not, we could exchange the positions of u and v to achieve a tree with smaller average code length.

25. Construct a function f from permutations of $\{1, 2, \ldots, n\}$ to binary trees: Given permutation π, construct a binary search tree by inserting the symbols of π, in order; $f(\pi)$ is the underlying binary tree. This function is onto the set of binary trees with n vertices, since given any such binary tree, we can label its vertices with the set $\{1, 2, \ldots, n\}$ in inorder and let π be the permutation obtained by listing the labels in level order—clearly $f(\pi)$ is the given tree. Furthermore, any binary tree in which the root has two children (and for each $n \geq 3$ there is at least one such tree) can be obtained from at least two permutations, the left-to-right level order and the right-to-left level order. The conclusion follows.

SECTION 9.5   GAME TREES

1. (a) and (b)

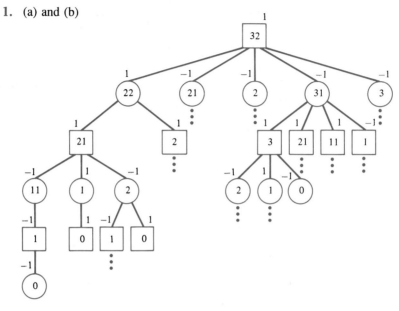

(c) first player

3. (a) and (b)

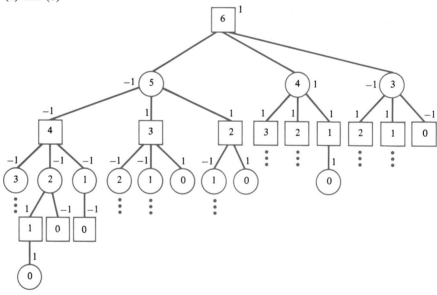

(c) first player

5. (a) and (b)

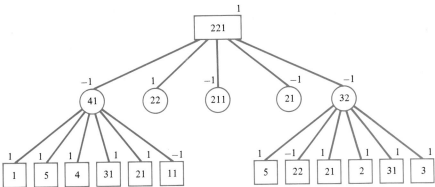

(c) first player

7. If $n = 0$ or 1, the statement is obvious. Now assume that n is odd and greater than 1. Blue takes one stone and by induction wins (he is now the second player in the game with a smaller, even number of stones). Finally, assume that n is even and greater than 0. If Blue takes one stone, Red wins by the inductive hypothesis (she is now the first player in the game with a smaller, odd number of stones). If Blue combines two piles into one, Red removes the new pile and wins by the inductive hypothesis (there are now $n - 2$ stones, and $n - 2$ is a smaller, even number).

9.

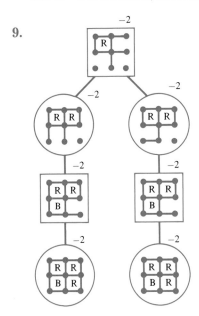

11. (a)   400   (b)   49   (c) $n(n-1)$, where $n = I(J-1) + J(I-1)$

13.

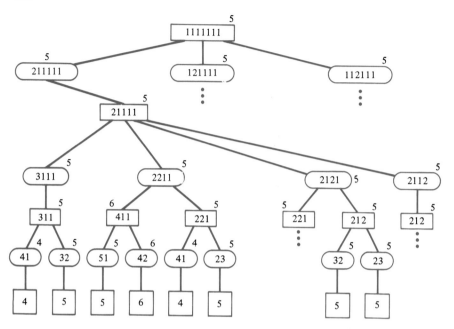

15. (a) and (b)

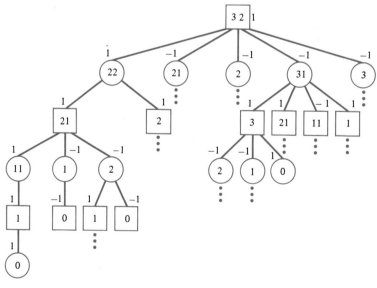

(c) first player

17. (a) and (b)

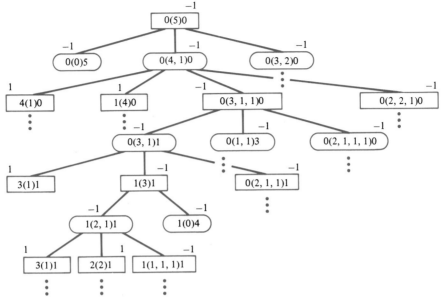

(c) second player

19. Second player wins if and only if the two piles have equal size.

21. (a) Maintain a global list of positions (including player to move) already analyzed, and their values. As a first alternative when P is not a leaf, check the list and return the appropriate value if the position is on the list. Otherwise, proceed with the recursive algorithm, but make sure to update the list before returning the value.

(b) Add another parameter, *depth_to_go*. The initial call to *game_value* uses a predetermined value of *depth_to_go*. Each recursive call to *game_value* passes on *depth_to_go* − 1. As a first alternative when P is not a leaf, if *depth_to_go* = 0, then return the value determined by the heuristic evaluation function.

(c) Add another parameter, *goal*. The initial call to *game_value* uses *goal* = ∞. Modify the first innermost **begin**...**end** block of the procedure to read as follows.

```
begin
    Q ← (P followed by move m)
    v' ← game_value(Q, Red, v)
    if v' > v then
        begin
            v ← v'
```

$$\textbf{if } v \geq goal \textbf{ then return}(goal)$$
$$\textbf{end}$$
$$\textbf{end}$$

Modify the second innermost **begin**. . . **end** block in a similar way (with each inequality reversed).

(d) The first outermost **begin**. . . **end** block is changed to read as follows (and the second in a dual way).

> **begin**
>   **for** each legal move m for Blue **do**
>   **if** *game_value*(P followed by move m, Red) $= 1$
>   **then return**(1)
>   **return**(-1)
> **end**

23. (a) The initial position is $P = \mathbf{N}$. A move consists of replacing P by a nonempty proper subset of P if P is infinite, or any proper subset of P if P is finite.

(b) The initial position is $P = \mathbf{N}$. A move consists of replacing P by a proper subset of P.

(c) The initial position is $P = \mathbf{N}$. A move consists of replacing P by a finite proper subset of P.

CHAPTER 10: GRAPHS AND DIGRAPHS WITH ADDITIONAL STRUCTURE

SECTION 10.1 SHORTEST PATHS AND LONGEST PATHS

1. a, f (weight 5); a, b, f (weight 4); a, e, f (weight 8); a, b, c, f (weight 9)

3.

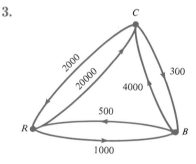

5. b, a, e

7. If $l = 1$, then the (i,j)th entry of $A = A^l$, i.e., a_{ij}, is the weight of the minimum weight walk of length 1 from vertex i to vertex j: Either there is a walk along the edge from i to j, or else there is no edge and $a_{ij} = \infty$. Assume that the statement is true for l. Then a minimum weight walk of length at most $l + 1$ from i to j is either a minimum weight walk of length at most l, or consists of a minimum weight walk of length l to some vertex k, followed by edge kj. Let $a_{ij}^{\diamond l}$ denote the (i,j)th entry of $A^{\diamond l}$. The weight of a minimum weight walk of length at most l from i to j is thus $a_{ij}^{\diamond l}$, which can be written as $a_{ij}^{\diamond l} + 0 = a_{ij}^{\diamond l} + a_{jj}$, and the weight of a minimum weight walk of length l from i to k followed by the edge kj is $a_{ik}^{\diamond l} + a_{kj}$. Thus the (i,j)th entry of $A^{\diamond(l+1)} = A^{\diamond l} \diamond A$, namely $\min_{1 \le k \le n}(a_{ik}^{\diamond l} + a_{kj})$, is the weight of a minimum weight walk of length at most $l + 1$, from i to j.

9. Consider the digraph consisting of edge ab with weight 10, edge bc with weight -6, and edge ca with weight -7. There is no minimum weight walk from a to b, since we can make the weight of a walk as small as desired simply by traveling around the triangle sufficiently often before stopping at b.

11. a, c, e, i, l, z

13. **procedure** $critical\_path(G$: weighted acyclic digraph)
    {assume that vertices of G are $0, 1, \ldots, n$, where 0 is the source and n is the sink; the weights are $w(i,j)$; we compute completion times $T(v)$ and the pointers $previous(v)$ needed to construct a critical path}
    for $i \leftarrow 1$ **to** n **do**
        $T(i) \leftarrow -1$
    $T(0) \leftarrow 0$
    for $i \leftarrow 1$ **to** n **do** {compute T for (i.e., label) another vertex}
        begin
            $v \leftarrow n$
            $v' \leftarrow 0$
            while $v' \ne v$ **do** {find vertex all of whose predecessors
                                are labeled}
            begin
                $v' \leftarrow v$
                for $j \leftarrow 1$ **to** $n - 1$ **do**
                    if jv is an edge and $T(j) = -1$ **then** $v \leftarrow j$
            end
            for $j \leftarrow 1$ **to** $n - 1$ **do** {label v}
                if jv is an edge and $T(j) + w(j,v) > T(v)$ **then**
                    begin
                        $T(v) \leftarrow T(j) + w(j,v)$

$$previous(v) \leftarrow j$$
          end
       end
       return$(T, previous)$

15. b, c, e, d, a, b

17. (a)   about 770,000 years   (b)   11

19. (a)   It will never halt.
     (b)   Insert before the statement $final(v) \leftarrow$ **true** the following statement.
           if $m = \infty$ **then return**$(distance, previous)$

21. Use Dijkstra's algorithm $n - 1$ times (without the stopping condition, as in Exercise 12, so that all distances from the source are computed). This will determine the distances between all pairs of vertices in $O(n^3)$ steps. Take the smallest value found.

23. Work backward from the sink, labeling vertices. The label $L(v)$ is intended to be the latest time at which the subproject represented by v may be started in order to finish the entire project by time $T(t)$. We start by setting $L(t) = T(t)$. Successively, find a vertex u all of whose successors have been labeled (this is possible because the digraph is acyclic), and set $L(u) = \min(L(v) - w(u, v))$, where the minimum is taken over all successors v of u.

SECTION 10.2   MINIMUM SPANNING TREES

1.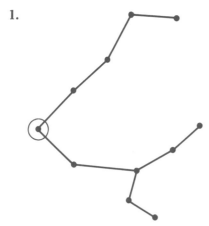

3. (a) bf, ae, cd, de, ab, fg
    (b) bd, ad, cd, ae
    (c) $ac, de, fg, hi, bc, ce, cf, eh, ij, jl, lz, jk$
    (d) ad, bc, ae, be

5. (a) When applying Prim's algorithm, we can start with edge e and vertex v.

   (b) no

7. (a) **procedure** *greedy_hamilton*$(G :$ complete weighted graph)

        {assume that vertices of G are $1, \ldots, n$; path is constructed as a list of vertices}

          $path \leftarrow (1)$

          $visited(1) \leftarrow$ **true**

          for $j \leftarrow 2$ **to** n **do**

            $visited(j) \leftarrow$ **false**

          $v \leftarrow 1$

          for $i \leftarrow 1$ **to** $n - 1$ **do**

            begin

              $m \leftarrow \infty$

              for $j \leftarrow 1$ **to** n **do**

                if $\overline{visited(j)} \wedge w(v, j) < m$ **then**

                  begin

                    $w \leftarrow j$

                    $m \leftarrow w(v, j)$

                  end

              $path \leftarrow path$ followed by w

              $visited(w) \leftarrow$ **true**

              $v \leftarrow w$

             end

          $path \leftarrow path$ followed by 1

          return$(path)$

   (b) a, d, b, e, c, a; no

9. **procedure** *sorted_kruskal*$(G :$ connected weighted graph)

      {same setting as Algorithm 2; edges denoted by $e_1, e_2, \ldots, e_m$}

        sort the edges of G so that $w(e_1) \leq w(e_2) \leq \cdots \leq w(e_m)$

        $j \leftarrow 1$

        $F \leftarrow$ the empty tree

        for $k \leftarrow 1$ **to** $n - 1$ **do**

          begin

            while $F \cup \{e_j\}$ contains a cycle **do**

              $j \leftarrow j + 1$

            $F \leftarrow F \cup \{e_j\}$

            $j \leftarrow j + 1$

          end

        return(F)

11. (a) Follow Prim's or Kruskal's algorithm, but add the largest weight edge at each stage, rather than the smallest weight edge (consistent with the rules of the algorithm).

(b) bc, df, ab, be, de

13. We will show that after each deletion, the resulting graph still contains a minimum spanning tree of G. Then since we end up with a tree, it must be a minimum spanning tree. We proceed by induction, the base case being $T = G$: Clearly, G contains a minimum spanning tree. Suppose that T contains a minimum spanning tree S. Suppose that edge e is deleted to form T'. If e is not in S, there is nothing to prove. If e is in S, consider $S - \{e\}$. It is a forest with two components, C_1 and C_2. Since e was in a cycle in T, there must be an edge e' in T joining a vertex in C_1 and a vertex in C_2. Since e' was not deleted from T to form T', we know that $w(e') \le w(e)$. Furthermore, $S' = S - \{e\} \cup \{e'\}$ is a tree. The weight of S' is at most the weight of S. Thus S' is a minimum spanning tree of G, and S' is a subgraph of T', since e' is in T'.

15. Let T be the minimum spanning tree of G produced by Prim's algorithm, containing edges $e_1, e_2, \ldots, e_{n-1}$. If G has another minimum spanning tree, let T' be one that agrees with T on $e_1, e_2, \ldots, e_k$ for the largest possible k. Proceed as in the proof of the correctness of Prim's algorithm. The tree T'' constructed there has smaller weight than T', since $w(e_{k+1}) < w(e')$. This contradicts the choice of T'.

17. Let T_k be the tree constructed after the kth pass through the outer loop of Prim's algorithm. We will show by induction that each T_k is contained in some minimum spanning tree of G; Theorem 1 follows by taking $k = n - 1$. The assertion is trivial for $k = 0$. Suppose that it is true for T_k. Let S be a minimum spanning tree containing T_k. We must show that $T_k \cup \{e_{k+1}\}$ is contained in a minimum spanning tree. If e_{k+1} is in S, we are done. If not, look at $S \cup \{e_{k+1}\}$. It contains a cycle. Since T_{k+1} does not contain a cycle, there is some edge in $S - T_{k+1}$ on the cycle. We follow the cycle around until we come to the first such edge; call it e. Thus e was a candidate when e_{k+1} was added to T_k, so $w(e) \ge w(e_{k+1})$. But $S' = S \cup \{e_{k+1}\} - \{e\}$ is a tree, and its weight does not exceed the weight of S. Thus S' is the desired minimum spanning tree containing T_{k+1}.

SECTION 10.3   FLOWS

1.

su	sv	uv	ut	vt
1	1	0	1	1
2	0	2	0	2
3	0	2	1	2
3	1	2	1	3
2	1	2	0	3

3.

su	sv	uv	ut	vt
4	1	1	3	2

5.

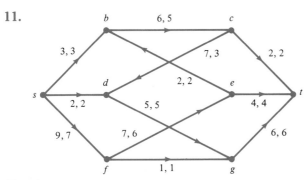

7. See answer to Exercise 11.

9. See answer to Exercise 11.

11.

13. 6

15. Add a new vertex S to be the new source, and put an edge of capacity ∞ from S to each source.

17. 24

19. 10

21. Replace each undirected edge by a pair of antiparallel edges, each with the capacity of the undirected edge. After each new flow is found, combine the flows on all pairs of antiparallel edges into a net flow in one of each pair and no flow in the other.

23. 5

INDEX

Numbers followed by a sharped number in parentheses [e.g., 356(#42)] refer to an exercise (following the #) on the given page. Numbers in bold type [e.g., **276**] give the most complete initial explanation of a topic; when no number is given in bold type, the first nonexercise reference will generally be the primary one. Numbers in italic type [e.g., *730*] refer to formal definitions in the end-of-section glossaries.

A

Absolute value ($|x|$), 53

Abstract algebra, 45, 105(#38)

Abstraction, levels of, xvi, 17, 76, 127, 132, 195, 463

Abundant number, 105(#35)

Ackermann, W., 304

Ackermann's function (φ), 302, **305**, *307*, 310(#15–16), 324, 333(#33), 344, 353(#28)

   Robinson's version (A), 310(#17), 330(#4), 354(#29), 356(#42)

Acquaintances, 417, 425(#28–31), 525

Acyclic graph or digraph, 726, *730*

Addition (+), *see also* Sum

   Boolean, 61, *71*, 171

   of functions, 153, *159*

   of matrices, 286, 291(#21)

   of natural numbers, 132, **276**, 362

recursive definition of, 303

   of polynomials, 282

   of rational numbers, 222(#2)

Addition principle, 361, **364**, *373*, 375(#10)

Adjacency lists, 572, *582*, 586(#20–21)

Adjacency matrix (A_G), 566, 570, 576, *582*, 585(#18), 586(#19), 720

Adjacent vertices, 522, 523, 527, 528, *536*

Admissions to college, 545(#49)

Airline flights, 377(#18), 527, 538(#2), 564(#35), 719, 731(#2)

Aleph ($\aleph$), 84

   aleph nought ($\aleph_0$), 84

Algebra

   abstract, 45, 105(#38)

   Boolean, 61, 75(#25), 106

   on a computer, 288, 292(#29)

   fundamental theorem of, 471

   linear, 105(#39), 286, 467

propositional, 6

Algebraic expression (FPAE), 300, *307*, 313(#32), 332(#24), 346

   evaluating, 301, 332(#25), 669

Algorithm(s), 128, **210**, *220*, 226(#43), *see also* Procedure, specific algorithms

   ambiguity, 219

   analysis using recurrence relations, 451

   arguments for, 215, **226**, *240*

      passing, 236

   bad, 265, *272*

   basic step of, 246

   characteristics, 218

   complexity, 247, 265

      average case, 247, 685

      exponential, 265

      linear ($O(n)$), 255

      logarithmic ($O(\log n)$), 255

      polynomial ($O(n^k)$), 255, **265**

      space, 245

889

List of Important Symbols

$x \bmod y$	the nonnegative remainder when the integer x is divided by the positive integer y, p. 137
$x \equiv y \pmod{m}$	x is congruent to y modulo m ($x \bmod m = y \bmod m$), p. 137
$n!$	n factorial $(1 \cdot 2 \cdot 3 \cdots (n-1) \cdot n)$, p. 138
$f(S)$	the image of the set S under the function f, p. 151
$f^{-1}(T)$	the inverse image of the set T under the function f, p. 151
$g \circ f$	g composed with f (the composition of the functions f and g), p. 154
f^{-1}	the inverse of the bijective function f, p. 157
i_A	the identity function on the set A, p. 158
xRy	x is related to y under the relation R, p. 165
$x\mathbb{R}y$	x is not related to y under the relation R, p. 165
$x \mid y$	the integer x divides the integer y, p. 166
M_R	the Boolean matrix representing the relation R, p. 168
Δ_A	the identity relation on the set A, p. 170
I_n	the n by n identity matrix, p. 170
$\overline{R}$	the complementary relation to the relation R, p. 170
R^{-1}	the inverse relation to the relation R, p. 171
$S \circ R$	S composed with R (the composition of the relations R and S), p. 172
R^n	the composition $R \circ R \circ \cdots \circ R$ (n R's), p. 172
$x \preceq y$	x precedes or is equal to y (in a poset), p. 186
$x \prec y$	x precedes y (in a poset), p. 186
$[a]_R$ or $[a]$	the equivalence class of a (under the equivalence relation R), p. 194
A/R	the set of equivalence classes of the set A under the relation R, p. 194
$\gcd(x, y)$	the greatest common divisor of x and y, p. 211
$x \leftarrow E$	the value of E is assigned to x, p. 227
T_A or T	the time complexity of the algorithm A, p. 247
$O(f)$	the big-oh class of the function f, p. 252
P	the class of polynomial problems, p. 267
NP	the class of nondeterministic polynomial problems, p. 267
$d_{n-1}d_{n-2}\ldots d_2 d_1 d_0$	the decimal numeral representing the number $\sum_{i=0}^{n-1} d_i \cdot 10^i$, p. 276
$A + B$	the sum of the matrices A and B, p. 286
$A \times B$	the matrix product of the matrices A and B, p. 286